1701–
(1531)
KV-038-392

ENCYCLOPAEDIC DICTIONARY OF PHY

Other Volumes of the Dictionar

QUEEN MARY COLLEGE
(University of London)
LIBRARY

CLASSIFICATION
QC 5

AUTHOR

TITLE
ENCYCLOPAEDIC dictionary of physics.
Supplementary Vol. 3.

LOCATION & STOCK No.
SCIENCE REFERENCE 138710

Confined to the Library

QMC 312558 3
a30213 003125583b

ENCYCLOPAEDIC DICTIONARY OF PHYSICS

WITHDRAWN
FROM STOCK
QMUL LIBRARY

SUPPLEMENTARY VOLUME 3

DATE DUE FOR RETURN

SUPPLEMENTARY VOLUME 3

ENCYCLOPAEDIC DICTIONARY OF PHYSICS

GENERAL, NUCLEAR, SOLID STATE, MOLECULAR
CHEMICAL, METAL AND VACUUM PHYSICS
ASTRONOMY, GEOPHYSICS, BIOPHYSICS
AND RELATED SUBJECTS

EDITOR-IN-CHIEF

J. THEWLIS

FORMERLY HARWELL

Associate Editors

R. C. GLASS
LONDON

A. R. MEETHAM
TEDDINGTON

THE QUEEN'S AWARD TO INDUSTRY 1966

PERGAMON PRESS

OXFORD · LONDON · EDINBURGH · NEW YORK
TORONTO · SYDNEY · PARIS · BRAUNSCHWEIG

138710
QC 5

Pergamon Press Ltd., Headington Hill Hall, Oxford
4 & 5 Fitzroy Square, London W.1

Pergamon Press (Scotland) Ltd., 2 & 3 Teviot Place, Edinburgh 1

Pergamon Press Inc., Maxwell House, Fairview Park, Elmsford, New York 10523

Pergamon of Canada Ltd., 207 Queen's Quay West, Toronto 1

Pergamon Press (Aust.) Pty., Ltd., 19a Boundary Street, Rushcutters Bay, N.S.W. 2011, Australia

Pergamon Press S.A.R.L., 24 rue des Écoles, Paris 5^{e}

Vieweg & Sohn GmbH, Burgplatz 1, Braunschweig

Copyright © 1969
Pergamon Press Ltd.

First published 1969

Library of Congress Catalog Card No. 60-7069

Printed in Germany

08 012447 X

QUEEN MARY
COLLEGE
LIBRARY

ARTICLES CONTAINED IN THIS VOLUME

LIST OF CONTRIBUTORS TO THIS VOLUME

APPLETON A. D. (*Newcastle upon Tyne*)
APTACK E. (*Burton-on-Trent, Staffs.*)
AXTON E. J. (*Teddington*)

BELL R. E. (*Montreal, Canada*)
BERGER H. (*Argonne, Illinois*)
BERKTAY H. O. (*Birmingham*)
BERMAN R. (*Oxford*)
BERREMAN D. W. (*Murray Hill, N. J.*)
BHAGAVANTAM S. (*New Delhi, India*)
BIKERMAN J. J. (*Cleveland, Ohio*)
BIRCH K. G. (*Teddington*)
BOES D. J. (*Pittsburgh, Pa.*)
BRADLEY R. S. (*Leeds*)
BRANS C. H. (*Louisiana*)
BRUCK H. (*Gif-sur- Yvette, France*)
BUCKMASTER, H. A. (*Calgary, Canada*)
BURKE, J. D. (*Pasadena, Calif.*)
BURRILL E. A. (*Burlington, Mass.*)
BURSON J. H., III (*Atlanta, Georgia*)
BUTCHER P. N. (*Great Malvern, Worcs.*)

CHADDERTON L. T. (*Thousand Oaks, Calif.*)
CHAMBERLAIN A. C. (*Harwell*)
CHRISTIAN J. W. (*Oxford*)
CHRISTIAN R. H. (*Aldermaston*)
CLARK R. J. H. (*London*)
COBLEY D. N. (*Burton-on-Trent, Staffs.*)
CROMPTON R. W. (*Canberra, Australia*)

DAWSON B. (*Victoria, Australia*)
DAY J. P. (*Keele, Staffs.*)
DE BARR A. E. (*Macclesfield, Cheshire*)
DEAN K. J. (*Letchworth, Herts.*)
DELANY M. E. (*Teddington*)
DER AGOBIAN R. (*Paris*)
DERING J. C. (*Calgary, Canada*)
DEW G. D. (*Teddington*)
DUGDALE J. S. (*Leeds*)

EICKHOFF K. G. (*Warrington, Lancs.*)
ELFORD M. T. (*Canberra, Australia*)
EMMONY D. C. (*Glasgow*)

FENDER D. H. (*Pasadena, Calif.*)
FENSHAM P. J. (*Victoria, Australia*)
FERRARI R. L. (*Cambridge*)
FLEISCHER R. L. (*Schenectady, N. Y.*)
FRANKS A. (*Teddington*)
FREEMAN J. M. (*Harwell*)
FRIEDLANDER H. Z. (*Tarrytown, N. Y.*)
FROST H. J. (*Victoria, Australia*)

GINSBERG I. W. (*Ann Arbor, Michigan*)
GLASER P. E. (*Cambridge, Mass.*)
GOLDSMID H. J. (*Bristol*)
GRAFF R. A. (*Tarrytown, N. Y.*)

HALMSHAW R. (*Sevenoaks, Kent*)
HARGREAVES J. K. (*Boulder, Colorado*)
HEAVENS O. S. (*York*)
HEYLEN A. E. D. (*Leeds*)
HILL C. R. (*Sutton, Surrey*)
HODGKINSON W. L. (*Harwell*)
HUGILL J. (*Culham*)
HUME-ROTHERY W. (*Oxford*)

JAMES G. D. (*Harwell*)
JEFFREY A. (*Newcastle upon Tyne*)
JONES S. S. D. (*Farnborough, Hants.*)

KHANNA M. L. (*New Delhi, India*)
KINSLOW R. (*Cookeville, Tennessee*)

LAITHWAITE E. R. (*London*)
LAST F. H. (*London*)
LAWSON J. D. (*Didcot, Berks.*)
LAWTON J. (*London*)
LEWIS E. (*Long Island, N. Y.*)
LIEBHAFSKY H. A. (*Schenectady, N. Y.*)
LIMPERIS T. (*Ann Arbor, Michigan*)
LONGSTAFF R. M. (*Wantage, Berks.*)

MANLEY G. (*Lancaster*)
MAYNE J. E. O. (*Cambridge*)
MCCARTY R. S. (*Tustin, Calif.*)
MEGAW W. J. (*Harwell*)
MELCHIOR P. (*Brussels*)
MICHEL F. C. (*Houston, Texas*)
MIELE A. (*Houston, Texas*)
MORLEY F. (*Harwell*)
MYNOTT P. (*Leicester*)

OWEN R. B. (*Harwell*)

PERRING J. (*Harwell*)
POLLOCK A. A. (*London*)
POOLE D. M. (*Harwell*)
PRICE P. B. (*Schenectady, N. Y.*)

REES D. J. (*Calgary, Canada*)
REES J. A. (*Liverpool*)
RENNIE C. (*Winfrith, Dorset*)
REYNOLDS J. A. (*Culham*)
ROBERTS W. T. (*Birmingham*)
RUNCORN S. K. (*Newcastle upon Tyne*)
RUNNELS L. K. (*Louisiana*)
RYMER T. B. (*Reading*)

SAYERS J. B. (*Harwell*)
SHARPE R. S. (*Harwell*)
SMART D. (*Newcastle upon Tyne*)

SMITH D. B. (*Wantage, Berks.*)
SPRING K. H. (*London*)
STEWARDSON E. A. (*Leicester*)
STROKE G. W. (*Ann Arbor, Michigan*)
SULLIVAN J. V. (*Melbourne, Australia*)

THEWLIS J. (*formerly Harwell*)
TOWNSEND P. (*Colchester, Essex*)

WALDRON M. B. (*London*)
WALKER R. M. (*Schenectady, N. Y.*)
WALKER T. G. (*Didcot, Berks.*)
WALLACE P. R. (*Montreal, Canada*)
WALTERS K. (*Aberystwyth*)
WEINBERG F. J. (*London*)
WELLS P. B. (*Hull*)
WHIFFIN A. C. (*Crowthorne, Berks.*)
WILLIS B. T. M. (*Harwell*)
WILLIS J. B. (*Melbourne, Australia*)
WILSON A. J. C. (*Birmingham*)
WILSON J. A. (*Bristol*)
WROE H. (*Didcot, Berks.*)

FOREWORD

I write this Foreword at a time when Volume 2 has just been published and when no reviews have appeared. The reception accorded to that volume remains, therefore, to be seen; but if it is comparable with that given to Volume 1 it would seem that we are succeeding in fulfilling our declared aims. These were set out in the Foreword to the first supplementary volume in which I wrote:

"Of its very nature, a work of the size and scope of the *Encyclopaedic Dictionary of Physics* can never reach ultimate completion. However, by issuing a continuous series of supplementary volumes, we shall strive to keep it as up to date and comprehensive as we can (having regard to the inevitable time lapse between writing and publication), and as free from errors as may be.

The volumes in this series are intended to form part of a unified whole, and are numbered accordingly. They are designed to deal with new topics in physics and related subjects, new development in topics previously covered and topics which have been left out of earlier volumes for various reasons. They will also contain survey articles covering particularly important fields falling within the scope of the Dictionary.

The contents of these volumes will be arranged alphabetically, as in the previous volumes. Articles will be reasonably short and will be signed. Cross references to other articles will be incorporated as necessary, and bibliographies will be included as a guide to further study. Each volume will have its own index, prepared on the same generous scale as before; and, in addition, it is intended to issue a cumulative index every five years. Errata and addenda lists will be published, referring to the original *Encyclopaedic Dictionary of Physics* and to those supplementary volumes which will already have been published.

In preparing the Supplementary Volumes regard will be had to the changing emphasis in many branches of physics—the invasion of the biological sciences by physics, the possibilities opened out by the increasing use of computers in all branches of science and technology, the ever-increasing scope of theoretical physics, the progress in high energy physics, the emergence of new instrumental techniques etc.; and, at the same time the authors of previously published articles will be given the opportunity of bringing those articles up to date. Naturally there are many articles for which this will not be necessary, and it is certainly not intended that new articles shall be written if there is no need for them. In short, it is our intention to produce a series of volumes in which the high standard already achieved in the *Encyclopaedic Dictionary of Physics* is fully maintained."

Once again I should like to express my gratitude to the publishers, and particularly to Mr. Robert Maxwell, for their constant support; to my Associate Editors, Dr. A. R. Meetham and Mr. R. C. Glass for their help and counsel; and to Mr. S. Crimmin, the Assistant Editor at the London Office of the Pergamon Press, for another year of solid, reliable and painstaking work. I should also like to thank the referees who, as in the past, have helped to ensure the quality of the published articles. Last but not least it is a pleasure to acknowledge the debt I owe to my wife, who has been of invaluable assistance.

J. THEWLIS
Editor-in-Chief

ERRATA: VOLUMES I—VIII AND SUPPLEMENTARY VOLUME I

Volume I

p. 419, Col. II, article Binomial distribution, last line.
For: $\frac{n!}{r!}(n - r)!$ *read:* $\frac{n!}{r!\,(n - r)!}$.

p. 454, Col. I, article Bohr magneton, line 2. *For:* $\frac{e\hbar}{4\pi m}$ *read:* $\frac{e\hbar}{4\pi mc}$.

p. 480, Col. II, line 32. *For:* $\left(n + \frac{1}{2}\right)h$ *read:* nh.

Volume VIII

p. 333, Col. I, entry "Prism", resolving power of. *For:* (4) *read:* (5).
p. 363, Col. II, entry "Resolving power of a prism". *For:* (4) *read:* (5).

Supplementary Volume I

p. 241, Col. I, line 4. *For:* sort *read:* sought.
p. 243, Fig. 3, right hand ordinate. *For:* Watt hours/CF *read:* Wh/ft^3.

A

ABLATION. The word ablation in the sense used in this article has been applied for many years by astronomers and astrophysicists to describe the erosion and disintegration of meteors entering the atmosphere of the Earth. The familiar, transient, incandescent trails, called shooting stars, are evidence of the intense heat generated by the interaction of these high velocity particles with the atmosphere. The phenomenon of ablation has recently acquired technological significance in the development of long-range missiles, Earth satellites and interplanetary vehicles which travel outside of the atmosphere and must re-enter to complete their mission.

The temperatures that are developed in the atmosphere near the surface of re-entering bodies, or indeed any bodies moving at very high speeds, depend on the velocity and may exceed those at which the most refractory solids are stable. Under these conditions the complete disintegration of such bodies requires a finite time which can be sufficient for the completion of the mission. Selected materials, notably organic plastics, erode remarkably smoothly and slowly in these severe environments. These so-called ablation materials can be used to shield payloads from thermal damage. More or less similar materials are also used for parts of rocket motors which are exposed to the hot exhaust gases.

The general behaviour of heat shields can be discussed in terms of the theory of thermal diffusion. However, it must be recognized that the effective values of the heat capacity, the thermal conductivity, and the density depend on temperature, time and position in the material and that phase changes and reactions may occur. In addition, the thermal boundary conditions depend not only on the velocity of the body and ambient pressure, but also on the body shape and the body station under consideration. The attitude, altitude, path angle and other details of its motion must also be taken into account. At the high temperatures that are encountered in practice, radiative transfer of heat into and out of the body can be important.

Not only is the detailed theory of the ablation phenomena complicated but laboratory simulations of the high temperatures and heating rates are also difficult. In connexion with development of ablation technology very elaborate computational schemes have been generated to correlate and predict the behaviour of the materials. Ingeneous and costly laboratory devices have also been developed for simulating the re-entry environment in order to provide an empirical base for the computations. The details of the large scale analytical and experimental studies relating to ablation technology are beyond the scope of this article. However, a general review of some experimental results and an approach that has been found to be useful in the analysis is given below.

Table 1. Relative weight loss from diverse materials exposed to gas at 7000 °K

Material	Relative weight loss
Graphite	0.81
Nylon-phenolic composite	1.2
Silicon carbide	1.7–6.3
Silica-phenolic composite	2.2
Glass-phenolic composite	2.2
Silica	2.3
Alumina	6.9–13.7
Mullite	8.22
Zirconia	12.9
Copper	60

Table 1 shows relative gravimetric erosion rates which have been observed in laboratory experiments using an electric arc heater delivering gas at about 7000 °K. The table entries were obtained from the weight loss during ten second exposures. The cold calorimeter heating rates were about 2000 Btu per square foot per second. The durability of the all-organic material is at first glance somewhat surprising. However, this result can be understood in terms of the high heat absorbing capacity, that is, the value of the integral of the specific heat of such materials up to high temperatures. Typical values of this integral up to 5000 °K are shown in Table 2. The durability of these materials is also enhanced by the relatively large volumes of gas that are produced when these materials are heated which interfere with the transfer of heat from the environment. Typical values of this gas volume are also shown in Table 2.

The high values of heat absorption for organic materials shown in Table 2 depend on the highly endo-

Table 2. Estimated integrated specific heats from 300 to 5000 °K and equilibrium volume of gas generated per gramme at 5000 °K for various substances

Substance	Heat absorbed (cal/g)	Relative gas volume (Mole)
H_2 gas	67,000	1·0
$(CH_2)_n$ organic plastic	24,000	0·21
$(CH)_n$ organic plastic	20,600	0·15
C graphite	16,670	0·08
$(C_6H_{12}O_6)_n$ cellulose	5760	0·10
BeO beryllia	7080	0·08
MgO magnesia	5500	0.05
SiO_2 silica	2800	0·05
Be	9876	0·11
Cu	1600	0·016
$(C_2F_4)_n$ Teflon	6300	0·06
H_2O	14,500	0·16
He	3525	0·25

thermic nature of the hydrogen dissociation process which only occurs at high temperatures. When the temperature is high, exothermic reactions of the material with the atmosphere are likely to be unimportant. When the environment is not very hot the advantages of these materials are lost as is shown by the data given in Table 3, which show relative erosion rates of selected materials at a range of thermal environments.

Table 3. Relative erosion rates of various materials vs. temperature of exposure

Material	Resin (%)	Temperature (°C) 1800	2500	7000
Phenolic-glass cloth	27	1·0	2·7	2·5
	37	1·2	2·5	2·0
	44	1·6	2·2	2·0
	65	1·7	1·5	1·4
Phenolic-silica cloth	41	1·4	1·0	2·1
Phenolic-nylon cloth	57	4·7	2·5	1·0

It may be seen that the durability of the material in a hot environment depends on the environment itself. For example, the high melting point of silica contributes to its stability in the 2000 °C range. When, however, the silica vaporizes its advantage is lost.

The usefulness of the organic materials can also depend on the fact that when they decompose one of the products may be carbon which is the most refractory material known. The carbon residue can become very hot, reducing the temperature gradient which drives heat from the environment and also radiating significant amounts of the convected heat back into the gas.

The rather gross description of the phenomena which may occur when selected organic ablating materials are heated is, of course, not adequate for the design of heat shields. Models of the process on which design computations can be based have been developed. These take into account the fact that as gas and carbon are produced at the heated surface, the interface between the virgin material and char moves into the material. The char layer thickness and its resistance to the passage of the gas increases. The gas pressure can cause the spallation of the char layer and reduce the total effectiveness of the system. When the heating is very severe the char itself ablates and may never get thick enough to permit dangerous pressures to develop. As the char layer develops, the primary gaseous products decompose further as they percolate through the pores and greatly complicate the analysis. Char integrity can sometimes be enhanced by the addition of strong refractory fibres.

At very high temperatures radiative processes contribute more significantly to the heat load on the material and photon capture processes become important. These phenomena are less well understood from a quantitative point of view than the equilibrium thermodynamics of the materials.

In practical situations the ablating material is part of a larger system and, for example, its mechanical and electrical properties can be as important as its heat resistance. These matters are not, however, relevant to this article, in which the object has been to provide orientation in a relatively new field that is likely to develop more technological importance as time goes on.

I. J. Gruntfest

ACOUSTIC EMISSION FROM MATERIALS. During the mechanical deformation of a number of materials, audible sounds are generated. In very many deformation processes, a small fraction of the energy of deformation is released in the form of mechanical vibration of the medium being deformed. In most cases the amplitude of the vibration is too low, or its frequency too high for it to be heard without special apparatus. In recent years increasing attention has been paid to this phenomenon, which is known as acoustic emission.

Acoustic emission would seem to be a likely concomitant of any abrupt deformation process. Consider a material in equilibrium with applied stresses. A deformation event such as slip, or the sudden growth of a crack, takes place and the equilibrium is locally disturbed (or in acoustic terminology, the boundary conditions change). The restoration of local equilibrium is achieved with the propagation of a stress wave away from the source of the disturbance (in acoustic terminology, the system adjusts itself to the new boundary conditions). The subsequent propagation of the stress wave is more or less complicated, and its form is usually drastically modified, according to the physical characteristics of the subject body and the observing system. Thus the acoustic emission finally observed does not have the same waveform as the stress wave at source.

The observation of acoustic emission can yield information about material behaviour and assist in the testing materials. Many emission sources have been identified. A particularly energetic source of emission is the Portevin-Le Chatelier effect (discontinuous or jerky slip). The "cry of tin" is known to be due to twinning. Schofield has found rather good correlation between emission count and slip line count under certain conditions in aluminium single crystals. The vibrations following fracture, when large amounts of stored elastic energy are released, constitute a special case of acoustic emission.

Three examples of the useful application of acoustic emission research are given here; other applications are being developed.

First, large-scale acoustic emissions in mines often herald the approach of rock-bursts which could endanger life. Acoustic techniques for detection and warning have been used in the United States and in the Soviet Union.

Second, in the testing of high-strength alloys, emission has been used as a criterion for crack initiation in notched sheets (Jones and Brown). Crack inception is often accompanied by an audible "click" or "pop" and there is quite good agreement between acoustic and other methods of identifying this critical point.

The third application relates to the measurement of peak pressures, and relies on the so-called "Kaiser effect". Emission behaviour is history-sensitive. Suppose that a material is stressed to a certain maximum level and then retested after relaxing the stress. During the second test, emission activity is generally much lower than in the first test, until the previous maximum stress level is attained. Above this stress level, full emission activity is restored. This effect is used in a passive peak pressure measuring transducer reported at the 1966 IEEE Ultrasonics Symposium in Cleveland, Ohio. An anodized aluminium diaphragm is used as the active element of this transducer. One side of the diaphragm is exposed to the test environment, in which the peak pressure is to be determined. In calibration and demonstration experiments, and in the controlled retest, the other side is fluid-coupled to a transducer which detects acoustic emissions from the diaphragm. After exposure to the test environment, the unit is removed and retested in controlled conditions. From observation of the point at which emission commences, the maximum pressure encountered in the test environment can be determined to within 3 per cent.

The apparatus for observing acoustic emission is simple. The central element is the transducer for converting the stress waves into electrical signals. Electrodynamic methods have been used but piezoelectric devices (accelerometers, gramophone styli or crystals of piezoelectric materials) are more common since they can be mounted directly on the test material and the energy transfer is reasonably efficient. The signal from the transducer is then amplified and fed into suitable recording, displaying or counting equipment. It is usually necessary to take some care to reduce background noise. Mechanical vibrations entering the specimen from the testing machine are a common source of noise in laboratory experiments, and special systems have used hydrostatic loading or thermal expansion for the silent application of stress. When precautions of this kind are taken, the limiting factor is usually the first stage of the electronics.

Acoustic emissions are commonly observed as transient signals showing the "ringing-down" of the system after shock excitation. Some observers have also reported a second kind of emission signal which has been described as a high-frequency continuous type (Schofield). Care must be taken to avoid confusing acoustic emissions with other signals. Emissions often occur discretely although in periods of high activity the signals may overlap and be difficult to count. During testing of a single specimen, up to tens of thousands of emission signals may be counted.

The transducer output may be recorded directly, or the signal may be processed in some way before a recording is made. In many applications the shape of the emission signal is of no interest and the transducer output can be fed into pulse shaping and counting circuits. In cases where it is desired to record the whole of the emission signal, the frequency range of the recording system should be chosen with the resonant frequencies of the transducer and of the test body in mind, since these frequencies are liable to be excited during the pulse shaping and filtering processes that intrude between source event and detection.

Acoustic emission from metals. A wide range of materials has been employed in acoustic emission research. Emission during tensile deformation has been studied using both engineering materials for applications research and others, for example, single crystals, for more fundamental studies. Kaiser has also observed acoustic emissions generated during melting and solidification, and during phase changes in metallic systems; these emissions were related to the volume changes during the changes of state.

B. H. Schofield of Lessells and Associates, and a group at Michigan State University, have studied acoustic emission from aluminium single crystals. It was found that emission behaviour varied widely according to such variables as crystal orientation, oxide film thickness, strain rate, previous heat treatment and stress history. Acoustic emission is bound up with the actual structural changes during deformation, and as such it is subject to the same complex of influencing factors. The only variable which showed a good simple correlation with emission rate was strain rate, which evidently gives an average measure of the damage going on.

During a typical test, emission count rate varies widely, but within these statistical fluctuations certain general patterns can be seen. Where the deformation process has already been classified into several stages, on the basis of metallurgical evidence, different emission behaviour may be exhibited in each stage. Another feature of emission behaviour is that the rate often increases very sharply in the final stages before fracture. Warning of fracture thus obtained is one of the promising uses of acoustic emission.

Schofield observed emissions which he believed to be generated by groups of dislocations bursting through the oxide film on the surface of the specimen. This illustrates the extreme fineness of observation possible when proper measures are taken to eliminate noise.

Emission during fatigue testing is relatively difficult to observe, owing to the high level of background noise during fatigue testing. However, Schofield has succeeded in identifying emission from growing fatigue cracks.

Acoustic emission from rocks. The largest emissions are earthquakes. Some of the largest earthquakes have been known to set the whole earth "ringing like a bell", an analogue of the laboratory situation where specimen resonances may be excited by localized deformation events. In Japan, where earthquakes have been extensively studied, earthquake situations have been simulated in the laboratory and techniques of analysis have been evolved which may be relevant to other fields of emission research.

The amplitude and time distributions of emissions have been studied in some detail. Logarithmic plots of the amplitude distribution were straight lines whose gradient depended on the material. In particular, this parameter of the amplitude distribution was found to characterize the degree of heterogeneity and also the strength of the material. This observation was made both with various rock types and with a model material comprising a mixture of resin and pumice.

The time distribution of acoustic emissions (and of earthquakes) is compared with predictions made on the basis of a stochastic theory in which it is assumed that the events take place independently of one another and that the occurrence of an event is to be described in terms of a transition probability which is a function of the stress state. In these tests the experimental data fitted quite well the predictions of stochastic theory, and this would seem to be a good starting point for analysis of the behaviour of other systems. Acoustic emission provides a unique method of registering the individual steps by which a material under stress moves towards failure. Time distribution analysis can show whether, at different stages of the deformation process, these steps occur in groups or individually.

Acoustic emission from other materials. Some work on emission from concrete has been carried out in connexion with the establishing of failure criteria. Emission from heated rock-salt under shear stress was observed in the late 1920's, twenty-five years before the phenomenon attracted serious attention in its own right. Emission during crack formation in ice has also been described.

Emission and fracture. Modern technology often demands that materials be used near to the limits of their performance, and acoustic emission is being studied in the hope that it may provide useful information about material behaviour near those limits. For instance, it has already been remarked that emission activity often shows a sharp increase just prior to failure. Also, it may vary according to the kind of failure. Working with soft steel, Crussard found that ductile fracture was preceded by emission while brittle fracture was not. Obert, investigating rocks under pressure, found that rocks which tended to shatter and rocks which tended to crush exhibited different emission characteristics during the later stages of deformation. Thick epoxy resin bonds tend to fail in a brittle manner without prior emission; thinner bonds which sustain higher stress give rise, before fracture, to emissions which are presumably caused by cracks which do not have space to propagate. Information of this kind is accumulating, but it remains to be seen whether study of the phenomenon of acoustic emission can lead to any major developments in fracture technology.

Bibliography

Barry D. L. (1967) IDO–17230 USAEC R & D Report

Crussard C. *et al.* (1958) *Comptes Rendus* **246**, 2845.

Jones M. H. and Brown W. F. Jr. (1964) *Mat. Res. and Stands.*, ASTM **4**, 3, 120.

Mogi K. (1962) *Bull. Earthqu. Res. Inst. Tokyo* **40**, 107.

Pollock A. A (1968) *Ultrasonics* **6**, 2

Schofield B. H. (1963) *4th Symp. on Physics and NDT*, San Antonio, Texas.

Schofield B. H. (1966) in *Encyclopaedic Dictionary of Physics* (J. Thewlis Ed.), Suppl. Vol. 1, 196, Oxford: Pergamon Press.

A. A. Pollock

AITKEN NUCLEI. The existence of submicron particles in the atmosphere was first demonstrated by Coulier (1875) and Aitken (1881) who showed that fogs could be generated at lower supersaturations in unfiltered than in filtered air and ascribed this to the presence of particles. Aitken nuclei, as now defined, occupy the lower end of the size spectrum of condensation nuclei and range in radius from 5×10^{-8} to 2×10^{-5} cm. Because of their small size they do not participate directly in normal condensation processes in the atmosphere, but they are nevertheless important in atmospheric electricity and air pollution. At the lower end of the size range small ions are responsible for the important electrical properties of the atmosphere while at the upper it is generally accepted that most of the naturally occurring radon and thoron decay products in the atmosphere are carried by the Aitken nuclei. The nuclei are so small that their rate of sedimentation under gravity is almost negligible (the terminal velocity of a 0·1 micron diameter particle is 10^{-4} cm/sec) but on a molecular scale the bigger particles are so large that they diffuse only very slowly. Such particles therefore tend to remain airborne for long periods and the deposition behaviour of gases and vapours which adsorb on them will be considerably modified and their presence may have an effect on the physiological properties of toxic gases and vapours.

Aitken nuclei are formed in all combustion processes, are emitted by hot metallic surfaces, and are formed by the action of alpha, beta, gamma and X radiation on particle free air. Very little is known of the chemical composition of the nuclei in the atmosphere, but it has been shown by neutron activation that the particles emitted by a hot platinum filament (around 1200°C) contain platinum and those from a nickel chrome wire consist almost entirely of mixed chromium oxides. Ammonium sulphate has been identified by means of electron diffraction in the nuclei formed by radiolytic reactions in air.

Electron microphotographs of Aitken nuclei were first obtained by Linke in 1943 and techniques have recently been developed which combine fine grain autoradiography and electron microscopy to make possible the identification of radioactive particles down to 10^{-5} cm diameter by means of the alpha tracks emanating from them. Concentrations of Aitken nuclei are usually measured by causing water to condense on the particles by an adiabatic expansion of a saturated sample of air. In the earlier instruments individual droplets were counted and the technique was difficult and time consuming. Concentrations are now generally measured in photoelectric instruments of the type developed by Nolan and Pollak in which the density of the fog formed on expansion gives a measure of the concentration. The useful range of such instruments is from 25 to $2{\cdot}5 \times 10^{-5}$ nuclei/cm³.

The size of Aitken nuclei can be measured directly from electron microphotographs but this inevitably involves some delay and it is more usual to use indirect methods to give a more approximate but rapid size estimation. In the diffusion method the aerosol is passed through a diffusion battery and the nucleus concentration measured at the entrance and exit of the battery. From the ratio of these concentrations and the dimensions of the battery an "average" diffusion coefficient for the nuclei can be calculated and hence an average radius estimated. It has been shown experimentally that where the particles are spherical and monodisperse this method gives the correct diffusion coefficient and particle radius. Diffusion batteries are useful up to a nucleus radius of about 10^{-5} cm.

An alternative method of obtaining a rapid estimate of the average size of Aitken nuclei is to measure the fraction of the nuclei which is uncharged when the nuclei are at charge equilibrium. This is done by measuring the concentration of nuclei before and after they have passed through an ion trap designed to remove all charged particles. The method depends on the application of the Boltzmann energy distribution to the electrical energy possessed by the particles by virtue of their electric charge and while it is not strictly applicable theoretically it has been shown to give good results with nuclei up to 2×10^{-5} cm radius. The variation of fraction uncharged with particle radius is shown in the figure.
See also Diffusion battery.

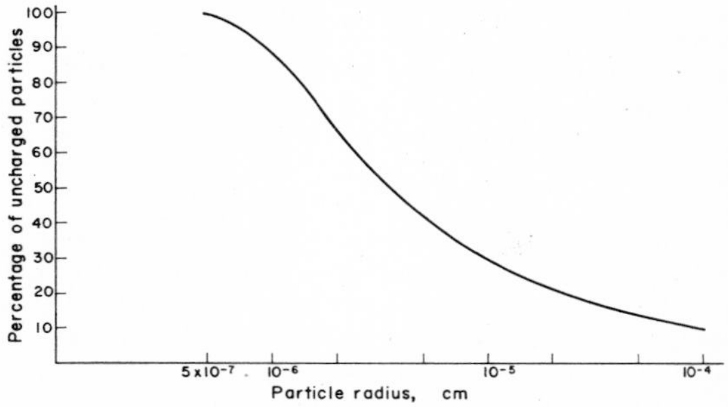

Percentage of uncharged particles at charge equilibrium vs. particle radius (assuming application of Boltzmann law).

Bibliography

Fuchs N. A. (1964) *The Mechanics of Aerosols*, Oxford: Pergamon Press.

Keefe D. Nolan P. J. and Rich T. A. (1959) *Proc. Roy. Irish Acad.* **60**, A 4, 27.

Megaw W. J. (1966) *J. de Recherches Atmosphériques* **2**, 2–3, 53.

Metnieks A. L. and Pollak L. W. (1959) *Inst. Adv. Studies Dublin, Geophysical Bulletin* No. 16.

W. J. Megaw

ANOMALOUS DISPERSION TECHNIQUES IN STRUCTURE ANALYSIS. Anomalous dispersion effects occur in both X-ray and neutron scattering, but are more important in the X-ray case where they have been used, both for the determination of the absolute configuration of optically active compounds and for the solution of the "phase problem". In a diffraction experiment, the intensities of the Bragg reflections are readily recorded, whereas the phases of the reflections are lost: anomalous dispersion provides the possibility of recovering the phases, and so arriving at a unique solution of the crystal structure by direct Fourier analysis.

When the wave-length of an incident X-ray beam is close to an absorption edge of an atom, the atomic scattering factor must be written as

$$f = f_0 + \Delta f' + i\Delta f''. \quad (1)$$

Here f_0 is the scattering factor for wave-lengths far from the absorption edge, and $\Delta f'$, $\Delta f''$ are corrections to the scattering factor arising from anomalous dispersion. The variation with wavelength of the real part, $\Delta f'$, of the correction and of the imaginary part, $\Delta f''$, are illustrated in Fig. 1. These correction terms are often small and can be ignored, but when the wave-length of the primary radiation is just below that corresponding to an absorption edge of the scattering atom, the term $\Delta f''$ can lead to important intensity effects, discussed below. The dispersion effect is associated with the resonant motion of the innermost electrons around the nucleus and so, to a very good approximation, both $\Delta f'$ and $\Delta f''$ are independent of

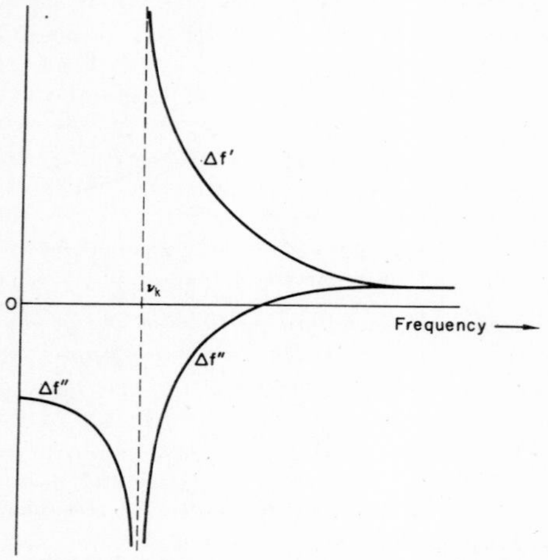

Fig. 1. Anomalous absorption effects near the K absorption edge, ν_K.

scattering angle θ. On the other hand, f_0, which is the Fourier transform of the total electron distribution, shows a form-factor dependence arising from the interference between the waves scattered by different parts of the total distribution.

If a crystal contains one set of atoms (N) which scatters X-radiation normally and one atom alone which scatters it anomalously, the amplitude-phase diagram for the hkl and $\overline{hkl}$ reflexions has the appearance shown in Fig. 2. $\Delta f''$ is always 90° ahead of the phase of $\Delta f'$, so that the resultant intensities F^2_{hkl}, $F^2_{\overline{hkl}}$ are different if two conditions are satisfied:

(i) the N atoms are distributed non-centrosymmetrically ($\alpha_N \neq 0, \pi$) and
(ii) $\Delta f'' \neq 0$.

This breakdown of "Friedel's law" was first demonstrated by Coster, Knol and Prins in the case of ZnS: by using AuLα radiation to excite anomalous scattering from Zn, they showed that the intensities of the 111 and $\overline{111}$ reflections were unequal. The inequality is related to the asymmetry of the sequence of Zn and S layers in the zinc blende structure. The layers are alternately close together and further apart, so that the structure can be envisaged as a succession of double layers, Zn—S. The Bragg intensity is larger if the X-radiation strikes the S-half of the double layer first, and smaller if it is incident on the Zn-half first (see Fig. 3).

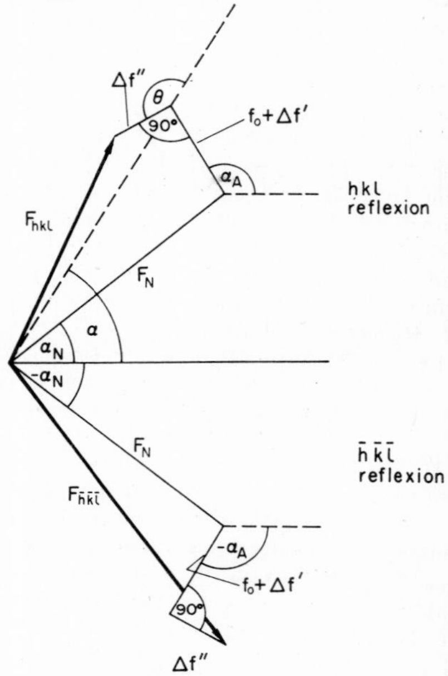

Fig. 2. Amplitude-phase diagram for scattering into hkl and $\overline{hkl}$ reflections, from a crystal containing normal atoms (N) and an anomalous scatterer (A).

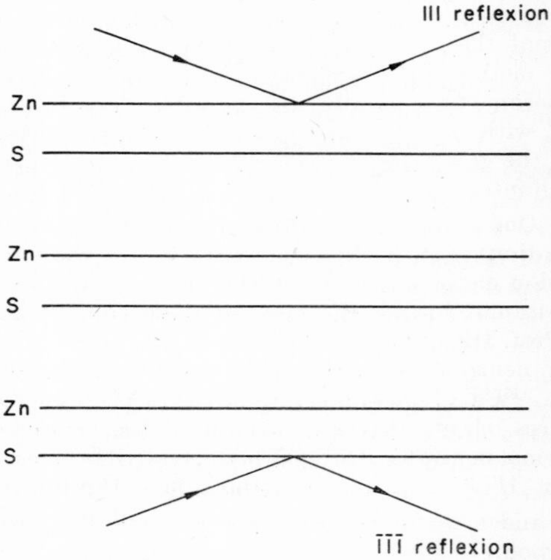

Fig. 3. Reflection of AuLα radiation from the (111) and ($\overline{111}$) faces of zinc blende. The intensity of the 111 reflection is less than that of $\overline{111}$.

The deeper implication that anomalous scattering leads to the possibility of distinguishing the enantiomorphous pairs of a non-centrosymmetrical molecule was not realized until many years later. The principle involved is the same as that used in the examination of ZnS, although zinc sulphide itself is its own enantiomorph. In 1949 Bijvoet determined the absolute configuration of dextro-rotatory tartaric acid from differences in the hkl and $\overline{\text{hkl}}$ intensities. He showed that the chemical convention for the configuration of the acid, first proposed by Emil Fischer in the nineteenth century, was the correct one. (There was a 50–50 chance that Fischer's convention was incorrect.) The diffraction method is the only effective means of establishing unambiguously the absolute configuration of a molecule.

To understand how the phases of reflections can be determined from anomalous scattering, we return to Fig. 2. By straightforward geometry, the angle θ is given by

$$\cos\theta \approx \frac{F^2_{\text{hkl}} - F^2_{\overline{\text{hkl}}}}{4\,|F_{\text{hkl}}|\,\Delta f''}$$

and the phase angle α by

$$\alpha = \alpha_{\text{A}} + \frac{\pi}{2} \pm \theta .$$

It is assumed in deriving these expressions that the intensities F^2_{hkl}, $F^2_{\overline{\text{hkl}}}$ differ only slightly so that α is equivalent to the required phase angle of the hkl reflexion. To determine the phase, α, it is necessary to know the intensities of hkl, $\overline{\text{hkl}}$ and the location of the anomalously scattering atom in the unit cell (i.e. α_{A}). The ambiguity in the phase angle, giving the choice of sign $\pm\theta$, can be resolved by various methods. In one method α is redetermined using a different incident wave-length: the two wave-lengths give four estimates of α, two of which should be equal and represent the correct estimate. The anomalous dispersion method for solving the phase problem is likely to become of increasing importance in structure analysis, with the development of counter diffractometers and the greater precision they allow in measuring the small intensity differences from Bijvoet pairs, hkl and $\overline{\text{hkl}}$. One of the most notable recent structure determinations, which relied entirely on the anomalous dispersion method, was that of Factor VIa, a derivative of vitamin B_{12}, carried out in Oxford by the school of Prof. Dorothy Hodgkin.

In neutron diffraction, there are only a few isotopes (^{6}Li, ^{113}Cd, ^{149}Sm, ^{157}Gd) which possess nuclear resonances close to the 1–2 Å wave-length range of thermal neutrons. For all other isotopes, the imaginary term $\Delta f''$ in (1) is zero, and the Bragg intensities from hkl and $\overline{\text{hkl}}$ are the same. The anomalous dispersion method, therefore, has much less scope in neutron diffraction as a technique for structure determination. It is possible that neutron anomalous dispersion could be used to determine the phases of the reflections from crystals of large biochemical molecules.

Bibliography

LIPSON H. and COCHRAN W. (1966) (3rd Edn) *The Determination of Crystal Structures*, London: Bell.

RAMACHANDRAN G. N. (1964) *Advanced Methods of Crystallography*, New York: Academic Press.

B. T. M. WILLIS

ARTIFICIAL EAR. An artificial ear is a device employed to calibrate earphones objectively for use in *audiometry* and *telephonometry*.

As a measure of sensitivity of the human ear it would be convenient to determine the minimum sound pressure at the eardrum which results in a sensation of hearing. However, it is virtually impossible to measure the sound pressure actually at the eardrum and as next best it is desirable to determine the minimum audible pressure at the entrance to the external auditory meatus and to use this as a measure of hearing acuity. Even so, only in sophisticated laboratory experiments on standard earphones is this practicable. In most cases hearing sensitivity is measured by applying an earphone to the ear under test; when the electroacoustical sensitivity of the earphone is known the sound pressure corresponding to threshold may be estimated from the measured threshold voltage. Thus an objective method of earphone calibration is indispensable in audiometry and this has, for many years, been a factor limiting the precision of measurements of hearing. Numerous attempts have been made to develop and standardize a procedure for determining the acoustical output of earphones using so-called artificial ears but the design of these objects has not been firmly based on valid data and few, if any, of these devices have been adequately tested or validated, although some may be adequate for comparing the electroacoustical sensitivities of earphones of one type. Extreme simplifications, e.g. a simple 6 cm³ cylindrical volume, are usually referred to as "*couplers*".

The acoustic source impedance of current earphones is of the same order as, or greater than, the load impedance presented by the human ear. Consequently for given electrical excitation the sound pressure generated by an earphone is strongly dependent on the *acoustic load*. Thus a fundamental factor in the design of an accurate artificial ear is a knowledge of the acoustical impedance presented to earphones by the average human ear. Although this was recognized many years ago, most of the data obtained on ear impedance have been confined to the restricted range of frequencies important for telephone communication and wide disparity of results precluded the extraction of reliable impedance values even over this range. Much of the discrepancy may be ascribed to leakage between earcap and ear, which particularly affects the impedance at frequencies below, kc/s. In audiometry one endeavours to minimize variability in sound pressure generated for given earphone excitation by controlling headband force and by careful positioning of earphone on the ear, thus logically it is the impedance of the human ear under audiometric (no leakage) conditions that is required. Only during the past few years have such data become available and been used as the basis of design of an artificial ear for use in audiometry.

Typical results for the impedance measured through the central aperture of an earcap made of hard material (type of earcap to the S.T.C. type 4026 A earphone) are shown in Fig. 1 as a function of frequency. Data on each of a range of practical earcaps have been obtained and the mean impedance contour below approximately 5000 c/s is conveniently represented in terms of the component values of an eight-element lumped-parameter electrical network analogue (Fig. 2). Basically a volume $V_1 = \gamma P C_1$ is shunted by a high resistance R_1, but below approximately 500 c/s non-rigidity of the pinna and surrounding flesh produces an increase in the effective enclosed volume of the ear cavity by an amount $V_2 = \gamma P C_2$. The resonance between 2 and 3 kc/s (L_3, C_3, R_3, in analogue) corresponds to a decrease in effective volume at higher frequencies due to "uncoupling" of the ear canal volume. Only with very soft rubber earcaps does the impedance contour become more complex.

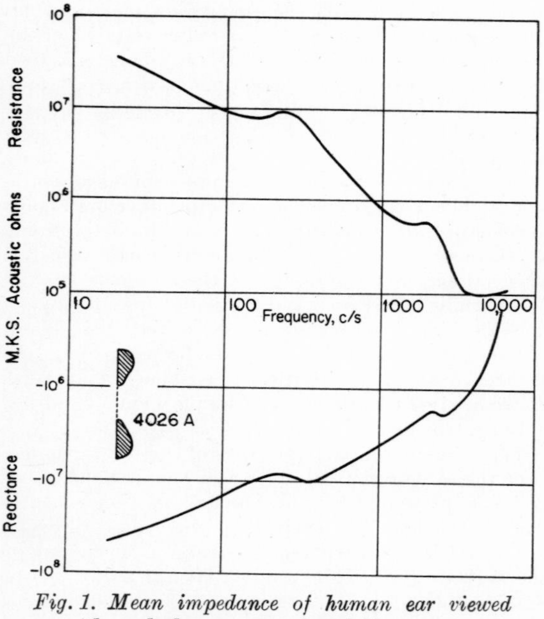

Fig. 1. Mean impedance of human ear viewed through the aperture of a rigid earcap.

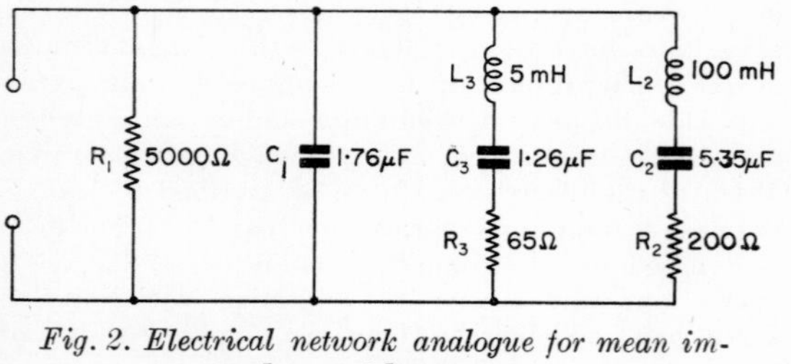

Fig. 2. Electrical network analogue for mean impedance of human ear.

An artificial ear must include a microphone to measure the acoustic pressure generated by the earphone under test, and a capacitor-type microphone is generally employed in view of its relatively uniform response over a wide range of frequencies. An example of a recently designed artificial ear, based on the analogue network shown above, is illustrated in Fig. 3. Validation studies have shown that this device calibrates various types of earphone commonly used in audiometry more nearly in accordance with the best subjective comparisons of sensitivity than earlier types.

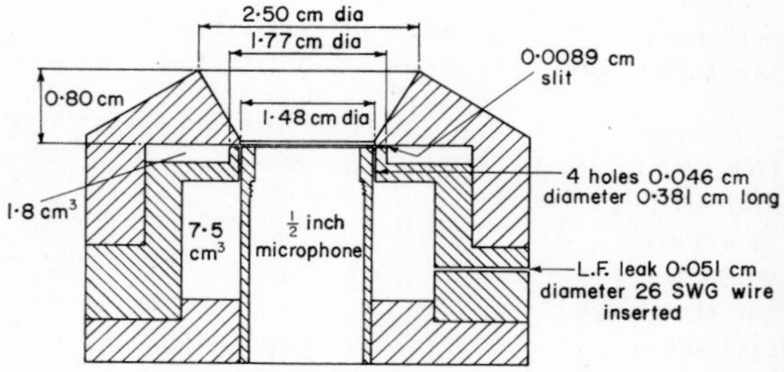

Fig. 3. Cross section through an artificial ear designed for use in audiometry.

Measurement of earphone sensitivity involves determining the voltage which must be applied to generate threshold sound pressure. Thus before an artificial ear can fulfil its purpose it is necessary to establish the sound pressure in the artificial ear which corresponds to the threshold of hearing in the average human ear; this is denoted the RETSPL (reference equivalent threshold sound pressure level). Values of RETSPL relating to five different combinations of reference earphone and artificial ear have been standardized by the International Organization for Standardization (R 389 : 1964) and further combinations are being added as values become available. The multiplicity of threshold data is a manifestation of the inadequacy of existing artificial ear types and when an accurate artificial ear is standardized a single set of reference threshold data will suffice for all types of earphone.

A precision artificial ear for use in audiometry should be ideally suited for the needs of telephonometry. For some applications it may prove necessary to incorporate an additional acoustic element to simulate the acoustic leakage associated with telephone usage and to apply a frequency-dependent correction in the associated electronic circuitry to obtain a correct representation of the subjective sensitivity of the earphone in question.

Increasing use is being made of circumaural earphones, which consist essentially of an acoustic source unit inside an earmuff which completely covers the pinna. No fully satisfactory calibration method has yet been devised for these and objective measurements are usually effected by sealing these earphones to a rigid plane in which a capacitor microphone has been arranged flush with the surface.

Earphones which are connected to an earmould to be inserted into the auditory meatus, as used in many air-conduction hearing aids, are usually calibrated according to a procedure recommended by the International Electrotechnical Commission. The earphone under test is connected by a tube 1·8 cm long and 0·3 cm diameter to a 2 cm³ coupler, one wall of which is formed by the diaphragm of a capacitor microphone. The use of this coupler does not allow the actual performance of the insert earphone on a person to be obtained but it does load the earphone with a specified acoustic impedance and permits the relative physical performance characteristics to be determined over the range 200–5000 c/s.

Reference equivalent threshold sound pressure level (*RETSPL*) for a specified combination of earphone and artificial ear. The sound pressure level developed in an artificial ear by an earphone of specified type when the latter is energized by the modal value of the threshold voltage for a group of otologically normal subjects, for that particular earphone.

Bibliography

Standard reference zero for the calibration of pure-tone audiometers (1964) *International Organization for Standardization*, Recommendation 389.

DELANY M. E. (1964) The acoustical impedance of human ears, *J. Sound Vib.* **1**, 455.

I.E.C. reference coupler for the measurement of hearing aids using earphones coupled to the ear by means of ear inserts (1961) I.E.C. Publication 126.

M. E. DELANY

ATOMIC ABSORPTION SPECTROSCOPY AND ITS APPLICATION TO CHEMICAL ANALYSIS. Bunsen and Kirchhoff showed in the middle of the 19th century that atomic spectra, either in emission or absorption, provided a powerful means of chemical analysis. Since that time emission methods have been steadily developed (see Wagner 1962; Harrison 1962), and have culminated in the direct-reading spectrographs now in widespread use, particularly for metallurgical analysis. The possibility of analytical methods based on atomic absorption spectra was almost completely ignored until the early 1950's, when A. Walsh pointed out the potential advantages of absorption over emission methods and devised simple and versatile apparatus applicable to the routine analysis of a wide range of elements.

A. Theoretical

In analytical methods based on atomic emission the sample for analysis is atomized and excited (in practice simultaneously), the emitted radiation dispersed, and the intensities of selected lines in the emission spectrum measured.

Neglecting self-absorption and induced emission, the integrated intensity of emission of a line is given by

$$\int I_\nu \mathrm{d}\nu = CN_j f, \tag{1}$$

where N_j is the number of atoms in the upper state involved in the transition responsible for the line, f is the oscillator strength of the line, and the proportionality constant C depends on the dispersing and detecting systems. Assuming that the atomic vapour is in thermal equilibrium at temperature T, then the number of atoms in an excited state j, of excitation energy E_j, is given by

$$N_j = N_0 \frac{P_j}{P_0} \exp(-E_j/\mathbf{k}T), \tag{2}$$

where N_0 is the number of atoms in the ground state and P_j and P_0 are the statistical weights of the excited and ground states respectively. Thus the emitted intensity depends on T and E_j, and examples of the variation of N_j/N_0 with these quantities are given in Table 1 for the resonance lines of various elements.

Consider now the case of absorption of a parallel beam of radiation of intensity $I_{0\nu}$ at frequency ν incident on atomic vapour of thickness l cm. If I_ν is the intensity of the transmitted beam, the absorption coefficient K_ν of the vapour at frequency ν is defined by

$$I_\nu = I_{0\nu} \exp(-K_\nu l), \tag{3}$$

Table 1. Values of N_j/N_0

Resonance Line, Å	Transition	P_j/P_0	N_j/N_0			
			T = 2000°K	T = 3000°K	T = 4000°K	T = 5000°K
Cs 8521	$^2S_{1/2} - {}^2P_{3/2}$	2	4.4×10^{-4}	7.2×10^{-3}	3.0×10^{-2}	6.8×10^{-2}
Na 5890	$^2S_{1/2} - {}^2P_{3/2}$	2	9.9×10^{-6}	5.9×10^{-4}	4.4×10^{-3}	1.5×10^{-2}
Ca 4227	$^1S_0 - {}^1P_1$	3	1.2×10^{-7}	3.7×10^{-5}	6.0×10^{-4}	3.3×10^{-3}
Zn 2139	$^1S_0 - {}^1P_1$	3	7.3×10^{-15}	5.6×10^{-10}	1.5×10^{-7}	4.3×10^{-6}

and the relationship between integrated absorption and concentration is given by

$$\int K_\nu d\nu = \frac{\pi e^2}{mc} N_\nu f, \tag{4}$$

where e and m are the electronic charge and mass respectively, c the velocity of light, and N_ν the number of atoms per cm^3 capable of absorbing in the range ν to $\nu + d\nu$. Table 1 shows that at temperatures up to 5000°K the number of excited atoms is in most cases negligible, and thus in equation 4, N_ν can be replaced by N, the total number of atoms per cm^3.

There is no satisfactory way of determining the integrated absorption coefficient by conventional scanning methods, owing to the small width (0·02–0·1 Å) of atomic absorption lines under normal conditions, and it is usual to measure the absorption coefficient at the centre of the line, using an atomic line-source which emits lines having a much smaller half-width than the absorption line. Under these conditions the peak absorption coefficient is linearly proportional to concentration, providing the shape and width of the line are independent of concentration. It is not generally practicable to use a continuous source of radiation, as the resolution required to resolve a sufficiently narrow wave-length band (width ~0·01 Å) is beyond that achievable with an ordinary monochromator.

B. Apparatus

An atomic absorption spectrophotometer thus consists of:

(1) A light source emitting, under conditions which ensure the production of extremely sharp lines, the spectrum of the element to be determined.

(2) A means of producing an atomic vapour of the sample to be analysed.

(3) A wave-length selector to separate the resonance line required from any other lines which may be emitted by the source.

(4) A detector, amplifier, and readout system.

Light sources. For the alkali metals and a few others spectral vapour lamps, normally under-run, are suitable, but for most metals hollow-cathode lamps are required. With a few metals, such as iron, nickel and cobalt, the presence of another neutral atom or ion line close to the resonance line may decrease the sensitivity somewhat and lead to undue curvature of the absorbance-concentration graph. In such cases use of a high-intensity hollow-cathode lamp is advantageous.

Production of an atomic vapour of the sample. Although solid materials can be directly vaporized by electric arc or cathodic sputtering techniques, the usual method involves spraying a solution of the sample into a flame, in a manner somewhat similar to that employed in flame photometry. For alkali metals and many heavy metals (Pb, Ag, Cu, Zn, Sb, etc.) an air-coal gas or air-propane flame (temperature ca. 1800°C) is suitable, but the alkaline earth metals, chromium, and molybdenum need an air-acetylene flame (2200°C), while metals forming highly refractory oxides (Be, Al, V, Ti, Si, Zr, W, etc.) require a nitrous oxide-acetylene flame (2950 °C). The solution is aspirated by the air or nitrous oxide and the resultant spray mixed with the combustible gas in a spray-chamber, where the larger droplets settle out, leaving only the finest ones to be carried forward to the burner where the mixture is burned. The burner port is normally 5–10 cm long and 0·05–0·15 cm wide.

Wave-length selector. For the alkali metals interference filters are adequate, but for other metals a monochromator is necessary, though the resolving power need not usually be better than 2–5 Å except where metals with very complex spectra are to be determined. For many elements the monochromator can be replaced by a resonance detector (see *High-intensity hollow cathode lamps*).

Detector, amplifier, and readout system. For the alkali metals a conventional photocell is a satisfactory detector, but in general a photomultiplier is required. The radiation from the sharp-line source is modulated and the signal from the detector is fed to an a.c. amplifier (preferably tuned to the modulation frequency of the light source), so that any radiation emitted by the flame gives rise at the output of the detector to a d.c. signal, which is not amplified. The readout consists of a microammeter, galvanometer, or recorder.

The figure shows schematically a single-beam atomic absorption spectrophotometer.

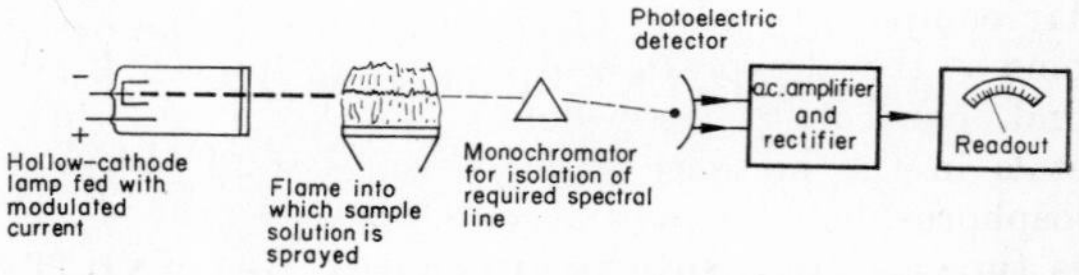

Fig. 1.

C. Application to Chemical Analysis

Scope and limitations. The atomic absorption method can be applied in principle to the determination of any element for which a source of resonance radiation is available and for which an atomic vapour can be produced from the analytical sample. Since most non-metals have their resonance lines in the vacuum ultraviolet (i.e. below 2000 Å), special difficulties arise owing to the strong absorption of these wave-lengths both by the oxygen of the air and by the flame gases, and as yet few non-metals can be conveniently determined by the atomic absorption technique. The very high resolution afforded by the use of sharp-line sources makes it possible to carry out isotopic analysis for metals, such as lithium and uranium, where the separation between the resonance lines of the isotopes is greater than their width in absorption.

The limit of detection for a given metal is determined by the ability to measure the small change in the intensity of the line from the sharp-line source caused by absorption by the metal atoms in the flame, which in practice is set by the noise level of the sharp-line source and of the detector-amplifier system. The sensitivity of the method is conventionally defined as the concentration of metal required to produce one per cent absorption when the solution is sprayed into a flame of standard length, usually 10 cm. This sensitivity varies slightly from one instrument to another, depending chiefly on the efficiency of conversion of the metal in solution to an atomic vapour in the flame.

Interferences. Spectral interference, due to overlapping of the lines of different elements, is very rare in atomic absorption spectroscopy, since the absorption line of the metal in the flame is only 0·02–0·1 Å wide, and the absorption at the peak of this line is measured with a line not more than about 0·01 Å wide emitted by the hollow-cathode lamp.

Ionization interference, i.e. loss of neutral atoms by ionization, to an extent which depends on the presence of other easily ionized metals, is found when determining alkali metals even in low-temperature flames, and with any metals having ionization potentials less than about 6·5 eV when using the nitrous oxide-acetylene flame. This interference is readily eliminated by the addition to the analytical sample of a high concentration of the salt of a readily ionized metal such as potassium.

Chemical interference, due to the formation of molecular compounds which are not fully dissociated to atoms at the temperature of the flame, is sometimes found, particularly when determining alkaline earth metals in the presence of elements such as silicon, phosphorus, sulphur, and aluminium. Interference of this type can frequently be overcome by the addition of a large excess of a metal which can compete with the metal being determined for combination with the interfering material, e.g. in the determination of calcium in the presence of phosphorus addition of a large excess of strontium liberates the calcium from combination with the phosphorus and enables its concentration to be measured accurately. Many such interferences can be reduced or eliminated by use of a higher temperature flame.

In the measurement of very low concentrations of one metal in the presence of a high concentration of other material a small background effect may arise from molecular absorption or scattering of light by the bulk of the dissolved material. Such losses are corrected for where necessary by measurement of the background absorption at an adjacent non-resonance line.

Preparation of analytical samples. Aqueous solutions may be sprayed as such or after appropriate dilution with water. Solutions in oils or other non-aqueous liquids may usually be sprayed after dilution with a solvent such as methyl isobutyl ketone. Plant and animal tissues are normally ashed by one of the standard wet- or dry-ashing techniques and a solution of the ash in hydrochloric acid is made to volume with water. Metals and alloys are dissolved in acid or alkali and made to volume with water. Trace elements in aqueous solutions can be concentrated by extraction with a chelating agent into an immiscible solvent and this solution sprayed into the flame. The usual microchemical precautions against contamination must of course be observed.

Calibrating solutions. When the metal being determined forms a significant fraction of the whole sample being analysed it is frequently sufficient to make up calibrating solutions containing only this metal, usually as a simple salt dissolved in water or dilute acid. When, however, the metal to be determined occurs as a minor or trace constituent, the viscosity and efficiency of atomization may be different for the sample and calibrating solutions. In this latter case it is usually desirable to add to the calibrating solutions the appropriate concentration of the major constituent.

The most usual method of calibration is to measure three or four calibrating solutions containing known concentrations of the metal to be determined and covering the concentration range expected in the sample solutions. The absorbances of these solutions are plotted against their concentrations and the sample concentrations read off by interpolation.

Sensitivity. In Table 2 are shown the wave-lengths and approximate sensitivities for a number of metals using a typical commercial single-beam instrument. The concentrations quoted are those giving one per cent absorption when sprayed as aqueous solutions into the flames indicated.

Applications. The principal fields of application to date have been:

Biology and Medicine: determination of sodium, potassium, calcium, magnesium, and trace heavy metals in serum, urine, and tissue.

Agriculture: determination of major, minor, and trace constituents in soils, fertilizers, and plant materials.

Geology and geochemistry: determination of trace metals in rocks, soils, and stream sediments.

Mining and mineral processing: analysis of drill samples, ores, and concentrates.

Metallurgy: determination of minor and trace constituents in ferrous and non-ferrous alloys.

Petroleum industry: determination of trace metals in crude oils, feedstocks, and petroleum, and of wear metals in lubricating oil.

Miscellaneous: determination of trace contaminants in beverages; analysis of electroplating solutions.

Table 2. Sensitivities for metals determined by atomic absorption spectroscopy

Metal	Spectral line, Å	Flame*	Concentration μg/cm³	Metal	Spectral line, Å	Flame*	Concentration μg/cm³
Li	6707·8	C	0·02	Yb	3988·0	N	0·25
Na	5890·0	C	0·02	Lu	3359·6	N	12
K	7664·9	C	0·03	Si	2516·1	N	2·5
Rb	7800·2	C	0·04	Ge	2651·6	N	1·5
Cs	8521·1	C	0·15	Sn	2246·1	A	2
Cu	3247·5	C	0·07	Pb	2170·0	C	0·4
Ag	3280·7	C	0·06	Ti	3642·7	N	3·5
Au	2428·0	C	0·2	Zr	3601·2	N	15
Be	2348·6	N	0·02	Hf	3072·9	N	14
Mg	2852·1	A	0·008	As	1937·0	A	1·5
Ca	4226·7	N	0·03†	Sb	2175·8	C	1
Sr	4607·3	N	0·06†	Bi	2230·6	C	0·4
Ba	5535·5	N	0·4†	V	3184·0	N	1·5
Zn	2138·6	C	0·02	Nb	3580·3	N	24
Cd	2288·0	C	0·02	Ta	2714·0	N	11
Hg	2536·5	C	5	Se	1960·3	C	0·6
B	2497·7	N	50	Te	2142·8	C	0·3
Al	3092·7	N	1	Cr	3578·7	A	0·15
Ga	2874·2	A	2	Mo	3132·6	N	0·4
In	3039·4	C	0·4	U	3584·9	N	120
Tl	2767·9	C	0·5	W	2551·4	N	5
Sc	3911·8	N	0·8	Re	3460·5	N	12
Y	4102·4	N	5	Fe	2483·3	A	0·1
La	3574·4	N	110	Co	2407·3	A	0·1
Pr	4951·4	N	72	Ni	2320·0	C	0·1
Nd	4634·2	N	35	Ru	3498·9	A	0·9
Sm	4296·7	N	21	Rh	3434·9	A	0·4
Eu	4594·0	N	2	Pd	2476·4	C	0·2
Gd	3684·1	N	38	Os	2909·1	N	1
Dy	4211·7	N	1·5	Ir	2088·8	A	3
Ho	4103·8	N	2	Pt	2659·5	A	2·5
Er	4008·0	N	1·5				

* C — Air-coal gas flame, 10 cm burner.
A — Air-acetylene flame, 10 cm burner.
N — Nitrous oxide-acetylene flame, 5 cm burner.
† Ionization suppressed with potassium chloride.

Bibliography

ELWELL W. T. and GIDLEY J. A. F. (1966) (2nd Edn.) *Atomic Absorption Spectrophotometry*, Oxford: Pergamon Press.

HARRISON J. A. (1962) in *Encyclopaedic Dictionary of Physics* (J. Thewlis Ed.), **6**, 641, Oxford: Pergamon Press.

ROBINSON J. W. (1966) *Atomic Absorption Spectroscopy*, New York: Marcel Dekker.

WAGNER W. F. (1962) in *Encyclopaedic Dictionary of Physics* (J. Thewlis Ed.), **6**, 707, Oxford: Pergamon Press.

WALSH A. (1955) *Spectrochim. Acta* **7**, 108.

WALSH A. and WILLIS J. B. (1966) *Atomic Absorption Spectrometry*, in Vol. III A of *Standard Methods of Chemical Analysis* (Ed. F. J. Welcher), Princeton: Van Nostrand.

J. B. WILLIS

ATOMIC CHARGE DENSITY, ANTISYMMETRIC FEATURES OF. In using the methods of X-ray and neutron diffraction to study properties of crystalline solids, the information inherent in Bragg scattering falls broadly into two categories: (a), where the constituent atoms are located in the space- and time-averaged version of the unit cell representative of a bulk experiment on a system displaying both zero-point and thermal vibration behaviour; and, (b), information on more subtle aspects of the individual atoms which relates either to the nature of their electronic charge distributions or to the nature of the vibrational excursions which they make about the equilibrium positions that would typify a static crystal structure. For non-magnetic structures, the standard kinematical theory of X-ray structure analysis can be applied with minor modifications to neutron structure analysis, and the first aim in either case is usually with matters such as molecular dimensions (bond lengths and angles) which belong to (a). Achieving this aim involves making simplifying assumptions about the atomic details of (b) which are relevant to each diffraction method, and the actual study of these details, which can be termed details of atomic charge density, is usually regarded as a quite separate, and subsequent, sphere of investigation to be undertaken only in cases where experimentation on suitable systems has yielded Bragg data of very high quality. This sort of tactical separation of aims is very convenient but it is now emerging that, if carried too far, neither aim can be prosecuted successfully to the point where the accuracy of the final information is commensurate with modern developments of experimental practice. The relevance of such information to the fields of theoretical chemistry and solid state physics requires that possible interactions between (a) and (b) be viewed as an integral part of attempts to extract maximum information from these powerful diffraction methods. The basic problems lie in (b), and present evidence is that an appreciation of antisymmetric features of atomic charge density is intimately involved with such diverse phenomena as covalent bonding and vibrational anharmonicity.

The general situation is shown by the structure factor for Bragg X-ray scattering. With radiation whose wave-length (λ) is sufficiently short to avoid complications of anomalous dispersion which arise if $\lambda \to \lambda_i$, where the latter define the absorption edges of an atom, the amplitude of coherent scattering $F(\mathbf{S})$ observed in the directions permitted by Bragg's law is related to the distribution of electron density, $\varrho(\mathbf{r})$, at all points in the unit cell by the simple Fourier transform

$$F(\mathbf{S}) = \int \varrho(\mathbf{r}) \exp\{2\pi i\, \mathbf{S}\cdot\mathbf{r}\}\, d\mathbf{r}. \qquad (1)$$

Here, $\mathbf{S}$ is the scattering vector of magnitude $S = |\mathbf{S}| = 2\sin\theta/\lambda$, where 2θ is the angle between the incident and scattered beams, so that $\mathbf{S}$ defines the directions (in reciprocal space) that correspond to the normals to the planes of spacing d (in real space) that enter Bragg's law, $n\lambda = 2d\sin\theta$. Equation (1) assumes a more useful form on expressing $\varrho(\mathbf{r})$ in terms of nuclear positions, $\mathbf{r}_j$, in the unit cell as

$$\varrho(\mathbf{r}) = \sum_j \varrho_j'(\mathbf{r} - \mathbf{r}_j), \qquad (2)$$

i.e. as a series of "atomic" distributions, in which case

$$F(\mathbf{S}) = \sum_j f_j'(\mathbf{S}) \exp\{2\pi i\, \mathbf{S}\cdot\mathbf{r}_j\} \qquad (3)$$

where

$$f_j'(\mathbf{S}) = \int \varrho_j'(\mathbf{r}') \exp\{2\pi i\, \mathbf{S}\cdot\mathbf{r}'\}\, d\mathbf{r}', \qquad (3a)$$

the vectors $\mathbf{r}'$ being referred to the different origins defined by the $\mathbf{r}_j$. In the general case where the $\varrho_j'(\mathbf{r}')$ do not possess the property of centrosymmetry about the reference $\mathbf{r}_j$, their expression in the general form

$$\varrho_j'(\mathbf{r}') = \varrho_{c,j}'(\mathbf{r}') + \varrho_{a,j}'(\mathbf{r}') \qquad (4)$$

implies that, in (3),

$$f_j'(\mathbf{S}) = f_{c,j}'(\mathbf{S}) + i f_{a,j}'(\mathbf{S}). \qquad (5)$$

The subscripts c, a in (4) define components which possess, respectively, centrosymmetry or antisymmetry about the $\mathbf{r}_j$: the simplest illustration of these terms is provided by the behaviour of the functions $\cos x$ and $\sin x$ about $x = 0$, the first showing even (centrosymmetric) behaviour whereas the second shows odd (antisymmetric) behaviour.

The situation described by (3) and (5) refers to vibrating atoms which, for Bragg scattering, can be viewed as static atoms whose intrinsic features have been smeared out over a greater volume in space because of zero-point and thermal motion phenomena: subtleties of such phenomena concerned with the degree of coupling between neighbouring atoms are not immediately relevant here, and each atom can be treated in terms of smearing functions which define the probability of atom-displacement away from the equilibrium position. In the customary "rigid-atom" approximation, each $\varrho_j'(\mathbf{r}')$ of (3a), (4) can then be regarded as the convolution of the static atom, $\varrho_j(\mathbf{r}')$, with the nuclear smearing function, $t_j(\mathbf{r}')$, and thus expressed as

$$\varrho_j'(\mathbf{r}') = \varrho_j(\mathbf{r}') * t_j(\mathbf{r}'),$$

where $*$ denotes convolution. The first consequence of this step is that the $\mathbf{r}_j$ are to be taken as defining equilibrium atomic positions, and also that (3) can be re-written as

$$F(\mathbf{S}) = \sum_j f_j(\mathbf{S})\, T_j(\mathbf{S}) \exp\{2\pi i\, \mathbf{S}\cdot\mathbf{r}_j\} \qquad (6)$$

where $f_j(\mathbf{S})$ is the unprimed equivalent of (3a) and $T_j(\mathbf{S})$ is the analogous transform involving $t_j(\mathbf{r}')$. The quantities $f_j(\mathbf{S})$ and $T_j(\mathbf{S})$ are, respectively, the at-rest atomic scattering powers and thermal vibration factors. Following (4) and (5), separation of centrosymmetric and antisymmetric components next leads to

$$f_j(S) = f_{c,j}(\mathbf{S}) + i f_{a,j}(\mathbf{S}), \quad T_j(\mathbf{S}) = T_{c,j}(\mathbf{S}) + i T_{a,j}(\mathbf{S}), \qquad (7)$$

so that, even for a centrosymmetric space group where $F(\mathbf{S}) = A(\mathbf{S}) + iB(\mathbf{S})$ with $B(\mathbf{S}) = 0$, the X-ray structure factor becomes

$$F(\mathbf{S}) = \sum_j [\{f_{c,j}(\mathbf{S})\, T_{c,j}(\mathbf{S}) - f_{a,j}(\mathbf{S})\, T_{a,j}(\mathbf{S})\} \cos 2\pi \mathbf{S} \cdot \mathbf{r}_j - \{f_{c,j}(\mathbf{S})\, T_{a,j}(\mathbf{S}) + f_{a,j}(\mathbf{S})\, T_{c,j}(\mathbf{S})\} \sin 2\pi \mathbf{S} \cdot \mathbf{r}_j]. \quad (8)$$

This shows that, in general, two types of antisymmetric charge density phenomena (obvious adjectives for which are electronic and nuclear (vibrational)) may be manifested in an accurate X-ray Bragg experiment. For the corresponding neutron experiment on a non-magnetic structure, the situation is simpler since now the complicated f-terms of (8), which are a function of **S** and thus angle- and orientation-dependent, are replaced by scattering lengths ($\bar{b}_j$) which are isotropic and angle-independent. The neutron analogue of (8) is thus

$$F(\mathbf{S}) = \sum_j \bar{b}_j\{T_{c,j}(\mathbf{S}) \cos 2\pi \mathbf{S} \cdot \mathbf{r}_j - T_{a,j}(\mathbf{S}) \sin 2\pi \mathbf{S} \cdot \mathbf{r}_j\}, \quad (9)$$

so that here features of vibrational antisymmetry can be studied in a more clear-cut fashion. In both types of experiment, however, it is clear that the information extracted on a particular parameter will depend on the treatment accorded to other parameters with which it is intimately associated in the $F(\mathbf{S})$ data.

The point of this discussion emerges when the customary conduct of structure analysis is compared with two quite basic properties of assemblages of bonded atoms. In normal practice, the implementation of aim (a) with X-ray data involves two assumptions: (i), that $\bar{f}$-values from theoretical (e.g. Hartree–Fock) models of isolated atoms, taken in spherically averaged form so that the values depend only on S, can be used as an adequate substitute for the $f_j(\mathbf{S})$ of (6); and (ii), that the $T_j(\mathbf{S})$ of (6) can be approximated by the Gaussian ellipsoids which correspond to the Gaussian forms of $t_j(\mathbf{r}')$ that arise on assuming that the atomic vibrations are harmonic in character. The simplest example of (ii) occurs when isotropic harmonic vibration is assumed, in which case

$$T(\mathbf{S}) = \exp(-\bar{B}S^2/4), \quad \bar{B} = 8\pi^2 \langle u_S^2 \rangle \quad (10)$$

where $\langle u_s^2 \rangle$ is the mean-square vibrational displacement in any direction. Both (i) and (ii) are legitimate starting points for investigating the $\mathbf{r}_j$, but their limitations are evident in the fact that they correspond to writing (8) and (9) as

$$\begin{aligned} F(\mathbf{S}) &= \sum_j \bar{f}_{c,j}(\mathbf{S})\, T^h_{c,j}(\mathbf{S}) \cos 2\pi \mathbf{S} \cdot \mathbf{r}_j \\ F(S) &= \sum_j \bar{b}_j T^h_{c,j}(\mathbf{S}) \cos 2\pi \mathbf{S} \cdot \mathbf{r}_j, \end{aligned} \quad (11)$$

where the h-superscript denotes the harmonic model.

All antisymmetric atomic characteristics are thus being ignored, and the attraction of this simplification must be balanced against its real relevance to scattering from an aggregate of bonded atoms. In cases where covalent bonding phenomena predominate the environment of each bonded atom is, except in rare instances, non-centrosymmetric. In consequence, the centrosymmetric restrictions of (i) and (ii) must ignore here the very essence of the situation they are attempting to treat. For cases where ionic bonding phenoma predominate, more instances exist where an atomic environment is centrosymmetric, but otherwise the same sentiments apply. The general situation that must be recognized is one of atomic non-centrosymmetry so that, unless special environments indicate otherwise, antisymmetric components of atomic charge density must be regarded as an integral part of detailed analysis of accurate diffraction data.

Although intensive study of antisymmetric phenomena has only commenced recently, it is already clear that such phenomena provide the key to extracting more information from structural investigations. The greater activity to date has been concerned with the X-ray method, not merely because this method is the more traditional of the two but rather because of the unique opportunities it offers for extended electron distribution studies, and attention has been focused on the f-terms of (8) and the $\bar{f}$-approximation of (i) above. Two distinct questions arise here: firstly, how does the spherical assumption for the isolated atom accord with the "prepared" version of the atom which we can imagine exists immediately prior to its transformation to a bonded atom?; and secondly, how does the transformation itself modify conclusions that may be drawn from examining the first question? Each question is relevant to theoretical chemistry, and both must be asked if antisymmetric phenomena at the "prepared" level are to be separated from those at the ultimate bonded level. (For example, the "prepared" version of the quadrivalent carbon atom would be regarded as involving the electronic configuration $(1s)^2\,(2s)\,(2p)^3$ rather than the ground configuration $(1s)^2\,(2s)^2\,(2p)^2$.)

The first question has been studied for the atoms N, O and F which possess lone pairs of electrons as well as bonding electrons in their valence shells and are thus important participants in hydrogen-bonding phenomena. Simple calculations based on the concept of orbital hybridization show that, for all three atoms, the charge distributions associated with sp^3, sp^2 or sp "prepared" states are usually markedly non-spherical and, in many cases, non-centrosymmetric so that the general description of their scattering power is of the form (7) above. In these circumstances, persistence with the $\bar{f}$-approximation and the presumption that (11) is an adequate replacement for (8) leads to the type of situation shown in Fig. 1. This illustrates features of the sp^3 "prepared" version of N, which we can envisage as the precursor to the bonded form of this atom as it exists in molecules such as ammonia.

Figure 1a shows the general disposition of non-spherical "prepared" features which could be manifested in a typical X-ray experiment with CuKα radiation when the atom is undergoing isotropic harmonic vibration, with $\overline{B} = 2{\cdot}0$ Å^{-2} in (10), about the equilibrium nuclear position given by $x = y = 0$ in the figure: the

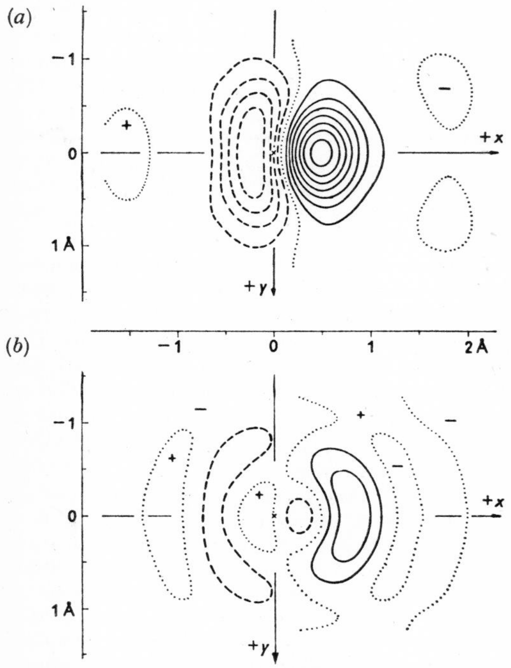

Fig. 1. Non-spherical features of electron density typical of the sp^3 "prepared" N atom discussed in the text. Contours are at 0·1 eÅ^{-2} intervals, with excess density shown as full lines and deficits as broken lines. The upper half (a) shows the disposition about the nuclear centre while the lower half (b) shows how these features can be eliminated by unsuitable structure refinement.

atom is oriented with the lone pair of electrons projecting outwards along the $+x$ direction, away from the bonded neighbours (H in the simple case of NH_3) which lie on the $-x$ side, so that the view corresponds to that along the pseudo-trigonal axis possessed by N in the more general case when it is trivalently bonded in geometry essentially similar to that in NH_3. The non-spherical features of Fig. 1a arise principally from the antisymmetric component in the charge density of this "prepared" atom, and there is an appreciable excess of charge density on the side of the atom away from its bonded neighbours. In such a situation the normal process of $\overline{f}$-refinement, either by Fourier or least squares processes, must lead ultimately to the stage illustrated by Fig. 1b. The "true" features of Fig. 1a are now markedly attenuated, and the residual features of Fig. 1b are all that are left when the spherical assumption is allowed to shift the atom outwards in the $+x$ direction by ca. 0·02–0·03 Å and to modify its original isotropic $\overline{B}$-parameter into an anisotropic harmonic version corresponding to a greater amplitude of vibration parallel to the x-axis than normal to it. Both these effects—the atomic shift arising from neglecting the antisymmetric component and the vibration modification arising from the fact that the centrosymmetric component of the "prepared" atom is prolately spheroidal rather than spherical—would be judged as legitimate features of the $\overline{f}$-refinement in terms of the statistical criteria which are used to judge parameter-reliability, and yet it is patently clear that both effects are false. In such a circumstance, neither of the aims (a) or (b) noted at the outset would be realized, and this simple example enables the general conclusion to be drawn that exclusive use of the $\overline{f}$-approach may in fact conceal the very structural features one wishes to define.

How this conclusion is modified by the fact that diffraction data involve bonded rather than "prepared" atoms is the second question earlier, and there is no answer with the overall simplicity of that above. Nevertheless, sufficient evidence is now available to permit the general form of this answer to be at least defined. In the field of covalent (organic) structures, the ubiquitous carbon atom has the invaluable attribute that its "prepared" distribution is spherically symmetric, except in rare instances, so that antisymmetric components associated with its non-centrosymmetric environment must be ascribed solely to bonding redistribution phenomena. The classic covalent structure is diamond, where the environment of each atom is tetrahedral (T_d) and the equilibrium atomic positions in the unit cell are known, so that detailed analysis of accurate X-ray data conducted in terms of (8) should reveal how this bonded atom differs from its "prepared" counterpart. At room temperature (where vibrational effects are so small that $\overline{B} = 0{\cdot}200$ Å^{-2} and the T_a-components of (8) can be ignored), it emerges that the bonded atom has the form of (4) which, on omitting the prime from $\mathbf{r}'$, can be written in further detail as

$$\varrho'(\mathbf{r}) = \overline{\varrho}_c(r) + \delta\varrho_c(\mathbf{r}) \pm \varrho_a(\mathbf{r})$$

with

$$\delta\varrho_c(\mathbf{r}) = Lr^2 \exp(-\alpha r^2)\,\{[(x^4 + y^4 + z^4)/r^4] - (3/5)\} \qquad (12)$$

$$\varrho_a(\mathbf{r}) = Kr^2 \exp(-\alpha r^2)\,\{xyz/r^3\}$$

$$K = 7{\cdot}5, \quad L = -2{\cdot}0, \quad \alpha = 2{\cdot}2\,\text{Å}^{-2},$$

where the terms in curly brackets are angular factors (Kubic Harmonics) satisfying the overall T_d symmetry, x, y and z being the components of r along the cubic axes of the unit cell: the $\pm$ alternatives apply to the two tetrahedral orientations which occur in the diamond structure. The "prepared" atom is simply the $\overline{\varrho}_c$-

component of (12), which is found to be described accurately by Hartree–Fock results for the lowest ground state of the isolated atom, and the non-spherical components provide a detailed account of all bonding phenomena evident in the X-ray data, including the "forbidden" reflection 222. The resultant picture to emerge of covalent bonding along each C—C bond is shown in Fig. 2, and further detail is given in Fig. 3. Both show the dominant part played by the antisymmetric atomic components in producing a build-up of electron density around the bond centre.

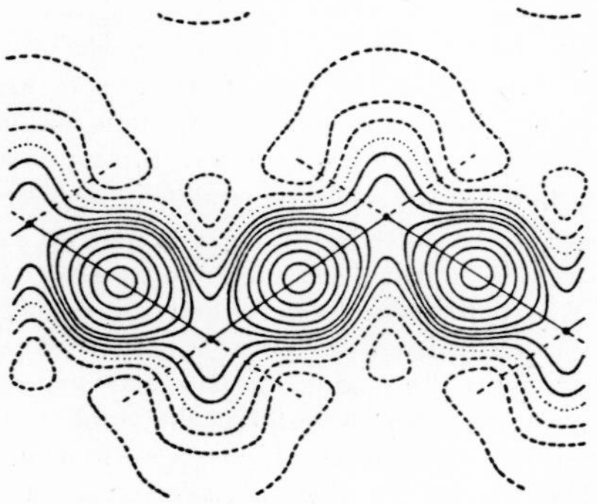

Fig. 2. The non-spherical features of electron density, in ($\bar{1}10$) of diamond, associated with covalent bonding in this structure at room temperature. Excess density is shown by full lines drawn at contours corresponding to 0·05, 0·10, 0·20, 0·30 etc. eÅ^{-3} above the zero level denoted by the dotted line. Deficits of density (compared with what applies to the spherical unbonded atoms) are shown by broken lines at the intervals —0·05, —0·10 eÅ^{-3}.

Similarly detailed evidence of electron redistribution phenomena has also been obtained for silicon, where (12) again applies with $K = 1{\cdot}11$, $L = -0{\cdot}32$, $\alpha = 0{\cdot}88$ Å^{-2} at room temperature (where now $\bar{B} = 0{\cdot}444$ Å^{-2}), so that both studies demonstrate unequivocally the sort of information required in answering the second question above. Supplementary information has also long been implicit in the fact that X-ray $\bar{f}$-studies of H-atoms invariably place them closer to their bonded neighbours (C, N or O) than do neutron studies, the different results being essentially a consequence of the $\bar{f}$-approach failing to recognize the antisymmetric component which is impressed on the spherical "prepared" H-atom by purely bonding phenomena. The general evidence is thus that antisymmetric atomic components of the "unprepared" type will generate excess features between bonded atoms possessing non-centrosymmetric environments. This conclusion, when used as the basis for prediction governed by symmetry considerations alone, has already proved to be a valuable means of interpreting non-spherical features of electronic charge density that have been evident for some time in accurate X-ray studies of quite complex organic molecules.

It is thus clear that, in an atom like N discussed earlier, situations can exist where antisymmetric phenomena at the "prepared" level may be opposed by similar phenomena generated at the ultimate bonded level: whereas the example provided by Fig. 1 shows that the $\bar{f}$-approach will shift the atom away from its neighbours, the stage of proceeding from the "prepared" to the bonded atom will tend to reverse this trend. Beyond saying this, however, it is not yet possible to be more precise since the details of each case will depend on specific factors such as the environmental geometry involved, the angular variations (about the nuclear centre) of the non-spherical features at the "prepared" and bonding levels, and the actual amount of electron redistribution that is responsible for each set of features. These are factors which still need to be defined, and one most promising way to avoid $\bar{f}$-ambiguities in the X-ray method is by a parallel neutron study where the definition of atomic positions depends on the $\bar{b}_j$ instead. Detailed correlation of position and thermal vibration parameters determined by the two methods can then be used to resolve the added subtleties of non-spherical electronic charge density that are present in X-ray data.

The role of neutron diffraction is by no means a subordinate one, however: instead, the non-intervention of f-considerations means that attention can be focused here on the adequacy of assumption (ii) concerned with the T-factors of (8), especially since the point-atom character of the $\bar{b}_j$ also means that neutron observations at large S-values suffer none of the attenuation of X-ray data produced by the general nature of $f_j(S)$. In approaching the vibrational problem, the obvious point to bear in mind is the fact that the potential energy curve of a diatomic molecule is markedly asymmetric about the minimum denoting the equilibrium atomic separation, and that vibrational excursions which increase the separation are energetically easier than those which diminish it. In other words, the smearing function $t_j(\mathbf{r}')$ of each atom in the vibrating molecule is non-centrosymmetric, in accord with the similar character of its environment, so that an antisymmetric component of nuclear charge density is present. This example cannot be pushed too far in contemplating solids, since the alkali halides are the classical instance of solids with atoms in centrosymmetric environments: here the final outcome for the $t_j(\mathbf{r}')$ must also be centrosymmetric, and this will now be the salient feature (for Bragg scattering) if the grand summation over the crystal is performed in the usual way over individual pairs of atoms whose interatomic potentials have similar form to that of the isolated diatomic molecule. However, these considerations do allow one to anticipate that, in simple solids containing atoms in non-centrosymmetric sites, situa-

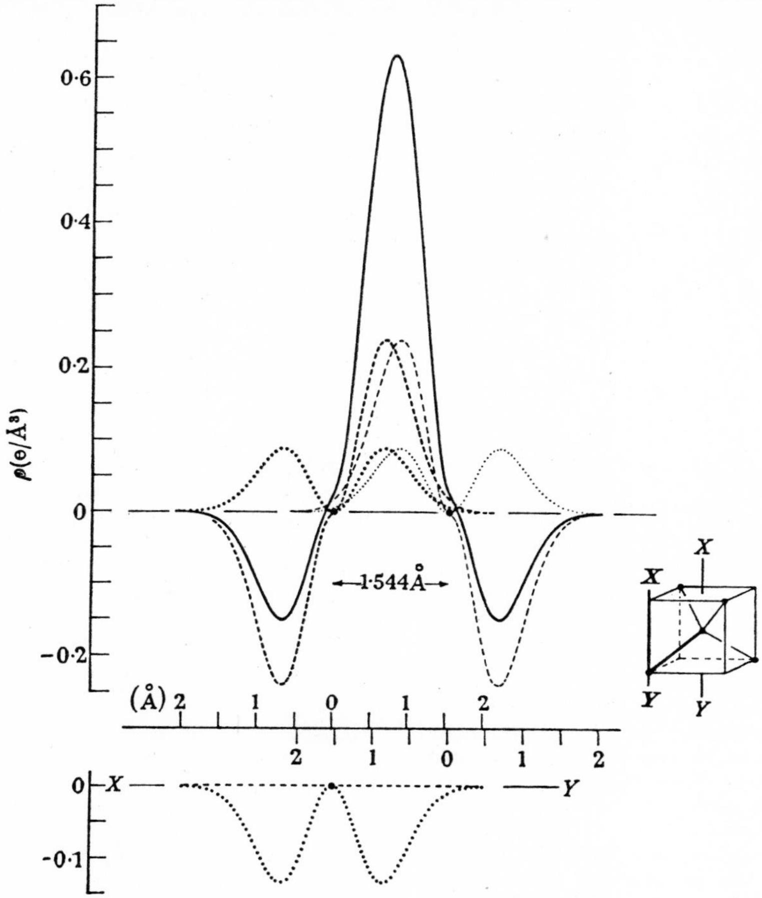

Fig. 3. Detail of the covalent bond in diamond between any pair of atoms. The major part of the figures shows the assemblage of non-spherical atomic components along the bond directions (⟨111⟩ in the structure). Broken lines are the antisymmetric atomic components and dotted lines are cubocentrosymmetric components, heavier and lighter lines differentiating from which atom they originate. The heavy full line shows the total resultant along the bond that appears in Fig. 2. The lower part of the figure shows that only the cubocentrosymmetric component is a feature of the bonded atom along directions corresponding to ⟨100⟩ in the structure.

tions should arise in suitable neutron Bragg experiments where the assumption (ii) of harmonic vibrations is patently inadequate to interpret the diffraction evidence in proper detail.

Such situations arise in fact in members of the fluorite system, whose parent member is CaF_2, where the cation sites have cubic (O_h) symmetry but the anion sites have T_d symmetry. The predictions of assumption (ii) here require that the T-factors of (11) have the simple isotropic from in (10), the main consequence of which is that, for odd-index reflections where (11) demands that the scattering be a function of the cations only, independent reflections possessing a common S-value should display identical amplitudes of scattering. A quite contrary situation has been found in accurate neutron experiments on UO_2, ThO_2 and CaF_2 made at elevated temperatures, typical results of which are given in Fig. 4. These show results found for three reflections possessing a common Bragg angle, and there are marked departures from harmonic "equivalence" with rise of temperature. Similar effects are manifest in other odd-index reflections, but

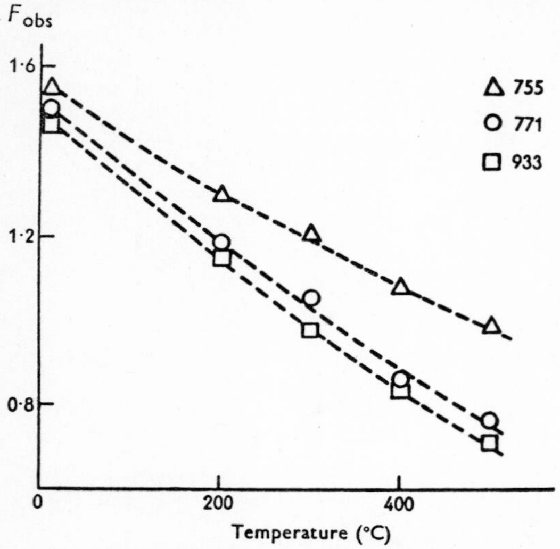

Fig. 4. Observed neutron structure factors for the odd-index reflexions 755, 771 and 933 of CaF_2, all of which possess a common Bragg angle, as a function of temperature.

extended analysis with the appropriate form of (8) shows that all features of the experiments become entirely consistent when anharmonic phenomena associated with the antisymmetric components of the anionic $t(\mathbf{r}')$ functions are recognized.

These neutron experiments provide the clearest evidence of vibrational anharmonicity yet available from studies of Bragg scattering, and their extension to other systems holds promise that the information thus obtained will ultimately provide valuable subsidiary criteria for judging the importance of the T_a-factors in (8) for accurate X-ray studies of complex systems. The need for such criteria is already indirectly evident in molecular systems undergoing rigid-body librational as well as conventional translational motions, where failure of the harmonic assumption (ii) shows itself in atomic positional parameters which require extensive correction before they can be judged suitable for estimating equilibrium bond lengths. In other molecular systems where the rigid-body assumption may be inadequate, it is likely that Bragg studies will have to be supplemented by others which permit more-detailed examination of the dynamics of a vibrating system in the sense of defining further subtleties concerned with correlations between the motions of individual atoms. Even here, however, the value of this supplementary information will depend on the extent to which structure analysis has exploited the flexibilities of data-treatment inherent in the form of (8).

The present position is thus one of considerable importance for future accurate structural studies. With the role of antisymmetric features of charge density, both electronic and nuclear, now being defined, the prospects for further exploitation of these powerful, yet simple, diffraction techniques are very attractive.

Bibliography

Dawson B. (1964a) *Acta Cryst.* **17**, 990.
Dawson B. (1964b) *Acta Cryst.* **17**, 997.
Dawson B. (1965) *Aust. J. Chem.* **18**, 595.
Dawson B. (1967a) *Proc. Roy. Soc.* A **298**, 255.
Dawson B. (1967b) *Proc. Roy. Soc.* A **298**, 264.
Dawson B. Hurley A. C. and Maslen V. W. (1967) *Proc. Roy. Soc.*, A **298**, 289.
Willis B. T. M. (1963) *Proc. Roy. Soc.* A, **274**, 134.
Willis B. T. M. (1965) *Acta Cryst.* **18**, 75.

B. Dawson

ATTACHMENT COEFFICIENTS. When electrons move through electronegative gases they may form negative ions by electron attachment (see Hasted 1961; Reynolds 1962). Under steady state conditions, for a swarm of electrons drifting through an attaching gas under the influence of a uniform electric field the number $\mathrm{d}n_e$ of electrons forming negative ions while drifting a distance $\mathrm{d}x$ in the direction of the field is given by

$$\mathrm{d}n_e = -n_e\alpha_a\,\mathrm{d}x$$

where n_e is the number of free electrons at $x = x$ and α_a is defined as the attachment coefficient. The minus sign indicates that the process of attachment results in a decrease in the number of free electrons. The value of α_a depends on the nature of the gas and on the ratio E/p of the electric field strength E to the gas pressure p (more correctly, α_a depends on the ratio E/N where N is the gas molecular number density).

The attachment coefficient is a swarm coefficient (i.e. it is a quantity describing the behaviour of a swarm of electrons drifting through a gas, the agitational energies of the electrons in the swarm being distributed over a wide range of values) and is related to σ_a, the attachment cross-section for electrons of speed c and energy ε through the expression

$$n_0\alpha_a W = N\int \sigma_{a_\varepsilon}\cdot c_\varepsilon\cdot \mathrm{d}n_\varepsilon$$

where the subscript ε denotes that each value is appropriate to an electron energy ε, and n_0 $(=\int \mathrm{d}n_\varepsilon)$ is the number density of electrons in the swarm. W is the drift velocity of the electrons. (Davies 1961)

Experimental determination of attachment coefficients. The more important methods now in use have been described in the books by McDaniel, Raether and Bates. They may be classified into steady-state and dynamic methods.

Among the steady-state methods, the *electron filter method* developed by Loeb and Bradbury is still useful particularly at low electron energies where ionization of the gas is negligible. The *diffusion method* developed by Huxley and his colleagues is based on the analysis of the lateral diffusion of a mixed swarm of electrons and negative ions and yields values of the Townsend energy ratio in addition to values of α_a. The method is applicable over a wide range of E/p but in analysing data obtained for electron energies at which ionization is important appropriate values of the primary ionization coefficient, α_i (Thewlis 1962) have to be adopted. At high values of E/p the *Townsend growth-of-current method* (Llewellyn Jones 1961; Harrison 1967) has been widely used to obtain values of α_a in addition to values of the ionization coefficients. The interpretation of the data from this method is not in general easy for attaching gases, particularly if detachment is also occurring.

The largest group of methods in the dynamic class are the *pulsed drift tube methods* in which a variety of pulse techniques have been used to examine the processes occurring in a drift tube after a burst of free electrons has been generated either at a photocathode or in the gas. The current flowing in the anode circuit has electronic and ionic components and the examination of the time-dependence of this current leads to values of α_a and of the electron and ion drift velocities. As for the static methods the simultaneous occurrence of ionization and attachment complicates the analysis of the data. A well-defined group of time-dependent studies are the *microwave methods* in which the deionization period following a pulsed discharge in a gas is studied using microwave probing signals. The methods apply at thermal electron energies. In general, the decay of the electron concentration after the discharge may be controlled by processes such as ambipolar diffusion (Sayers 1961) and electron/ion recombination (Llewellyn Jones 1961) in addition to electron attachment but it is sometimes possible to separate the contributions of the various decay mechanisms. The analysis of the data can be further complicated by, for example, the creation of excited molecules during the discharge. Finally, for relatively high values of E/p, the *avalanche methods* of Raether and his colleagues (Harrison 1967) can be utilized to determine α_a. The electron and ion components of single avalanches passing between cathode and anode of a discharge gap are studied through the transient voltages they produce across a resistor placed in series with the gap and values are deduced for α_a, ionization coefficients, and electron and ion drift velocities. Electron detachment can again be a complicating factor.

Bibliography

BATES D. R. (Ed.) (1962) *Atomic and Molecular Processes*, New York: Academic Press.

DAVIES D. E. (1961) in *Encyclopaedic Dictionary of Physics* (J. Thewlis Ed.), **2**, **523**, Oxford: Pergamon Press.

HARRISON J. A. (1967) in *Encyclopaedic Dictionary of Physics* (J. Thewlis Ed.), Suppl. Vol 2, 407, Oxford: Pergamon Press.

HASTED J. B. (1961) in *Encyclopaedic Dictionary of Physics* (J. Thewlis Ed.), **2**, **744**, Oxford: Pergamon Press.

LLEWELLYN JONES F. (1961) in *Encyclopaedic Dictionary of Physics* (J. Thewlis Ed.), **2**, 452, 809, Oxford: Pergamon Press.

McDANIEL E. W. (1964) *Collision Phenomena in Ionized Gases*, New York: Wiley.

RAETHER H. (1964) *Electron Avalanches and Breakdown in Gases*, London: Butterworths.

Reynolds J. A. (1962) in *Encyclopaedic Dictionary of Physics* (J. Thewlis Ed.), **4**,**789**, Oxford: Pergamon Press.

SAYERS J. (1961) in *Encyclopaedic Dictionary of Physics* (J. Thewlis Ed.), **1**, **147**, Oxford: Pergamon Press.

THEWLIS J. (1962) *Encyclopaedic Dictionary of Physics*, **3**, 397, Oxford: Pergamon Press.

J. A. REES

B

BALLOON TECHNOLOGY The balloon, invented in 1782, served for more than a century as the sole vehicle by which men could interpenetrate the atmosphere for direct observations and measurements, and many scientific *firsts* were achieved through the use of balloons. Today the airplane and the space vehicle tower above the balloon in overall utility. Nonetheless, for many roles the balloon remains the best, or frequently the only means of performing crucial experiments. Aided by advances in technology, balloons are being flown in increasing numbers, at higher altitudes, with heavier loads, and for longer durations than thought possible a few years ago.

The first balloon conceived and fabricated by Joseph and Etienne Montgolfier depended upon heated air for buoyancy. The physicist J. A. C. Charles quickly followed with a demonstration of the hydrogen balloon. Hot air, hydrogen, and natural gas are used to some extent but the bulk of today's flights depend upon helium for lift.

The majority of research balloons flown range between 85,000 and 300,000 cubic metres capacity. However, July, 1966 marked the beginning of a series of flights of a 740,000 cubic metre design fabricated from a lamination of Mylar film and Dacron scrim. The initial success of these flights suggests many new experimental possibilities.

Lift. The gross lift of a balloon may be derived from Archimedes law which states that the buoyant force exerted on a body immersed in a fluid equals the weight of the displaced fluid. However, a much greater insight into the behavior of a balloon whose skin is unstretched can be achieved by combining the ideal gas law

$$P = \varrho \boldsymbol{k} T/m$$

with the differential form of the barometric equation

$$\mathrm{d}P/\mathrm{d}h = -\varrho \boldsymbol{g}$$

to yield

$$\mathrm{d}P/P = -(m\boldsymbol{g}/\boldsymbol{k}T)\,\mathrm{d}h$$

where

P = pressure (dyn/cm²)
ϱ = density (g/cm³)
$\boldsymbol{k}$ = Boltzmann's constant
T = temperature (°K)
m = molecular mass (g)
h = vertical distance or altitude.

Upon integration (assuming constant temperature) we obtain

$$P = P_0 \exp(-m\boldsymbol{g}h/\boldsymbol{k}T)$$

where P_0 is the pressure at $h = 0$. This result shows that if two gases have the same pressure at some reference point $h = 0$, then the gas with the lower molecular weight will experience a lower decrease in pressure with altitude than the gas with the higher molecular weight. A lighter-than-air gas enclosed in a balloon envelope therefore manifests a pressure head which is zero at the base but which becomes significant toward the top of the balloon. That pressure provides the force to extend the balloon envelope; and the resultant force is vertical and just equal to the gross buoyancy predicted by Archimedes law. Furthermore, as we will note later, it is this distribution of pressure combined with the weight of the envelope which generates the onion shape which is characteristic of the modern balloon.

Flight. Balloon flight behaviour is complex but some success in modelling has been achieved through the use of digital computers. Such analysis requires considerable knowledge of the variability of air density with altitude plus information on the intensity of the local infra-red radiation field. Since these are seldom known well in advance of a flight, it is customary to provide limited velocity control by means of disposable ballast plus, in some instances, a means of releasing controlled amounts of helium.

The intricate web of forces governing the flight of a balloon can best be visualized by following a typical flight formulated as a composite of many actual flights. The balloon has a 265,000 cubic metre volume. The envelope has a nominal polyethylene film thickness of 0·0019 cm. It will weigh 430 kg and lift a load consisting of 91 kg of scientific instrumentation, 41 kg of ballast, 36 kg of tracking and control instrumentation, and a 7 kg payload recovery parachute. Prior to launch some 650 cubic metres of helium are metered into the balloon. With a nominal ground-level lifting force of 1·05 kg/m³, the gross lift will be 675 kg or 11 percent more than the system gross weight. The difference between gross lift and gross weight is called the net or free lift. It provides the motive force to the system.

Eleven per cent free lift will normally cause the balloon to rise initially at approximately 5 m/sec.

There will be two retarding mechanisms. The first is the familiar aerodynamic drag. The aerodynamic drag coefficient is slightly complicated by the fact that the balloon volume changes constantly during ascent but is otherwise fairly normal. Of more interest is an effect which has been labeled the *thermodynamic drag.*

Thermodynamic drag arises because the helium cools as a result of expansion at a rate which is greater than the atmospheric lapse rate. To be more explicit, the *normal atmosphere* (which the real atmosphere only approximates) experiences a *lapse rate* of 6·4 °C/km from the ground to the *tropopause* at 11 km; is isothermal from 11 km to 25 km; and has a negative lapse rate (i.e. warms) of 3°C/km above 25 km. Helium if allowed to expand adiabatically would have a lapse rate of 12·7°C/km.

In actual flights the gas temperature drops much below the temperature of the air surrounding the balloon but some heat enters the balloon through conduction and radiative absorption. The difference between helium and air temperature is called supertemperature (or sometimes superheat). When the helium is cooler than the outside air, which is always the case for a rising balloon, there will be an apparent loss of free lift. Returning to the equation

$$P = P_0 \exp(-m\mathbf{g}h/\mathbf{k}T)$$

we see that a low lifting-gas temperature causes a reduction in pressure head and hence lift. The name thermodynamic drag has been given to this loss of lift or rising (and to a corresponding loss of negative lift when a balloon is falling). Since an increase in upward velocity causes the gas to expand and cool more rapidly than can be compensated by heat flow there will be an increase in thermodynamic drag, which is to say that the system is inherently stable.

Balloons tend to rise with an average velocity of 4·5 m/sec below the tropopause and an average of perhaps 3·0 m/sec thereafter. A reduction in velocity upon entering the tropopause is considered desirable. It is there that the atmospheric temperatures are a minimum and balloon films tend to embrittle at low temperatures. Winds and wind gusts tend to maximize in the tropopause. Finally the greatest unfolding of material takes place at tropopause altitudes. As a consequence most balloon failures occur between 11 km and 25 km and caution is advisable.

If it was not for the thermal effects, the free lift of a rising balloon would remain essentially constant. The volume of the balloon would expand at just the rate necessary to compensate for the decreasing air density.

When the expanding gas attains a volume equal to the design volume of the balloon it becomes necessary to vent off the excess gas. As the gas is vented through a duct in the bottom of the balloon the free lift will gradually diminish. As the lift diminishes, the rate of rise will also diminish and heating will gradually warm the balloon. A properly designed *system* will round off to the equilibrium float altitude with little undershoot or overshoot. Overshoot is particularly bad since gas over and beyond that originally required for free lift will be vented and a negative net lift will be generated. To keep the balloon from descending, it will be necessary to expend ballast.

The *equilibrium altitude* may not be exactly that which was predicted for a number of reasons. First the atmosphere at that altitude may be more or less dense than expected. The balloon may have a slightly smaller or larger volume than planned if the payload weight is *off-design*. And of course payloads tend to grow in weight between design and operations. Then too, the radiation environment may be unexpected. This latter effect is one of the most interesting aspects of balloon technology.

Radiation effects. Balloons made of either polyethylene or Mylar tend to be quite transparent to visible radiation but possess many infra-red absorption bands. They therefore tend to radiate energy to the sky from their upper quadrants and to receive energy from the Earth and/or lower atmosphere in the lower quadrants. Thermal energy is transferred to the helium from the envelope through conduction and convection. Free space represents a constant heat sink. The Earth, on the other hand, is a variable heat source. Its apparent temperature depends upon seasonal effects, diurnal effects, and weather (especially cloud cover).

The most serious thermal effect is encountered when a balloon in equilibrium at altitude experiences a cooling trend as, for example, when passing over a high cold cloud deck or the cooling of the Earth due to sunset. The net lift of a balloon at equilibrium is of course zero. A loss of gross lift due to cooling produces a negative net lift and the balloon will begin to descend. To maintain the desired float altitude ballast must be dropped. A subsequent warming trend will cause the balloon to rise and gas will be vented. Repetitions of such alternate heating and cooling trends are common and the time at which the balloon can be held at a constant altitude depends upon the amount of the original gross weight which was apportioned to expendable ballast.

A means of achieving long flights without the use of expendable ballast is through the use of *superpressure balloons.* A superpressure balloon is one which is sealed so that the initial free lift will not be lost through venting of the excess lifting gas. Instead the free lift is converted to superpressure within the envelope—10 per cent free lift converts to approximately 10 per cent overpressure at the float altitude. Now when a superpressure balloon experiences cooling the internal pressure will diminish but so long as it does not drop below the outside pressure the volume will remain constant and so will the float altitude. These balloons are sometimes called constant-volume balloons or constant-altitude balloons. Due to the skin stresses which are introduced by pressurization they are normally spherical in shape and carry extremely light loads. Superpressure balloons have flown for more than 75 days.

Design. The normal zero-pressure balloon as we noted earlier has an inverted onion or tear drop shape. In the balloon field it is called the *natural* shape. A deviation from the older more spherical shapes was necessitated by the introduction of lightweight plastic films and made possible by the development of the analogue computer.

Experimentation with plastic films began prior to World War II but where not used extensively until after the war. They offered the opportunity to design a balloon which would be extremely light and yet strong enough to carry instrumentation to altitudes in excess of 30 km. The strength of plastic films and especially of seams was (and is) such that little safety factor exists in the structure. It was therefore important to analyse the stresses in a balloon and to develop designs which reduced localized stresses. Differential equations relating the curvature of the envelope to the internal pressure and the suspended weight below an elemental area are rather easily written. However, these equations cannot be solved in a closed form. The analogue computer following the war and the digital computer in more recent times have been used to provide working solutions. These results describe the shape of the balloon subject to boundary conditions.

The vast majority of balloons flown today are designed to have zero-superpressure under equilibrium conditions, to have flat tops, and to have zero circumferential stress under equilibrium conditions. Design tables have been developed which provide gore pattern relationships (length to width ratios) for a family of natural balloon shapes of differing oblateness. To determine the correct shape (designated by a *Sigma* number) it is necessary to specify the design altitude, the areal weight of the proposed film, and the load to be carried. Manufacturers have developed a set of balloon designs of varying material thickness, oblateness, and size which will meet, within reasonable tolerances, most requirements. New balloons are designed as requirements or materials change.

Design innovations have been introduced to handle problems that appear operationally and are outside the realm of analysis. For example, it is customary to use so-called *load tapes* on polyethylene balloons that are to be used with heavy loads. The load tapes are made from glass or Fortisan fibres sealed between polyethylene ribbons and are incorporated into the balloon as the gores are sealed together. The tapes are gathered together at the top and bottom fittings and serve to carry much of the meridional load. A number of schemes have been used to provide additional strength near the top of the balloon during pre-launch and the early flight when the total lift is concentrated near the center.

The relatively low cost of polyethylene makes it the most favoured fabric. Small to medium size polyethylene balloons are quite reliable when launched in fair to moderate weather. As the size increases or the weather deteriorates, the failure rate of polyethylene balloons rises to 20 per cent or more. The load limit on a polyethylene balloon is in the neighbourhood of 1300 kg.

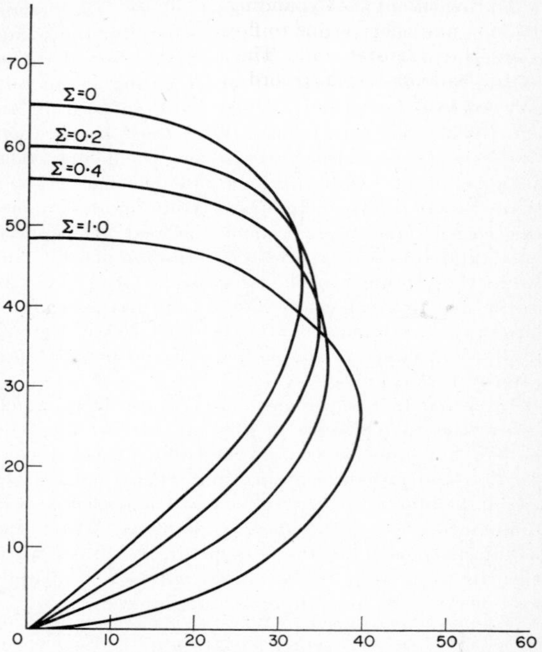

Balloon shapes—all the same gore length.

Greater reliability and greater load carrying capacity may be realized with a lamination of Mylar film with Dacron scrim. Polyethylene films have ultimate strengths of approximately 2×10^6 kg/m^2 and the Mylar–Dacron scrim is about seven times as strong. The laminate does have the disadvantage that very light weight composites are not yet available and the cost is nearly 10 times that of an equivalent polyethylene balloon. Mylar itself is much stronger than polyethylene but its resistance to tear propagation is low.

The largest Mylar–Dacron scrim balloon built and flown is a 740,000 cubic metre balloon designed to carry an assortment of experimental equipment to 45 km. It and most other Mylar–Dacron scrim balloons have been fabricated as dual-cell balloons. A small balloon, the launch balloon, sets on top of the main balloon. All of the helium is injected into the launch balloon which will not be fully inflated on the ground. The main balloon is protected by a reefing sleeve prior to launch. In this way the profile drag of the system is minimized, the main balloon is not committed prior to actual launch, and the launch balloon may be fabricated of a heavier material than the main balloon.

During ascent the expanding gas first fills the launch balloon and then begins to flow into the main balloon through a transfer duct. The two-cell Mylar–Dacron scrim balloon has a record of success approaching 100 per cent.

R. S. McCarty

BLACKBODY INSTRUMENTS FOR INFRA-RED. Commercial instruments are available with spectral energy densities comparable to a blackbody. These instruments, which are used for calibrating infra-red sensors are usually made of a corroded metal—the shape of a conical cavity which can be varied in temperature from 300°K to as high as 1500°K. The temperature can be set at any value in this range.

I. W. Ginsberg and T. Limperis

C

CERENKOV DETECTORS. When a charged particle passes through a dielectric medium with a velocity greater than that of light in the medium, then an electromagnetic shock wave, Čerenkov radiation, is produced. The velocity dependence and directional properties of the radiation have been used in the detection and identification of elementary particles for research in high energy nuclear and cosmic-ray physics. Essentially, a Čerenkov counter consists of a transparent medium viewed by a photomultiplier tube, the medium having a refractive index greater than 1 for light close to the visible region.

From each element of the particle track Čerenkov light is emitted uniformly along the surface of a cone whose apex is the instantaneous position of the particle. The angle, θ, which the conical surfaces make with the particle direction is related to β, the ratio of the velocity of the particle to the velocity of light *in vacuo*, and n, the refractive index of the medium by the Čerenkov relation,

$$\cos\theta = \frac{1}{\beta n}.$$

The number of photons radiated by a particle of charge Ze in a spectrum interval defined by the wavelengths λ_1 and λ_2 is given by

$$N = 2\pi\alpha Z^2 l \sin^2\theta (1/\lambda_1 - 1/\lambda_2),$$

where l is the length of track of the particle in the medium, α is the fine structure constant $\left(\alpha = \frac{1}{137}\right)$ and θ is the Čerenkov angle. The number emitted between 3500 and 5000 Å (the range of response of a typical photomultiplier) is

$$N \cong 400\, l \sin^2\theta.$$

Since the photocathode efficiency is about 10 per cent, a practical Čerenkov detector should be designed to detect at least 200 photons to eliminate loss of detection efficiency due to statistical fluctuations in the number of electrons emitted from the cathode.

The principal use of Čerenkov counters is to restrict the particles detected to those within a defined velocty interval. This can be achieved in two distinct ways.

1. The Čerenkov relation implies a threshold condition on the velocity of the particle, given by $\beta > 1/n$. A threshold Čerenkov counter is designed to collect as many photons as necessary for high efficiency on to the photomultipliers.

2. A measurement of the angular range of the Čerenkov cone determines the velocity within limits $\beta_1 < \beta < \beta_2$ where both β_1 and β_2 are above the threshold value. The range selected by a differential Čerenkov counter is determined by the refractive index of the medium and the details of the optical focusing system.

Two typical designs for threshold counters using liquid radiators are shown in Figs. 1 and 2. The "end-on" type of Fig. 1 collects the Čerenkov light directly on the photocathode surface of the photomultiplier. Such a counter is useful as the last unit of a detection system where it can be made fairly thick for high efficiency. For an intermediate position in a particle detection system the relatively thin "transmission" type of counter (Fig. 2) is used. In this case the collection of light depends on reflections from the walls of the liquid container which is coated with white reflecting paint. Efficiencies greater than 98 per cent, for example, can be achieved for relativistic particles in a 4″ water transmission counter. For both types it is normal to immerse the ends of the phototubes directly into the liquid, with O-ring seals round the barrels of the tubes, for greater optical efficiency.

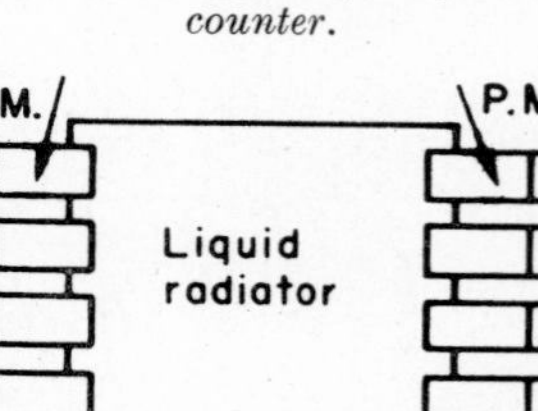

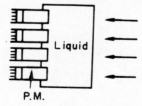

Fig. 1. "*End-on*" *type of liquid Čerenkov counter.*

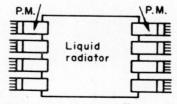

Fig. 2. "*Transmission*" *liquid Čerenkov counter.*

Since the number of photons produced by the scintillation process is about a factor of 40 greater than that produced by the Čerenkov effect, care has to be taken in the design of Čerenkov counters to avoid using materials which have scintillation properties. A range of refractive indices from 1·27 to 1·7 can be achieved by the appropriate choice of liquids. The refractive indices and corresponding threshold velocities of some commonly used liquids and solids is given in the table. To cover the range of refractive indices 1·0–1·2 a variable pressure gas counter has to be used. In this case the photomultiplier tubes view the radiator through lucite or quartz windows. The threshold velocity can be selected by varying the gas pressure. Commonly used gases are nitrogen, ethylene, methane, carbon dioxide and freon.

Material	Refractive index n	Threshold velocity
Solids		
quartz	1·46	0·685
lucite	1·49	0·67
perspex	1·49	0·67
crown glass	1·5–1·6	0·62–0·66
flint glass	1·5–1·7	0·59–0·66
thallous chloride	2·2	0·455
Liquids		
FC 75	1·276	0·78
water	1·333	0·75
carbon tetrachloride	1·46	0·685
glycerol	1·47	0·68
benzene	1·50	0·67

A threshold counter using only total internal reflections can be made using a slab of solid radiator painted black and viewed at the ends (Fig. 3). The threshold for a normal counter using perspex as radiator is 0·67 while the same material used in such a way as to observe only the internally reflected light gives an effective threshold of about 0·90. The condition for total internal reflection is $\sin\theta > 1/n$ and $\sin\theta > \cos\theta$ to satisfy the Čerenkov threshold relation.

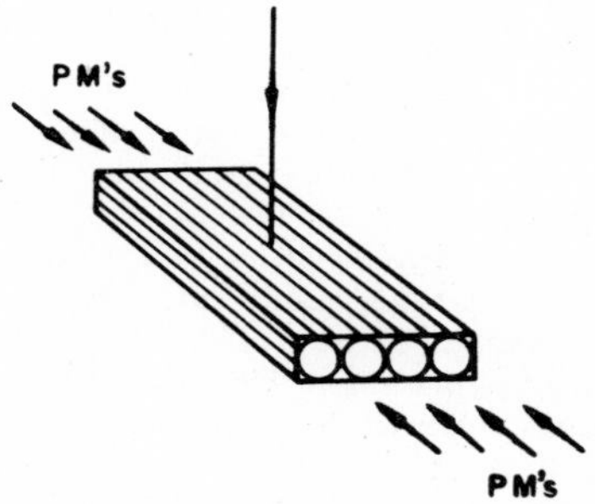

Fig. 3. Total internal Čerenkov counter using solid radiator.

The most common type of differential counter uses a convex lens or spherical mirror to focus the Čerenkov light from a gas radiator into a ring image. The angular range over which the focused light is detected and consequently the range in velocity of the selected particle can be chosen by placing an annular slit at the focal plane (Fig. 4). The upper arm of the counter of Fig. 4 operates in an integral mode and the requirement of a coincidence between the integral arm and the differential arm gives a marked reduction in false counts. The refractometer serves to measure the refractive index of the gas.

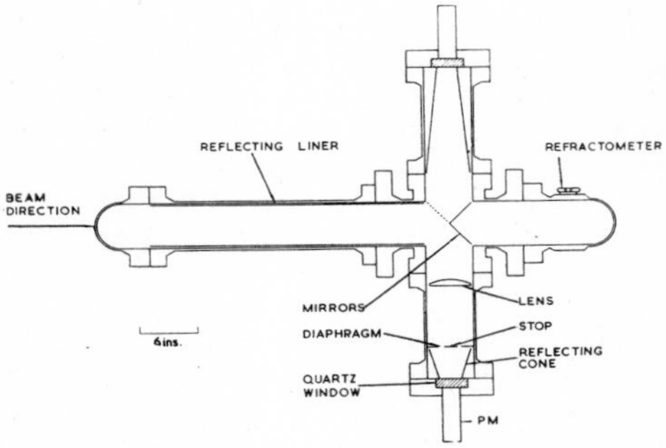

Fig. 4. High pressure gas Čerenkov counter with integral and differential arms.

Differential Čerenkov counters have been extensively used for the mass analysis of fixed momentum particle beams from high energy accelerators. For a fixed momentum the mass relation is related to the angular resolution by

$$\tan\theta\, d\theta = m\,dm/E^2$$

where E is the total energy of the particle and θ is the angle at which the Čerenkov light is detected. Figure 5 shows the velocity spectrum of a high reso-

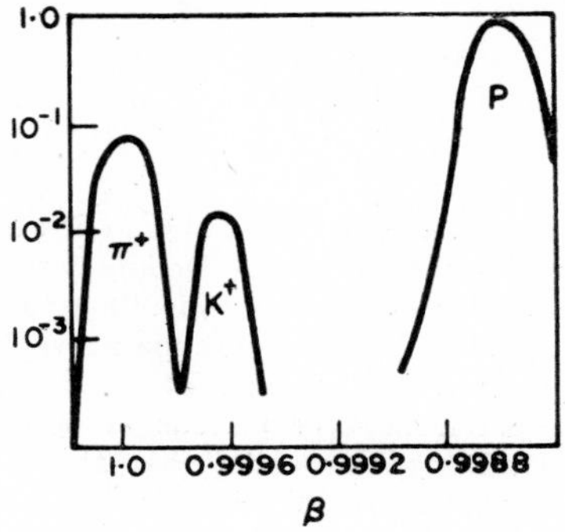

Fig. 5. Velocity spectrum of particles using a differential gas counter indicating separation of π-mesons, K-mesons and protons at 18 GeV/c.

lution counter operating in a positive particle beam of 18 GeV/c. The velocity resolution $\Delta\beta/\beta$ was 2×10^{-4} and K^+ mesons could be identified with a certainty in excess of 99 per cent.

Čerenkov counters have also been used to detect high energy photon electron cascades. A block of fairly transparent material of short radiation length, such as lead glass, with dimensions of many radiation lengths is used to produce and contain the shower initiated by a single incident photon or electron. The number of photoelectrons produced in the photomultipliers of this total absorption Čerenkov counter is roughly proportional to the energy of the incoming photon or electron.

The Čerenkov counter, like the scintillation counter, has a short response time and high efficiency and is capable of operating in high particle fluxes. The unique directional and velocity properties of the Čerenkov radiation can be used to select a group of particles out of a high background of other particles by a suitably designed optical system. Discrimination can be made between particles of different mass and same momentum using differential counters or two threshold devices operating with different refractive indices. The energies of high energy photons and electrons ($E > 20$ MeV) can be measured using total absorption counters with a Čerenkov radiator. The associated use of scintillation and Čerenkov counters in elementary particles physics has played an important role in the detection of high energy particles.

Bibliography

Hazen W. E. (1961) in *Encyclopaedic Dictionary of Physics* (J. Thewlis Ed.), **3**, 327, Oxford: Pergamon Press.

Hutchinson G. W. (1960) *Progress in Nuclear Physics*, Vol. 8, Oxford: Pergamon Press.

Jelley J. V. (1958) *Cerenkov Radiation and Its Applications*, Oxford: Pergamon Press.

Jelley J. V. (1961) in *Encyclopaedic Dictionary of Physics* (J. Thewlis Ed.), **1**, 618, Oxford: Pergamon Press.

Lindenbaum S. J. and Yuan L. C. L. (1961) in *Methods of Experimental Physics*, Vol. 5A, New York: Academic Press. Chap. 1.5.

T. G. Walker

CLIMATIC VARIATIONS (CLIMATIC CHANGE). Climate varies from place to place, country to country; and it likewise varies through time. Regarded as the sum total of the weather conditions prevailing, it is customarily expressed by a series of averages and of departures therefrom over periods of 30 years. In temperate lands popular belief that the climate is not what it was, even in a lifetime, is widespread; but in such matters the unaided human memory is often selective and fallible. Throughout Europe and North America migration from country to city has long prevailed, where the effects of weather are less keenly felt. Agricultural difficulties arising from cold springs and wet summers are less noticed; severe winters are more easily borne. Yet outside the cities the yield of crops and gardens remains much the same, and meteorological statistics confirm that there has been but little change. Nevertheless, within a century there can be fluctuations of sufficient magnitude to ensure that the accumulated experience of countrymen may differ appreciably in one generation from that of its predecessor, especially in those regions where the mean temperature of winter or spring lies close to the freezing point. There, it requires but a small change in temperature over a decade or two to ensure considerable differences in the frequency of such noticeable phenomena as snowfall, frost and ice on open waters. Between 1895 and 1940, for example, in England there were only two winters, 1917 and 1929 that gave sufficiently lasting cold for the majority of the smaller rivers to freeze. By contrast, since 1940 there have been four, and in the last half of the 17th century, more than ten. Variations in the proportion of wet to dry summers likewise occur. In England, the 1850's and 1740's were relatively dry, the 1870's and 1920's mainly wet. The droughts of the 1930's in America, with the three successive severe "war winters" 1940–42 in Europe, have called attention to the detailed study of older records.

The period of instrumental recording and historical accounts. Official meteorological services began to be set up after 1850; few countries possess extensive networks of reliable observations for more than 80 years. Here and there in Europe instruments came into use during the later 17th century and some of the scattered journals of daily readings can, with care, be integrated to provide a summary of the behaviour of the climate of the lands bordering the North Atlantic, including a small part of coastal North America, for about 200 years. For a very few localities in England, Holland, France and Germany they can be carried back before 1700.

Throughout Europe and much of North America the extent and character of the amelioration of the early 20th century is shown by the meteorological statistics. It has been accompanied by diminution of the mountain glaciers, better tree-growth in Scandinavia, the ripening of corn in Iceland, improved fishing off Greenland and greater freedom of Arctic navigation, of much importance to Northern Russia. Elsewhere in the world the rise of temperature has not been so marked, but a significant change in the seasonal incidence of Australian rainfall has been noted. Accordingly it appears that these small but significant fluctuations result from slight variations in the predominant pattern of the atmospheric circulation; for example slight displacement of the prevailing tracks followed by depressions across North America and the North Atlantic. There is evidence from the instrumental record that similar fluctuations have occurred during the past three centuries but without any clearly-defined regularity. At times they can be linked with changes in the extent of the Arctic sea-ice, or with those groups of years in which, since 1670, the mountain glaciers of the Alps, Scandinavia, Iceland and also the Canadian Rockies have been actively advancing. Further culminations were attained about 1750, 1820, 1850. Instrumental readings in N.W. Europe show that the 1690's were exceptionally cool and 1726–1739 generally warm. Economic effects in Scotland and Sweden, for example, were notable.

In the 19th century, recognition of the former existence of an Ice Age over much of the northern hemisphere was followed by much study of documentary material on the occurrence of floods and droughts, bad harvests, the date of the vintage in France, dates of freezing and thawing of the Russian rivers, of ice in the entrance to the Baltic, of the level of the Caspian Sea. Accounts of the agricultural practices of ancient Rome and Greece indicate a climate essentially similar to that of the present, and Tacitus's remarks on the climate of Britain might be echoed by thoughtful Italians today. Evidence, however, supports the view that the sixth and seventh centuries A.D. may have been drier and less stormy, and the decades around 1200 were characterized by mild winters, earlier springs and warm summers. The magnitude of this "Little Climatic Optimum" should not be exaggerated; it may have amounted to a shift during 40 or 50 years of the present summer warmth of Oxford to York, i.e. a rise in the average temperature of about 1°C. But this sufficed to encourage cultivation of a few Southern English vineyards. In Germany wine was made in localities up to 800 feet above today's profitable limit.

Climate became more disturbed around the early 14th century and some Alpine glaciers began to advance. The 15th century saw an increase in the extent of permanently frozen ground in South Greenland, and decline of the Norse colony. About the time of the Reformation, summers in Western Europe were often favourable. Towards the end of the 16th century there was a general cooling, followed by a sharp increase in the frequency of severe winters and widespread advance of the Alpine glaciers, sometimes beyond the limits of any previous advance since the final stages of the Ice Age 10,000 years ago. Tree-growth studies in Lapland indicate markedly cooler summers, and the frequency of arrival of sea-ice on the coast of Iceland increased markedly after 1590.

In North America, where historical records are lacking, much has been learnt from examination of long series of tree-rings in critical areas; these indicate widespread drought in the South-West in the late 16th century. Dating of glacier advances has been effected through ring-counts on trees that were pushed over and buried beneath morainic material; there is fair agreement with the European dates. Much evidence has been assembled in Asia from the Chinese Annals, from records of the blooming of the cherry in Japan, and from changes in the level of lakes without outlet; while in North Africa the height of the Nile flood has been recorded over many centuries. That the course of history has been affected by the great invasions of pastoral peoples, Huns, Magyars, Mongols, Turks, from Central Asia is evident; but Huntington's argument that these migrations were caused by fluctuations in the rainfall is no longer tenable.

Prehistoric variations of climate since the Ice Age. The last great phase of the Ice Age, when Scandinavia, the Baltic, much of Britain and all of north-eastern North America were covered by ice, began to wane about 18000 B.C. Melting exceeded accumulation and from North Germany, Yorkshire, New England and Wisconsin the margin of the ice retreated northward, but not uniformly. Soon after 9000 B.C. an appreciable readvance set in, leaving its end-moraines through Central Sweden and northern Michigan. South of the ice, tundra vegetation with many bogs and small lakes prevailed. As the ice retreated, trees spread northward. Their pollen, falling upon the surface of bogs and lakes, has been preserved in the peat or the sediments beneath. Examination of successive layers thus reveals the changes in the prevailing tree-cover and that of other vegetation, from which changes in climate can be deduced; although in recent centuries human interference with the vegetation must be allowed for. The climate of temperate Europe and North America attained an "optimum", characterized by milder winters and warmer summers than today, somewhere between 5000 and 3000 B.C. In Denmark summers were probably 2°C (3·6°F) warmer than today, winters 0·5°C (1°F) warmer. From raised beaches of this age in Britain the remains of shellfish indicate that the sea was probably 2°C warmer. Spitsbergen was relatively mild, and it is probable that the Arctic seas were much more free from ice than now. There is evidence that subsequently the climate underwent several successive recessions, followed by some recovery. Each of these may have been marked by up to half a century of mainly unfavourable seasons, with notably cooler and wetter summers; in Europe one of the most notable is dated around 600 B.C. These post-glacial climatic oscillations appear to be similar in type to the smaller episodes of historical time; they may be susceptible to similar explanation. Radiocarbon dating has provided since 1948 a method by which the timing of climatic events, such as advances of the Andean and Patagonian glaciers, can sometimes be compared with that of events in the tropics or in the northern hemisphere. Since the effect of these small declines appears to have exceeded that of recoveries since the post-glacial "optimum", and as 3 per cent of the Earth's surface is still covered by ice, some think that after 10,000 years or so another glacial phase may have developed in northern lands.

The Pleistocene Ice Age, and earlier variations in geological time. Early last century it was recognized that the very extensive surface deposits, and notably the erratic boulders so widely found in N. Europe and N. America could only have resulted from the spread of great ice sheets from the highland areas in quite late ("Pleistocene") geological time. Later it became evident that there had been several such episodes, of varying length and intensity, within the last one million years, and that the last culminated between 40,000 and 20,000 years ago. The amount by which the mountain snowline would need to be lowered to provide for so much ice has been calculated; about 1200 m in the Alps, perhaps 1500 m in Scotland and W. Canada, rather less towards the tropics. More recently, stratified cores have been obtained by boring into the deep ocean sediments. These derive largely from the skeletons of minute creatures living at or near the surface; they accumulate excessively slowly. From their characteristics, an index of the variation of the surface water temperature can, with care, be derived. At the Ice Age maxima, equatorial waters were probably 6°C cooler than at present, and within the last million years a number of shorter and longer cool phases can be recognized. Scandinavian ice then spread to Holland and the mouth of the Thames. Canadian ice spread as far as New York and Long Island, where the terminal moraine now forms the small hills. Adjacent to the ice, the mean summer temperature at New York was probably at least 15°C below that of today; at London, 11°C. But in drier lands such as Siberia where, on account of deficiency of snowfall, ice sheets were not widely extended, summer temperatures were not lowered by so large an amount. They have been deduced from the characteristics of fossil vegetation, animals, insects and shellfish.

It is evident that the temperature gradient in the surface air, between equator and pole, was much greater than today. Depressions moving along the Southern margin of the ice sheets were probably more vigorous; one result was that the Northern Sahara was for long periods considerably wetter than now, with extensive grassland capable of supporting animals. Indeed during the glacial phases when up to 10 per cent of the Earth's surface lay under ice compared with 3 per cent today, the desert belt was in all probability considerably narrower than we now see. The North Atlantic surface waters were much colder, and with so much water locked up in the ice sheets, sea level is estimated to have been lowered by 90 m (300 ft).

Before the Pleistocene ice age, that is, the sequence of cold phases within the last million years with whose effects we are familiar, deducations with regard to variation of climate in geological time, for upwards of 600 million years, must be derived from the nature of the rocks and their fossils. It is evident that the temperature of the equatorial tropics has differed little from that of today since Cambrian times, if not earlier, and that during the greater part of the Earth's history the climate of higher latitudes has been relatively warm, with no permanent ice sheets on land and little or no sea-ice even at the poles. In more than one widely separated and relatively brief period of geological time, however, certain parts of the world lay under ice sheets. For example, at the end of Carboniferous times about 280 million years ago, ice covered much of South Africa, India and Argentina. No satisfactory explanation of such a distribution can be provided without postulating continental drift and consequent change of latitude. Together with this, we must assume that high land was then to be found in high latitudes; we see today that this provides the base for the two largest remaining ice sheets, in Greenland and Antarctica.

Possible causes of climatic change. Theories are many. Changes in the amount of solar radiation received at the Earth's surface might be a prime cause, and might occur through extraterrestrial events such as a change in the output of solar energy, or the interposition of matter in outer space. On the other hand change might result from occurrences on the Earth itself. Long-term periodic variations in the elements of the Earth's orbit and the tilt of the axis occur, and undoubtedly they will affect the receipt of solar energy in higher latitudes; but on quantitative grounds they do not appear to afford a sufficient explanation of ice ages.

Sunspots have been observed and counted for over 200 years. Their number rises and falls over a somewhat irregular interval averaging 11 years. As they represent disturbances whose appearance is associated with auroral displays and disturbance of the Earth's magnetic field, their number has been presumed to be related to the output of energy; but it is far from certain that that part of the energy that goes to warm the Earth is affected. Very many attempts have been made to establish correlations between sunspots and weather, too often with unconvincing results as the period of available records grows longer.

Changes in the composition of the atmosphere itself can seriously affect the energy balance. Increased cloud leads to reflection of a greater proportion of the incident radiation. Small changes in the variable gaseous constituents (water vapour, carbon dioxide and ozone, the latter at high altitudes) can be shown to lead to greater absorption of radiation by the atmosphere itself. This screening or "greenhouse effect" means that the Earth will become warmer. It has been argued that the enormous increase in the burning of fossil fuels, coal and oil, since the Industrial Revolution began has produced sufficient CO_2 to lead to a world-wide warming. But there are objections to this theory; many fluctuations of similar magnitude occurred before 1800, and for large parts of the world the reliability of meteorological data can be questioned when it is a matter of fractions of a degree.

In many large-scale volcanic eruptions fine ash is driven high into the atmosphere, whence it spreads to form a veil whose effects may remain for two or three years. Measurements showed that the receipt of solar radiation was for a time appreciably curtailed after the eruptions of Krakatoa in 1883 and Katmai in 1912; each eruption was followed after an interval by a sequence of unusually cool months in Europe. The cool summers of 1784 and 1816, and recently the very cool August of 1956 can be similarly attributed. But it does not appear that volcanic eruptions occur with sufficient frequency to produce more lasting effects.

Poleward transfer of heat by surface ocean currents is considerable, serving to reinforce that resulting from the prevailing atmospheric circulation, for example in the North Atlantic where it affects the area covered by sea-ice. Diversions and modifications of these ocean movements, both on the surface and at depth may affect climate, but it is not easy to disentangle their effects from those produced by the prevailing winds.

For some other theories reference should be made to standard texts. No single theory yet put forward receives universal support, beyond the recognition that for the genesis of an ice age the initial presence of high land in high latitudes upon which snow can accumulate appears to be necessary. Yet this does not readily explain the fact that the great ice sheets of the northern hemisphere waxed and waned several times in the course of the Pleistocene, i.e. the last million years.

For the fluctuations of smaller magnitude whose effects are manifest in the historic record, and through pollen-analysis, archeological studies, glacier behaviour and the study of core-samples from the deep ocean sediments, the cumulative effects of small variations in the receipt, or reflection, of solar radiation coupled with lag effects in the ocean, that in turn influence the prevailing pattern of the world's atmospheric circulation, may provide an explanation. But we need more precise estimates of the changes in temperature and rainfall, at least; and these must be better dated. Climatic variation provides a field of study in which there is still much to learn.

Bibliography

Brooks C. E. P. (1949) *Climate through the Ages*, London: Benn.

Flint R. F. (1957) *Glacial and Pleistocene Geology*, New York: Wiley.

Poole J. H. J. (1961) in *Encyclopaedic Dictionary of Physics* (J. Thewlis Ed.), **1**, 95, Oxford: Pergamon Press.

Schwarzbach, M. (1963) *Climates of the Past* (English translation) New York: Van Nostrand.

Changes of Climate (1963) UNESCO Rome Symposium, Paris.

Symposium on Climate, 8000–0 B.C. (1966) London: Royal Meteorological Society.

G. Manley

COATED PARTICLE NUCLEAR FUELS. Coated particle fuels are now being used in all the existing high temperature gas-cooled reactor (HTR) projects (namely Dragon (OECD Winfrith), Peach Bottom and UHTREX (U.S.A.), and AVR (Germany)). In their conceptual form, however, they contained uncoated fuel which emitted a large proportion of the volatile fission products, therefore the original reactor design incorporated a complicated and costly fission product purge circuit and large clean-up plant to maintain low radiation levels in the primary coolant circuit. Attempts to reduce fission product emission by cladding the compacts into which the fuel particles were embedded with an impermeable coating proved unsuccessful due to thermal stress cracking, while the use of a metal can in a high temperature reactor results in severe restrictions on operating temperature and excessive neutron absorption. The coated particle nuclear fuel, in which each individual particle is clad with an impermeable ceramic coating, of low neutron cross-section, was thus conceived as a means of minimizing the release of fission products, i.e. a canned fuel on a micro-scale.

The earliest references to the application of ceramic coatings to nuclear fuel particles appeared in 1959. The first irradiation results on coated particles some two years later were so encouraging that large development programmes were initiated both in the U.S.A. and in Europe.

The most favoured fuel cycle for HTR's is that based on $^{235}U-Th-^{233}U$ and hence the emphasis on fuel development has been on mixed Th—U fuels either as carbides or oxides. A low uranium enrichment fuel cycle and fuel cycles incorporating plutonium are only slightly less economic and fuels appropriate to these concepts have also been developed. Substantial savings in fuel cycle cost are claimed if the fissile and fertile materials are partially segregated—designated "feed" and "breed". The breed fuel element, which is retained in the reactor for periods up to about 10 years, contains a high fertile to fissile ratio, while the feed elements, which are replaced at fairly frequent intervals ($\sim$2 years), contain all fissile material. As the HTR is essentially a converter system, optimum fuel cycle costs are obtained with recycled fuel, hence the choice of fabrication route for the fuel must consider its suitability for remote or semi-remote operation in addition to the normal metallurgical requirements.

Three main methods have been developed for fabricating suitable fuel kernels, namely: melting in a plasma torch or in a carbon bed, a powder metallurgy agglomeration process, and by the gelation of spherical droplets (sol-gel process). It is desirable for the fuel kernel to be fairly spherical in order that the mechanical strength of the coating is at an optimum. The powder metallurgical route has particular advantages in the flexibility of composition obtainable, high yield, and the ability to incorporate controlled amounts of porosity into the fuel kernel.

One advantage of the sol-gel process is that it requires a much lower densification temperature, but with recycled fuels the disposal of large quantities of active fluids is a problem. The initial specification for fuel kernels was in the size range 150–250 μ diameter, but for economic reasons and to enable higher fissile loadings, fuel kernels of 500 μ diameter and above are now common.

The fuel particles thus produced are then coated to a thickness of $\sim$100–150 μ, the coatings normally being applied by the pyrolysis of a gas while the particles are suspended or agitated in some way, generally in a fluidized bed. Pyrocarbon coatings have been the most extensively studied. Hydrocarbon gas mixed with an inert diluent is introduced into the heated zone of the fluidized bed and the resulting carbon is deposited on the particle. Methane is most commonly used but other carbonaceous gases that have been employed include propane, butane, acetylene and carbon tetrachloride. The behaviour of pyrocarbon coatings under irradiation is particularly sensitive to the density, preferred orientation, and apparent crystallite size of the coatings. The factors that affect these properties are the temperature and rate of deposition of the coatings, the latter being controlled by the size of bed and both partial pressure and flow rate of the hydrocarbon gas. In the temperature range 1300–2200°C, coating density can vary over the range 1·3–2·1 g/cm^3 (theoretical density = 2·27 g/cm^3). Preferred orientation, characterized by Bacon's "anisotropy factor", which is the calculated ratio of thermal expansion coefficients in the parallel and transverse directions, is obtained by X-ray techniques, typical values ranging from unity (isotropic) to values >3. The crystallite size is primarily a function of coating temperature and varies from $\sim$40 Å to 150 Å.

Other coating materials, which have better retention properties for certain of the volatile metal fission products, have also been investigated, e.g. SiC deposited from methyltrichlorosilane, Al_2O_3 from $AlCl_3$ and BeO from $BeCl_2$. A coating containing a SiC layer of $\sim$30 μ thickness sandwiched between two pyrocarbon layers has proved successful and is used by the Dragon Project as their reference coating. Al_2O_3 coatings have performed satisfactorily but have a higher neutron absorption and are more expensive to produce than pyrocarbon, and thus have not been considered for reactor use; BeO coatings have only been partially successful due to the difficulty in producing fine-grain structures.

There are essentially two forms of high-temperature gas-cooled reactor fuel element utilizing coated particles. In the type typified by Dragon and Peach Bottom the core consists of fixed prismatic fuel elements. The particles are normally fabricated into a graphite matrix to form a compact, these being stacked in graphite tubes (6–8 cm diam.) running the length of the reactor core. These graphite tubes form the structural members of the fuel element. This form of fuel element can be purged if desired to reduce the release of volatile fission products into the coolant, however, it is now generally accepted that with coated particle fuels purging is no longer necessary, thus giving greater flexibility of design for power reactors. In the second type, as in the AVR, the fuel particles are incorporated into graphite spheres (6 cm diam.); the reactor core simply consists of these spheres loosely bedded in a cylindrical graphite cavity.

A further concept that has also been considered is the SiC matrix fuel, this fuel being considered as a direct replacement for a stainless-steel clad UO_2 fuel pin in an AGR (advanced gas-cooled reactor). The essential difference is that this fuel could operate in the CO_2 used for cooling in the AGR, while graphite fuels require a high purity inert coolant, i.e. helium, to avoid excessive corrosion of the fuel element graphite.

The main requirements for the matrix are to give an even distribution of fuel and to provide good thermal conductivity; on the other hand particles must not be broken during compaction. The compacts in the prismatic type of fuel elements are normally formed by hot-pressing a mix of essentially graphite powder and coated particles. Breakage can be minimized by first overcoating the particles with graphite powder. Warm extrusion is an alternative fabrication method. In both methods the finished compact requires thorough degassing. The normal method for fabricating the sphere type of fuel element is to injection-mould a graphite-particle mix into a machined hollow graphite sphere, the filling hole being sealed with a graphite plug.

To avoid the compacting stage, which adds to the fuel element cost, and also to reduce the amount of graphite that needs to be reprocessed after irradiation, consideration has been given to ways of incorporating loose particles into fuel elements, e.g. the fuel element for the proposed Colorado power reactor. In the form of a loose packed bed, however, they have a low bulk thermal conductivity, added to which there could be serious problems in the event of fracture of the fuel element due to particles being deposited all round the coolant circuit, hence some form of bonding between the particles is desirable.

The operational requirements for coated particle fuels will depend both on the fuel cycle to be adopted and the design of the fuel element. In the initial prismatic fuel element designs calculated fuel temperatures ranged up to $\sim$1600°C, due mainly to the large clearances between the fuel compact and the graphite tube. Recent designs have resulted in a reduction in the maximum fuel temperatures to 1000–1200°C in order to reduce the rate of movement of fission products from the fuel into the reactor circuit. The burn-up requirements are coupled to the fuel cycle. In a homogeneously loaded reactor using the $^{235}U-Th-^{233}U$ cycle, where the Th/U ratio is around $^{10}/_{1}$, the optimum burn-up approaches

200% fifa (fissions per initial fissile atom) which is equivalent to ~20% fima (fissions per initial metal atom). For Pu/Th and Pu/^{238}U cycles higher fissile/fertile ratios are required, but the fifa burn-up is lower so the burn-up levels required are in the same range. In the feed-breed systems it is desirable to obtain almost complete burn-up of the fissile feed particles (in practice 70–80%) whilst the breed elements with Th/U ratios of $\sim^{40}/_{1}$ will have relatively low fima burn-ups despite their residence time in the reactor being about 3 times that of the feed fuel.

For the particle to fulfil its function the coating must remain essentially impervious to fission products during its service in the reactor. In practice the quality of a coating can be judged by obtaining the ratio of the release rate of a specific isotope to the birth-rate of the same isotope (R/B) and upper limits can be specified. In order to obtain low R/B values the rate of failure of particles must obviously be low (i.e. below 1 in 10^4). The experimental programmes to develop these fuels have therefore concentrated on two aspects, one, the factors that influence the failure rate of the particles, and secondly, studies relating to the migration of fission products through coatings and their subsequent movement in the reactor circuit.

It has been shown that failure of coated particles can result from four main mechanisms:

1. Pressure exerted on to the coating both from swelling of the fuel due to retained fission products, and from volatile fission products released from the fuel kernel but retained by the coating.

To minimize this pressure adequate voidage must be incorporated into the particle. In practice this is achieved either by the use of low density fuel kernels, or by using a porous inner pyrocarbon layer. Both methods have proved successful providing there is an adequate thickness of coating to give the necessary restraint. To attain the very high burn-ups required in a feed fuel the fissile material requires to be diluted; this led to the development, by the Dragon Project, of a carbide fuel diluted in excess carbon, thus reducing parasitic neutron capture to a minimum.

2. Dimensional changes to the coating caused by fast neutron damage. The failure of coatings from this cause has only recently been investigated, since most of the high burn-up tests have been carried out in test reactors with essentially thermal fluxes. In HTR's there is, however, a large fast flux component giving integrated doses of the order of 10^{22}nvt (>0·18 MeV) in a four-year residence time. It has been shown that pyrocarbon coatings deposited in excess of 1800°C having a high density (>2·0 g/cm^3) and low anisotropy are the most dimensionally stable under irradiation and there is every indication that they will achieve this dose without failure. It has also been shown that pyrocarbon can accommodate at least 3% creep strain without cracking. SiC appears to undergo only very small dimensional changes under irradiation at high temperatures and hence should also withstand these fast neutron doses.

3. Fission fragment damage to the inner coating layers. Qualitatively the effects are similar to fast neutron damage; the shrinkage of some low density pyrocarbon layers can result in the initiation of cracks which together with a high internal pressure can cause complete penetration of the coating. This can be overcome either by incorporating crack-arresting interrruptions in the coating, or by using a coating that is relatively immune to damage (e.g. a very porous coating).

4. Diffusion of fuel through the coating. Although this is predominately a thermal rather than an irradiation effect it determines the upper temperature limit for operation of a specific type of coated particle. For pyrocarbon coated dicarbide fuel kernels this temperature limit is around 1600°C. The use of a SiC layer and/or an oxide fuel kernel appears to raise this temperature.

There is now an accumulation of irradiation data to show that particles can be designed to achieve burn-ups well in excess of 20% fima at temperatures up to 1400°C with extremely low failure rates.

Fission product release into the HTR circuit is of concern in two ways. The first relates to leakage or accidents, in which one must consider the permissible release of activity, and here the biologically important isotopes ^{131}I and ^{90}Sr are of most concern. The second problem concerns maintenance of the active reactor circuit. The isotopes ^{140}Ba and ^{144}Ce provide the most serious problems due to their high γ-energy; ^{137}Cs is a much less serious hazard.

Fission gas release from pyrocarbon coated particles can be related either to contamination of the outer coating surface by fissile material or to broken particles, as in the temperature range up to ~1500°C pyrocarbon is essentially impermeable to fission gases. The inclusion of a SiC layer is only of importance in that such coatings are normally less contaminated. ^{131}I behaves similarly though the rate of release is only about 10–20 per cent of that of ^{133}Xe. Plate-out round the circuit reduces the iodine in the coolant to extremely low levels.

Pyrocarbon coatings release substantial quantities of the relatively volatile carbide-forming fission products. The order of release is $^{89}Sr \gg {}^{137}Cs > {}^{140}Ba > {}^{144}Ce$. The addition of SiC reduces the release of strontium, barium and cerium by up to two orders of magnitude as shown by the analyses below which relate to Pu/Th fuel.

The subsequent movement of the fission products depends both on the temperature and geometry of the fuel element. For fuel tube temperatures of 1000°C caesium reaches diffusion equilibrium rapidly, hence the rate of ^{137}Cs entry into the coolant is effectively that at which it leaves the fuel compact. Strontium migrates more slowly and its rate of entry will not approach its rate of release from the fuel for about one year. Barium and cerium isotopes should never

	Max. irrad. Temp. (°C)	Burn-up (fifa)	Fractional release			
			^{137}Cs	^{89}Sr	^{140}Ba	^{141}Ce
Pyrocarbon coated	1400	22	4×10^{-2}	2×10^{-1}	4×10^{-4}	1×10^{-4}
PyC/SiC/PyC coated	1400	22	1×10^{-3}	4×10^{-4}	4×10^{-5}	6×10^{-5}

Al_2O_3 coatings are also effective in reducing the release of metallic fission products.

enter the coolant as metals. The main source of ^{140}Ba in the coolant is from the decay of ^{140}Xe released from exposed fuel.

Coated particle fuels have thus rapidly attained a high state of development and appear fully capable of achieving the requirements of the present HTR concepts. There is also every indication that higher fuel operating temperatures are possible, allowing greater flexibility in design. Coated particle fuels have so far only been considered for high-temperature gas-cooled reactors but it may be possible to take advantage of their excellent high burn-up performance in other systems where high fissile densities are not at a premium.

Bibliography

Journal of British Nuclear Energy Society, July 1966, Vol. 5, No. 3, (containing the Papers of the Symposium on High Temperature Reactors and the Dragon Project).

Proceedings of the Symposium on Ceramic Matrix Fuels containing Coated Particles, November 1962, USAEC TID-7654.

Development of Ceramic Coated Particles—Nuclear Fuels Summary Report, April 1966, BMI-1768.

J. B. Sayers

CONDENSER MICROPHONE. In a condenser microphone the displacement of a diaphragm, under the action of a sound wave incident on one side, is detected capacitatively using a fixed rigid electrode or back-plate near and parallel to the diaphragm. Usually the diaphragm is made so thin and flexible that the dominant force opposing motion is due to tension rather than the stiffness of the diaphragm material although, for use at very high sound pressures, stiff diaphragms or plates under no radial tension have been used. Laboratory condenser microphones generally have nominal outside diameters equal to 1, 1/2, 1/4, 1/8, inch; the smaller diameter types, while being less sensitive, extend the limit of substantially flat response to higher frequencies.

For an elementary analysis of the electroacoustical performance, the system is treated as having a single degree of freedom—the mean spacing of the membrane relative to the back-plate—and in this representation the diaphragm can be replaced by an equivalent plane piston with compliant mounting. Damping is provided by viscous motion of the gas molecules in the gap between diaphragm and back-plate, the latter having holes or annular grooves to reduce the damping to near critical.

The system most widely used for measuring the mean diaphragm displacement is that in which a d.c. polarizing voltage is applied to the microphone from a high-impedance source so that the capacitor operates under approximately constant charge conditions, capacitance change due to diaphragm displacement being converted directly into voltage change. Other detection systems include direct measurement of capacitance using amplitude or frequency modulation of a radio frequency signal.

Basic equations. In the simplest analysis the mechanical configuration consists of an equivalent plane piston of area S with mechanical compliance c_m, mass m_m, resistance r_m and at distance x from a parallel back-plate. Electrically the microphone consists of a capacitance C with charge q in series with a polarizing voltage E_0 and an alternating emf e. Assuming sinusoidal time dependence, with the acoustical impedance of the diaphragm Z_a given by

$$Z_a = j\omega m_a + r_a + \frac{1}{j\omega c_a},$$

where $m_a = m_m/S^2$, $r_a = r_m/S^2$, and $c_a = S^2 c_m$, and with $Z_e = 1/j\omega C_0$ the two basic equations of operation are

$$p = Z_a U + \frac{-\Phi}{j\omega C_0} i$$

$$e = \frac{-\Phi}{j\omega C_0} U + Z_e i,$$

where p = sound pressure, $U = -S\dot{x}$ = volume velocity at the diaphragm, $i = \dot{q}(t)$ = electrical current, and the transduction factor $\Phi = q_0/Sx_0 = \varepsilon_0 E_0/x_0^2$ where subscript zero denotes equilibrium values when polarized and ε_0 is the permittivity of free space. These equations suggest the equivalent network shown:

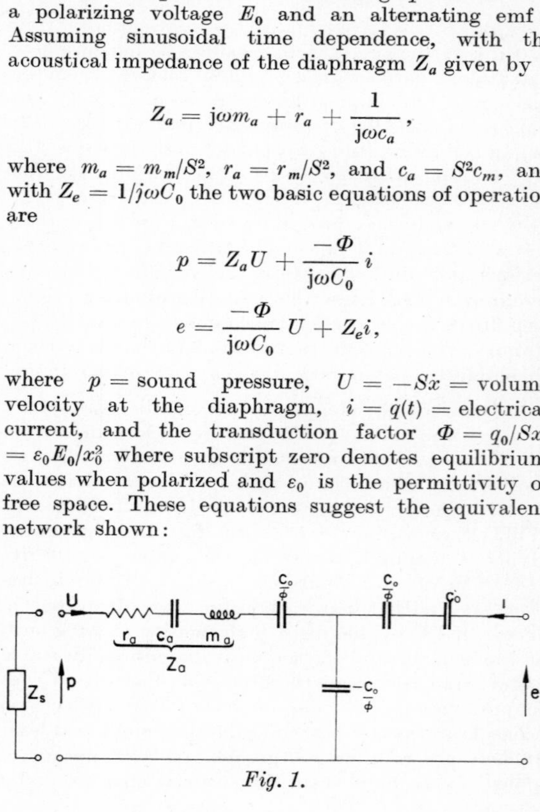

Fig. 1.

Pressure sensitivity. The pressure sensitivity of a microphone M_p is defined as $M_p = \left(\frac{e}{p}\right)_{i=0}$ where p is the sound pressure uniformly applied to the face of the microphone. From the above equations

$$M_p = \frac{-\Phi}{j\omega C_0} \frac{1}{Z_a}.$$

Thus at frequencies where the system is stiffness-controlled $Z_a \cong 1/j\omega c_a$ and the response is independent of frequency, while above resonance $Z_a \cong j\omega m_a$ and the sensitivity decreases at approximately 12dB/octave.

If Y is the electrical input admittance measured with the acoustical terminals short-circuited ($Z_s = 0$), achieved practically using a $\lambda/4$ termination

$$M_p = -\frac{1}{\Phi}\left\{1 - \frac{j\omega C_0}{Y}\right\}.$$

While Φ is not easily calculated it is, to first order at least, independent of frequency and this provides a method of determining the relative response as a function of frequency from purely electrical measurements. C_0 may be determined from an admittance measurement at very high frequencies (>100 kc/s) where diaphragm inertia effectively prevents appreciable motion.

Acoustical impedance. It is difficult to measure directly the acoustical impedance of the microphone diaphragm as a function of frequency, but using the above analysis it may be derived from measurements of electrical admittance with the microphone terminated acoustically in a $\lambda/4$ shunt when

$$Z_a = \Phi^2 \frac{Y}{j\omega C_0(Y - j\omega C_0)},$$

in which the constant Φ may be deduced from an approximate value of microphone sensitivity at any given frequency. At low and middle frequencies the impedance is primarily that of a compliance and the diaphragm impedance is often expressed in terms of an equivalent air volume.

Harmonic distortion. For an electrically charged but unloaded capacitor there is no distortion when the electrode spacing varies sinusoidally. However, loading the microphone capsule with resistance or capacitance does produce harmonic distortion at high amplitudes. For example, loading with resistance R produces a second harmonic distortion of

$$D_2 = \frac{e}{E_0} \frac{1}{2\omega RC_0}$$

for $R \gg 1/\omega C_0$ where e is the output at fundamental frequency. (The inequality $\omega RC_0 \gg 1$ is usually satisfied in practice over the frequency range of interest as otherwise the low-frequency sensitivity decreases due to the time-constant of the electrical circuit.)

Noise in condenser microphones. From the purely electrical noise source associated with the input resistance R of the microphone preamplifier and polarizing source, an r.m.s. noise voltage e_g is generated where

$$\overline{e_g^2}\Big|_f^{f+df} = \frac{4kTR\,df}{1 + \omega^2 R^2 C^2},$$

where k is Boltzmann's constant and T the absolute temperature. Since ωRC is usually much greater than 1, the spectrum level generally decreases with increasing frequency over the major part of the audio range. The need for very high-impedance electrical circuitry is thus sometimes a disadvantage of the polarizing system and amplitude or frequency modulation of radiofrequency signals is employed to overcome this difficulty. Such r.f. systems may operate at an impedance of tens of ohms instead of hundreds of megohms with consequent reduction in electrically generated noise.

However, the lower noise limit is ultimately set by the Brownian motion of gas molecules against the diaphragm of the microphone. With r_a the acoustical resistance acting on the diaphragm the mean square noise pressure is given by

$$\overline{p^2}\Big|_f^{+df} = 4kTr_a\,df.$$

The output spectrum of the noise voltage will thus have precisely the same shape as the sensitivity curve of the microphone. Typically $r_a \sim 10^7$ N.s.m^{-5}.

See also: Electrostatic actuator. Pistonphone.

Bibliography

ZAALBERG VAN ZELST J. J. (1947/48) *Circuit for condenser microphones with low noise level, Philips Tech. Rev.* **9**, 357.

HUNT F. V. (1954) *Electroacoustics*, New York: Wiley.

M. E. DELANY

CRYSTAL SYMMETRY, MAGNETIC. The conventional symmetry operations of a crystal, namely, rotations, rotation-inversions, screws, glides and lattice-translations, bring the spatial arrangement of atoms into self-coincidence. This self-coincidence is achieved both in respect of density and charge distributions. If, however, the crystal is ferromagnetic or antiferromagnetic in which spontaneous magnetic moments are associated with the atoms, the conventional symmetry operations of the crystal may not bring about self-coincidence in respect of the magnetic moment orientations of the atoms. This point may be illustrated by considering a simple system comprising of two identical spherical atoms with magnetic moments aligned parallel to each other as shown in the figure.

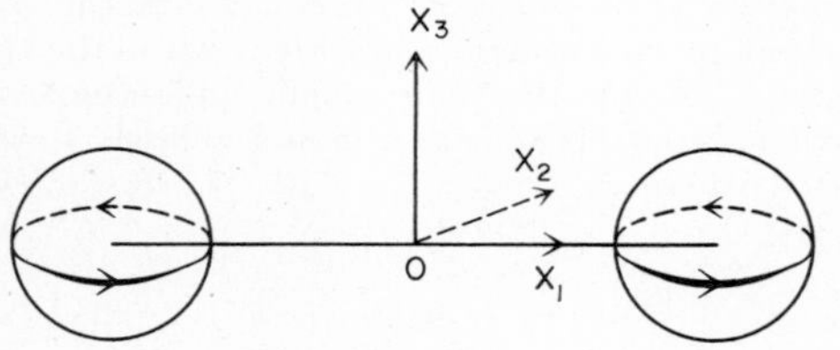

Fig. 1.

The magnetic moment of each atom can be considered to be generated by a circular electric current in X_1X_2 plane. The direction of the current is indicated by an arrow head. The orientation of the magnetic moment vector is fixed by the sense or direction of the current. For this system, a two-fold rotation about X_3 is a symmetry operation in every respect including the orientation of magnetic moment because it does not alter the current direction. A two-fold rotation about X_1 or X_2 brings the atoms into coincidence but reverses the current thereby reversing the magnetic moments. Such a rotation cannot be taken as a symmetry operation of the system if self-coincidence is demanded in all respects. Thus we find that conventional notions are inadequate to describe the symmetry of crystals which possess magnetic moment distributions. To meet this situation, the concept of symmetry operation has been generalized. The operation of reversal of magnetic moment is defined as a possible symmetry operation and is denoted by $\mathscr{R}$. It is also referred to as a time-reversal operation because reversal of magnetic moments is equivalent to the reversal of an electric current which in turn is equivalent to time-reversal. The products of $\mathscr{R}$ with the conventional symmetry operations, where product means consecutive application, are also possible symmetry operations. They are called *complementary operations* or *anti-operations*. The product of a conventional operation S with $\mathscr{R}$, i.e. the complement of S is denoted by $\underline{S}$. The operations complementary to rotations 2, 3, 4, 6 are respectively denoted by $\underline{2}$, $\underline{3}$, $\underline{4}$, $\underline{6}$; those complementary to rotation-inversions $\bar{3}$, $\bar{4}$, $\bar{6}$ are $\underline{\bar{3}}$, $\underline{\bar{4}}$, $\underline{\bar{6}}$ respectively. $\bar{2}$ is a simple reflection and $\underline{\bar{2}}$ is complementary to it. $\bar{1}$ is inversion and $\underline{\bar{1}}$ is complementary to it. The operation $\mathscr{R}$ itself is considered as complementary to the identity operation 1 and it is also denoted by $\underline{1}$. A similar adaptation of notation is extended to glides, screws and lattice translations.

It is readily seen that $\mathscr{R}^2 = 1$. If S_1, S_2, S_3 ..., denote conventional operations, we have $\mathscr{R} \cdot S_i = S_i \cdot \mathscr{R} = \underline{S_i}$.

If $S_i \cdot S_j = S_k$, we have $\underline{S_i} \cdot S_j = S_i \cdot \underline{S_j} = \underline{S_k}$. Using these product rules, all possible space groups and point groups that can be built out of the set of conventional and complementary operations can be constructed.

A straightforward method of deriving the generalized point groups from the 32 conventional ones is as follows. The two elements 1 and $\mathscr{R}$ form a group of order two. By taking the direct product of each of the 32 conventional groups with the group (1, $\mathscr{R}$), we get 32 groups known as *grey groups*. We then pick out the distinct sub-groups of the grey groups. They are 90 in number. 32 of these are the conventional groups themselves and they are called single-colour groups. The remaining 58 are called double-colour groups. They are also referred to as variants of the 32 conventional groups. There are thus 122 groups in all. This generalization was originally introduced by Shubnikov. He called them *colour groups* because he took a change of colour of the motif from black to white and white to black as his operation $\mathscr{R}$.

As an illustration, the generalized point groups related to the conventional point group 2/m will be derived here. $2/m$ consists of the elements (1, 2, $\bar{1}$, $\bar{2}$). The direct product with (1, $\mathscr{R}$) yields the grey group (1, 2, $\bar{1}$, $\bar{2}$, $\mathscr{R}$, $\underline{2}$, $\underline{\bar{1}}$, $\underline{\bar{2}}$). The distinct sub-groups are (1, 2, $\bar{1}$, $\bar{2}$) which is $2/m$ itself; (1, $\underline{2}$, $\underline{\bar{1}}$, $\bar{2}$), (1, 2, $\underline{\bar{1}}$, $\underline{\bar{2}}$), (1, $\underline{2}$, $\bar{1}$, $\underline{\bar{2}}$) which are double-colour groups denoted by $\underline{2}/m$, $2/\underline{m}$, $\underline{2}/\underline{m}$ respectively.

The grey point groups contain the operation $\mathscr{R}$ explicitly. They cannot therefore describe the symmetry of magnetic crystals because moment reversal cannot be a symmetry operation in such crystals. In paramagnetic crystals, the time-average magnetic moment at every lattice point is zero whereas in diamagnetic crystals there is no spontaneous magnetic moment at all for the atoms. For these two categories of crystals, the moment reversal is a symmetry operation in a trivial way. The symmetry of the paramagnetic and diamagnetic crystals is therefore classified under the grey groups.

The single-colour and double-colour groups do not contain $\mathscr{R}$ explicitly. The ferro and antiferromagnetic crystals are classified under these groups. These 90 groups are for this reason called *magnetic point groups* and are listed in the Table.

We may refer to Fig. 1 for an illustration of a system possessing the symmetry of a double-colour group. If we ignore the magnetic moments of the atoms, the system has for its symmetry elements three two-fold rotations about X_1, X_2, X_3 denoted by 2, 2′, 2″ respectively, the inversion $\bar{1}$ about the origin O, three reflections in the coordinate planes X_2X_3, X_3X_1, X_1X_2 denoted by $\bar{2}$, $\bar{2}'$, $\bar{2}''$ respectively. This group is denoted by the symbol *mmm*. We note that operations 2, 2′, $\bar{2}$ and $\bar{2}'$ reverse the current and thereby reverse the magnetic moment. Therefore, if self-coincidence in respect of magnetic moment orientation is also demanded, operations complementary to the above have to be taken as symmetry operations; and the

Magnetic point groups

No.	Class symbol	No.	Class symbol	No.	Class symbol
1	1	29	$4mm$	60	$6/m$
2	$\bar{1}$	30	$4\underline{mm}$	61	$6/\underline{m}$
3	$\underline{\bar{1}}$	31	$\underline{4}m\underline{m}$	62	$\underline{6}/m$
4	m	32	$\bar{4}2m$	63	$\underline{6}/\underline{m}$
5	$\underline{m}$	33	$\underline{\bar{4}}2\underline{m}$	64	$\bar{6}m2$
6	2	34	$\underline{\bar{4}2}m$	65	$\bar{6}\underline{m2}$
7	$\underline{2}$	35	$\bar{4}\underline{2m}$	66	$\underline{\bar{6}}m\underline{2}$
8	$2/m$	36	422	67	$\underline{\bar{6}m}2$
9	$\underline{2}/m$	37	$4\underline{22}$	68	$6mm$
10	$2/\underline{m}$	38	$\underline{4}2\underline{2}$	69	$6\underline{mm}$
11	$\underline{2}/\underline{m}$	39	$4/mmm$	70	$\underline{6m}m$
12	$2mm$	40	$\underline{4}/\underline{m}m\underline{m}$	71	622
13	$\underline{2}m\underline{m}$	41	$4/\underline{m}mm$	72	$6\underline{22}$
14	$2\underline{mm}$	42	$4/m\underline{mm}$	73	$\underline{6}2\underline{2}$
15	222	43	$4/\underline{mmm}$	74	$6/mmm$
16	$\underline{22}2$	44	$\underline{4}/mm\underline{m}$	75	$6/\underline{m}mm$
17	mmm	45	3	76	$6/m\underline{mm}$
18	$\underline{m}mm$	46	$\bar{3}$	77	$6/\underline{mmm}$
19	$\underline{mmm}$	47	$\underline{\bar{3}}$	78	$\underline{6}/mm\underline{m}$
20	$\underline{mm}m$	48	$3m$	79	$\underline{6}/\underline{m}m\underline{m}$
21	4	49	$3\underline{m}$	80	23
22	$\underline{4}$	50	32	81	$m3$
23	$\bar{4}$	51	$3\underline{2}$	82	$\underline{m}3$
24	$\underline{\bar{4}}$	52	$\bar{3}m$	83	$\bar{4}3m$
25	$4/m$	53	$\bar{3}\underline{m}$	84	$\underline{\bar{4}}3\underline{m}$
26	$\underline{4}/\underline{m}$	54	$\underline{\bar{3}}m$	85	432
27	$4/\underline{m}$	55	$\underline{\bar{3}m}$	86	$\underline{4}3\underline{2}$
28	$\underline{4}/m$	56	$\bar{6}$	87	$m3m$
		57	$\underline{\bar{6}}$	88	$\underline{m}3\underline{m}$
		58	6	89	$m3\underline{m}$
		59	$\underline{6}$	90	$\underline{m}3m$

symmetry of the system is given by the double-colour group $\underline{mm}m$.

The *generalized space groups,* also known as *Shubnikov groups,* may be derived from the 230 conventional space groups in a similar manner. A convenient procedure would be to first derive what are called the *magnetic lattices.* These are obtained by generalizing the 14 Bravais lattice. One of the ways in which a graphic description of the generalized lattices may be obtained is to regard the single-colour lattices as consisting of lattice points of a single colour—say white, and the double-colour lattices as consisting of lattice points of two colours—say black and white. A complementary translation is hence a lattice translation followed by change of colour. The double-colour lattices derivable from each Bravais lattice are obtained by considering all distinct possibilities of introducing black lattice points at the edge centres or side centre or face centres or the body centre of the crystallographic unit cell of the lattice in a consistent manner such that the system of points still forms a lattice and retains the original symmetry. The distinct double-colour lattices thus generated from all the 14 Bravais lattices are 22 in number and are given below. The change of colour is synonymous with reversal of magnetic moment. Accordingly the single-colour and double-colour lattices are together known as magnetic lattices. In the description given below, P stands for primitive; R for rhombohedral primitive; A, B, C for the respective side-centred; F for face-centred; I for body-centred. Suffixes a, b, c indicate that the respective edges are centred with black lattice point; suffix s indicates that edge a or b or c is similarly centred; suffixes A, B, C indicate that the respective side is centred similarly; and suffix I indicates that the cell body is centred similarly. The international convention is adopted regarding the choice of vectors a, b, c.

Triclinic:	P, P_s
Monoclinic:	P, P_a, P_b, P_C; C, C_c, C_a
Orthorhombic:	P, P_a, P_C, P_I; C, C_c, C_a, C_A; F, F_s, I, I_c
Tetragonal:	P, P_c, P_C, P_I; I, I_c
Trigonal:	R, R_I
Hexagonal and Trigonal	P, P_c
Cubic:	P, P_I; F, F_s; I

To derive the generalized space groups, we proceed by dealing with the conventional space groups one by one. The conventional group under consideration will itself be the single-colour group. The conventional group, augmented by the complements of all its elements, is the associated grey group. The distinct variants of the factor group are derived in a manner analogous to the variants of the point group. The number of distinct variants, however, may be more than those of the point group. The double-colour factor groups on the single-colour lattice give rise to some space groups. The single-colour factor group on the double-colour lattices give some more space groups. Whenever there is arbitrariness regarding the direction of the screw, glide or complementary translation all possible combinations should be investigated and distinct groups generated taken.

As an illustration, we may derive the Shubnikov generalizations of the conventional space group orthorhombic Pma2. The grey group is $Pma2\underline{1}$. The orthorhombic primitive lattice has 3 distinct variants P_c, P_C, P_I. The variants of the factor group ma2 are 3. They are $\underline{m}a\underline{2}$, $m\underline{a2}$, $\underline{ma}2$. Some of the double-colour space groups which can be readily written down are $P\underline{m}a\underline{2}$, $Pm\underline{a2}$, $P\underline{ma}2$, P_cma2, P_Cma2, P_Ima2. However, the two-fold axis in this group is along c whereas the glide is along a. A close examination shows that the magnetic translation being along a or along b

or along c, each generates a distinct group. Thus, $P_a ma2$ and $P_b ma2$ are distinct from each other and from $P_c ma2$. In a similar manner, $P_A ma2$ and $P_B ma2$ are distinct from each other and from $P_C ma2$. Thus we have 10 double-colour groups derivable from the conventional space group $Pma2$.

In all there are 230 single-colour space groups, 230 grey space groups and 1191 double-colour space groups, making a total of 1651 generalized space groups.

Bibliography

BACON G. E. (1961) in *Encyclopaedic Dictionary of Physics* (J. Thewlis Ed.) **2**, 235, Oxford: Pergamon Press.

BHAGAVANTAM S. (1966) *Crystal Symmetry and Physical Properties*, New York: Academic Press.

PHILLIPS F. C. (1961) *in Encyclopaedic Dictionary of Physics* (J. Thewlis Ed.) **1**, 493; **2**, 232, 236, Oxford: Pergamon Press.

SHUBNIKOV A. V. and BELOV N. V. (1964) *Coloured Symmetry*, Oxford: Pergamon Press.

S. BHAGAVANTAM

CRYSTAL SYMMETRY AND PHYSICAL PROPERTIES. *1. Non-magnetic properties.* We are concerned with single crystals and their macroscopic physical properties. From this point of view, a crystal may be considered as homogeneous and continuous. Its properties depend only on direction, the translational symmetry playing no part. Thus we need consider only the point group symmetry of a crystal while studying its physical properties. In this part of the article. we concern ourselves only with non-magnetic properties. A unified treatment including magnetic properties as well would have been more logical but it is not preferred here.

Any physical property that a crystal may exhibit must possess at least the symmetry of the crystal. This principle enunciated by Neumann is axiomatic in crystal physics. It means that the symmetry elements of any physical property observed in a crystal must include all the symmetry elements of the point group of the crystal. By the symmetry elements of a physical property, we mean those coordinate transformations that make the set of functions defining the property invariant. The principle implies that any particular physical property may possess a higher symmetry than that of the crystal. For example, properties such as optical refraction and elasticity are centrosymmetric. Such properties will continue to be centrosymmetric even in crystals that are not centrosymmetric.

All physical properties may be interpreted as relations between two measurable physical quantities of which one may be called the influence, the other the effect. For example, the physical property of electric polarizability of a crystal relates the electric field imposed on it and the electric moment induced in it. Similarly, the physical property of elasticity relates the physical quantities, stress and strain.

The appropriate mathematical entities to represent physical quantities are tensors and the relations between them are expressed as tensor relations. To a first order of approximation, the relations are linear and their general form will be as in (1).

$$B_{ijk\ldots} = a_{ijk\ldots\,lmn\ldots} A_{lmn\ldots} \tag{1}$$

The tensors $A_{lmn\ldots}$, $B_{ijk\ldots}$ represent the influence and effect respectively. The tensor $a_{ijk\ldots lmn\ldots}$, the components of which form the coefficients in the relation, represents the physical property. If one of the physical quantity tensors and not both in (1) is an axial tensor, the tensor $a_{ijk\ldots lmn\ldots}$ is also axial.

Neumann's principle lays down that a physical property tensor acquires the symmetry of the crystal in which the property is observed. That is, the tensor will have an appropriate form for each crystal class such that it is invariant under all the symmetry operations of that class. To determine the form of a tensor in a particular class, we perform the coordinate transformations appropriate to each symmetry operation of that class on the tensor components and demand invariance. Each such transformation gives rise to as many linear equations as there are components. In the final analysis, we shall be left with the result that certain components of the tensor vanish and certain components are only linear combinations of others. The generality in the situation is apparent. All physical properties represented by tensors of the same rank and kind are affected in the same manner by the symmetry of a class. Once the forms of a tensor appropriate to the 32 classes were determined, they would serve as a permanent reference for all properties represented by tensors of the same rank and kind. The study of the effect of symmetry on physical properties is thus simplified.

When we say tensors of the same kind, we not only refer to their polar or axial nature but also to their intrinsic symmetry. The intrinsic symmetry of a tensor is given by the group of permutations of its suffixes under which the tensor remains invariant or changes sign. In the case of a tensor representing a physical quantity or property, such intrinsic symmetry is due to considerations that are either mechanical or thermodynamic. We shall consider the elasticity tensor as an illustration. This tensor c_{ijkl} is defined by the relation $\varepsilon_{ij} = c_{ijkl}\tau_{kl}$ where ε_{ij} is the strain tensor and τ_{kl} is the stress tensor ($i, j, k, l = 1, 2, 3$). We have $\varepsilon_{ij} = \varepsilon_{ji}$ from the very definition of strain and $\tau_{kl} = \tau_{lk}$ from certain mechanical considerations. c_{ijkl} automatically acquires the intrinsic symmetries of the influence and effect. That is, c_{ijkl} is invariant under the interchange of suffixes 1 with 2 and suffixes 3 with 4. In addition, we also know that the c_{ijkl} are second order partial derivatives of the strain energy function expressed as a quadratic from in ε_{ij}. From the fact that the order of differentiation should not matter, it follows that the tensor is invariant for interchange of the set of suffixes ij with kl.

In the case of transport properties, the intrinsic symmetry is determined by Onsager's principle. For example, it follows from this principle that the thermal conductivity and electrical conductivity tensors which are of second rank are symmetric with respect to interchange of suffixes.

Thus the intrinsic symmetry of a physical property tensor depends on the nature of the property. It acquires the intrinsic symmetry of the physical quantity tensors it relates. In relation (1), the tensor $a_{ijk...lmn...}$ remains invariant or changes sign under the permutations of the suffixes $ijk...$ and $lmn...$ under which the tensors $B_{ijk...}$ and $A_{lmn...}$ remain invariant or change sign. Once the intrinsic symmetry is established, it is a straightforward matter to determine which components of the tensor are related to each other and in what manner.

The procedure of applying the symmetry transformations of a crystal class to the components of a tensor is simplified due to a method developed by *Fumi*. It uses the correspondence between polar tensor components and coordinate products. For example, a second rank polar tensor has the same transformation properties as the 9 coordinate products x_1x_1, x_1x_2, x_1x_3, x_2x_1, x_2x_2, x_2x_3, x_3x_1, x_3x_2, x_3x_3. It is easy to see by direct inspection how the coordinate products transform. As an illustration, let us obtain the form of second rank polar tensor α_{ij} appropriate to the point group 2. If we take the 2 axis along X_3, the operation transforms the coordinates as $x_1 \to -x_1$; $x_2 \to -x_2$; $x_3 \to x_3$. Under this transformation, all products in which x_3 occurs only once will be transformed into their negatives and since this is a symmetry transformation which should leave the components invariant, all such products must vanish. We conclude that the corresponding components α_{13}, α_{23}, α_{31}, α_{32} must vanish. The rest of the components, namely, α_{11}, α_{12}, α_{21}. α_{22}, α_{33} remain. If the tensor α_{ij} has the intrinsic symmetry $\alpha_{ij} = \alpha_{ji}$, it follows that amongst the non-vanishing components $\alpha_{12} = \alpha_{21}$. Thus only 4 components are independent. This method can be easily extended to tensors of higher rank and also to axial tensors. The direct inspection, however, is possible only for triclinic, monoclinic, orthorhombic, tetragonal and cubic classes in which the conventional orthogonal coordinates do not transform into linear combinations of more than one coordinate. The trigonal and hexagonal classes can be treated directly only by using special coordinate frames. Fumi has developed the necessary techniques to handle these cases but they are not given here. In such cases, the elaborate method of writing down the transformation equations of the tensor components may be adopted. The forms of the general polar tensors up to sixth rank appropriate to the 32 classes are available in the literature.

A group theoretical method developed by *Bhagavantam* enables one to find the number of non-vanishing independent components of a physical property tensor subject to a point group symmetry. A determination of the number of non-vanishing independent components serves as a useful check on the results obtained by direct methods. Further, the knowledge of these numbers itself is quite useful in considering the physical properties of crystals. This method is given below.

Consider a tensor subjected to a particular point group. To each symmetry operation of the group corresponds a matrix which transforms the components of the tensor and the dimension of this matrix will be equal to the number of components of the tensor. All such matrices form a reducible representation of the point group called the tensor representation on the linear space of the tensor components. The number n of non-vanishing independent tensor components we are seeking is equal to the number of times the total symmetric irreducible representation occurs in the tensor representation. We have

$$n = \frac{1}{N} \sum_{\varrho} h_\varrho \chi_\varrho(R) . \qquad (2)$$

In (2), N is the order of the group, h_ϱ is the number of elements in class ϱ, $\chi_\varrho(R)$ is the character of an element R of the group belonging to the ϱth class in the tensor representation. Thus to find n, all that we need to know is the character of each of the group elements in the tensor representation. The character of an element in a representation is the sum of the diagonal elements of the corresponding matrix. The routine procedure of calculating the characters in a tensor representation is the following. The character of a pure rotation through φ in a polar vector representation is evidently $(2c + 1)$ where c stands for $\cos\varphi$. The character of a rotation-reflection is $(2c - 1)$. In an axial vector representation the characters are $(\pm 2c + 1)$ respectively. A tensor of rank s can be expressed as an outer product of s vectors and the character in the tensor representation is the product of the characters in the representations due to the s vectors. Also, the characters in a representation due to a physical property tensor are the products of the characters in the influence and effect tensor representations. The characters in higher rank tensor representations are thus evaluated from those of the lower rank.

The diagonal elements of the matrices of the tensor representation are easily computed to find the characters directly by noting that the matrices in the tensor representation are Kronecker products of the matrices of the vector representations. In this process, it is possible to take account of the intrinsic symmetry of the tensor by direct inspection. For example, the matrices in a polar vector representation have the form $\begin{pmatrix} c & s & 0 \\ -s & c & 0 \\ 0 & 0 & \pm 1 \end{pmatrix}$ where $c = \cos\varphi$; $s = \sin\varphi$; $+1$ is taken for pure rotation through φ, and -1 is taken for rotation-reflection. A second rank polar tensor representation consists of 9×9 matrices. They are obtained by the following procedure. In the first row first column position of the above 3×3 matrix, put the entire 3×3 matrix after multiplying every one of its elements with the first row first column element. Repeat this process for each of the elements of the 3×3 matrix. It is easy to see that the diagonal elements of the 9×9 matrices are c^2, c^2, 1, $\pm c$, $\pm c$, $\pm c$, $\pm c$, c^2, c^2 and the character is $(4c^2 \pm 4c + 1)$. This is equal to $(2c \pm 1)^2$, that is, square of the character in the vector representation. If the tensor is symmetric with respect to interchange of indices, we note that in the 9×9 matrix we need take only the rows corresponding to the tensor components with suffixes 23, 31, 12 and add the elements of columns corresponding to 12 and 21, 13 and 31, 23 and 32. We thus find that the character of a symmetry element in a second rank symmetric polar tensor representation is $(4c^2 \pm 2c)$. The results of investigations in respect of 32 crystal classes for several particular physical properties are available in the literature.

It may be noted that crystal symmetry by itself does not forbid any physical property tensor in all the crystal classes. All polar tensors of odd rank vanish identically in all the eleven centrosymmetric classes. The fact that piezoelectricity which is a third rank polar tensor is not exhibited by centrosymmetric crystals is an illustration of this statement. All axial tensors of even rank vanish identically in all the eleven centrosymmetric classes. The fact that optical activity which is a second rank axial tensor is not exhibited by centrosymmetric crystals is again an illustration of this statement.

2. *Magnetic properties.* For a long time, it was supposed that properties like piezomagnetism were forbidden. It was assumed that crystal structures are time symmetric whereas the above mentioned magnetic property is time asymmetric. For this reason, the property was assumed to identically vanish in all crystal classes. In recent years, it has been shown that in their ferromagnetic and antiferromagnetic phases, crystals do not possess a time symmetric structure and therefore, magnetic properties of the above type are not ruled out. A magnetic moment is equivalent to a circular electric current and a reversal of current reverses the moment. Reversal of current is equivalent to reversal of time. The *time inversion operation* has to be recognized as a possible symmetry operation and we shall designate it by $\mathscr{R}$. Its products with the conventional operations are also possible symmetry operations and are called complementary operations. They are denoted by an underscore on the symbol for the corresponding conventional operation. For example complementary diad is denoted by $\underline{2}$. Obviously $2\mathscr{R} = \mathscr{R}2 = \underline{2}$. Taking the new symmetry operations into account, the possible crystal classes have been constructed and they are 122 in all. Out of these, 32 classes are obtained by taking direct products of the conventional point groups with the group $(E, \mathscr{R})$. Obviously these groups contain $\mathscr{R}$ explicitly and include the complementary operation of every one of the conventional operations in it. They are called the grey classes. The remaining classes are merely the distinct subgroups of these and may be arrived at by identifying them in that manner. Of these, 32 classes do not include $\mathscr{R}$ or any complementary operation. They are the same as the conventional classes and are called the *single coloured classes.* The remaining 58 classes do not contain $\mathscr{R}$ explicitly but contain complementary operations. They are called the *double coloured classes.*

The paramagnetic and diamagnetic crystals do not possess magnetic structures and are considered to be time symmetric. We may note that $\mathscr{R}$ is a symmetry operation for these crystals and they have to be within the 32 grey classes. The ferromagnetic and antiferromagnetic crystals come from the remaining 90 classes which are known as *magnetic classes.*

The magnetic field, moment, and intensity are axial vectors and are time asymmetric. That is, their components get multiplied by -1 on the application of $\mathscr{R}$. It follows that every physical quantity tensor which contains as a factor an odd number of these magnetic vectors is also time asymmetric. All such properties are called magnetic properties and the tensors representing them are called magnetic tensors. Under conventional symmetry operations, the magnetic tensors transform in the usual way. Under the operation $\mathscr{R}$, their components get multiplied by -1. To apply a complementary operation to a magnetic tensor, we first perform the corresponding conventional operation and then multiply the transformed components by -1. Physical property tensors other than the magnetic ones transform as under the identity operation when $\mathscr{R}$ is applied. Similarly, under complementary operations, they transform as under corresponding conventional operations. In relation (1), if one of the physical quantity tensors is magnetic, it follows that the physical property tensor is also magnetic. Neumann's principle applies to the case of magnetic tensors as well with the understanding that the crystal symmetry is that of one of the 122 classes.

The following results emerge. All magnetic tensors vanish in all the grey classes. For non-magnetic properties, all the 122 classes merge into the 32 conventional ones and for this reason, the generalization effected does not alter the results obtained in section 1. The operation of inversion i leaves a magnetic vector invariant because it is an axial vector. $\underline{i}$ which is the complement of inversion is therefore equivalent to $\mathscr{R}$ for magnetic vectors. It follows that all magnetic axial tensors of even rank and magnetic polar tensors of odd rank vanish in the 21 magnetic classes that contain i. They are $\bar{1}$, $2/m$, $\underline{2}/\underline{m}$, mmm, $m\underline{mm}$, $4/m$, $\underline{4}/m$, $4/mmm$, $\underline{4}/mm\underline{m}$, $4/m\underline{mm}$, $\bar{3}$, $\bar{3}m$, $\bar{3}\underline{m}$, $6/m$, $\underline{6}/\underline{m}$, $6/mmm$, $\underline{6}/\underline{m}m\underline{m}$, $6/m\underline{mm}$, $m3$, $m3m$, $m3\underline{m}$. Magnetoelectric polarizability is a second rank magnetic axial tensor which relates the axial magnetic moment vector with the polar electric field vector. It vanishes in the above 21 classes. It also follows that magnetic axial tensors of odd rank and magnetic polar tensors of even rank vanish in the 21 magnetic classes that contain *i*. They are $\underline{\bar{1}}$, $2/\underline{m}$, $\underline{2}/m$, $\underline{mmm}$. $\underline{m}mm$, $4/\underline{m}$, $\underline{4}/\underline{m}$, $4/\underline{mmm}$, $4/\underline{m}mm$, $\underline{4}/\underline{m}m\underline{m}$, $\underline{\bar{3}}$, $\underline{\bar{3}m}$, $\underline{3}m$, $6/\underline{mmm}$, $6/\underline{m}mm$, $6/\underline{m}mm$, $\underline{6}/mm\underline{m}$, $\underline{m}3$, $\underline{m}3\underline{m}$, $\underline{m}3m$. Piezomagnetism is a third rank magnetic axial tensor which relates the magnetic moment vector with the stress tensor. It vanishes in the above 21 classes.

Using the rules given above for the transformation of magnetic tensors, one can proceed to apply the various techniques developed to find the forms of any magnetic tensor in the 90 magnetic classes. In applying the group theoretical method to obtain the number of independent non-vanishing components, we note that the character of a complementary operation in a magnetic tensor representation is obtained by multiplying the conventional character by -1. An important result relating to ferromagnetism is that it is possible in those magnetic classes in which the magnetic moment vector does not vanish. These ferromagnetic classes are readily obtained as 31 in number and they are 1, $\bar{1}$, 2, $\underline{2}$, m, $\underline{m}$, $2/m$, $\underline{2}/\underline{m}$, $2\underline{22}$, $2\underline{mm}$, $\underline{2}m\underline{m}$, $m\underline{mm}$, 4, $4\underline{22}$, $4/m$, $4\underline{mm}$, $4/m\underline{mm}$, $\bar{4}$, $\bar{4}\underline{2m}$, 3, $3\underline{2}$, $3\underline{m}$, $\bar{3}$, $\bar{3}\underline{m}$, $\bar{6}$, $\bar{6}\underline{m2}$, 6, $6\underline{22}$, $6/m$, $6\underline{mm}$, $6m\underline{mm}$. Results obtained for pyromagnetism, magnetoelectric polarizability and piezomagnetism are available in the literature. The existence of magnetoelectric polarizability and piezomagnetism in some antiferromagnetic crystals has been experimentally demonstrated.

Bibliography

BHAGAVANTAM S. (1966) *Crystal Symmetry and Physical Properties*, New York: Academic Press.

NYE J. F. (1960) *Physical Properties of Crystals*, Oxford: The University Press.

PHILIPS F. C. (1961) in *Encyclopaedic Dictionary of Physics* (J. Thewlis Ed.), **2**, 232, Oxford: Pergamon Press.

S. BHAGAVANTAM

CRYSTAL WHISKERS. The term is applied to filaments of metals and compounds which normally are found to be single crystals and to possess abnormal properties, often approaching those of a perfect solid. Whiskers have now been prepared in more than 30 substances by a variety of methods which may be grouped under several heads.

1. From electrodeposits and alloys containing a low melting point component. This often happens spontaneously or in special atmospheres. In many cases, such as cadmium, zinc and tin, the whiskers have been shown to be single crystals of the pure metal, but in other cases they may be oxides or sulphides. In some cases, impurities or contaminants or plating faults are essential to whisker formation, while pressure on the metal base has been shown to increase whisker formation.

2. From the vapour. Metals having a high vapour pressure may be successfully grown in whisker form in a chamber in which a suitable temperature gradient is induced so that progressive condensation occurs. The nature of the gas present and its pressure, or the existence of a vacuum have a marked influence on whisker growth.

3. Reduction of salts. Suitable salts are held in a container or boat near their melting point in a reducing atmosphere such as hydrogen, and whiskers are grown from the container, which often is made of the same material as that of the desired whisker.

4. Electrolysis. By careful control of a well-known tendency for dendritic growth under certain conditions in electrolysis, such as high current density, it is possible to produce whiskers of a variety of metals.

5. Controlled oxidation. Oxide whiskers may be produced by heating in suitable atmospheres at high temperature.

6. From solution. Rapid cooling of a solution encourages dendritic growth often in the form of whiskers.

Other special methods have been used for specific materials, notably the use of a d.c. arc for graphite whiskers or decomposition processes for silicon and germanium.

The mechanism by which whiskers form depends very much on the method of preparation. Whiskers grown from plated deposits, for example, are believed to grow from the base by some type of dislocation motion accompanied by vacancy diffusion away from the base of the whisker. On the other hand, whiskers grown from the vapour, by reduction, or from solution, more probably occur by progressive addition of deposited atoms at the tip and side, probably in many cases involving an axial screw dislocation to provide continuous growth steps although the presence of this has been difficult to establish. Side growth is often thought to be influenced by impurities, possibly by the formation of an adsorbed layer which prevents nucleation.

The properties of the whiskers often reflect either a close approach to crystalline perfection, free from dislocations or vacancies, or else the high surface/volume ratio. Most attention has been paid to the exceptional mechanical strength of the more nearly perfect whiskers since these offer promise for fibre-reinforcement of composite materials. It has been shown that the tensile strength of a perfect solid should be approximately one tenth of the elastic modulus, E and values of 0·002–0·066 E have been reported for metal whiskers and from 0·01 to 0·099 E for ceramics, the highest values being found for iron and sapphire respectively. The association of high strength with crystal perfection is consistent with the observation that strength rapidly increases with decreasing diameter below 10 μ and by the development of a very pronounced yield as soon as plastic deformation occurs. It is necessary to discriminate between strengthening attributable to crystal perfection and that caused by surface films which might pin dislocations. Perfection of the crystal surface influences not only nucleation of slip but also imparts enhanced resistance to oxidation or chemical attack. The high surface/volume ratio affects electrical and magnetic properties where the thickness is small compared with electron free mean paths, flux penetration depth, magnetic domain size, etc.

Bibliography

COLEMAN R. V. (1964) *The Growth and Properties of Whiskers, Metall. Rev.* **9**, 261.

DOREMUS, R. H. ROBERTS B. W. and TURNBULL D. (1958) *Growth and Perfection of Crystals,* London: Chapman & Hall.

HARDY H. K. (1956) The Filamentary Growth of Metals, *Progress in Metal Physics,* **6**, 45.

NADGORNY E. M. (1962) The Properties of Whiskers, *Soviet Physics Uspekhi* **5**, (3), 462.

M. B. WALDRON

D

DELAYED ALPHA EMISSION. This is a form of alpha-radioactivity in which a preceding beta decay populates a nuclear excited state that is unstable against relatively prompt alpha emission. The alpha particles then appear to follow the half life of the parent beta decay. The nucleide ^{20}Na is an example of a delayed-alpha precursor.

R. E. Bell

DELAYED NEUTRON EMISSION. This occurs when a nucleus undergoes negative beta decay to a state of the daughter nucleus that is excited by more than its own neutron separation energy. Since the actual neutron emission is relatively prompt, the neutrons appear to follow the half life of the parent beta decay.

Keepin (1961) has discussed delayed neutrons in nuclear fission but it should be noted that the process is not confined to fission products; for example, ^{17}N is a well known delayed neutron precursor with a half-life of 4·2 seconds.

Bibliography

Keepin G. R. (1961) in *Encyclopaedic Dictionary of Physics* (J. Thewlis Ed.) 2, 280, Oxford: Pergamon Press.

R. E. Bell

DELAYED PROTON EMISSION. This occurs when a nucleus undergoes positive beta decay to a state of the daughter nucleus that is excited by more than its own proton separation energy. Since the actual proton emission is usually relatively prompt, the protons appear to follow the half-life of the parent beta decay.

See also: Proton radioactivity.

R. E. Bell

DEMODULATION. Demodulation is the process of extracting a modulating signal frequency from a modulated carrier frequency. Demodulation is a specific process which can be classified under the more general heading of detection. It is common practice to use "detection" to describe the demodulation process but this term is misleading.

Demodulation procedures are outlined below.

1. Quadratic demodulation. Also known as *square law* demodulation. The relation describing quadratic demodulation is

$$V_{out}(t) \propto V_{in}^2(t)$$

where $V_{out}(t)$ is the output voltage and
$V_{in}(t)$ is the input voltage at the modulated carrier frequency.

Typical devices which may be used as quadratic demodulators are semiconductor junctions (diodes, transistors), electronic valves, thermistors, bolometers and thermocouples.

2. Linear demodulation. The relation describing linear demodulation is

$$V_{out}(t) \propto V_{in}(t)$$

where $V_{out}(t)$ is the output voltage.

Semiconductor junctions and electronic valves may be used as linear demodulators.

3. Synchrodyne demodulation. Also known as phase-sensitive, coherent synchronous or homodyne demodulation. The relation describing synchrodyne demodulation is

$$V_{out} \propto V_{in} \times V_{ref} \times \cos\theta$$

where V_{in} is the input voltage, V_{ref} is the reference voltage and θ is the phase angle between V_{in} and V_{ref}. This signal processing technique provides baseline stabilization since it eliminates the necessity for stable d.c. gain and replaces it by stable a.c. gain which is achieved more easily.

Synchrodyne demodulation is useful in laboratory experiments where a physical parameter may be modulated. The modulating frequency serves as the coherent reference frequency for the synchrodyne demodulator. The noise response of a synchrodyne demodulator is less than the noise response of either a quadratic or a linear demodulator.

4. Superheterodyne demodulation. This technique is widely used in microwave and radio frequency receivers. A locally generated frequency is mixed with the modulated carrier frequency in order to produce a modulated difference frequency at which selective, high gain amplification can be obtained. Two successive difference frequencies may be employed where image rejection is important. The modulation is then extracted using a discriminator for frequency modulated signals and quadratic or linear demodulation for amplitude modulated signals. In the laboratory, synchrodyne demodulation can be employed since coherent local oscillator power can be derived from the signal carrier power. In this case, synchrodyne demodulation should also be employed at the intermediate frequency when maximum stability of the demodulation system noise figure is required.

Bibliography

TUCKER D. G. (1953) *Modulators and Frequency Changers,* London: Macdonald.
VAN DER ZIEL (1954) *Noise,* New York: Prentice Hall.

H. A. BUCKMASTER and J. C. DERING

DETECTIVITY. This is a measure of a radiation detector's performance. It is defined as the r.m.s. signal-to-noise ratio divided by the r.m.s. radiation power required to produce that ratio, and it has units of reciprocal watts. Detectivity is the inverse of NEP—another commonly used figure of merit for detectors. The reason for introducing it in addition to NEP (noise equivalent power) is purely a psychological one. Increasing values of a figure of merit should mean better performance, however, for NEP the opposite is true—lower values of NEP imply better detectors. Detectivity alleviates this problem.

I. W. GINSBERG and T. LIMPERIS

DIAMOND, THERMAL PROPERTIES OF. At normal temperatures and pressures diamond is the metastable crystalline form of carbon, graphite being the stable form. As diamond graphitizes when heated to between 1000 and 2000°C (the temperature at which graphitization begins depends on the atmosphere in which heating is carried out), it is not possible to start discussing the thermal properties of diamond by listing its melting and boiling points at atmospheric pressure.

At any temperature there is a minimum pressure above which diamond is stable and a line can be drawn on a temperature-pressure plot which separates the diamond-stable and graphite-stable regions. The calculation of this equilibrium line and its experimental determination will be discussed later. The upper limit to the region of diamond crystal stability is the melting curve of diamond. For pressures less than about 100,000 to 150,000 atm graphite is the stable form just below the melting temperature (near 4000°K for all pressures), but if diamond is heated while maintained at still higher pressures it will remain the stable form until it melts. There is only slight experimental evidence for the shape of the diamond melting curve.

The phase diagram of carbon is sketched in Fig. 1. The two continuous lines are not known with great certainty, but experiments have indicated their general shapes; the diamond melting line which has been

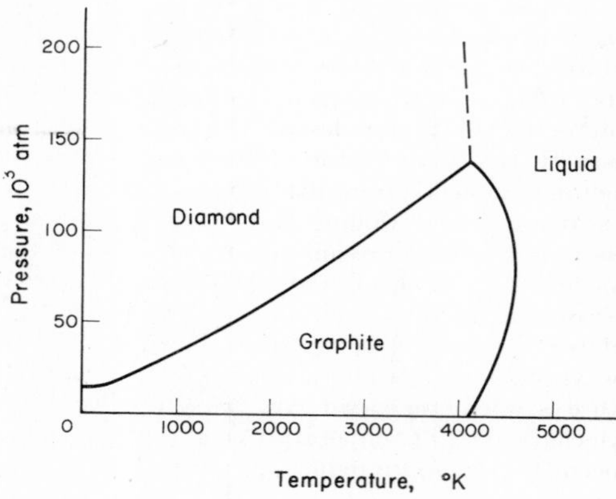

Fig. 1. Phase diagram of carbon. Evidence for the equilibrium conditions represented by the full lines is much more reliable than for the dashed section.

drawn is only meant to suggest the initial trend of this boundary. The vapour pressure curve of graphite has not been shown since it would scarcely be visible above the horizontal axis.

Subjecting one form of carbon to the conditions under which the other form is stable does not necessarily ensure that the transition will take place, and the existence of diamonds in everyday life is evidence of this fact. However, the transition can be made to proceed in either direction under suitable conditions. As mentioned already, diamond can easily be converted into graphite by heating, and graphite has been converted directly into diamond by the simultaneous application of high pressure and high temperature and indirectly by solution in a molten metal or alloy under less extreme conditions.

The heat of transformation of graphite into diamond has been determined from the difference between the heats of combustion to CO_2. Since the individual heats of combustion are both nearly 100,000 cal/mole, the difference, 453 cal/mole at 25°C and zero pressure, cannot be considered known to any great accuracy and there is little prospect of a direct determination being possible.

Specific heat, thermal expansion and thermal conductivity. The thermal properties of diamond correspond to those of a Debye solid with a characteristic temperature, θ, of about 2000°K. This means that a given temperature corresponds to a lower reduced temperature, T/θ, for diamond than for any other solid and it has the smallest specific heat (per g atom). At normal temperatures its thermal expansion is very small and is still varying appreciably with temperature, while its thermal conductivity is greater than that of any metal or other solid.

A more detailed consideration requires a knowledge of the frequency distribution function for the normal modes of vibration of the lattice and their dispersion relations. The former relates the number of modes, $g(\nu)\,d\nu$, within a small frequency range $d\nu$ to the frequency ν; the latter, $\nu(q)$, describes how the frequency varies with the wave vector q. Warren, Wenzel and Yarnell determined $\nu(q)$ for diamond by inelastic neutron scattering, and Dolling and Cowley fitted these results to a dipole approximation model of the interatomic forces. From this fit the frequency distribution $g(\nu)$ shown in Fig. 2 is derived and the specific heat calculated as a function of temperature.

The variation of specific heat of a solid with temperature is often translated into the variation in the Debye characteristic temperature. A given value of C_v will occur for a definite reduced temperature, T/θ, on

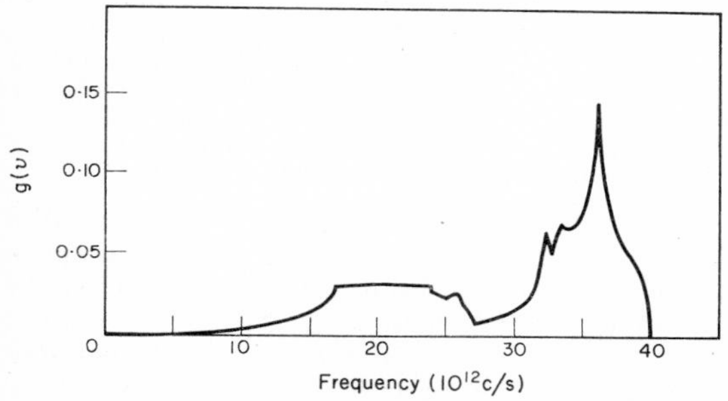

Fig. 2. Frequency distribution for lattice vibrations of diamond.

the Debye theory, so that an effective θ can be ascribed to each pair of C_v and T values. If the variation of specific heat with temperature were given exactly by the Debye theory, then the effective θ would be independent of temperature, but for all real solids there are considerable departures from this behaviour. The difference between the calculated and the experimentally determined specific heat of diamond is small, but is seen clearly by comparing the effective θ's as is shown in Fig. 3. The specific heat at constant volume, C_v, itself is shown in Fig. 4. The difference between the specific heat at constant pressure, C_p, and C_v is small at normal temperatures; at high temperatures it is almost proportional to the absolute temperature and can be estimated to be between 1 and $1{\cdot}5 \times 10^{-4}$ cal/mole-deg above 1000°K.

There have been several series of measurements of the expansion coefficient of diamond within the temperature range from 25 to 2000°K and the linear coeffi-

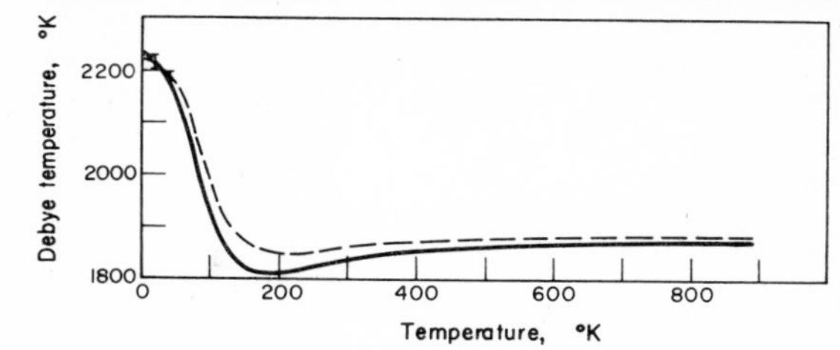

Fig. 3. Comparison of the calculated and experimental curves of the temperature variation of Debye temperature, θ, for diamond: — calculated; – – – experimental.

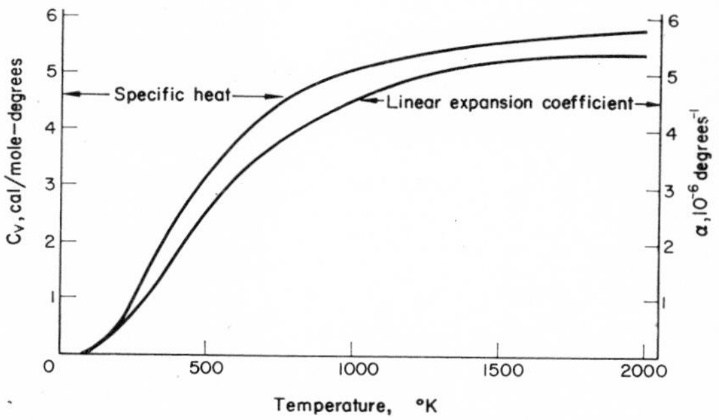

Fig. 4. Specific heat and thermal expansion coefficient of diamond.

cient, α, is shown in Fig. 4. It is about zero at 90°K and increases to a limiting value of $5{\cdot}4 \times 10^{-6}\,\mathrm{deg}^{-1}$. It is possible that it becomes negative at lower temperatures, as is found for silicon and germanium, but the coefficient has become so small as to be unmeasurable.

Thermal expansion of a solid is determined by the variation in normal mode frequencies with specific volume which results from the anharmonic terms in the potential energy of the atoms. Dolling and Cowley were able to reproduce the expansion coefficients of diamond, silicon and germanium by choosing for each element appropriate values for two parameters which represent the anharmonicity.

The results of thermal expansion measurements are often discussed in terms of the Grüneisen "constant" γ, which is given by

$$\gamma = \frac{3\alpha V\beta}{C_v}$$

where V is the molar volume and β the isothermal bulk modulus. Since at normal pressures V and β change slowly in opposite directions as the temperature varies, γ is essentially proportional to α/C_v. It is evident from Fig. 4 that the calculated γ will decrease with decreasing temperature to a minimum and then rise again. The minimum occurs at about room temperature and is between 0·8 and 0·9. A similar variation has also been found for silicon and germanium, although they have minima which are negative, corresponding to the negative expansion coefficients at low temperatures. For diamond the high temperature limiting value of γ is 1·3, which is considerably greater than for Si (0·8) and Ge (0·4).

The thermal conductivities of pure non-metallic crystals reach very high values with a maximum usually occurring at about $\theta/30$. On either side of the maximum the conductivity varies quite rapidly with temperature, so that at room temperature, which is not very different from θ for most solids, the conductivity has decreased to a rather small value, giving rise to the generalization that "metals conduct heat better than non-metals". For diamond, however, room temperature is less than $\theta/6$ and the conductivity has not fallen off so much from its value at the maximum and Fig. 5 shows that even the least perfect specimen

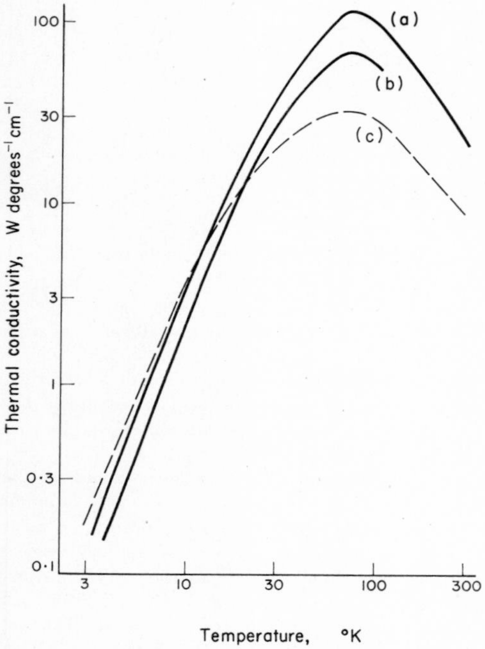

Fig. 5. Thermal conductivity of three gem quality diamonds (a) type IIa, (b) type IIb, (c) type I. All specimens had approximately square cross sections, 1 × 1 mm.

measured conducts heat better than copper at room temperature (4 W/cm deg). The three specimens were all of gem quality: type I contains appreciable nitrogen, type IIa is semiconducting and type IIa is probably the purest. It is unlikely that even more perfect diamonds would have appreciably higher conductivities, since the presence of 1 per cent ^{13}C in natural carbon introduces sufficient irregularity in the distribution of atomic mass in a diamond to restrict the conductivity. Of course, most diamonds, such as would be used in grinding, would have smaller conductivities than shown in Fig. 5. but still probably larger than any other common non-metallic solid.

The graphite-diamond equilibrium. The relative stability of two mechanical situations depends on a difference in potential energy. If there are two valleys in a switchback railway, the car will be in stable equilibrium in the lower and in metastable equilibrium in the upper valley, since it has a lower gravitational potential energy in the former. The ease of transition between the two states is determined by the height of the hill between the valleys and a car in the upper valley can only pass to the lower if it is given sufficient kinetic energy. For a thermodynamic system the relative stability of two phases is determined by the values of the appropriate thermodynamic potential, and for fixed pressure and temperature this is the Gibbs free energy, $G = H - TS$, where H is the enthalpy and S the entropy. The ease of transition is determined by the activation energy for the process.

For graphite and diamond to be in equilibrium with one another their Gibbs free energies must be equal. If conditions are such that they are not equal, then the phase with the lower value of G will be stable and the other metastable (or unstable). If we use the symbol $\triangle$ to denote the difference between the values of a property for diamond and for graphite, then the condition for equilibrium can be written:

$$\triangle G = \triangle H - T\triangle S = 0.$$

$\triangle H$ is the heat of transition from graphite to diamond and its variation with temperature is given by $\left(\frac{\partial \triangle H}{\partial T}\right)_p = \triangle C_p \cdot \triangle S$, the entropy difference between diamond and graphite, at temperature T is found from

$$\triangle S = \int_0^T \frac{\triangle C_p}{T}\, \mathrm{d}T.$$

Since $\triangle H$ is known at 25°C and zero pressure, measurements of the specific heats over as wide a temperature range as possible, with suitable extrapolations, enable $\triangle G(T, 0)$, the value of $\triangle G$ at temperature T and zero pressure, to be calculated. These values are given in the table and as they are all positive, graphite is the stable solid form of carbon at all temperatures for low pressures.

$\triangle G$ is, however, a function of pressure: $\left(\frac{\partial \triangle G}{\partial P}\right)_T = \triangle V$, where $\triangle V$ is the difference in molar volumes of diamond and graphite. Since diamond is always denser than graphite under the same conditions, $\triangle V$ is negative, so that $\triangle G$ can be reduced to zero or made negative by the application of sufficiently high pressure. The pressure required to reduce $\triangle G$ to zero is the equilibrium pressure for a given temperature.

Temperature (°K)	$\Delta G(T, 0)$ (cal/mole)	$V(T, 0)$ (cm^3/mole)		Equilibrium pressure P_{eq} (Atm)	$V(T, P_{eq})$ (cm^3/mole)		$\Delta V(T, P_{eq})$ (cm^3/mole)
		Graphite	Diamond		Graphite	Diamond	
298	694	5·299	3·416	16,100	5·1	3·4	1·7
400	785	5·313	3·418	18,200	5·1	3·4	1·7
500	885	5·327	3·420	20,500	5·1	3·4	1·7
700	1090	5·356	3·427	25,500	5·0	3·4	1·6
900	1310	5·387	3·434	31,000	5·0	3·4	1·6
1100	1530	5·421	3·443	36,700	5·0	3·4	1·6
1500	1970	5·493	3·461	48,000	5·0	3·4	1·5
2000	2530	5·584	3·485	64,000	4·9	3·4	1·4
2500	3080	5·678	3·510	80,000	4·8	3·5	1·3
3000	3640	5·779	3·535	97,000	4·7	3·5	1·3

The graphite-diamond equilibrium conditions calculated according to the assumptions discussed.

The molar volumes of graphite and diamond at zero pressure shown in the table are obtained from the thermal expansions. The total change in ΔG produced by pressure P is $\int_0^P \Delta V \, dP$, so that for each temperature the pressure-volume relations must be known. Diamond is very incompressible and it is sufficiently accurate to assume that the compressibility is independent of pressure and temperature and equal to the initial compressibility at room temperature, $1{\cdot}8 \times 10^{-7}$ cm^2/kg. Several assumptions have been made in previous equilibrium calculations about the compressibility of graphite at high temperatures, and as the molar volume of graphite is appreciably reduced by the pressures concerned, the calculated equilibrium pressure is sensitive to the assumed $P - V$ relation. It now seems best to assume that the bulk modulus at a particular temperature is linear in the pressure and that it is only a function of the molar volume, i.e.

$$B(T,P) = B(T, 0) + nP \qquad (1)$$

and

$$B(T,V) = B(V) \quad \text{or} \quad \left(\frac{\partial B}{\partial T}\right)_v = 0. \qquad (2)$$

The initial, zero pressure, bulk modulus at any other temperature is obtained from the ratio of the zero pressure volume at this temperature to the room temperature zero pressure volume, using the relation

$$\frac{B(T_1, 0)}{B(T_2, 0)} = \left\{\frac{V(T_2, 0)}{V(T_1, 0)}\right\}^n$$

which can be derived from equations 1 and 2. $\int_0^P V \, dP$ for graphite is then calculated.

The table shows the values of pressure which at each temperature satisfy the relation $\int_0^P \Delta V \, dP = -\Delta G(T, 0)$, and these calculated equilibrium pressures are shown in Fig. 1. The molar volumes of graphite and diamond under equilibrium conditions, calculated as described, and the volume changes, $\Delta V(T, P_{eq})$ at the transition are also given. It is useful to know the volume change since it is then easy to estimate how the calculated equilibrium pressure would be affected by any change in the measured thermodynamic properties. At all equilibrium pressures graphite has become very incompressible, so that for a change δP in P of a few thousand atmospheres on either side of the equilibrium pressure, the changes in $\int_0^P \Delta V dP$ are roughly $\Delta V(T, P_{eq})\, \delta P$. A change δG in G therefore corresponds to a change in equilibrium pressure of $\dfrac{a \times \delta G}{\delta V(T, P_{eq})}$, where the constant a is 41·3 if G is measured in calories/mole and P in atmospheres. If, for example, the heat of transition were 100 cal/mole higher than assumed, the equilibrium pressure at, say, 2000°K would be increased by $\dfrac{100 \times 41{\cdot}3}{1{\cdot}41} \approx 3000$ atm.

Diamonds for industrial use are now synthesized in large quantities, the usual method consisting in subjecting graphite and a metal or alloy, which dissolves carbon when molten, to a temperature such that the metal melts and to a pressure such that diamond is the stable form of carbon. After a short time the temperature is reduced, followed by the pressure. Bundy and his co-workers at G.E. in Schenectady have determined a section of the equilibrium line by observing the conditions for the transition to take place. These results agree as well as can be expected with the calculated line, considering the difficulties in measuring the high pressures and high temperatures involved and also the uncertainties in the relevant thermodynamic properties. The activation energy is too great for a direct transition to be possible under equilibrium conditions below about 3000°K, but Bundy has produced diamond by subjecting graphite to pressures above about 125,000 atm and flash heating to the neighbourhood of 3000°K. The $P - T$ values for which diamond was produced were all well within the calculated diamond-stable region. Diamonds have also been produced from graphite directly by shock wave heating and compression, but it seems that for this method also, the conditions must again be well inside the diamond-stable region.

Bibliography

BERMAN R. (Ed.) (1965) *Physical Properties of Diamond*, Oxford: The University Press.

BUNDY F. P. (1963) *J. Chem. Phys.* **38**, 618, 631.

BUNDY F. P. BOVENKERK H. P. STRONG H. M. and WENTORF R. H. (1961) *J. Chem. Phys.* **35**, 383.

DOLLING G. and COWLEY R. A (1966) *Proc. Phys. Soc.* **88**, 463.

PEARLMAN N. (1962) in *Encyclopaedic Dictionary of Physics* (J. Thewlis Ed.) **6**, 620, 626, Oxford: Pergamon Press.

SLACK G. A. (1961) in *Encyclopaedic Dictionary of Physics* (J. Thewlis Ed.) **3**, 606, Oxford: Pergamon Press.

WEINTROUB S. (1961) in *Encyclopaedic Dictionary of Physics* (J. Thewlis Ed.) **3**, 544, Oxford: Pergamon Press.

R. BERMAN

DIFFUSION BATTERY. A diffusion battery consists of a number of parallel rectangular or circular channels through which an aerosol may be passed. While passing through the channels a proportion of the particles will diffuse to the channel walls and be removed from the gas flow. The diffusion coefficient of the particles of the aerosol may be determined by measuring the fraction of the particles which penetrate the battery. The method is of practical use up to particle radii of 10^{-5} cm. For a battery with rectangular channels.

$$F = 0{\cdot}91 \exp(-x) + 0{\cdot}053 \exp(-11{\cdot}4x)$$

(Nolan *et al.*), where $x = 3{\cdot}77\, bDl/aq$

In this equation F is the fraction of particles of diffusion coefficient D (cm²/sec) penetrating a battery of channels l (cm) long, b (cm) deep and $2a$ (cm) wide and q (cm³/sec) is the gas flow in one channel.

For circular channels

$F = 0{\cdot}82 \exp(-3{\cdot}66y) + 0{\cdot}097 \exp(-22{\cdot}3y) + 0{\cdot}032 \exp(-57y)$ (Gormley and Kennedy). $y = Dl/r^2u$ and F is the fraction of particles of diffusion coefficient D (cm²/sec) penetrating a channel of radius r (cm) and length l (cm) at a velocity u (cm/sec). Experimental results agree very well for the spherical monodisperse particles for which the theory was derived. For aerosols consisting of particles of different sizes the fraction penetrating the battery is a function of gas velocity as well as of diffusion coefficient. Many freshly formed aerosols follow a logarithmic normal size distribution and it is possible to determine completely the size distribution of these by making three measurements with a diffusion battery at different flow rates in the method of Fuchs *et al.*

Bibliography

FUCHS N. A. STECHKINA I. B. and STAROSSELSKII V. I. (1962) *Brit. J. Appl. Phys.* **13**, 280.

GORMLEY P. G. and KENNEDY M. (1949) *Proc. Roy. Irish Acad.* **52** A, 163.

METNIEKS A. L. and POLLAK L. W. (1959) Inst. Adv. Studies, Dublin, *Geophysical Bulletin* No. **16**.

NOLAN J. J. NOLAN P. J. and GORMLEY P. G. (1938) *Proc. Roy. Irish Acad.* **45**-A4, 47.

W. J. MEGAW

DIRECT CONVERSION OF HEAT TO ELECTRICITY. *Introduction.* There are three classical effects of current electricity—magnetic, thermal (Joule heating) and chemical. All can be reversed, though only the first has been used to any extent for large scale generation, by the rather indirect process of using mechanical energy to drive a dynamo, which acts by cutting magnetic lines of flux with a coil, in which current is induced. The mechanical energy, in turn, is supplied by water power, or from thermal energy in the form of fossil or atomic fuel. If a more direct process could be used it might be possible both to improve efficiency (for example, by removing the Carnot cycle limitations associated with heat engines), and to simplify and hence reduce the capital cost of equipment.

Quite apart from these possible economic advantages, there are prospects with some processes of saving weight, and of increasing reliability and reducing noise and vibration by the absence of moving parts. There are thus considerable military and space applications developing.

This article therefore examines the possibilities of using some of these processes more directly. The inverse chemical process, which is the basis of the *Fuel cell*, is described in a separate article (see Liebhafsky 1966).

Technical and economic background. An ideal heat engine, taking in all its heat at an upper (absolute) temperature T_H and rejecting it all at a lower temperature T_C has an efficiency given by the well-known Carnot expression

$$\eta = 1 - \frac{T_c}{T_H}.$$

Practical heat engines always fall short of this ideal. They may not be able to use all the temperature range available, as illustrated in Fig. 1. At the upper end, materials are not yet available which will enable boilers, turbine blades, etc. to operate at the flame temperature which can be realized. At the lower end there are economic limits (e.g. to the size of condensers) which determine the "throwaway" temperature.

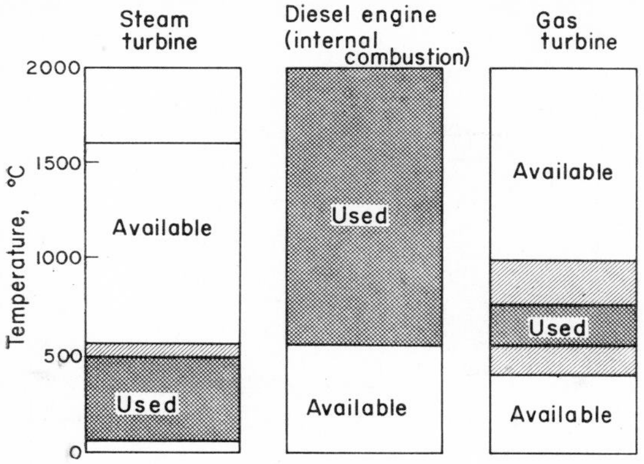

Fig. 1. The temperature ranges available, and normally used, in various types of heat engine.

A more subtle part of the reason is that not *all* the heat can be taken in and rejected at the maximum and minimum temperatures respectively. This is illustrated in Fig. 2, which shows a typical power station temperature–entropy diagram. Because of the thermodynamic properties of water/steam, the figure departs markedly from the rectangular temperature–entropy diagram of the ideal Carnot engine. Much ingenuity has been exercised, by superheat, reheat and feed-water heating, to improve the shape.

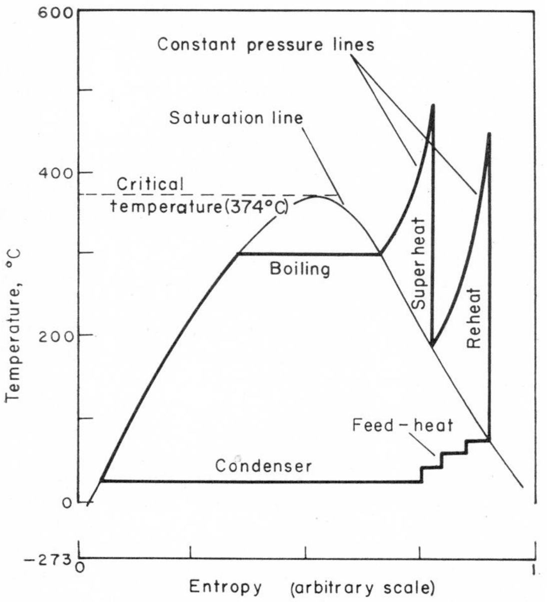

Fig. 2. Temperature-entropy diagram for a steam-turbine power-station heat-cycle.

A considerable improvement can be made by using mercury for the upper part (only) of the temperature range, as illustrated in Fig. 3. This combination is known as a *binary cycle,* and it provides a simple example of the concept often called *thermodynamic topping* to improve the efficiency of a simple cycle. Mercury binary plants have an efficiency about 10 percentage points above simple steam plants with the same top temperature. They pose severe metallurgical problems, and, in particular. do not lend themselves to the scaling up in size typical of modern central-station plants.

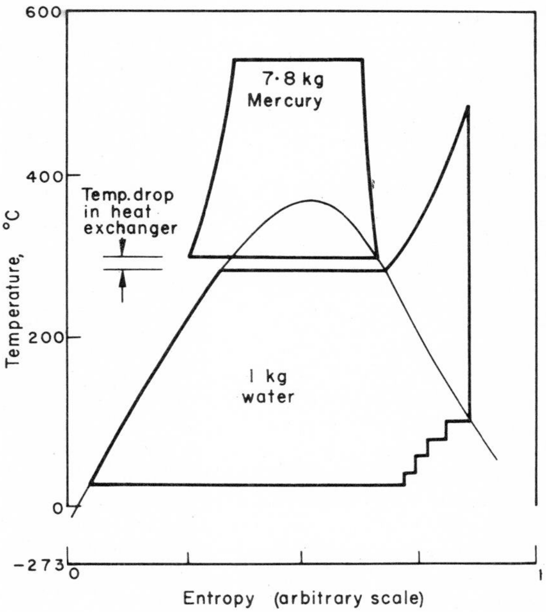

Fig. 3. Temperature-entropy cycle for a mercury-water binary plant heat-cycle.

Another idea being pursued—and this could be called *thermodynamic tailing*—is the use of an organic liquid instead of water for the lower end of the cycle. The object here is primarily to reduce the size and blade erosion problems of large low-pressure steam turbines, rather than to reduce the heat rejection temperature.

Generators intended for use in space must reject all their waste heat as some form of radiant energy—and the same is, in fact, ultimately true for all the power generated within the space-craft. In such vehicles it is the weight and volume of the power supply which matters most, and not the cost of power, as in a land-based station. Because of Stefan's fourth-power radiation law, a small radiator area implies a fairly high rejection temperature. In an overall optimization, these factors, as well as efficiency, must be considered. It has been found that, to keep the overall weight as low as possible, a ratio of T_C/T_H of about

$\frac{2}{3}$ to $\frac{3}{4}$ is best, with the actual values as high as materials permit. Heat cycles based on mercury or the alkali metals have been used or considered for the SNAP (System for Nuclear Auxiliary Power) generators.

The cost of supplying electric power is mainly divided between the capital value of the plant (expressed as an annual charge) and the fuel costs, with a small addition for maintenance and other running costs. In typical coal- or oil-fired plants, the fuel represents nearly two-thirds of the total costs, but in nuclear plants the balance is rather different and the cost of the fuel is often only about one third of the total cost of power. Extra capital cost can be afforded to the extent that fuel costs can be reduced as a consequence of higher efficiency.

In the concept of "topping", already briefly mentioned, two devices are used thermally in series. The topper could be, for example, an MHD or thermionic generator, preceding the main boiler/furnace arrangement. In the ideal topper the energy in the fuel is used with 100 per cent efficiency, i.e. all the heat energy is either converted into electricity, or passed as heat into the next stage. Thus, if the topper has an efficiency η_0, and the next stage (steam plant for example) η_1, the combined efficiency is

$$\eta = \eta_0 + (1 - \eta_0)\,\eta_1.$$

If η_0 is small the efficiencies of the two parts are approximately additive.

The capital available for a given overall efficiency can easily be evaluated algebraically. Topping is most worthwhile when the fuel costs are high, and the efficiency of the main generating plant is low. When making calculations of this sort, however, it has to be remembered that one cannot usually add a topper to an existing device in a simple fashion. For example, thermoelectric materials have low thermal transmission by comparison with normal boiler materials, and even if they had acceptable efficiency as toppers, would force a complete redesign of the conventional parts of the generator. MHD toppers are likely to need magnet power supplies which can reduce the gross efficiency, and the exhaust may contain corrosive seeding materials which could shorten the life of the conventional furnace.

In a "tailing" device, which thermally is in series with, and follows, the main plant, use is made of heat which would otherwise be rejected to exhaust and in this sense it is therefore free. For this reason the allowable capital cost can be higher than for a comparable topper. If the tailer has an efficiency of η_2, the overall efficiency is

$$\eta = \eta_1 + (1 - \eta_1)\,\eta_2.$$

A good deal of effort has been devoted to finding an effective way of using the rejected heat from a gas turbine or internal combustion engine, where the sink temperature may be around 500 °C. However, quite large heat exchangers are needed to extract heat from *gases*, and this, coupled with the poor thermal and mechanical properties of current thermoelectric materials, has made the search rather unrewarding. Where small amounts of power are wanted for some special purpose, a tailer to convert some of the exhaust gas heat can be justified.

Magnetohydrodynamic (*MHD*) *generation* (see also Swift-Hook). When an electrical conductor, such as a piece of copper, is moved through a magnetic field, a potential difference is created between its ends, and if joined into an external circuit a current can flow. The same is true of a liquid or gas if it can be rendered sufficiently conducting. With a gas this can be done by heating it, and the conductivity can be greatly enhanced by "seeding" it with a readily-ionizable material such as a caesium or potassium salt. Even so, the conductivity ϱ so attained is many orders of magnitude below that of a metal, and a very high temperature, certainly in the region of 2000°C, is needed for a practicable generator.

The hot gas is forced at a velocity V, between the poles of a magnet producing a field H, and between electrodes separated by a distance l. The open circuit voltage is

$$V_0 = VlH$$

which is independent of ϱ. However, when the electrodes are connected to an external circuit, the gas is slowed because energy is extracted from it, part being dissipated in the load and part as ohmic losses in the gas itself. Whereas with a conventional dynamo the internal electrical losses are extremely small (say 1 per cent), with an MHD generator they can be very high. These losses, however, appear as heat in the gas, and can be partly re-converted into electricity farther along the duct. The generator behaves rather like a turbine (either impulse or reaction) with "magnetic" blades in place of the conventional metallic ones of a gas turbine. It is because material blades are not needed that the MHD generator can operate, in principle, at high temperatures, and hence with high Carnot efficiency.

It is important to minimize the size of the MHD generator to reduce wall friction losses, heat dissipation to the outside, and to keep down the cost of providing electrodes and the magnetic field. Various proposals have therefore been made to attain higher conductivity by means of non-equilibrium ionization, achieving an electron temperature higher than the mean gas temperature. Consideration of the power required appears to show that thermal ionization, by preferential ohmic heating of the electrons, is more promising than electrical injection methods, and may be specially useful for monatomic gases in closed-cycles.

The MHD generator may operate on either a direct or an indirect cycle. In the former case, the hot gas is provided by combustion of a fuel. To attain a higher temperature the air may be pre-heated (by the exhaust gas) and/or enriched with oxygen. Seeding is then usually carried out with a potassium salt; recovery of the major part being necessary for reasons of both economy and avoidance of corrosion in the exhaust system. The indirect cycle is usually designed to operate in conjunction with a nuclear reactor. The choice of working fluid is then primarily a matter of technical merit, rather than economics, since losses are small: helium seeded with caesium has been proposed.

Conduction in a gas in a magnetic field is more complex than has been suggested so far. The electrons tend to move in circular paths about the lines of force, until they collide with gas molecules. They therefore tend to lag behind the heavy ions. Thus in addition to the normal (Faraday) current at right angles to the field and to the gas velocity, there is a *Hall* current component along the gas-flow direction. This has a great influence on the design of a generator, but cannot be discussed in detail here.

There are immense practical difficulties to be faced in designing a generator for long life. Apart from the combustion cycle, and seed injection and recovery already mentioned, the duct walls must be resistant to heat, corrosion and erosion, and so too must the electrodes with the additional proviso that they must emit electrically. Some form of cooling is usually required, though the tips of the electrodes must remain above emission temperature. Usually the electrodes must be mounted in independent groups down the duct, to overcome Hall-effect problems. Provision of a strong magnetic field over the large volume of the duct is likely to prove expensive both in installation cost and operating power, and a good deal of attention has been given to the possibilities of superconducting magnets.

It is possible, as a long shot, that the need for a magnet may be avoided by the so-called "electrostatic" or plasmadynamic method. This is based on the observation that when two identical electrodes at different temperatures are placed in a flame or other ionized gas, a potential difference is generated. The method has been developed; and power has been extracted from a flame laden with electrostatically charged dust, directed down a duct.

A more detailed treatment of MHD generation is given by Swift-Hook (1966).

Thermionic generation. In a simple vacuum diode valve, a stream of electrons flows from the hot filament to the anode, if the latter is biased slightly positive. This is the simple Edison effect discovered in 1883. However, if the anode and filament are sufficiently close together, a small current flows even with no applied anode voltage. This is true thermionic emission and can be explained by reference to Fig. 4. The electrons in the cathode are at an average energy corresponding to the Fermi level Φ and to release them into the vacuum thermal energy equivalent to the cathode work function Φ_C must be supplied. In dropping to the level Φ in the anode, heat corresponding to Φ_A is released, but a further potential V_L remains to do useful work, against an external load resistance.

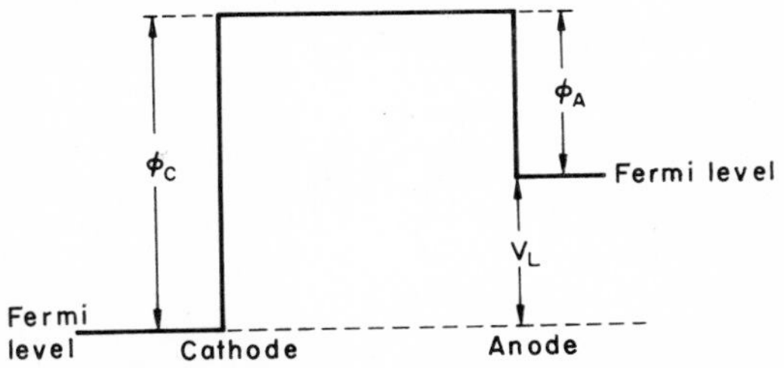

Fig. 4. Potential diagram for the thermionic converter: ideal case, matched load.

However, as soon as an appreciable number of electrons is released into the vacuum, they form a repulsive cloud, which tends to inhibit the release of further electrons. They have to surmount a potential barrier (Fig. 5) caused by the space-charge before they can reach the anode. The barrier does not of itself reduce the voltage V_L, but it limits the current to a small value, and in any practical device ways have to be found to reduce or eliminate space-charge effects.

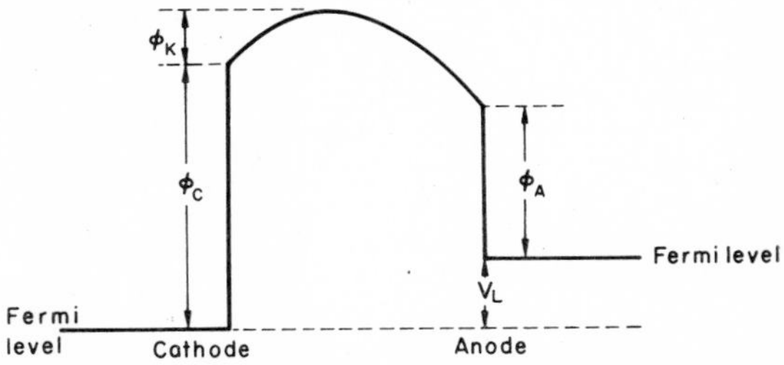

Fig. 5. Potential diagram for the thermionic converter, with space-charge potential barrier, Φ_K

A thermionic converter can be regarded as a heat engine, in which the cathode and anode are source and sink respectively, at temperatures T_C and T_A. The electron "gas" is the working fluid. The Carnot efficiency is thus

$$\eta = \frac{T_C - T_A}{T_C}.$$

Under idealized conditions the work functions are almost proportional to the absolute temperatures. Then, to a very simple approximation, the efficiency is given by

$$\eta = \frac{\Phi_C - \Phi_A}{\Phi_C + P_R/I_A},$$

where I_A is the anode current and P_R the cathode heat loss,. most of which occurs by radiation. Φ_A and P_R

must be kept small: the cathode temperature must be kept high.

There are three available methods of neutralizing space-charge: using a very small anode-cathode spacing, using crossed electric and magnetic fields, and introducing positive ions. In practice, the first two methods can be disposed of fairly quickly. In the "vacuum converter" the electrode spacing must be less than 0·01 mm to attain a power density of 1 W cm^{-2}. In the "magnetic triode" an additional accelerator plate is used in conjunction with a magnetic field to assist electrons across to the collecting anode. It has been found impossible to prevent a large part of the cathode emission being lost to the auxiliary electrode, and in addition, power is required to provide the field.

The main method of space charge neutralization which has been pursued is therefore that of providing positive ions, usually by introducing caesium vapour into the electrode space. Caesium is readily ionized and a single, heavy ion is effective in neutralizing the effect of a considerable cloud of electrons. At low pressure (say, 10^{-4} mm Hg) the electron mean free path is about 10 cm. The potential distribution is restored to approximately that of Fig. 4 but it is not possible to attain practical efficiencies above a few per cent with a plain tungsten emitter, except at temperatures so high that evaporation of the metal becomes troublesome. One way out of this difficulty is to use an impregnated or *dispenser* cathode; it acts by allowing barium (or thorium) to diffuse through porous tungsten to the surface.

An alternative is to use caesium at a higher pressure, in the region of 0·1–1 mm Hg. In this region rather complex potential sheaths form at the electrodes, but the plasma resistance as a whole falls, allowing large currents to flow, despite the fact that the electron mean free path is now less than the electrode spacing. Even more important, however, a monomolecular layer of caesium forms on the surface of the tungsten cathode, and reduces its work function from about 4 eV to 2·0 eV. Efficiencies in the region of 10–20 per cent then become possible.

Caesium vapour is highly aggressive to envelope and sealing materials, and considerable ingenuity in design is needed. The other main practical difficulty is to heat the cathode. With a thermionic diode immersed in a nuclear reactor this can be done directly, using uranium carbide as a fissile, emitting cathode material. Recently, so-called "heat-pipes" have been developed to convey heat from flames or radioactive isotopes to the cathode at temperatures around 1500°C.

Thermoelectric generation. Thermoelectric effects were first noticed by Seebeck in the early 1820's, but the efficiency attainable with metallic couples, even the familiar favourites antimony and bismuth, is rather low. Not much progress was made until the 1950's, when the semiconductor era arrived. Some of these materials can be "doped" in such a way that they give a much more favourable combination of properties.

The important properties are the Seebeck coefficient α, and the thermal and electrical conductivities, K and σ respectively. The voltage V developed by the junction is given by

$$V = \alpha(T_H - T_C),$$

where T_H and T_C are the hot and cold junction temperatures. V is proportional to the temperature difference if it is small.

It is possible to define a figure of merit Z for a junction, and ultimately for a *material*, in the form

$$Z = \frac{\alpha^2\sigma}{K}.$$

In a semiconductor suitably doped, the value of α can be very high. However, the ratio of σ/K is not uniquely determined by the Wiedemann-Franz law as it is (because of electronic movement) in metals, and σ is usually rather low, especially when α is high.

Semiconductors give a good thermoelectric performance only when they are "extrinsic", i.e. when the properties are determined by the impurity atoms and not by the basic properties of the material. To remain extrinsic at any operating temperature T requires an energy gap E_g of not less than about $4kT$ (where k is Boltzmann's constant) coupled with the correct level of doping. For room temperature operation the minimum energy gap is around 0·1 eV.

Simple semiconductor theory shows that the optimum value of α is 172 μV/°C, and the doping should be adjusted to give this value. More detailed theories take into account the interaction of the electrons with the crystal lattice, and the optimum α is then found to be somewhat over 200 μV/°C. This has all been confirmed by experiment to give the best value for $\alpha^2\sigma$. However, K is the sum of both electronic and lattice contributions to the thermal conductivity, the former also depending on doping. Thus further optimization is needed to get the highest value of Z for a particular operating temperature and material, and this is usually done empirically, over a range of α from say 150 to 300 μV/°C.

It is found that the maximum efficiency of a junction is given by

$$\eta_{\max} = \frac{(1 + ZT_m)^{1/2} - 1}{(1 + ZT_m)^{1/2} + T_C/T_H} \cdot \frac{T_H - T_C}{T_H},$$

where T_m is the mean operating temperature. The first factor is less than unity, and is highest for large values of ZT_m. The second factor is just the Carnot efficiency.

Work over the last few years has tended to show that the highest value of ZT_m that can be produced is close to or a little over unity. Low temperature materials have relatively high values of Z and vice versa. As shown in Fig. 6 this corresponds to theoretical efficiencies of around 20 per cent of the ideal Carnot.

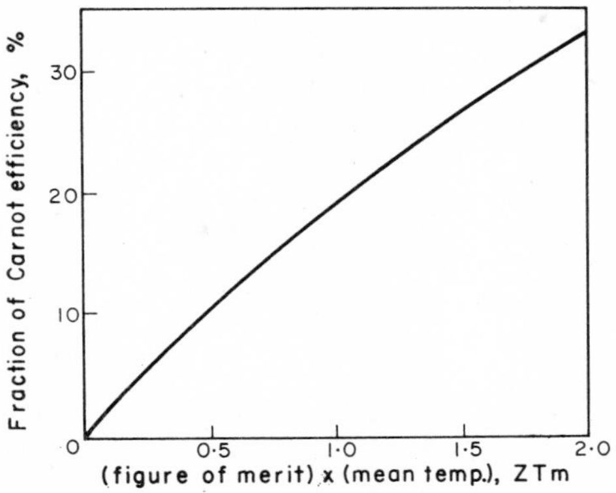

Fig. 6. Efficiency of a thermojunction as a function of its figure merit.

Unfortunately materials give a good performance over only a small temperature range, and to cover a wide range a series of materials must operate in cascade. It is difficult to arrange an efficient heat source and sink for a single set of junctions; it is still more difficult with a cascaded series to arrange this, together with good thermal contact and electrical insulation between neighbouring sets. Carnot efficiencies comparable for example with those of a steam or diesel engine cannot therefore so far be attained.

Among good thermoelectric materials are the tellurides of bismuth, lead and germanium; indium arsenide; iron disilicide; nickel oxide; and alloys of germanium/silicon for high temperature (~1000°C).

When a semiconductor is placed in a transverse magnetic field, the electrical conductivity, and the electronic component of the thermal conductivity tend to decrease. Under some circumstances α and also Z can increase, particularly with intrinsic materials. Even then the effect is marked only if the mobilities of both the electrons and holes are high. There is also the question of the power needed to provide the high magnetic field. In due course this may be partially solved by superconducting magnets, but on the whole it seems unlikely that overall efficiencies higher than obtainable by normal thermoelectric generation will be attainable.

Although the efficiency of thermoelectric generators is only 10 per cent or less, they are attractive for many applications because of portability, silence, absence of moving parts, and consequently high reliability in unmanned operation. One particular "fuel" which is attractive for satellite power supplies, submarine repeater stations, remote buoy-lighting and the like, is a radioactive isotope. Some of the U.S. SNAP generators, for example, are powered with Po-210 or Sr-90.

The solar cell. Today's photo-voltaic or solar cell provides a convenient, though expensive, way of converting radiant energy from a hot body (usually the Sun) into electricity. It can do so at efficiencies approaching 20 per cent and is widely used to charge satellite batteries.

The usual type of cell consists of a base of semiconductor material (in the early days generally n-type silicon) on the surface of which a thin layer is doped (usually with boron) to be p-type. Its action depends on quanta of radiation striking the barrier layer between the p and n type materials. These photons, if they have an energy larger than the energy gap E_g of the material, can raise electrons from the valence band to the conduction band, leading to the formation of electron-hole pairs (Fig. 7). Provided the carriers do not recombine too quickly—and there is always a tendency for this to happen—some can cross the p–n junction, giving rise to a potential drop which can be measured. On open circuit it is about half the energy gap, but when a circuit is completed external current can flow, and the voltage falls.

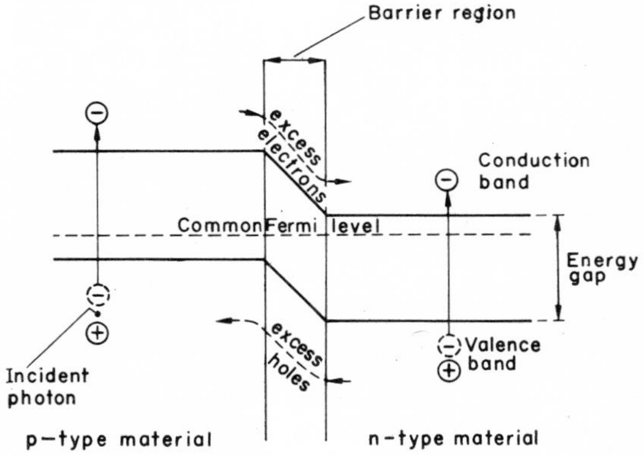

Fig. 7. Physical principles of the operation of a photovoltaic barrier-layer cell.

Only photons with energy $>E_g$ are effective in liberating electrons—but excess photon energy above this value is largely wasted. Thus, with a broad radiant energy spectrum, such as that from the Sun, it is inevitable that the efficiency is fairly low, even when the quantum efficiency is high and the cell is properly matched to the output load for maximum power generation. With silicon cells exposed to sunlight, the maximum efficiency attainable in practice is about 15 per cent.

The efficiency varies with energy gap in the manner shown in Fig. 8. Silicon is close to the optimum for use at sea level, but a higher value is better for "space" applications. This figure is based on simple theory, but practical efficiencies are lower for two main reasons—optical and electrical.

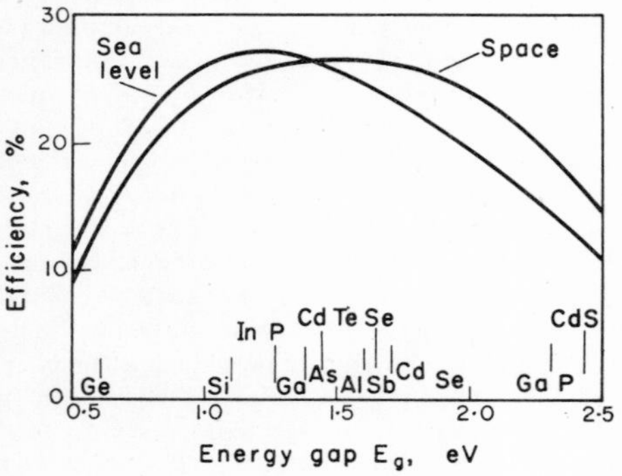

Fig. 8. Maximum theoretical efficiency of a photovoltaic cell as a function of energy gap, under space and sea-level conditions: also, energy gaps of typical semiconductor materials.

The proportion of radiation usefully absorbed near the critical junction area depends on the variation of absorption coefficient with wave-length. Once carriers have been generated, their effectiveness depends on their mobility and lifetime. The external current ultimately depends on the *transverse* movement of majority carriers across the cell surface to contacts at the edge. The provision of a highly-conducting grid on the surface is helpful.

Because its optical and electrical properties are better than those of silicon, gallium arsenide may ultimately prove to be a better material, yielding efficiencies of about 20 per cent. However, there are difficult technological problems to be solved before the theoretical ideal can be approached.

Cell temperature has a pronounced effect on performance. The diffusion length may alter slightly, though the lifetimes are practically unchanged. The energy gap has a small negative coefficient. The combined effect is that the solar collection efficiency and hence the cell current rise slightly with temperature. The cell voltage, however, which depends on the intrinsic carrier concentration, falls almost exponentially. The overall effect is given in Fig. 9. The temperature must be kept down by reflecting all unwanted radiation, and a high energy gap is advantageous.

A lot of thought has been given to the problems of permanence and efficiency under space conditions. Cells of reversed (n on p) type are more radiation-resistant, because they have a better short-wave response, and improved transverse current conduction. An extremely thin surface layer permits the collection of carriers, whose lifetime, and hence diffusion length, have been degraded by radiation. Aluminium-doped silicon is used in the new "super blue" cells. A thin cover glass gives protection from proton damage.

Many attempts have been made to prepare large-area cells, but not so far with complete success, and

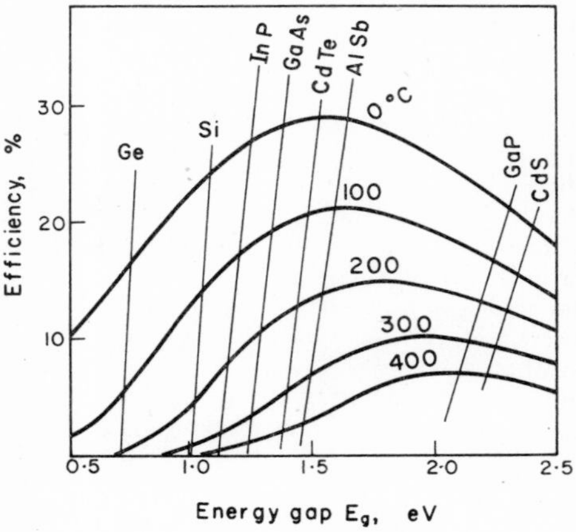

Fig. 9. Variation of maximum efficiency and optimum energy gap of a photovoltaic cell with temperature of cell surface: solar spectrum in space.

the usual arrangement is a shingled array. Attempts to improve the efficiency of individual cells have also been made, based on the idea of combining two different energy gap materials to increase the useful photon absorption. Parasitic losses in such multilayer cells, however, always seem to prevent a worthwhile improvement.

There are a number of variants on the standard photovoltaic cell. The *photoelectromagnetic* (PEM) cell depends on applying a magnetic field across a semiconductor, illuminated by radiant energy, the consequence of which is to produce a lateral charge separation. In the *photoemissive* converter, radiant energy causes electrons to be emitted from a photosensitive cathode on to a closely-spaced anode. In physical principle it depends on the photoelectric effect and therefore differs from the thermionic converter which is a true heat engine. The *thermal photovoltaic* (TPV) converter is essentially a device based on photovoltaic principles, but matched to a terrestrial thermal source. This source can itself be doped, e.g. with rare-earth oxides, to enhance its output at some specific (usually infra-red) wavelength, at which the cell is most sensitive.

Other types of generator. Various attempts have been made to use some variation of material properties with temperature to provide a form of heat engine. Some of these devices are highly ingenious, but the efficiency is usually disappointing. At first sight it seems that quite an effective generator could be made by thermally cycling a magnetic material around its Curie point, so that it was alternately *ferromagnetic* and diamagnetic. Close inspection shows that even under ideal conditions the efficiency is likely to be very low (not more than 1 per cent), and in practice it is much less, because of considerations of thermal inertia.

A rather similar generator may be made using the sudden change in polarization of a *ferroelectric* material (such as barium titanate) at its Curie point. Although the efficiency of such devices is also very low, they provide a convenient way of generating very high AC voltages and this may be used for some satellite applications.

Although not a form of direct generation from *heat*, it is worth noting that a useful form of high voltage can be obtained from alternating mechanical forces applied to *piezoelectric* materials. Petrol engine ignition systems have been developed, based on this principle.

Bibliography

DEBIESSE J. and KLEIN S. (1966) Flame Ionization and Magnetohydrodynamics, *Nature*, **212**, 1405.

LIEBHAFSKY H. A. (1966) in *Encyclopaedic Dictionary of Physics* (J. Thewlis Ed.), Suppl. Vol. I, 110, Oxford: Pergamon Press.

PEDERSEN E. S. (1967) Heat Pipe Thermionic reactor Concept, *Nuclear Eng.* **12**, 112.

SNYDER N. W. (Ed.) (1961) Energy Conversion for Space Power, *Progress in Astronautics and Rocketry*, Vol. 3, New York: Academic Press.

SPRING K. H. (Ed.) (1965) *Direct Generation of Electricity*, New York: Academic Press.

SPRING K. H. and SWIFT-HOOK D. T. (1962) Prospects for Large-scale Thermoelectric Power Generation, *B. J. Appl. Phys.* **13**, 159.

SWIFT-HOOK D. T. (1966) in *Encyclopaedic Dictionary of Physics* (J. Thewlis Ed.), Suppl. Vol. 1, 62, Oxford: Pergamon Press.

K. H. SPRING

DIRECT CURRENT TRANSMISSION OF POWER. In order to study the application of direct current transmission in electricity supply systems, it is necessary to know a little of the historical background of electricity supply development to realize why transmission of large quantities of electrical power is necessary and why direct current has certain advantages over the more usual alternating current systems.

When electricity became available for domestic and industrial use, the generating plant was built in the immediate vicinity of the demand for it. As the years passed, the need for electricity spread over a wider area and extensive distribution systems began to develop.

It soon became clear that, better economy of operation could be achieved by pooling the generating capacity whereby the modern and more efficient plant could be used to produce electricity to be transmitted to consumers many of whom were remotely situated from the site of generation.

It soon became economic to consider the national requirement for new generating stations to be based on the best site to produce it most cheaply. i.e. on or near coal fields to keep overland transport costs to a minimum, on river banks to utilize river water for cooling purposes, or on the coast to take advantage of cheap seaborne transport of coal and oil or on remote sites which are at present required for generation of electricity by nuclear power and finally on sites suitable for the utilization of cheap water power.

The development in design of generating plant has resulted in greater efficiency and the bigger the total capacity of the machine the less is the capital cost per kW installed; abroad it also became more economic for networks of different nations to be interconnected for the same reasons.

In the early 1900's, M. Thury designed a d.c. transmission system which transmitted 20 MW over a distance of nearly 225 km from Moutiers to Lyons in France at a voltage of 125 kV. At one end it consisted of several d.c. commutating generators each operating at 3000 V, connected in series and feeding into the transmission circuit. At the receiving end similar commutating generators were connected in series with the transmission circuit and each provided the driving power for generators of alternating current.

Several systems of this type were constructed but at the time, the major controversy over the merits of a.c. versus d.c. was at its height. On the one hand, with increasing demand for power transmission, there were commutation difficulties in generating d.c. with rotating machines. On the other hand, there were big developments being made in the design and construction of turbo alternators, transformers made it easier to change from one voltage level to another, there were rapid developments in a.c. induction motor design and the interruption of alternating current was made much easier by virtue of its passage through zero twice every cycle. The latter made the design of circuit breakers much more simple.

It was with this background that a.c. gradually replaced d.c. in generation and utilization and its application to transmission did not develop until later, when it became possible to replace the rotary plant by static equipment.

It is appropriate that the problems of a.c. transmission should be examined to see how d.c. is able to overcome many of the difficulties and indeed provide a much cheaper alternative in some cases.

The power that is to be transmitted by any transmission circuit must be raised to the highest practicable value. This is because routes are scarce and only by the exercise of efficient design can the interests of economy and amenity be met. However, in the case of an overhead line transmitting alternating current,

 Direct current transmission of power

both the length and the power transmitted have a direct effect on the stability of the system as a whole. For a given voltage there is a maximum distance of transmission of the order of 300 miles which cannot be exceeded without introducing series capacitors or shunt synchronous condensers and these require switching stations at intermediate points and increase the complication and cost. Raising the cross-section of the conductor does not increase the transmission capacity of an overhead line because the stability limits are set by conductor spacing one from another and from earth.

The transmission capacity can of course be raised by increasing the operating voltage but this again has a considerable bearing on cost, interference with communications, reliability in bad weather and on amenity by virtue of the need for higher towers.

It must be said that the use of cables for a.c. transmission has fundamental technical restrictions as well as economic ones. The main technical difficulty arises from the current required by the capacitance of the cable; for a 400 kV circuit this can be as much as 15,000 kVA per mile. It is only possible to supply this charging kVA from the terminal stations if the feeders are short, otherwise intermediate compensating equipment must be installed. For this reason it is virtually technically impossible to use alternating current for submarine circuits of any appreciable distance.

Earlier in this text reference was made to linking systems in order to share generating capacity. If a.c. is used for this purpose, the systems must each operate at exactly the same frequency. This is one of the main reasons why direct current was used to couple the British and French systems by the Cross-Channel d.c. link. Each system can operate at any desired frequency and without affecting the interchange of power between the two countries.

Let us now look to see where d.c. transmission can contribute to these problems of system development. The greatest of all is its ability to transmit power with no technical limit upon the distance. No intermediate compensating plant is required whatever the length of the feeder. This is particularly significant where underground cables are concerned, their capacitance is immaterial because normally no capacitance current flows for a direct current system.

On the subject of insulation stress, d.c. also has a contribution to make in using insulating materials to better advantage. It is possible to increase the working voltage of a d.c. system to $\sqrt{2}$ of the RMS voltage of the equivalent a.c. system without exceeding the maximum a.c. voltage stress. In an overhead system, where thermal rating is not always the factor that limits the current, this can also be increased by $\sqrt{2}$ without exceeding the same percentage losses of its a.c. counterpart. In other words, a 2-circuit 3-phase a.c. line can be converted into a 3-circuit d.c. line and transmit twice the power with the same size of conductors and clearance distance.

The savings in insulation in a power cable are even more significant. The radial voltage distribution in a cable under a.c. voltage conditions is non-uniform because it is controlled by the capacitance effects of the dielectric. The use of a direct voltage produces a much more uniform voltage gradient and working stresses in the dielectric and hence the operating voltage for a direct current system can be two or three times that for the same system working at an alternating voltage.

There is obviously now a technical and economic case for using d.c. transmission systems in the future. It is fortunate that the need for high voltage high current transmission schemes using static conversion equipment was foreseen. In 1929 the A.S.E.A. Company of Sweden patented the multi-grid-controlled mercury arc valve and this Company has pioneered the design of this type of equipment. In conjunction with the Swedish State Power Board they designed and constructed the first high voltage d.c. link in the world consisting of a connexion between the Swedish Mainland and the Island of Gotland, which lies in the Baltic Sea approximately 100 km from the mainland. This is rated at 20 MW, 100 kV and has given excellent service since it was first commissioned in March 1954.

On the Swedish side the power is taken from the 132 kV a.c. system and two mercury arc rectifiers each having a rating of 50 kV convert it to direct current, one pole being taken at each terminal and connected to electrodes in the sea. A similar arrangement on the Island of Gotland inverts the d.c. system back into the a.c. system at 30 kV. Thus the two systems are coupled together by means of a single core cable, earth return being provided by the sea.

Considerable advances have been made since 1954 in the design of mercury arc rectifiers in Sweden and the table illustrates the schemes which are operating, which are under construction and which are under consideration.

Circuit arrangements. The common arrangement of valves is in the form of a 3-phase bridge as shown in Fig. 1.

The current flows from the phase terminals of the transformer through two valves in series with the load. For example, the current due to the R-Y voltage flows through valves 1 and 6 for the positive half

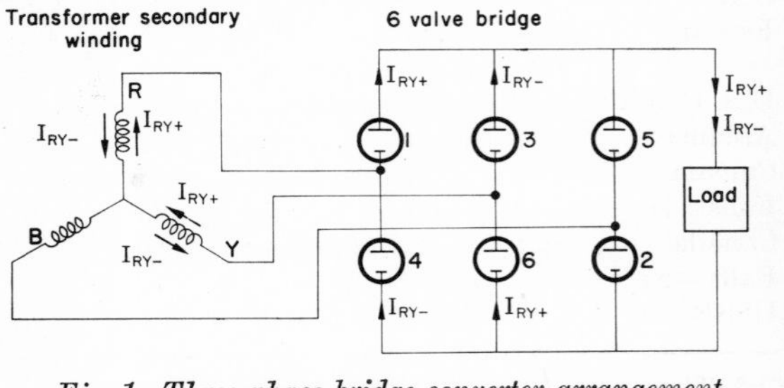

Fig. 1. Three-phase bridge converter arrangement.

cycle and through valves 3 and 4 for the negative half cycle.

Figure 2 shows how the currents flow in the valves for all the 3 phases, the pairs of valves concerned for each half cycle being indicated. It is of interest to note the transfer of current which takes place from valve to valve as time passes, a process called "commutation" and, starting with the *R-Y* voltage positive half cycle, during which valves 1 and 6 conduct, the commutation of current takes place between valves in the following order:

6 to 2, 1 to 3, 2 to 4, 3 to 5, 4 to 6, 5 to 1 and again 6 to 2 and so on.

Figure 3 shows the voltage wave form at the terminals of the load and it will be seen to contain 6 pulses per cycle, which must of course be separated by 60 electrical degrees.

If a similar wave form is superimposed but displaced relative to it by 30 electrical degrees, a smoother

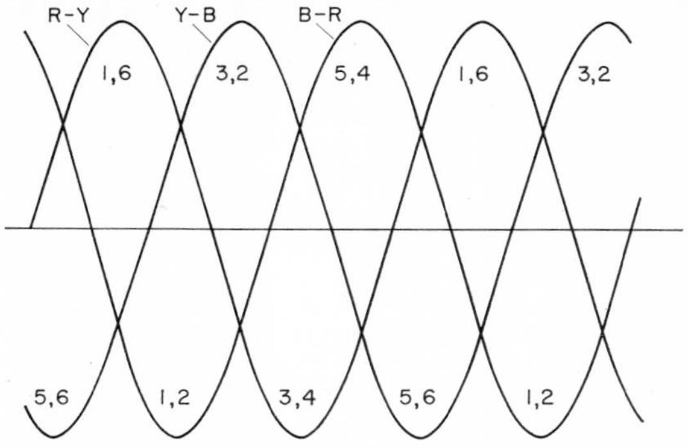

Fig. 2. Flow of currents in valves with respect to time. Pairs of conducting valves shown for each half cycle.

12 pulse output voltage is produced. This principle is adopted in large power transmission systems and is achieved by connecting two bridges in series with

High voltage d.c. transmission schemes

Scheme	Commis-sioning date	Capacity MW	tage kV d.c.	Route length km		Remarks
				Over-head	Under-ground	
Swedish Mainland—Gotland	1954	20	*100	—	100 sea	A link between asynchronous systems
England—France	1961	160	±100	—	56 sea	Asynchronous link
USSR Volgograd—Donbass	1962	750	±400	473	—	A bulk transmission system
Japan Sakuma Frequency Changer	1965	300	±125	—	—	A self contained link between 50 and 60 c.p.s. systems
New Zealand	1965	600	±250	570	40 sea	Asynchronous link between N & S Islands
Sweden—Denmark	1966	250	*250	85	85 sea	Crossing via Laesö
Italian Mainland—Sardinia	1966	200	*200	290	115 sea	Asynchronous link via Corsica
Canadian Mainland—Vancouver Island	1967–1969	310	*260	42	29 sea	Asynchronous link
U.S.Columbia River—Los Angeles	1968	1440	±400	1300	—	Bulk transmission
G.B.—Thames Estuary—London	1970	640	±266	—	85 land	Link between main transmission and distribution systems
U.S. Columbia River—Arizona	1971	1440	±400	1350	—	Bulk transmission
Canada—Churchill Falls—Montreal	Proposed	2000	900	1100	—	Bulk transmission
Canada—Churchill Falls—New York	Proposed	4500	1000	2500	150 sea	Bulk transmission via Newfoundland
USSR—Siberia—Urals	Proposed	12000	750	2500	—	Bulk transmission

* These schemes use the sea for the return path.

one of them electrically displaced on the a.c. side by 30 electrical degrees, as shown in Fig. 4.

The development of the grid controlled mercury arc valve has provided very precise control of the flow of current in these converter circuits and they now form the basis of the design of the terminal stations, which are used to connect the direct current transmission circuits into the a.c. system or systems.

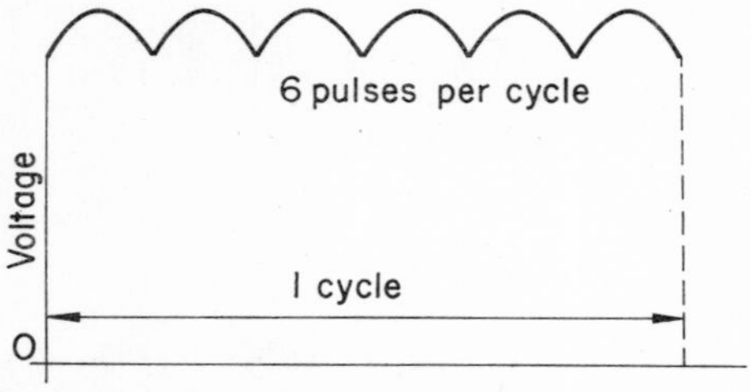

Fig. 3. Voltage output of the three-phase bridge shown in Fig. 1.

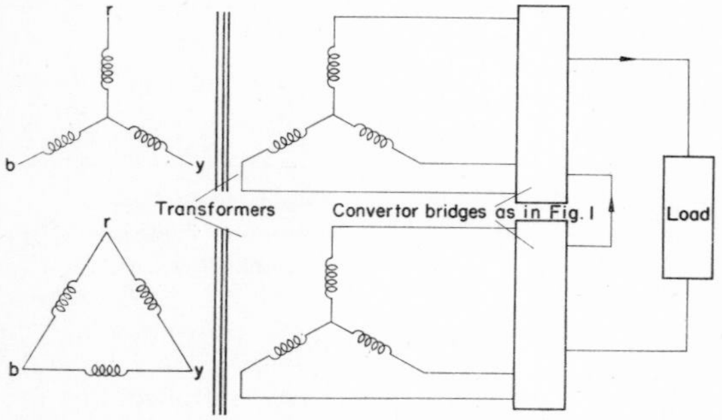

Fig. 4. Six-pulse bridges connected in series to give a twelve-pulse arrangement.

Many fascinating aspects arise among which are control and protection techniques, harmonic voltage generation and suppression and transient behaviour. Readers who wish to study the subject further are referred to the bibliography. The most comprehensive record anywhere in the world at the present time is the first-named I.E.E. Conference Publication. This Conference was attended by over 450 international specialists and every aspect of the subject was discussed. Part 1 is a complete record of the written contributions and Part 2, a record of the discussion.

New development. The striking achievement to date has been the development of the mercury arc valve for higher currents and voltages. There are now rapid developments taking place in commercial solid state devices and it is expected that thyristor type converters will soon become vigorous competitors. Big savings in terminal station ground area and hence cost are possible.

Bibliography

High voltage d.c. transmission, Conference publication No. 22, Parts 1 and 2, 19–23 September 1966, London: Institution of Electrical Engineers.

Direct current transmission, *Electrical Review,* 25 February 1966, 289.

KAISER F. D. (1966) Thyristor converters change high voltage d.c. economics, *Electrical World,* 21 February

FUKADA S. and TAKEI I. (1964). The Sakuma d.c. frequency converter project, *Direct Current,* **9**, No. 1, January, 1.

ADAMSON C. and HINGORANI N. G. (1960) *High voltage d.c. power transmission,* London: Garraway.

UHLMANN E. (1953). The operation of several phase displaced inverters on the same receiving network, *Direct Current,* **1**, No. 5, June 106.

RATHSMAN B. G. and LAMM U. (1952). The Gotland high voltage d.c. link, *Direct Current,* **1**, No. 1, June, 2.

LAMM U. (1946) *Mercury arc converter stations for high voltage d.c. power transmission.* C.I.G.R.E. Report No. 133.

DISPERSING CRYSTALS FOR X-RAY SPECTROMETRY. The dispersion of X-rays for identification or intensity measurement can be carried out using gratings, but is more usually performed by selective reflection of specific wave-lengths from planes (spacing d) of a suitable crystalline material; the wave-length (λ) of the reflected X-rays satisfies the Bragg relation $n\lambda = 2d \sin \theta$ where n is the order of the reflection and θ is the angle of diffraction. The range of wave-lengths normally encountered in X-ray analytical studies is from a few tens to a few tenths of angstroms; to accept such a wave-length range in conventional spectrometers with a θ range of perhaps 10–60° requires crystals with d values covering a range from a few tens to a few angstroms. It should be noted that the limiting of the accessible wave-length range with a given crystal and geometry is absolute at the high end, whereas at the low end radiations which cannot be accepted when $n = 1$ may be looked at in higher order ($n = 2, 3, 5$) thus effectively extending the working range considerably.

The table shows crystals available at present and their d spacings (taken largely from Elion).

The physical properties of the crystals are important in that they determine the type of spectrometer geometry in which the crystals can be used. (See Blokhin for a full discussion of the possible geometries.) Quartz, KAP, ADP, EDT, LiF and others can be cut, ground and bent to satisfy the fully focusing Johansson arrangement; mica can be used cleaved and bent to the semi-focusing Johann condition; the stearate family of pseudo-crystals, preparation of which is described by Henke, must be laid down on a surface which is either curved or can be bent subsequently to the Johann geometry.

The efficiency with which a given crystal reflects a specific wave-length can be considered from the two viewpoints of the intensity of the reflected line and the

Material	hkl Planes	Approx-imate *d* spacing	Relative efficiency—Arbitrary units (ref. Blokhin)				Relative efficiency—Arbitrary units (ref. Chan)	
			$n = I$	$n = II$	$n = III$	$n = IV$	Si-K rad$^{n.}$	Al-K rad$^{n.}$
Topaz	303	1·353	30	2				
Quartz	$20\bar{2}3$	1·375						
Quartz	$13\bar{4}0$	1·178	20	0 5				
Lithium Fluoride	200	2·01	110	18				
Aluminium	111	2·31	140					
Quartz	$11\bar{2}0$	2·45	10	10				
Sodium Chloride	200	2·814	120	25				
Calcite	211	3·05	35	12	17	18		
Silicon	111	3·135						
Potassium Chloride	200	3·14						
Fluorite	111	3·16	30					
Germanium	111	3·26						
Potassium Bromide	200	3·29						
Quartz	$10\bar{1}1$	3·343	35	5·5	0·3			
Quartz	$10\bar{1}0$	4·255	10	10	3·5	4	58	
†Pet	002	4·371					224	11
†Edt	020	4·404					86	3
†Adp	101	5·325						
Gypsum	020	7·60	17					
Mica	002	*~9·85	7				26	1
†Kap	$10\bar{1}0$	13·30						
†Ba or Pb My	—	39·5						
†Ba or Pb St	—	~50·0						
†Pblg	—	~65·0						

†Pet Pentaerythretol
†Edt Ethylene diamine d-tartrate
†Adp Ammonium dihydrogen phosphate
†Kap Potassium acid pthallate

†My Myristate
†St Stearate
†Lg Lignocerate
* Dependent on type.

angular width of the line. The intensity of reflection may sometimes be increased by surface abrasion of the crystal although an increase in line width generally results from this treatment; some discussion of this topic can be found in the references by Elion and Birks.

The intrinsic reflection efficiencies of the crystals vary considerably but assessment of this is complicated by the variation with X-ray wave-length for any one crystal. The table includes comparisons of efficiency given by Blokhin and by Chan; further performance figures are to be found in the papers by Ehlert *et al.*, Birks *et al.* and Jones *et al.*

Bibliography

BIRKS L. S. (1963) *Electron-Probe Microanalysis*, New York: Interscience.

BIRKS L. S. and SIOMKAJLO J. M. (1962) *Spectrochim. Acta*, **18**, 363.

BLOKHIN M. A. (1965) *Methods of X-Ray Spectroscopic Research*, Oxford: Pergamon Press 1965.

CHAN F. L. (1965) *Advances in X-Ray Analysis*, **9**, 515.

EHLERT R. C. and MATTSON R. A. (1966) *Advances in X-Ray Analysis*, **10**, 389.

ELION H. A. (1966) *Progress in Nuclear Energy, Series IX, Analytical Chemistry*, Vol. 5, Oxford: Pergamon Press.

HENKE B. L. (1963) *Advances in X-Ray Analysis*, **7**, 465.

HENKE B. L. (1964) *Advances in X-Ray Analysis*, **8**, 269.

JONES J. L., PASCHEN K. W. and NICHOLSON J. B. (1963) *Appl. Optics* **2**, 955.

D. M. POOLE

DIVERGENT BEAM X-RAY DIFFRACTION—PRESENT APPLICATIONS. When a highly diverging beam of characteristic X-rays penetrates into a single crystal, Kossel diffraction occurs in specific directions governed by the crystal orientation and the lattice symmetry. For a particular set of lattice planes in the crystal, the directions within the divergent beam for which the X-rays are incident on the crystal at the Bragg diffraction angle lie on the surface, or part surface, of a cone. The axis of this cone is normal to the planes, its apex is the source of the divergent beam, and its semi-angle the complement of the Bragg angle. The attenuation of the X-ray beam within the crystal is considerably enhanced in these directions, and a pattern of absorption, or deficiency, cones is consequently present in the radiation emerging from the crystal, each cone being associated with its appropriate set of diffracting planes. The diffracted X rays themselves radiate out from the lattices as reflection cones, but in these directions the local intensity is higher than in the adjacent area of the radiograph, so that they appear with reverse contrast. However, the contrast and resolution of the reflection cones is, in general, not as great as for the absorption cones.

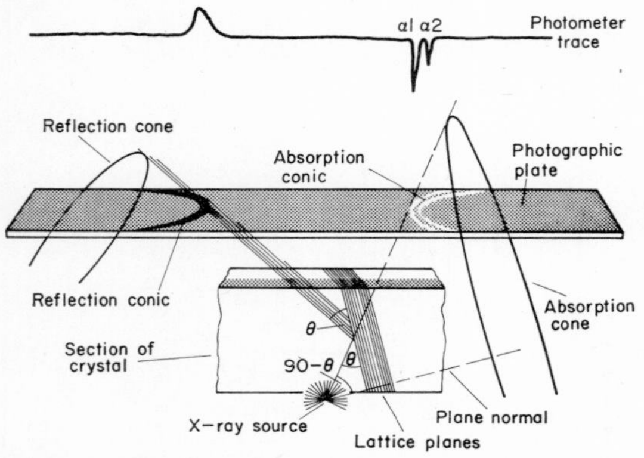

Fig. 1. The formation of absorption and reflection conics and their geometrical relationships.

The absorption and reflection cones that characterize divergent beam diffraction can be projected forward 2–3 in., and then photographically recorded, or alternatively recorded by a back reflection arrangement. They appear as a tracery of dark and light arcs contrasted against the radiographic image of the crystal. If a fine grained photographic emulsion is used, magnification to X 200–X 500 is possible and from a study of the relative positions and points of coincidence in the enlargements accurate determinations of lattice constants can be deduced. If the crystal gives cones with well resolved $\alpha_1\alpha_2$ components, a precision in lattice constant determination of $\pm 0{\cdot}001$ per cent is possible with cubic crystals. Measurements have been reported on diamond, nickel and magnesia with an accuracy approaching this precision. With hexagonal beryllium a reported accuracy of the measured lattice constants is $\pm 0{\cdot}01$ per cent. This very high accuracy can be achieved without any ancillary equipment requiring precision engineering, without the need for any precise film measurements, and without the need for the extrapolation corrections associated with other techniques. The method is essentially non-destructive as regards the crystal, and sequential measurements have been made on the same crystal during thermal and stressing sequences. The thickness of the crystal must be within certain limits, but the other dimensions are immaterial. Any observation of conic coincidence leading to a parameter measurement naturally refers only to the small cross-sectional element of crystal lying in the appropriate direction. This can be a disadvantage if an average lattice constant is required in a crystal in which there is a lack of overall homogeneity due, for example, to local variations in the level of impurity content. It may, however, be used to advantage if one requires some statistical assessment of the non-homogeneity in the lattice constants of a particular crystal. Apart from the measurement of lattice constants, divergent beam X-ray diffraction has been used to locate sub-structure boundaries in MgO. The lattice misorientation at such boundaries was found to be as small as 0·2°, but this gave a pronounced and measurable displacement of the conic.

Fig. 2. Part of divergent beam X-ray diffraction pattern from an MgO crystal.

Measurement of the broadening of the absorption conics has been used to study deformation mechanisms in aluminium and copper single crystals; to measure lattice strain; to estimate the density of edge dislocations; to study the substructure modifications in a fatiguing crystal of silver and to measure the distribution of point defect clusters in irradiated quartz.

Now that suitable X-ray microscopes are commercially available for both transmission and reflection divergent beam crystallography with X-ray source sizes as small as 1 μm diameter, the potentialities of the techniques can be examined in far more detail and more applications will doubtless be found.

Bibliography

Milledge H. J. (1962) in *Encyclopaedic Dictionary of Physics* (J. Thewlis Ed.) **7**, 796, Oxford: Pergamon Press.

SCHWARZENBERGER D. R. (1959) *Phil. Mag.* **4** (47), 1242.

SHARPE R. S. (1965) *Appl. Mat. Res.*, **4**, (2), 74.

R. S. SHARPE

DRUYVESTEYN DISTRIBUTION. When electrons are introduced into a gas by natural radioactivity, cosmic radiation, ultra-violet illumination or by some other means such as an electron gun, the electrons collide elastically with the gas particles so that momentum and energy are conserved. The electrons soon attain thermal equilibrium with the gas particles and due to the innumerable and incessant collisions occurring, the energy is spead over a wide range of values. The way the energy varies among the members of the electron population is described by the energy distribution function $F(\varepsilon)$ in terms of the energy ε. The number of electrons in a total population n which have an energy between ε and $\varepsilon + d\varepsilon$ is then $nF(\varepsilon)\,d\varepsilon$. If one electron only is considered by substituting $n = 1$, then the function $F(\varepsilon)\,d\varepsilon$ expresses the probability of the electron having an energy within the interval $d\varepsilon$, as compared with its probability $\int_0^\infty F(\varepsilon)\,d\varepsilon = 1$ of having an energy in the range $0 \leqq \varepsilon \leqq \infty$. When the latter condition is fulfilled the distribution function is said to be "normalized". It is often convenient to consider the mean energy $\bar{\varepsilon}$ of a normalized distribution as given by

$$\bar{\varepsilon} = \int_0^\infty \varepsilon F(\varepsilon)\,d\varepsilon. \tag{1}$$

The most prevailing type of energy distribution encountered in practice is the Maxwellian one, illustrated in the figure. It is derived making use of the equipartition of energy principle and can be

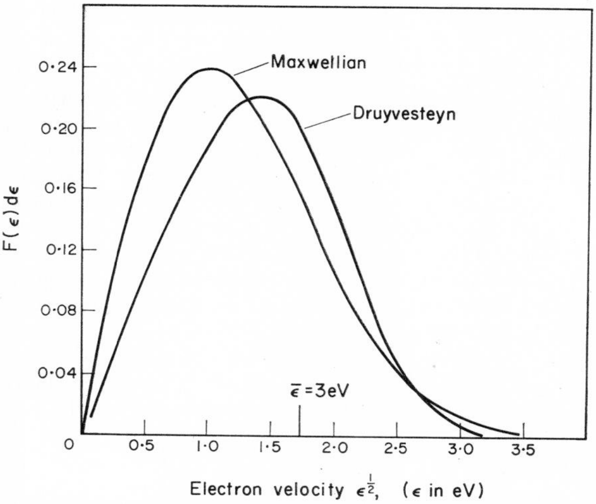

Fig. 1. Normalized Maxwellian and Druyvesteyn distribution functions for a mean energy of $\bar{\varepsilon} = 3eV$

represented by the equation

$$nF(\varepsilon)\,d\varepsilon = 2{\cdot}08 \cdot n/(\bar{\varepsilon})^{3/2} \cdot \varepsilon^{1/2} \exp \frac{3\varepsilon}{2\bar{\varepsilon}}\,d\varepsilon. \tag{2}$$

It can be shown from the formula that the mean energy is three times the most probable value and this is illustrated in the figure. For such a distribution it is possible to speak of a temperature of the particle population and this is given by

$$\boldsymbol{e}\bar{\varepsilon} = \frac{3}{2}\boldsymbol{k}T \tag{3}$$

in which $\boldsymbol{e}$ is the electronic charge ($1{\cdot}59 \times 10^{-19}$ coulomb), $\bar{\varepsilon}$ is the mean energy in eV, $\boldsymbol{k}$ is Boltzmann's constant ($1{\cdot}37 \times 10^{-23}$ Joules/T°) and T is the temperature in degrees absolute (Kelvin). At room temperature ($T = 293$°K), $\bar{\varepsilon} \cong 1/25$ eV.

When an electric field is applied the electrons are deflected in between collisions in the field direction and thus acquire a drift motion superimposed on their random movement. The electrons drifting towards the anode pick up energy from the electric field and lose some of this during collisions. At low voltage, though the energy gain per mean free path is small, the electron traverses many free paths and thus picks up energy continuously whilst the energy lost at each elastic collision is small, the fraction being $2\boldsymbol{m}/M$ where $\boldsymbol{m}$ is the electron mass and M is the mass of the gas particle; even for the lightest gas atom hydrogen, $\boldsymbol{m}/M = 1/1836$. The electron mean energy is thus continuously increased and quickly exceeds by far its initial thermal value equivalent to room temperature.

It is the great merit of Druyvesteyn, a Dutch physicist, that around 1935 he was able theoretically to account not only for the new value of the mean energy possessed by the electrons but also for the changed form of the electron energy distribution function. By carefully balancing the energy gained by the electrons from the electric field against the energy lost in elastic collisions and assuming the mean free path to be constant and independent of electron energy (see *Ramsauer effect*), Druyvesteyn derived an energy distribution which now carries his name and can be represented by

$$F(\varepsilon)\,d\varepsilon = 1{\cdot}05 \frac{\varepsilon^{1/2}}{(\bar{\varepsilon})^{3/2}} \exp - 0{\cdot}55(\varepsilon/\bar{\varepsilon})^2\,d\varepsilon. \tag{4}$$

This is illustrated in the figure for the same mean energy $\bar{\varepsilon} = 3$ eV and differs markedly from the Maxwellian form. Neglecting thermal energy, the mean energy is related to the applied voltage by $\bar{\varepsilon} = \left(\frac{0{\cdot}55M}{\boldsymbol{m}}\right)^{1/2} \boldsymbol{e}\lambda_0 \frac{E}{p}$ in which p is the gas pressure and λ_0 the mean free path at 1 torr. This was one of the first standard forms of distribution derived specifically for electrons drifting elastically in the field direction and it thus created great interest. It was used for detailed calculation of the excitation and ionization capability of electron swarms in the Townsend discharge and is also applied to obtain the attachment coefficient in electronegative gases such as air, although attaching type collisions were ignored by Druyvesteyn in his derivation. Its popularity at the time was due to the lack of any other alternative to the Maxwellian form. Lately a number of standard functions have become available which reflect the change in electron collision cross-section with electron energy. A difficulty brought about by the use of the Druyvesteyn distribution is that the once well-established Einstein expression $\bar{\varepsilon} = \frac{3}{2} D/\mu$, which related the mean energy to the ratio of the diffusion coefficient D to the mobility of electrons, is no longer accurately obeyed unless the time between collisions is constant. When a constant mean free path is assumed, as Druyvesteyn did, the relation is 87 per cent accurate and when a Ramsauer effect is present in the gas the Einstein expression may be different by a factor of three.

The Druyvesteyn distribution is now regarded as one of a number of standard distribution forms. It serves a useful purpose in that it can be used as a reference and yardstick with which other more accurate and intricate forms obtained by computer techniques can be compared.

A. E. D. Heylen

E

EARTH TIDES.

1. Description of the Phenomenon

The phenomenon of Earth tides is an elastic deformation of the Earth under stresses caused by the attraction of the Moon and the Sun.

If the Earth were completely rigid it would be possible to observe, by means of very sensitive instruments, small periodic deviations of the vertical (amplitude about $0''\cdot 02$) and small variations of the force of gravity (amplitude about 2×10^{-7}). The law of variation of these perturbations and their amplitude at any moment can be calculated with any desired accuracy from the orbital elements of the Earth and the Moon and the masses of the Moon and the Sun.

The Earth, however, of course undergoes deformation; it is not at all an ideal body, and has physical properties which are governed by very complicated and still largely unknown laws. The study of these laws is in the field of rheology; they involve the concepts of elasticity, viscosity, plasticity, and so on.

These deformations of the entire Earth, whose law of variation is naturally the same as that of the luni-solar forces which cause them, will influence the amplitude, and probably also the phase, of the effects to be measured; they will also bring about variable internal stresses and periodic changes of volume.

The significance of the measurements therefore lies in a comparison of the observed effect with the corresponding effect calculated for the case of an ideal rigid Earth. The ratio of the amplitudes and the difference of phases for each of the principal waves which the instrumental accuracy enables us to detect form the basis of geophysical study.

The importance of the earth tides is that they are the only deformations of the Earth for which we are able to calculate exactly and in advance the forces concerned (the attraction of the Moon and the Sun) and their variation with time.

However, the amplitude of these deformations itself varies from point to point as a function of latitude and also of the local or regional geological structure.

The potential of the tides at a given point is found by considering the difference between the luni-solar attraction force at that point and the same force applied at the centre of mass of the Earth. It is given by an expansion in Legendre polynomials, only the second-order polynomial being significant:

$$W_2 = \tfrac{1}{2}\, Gm \frac{a^2}{r^3} (3\cos^2 z - 1), \qquad (1)$$

where z is the zenith distance of the Sun or Moon at the point, G the gravitational constant, m the ratio of the Sun's or Moon's mass to that of the Earth, a the Earth's equatorial radius, and r the distance of the Sun or Moon (see Doodson 1962).

The form given in (1) is unsuitable for discussing the importance of various kinds of tides; and using the formula of spherical astronomy

$$\cos z = \sin \Phi \sin \delta + \cos \Phi \cos \delta \cos H,$$

where Φ and λ are the geometric coordinate (latitude and longitude) of the point considered, α and δ

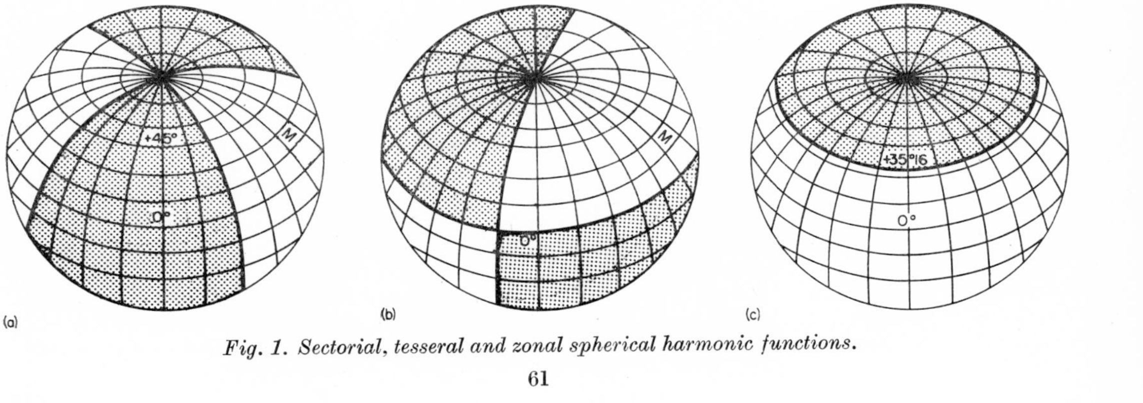

Fig. 1. Sectorial, tesseral and zonal spherical harmonic functions.

the equatorial coordinates (right ascension and declination) of the Sun or Moon, H is its hour angle ($= t_s - \alpha - \lambda$ where t_s is the sidereal time), Laplace obtained the potential as the sum of three terms:

$$W_2 = \frac{3}{4} Gm \frac{a^2}{r^3} \begin{cases} \{\cos^2 \Phi \cos^2 \delta \cos 2H \\ + \sin 2\Phi \sin 2\delta \cos H \\ + 3(\sin^2 \Phi - \tfrac{1}{3})(\sin^2 \delta - \tfrac{1}{3})\}, \end{cases} \quad (1')$$

which are the sectorial, tesseral and zonal spherical harmonic functions shown in Fig. 1.

The components of the tidal force at the point of observation are found by differentiating the potential with respect to a, Φ and λ:

vertical component: $\partial W_2/\partial a$, (2)

horizontal components:

$$\begin{cases} \text{meridian (North-South) } \xi = \partial W_2/a\, \partial \Phi, & (3) \\ \text{prime vertical (East-West) } \eta = \partial W_2/a \cos \Phi\, \partial \lambda. & (4) \end{cases}$$

These derivatives are shown in Figs. 2, 3 and 4 respectively.

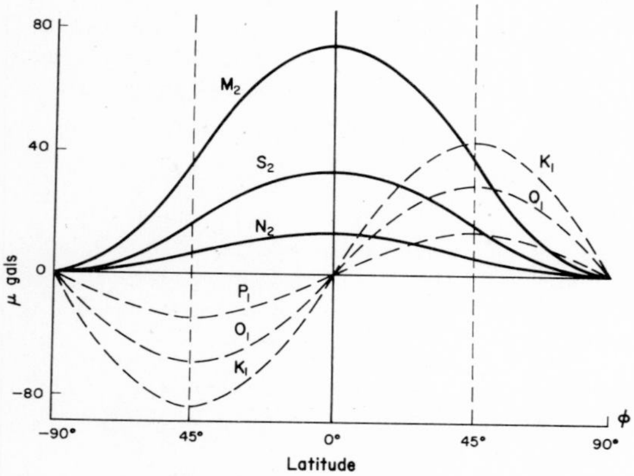

Fig. 2. Amplitude variations of the principal waves of the vertical component of tidal force, as a function of latitude. $K_1O_1P_1$: diurnal tesseral waves. $M_2S_2N_2$: semi-diurnal sectorial waves.

The *first function* has nodal lines (lines where the function is zero) only along the meridians 45° on either side of the meridian of the Sun or Moon; these lines divide the sphere into four sectors in which the function is alternately positive and negative. The regions of positive and negative W are those of high tide and low tide respectively. It is called a *sectorial function*; the corresponding tidal period is 12 hours, and the maximum amplitude occurs at the equator when the declination of the perturbing body is zero. Each of the three components is zero at the poles; the vertical and East–West components have maxima at the equator, while the North–South component is zero at the equator and maximum at 45°.

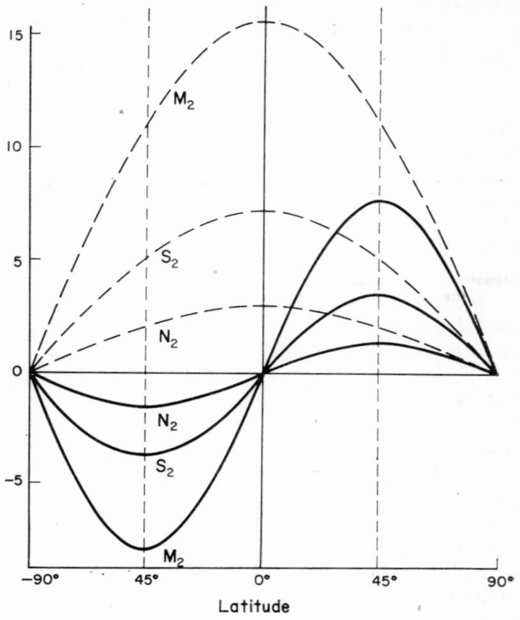

Fig. 3. Amplitude variations of the semi-diurnal sectorial waves ($M_2S_2N_2$) of the tidal force in the North-South (full lines) and East-West (broken lines) components, as a function of latitude.

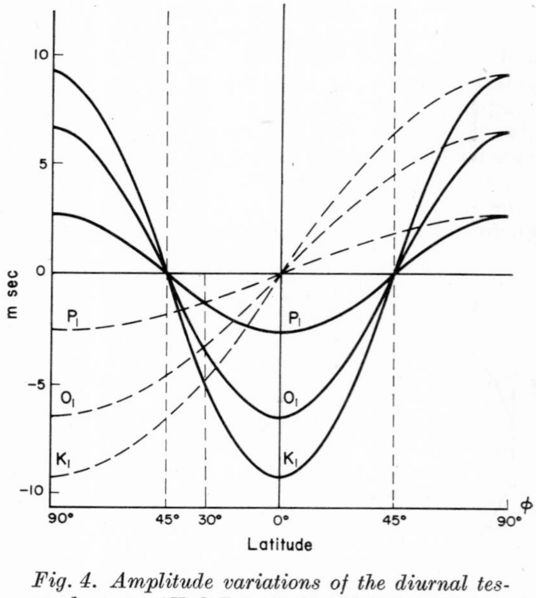

Fig. 4. Amplitude variations of the diurnal tesseral waves ($K_1O_1P_1$) of the tidal force in the North-South (full lines) and East-West (broken lines) components, as a function of latitude.

It may be noted that variations in the distribution of mass on the Earth's surface which give a sectorial distribution do not affect either the position of the pole of inertia or the largest moment of inertia C (which determines the rate of rotation of the Earth).

The *second function* has nodal lines along a meridian (90° from that of the perturbing body) and along the equator. It is a *tesseral function*, and the regions into which it divides the sphere change sign with the declination of the body. The corresponding tidal period is 24 hours. The vertical component has a maximum amplitude in latitudes 45° North and South and when the declination is greatest; it is always zero at the equator and the poles. The North–South component has zero amplitude at 45°, and maximum amplitudes of opposite sign at the equator and the poles. The East–West component is zero at the equator, and, of course, equal to the North–South component at the poles, where the foot of the vertical can only describe a circle, since the azimuth has no meaning there. Variations in the distribution of mass on the Earth's surface which give a tesseral distribution cause a change in the position of the pole of inertia but not in the largest moment of inertia C.

Figure 4 shows clearly that the North–South component of the diurnal tides gives rise to a couple acting on the Earth, whose resultant is not zero, owing to the oblateness of the Earth. The effect of this couple is to move the plane of the Earth's equator relative to the plane of the ecliptic. This is the precession—nutation effect in astronomy, as shown in Fig. 5. If the ecliptic and the plane of the lunar orbit coincide with the plane of the terrestrial equator ($\delta = 0$ for the Sun and for the Moon) the phenomena of precession-nutation disappear at the same time as the diurnal tesseral tidal waves. Thus it follows that one term in the expansion of the precession—nutation corresponds to each wave of the diurnal tide.

The *third function* is a *zonal function* depending only on the latitude; its nodal lines are the circles of latitude $\pm 35° 16'$. Since it otherwise depends only on the square of the sine of the declination, its period is fourteen days for the Moon and six months for the Sun. Variations in the distribution of mass on the Earth's surface giving a zonal distribution do not change the position of the pole of inertia but do affect the largest moment of inertia C. We should therefore expect the rate of rotation of the Earth to vary with the abovementioned periods. This has in fact been detected by time services using the most accurate instruments (photographic zenith tubes).

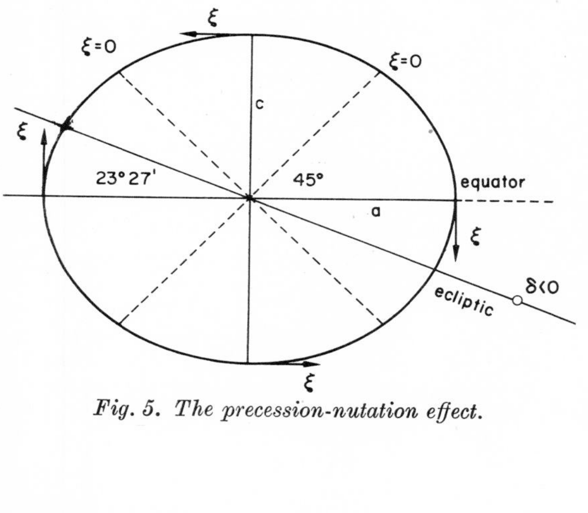

Fig. 5. The precession-nutation effect.

In order to derive a purely harmonic expansion of the tidal potential it is necessary, following Doodson, to define the following six independent variables: mean lunar time, mean longitude of the Moon, mean longitude of the Sun, longitude of the Moon's perigee, longitude of the Moon's ascending node, longitude of perihelion. This leads to a resolution into a very large number of waves: here we shall mention only the six principal ones which at present provide important geophysical information. For further details we refer to a more extensive work (Melchior 1962; Doodson 1962). These principal waves comprise three semidiurnal waves known as:

lunar	(M_2)	period $12^h25^m14^s$, speed 28°9841 per mean solar hour
solar	(S_2)	period 12^h, speed 30°000 per m.s.h.
lunar elliptical	(N_2)	period $12^h39^m30^s$, speed 28°4397 per m.s.h.

(i.e. due to the eccentricity of the Moon's orbit)

and three diurnal waves known as:

lunisolar	(K_1)	period 1 sidereal day exactly, $= 23^h56^m4^s$, speed 15°0411 per m.s.h.

(associated with the lunisolar precession)

lunar	(O_1)	period $25^h49^m10^s$, speed 13°9430 per m.s.h.

(associated with the fortnightly nutation)

solar	(P_1)	period $24^h3^m57^s$, speed 14°9589 per m.s.h.

(associated with the six-monthly nutation)

If we know the orbital elements of the Earth and the Moon and the ratios of the masses of the Sun and Moon to that of the Earth, we can calculate the amplitude of each of these waves as a function of the latitude of observation [see formula (1')]. Thus in Figs. 2, 3 and 4 the amplitudes are given in microgals for the vertical component and in milliseconds of arc for the horizontal components.

Then, knowing the theoretical amplitudes of the six principal waves in the three components of the force exerted on the Earth, we require suitable instruments to record the resulting deformations, followed by harmonic analysis of the results in order to derive the elasticity characteristics of the Earth.

2. *Measuring Instruments*

At the chosen point of observation, the three components of the tidal force have distinct effects. The resultant of the applied forces at any point on the Earth's surface is called the force of gravity; its direction is that of the vertical and its strength is denoted by $\boldsymbol{g}$. The forces which contribute to this resultant are

(1) the gravitational attraction of the Earth's mass,
(2) the centrifugal force of the Earth's rotation,
(3) the external luni-solar tidal forces.

The respective strengths of these forces are (1) about +981 gal, (2) −3 gal, (3) ±0·0002 gal. The first two forces are constant and the tidal effect is only a small periodic perturbation causing

(a) a variation in the force of gravity $\boldsymbol{g}$ which comes directly from the vertical component of the tidal force,

(b) an oscillation of the direction of the vertical which can be resolved in the North–South and East–West directions and whose angular value, n is given by the ratio of the horizontal components of the tidal force to the force of gravity g:

$$n_{\mathrm{NS}} \cong \tan n_{\mathrm{NS}} = \frac{1}{a\boldsymbol{g}} \frac{\partial W_2}{\partial \Phi}, \tag{5}$$

$$n_{\mathrm{EW}} \cong \tan n_{\mathrm{EW}} = \frac{1}{a\boldsymbol{g} \cos \Phi} \frac{\partial W_2}{\partial \lambda}. \tag{6}$$

In order to observe the three components of the tidal force it is necessary to use three instruments based on two quite different principles, owing to the large constant vertical acceleration of about 980 gal:

(1) a gravimeter to record the periodic variations of $\boldsymbol{g}$,

(2) two clinometers (generally horizontal pendulums) to record the oscillations of the vertical in the N–S and E–W directions.

(*a*) *Gravimeters* (Dooley 1961; Loncarevic 1961). Here we shall consider only differential apparatus which can measure very small variations of $\boldsymbol{g}$, but not determine the value of $\boldsymbol{g}$ itself. The main difficulty is to calibrate the recordings in terms of the unit of acceleration (1 gal). The calibration is often achieved by observing the movement of the gravimeter beam under the influence of a known measuring spring. More recent methods employ electrical techniques or involve raising the equipment (the vertical gradient of gravity).

The instruments used are all based on the change in the extension of a high-quality spring made of a special alloy and placed in a temperature-controlled enclosure. Two types have been designed.

(1) stable gravimeters, which are simple balances whose sensitivity is inherently low and which are equipped with an electronic amplification system (Askania).

(2) astatized gravimeters, so designed that the restoring force of the spring remains equal and opposite to the force of gravity over a fairly wide range. This is achieved by means of a spring placed at 45° to the balance beam and whose unstressed length would be zero (North American, LaCoste-Romberg, etc.) (Loncarevic 1961).

(*b*) *Horizontal pendulum.* This instrument consists of a small beam about 10 cm length and of mass approximately 10 grammes supported by two very thin threads as shown in Fig. 6 (Hengler-Zöllner type). The axis of rotation of this pendulum passes through the points at which the threads are attached to the supporting frame, and the angle i which it makes with the vertical can be made very small (in practice, $i \sim 5''$), so that the force of gravity is reduced to an effective value of $\boldsymbol{g} \sin i$ and the horizontal components of the tidal force become correspondingly more significant. The sensitivity of the instrument is therefore proportional to the reciprocal of $\sin i$ (~40,000).

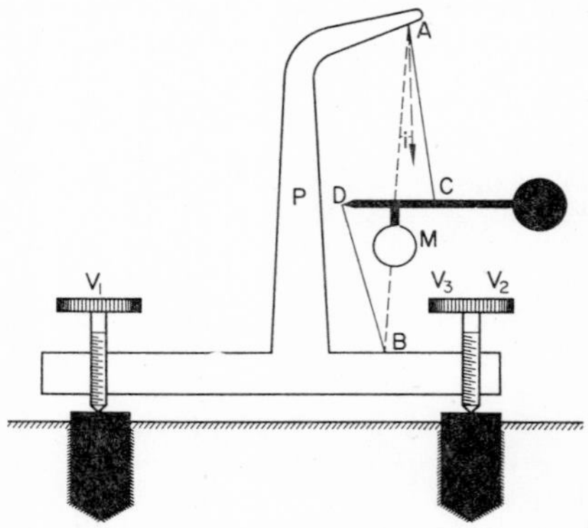

Fig. 6. The horizontal pendulum (Hengler-Zöllner type).

A mirror is fixed to the oscillating beam and reflects a spot of light on to a photographic recording system of the Poggendorff type. With a focal length of 5 m, a further amplification factor of 100 is obtained (2 × 500/10). There is, however, a decrease in sensitivity on account of the torsion couple exerted by the threads, and the final amplification factor of the best instruments is of the order of 2×10^6.

Two types of pendulum have been designed.

(1) the metal type, whose sensitivity is usually not better than 0″·01 per millimetre on the recording paper (Schweydar, Tomaschek).

(2) the fused-quartz or fused-silica type (Verbaandert-Melchior), where the points of attachment of the threads to the frame can be identified very accurately by autogenous welding of the quartz threads (diameter 50 μ) to the quartz frame. The pendulum is also of quartz. The sensitivity may be as high as 0″·0007 per millimetre.

These instruments can be calibrated in a rigorous and accurate manner by using the Verbaandert variable base-plate. This is a steel capsule into which mercury is forced under pressure and which is placed under the azimuth foot screw of the pendulum frame. When the mercury pressure is varied, the capsule changes shape and causes the pendulum to execute artificial oscillations. These oscillations are known because the bearing has been previously calibrated with an interferometer. The response of the pendulum to the oscillation provides the calibration of the instrument.

Some underground observing stations are now equipped with automatic means of carrying out regular calibrations of horizontal pendulums.

Observing stations. In order to eliminate thermal effects at the Earth's surface, it is desirable to place the instruments in underground stations at a depth of at least 40 m. This is essential for horizontal pendulums but less so for gravimeters, which must be placed in an enclosure at constant temperature and, if possible, in a sealed vessel. Permanent gravimetric stations are in operation and have given important results in France, Belgium, Germany, Luxembourg, Italy, Sweden, USSR, Iran, Japan, USA and Venezuela.

The pendulums must be set up in rock with the minimum of artificial connexions. Very detailed installation techniques have been worked out (see Melchior 1966).

Such stations now exist and have given important results in Belgium, Germany, Austria, Italy, Sweden, Czechoslovakia, Bulgaria and USSR.

3. *Interpretation of Observations*

The parameters derived from observation are not the same for the variations of gravity and the deviations of the vertical. The Earth's crust, under the Moon's attraction which reduces the value of g, is raised by the tidal effect, and the gravimeter upon it moves away from the centre of the Earth, so that the force of gravity is further reduced; thus there is an amplification effect, and the observed amplitude factors (ratios of observed and theoretical amplitudes) are greater than unity. On the other hand, the crust is tilted in the direction of the deviation of the vertical, in order to remain perpendicular to the vertical; this deformation partly compensates the amplitude of the deviation of a pendulum, and the observed amplitude factors for the horizontal components are less than unity.

Love has given simple expressions for these amplitude factors as functions of parameters (Love numbers) related to the internal distribution of the elastic properties and of the density. Theoretical calculations have been made for various models of the Earth. These parameters are defined as follows.

h = ratio of the amplitude of the radial deformation of the Earth to the theoretical amplitude of the corresponding deformation of the equipotential surface.

k = ratio of the change in the Earth's potential caused by the deformation of the Earth to the external potential causing this deformation.

It can then be shown that the amplitude of the deviations of the vertical is reduced by the deformation of the Earth in the ratio

$$\gamma = 1 + k - h,$$

while the variations of g are amplified in the ratio

$$\delta = 1 + h - \frac{3}{2}k.$$

Methods of calculation. Just as the instruments must be calibrated as accurately as possible, the theoretical values of the amplitudes of the various waves of the tides, with which the observed values are to be compared as with a standard, must be determined as accurately as possible. This may seem to be quite simple and to depend only on the accuracy with which the orbital elements of the Earth and Moon, and in particular the Moon's mass, are known.

We must, however, also take into account the fact that a month's or even a year's continuous observations do not suffice to give measurements from which even the most accurate possible selection process can define a wave free from all contributions from neighbouring waves, since the frequencies of the various waves are too close together. Consequently, a harmonic analysis, by whatsoever method, will always yield a resultant wave containing definite contributions from neighbouring waves. What is important is to calculate exactly the contribution from each wave and, for the time in question, the homologous theoretical wave corresponding to the composite result of analysing the observations; the latter is called the required principal wave, but it is never completely free from other waves.

This idea has been particularly well developed in Lecolazet's method, which takes account of 79 waves in each calculation and can be immediately extended to a much larger number of waves without any fundamental modification.

Numerous tests have confirmed these theoretical arguments and have shown that Lecolazet's method gives the smallest errors in the determination of the harmonic constants.

For this reason the method has been the most widely used, and in particular has been adopted by the International Centre for Earth Tides, which has used it on more than 3500 occasions.

The development of electronic computation. This work has been greatly extended by the use of electronic computation. Since 1958, the International Centre for Earth Tides in Brussels has used increasingly powerful computers to analyse observations of the deformations of the Earth. For several years, the classical methods of Doodson and Lecolazet, devised before the development of large computers, have been applied at this Centre.

In 1966, A. P. Venedikov at the same Centre worked out a new method making use of the full potential of a powerful computer, first in choosing selection methods and then in applying them. This method can be used for recordings of any length and with interruptions of any kind.

The comparison with theoretical values is based on Lecolazet's principle of homologous waves.

4. The Internal Structure of the Earth as shown by the Earth Tides

Theoretical study of deformations. The problem of the theory is to express the numbers h and k as functions of the internal distribution of the elastic properties and of the density. The problem was solved by Kelvin for the case of a uniform Earth. In 1905 Herglotz worked out the theory for a non-uniform but incompressible Earth, arriving at a sixth-order differential equation for h; this reduces to a Clairaut-type second-order equation if the rigidity is put equal to zero. In 1950 Takeuchi integrated numerically the equations for a non-uniform and compressible Earth, using various models of the Earth's interior based on seismological results. Pekeris, and later Molodensky, instead expressed the deformation equations as a linear set of six homogeneous first-order differential equations, which is convenient for applying the boundary conditions (the surface of discontinuity between the core and the mantle, and the outer surface) and carrying out a numerical calculation.

In 1957 Jeffreys, extending earlier work by Poincaré, showed that a resonance effect due to motions in the liquid core influences waves whose period is sufficiently close to the sidereal day (K_1, and to a lesser extent P_1). Jeffreys and Vicente described a series of liquid-core models (Bullen 1961). Molodensky carried out similar work in 1961 and obtained results very like those of Jeffreys and Vicente.

According to these theoretical calculations the three diurnal waves which one can expect to derive with fair accuracy from the observations will have the amplitude coefficients shown in the following table.

A relation $k \sim \frac{1}{2}h$ pointed out by Melchior in 1951 was also found by Molodensky, but does not occur in Jeffreys and Vicente's models. This is the reason for the quite large differences between the values of δ given by the two theories.

The amplitude coefficients for the semi-diurnal and longperiod waves should be practically the same as for O_1, which is not subject to resonance.

Experimental results. Table 1 shows that the effect of the core is more noticeable in the horizontal components (factor γ) than in the vertical component (factor δ), which is evident if we put $k \sim \frac{1}{2}h$:

$$\gamma = 1 + k - h \sim 1 - h/2$$
$$\delta = 1 + h - 3/2k \sim 1 + h/4$$

However, the calibration of horizontal pendulums is effected much more precisely than that of gravimeters. The horizontal pendulum is therefore the best instrument for detecting the dynamic effects of the liquid core.

To appreciate the quality of the instruments used and the accuracy of their calibration it may be noted that the amplitudes of the three waves are respectively as follows for the stations concerned (latitude about 50°):

Vertical component	*East-West component*
51 μ gal	0″0070 for K_1
17 μ gal	0″0020 for P_1
34 μ gal	0″0050 for O_1

The solar P_1 wave is very difficult to obtain: its amplitude is very small and it is very sensitive to diurnal perturbations of a thermal character.

The Bulletin d'Informations des Marées Terrestres No. 46 gives the results of 23 series of observations each long enough in duration (at least 300 days) to permit the achievement of sufficient precision for the P_1 wave. These are reproduced in Tables 2 and 3.

Table 1.

	Deviations of the vertical: factor $\gamma = 1 + k - h$				Variations of $\boldsymbol{g}$: factor $\delta = 1 + h - \frac{3}{2}k$			
	(1)	(2)	(3)	(4)	(1)	(2)	(3)	(4)
K_1	0·714	0·693	0·734	0·730	1·183	1·185	1·137	1·143
P_1	0·695	0·696	0·699	0·697	1·196	1·172	1·154	1·158
O_1	0·654	0·658	0·688	0·686	1·224	1·211	1·159	1·164
$\Delta(O_1 - K_1)$	−0·060	−0·035	−0·046	−0·044	+0·041	+0·026	+0·022	+0·021

(1) Jeffreys and Vicente's model, assuming Roche's law in the core;
(2) Jeffreys and Vicente's model with central particle;
(3) Molodensky's model with homogeneous fluid core;
(4) Molodensky's model with central kernel.

It can be seen that these results show good agreement among themselves: they also agree well with theoretical analysis. In Table 4 the results are extracted for comparison with theoretical models.

Table 2. Horizontal pendulums. Results obtained for the East-West component for the three principal diurnal waves (N = duration of observations in days).

K_1 WAVE

Stations	$\gamma = 1 + k - h$	$\varepsilon(\gamma)$	N
Dourbes 1	0·7578	0·0067	1350
Dourbes 2	0·7514	0·0081	988
Sclaigneaux 1	0·7428	0·0125	860
Sclaigneaux 3	0·7509	0·0228	588
Sclaigneaux 2	0·7269	0·0154	544
Kanne	0·7254	0·0191	364
Tiefenort	0·7330	0·0097	1048
Graz	0·6971	0·0203	286
Dannemora	0·7098	0·0153	898

P_1 WAVE

Stations	$\gamma = 1 + k - h$	$\varepsilon(\gamma)$	N
Dourbes 1	0·7327	0·0209	1350
Dourbes 2	0·7092	0·0255	988
Sclaigneaux 1	0·7204	0·0396	860
Sclaigneaux 3	0·7307	0·0700	588
Sclaigneaux 2	0·6540	0·0494	544
Kanne	0·6480	0·0624	364
Tiefenort	0·7933	0·0277	1048
Dannemora	0·6535	0·0489	898

O_1 WAVE

Stations	$\gamma = 1 + k - h$	$\varepsilon(\gamma)$	N
Dourbes 1	0·6583	0·0094	1350
Dourbes 2	0·6767	0·0114	988
Sclaigneaux 1	0·6899	0·0172	860
Sclaigneaux 3	0·6677	0·0323	588
Sclaigneaux 2	0·6884	0·0203	544
Kanne	0·6965	0·0242	364
Tiefenort	0·6745	0·0141	1048
Graz	0·6333	0·0275	286
Dannemora	0·6795	0·0212	898

All these stations, with the exception of Tiefenort, are equipped with Verbaandert-Melchior horizontal pendulums. The station at Tiefenort employs a Schweydar metallic pendulum.

Strasbourg and Frankfurt/M use North American astatized gravimeters. All the other stations employ Askania stable gravimeters.

The dynamic effects are in agreement with theory although a little higher than expected for the factor γ.

The following observations may be made regarding the (static) solution corresponding to the O_1 wave:

1. It gives for the factor δ a value equal to that in Central Asia.
2. The same factor δ is equal to that found by Lecolazet and Steinmetz for the zonal wave *Mf* ($\delta = 1{\cdot}16 \pm 0{\cdot}09$) for which static theory holds.

Table 3. Gravimeters. Results obtained for the vertical component for the three principal diurnal waves (N = duration of observations, in days).

K_1 WAVE

Stations	$\delta = 1 + h - 3/2k$	$\varepsilon(\delta)$	N
Brussels I	1·1386	0·0082	1136
Brussels II	1·1575	0·0088	916
Brussels I B	1·1411	0·0135	252
Luxembourg	1·1130	0·0062	1058
Bonn	1·1415	0·0077	266
Strasbourg	1·1592	0·0012	365
Frankfurt/M	1·1375	0·0029	458
Genoa	1·1535	0·0139	702
Trieste	1·1547	0·0125	338
Stockholm	1·1241	0·0078	664
Helsinki	1·1002	0·0052	354
Talgar	1·1630	0·0083	546
Teheran	1·1100	0·0187	552
Kyoto	1·1479	0·0198	372

P_1 WAVE

Stations	$\delta = 1 + h - 3/2k$	$\varepsilon(\delta)$	N
Brussels I	1·1693	0·0228	1136
Brussels II	1·2522	0·0239	916
Brussels I B	1·0514	0·0445	252
Luxembourg	1·1052	0·0196	1058
Bonn	1·1021	0·0246	266
Strasbourg	1·1705	0·0038	365
Frankfurt/M	1·1576	0·0088	458
Genoa	1·2014	0·0384	702
Trieste	1·0304	0·0330	338
Stockholm	1·0745	0·0235	664
Helsinki	1·1525	0·0171	354
Talgar	1·1590	0·0233	546
Teheran	1·1696	0·0596	552
Kyoto	1·1329	0·0522	372

O_1 WAVE			
Stations	$\delta = 1 + h - 3/2k$	$\varepsilon(\delta)$	N
Brussels I	1·1579	0·0124	1136
Brussels II	1·1616	0·0134	916
Brussels I B	1·1660	0·0174	252
Luxembourg	1·1615	0·0087	1058
Bonn	1·1537	0·0105	266
Strasbourg	1·1773	0·0017	365
Frankfurt/M	1·1406	0·0040	458
Genoa	1·1624	0·0205	702
Trieste	1·1636	0·0192	338
Stockholm	1·1456	0·0110	664
Helsinki	1·1254	0·0071	354
Talgar	1·1780	0·0123	546
Teheran	1·1382	0·0242	552
Kyoto	1·1646	0·0307	372

Table 4. Experimental results.

Waves	γ	δ	k	h	k/h
K_1	0·747	1·143	0·220	0·473	0·465
P_1	0·721	1·148	0·262	0·541	0·484
O_1	0·676	1·160	0·328	0·652	0·503
Q_1	0·654	1·167	0·358	0·704	(0·517)

3. The value of k agrees with that found by Markowitz from the variations of the speed of the Earth's rotation resulting from variations in flatness due to the same wave Mf ($k = 0{\cdot}34 \pm 0{\cdot}07$).
4. The value of k agrees perfectly with that deduced by R. Newton from the study of perturbations of satellite orbits by the solar S_2 wave:

$$k = 0{\cdot}327 \pm 0{\cdot}036.$$

5. The value found for the ratio k/h is almost the same as that obtained by Melchior (0·504).
6. The value of k obtained here would give to the movement of the pole a period (Chandler) of more than 450 days, which is excessive and should be examined further.

These results will undoubtedly need some modification when the corrections due to indirect effects are applied. But these are small for the diurnal waves and we do not think that the essential conclusions can be affected.

5. Indirect Effects and the Structure of the Earth's Crust

The semi-diurnal earth tides are more difficult to interpret, because the oceans, and in particular the coastal waters of the Atlantic, exhibit large semi-diurnal tides through belonging to systems having a resonance in this frequency range.

In consequence, the recording apparatus is subject to large perturbations at these frequencies, called indirect effects because they have all the properties of the tide but are transmitted to the instruments via the ocean tides, which cause (1) an attraction due to the moving mass of water itself, (2) a bending of the crust under the extra load, (3) a change in the Earth's potential due to this bending.

Near the coasts these effects are greater than the Earth tide itself, and they are quite perceptible at 100 to 200 km from the coast. They cause different perturbations of the amplitude and phase of each wave, and it has been possible to develop on this basis an empirical method of separating the direct and indirect effects.

It is obviously desirable to try to calculate the indirect effects in order to deduce the elastic properties of the crust and the upper mantle, thus demonstrating various nonuniformities. This calculation meets with great difficulties due to the lack of accurate data on the size of the tides in mid-ocean and also the problem of calculating the bending of the non-uniform crust of the Earth. It is therefore too soon to state values of γ for the semi-diurnal waves. We can at most say that the values obtained are of the order of 0·8 in Western Europe and appear to tend to 0·66 in the centre of the continent. The phase difference between the waves M_2 and S_2 appears to agree well also with the indirect effects of the oceans. There are, however, fairly large anomalies for the N–S component.

Series of measurements are now being made in networks of numerous stations to try to elucidate these phenomena, mainly aiming at a knowledge of the indirect effects and the structure of the crust rather than that of the general earth tides.

6. Earth Tides and the Retardation of the Speed of the Earth's Rotation. The part played by sectorial Earth tides.

The retardation of the speed of rotation of the Earth can only take place if a non-zero couple acts on the equatorial plane producing a rotation vector ω_1 parallel to the vector ω and in the opposite sense.

If the Earth revolved symmetrically about its axis, the resultant couple exerted on the equatorial plane by the Moon (or Sun) would be zero if the Earth were not deformed (Euclidean solid, Love number $h = 0$) or if its deformation were symmetrical with respect to the Earth-Moon direction (Hookean solid, no phase lag, no energy dissipation).

If the Earth were viscoelastic (on whatever rheological model) its deformation would occasion a retardation (phase lag 2ε) with respect to the stress imposed by the tidal force, and energy,

Table 5. Vertical component. Phase lag of the M_2 wave.

Atlantic Europe		*Central Europe*		*USSR*	
Brussells I	+0°72	Kieselbach	−4°00	Poulkovo	−2°09
Brussells II	+1°56	Potsdam	−0°98	Krasnaya P	−4°23
Dourbes	+0°96	Berggieshübel	−2°98	Moscou	−1°34
Vedrin	+0°03	Tihany	−0°37	Kiev	−2°82
Battice	+0°78	Borowiec	−1°46	Poltava	−1°30
Luxembourg	+0°71	*Northern Europe*		Tiflis	(+1°20)?
Strasbourg	+2°08	*Stockholm*	−2°63	*Asia*	
Karlsruhe	+0°34	*Helsinki*	−0°45	*Talgar*	−3°43
Frankfurt/M	+0°72	*Southern Europe*		Tashkent	−4°10
Bonn II	+0°89	Sofia	−1°45	Alma Ata	−3°51
Hannover	+1°37	*Genoa*	−0°20	Frounze	−3°60
Berlin	+0°68	*Trieste*	−0°80	Langchow	−2°88
Bad Salzungen	+0°60	Costozza	−2°18	*Teheran*	−3°67
Freiberg	+0°20	Resina	−2°72	*Kyoto*	−2°17

+: in advance
−: behind

The stations underlined were those for which a complete analysis has been made on a very series of observations (250–100 days).

abstracted from the kinetic energy of rotation of the Earth, would be dissipated. This is known as *secular retardation*. An expression for this may easily be obtained by calculating the couple, N, exercised by the horizontal East-West component of the sectorial tidal force on the Earth, which is radially deformed (vertical component) by the same sectorial force.

$$N = \frac{8\pi}{15} A^2 \cos^4 \delta \sin 2\varepsilon \int_0^a \varrho \frac{\partial}{\partial r} \left[\frac{H(r)\, r^6}{g} \right] dr$$

where $A = \frac{3Gm}{2\tau^3}$. The couple is proportional to the square of A since it expresses a tidal force exerted on a tidal deformation. In these conditions the effect of the Moon is about five times more pronounced than that of the Sun. The angle 2ε is given directly by the analysis of gravimetric records.

We know that oceanic tides give rise to perturbations (indirect effects) in earth tides. The angle 2ε must be corrected for these effects if it is to be interpreted as a parameter describing the rheological behaviour of the planet. However, if one wishes to calculate the couple responsible for the retardation of the terrestrial rotation as such it is necessary to consider the actual total deformation to which the East-West component of the sectorial tidal force applies. At the surface this deformation is characterized by the observed angle 2ε. But the indirect oceanic effects are essentially superficial and it is difficult in these conditions to estimate the value of 2ε in the interior of the Earth.

Table 5 gives the results obtained by all the gravimetric stations which have submitted their data to the International Centre for Earth Tides.

It may be seen that the geographical distribution of the angle 2ε is not smooth but is characterized by a definite number of zones:

Atlantic Europe: about 0°7 in advance
Central Europe and Italy: about 1°7 behind
USSR, Asia and Japan: about 3°5 behind

These phases refer to the entire tide, but we are interested in the phase lag of the deformation tide and this is six times greater if the amplitude factor δ is equal to 1·20.

We are here dealing with the existence of relative deformations between different regions of the Earth's crust, which will give rise to an additional dissipation of energy.

Comments

(1) Earth tides lead to other effects which have so far been less easy to measure than the three components of gravity:

(a) volume changes causing fluctuations in the water level in many wells and apparently also slight periodic variations in volcanic activity.

(b) variable stresses in the Earth's crust (measured with extensometers), which some workers have regarded as triggering earthquakes, but this is disputed.

(2) The experimental study of earth tides requires organization on a world scale. The International Association of Geodesy has a Permanent Commission on Earth Tides, whose work is based on a Centre (already mentioned) affiliated to the Federation of Astronomical and Geophysical Services. This Centre arranges for all reductions of observations, translates into French the Russian publications on this subject, publishes an Information Bulletin, receives and invites research workers, and arranges symposia.

Bibliography

BULLEN K. E. (1961) in *Encyclopaedic Dictionary of Physics* (J. Thewlis Ed.), **2**, 558, Oxford: Pergamon Press.

Bulletin d'Informations sur les Marées Terrestres No. 46, Brussels, 31 Dec. 1966.

DOODSON A. T. (1962) in *Encyclopaedic Dictionary of Physics* (J. Thewlis Ed.), **7**, 355, 358, Oxford: Pergamon Press.

DOOLEY J. C. (1961) in *Encyclopaedic Dictionary of Physics* (J. Thewlis Ed.), **3**, 498, Oxford: Pergamon Press.

LONCAREVIC B. D. (1961) in *Encyclopaedic Dictionary of Physics* (J. Thewlis Ed.), **3**, 516, Oxford: Pergamon Press.

MELCHIOR P. (1966) *The Earth Tides*, Oxford: Pergamon Press.

MELCHIOR P. and GEORIS B. (1967) *Earth tides, precession-nutation and retardation of the Earth's rotation*, Astronomical Union-Commission 19, Prague.

P. MELCHIOR

ELASTIC LIQUIDS. *1. Introduction.* In the development of classical mechanics, the distinction between solids and liquids was assumed to be quite sharp, and separate physical laws were formulated to account for the mechanical behaviour of each: the solid obeying Hooke's law and the liquid Newton's law of constant viscosity. It is now common knowledge that many materials cannot be classified as either simply Hookean or simply Newtonian. These materials exhibit both elastic and viscous properties and may be conveniently divided into two categories: they are called viscoelastic *solids* if they do not continually change their shape when subjected to small stresses; and elastico-viscous *liquids* (or simply elastic liquids) if they do change their shape continually when subjected to stresses, irrespective of how small these stresses may be.

Elastic liquids are best thought of as being materials that are predominantly fluid in behaviour but have some of the properties usually associated with solids. For example, they possess a memory of past deformation and have the ability to store energy when sheared. These "elastic" effects are present in a variety of liquids including emulsions, suspensions of elastic solids in purely viscous liquids, suspensions of rigid particles in viscous liquids (when the Van der Waals forces are significant), polymer solutions, and various materials associated with the food and pharmaceutical industries.

2. The linear behaviour of elastic liquids. The study of the behaviour of elastic liquids is greatly simplified if it can be assumed that non-linear effects in the viscous and elastic deformation processes are absent. Such an assumption limits the type of motion that may be studied, but is justified, for example, in the case of the small amplitude oscillatory motions which are so often used as a basis of systematic observations on small samples.

If attention is confined to those types of motion in which non-linear effects can be ignored, elastic liquids can be characterized in a number of different, but equivalent, ways (Gross 1953: Staverman and Schwarzl 1956: Alfrey and Gurney 1956). From the theoretician's standpoint, it is most convenient to characterize them by means of linear equations of state relating the stress tensor p_{ik} and the rate-of-strain tensor $e^{(1)}_{ik}$—the equations of state of the material. Using Boltzmann's superposition principle, the equations of state for an incompressible elastic liquid can be conveniently expressed in the form

$$p_{ik} = -pg_{ik} + p'_{ik}, \tag{1}$$

$$p'_{ik}(x, t) = 2 \int_{-\infty}^{t} \Psi(t - t')\, e^{(1)}_{ik}(x, t')\, dt', \tag{2}$$

where p is an arbitrary isotropic pressure, Ψ is a relaxation function, g_{ik} is the metric tensor of a fixed coordinate system x_i, and t is the current time.

When the motion is slow and *steady*, equation 2 reduces to

$$p'_{ik} = 2\eta_0 e^{(1)}_{ik}, \tag{3}$$

where $\eta_0 \left(= \int_0^\infty \Psi(\xi)\, d\xi\right)$ is a constant known as the limiting viscosity at small rates of shear. Equation 3 implies that all elastic liquids behave in the same way as purely viscous Newtonian liquids when the motion is steady and sufficiently slow.

Experimentalists usually investigate the linear time-dependent behaviour of elastic liquids by subjecting the liquids to a small-amplitude sinusoidal stress $p'_{ik} = p^{(0)}_{ik} e^{i\omega t}$, where ω is the angular frequency of oscillation and the real part is to be understood. In this type of motion, the equations of state for any linear elastic liquid can be written in the form

$$p'_{ik} = 2\eta^* e^{(1)}_{ik}, \tag{4}$$

where η^* is a function of ω and is in general complex. It is usual to express the complex dynamic viscosity η^* in the form

$$\eta^* = \eta' - \frac{iG'}{\omega}, \tag{5}$$

and η' is given the name "dynamic viscosity" and G' the name "dynamic rigidity". η' and G' are related to the relaxation function by the equation

$$\eta' - \frac{iG'}{\omega} = \int_0^\infty \Psi(\xi)\, e^{-i\omega\xi}\, d\xi. \tag{6}$$

For a purely viscous Newtonian liquid, η' is a constant and G' is zero. Typical graphs of η' and G' for an elastic liquid are given in Fig. 1.

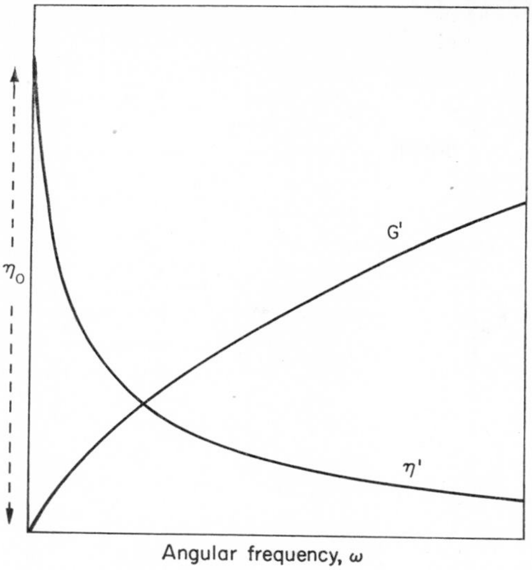

Fig. 1. Typical η' and G' curves for an elastic liquid.

Elastoviscometers with various geometries have been constructed to investigate the linear time-dependent behaviour of elastic liquids (see, for example, Oka 1960; Weissenberg 1964). Most of these instruments are confined to frequencies below 100 cycles per second. However, Lamb and his collaborators (see, for example, Lamb 1963) have developed instruments that make use of very high frequency oscillations. In this way, they have shown that some liquids which are popularly regarded as purely viscous are in fact slightly elastic.

3. The non-linear behaviour of elastic liquids. We now consider the behaviour of elastic liquids when the linearity restriction of the last section is removed.

Consider a steady simple shear in which the velocity components in a cartesian frame x_i are

$$v_1 = \gamma x_2, \quad v_2 = 0, \quad v_3 = 0, \tag{7}$$

where γ is a constant. In the case of a purely viscous Newtonian liquid, the corresponding stress distribution is

$$\left.\begin{array}{l} p_{11} - p_{33} = 0, \quad p_{22} - p_{33} = 0, \\ p_{12} = \eta_0 \gamma, \end{array}\right\} \tag{8}$$

where η_0 is the constant coefficient of viscosity. p_{11}, p_{22}, p_{33} are known as the normal stresses and p_{12} as the shearing stress.

It is well known (see, for example, Coleman *et al.* 1966: Lodge 1964: Fredrickson 1964), that when *elastic* liquids are subjected to a steady simple shear, normal-stress differences occur and the viscosity is not a constant but is in general a monotonic decreasing function of the velocity gradient γ. In this case, equation 8 has to be replaced by

$$\left.\begin{array}{l} p_{11} - p_{33} = \sigma_1(\gamma), \quad p_{22} - p_{33} = \sigma_2(\gamma), \\ p_{12} = \eta(\gamma)\, \gamma, \end{array}\right\} \tag{9}$$

where σ_1, σ_2 and η are even functions of γ which are such that when γ tends to zero, the normal-stress differences σ_1 and σ_2 also tend to zero, but the "apparent" viscosity η tends to a constant η_0. σ_1, σ_2 and η are sometimes referred to as the "material functions".

The experimental determination of σ_1 and σ_2 is a subject of controversy. Numerous researchers, including Roberts (1952), Philipoff (1957) and Kotaka *et al.* (1959) support a hypothesis of Weissenberg

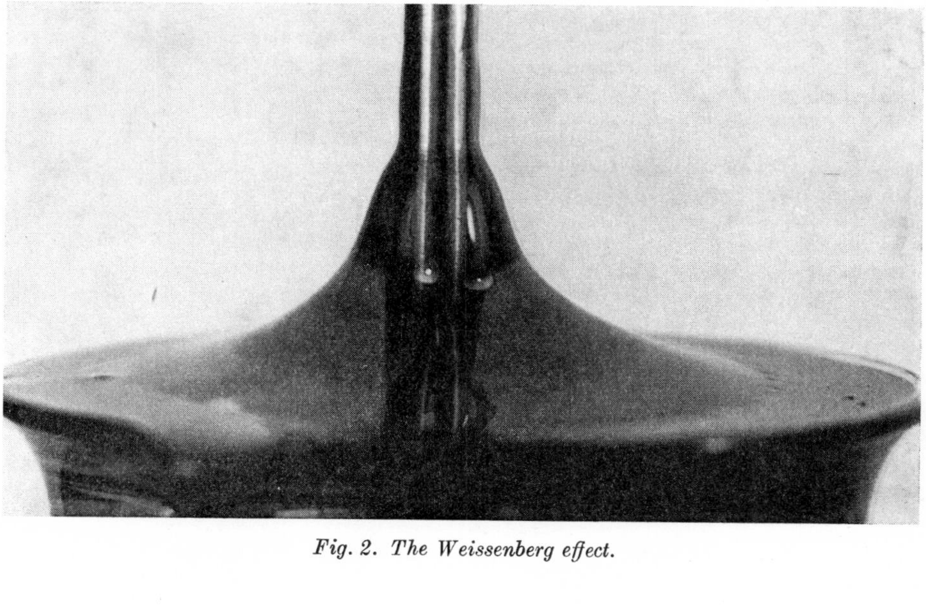

Fig. 2. The Weissenberg effect.

that σ_2 is zero, while others, including Markovitz and Brown (1963) and Adams and Lodge (1964) have given evidence to the contrary. However, it is generally agreed that for most elastico-viscous systems σ_1 is much bigger than σ_2. This essentially means that the normal stress in the direction of the streamlines is greater than those in planes normal to the streamlines.

The presence of normal-stress differences in the flow of elastic liquids gives rise to a number of startling effects which are not present in the flow of purely viscous Newtonian fluids. For example, when a cylindrical rod is rotated about its axis in an elastic liquid having a free surface, the liquid climbs the rod in a spectacular fashion, this being known as the "Weissenberg effect" (cf. Fig. 2).

Another normal-stress effect is the "die-swell" phenomenon associated with Poiseuille flow, sometimes referred to as the "Merrington" or "Barus" effect. When an elastic liquid flows out of a tube, the diameter of the emerging liquid stream increases, in some cases to as much as three or four times the diameter of the tube (cf. Fig. 3).

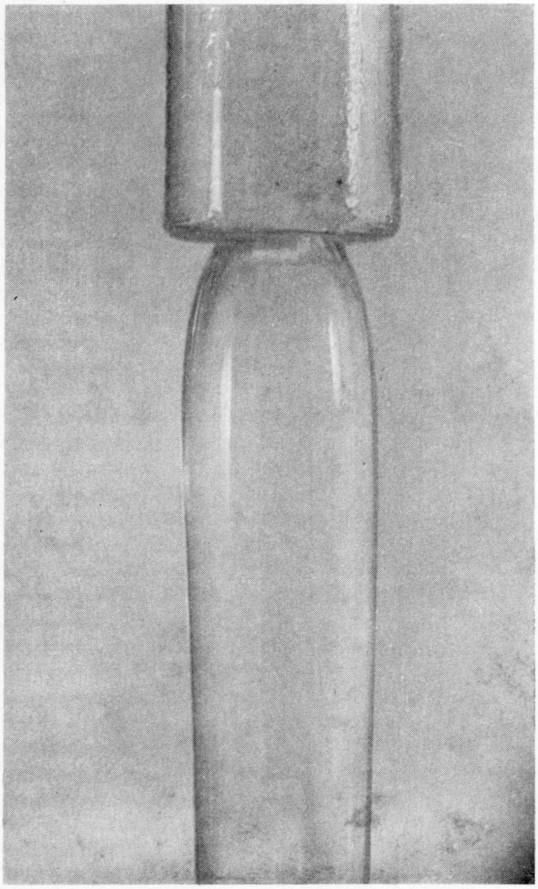

Fig. 3. The Barus effect.

Normal-stress differences in the flow of elastic liquids also give rise to many other interesting effects. For example, steady rectilinear flow of an elastic liquid down a straight pipe of non-circular cross-section is not possible in general, and some type of secondary flow in the cross-section of the pipe is to be expected (cf. Fig. 4).

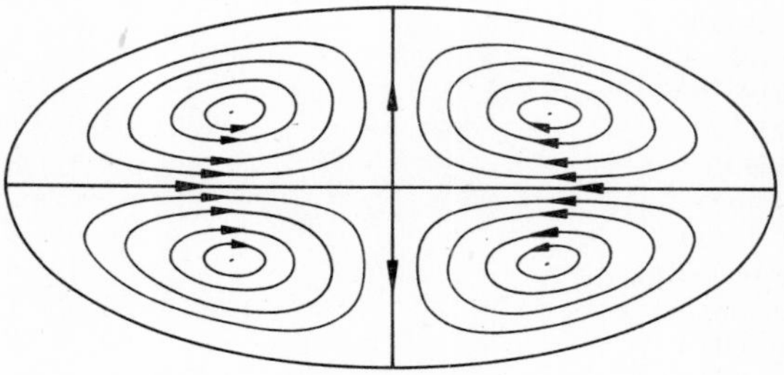

Fig. 4. The predicted secondary flow in the flow of an elastic liquid through a pipe of elliptic cross-section.

In many circumstances, the presence of elasticity in a liquid can significantly alter the secondary flow which usually results when a solid of revolution is rotated in a liquid. Figure 5 shows the type of secondary flow to be expected when a cone is rotated slowly in a purely viscous liquid and Fig. 6 illustrates the corresponding secondary flow for an elastic liquid (Walters and Waters 1967). In the viscous case, the direction of the open streamlines is outwards near the rotating cone and inwards near the stationary plate. For the elastic-liquid case, the streamlines are divided

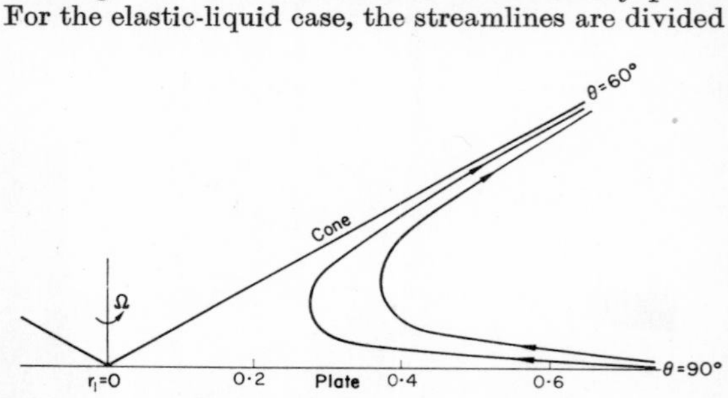

Fig. 5. Secondary flow for a viscous liquid.

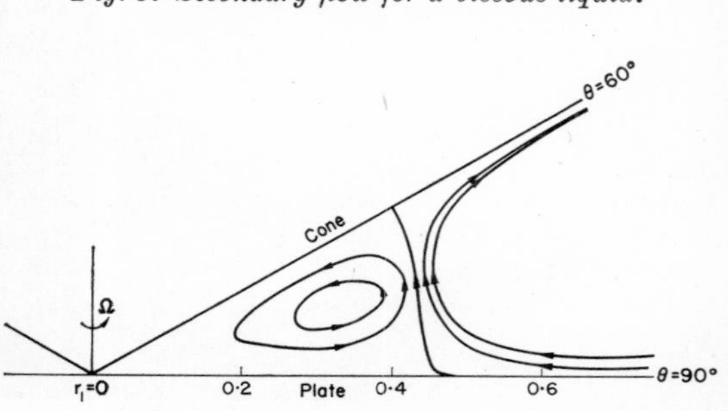

Fig. 6. Secondary flow for an elastic liquid.

into two regions. In the outer region, the streamlines behave similarly to those in the viscous case; but the inner region contains closed streamlines, the direction of which are in the opposite sense to those in the outer region.

4. The measurement of the material functions. The measurement of the apparent viscosity $\eta(\gamma)$ is fairly straightforward and a number of viscometers with various geometries have been built for this purpose (see, for example, Oka 1960; Coleman *et al.* 1966). The most popular geometries are those in which the elastic liquids are either sheared between coaxial cylinders in relative rotation (*Couette flow*) or in the gap between a cone and a plate. The latter geometry is especially convenient since the shear rate is constant in the gap if the angle between the cone and the plate is small ($\leqq 4°$). In this case, the measurement of the couple on either the moving or the stationary component leads to a direct measurement of the viscosity.

The measurement of the normal-stress differences σ_1 and σ_2 is less straightforward and is still the subject of some controversy. The most popular instrument for such measurements is the cone-and-plate *rheogoniometer* (Adams and Lodge 1964; Lodge 1964; Weissenberg 1964; Coleman *et al.* 1966) in which the steady shear is usually brought about by rotating the cone about its axis and keeping the plate stationary (cf. Fig. 7). Measurements of the radial distribution of pressure on the stationary plate can be used to

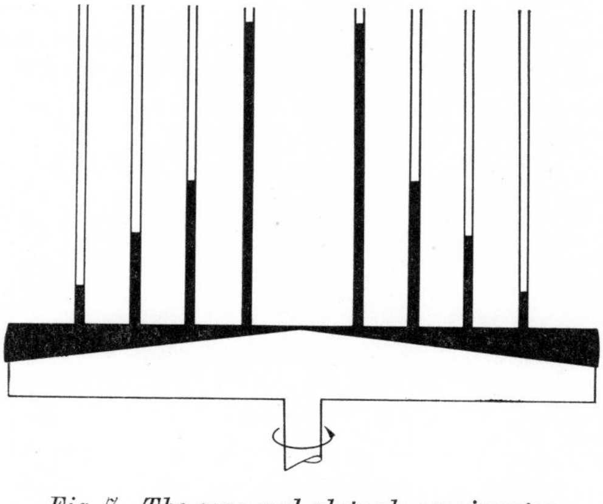

Fig. 7. The cone-and-plate rheogoniometer.

determine $\sigma_1 + \sigma_2$. A further measurement is needed to determine σ_1 and σ_2 separately. If it can be assumed that the free liquid boundary is spherical in shape, that surface tension effects are negligible, and that the state of steady shear exists up to the free boundary, the total thrust on the plate can be used to determine $\sigma_1 - \sigma_2$, while the pressure reading at the rim leads to a value for σ_2 (see, for example, Adams and Lodge 1964).

Radial pressure measurements made in the corresponding parallel-plate arrangements can also be used in conjunction with those in the cone-and-plate system to determine σ_1 and σ_2 separately (Adams and Lodge 1964). Further, a modification of the cone and plate system in which the apex of the cone does not touch the plate also provides a means of determining σ_1 and σ_2 (Jackson and Kaye 1966).

Experiments with the cone-and-plate and parallel-plate systems have the disadvantage of being confined to fairly low rates of shear. It appears that the only way of determining normal-stress information at high shear rates is by means of the "jet thrust" or "jet expansion" methods (Harris 1961; Metzner *et al.* 1962; White and Metzner 1963). These methods involve the extrusion of the elastic liquid from a tube at high speed. The relevant information is either obtained by observing the expansion of the emerging jet or by taking thrust measurements of the fluid stream as it emerges from the tube. In the case of a cylindrical tube it is necessary to assume that σ_2 is zero in order to determine σ_1. This assumption is unnecessary if one considers instead the corresponding situation when the jet emerges from between two parallel plates (White and Metzner 1963).

The above methods have all been used with varying degrees of success by experimenters in the determination of the material functions η, σ_1 and σ_2. This is clearly an important step in the characterization of elastic liquids and in the prediction of their behaviour in situations of technological importance; but it must be borne in mind that a knowledge of these functions alone is not in general sufficient to determine the behaviour of elastic liquids in flows other than those involving simple shearing. To predict the behaviour of elastic liquids in more complicated flows it is necessary to consider suitable equations of state for the liquids in conjunction with the familiar equations of motion and continuity. It is therefore necessary to give consideration to the important subject of the formulation of equations of state for elastic liquids.

5. The formulation of equations of state for elastic liquids. Since 1950, theoreticians have been remarkably successful in developing systematic methods for writing down equations of state for elastic liquids. These equations have to satisfy certain intuitive physical principles that one would associate with materials possessing memory. The two major principles may be stated as followed:

(i) The equations of state must be consistent with the requirement that the behaviour of a material element depends only on its previous deformation history and not on the state of neighbouring elements. This is sometimes called "*the principle of determinism*".

(ii) The equations of state must be consistent with the requirement that the behaviour of a material element does not depend on the translatory or rotatory motion of the material as a whole in space. This principle may also be stated in terms of indifference to an observer (identified with a moving coordinate frame) rather than indifference to the absolute motion of the material as a whole in space, but the two approaches are precisely equivalent (see, for example, Walters 1965). This principle is known by at least two names—"material objectivity" and "material indifference".

Oldroyd (1950) was the first to realize the importance of these principles, and in a comprehensive paper he showed how it was possible to incorporate them into a general mathematical theory of formulation. In order to do this, he introduced a convected coordinate system embedded in the material and deforming continuously with it. With this particular formulation the second principle is associated with indifference to absolute motion in space rather than indifference to an observer. Oldroyd's analysis led to simple rules for writing down generally valid equations of state and could have been used by later workers as the basis of their work on particular special forms for equations of state for elastic liquids (see, for example, Fredrickson 1964; Walters 1965). However, many later workers, especially those in the U.S.A. chose to regard the second principle as relating to indifference to an observer. This led to some simplification in that a set of cartesian coordinates could be used in place of the convected coordinates employed by Oldroyd, but at the same time, some of the "physics" of the situation was hidden in the process.

Notable amongst these mathematical approaches are the works of Rivlin and Ericksen (1955), Green and Rivlin (1957, 1960) and Coleman and Noll (1960, 1961a, 1961b, 1964). The latter contributions involve the introduction of "Functional Analysis" into the formulation of what are very general and therefore very useful equations of state. The form of the equations used by Green and Rivlin is more convenient for some situations but the formulation of Coleman and Noll is more elegant and has been extensively developed by these authors. These developments involve the derivation of very useful approximation schemes based on the type of memory possessed by the elastic liquid and the type of flow problem considered. An alternative approximation scheme for the general fluid model considered by Coleman and Noll has been suggested by Truesdell (1964) based on a suitable "characteristic time" of the fluid. In steady flow problems, the approximation schemes of Coleman and Noll, and Truesdell lead to relatively simple equations which are in fact the same as those obtained by Rivlin and Ericksen (1955) starting from a more restricted standpoint.

The mathematical theory of elastic liquids can now be said to be well-developed and in a form which can be readily used by experimenters who are faced with the problem of predicting the behaviour of elastic liquids in situations of technological importance. It must be pointed out however that this is a statement "in principle" and the solution of a specific flow problem can be extremely complicated.

Bibliography

Adams N. and Lodge A. S. (1964) *Phil. Trans. Roy. Soc.* London, A **256**, 149.

Alfrey T. and Gurney E. F. (1956) *Rheology*, Vol. 1 New York: Academic Press.

Coleman B. D., Markovitz H. and Noll W. (1966) *Visometric Flows of Non-Newtonian Fluids*, Berlin: Springer.

Coleman B. D. and Noll W. (1960) *Arch. Rat. Mech. Anal.* **6**, 355.

Coleman B. D. and Noll W. (1961a) *Annal. N.Y. Acad. Sci.* **89**, 672.

Coleman B. D. and Noll W. (1961b) *Rev. Mod. Phys.* **33**, 239.

Coleman B. D. and Noll W. (1964) *Second-order Effects in Elasticity, Plasticity and Fluid Dynamics*, Oxford: Pergamon Press.

Fredrickson A. G. (1964) *Principles and Applications of Rheology*, New York: Prentice-Hall.

Gross B. (1953) *Theories of Viscoelasticity*, Paris: Hermann.

Green A. E. and Rivlin R. S. (1957) *Arch. Rat. Mech. Anal.* **1**, 1.

Green A. E. and Rivlin R. S. (1960) *Arch. Rat. Mech. Anal.* **4**, 387.

Harris J. (1961) *Nature* **190**, 993.

Jackson R. A. and Kaye A. (1966) *Brit. J. Appl. Phys.* **17**, 1355.

Kotaka T., Kurata M. and Tamura M. (1959) *J. Appl. Phys.*, **30**, 1705.

Lamb J. (1963) *Proc. 4th Intern. Cong. Rheol.* Part 1, 317, New York: Interscience.

Lodge A. S. (1964) *Elastic Liquids*, New York: Academic Press.

Markovitz H. and Brown D. R. (1963) *Trans. Soc. Rheol.* **7**, 137.

Mason W. P. (1961) in *Encyclopaedic Dictionary of Physics* (J. Thewlis Ed.), **2**, 617, Oxford: Pergamon Press.

Metzner A. B., Houghton W. T., Sailor R. A. and White J. L. (1961) *Trans. Soc. Rheol.* **5**, 133.

Oka S. (1960) *Rheology*, Vol. 3, New York: Academic Press.

Oldroyd J. G. (1950) *Proc. Roy. Soc.* A **200**, 523.

Philippoff W. (1957) *Trans. Soc. Rheol.* **1**, 95.

Roberts J. E. (1962) Unpublished ADE report 13/52, Armament Design Establishment Knockhold, Kent.

Rivlin R. S. and Ericksen J. L. (1955) *J. Rat. Mech. Anal.* **4**, 323.

Staverman A. J. and Schwarzl F. (1956) *Die Physik der Hochpolymeren*, Berlin: Springer.

Truesdell C. (1964) *Phys. Fluids* **7**, 1134.

Walters K. (1965) *Nature* **207**, 826.

Walters K. and Waters N. D. (1967) *Deformation and flow of High-Polymer Systems*, 211, New York: Mac-Millan (to be published).

WEISSENBERG K. *The Testing of Materials by Means of a Rheogoniometer*, Sangamo Controls.
White J. L. and METZNER, A. B. (1963) *Trans. Soc. Rheol.* **7**, 295.

K. WALTERS

ELECTRICAL STANDARDS AND MEASUREMENTS, RECENT DEVELOPMENTS IN. The development of electrical standards is part of a continuing history which follows a common path. For a start, an idealized unit is conceived in terms of the mechanical units (length, mass and time) and the permeability of free space which is designated ($\mu_0 = 4\pi \times 10^{-7}$ H/m). A material standard is constructed to embody this unit and its value is measured, as well as possible, with mechanical standards. The unit is then defined by decreeing that the standard has exactly the measured value. Replicas of the standard are distributed so as to disseminate this value, and an attempt is made to define a standard which can be realized by following a specification. The development of electrical standards is the story of improvements in the measurements and definitions.

Although electrostatics is an ancient science, electromagnetism and hence electrical engineering was born in the year 1800 when Volta first produced a steady flow of electric current by means of the voltaic pile. In 1827 Ohm's law was enunciated, and then in 1831 Gauss and later Kohlrausch and Weber made magnetic and electrical measurements in terms of fundamental units. A step of major importance was taken in 1861 when the British Association for the Advancement of Science appointed a committee to establish a system of electrical standards and this included such notables as Maxwell, Joule, Kelvin and Wheatstone. After an extensive study, they decided to adopt Weber's system based on c.g.s. units. As these absolute units were not of a convenient size, however, they also selected practical ones of a suitable magnitude, these being the ohm of 10^9 e.m.u. and the volt of 10^8 e.m.u. thus giving an ampere of 10^{-1} e.m.u. They adopted a decimal system of scaling magnitudes above and below these units. This set the pattern upon which all subsequent work has been based. In 1864 they established the Standard Ohm, distributing copies to various organizations. Prior to this, workers, mainly telegraph engineers, used arbitrary standards such as "1 German mile of iron wire $^1/_{10}$ inch in diameter": or the Siemens unit "a column of mercury 1 metre long and 1 mm^2 in cross-section". The limitations on the accuracy of their measurements can be judged from the fact that in 1878 Rowland showed the B. A. Ohm to be in error by 1·5 per cent.

The Paris Electrical Congress in 1881 gave international status to the B. A. absolute units and to the parallel practical set to which it gave the names ohm, volt, ampere, coulomb and farad. It set up a commission which in 1884 recommended a legal ohm based on the mercury column of 106 cm length and 1 mm^2 cross-section, but this was not adopted. In 1893 the I.E.C. moved a step further and now recommended a lengthened mercury column of 106·3 cm of 1 mm^2 cross-section, a volt based on the Clark cell and an ampere based on a specified rate of deposition of silver, but still had no success, although the U.S. Government passed a bill recognizing both the practical and absolute units. In 1908 another International Conference attended by representatives of 24 countries agreed upon the need for units which could be established with the accuracy required by the engineering industry and which were to be called "International Electrical Units". The mercury ohm was generally agreed as being suitable and was now defined as 106·300 cm in length and a specified mass. After prolonged discussion as to which of the two remaining competing standards, the electrolytic ampere and the Weston cell, should be chosen, the former won out. So after nearly 30 years of discussion, the units adopted were ones which were soon found to be extraordinarily difficult to reproduce to the accuracy that had by that time become necessary.

In 1910, at a meeting at the National Bureau of Standards in Washington, it was decided that the international ohms as maintained by Britain at N.P.L. and by Germany at P.T.R. were the best available values, so the mean of these was accepted as the true value. By the mid 1920's it was clear that the standards of resistance in the various co-operating countries were drifting quite seriously. In 1935 the International Committee for Weights and Measures, which assumed responsibility for electrical standards in1921, adopted the mean of the values of the units of Germany, England, Japan, Russia and U.S.A. France, whose unit was excluded, had to adjust or "recapture" it by nearly 70 parts per million! As similar trouble had been experienced with the determination of the ampere, it was agreed that it was essential to improve determinations by absolute means. It was also decided that the m.k.s. system should be adopted for electrical measurements. The absolute determinations involved were very tedious and with the war intervening, it was 1946 when the International Committee decided on the date of effect of the changes. On 1st January 1948, the absolute units replaced the so-called international units based on the mercury ohm and silver ampere, and the ratios adopted were:

One international ohm = 1·000,49 absolute ohm
One international volt = 1·000,34 absolute volt.

In determining these values, resistance had been compared with calculable mutual or self inductors, and currents determined by means of current balances.

About that time techniques for determining physical properties of atoms and molecules were advancing apace and had been given a tremendous stimulus by the wartime development of techniques for isotope separation. The impetus given to microwave spectroscopy by radar was among the foremost of these and it was quickly recognized that the physical properties of atoms or molecules would serve as better standards if determined with sufficient accuracy. By 1950 this goal was well in sight, and Huntoon and Fano from U.S. National Bureau of Standards put forward a scheme of primary standards based on the mercury 198 radiation for length, the ammonia absorption clock for time, and the gyromagnetic ratio of the proton for current. These ideas were not completely new—in fact the proposal to use the wave-length of light as a standard of length was first made in 1827 by Biot.

In the standards field changes are made very slowly, and it was 1960 before the metre was defined as being so many wave-lengths *in vacuo* of a specified line of krypton-86.

The next unit to be established on such a basis was the second. In 1955 atomic beam experiments by Essen and Parry at the National Physical Laboratory, England, showed that a frequency or time standard could be established with a precision and accuracy better than that provided by the rotation of the Earth. In 1964 the International Committee adopted the atomic definition of time.

The International System of units adopted in 1960 (Système International) is based on the metre, kilogram, second, ampere, degree Kelvin and the candela. Omitting the last two as being of no interest here, two of the other four have been established on atomic standards. Of the remaining two, the ampere should be the next. For nearly twenty years, work has been in progress with the object of determining the ampere in terms of the gyromagnetic ratio of the proton and already a provisional value has been adopted by the International Committee.

In the mid 1950's, as a result of the work of Lampard and Thompson at the Australian National Standards Laboratory, a new possibility arose, namely, the use of the calculable capacitor as a means of determining the value of the ohm. N.S.L. has now adopted officially the calculable capacitor as the basis for its measurements of capacitance, inductance and resistance.

There now follows a brief account of methods used in determining these physical quantities together with some discussion of the newer measurement techniques.

Although frequency is not an electrical quantity, it is appropriate to include it, not only because its determination has been the province of those concerned with electrical measurements but also because it illustrates so well the principles of atomic standards. Mechanical clocks for keeping time accurately have been established for many hundreds of years, long before success was achieved in the accurate measurement of any other quantity. It was not until about 1920, however, that electrical means reached the stage where they could be considered important. At that time, tuned LC circuits with a stability of about 1 part in 10^4 were available as frequency standards. By 1930, the tuning fork control of oscillators had advanced to 1 in 10^5. Quartz crystal oscillators then took over and by the end of the war were approaching a stability of 1 in 10^8. By the early 1950's further improvements showed up the lack of uniformity in the rotation of the Earth, but of course, the quartz crystal is only a transfer device and not an absolute means of establishing frequency.

In 1945 Rabi, a scientist engaged in atomic beam research, pointed out that the hyperfine structure of certain atomic spectra gave rise to frequencies in the microwave region and that in principle it should be possible to use these as a means of establishing standards of frequency far better than any available at that time. It was not until 1955, however, that Essen and Parry achieved this.

The ground state of caesium comprises two components spaced by the energy corresponding to about 9192 Mc/s in accordance with the Bohr relationship $W_1 - W_2 = h\nu$. The splitting of the ground state resulting from the interaction between the spin of the valency electron and that of the nucleus gives rise to the hyperfine lines in the optical spectrum of caesium. These levels are subject to Zeeman splitting in the magnetic field which is necessary to sort out the desired transition from the others from which it differs by being nearly independent of the field.

Figure 1 shows schematically the arrangement of the evacuated beam tube. Caesium is heated in an

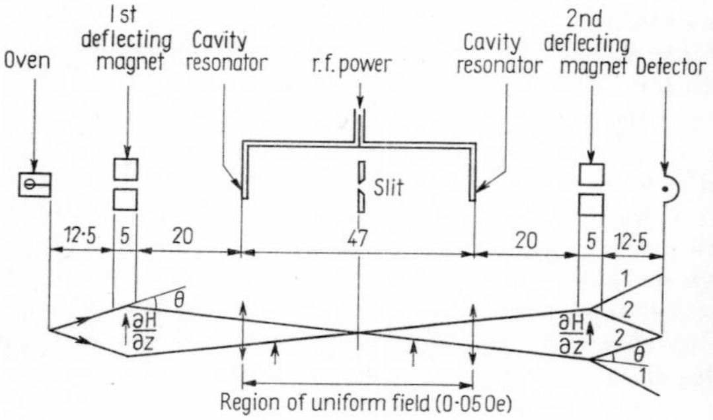

Fig. 1.

oven to about 200°C and atoms are emitted in a beam passing between the poles of the first deflecting or "A" magnet. The spins impart magnetic properties, so the atoms are deflected in the non-uniform field and then pass through a cavity resonator in which there is maintained a microwave field. If the frequency of this is appropriate, a change of state may occur. The atoms continue on through the slit and through the second cavity resonator, where more changes of state may occur, and on into the gap of the second deflecting or "B" magnet which is identical with "A". Those atoms which have undergone one change of state will now be deflected in the opposite direction and fall on the slit of the detector comprising a heated tungsten wire. As the work function of tungsten is higher than the ionization potential of caesium, the atoms will each give up an electron to the tungsten and be emitted as ions. The resulting current is measured by a vibrating reed electrometer.

In the space between the cavity resonators there is a weak magnetic or "C" field which induces Zeeman splitting. The efficiency of the excitation depends upon the path length in the microwave field. Clearly this should be as long as possible but the maintenance of a microwave field uniform in intensity and phase over the distances required would be very difficult. It has been shown by Ramsey, however, that having short regions of field at the two ends of the space in question is about 40 per cent better than having the field throughout its length. This not only greatly simplifies matters but also permits the use of standard waveguide. The deviation of the beam from a straight line is exaggerated in the diagram, being less than a millimetre. If the velocity of the atoms is about $2 \cdot 5 \times 10^4$ cm/s the time spent by the atom in traversing a field of 50 cm in length is about $^1/_{500}$ s. From the Uncertainty Principle ($\Delta\nu\Delta t = 1$), $\Delta\nu$ should be 500 c/s. With the Ramsey configuration, however, the width of the peak obtained by Essen and Parry with a path of 47 cm was about 350 c/s. They determined its centre to about 1 c/s. There is a small correction (about 1 c/s) to be made because of the effect of "C" field.

To avoid errors due to drift during these measurements, the quartz crystal oscillator had to have very high stability, preferably 1 part in 10^{10} per day or better. These are readily available nowadays but represented the peak of performance at that time. Those crystals in use for this experiment were the ring type developed by Essen himself.

Using this technique, Essen and Parry, in conjunction with Markowitz and Hall of the U.S. Naval Observatory, determined the frequency of the transition in terms of the ephemeris second as 9192631770 c/s and this value has now been adopted as the constant for the definition of the atomic second.

Improved caesium beam standards have been built since in several countries and, where they could be compared, found to agree to 1 or 2 parts in 10^{11}. They are also made commercially by the National Co. (Atomichron), Hewlett Packard and Pickard and Burns in U.S.A. The Hewlett Packard clock is claimed to be accurate to 1 part in 10^{11} and stable to 5 parts in 10^{12}.

Although the transition of the caesium atom has been used as a standard, in theory the thallium atom is better mainly because the correction for the "C" field is much less (about one fiftieth). A thallium atomic beam unit was made by Bonanomi at the Swiss Laboratory for Horological Research (L.S.R.H.) and another by Mockler at National Bureau of Standards was completed in September 1962. The frequency values referred to caesium determined with these two instruments are:

$$\nu_0 = 21{,}310{,}833{,}945 \cdot 1 \pm 1 \cdot 0 \text{ c/s L.S.R.H.}$$
$$\nu_0 = 21{,}310{,}833{,}945 \cdot 9 \pm 0 \cdot 2 \text{ c/s N.B.S.}$$

The hydrogen maser is a device with a stability approaching 1 in 10^{13} and Fig. 2 shows some details. It operates on the transition between the ground state hyperfine levels of atomic hydrogen. Molecular hydrogen is dissociated by an r.f. discharge passing through a fine hole into a state separator and then into the storage bulb where the atoms remain for a

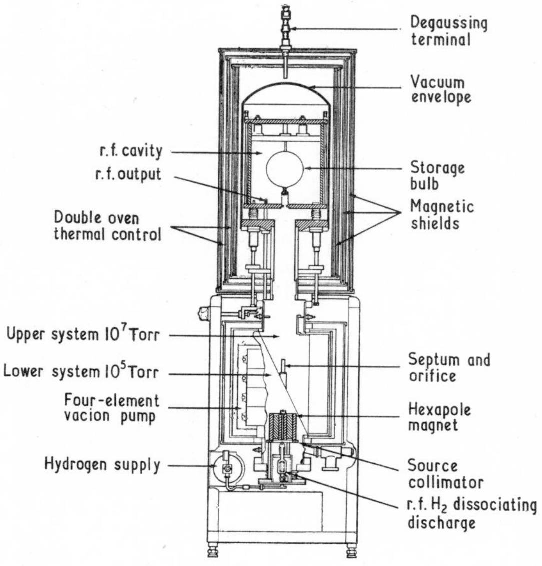

Fig. 2.

period of up to 1 second involving some 10^5 collisions. This bulb is located in a cavity tuned to the hyperfine transition frequency and if the beam flux is sufficiently high, stimulated emission takes place. The cavity is surrounded by a magnetic shield and a small Zeeman field is applied to separate out the desired transition. The storage bulb is of fused quartz and its inner wall is coated with an inert material such as Teflon to reduce chemical and other interactions at the wall. Due to long life in the bulb first order Doppler effects are virtually absent. For a time of 1 second the Uncertainty Principle gives a band width of 1 c/s which is the figure achieved in practice. Hence a greater precision is attainable than with the caesium beam device.

The most recent determination of the hydrogen maser with caesium beam atomic standards carried out in 1965 by U.S. National Bureau of Standards, Varian Associates and the Hewlett-Packard Co. gave the following result:

$$H_\nu = 1420405751 \cdot 7864 \pm 0 \cdot 0017 \text{ c/s}.$$

No discussion of this topic would be complete without reference to devices based on ammonia. The very first "atomic" clock built by the Bureau of Standards in 1948 used the absorption line in the microwave spectrum of ammonia at 23,870 Mc/s, and was stable to about 1 part in 10^7. Next in 1955 the ammonia maser was developed and in essence its operation depends upon the emission of radiation from stimulated ammonia molecules, i.e. the reverse of absorption. Inherently the ammonia maser is not as good a standard as the caesium beam device but by the use of various refinements including double beams to overcome Doppler effects, it can achieve a stability approaching 1 in 10^{11}.

The rubidium vapour frequency standard has reached an advanced stage of efficiency, having a stability approaching 1 part in 10^{12}. As it is not an absolute device it will not be discussed except to note that if the frequency is adjusted to a known value with a known magnetic field, it can then be used to measure unknown fields to a higher order of accuracy than any other device.

Turning to the first of the purely electrical units, the ampere, the determination of the gyromagnetic ratio of the proton is made by observing the precessional frequency of protons in a known magnetic field produced by a coil of measured dimensions and current. The frequency of precession $\omega = \gamma B$ where γ is the ratio of the magnetic moment to the angular momentum of the proton, i.e. the gyromagnetic ratio.

This technique enables the ampere to be reproduced with precision greater than can be realized by a current balance and with less difficulty and labour. The set-up used by Vigoreux of N.P.L. for this determination is shown in Fig. 3.

There is at the centre a sphere of water, the hydrogen atoms of which supply the protons. The container is

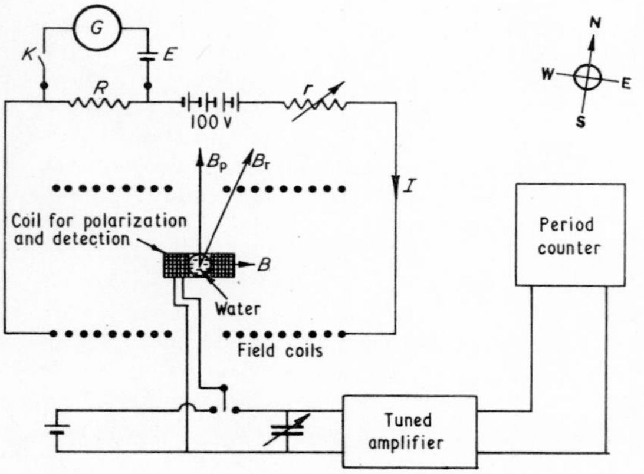

Fig. 3.

surrounded by a detector coil in which an e.m.f. is induced by the protons if their magnetic moment, on the average, has a component at right angles to the standard field B. Initially the magnetization is in the direction of B so there is no net induced e.m.f. A polarizing field B_p about 100 times as large as B is applied by means of the detector coil, or otherwise, at right angles to B. If the polarizing field is now suddenly removed, precession occurs about B, commencing at a polar angle of nearly 90° but reducing to zero again, the relaxation time being about 3 seconds during which the measurement is made.

As the field B is proportional to the current $I = E/R$, γ is obtained in terms of $\omega R/E$, where ω is the observed frequency of precession, E is one or more standard cells, and R a standard resistor. R/E can be determined by means of a current balance yielding γ in terms of the metre, kilogram and second. The field B is calculated from the dimensions of the coils—a very laborious job. Frequency is measured by counting with a 0·1 μs counter the time of 10,000 to 40,000 periods and turns out to be about 73 kc/s. The value of γ obtained by Vigoreux for protons in water is $2{\cdot}675171 \times 10^8$ Wb^{-1} m^2 s^{-1}.

The accuracy of this method is greater than that of the current balance because the field depends on the pitch of the helix, whereas in the balance the force of attraction depends on both the pitch and diameter.

Once having established the value of γ, it can be used to determine the magnetic field of a coil and hence the current or E/R.

Clearly, the adoption of this ratio as the basic constant would not confer the same advantages as the corresponding ones for length and time because of the relative complexity of the measurement and of the much lower accuracy.

The calculable capacitor is the most significant modern development in electrical measurements. Announced in 1956, Lampard's theorem in electrostatics states that, in a cross capacitor comprising four electrodes such as are shown in the inset Fig. 4, the mean capacitance per unit length of the opposing pairs of electrodes

$$\frac{\overline{C}}{L} = \frac{\ln 2}{4\pi^2} = 2 \text{ pF/m approximately}.$$

Thus a capacitance can be determined by one single measurement of length. If the electrodes are cylinders nearly in contact, the effect of the gaps between the cylinders is negligible. By using guard devices and simply measuring the change of capacitance brought about by a change in the effective working length of the cylinders, end effects are eliminated and the capacitance can be determined by measurement of the change of length by interferometry. Both N.S.L. and N.B.S. have set up such a standard of capacitance and from this have determined resistance in absolute units.

The form of the calculable capacitor used by Clothier at N.S.L. is shown in Fig. 4. In the vertical section, light from a Hg-198 source 0 passes through a monochromator which selects the desired green line. This passes into the capacitor and after collimation is reflected by prism T through the tube C which forms the lower guard rod. It continues up through the upper tube and guard rod B and is finally reflected into a telescope. In the upper and lower tubes are optical flats M and L which, being half-silvered, form a Fabry-Perot interferometer. C is adjusted electromagnetically so that L is parallel to M and a bright spot is seen in the centre of the field of view of the telescope. The

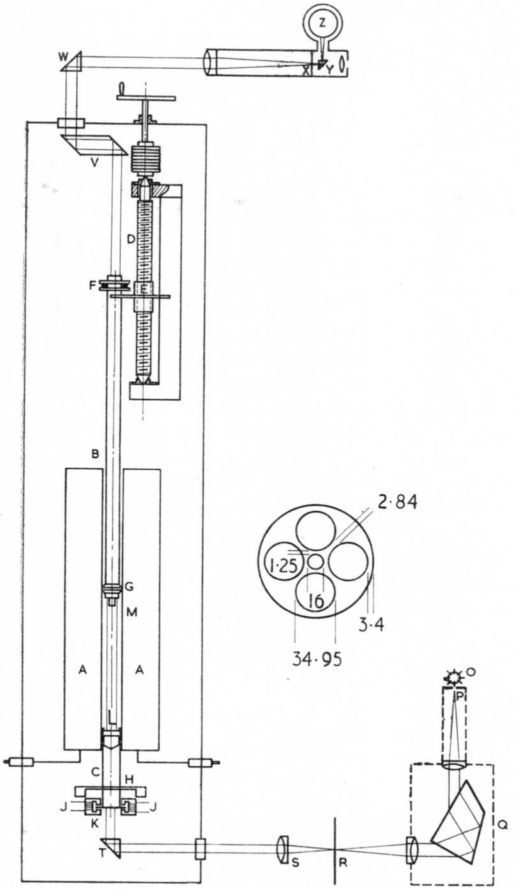

Fig. 4.

prism Y is then introduced to deflect the beam onto a photomultiplier Z. The capacitor is always used in such a way that the change in distance between the guard rods is a whole number of "fringes" or half wave-lengths. Once set, the photomultiplier allows the separation of the mirrors to be maintained with ease to 1/100th of a fringe or 10^{-7} in. As the capacitor is evacuated, the "fringe units" of capacity are absolute, there being no effects due to refractive index, permittivity or temperature. The "fringe unit" is approximately $5{\cdot}335405 \times 10^{-7}$ pF, i.e. about half an attofarad.

A standard capacitor of 0·002 pF is calibrated by means of the calculable capacitor and a transformer bridge. In turn, this is used to calibrate, by means of substitution techniques, a series of capacitors finishing with two 0·005 μF (or 5 nF) standard capacitors. The admittances of these are then compared in the "quad" bridge with those of two 20,000 ohm standard resistors R_2 and R_4 at 10^4 radians/second, at which frequency they are nominally equal (Fig. 5). The condition for balance $\omega^2 C_1 C_3 R_2 R_4 = 1 + \alpha$ gives the product $R_2 R_4$.

R_2 and R_4 are then connected in parallel and compared in a d.c. bridge essentially of the Kelvin type with conventional build-up resistors in two 100 to

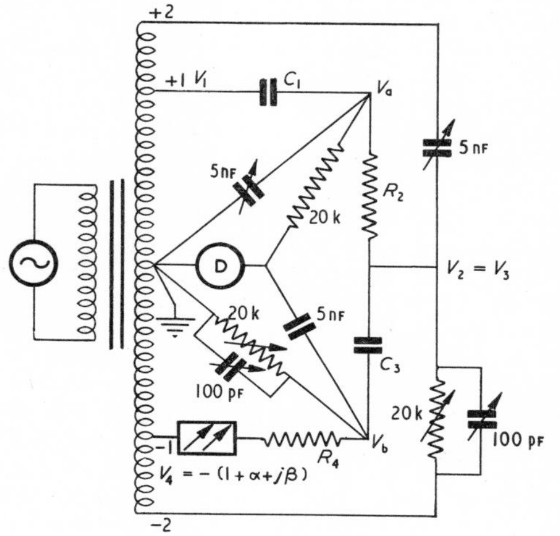

Fig. 5.

1 steps, so permitting reference to a standard 1 ohm resistor, thus yielding the value of the standard ohm in terms of capacitance. The a.c./d.c. transfer characteristics of R_2 and R_4 are determined by use of an intermediate resistor of calculable impedance and with a stability sufficient to cover the period of time required for this determination. The calculations involve the square of the speed of light which is known to about 3 parts in 10^7. N.S.L. have found that the complex series of measurements can be done quite quickly and that they are reproducible over many determinations to ± 1 in 10^7, so the greatest uncertainty involved is that in the speed of light.

N.S.L.'s contributions to this field are of great importance in two respects. The calculable capacitor enabled the accuracy of capacitance standards to be improved by more than an order of magnitude. This would, in itself, not have been of immediate value, however, if Thompson had not developed the transformer bridge techniques which made possible the measurement of the extremely small capacitance values to the required accuracy. These techniques have revolutionized the measurement of capacitance and inductance and it is now a comparatively simple matter to compare capacitors and inductors by means of commercially available ratio transformers to a part in a million.

In 1962 Hill and Miller at the National Physical Laboratory developed an inductively coupled bridge for use in resistance thermometry, where one is concerned with the precise measurement of resistance ratios rather than absolute values. This bridge is operated at 400 c/s and is accurate to about 4 parts in 10^7, equivalent to 0·0001°C change in temperature. A major virtue of such a bridge is that, because the inductive ratios of coils with high permeability cores depend only on the number of turns, their manufacture is a simple matter compared with that of precision resistances. The use of a.c. instead of d.c. avoids troubles due to contact potentials and improves detector efficiency.

A further development at N.P.L. has been the extension of this technique to the determination of the zero frequency resistance of standard resistors of values between 1 and 100 ohms to about 2 parts in 10^7 by extrapolating from measurements in the frequency band 40–600 c/s. It also uses a Kelvin double-bridge type network, the ratio arms comprising inductively coupled voltage dividers.

To try to overcome troubles in the precise comparison of resistance standards such as those due to contact potential, lack of detector sensitivity and the like, Miller at N.S.L. has used a very low frequency bridge operating in the vicinity of one cycle per second or less and has achieved a precision of about 1 in 10^8. Probably a.c. techniques will, in due course, supplant d.c. for much of the high precision measurement work but it is unlikely that a higher precision will be achieved until the design of standard resistors is improved.

If the ampere and the ohm are determined in absolute terms, the volt is automatically defined. Of course, there must be a practical means of maintaining it and the Weston cell has provided this since 1893. The standard cell is also unique in another respect because it has been, until very recently, the only means of providing an acceptable standard of e.m.f. for ordinary laboratory and industrial purposes. The *Zener diode*, however, now provides a source of e.m.f. as constant as the ordinary standard cell, i.e. to about a part in 10^4. It is not absolute, however, and must be calibrated.

The Zener diode is merely a means of making use of the constancy of reverse voltage breakdown of a silicon diode rectifier. The critical voltage value can be controlled but the most suitable values are near 5 volts, the point of change-over from a negative to positive temperature coefficient. The best of the Zener diodes have stabilities comparable with standard cells and are rugged devices. Tests on the stability of Zener diodes are being conducted under the auspices of the International Committee with the object of using them as transfer standards. In some cases a stability of a few parts in 10^6 per annum has been obtained. It is too soon yet to forecast their future because they have been in use only a short time in terms of standards practice, but clearly they have already a very wide field of application.

Reverting to the list of four fundamental units, mass is the only one either not yet established on the basis of an atomic physical constant or in the process thereof. Nor is there any prospect in sight. Masses of a kilogram can be compared to almost 1 in 10^9–a figure which has not changed significantly in the past 20 years so this is one area awaiting a completely new development.

Bibliography

BROWN R. (1961) in *Encyclopaedic Dictionary of Physics* (J. Thewlis Ed.), **2**, 673, 674, Oxford: Pergamon Press.

RICHARDSON J. M. (1966) in *Encyclopaedic Dictionary of Physics* (J. Thewlis Ed.), Suppl. Vol. **1**, 351, Oxford: Pergamon Press.

H. J. FROST

ELECTROCATALYSIS. *Catalysis* is generally used to denote the speeding-up of reactions under conditions that, hopefully, leave the active agent, the catalyst, not permanently changed. The great importance of catalysis is matched by its complexity and by the attendant confusion. One would be naive to believe that using *electrocatalysis* to denote the speeding-up of electrochemical reactions will explain this process. But the new word has gained acceptance in fuel-cell work because it is needed to differentiate the speeding-up of reaction at an electrode from that in non-electrochemical reactions. Justification for the use of *electrocatalysis* arises from the knowledge that it proceeds under conditions superimposed upon those present when an electrode is not the seat of reaction. Among these added conditions, the most important are (1) the presence of a difference in the electrochemical potential of electrons that makes them tend to move from anode to cathode through the external circuit; this difference promotes the loss of electrons at the anode and their capture at the cathode, (2) the presence of an electrical double layer at the electrode surface, (3) the opportunity for the electrolyte (solvent included) to affect the rate of reaction.

For fuel cells, the principal qualitative requirements of a good electrocatalyst are:

1. Invariance;
2. Adequate electrical conductivity;
3. High surface area;
4. Active morphology;
5. Transition-metal character;
6. Catalytic activity as needed.

The first requirement covers both poisoning of the electrocatalyst and its attack under operating conditions. The fifth is not intended to imply that the electrocatalyst must be a transition metal, though these as a class make the best electrocatalysts, but only that the adsorptive and reactive character associated with the transition metals is desirable. The sixth recognizes that the overall electrochemical reaction usually involves chemical reactions that

 Electrogasdynamic generation of power

must proceed at high enough rates if performance is not to suffer.

An electrocatalyst may be the entire electrode (platinum black confined in a "stocking") or it may be spread upon an electronically conducting substrate (platinum black on carbon or boron carbide). In the latter case, there is the possibility of favourable interaction between electrocatalyst and substrate.

The difference between catalysis and electrocatalysis is shown by absorbed hydrocarbons. When these are adsorbed on platinum, carbon-hydrogen bonds are often broken and hydrogen atoms form (dissociative adsorption). When the platinum is an anode electrocatalyst, electrons can leave through the external circuit for capture at the cathode, which can thus act as an electron sink. When the platinum acts as a (non-electrochemical) catalyst, this loss of electrons cannot occur. Often this means a much lower (even zero) concentration of hydrogen atoms in the case of electrocatalysis. As is well known, the concentration of hydrogen atoms can govern the rates and mechanism of reactions.

Electrocatalysis is the most important research problem of the fuel cell. Electrocatalysis is not understood. Neither is it understood why platinum should be the most important and the most versatile single material in catalysis and in electrocatalysis, but its great inertness though a transition metal may be a clue in light of the requirements listed above. The difficulty of the problem is illustrated by the results of Grubb, who found that platinum is a good electrocatalyst on a propane anode while iridium under his conditions was hopeless. Yet, as he points out, both metals have partially filled d-orbitals, both form face-centred cubic crystals in which the lattice parameters (a_0) differ by less than 3 per cent, and both are excellent catalysts for many non-electrochemical reactions. Present theories of the solid state seem helpless in this situation, which is complicated further because one must deal with the surface of the metal—not with its interior. For the present, the search for improved electrocatalysts must continue to be intuitive and empirical.

Bibliography

LIEBHAFSKY H. A. and CAIRNS E. J. (1968) *Fuel Cells and Fuel Batteries*, New York: Wiley.

H. A. LIEBHAFSKY

ELECTROGASDYNAMIC GENERATION OF POWER. There are considerable advantages to be gained from generating electricity directly from thermal sources, e.g. coal, gas and nuclear fuel, without the necessity of using gas or steam turbines in the thermodynamic cycle. The reason is that these devices set a limit to the maximum working temperature and thereby limit the overall thermodynamic efficiency. A number of different kinds of generator have been suggested to effect the direct generation of electricity. The electrogasdynamic generator (EGDG) is one of the various devices of this kind at present under consideration.

A typical EGD generator is shown in the Figure. There is a source of charge, S, which serves as one electrode and from which charge is removed by a gas stream and subsequently deposited upon a collector

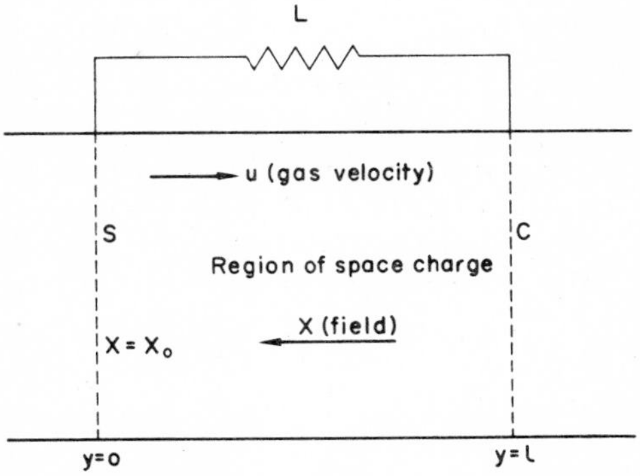

Fig. 1. Electrogasdynamic generator.

electrode, C. The circuit is completed by the external load L. Having passed through the first stage, the gas can then go on to similar stages until the required energy removal has been achieved.

A concrete example will be taken to illustrate the general features of these devices. The charge source S, and the collector electrode C, which are both circular grids of a diameter greatly exceeding their mutual separation, l, are set in a cylindrical, insulating container. The purpose of the large diameter and small separation is to increase the total current flowing— the current density, which can be a limiting factor, increases with decreasing separation, as will be shown later. A gas stream passes through the grids at u cm sec^{-1}. The space charge set up by the flow of gas tends to hinder further removal of ions from the source by retarding the ions, relative to the gas flow, by the velocity KX; where K is the ionic mobility and X the local field. The ion velocity at any point between the electrodes u_i is given by

$$u_i = u - KX \quad (1)$$

and the current density j by

$$j = u_i ne = (u - KX)\, ne \quad (2)$$

where n is the local ion concentration and e the ionic charge. In a general system equations 1 and 2 should relate vector quantities. Fortunately, however, because of favourable geometry the system is essentially one-dimensional and the quantities may be treated as scalars, as may readily be shown: the container, being an insulator, retains all the charge falling upon it. A time will be reached at which it has collected so much charge on its surface that additional charges in the gas stream are repelled. When this occurs, the lines of force run parallel to the surface of the container (and therefore to the gas stream) and deliver no more charge to it. The unidimensional field pattern is reinforced by the electrodes, which, as parallel, isopotential planes of large extent, tend to restrain the field to a direction at right angles to their surfaces.

For the unidimensional system, Gauss's law in e.s.u. reduces to

$$\frac{dX}{dy} = -4\pi ne \tag{3}$$

The negative sign arises because y and X point in opposite directions. Since j of equation 2 is constant, substitution into (3) and integration between $y = y_1$, $X = X_1$; $y = y_2$, $X = X_2$ yields

$$[uX - \tfrac{1}{2}KX^2]_{X_1}^{X_2} = -[4\pi jy]_{y_1}^{y_2} \tag{4}$$

the maximum field, X_0, occurring at the ion source, $y = 0$. Clearly, for a given value of u for any current density j the power will be a maximum when the potential drop is a maximum. This implies, firstly, that the generator should be as long as possible (i.e. $X = 0$ at $y = l$) and that the field, and hence X_0, should be as large as possible. Using the condition that $X = 0$, $y = l$, the field distribution is given by

$$X = \frac{u}{K}\left[1 - \sqrt{\left(1 - 8\pi j(l - y)\frac{K}{u}\right)}\right] \text{ e.s.u. V cm}^{-1} \tag{5}$$

The power generation, W, per cm³ of electrode is given by

$$W = j\int_0^l X\,dy = j\int_{X_0}^0 X\frac{dy}{dX}\,dX \text{ erg cm}^{-2}\text{ sec}^{-1} \tag{6}$$

Substituting for dy/dX from equation 2 into equation 6 and integrating:

$$W = \frac{uX_0^2}{8\pi}\left[1 - \frac{2}{3}\frac{KX_0}{u}\right] \text{ erg cm}^{-2}\text{ sec}^{-1}. \tag{7}$$

This is the maximum power that can be obtained in a single stage for given gas flow velocity, ion mobility and electric field at the source electrode. X_0 is ultimately limited by electrical breakdown of the gas to a value, X_b. The ion mobility, on the other hand, can be made as small as desired by attaching the charges to macroscopic particles or droplets, this can be achieved by using a dust or droplet laden gas, for example. Thus the absolute maximum power per stage for a given flow velocity occurs when $X_0 = X_b$ and $K = 0$, i.e.

$$W_{max} = \frac{uX_b^2}{8\pi} \text{ erg cm}^{-2}\text{ sec}^{-1}. \tag{8}$$

It is easy to show that under these conditions

$$j_{max} = \frac{uX_b}{4\pi l} \text{ e.s.u. sec}^{-1}\text{ cm}^{-2}.$$

$$V_{max} = \tfrac{1}{2}X_b l \text{ e.s.u. V.}$$

The operating conditions of the EGD generator are best illustrated by taking a specific case, for example, air at 10 atm and 900°K flowing at Mach 1. Under these circumstances, one finds $W_{max} = 23$ W cm⁻². $V_{max} = 5 \times 10^4\, l$ V and $j_{max} = 0{\cdot}43/l$ mA cm⁻². The total energy flux carried by the gas stream, including thermal, kinetic and pressure energy, is 24×10^4 W cm⁻². These figures point to important and generally valid conclusions. Thus, the main difficulty with this kind of generator is, clearly, the necessity to use a very large number of stages in view of the small fraction of the total power of the gas stream removed at each stage (in the example, 10^{-4}). However, on the credit side, these devices generate power at high potential and low current. In the example it was calculated that, at maximum power, 0·43 mA cm⁻² could be generated at 50,000 V. Potential of this order of magnitude are ideal for power transmission over long distances.

J. Lawton

ELECTROMAGNETIC LEVITATION. Levitation by the forces of electromagnetic induction is only one of several methods of suspending objects in space against the force of gravity without physical contact by mechanisms which make use of the known laws of physics. Among the others listed by Boerdijk in 1956 are included:

(1) Levitation by gravitational forces.
(2) Levitation by reaction forces (rockets).
(3) The use of forces due to radiation fields.
(4) Levitation of diamagnetics or superconductors in stationary magnetic fields.
(5) Levitation of permanent magnets by diamagnetics or superconductors.
(6) The use of quasi-stationary electromagnetic fields with feedback systems.

Electromagnetic levitation is the cheapest way to support objects, provided the distance between the suspended object and the levitation apparatus is not large. The object supported must be an electrical conductor which is fed with current by induction from a primary coil system. The forces required to provide both lift and stabilization are produced as the result of interaction between the primary currents and the currents which they induce in the suspended object. The art of designing induction levitation systems is that of devising the right shape of primary coil configuration to support the shape of object required. A thorough knowledge of electromagnetic induction theory is of considerable help in design although effective systems can be, and often have been, developed entirely by experimental "trial and error" methods. The fact that levitation apparatus is often of very simple construction, whereas the application of Maxwell's equations to the calculation of its performance is difficult encourages the experimental method of developing new systems.

The famous "jumping ring" experiment perhaps represents historically the beginning of the subject of levitation by induction. The apparatus is shown in Fig. 1 and consists of a single coil of wire, with a vertical axis, preferably (although not necessarily) containing a laminated iron core which projects upwards from it. When a conducting ring is dropped

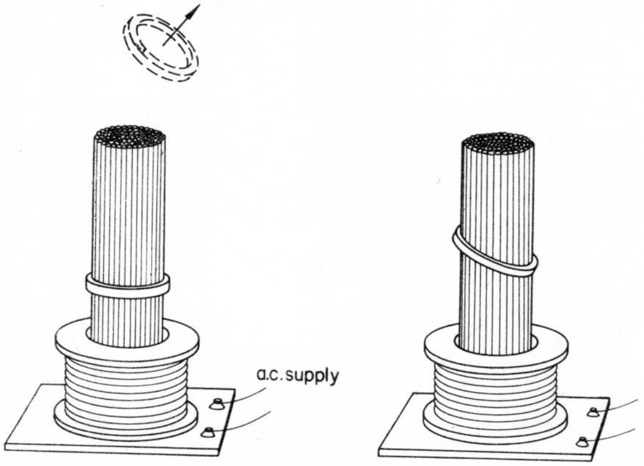

Fig. 1. The "jumping ring" experiment.

over the core and the coil is connected to a source of alternating current, the ring is thrown into the air. If the ring is replaced over the core whilst the current is still flowing it will remain "floating" some distance up the core. It is not, however, a levitated object, for it requires mechanical restraint laterally and "leans" on the core, provided the latter is long enough. If it is not, the ring drifts sideway and falls. Attempts to balance the ring exactly over the core are less likely to succeed than an attempt to balance a pencil on its point. The system as it stands is unstable.

The basic problems in designing levitation systems are as follows:

Apart from the necessity to provide sufficient ampere-turns in the primary coil system to produce the necessary lift to overcome gravity, the principal problem is that of stabilizing the supported mass. In most cases more than one primary coil is needed. The currents induced in the floating member are generally large and generate a lot of heat. Sometimes the problem is to limit the amount of heat so that the suspended body is not melted. Sometimes it is quite the reverse and the problem is to increase the heat in a specimen in order to melt it (for liquids as well as solids may be suspended) and to do so without overheating the primary coils.

Shaded pole action. A useful insight into the mechanism of levitation may be obtained from a study of the action of a shaded pole motor. Figure 2(a) shows the conventional method of shading the poles of a single phase induction motor. The part of a pole which is embraced by the closed conducting loop carries a flux whose phase is delayed in relation to that of the main part of the pole by the action of the current in the loop, so that a travelling magnetic field moves in a direction from the main part of the pole *towards* the shading ring. If the shading mechanism consists not of a loop in a slot, but of a solid slab of conductor on the pole face as shown in Fig. 2(b) the travelling field acts on the shading block itself and if not secured to

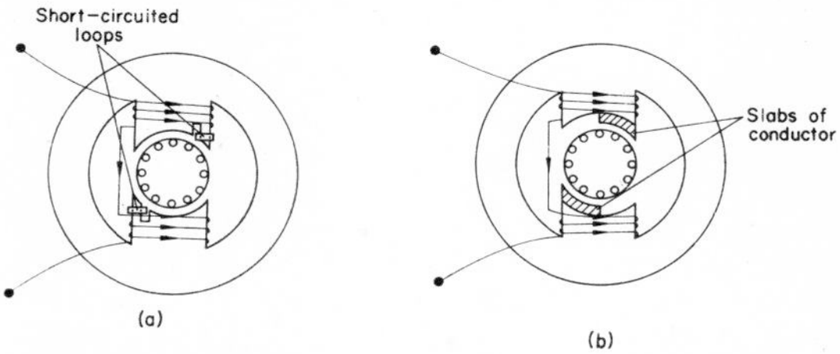

Fig. 2. Two methods of shading motor poles: (a) by conducting rings in slots; (b) by solid sheets on the pole faces.

the pole it will eject itself from the pole, whether or not there be rotor conductor or even rotor iron present.

Appreciation of this action provides an explanation of the basic lateral instability of the jumping ring, or of any sheet of conductor which is placed over a single coil carrying alternating current. Figure 3 shows that if a circular conducting plate is placed over a circular

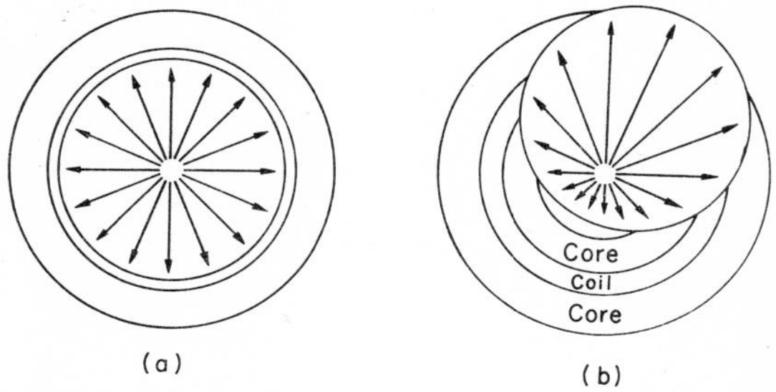

Fig. 3. Internal horizontal forces on a supported disk: (a) when the disk is symmetrically placed there is no net force; (b) in an asymmetric position the disk is expelled.

coil in any position other than of exact concentricity the plate will shade the coil, produce an outward-travelling field and eject itself along the radius of the coil on which its centre lies.

The mechanism of the lifting force can also be explained qualitatively in terms of shaded pole action. Figure 4(a) shows a cross-section through a jumping ring apparatus in which the ring has been extended into a long cylinder. Figure 4(b) shows an a.c. coil which is expelling an eccentric conducting plate laterally. Figure 4(c) is a cross-section through it. If the system of Fig. 4(c) is duplicated as shown in (d), it will be seen to be basically the same as that of (a) with the whole diagram turned through 90°, so that it can be argued that the jumping ring is lifted by the action of a travelling magnetic field which is in turn set up as a result of the fact that the phase of the total current opposite the base of the ring is different from that of the current around the top of the ring.

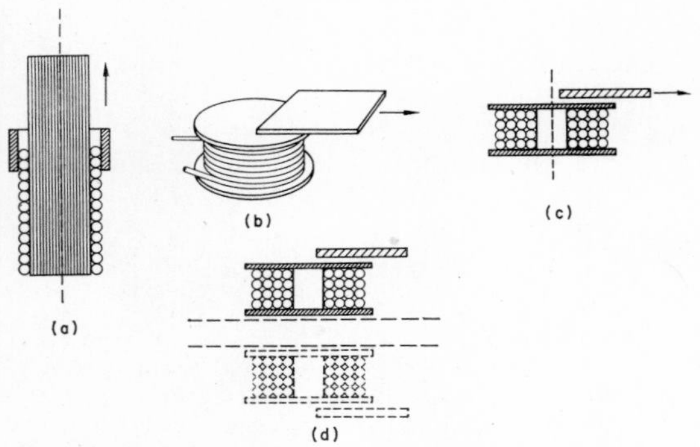

Fig. 4. The shaded-pole principle used to explain the jumping ring effect: (a) cross-section through jumping ring apparatus; (b) sheet of conductor being expelled from an a.c. coil; (c) cross-section through (b); (d) duplication of (c) gives the same system as at (a).

The most important result of the shaded pole study is that it suggests a possible primary coil arrangement for producing lateral stabilization. The inherently outward-travelling fields produced by the supported object itself can be countered by using a second coil inside the first, carrying a current which lags in phase behind that of the first, so as to produce an inwardly-travelling field, as shown in Fig. 5(a). The second coil may be simply short-circuited, or may even consist of a solid conducting tube as shown in (b) since the latter constitutes a shading ring for the centre of the primary

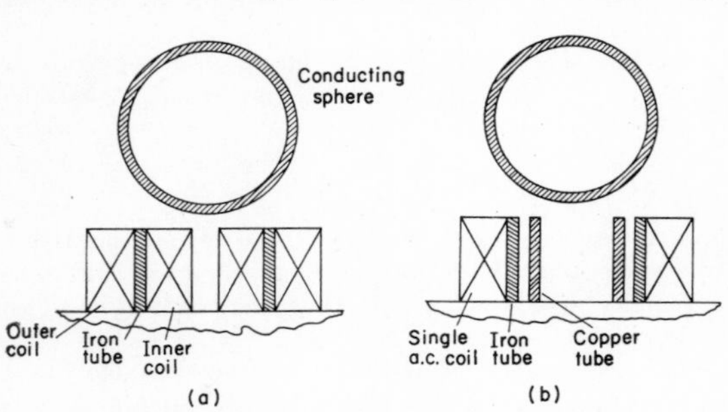

Fig. 5. Cross-section through power-frequency sphere levitators: (a) double coil system; (b) single coil and "shading" tube.

system. Levitation of spheres or of disks may be carried out with circular coil systems of the types shown in Fig. 5.

Dynamic instability. A second kind of instability may arise as the result of e.m.f.'s induced in the supported object due to a motional disturbance. For example, the coil systems of Fig. 5 have been designed to produce inwardly-travelling fields. Figure 6(a) shows an arrangement of two linear induction motors placed back-to-back which constitute a similar arrangement but one in which the travelling fields are more highly organized, so that for purposes of explanation it is reasonable to assume a fixed speed of field travel and to ascribe to each half of the system a speed/force relationship shown in Fig. 6(b) which is typical of induction motors in general.

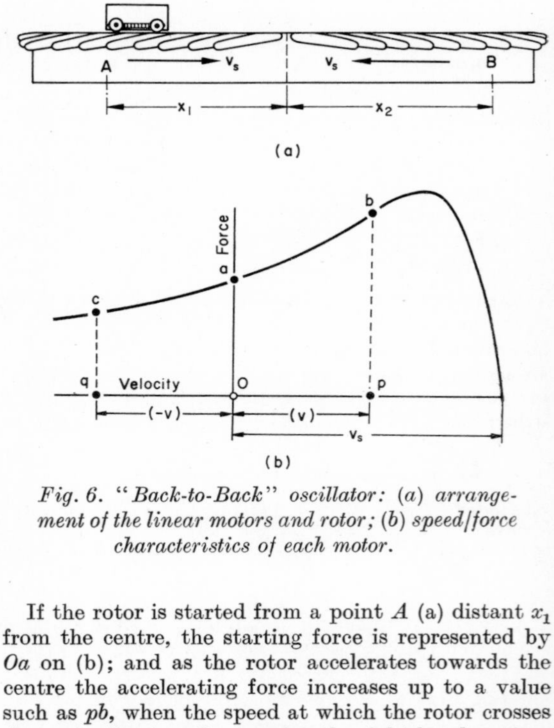

Fig. 6. "Back-to-Back" oscillator: (a) arrangement of the linear motors and rotor; (b) speed/force characteristics of each motor.

If the rotor is started from a point A (a) distant x_1 from the centre, the starting force is represented by Oa on (b); and as the rotor accelerates towards the centre the accelerating force increases up to a value such as pb, when the speed at which the rotor crosses the centre line is v. On entering the right-hand section, the rotor is travelling in a direction opposite to that of the field. With reference to that field its speed is $-v$, and the initial decelerating force is as represented by qc. This force rises to Oa as the rotor is brought to rest, but it can be seen that the decelerating forces are everywhere smaller than were those producing acceleration. Thus the point B (a) at which the rotor again comes to rest is at a distance x_2 from the centre where $x_2 > x_1$. The rotor now returns to the left, but (starting from a greater distance from the centre) it attains a velocity greater than v and as a result progresses to the left beyond point A before coming to rest again. Thus the oscillation continues to build up in amplitude until the maximum speed attained at the centre approaches the synchronous speed v_s.

This dynamic instability is a function of the effective "coupling" between primary coil system and levitated object which has been defined in terms of a factor of "goodness", G, in a recent paper. Oscillations will build up if $G > 1$ but dynamic disturbance will be damped if $G < 1$. The value of G is proportional to the square of a linear dimension for a given type of levitation, proportional to the supply frequency and to the conductivity of the material of the supported object.

Classes of levitator. Levitators may be divided into two classes:

(i) Low-frequency systems in which the primary coil system is often embedded in laminated steel with advantage. The range of frequencies used is 0–1000 c/s and the systems are usually designed to support solids, rather than both to support and melt the metal.

(ii) High-frequency systems (1000 c/s to 1 Megacycle) use air-cored coils, the primary conductors often consisting of water-cooled tubes. Their most common use is in supporting molten metal but they can also be used to support solids. Figure 7 shows a cross-section through a typical high frequency levitator.

Most levitators in both classes make use of two coils in the primary system, one of which is used primarily to provide the lift and the other to effect stabilization. Generally the latter actually produces *downward* forces on the supported specimen and is therefore often called the "attractor". In some cases a part of the supported object itself may perform the function of the stabilizing coil so that it is possible to support an object with a single primary coil. A variety of systems which have been used successfully are shown in Fig. 8.

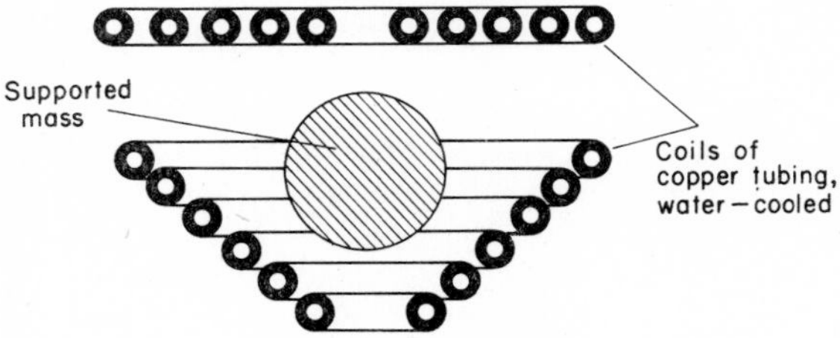

Fig. 7. Cross-section through a high frequency levitator.

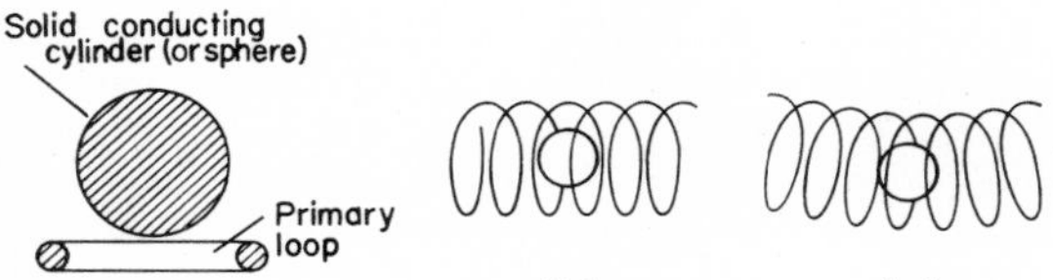

Fig. 8. Single coil systems: (a) cross-section through a system which can be used for spheres or cylinders; (b) Okress' system; (c) Okress' coil modified to prevent lateral drift.

The Cone of Attraction

When a 2-coil system has been set up to produce inwardly-travelling fields with a view to stabilization it can be shown that it also produces downward travelling fields within certain regions over the coils. Figure 9 shows a typical conical region within which small non-ferrous conductors are attracted but outside of which they are repelled. It appears to be a necessary condition of stability that the supported object should lie partly within and partly without the cone of attraction.

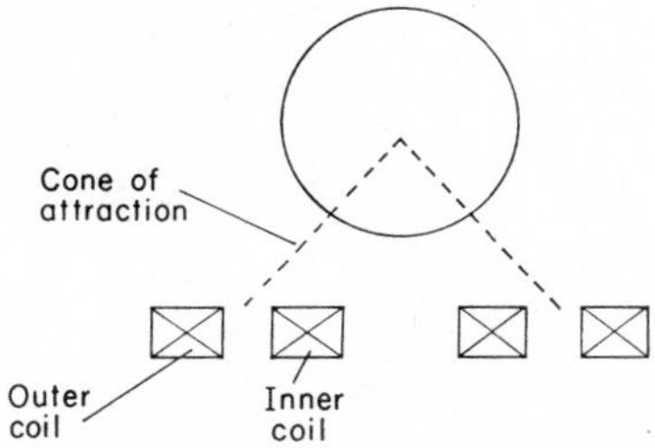

Fig. 9. Typical dimensions of the cone of attraction in a sphere levitator.

Supporting circular and rectangular plates. Spheres are possibly the easiest objects to levitate. Disks are also reasonably easy to stabilize even though there are several axes about which they may be unstable and several forms of instability which may occur. For example a disk may be statically laterally unstable, or it may oscillate laterally. It may oscillate in a vertical direction. About a vertical axis it is always free to rotate without generating any additional currents provided the whole system is symmetrical about that axis. Once set in motion in this way, only air friction will bring it to rest. It may be stable and level but eccentric and this is usually symptomatic of the disk

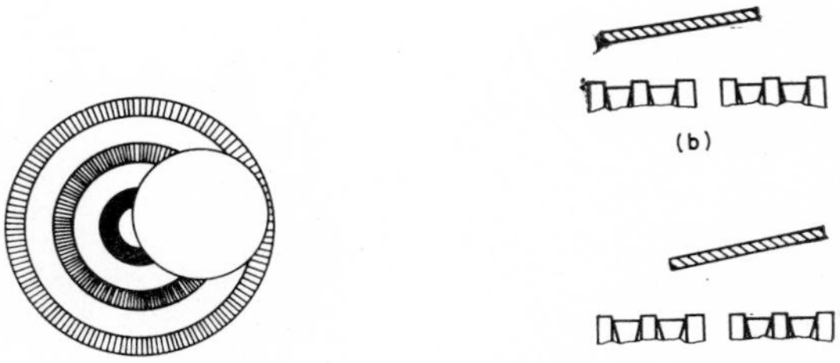

Fig. 10. Asymmetric stable positions for a disk: (a) plan view of eccentric disk; (b) side view of disk, tilted inwards; (c) side view of disk, tilted outwards.

being too small in diameter in relation to the primary coils. It may be stable but tilted. Some possible positions are shown in Fig. 10. The complexity of the system is illustrated in Fig. 11 which shows the various states of a circular plate system in terms of variation in outer coil current in magnitude and phase for one particular value of inner coil current.

Within the stable region the supported height and the degree of stability vary with coil current amplitudes and phase. Figure 12(a) shows the variation

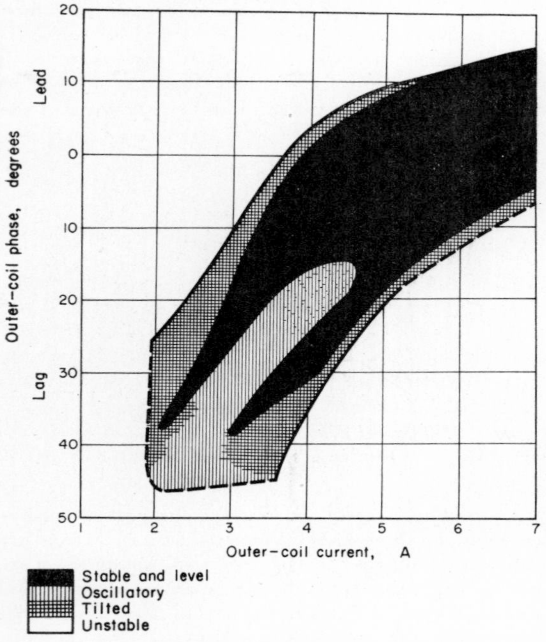

Fig. 11. A "behaviour" diagram for a plate levitator.

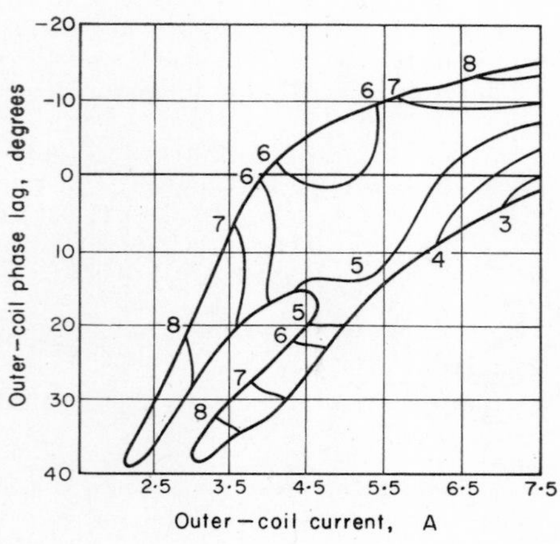

Fig. 12a.

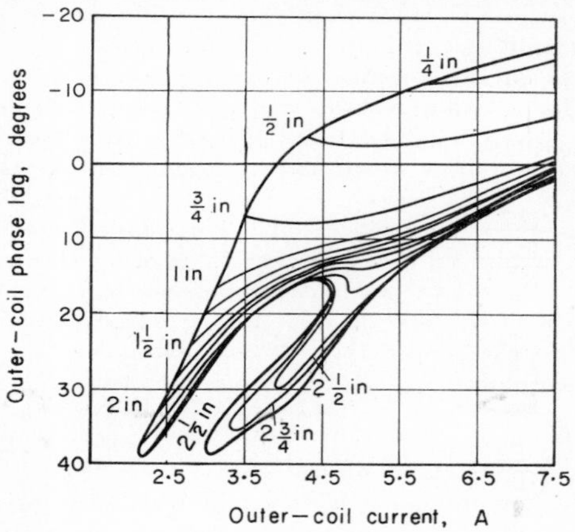

(b)

Fig. 12. Properties of a plate levitator in its stable region (see Fig. 11): (a) height lines (numbers indicate eighths of an inch); (b) maximum displacement lines.

in height for the system whose behaviour is displayed in Fig. 11. Figure 12(b) shows the degree of stability for the same example by indicating the maximum distance by which the plate can be laterally displaced and yet still be returned to the stable position on release. To a first approximation the lifting force for a given coil is inversely proportional to the fourth power of the suspended height.

Rectangles have more axes about which they may be unstable. In particular they may tilt about one or both of 2 horizontal axes or they may twist about a vertical axis. Figure 13 shows a plan view of a system of primary coils in which separate control of the conditions in the X and Y directions is possible. A simplified system results from the fact that stability is possible in either axis with pairs of primary currents

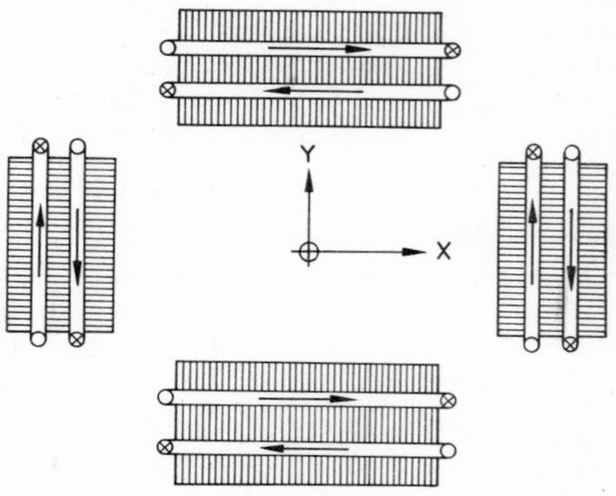

Fig. 13. Plan view of rectangular plate levitator primary consisting of 8 separate coils.

which are 180° out of phase but equal in magnitude. This leads to a simple 2-block construction as shown in Fig. 14. Variations in stable position with such a system are shown in Fig. 15 and of course in all cases the supported plates are free to move in the XX^1 direction with only air friction to resist the motion.

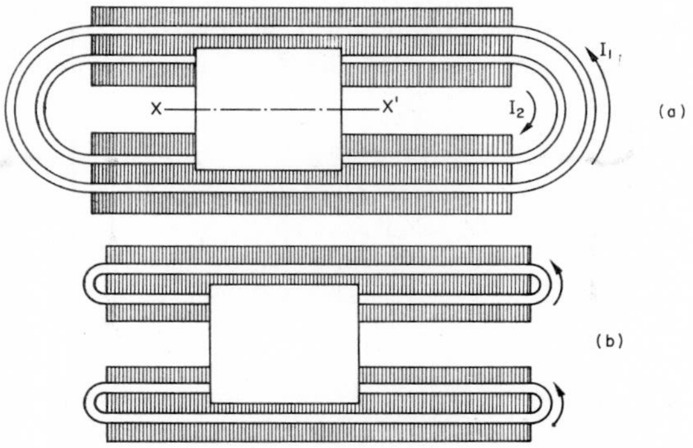

Fig. 14. Simplified rectangular plate levitator systems: (a) two coil, 2-block arrangement for $I_1 \neq I_2$; (b) If $I_1 = I_2$ blocks can be self-contained for adjustment to any width of plate.

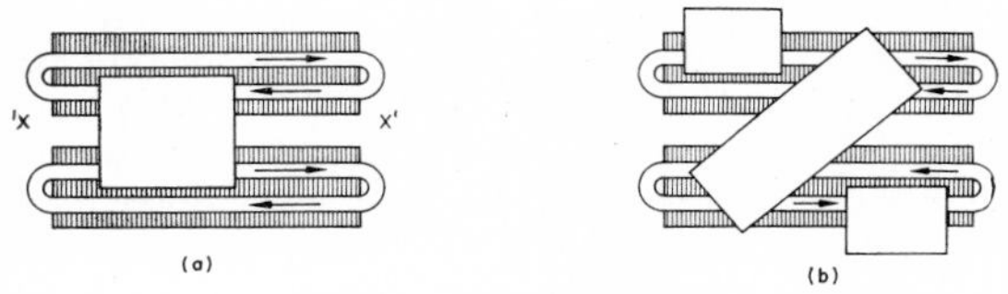

Fig. 15. Possible plate positions with a rectangular system: (a) normal symmetrical position; (b) 3 possible asymmetric positions.

Suspension of liquids. Liquids are generally only levitated using high-frequency systems. The reasons for this are twofold. Firstly the conductivities of liquids are generally much lower than those of copper or aluminium and secondly the suspension of liquid metals is usually only required in metallurgical research where only small quantities are to be supported. Both these features contribute to a very low value of G at power frequencies where it becomes extremely difficult to produce enough primary current to lift the specimens.

The same techniques can be used for stabilizing liquids in suspension as those used for solids but there are additional difficulties in that the internal stresses experienced by solids can now produce motion and the downward forces inside the cone of attraction tend to make the supported metal take up the shape of a pear or child's top and the metal often drips. The tendency to drip is greater in liquids with a lower surface tension. Liquids suspended *in vacuo* are more likely to drip than when a layer of oxide can form on the outside of the specimen.

One of the principal limitations of a high-frequency system is that the ultimate temperature of the metal is a function of its size. As the quantity of metal is reduced, the heat generated in it increases rapidly for a given suspended height and this is a quite fundamental limitation. The temperature of a melt can be increased at will simply by raising the frequency but it cannot necessarily be lowered by reducing frequency, for to reduce frequency beyond a certain limit is to reduce lift.

The following metals have been recorded as having been successfully levitated: aluminium, steel, gold, antimony, gallium, platinum and indium.

Advantages which are claimed for levitation melting include:

(a) The specimen touches no crucible during heating, melting and draining.
(b) The molten specimen can be protected by a suitable atmosphere or vacuum.
(c) Volatile impurities can be distilled or pumped away.
(d) The molten metal can be drained gradually, dropped as a whole or solidified in suspension.
(e) The melt is continually agitated by the internal travelling fields and therefore thoroughly mixed.
(f) The specimen can consist of metal powders and alloying constituents mixed and sintered together. The additions can be made during suspension.

Conclusions. Electromagnetic levitation, although the cheapest method of obtaining stable support is very expensive in terms of power input. The power required to support 1 lb of metal varies from 50 watts for a well-designed, low-frequency system where it is only required to lift the specimen by a fraction of a millimetre, to perhaps 50 kW or more for suspension of a rare metal by high-frequency currents. Electromagnetic levitation is not therefore likely to find large-scale application in transport systems where fluid cushion support is very much more economical.

Bibliography

Laithwaite E. R. (1966) *Induction Machines for Special Purposes,* London: Newnes.

Laithwaite E. R. (1966) *Propulsion Without Wheels,* London: English Universities Press.

Geary P. J. (1964) *Electric and Magnetic Suspensions,* Chislehurst, Kent: B.S.I.R.A.

Boerdijk A. H. (1956) Technical Aspects of Levitation, *Phillips Research Report,* **11**, 45.

Okress E. C., Wroughton D. M., Comenitz G., Brace P. H. and Kelly J. C. R. (1952) Electromagnetic Levitation of Solid and Molten Metals, *J. Appl. Phys.* **23**, 545.

Laithwaite E. R. (1965) Electromagnetic Levitation, *Proc. I.E.E.* **112**, 2361.

E. R. Laithwaite

ELECTRON-PROBE MICROANALYSIS OF LIGHT ELEMENTS. An article on Electron-probe microanalysis can be found in Supplementary Vol. 1 of the Dictionary: the term "light elements" is used in the present context to denote elements of low atomic number. Conventionally, elements below Na in the periodic table are said to fall in this category. Two principal factors give rise to the distinction between light element analysis and the analysis of the elements of higher atomic number. One is that the characteristic X-ray wave-lengths of the lighter elements are so large that they cannot be dispersed by the crystals normally used for the wave-lengths of the heavier elements. The other is that these long wave-length or soft X rays are easily absorbed, so that the X-ray path must be in vacuum and special precautions must be taken to enable the X-ray detector to function efficiently.

The dispersion of soft X rays. A fundamental step in microanalysis is to identify the characteristic X-ray wave-lengths emitted by the specimen and a crystal spectrometer is normally used for this purpose. An X-ray counter detects the X-ray beam diffracted by the analysing crystal when the Bragg relationship holds: $\lambda = 2d \sin \theta$, where λ is the characteristic wave-length, d is the spacing of the diffracting lattice planes and θ is the glancing angle of incidence made with these planes. This method of analysis fails when $\lambda > 2d$, since the Bragg condition can no longer be fulfilled ($\sin \theta$ cannot exceed unity).

The wave-length range of the characteristic X rays of the light elements extends from 18·3 Å (for fluorine) to 228 Å (for lithium), while the d spacings of analysing crystals commonly used vary between 1·3 Å (topaz) and 9·9 Å (mica). Theoretically it is therefore just possible to detect fluorine, although the practical limit is normally sodium. Extension of the range of elements capable of being analysed by means of a crystal spectrometer must depend on the existence of crystals having correspondingly greater d spacings. Large and relatively stable crystals have been grown of KAP (potassium acid phthalate), with a d spacing of 13·2 Å and the use of these has extended the range of elements down to oxygen. Other crystals are known to have larger spacings, but attempts to produce them in a suitable form have so far been unsuccessful.

Three methods are employed to analyse the radiations from the lighter elements. In one of these, the X rays are not dispersed and in the others, the X rays are dispersed by a multilayered "crystal" of large d spacing or by a diffraction grating.

In the *non-dispersive method* of Dolby, X-radiation from the specimen impinges directly on the window of a proportional counter. The amplitude of the output pulses of the counter should ideally be proportional to the energy of the incident quanta, but owing to the statistical nature of the mechanism of pulse formation, there is a considerable spread in the pulse height distribution for a given radiation Pulse height distributions from adjacent elements in the periodic table overlap, so that it is not possible to resolve their characteristic wave-lengths directly and the following technique was devised by Dolby to unfold these overlapping pulse height spectra. In Fig. 1, the three Gaussian distributions A, B and C, separated by distances equal to the standard deviation σ, represent idealized pulse height distri-

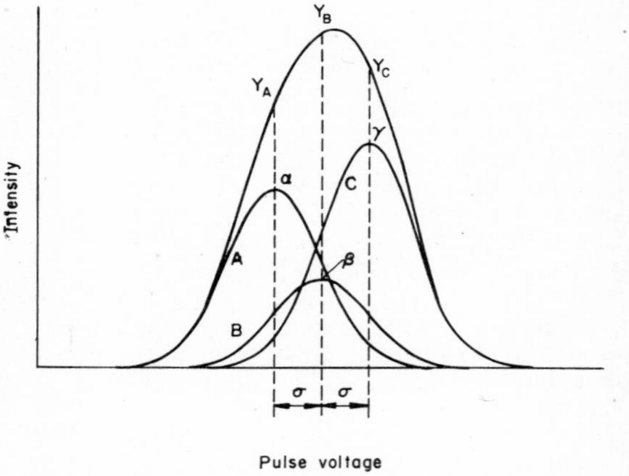

Fig. 1. Idealized proportional counter curves (after Dolby).

bution curves of X-ray lines from adjacent elements. The amplitudes α, β and γ of the constituent components must be derived from the composite curve, which is obtained directly from the output of the counter. Three linear simultaneous equations relate α, β and γ to the measured count rates Y_A, Y_B and Y_C. A single channel pulse height analyser is set on each constituent peak and the outputs are fed to the corresponding rate meters to give the values of Y. A network unit performs the arithmetical operations on the Y values according to the requirements of the simultaneous equations, thus yielding three voltages proportional to α, β and γ. In practice, the pulse height distributions do not have the idealized shapes shown in the figure, so that instead, the network is set up and calibrated by using pulses from three standard specimens consisting of the elements under investigation. This technique has been used to analyse elements down to beryllium ($\lambda = 114$ Å).

The *multilayered "crystal" method* of analysis was pioneered by Henke and resembles conventional crystal spectrometer analysis. The crystal is built up by the deposition of 50 to 100 monolayers of metal salts of straight-chained fatty acids on a plane or curved substrate. The effective d spacing normal to the surface of the crystal is typically 50 Å and crystals of barium or lead stearatedeconoate have been used to analyse elements down to boron ($\lambda = 67·8$ Å).

In the third method of analysis, *diffraction gratings* are employed as the dispersing element. X rays are diffracted by gratings if the glancing angle of incidence does not appreciably exceed the critical angle of total external reflection. In the range of wave-lengths considered here, the glancing angle should not exceed a few degrees. Adequate dispersion for all the light elements can be achieved with gratings having the same pitch (typically 600 lines per mm) as those used in the visible region of the spectrum. However, in order to diffract efficiently, X-ray gratings differ from optical ones in that the diffracting surfaces must be smoother and must not be marred by surface imperfections. The effects of surface roughness are accentuated by the small grazing angle and by the comparatively short wave-length of X rays. Blazed gratings, in which the blaze angle is about a degree, and unblazed ones are both suitable for light element analysis. A grating spectrometer differs from the conventional crystal spectrometer in that the 2 : 1 angular relationship between detector and crystal, arising from the Bragg condition, does not apply. The geometry of a grating spectrometer is illustrated in Fig. 2. All the spectra are diffracted simultaneously, provided the beam is incident on the grating at a sufficiently small angle. The relationship between the glancing angle of incidence i, the glancing angle of diffraction θ and the grating pitch d is $n\lambda = d(\cos i - \cos \theta)$, where n is the order of diffraction. As in a crystal spectrometer, it is advantageous if the source, grating and detector are positioned to conform to Rowland circle geometry.

Relative merits of the three analytical techniques. (*i*) *Sensitivity.* The non-dispersive method has an intrinsically high sensitivity, since the detector may subtend a large solid angle at the specimen. Count rates are typically two orders of magnitude greater than those obtained from stearate crystals or gratings.

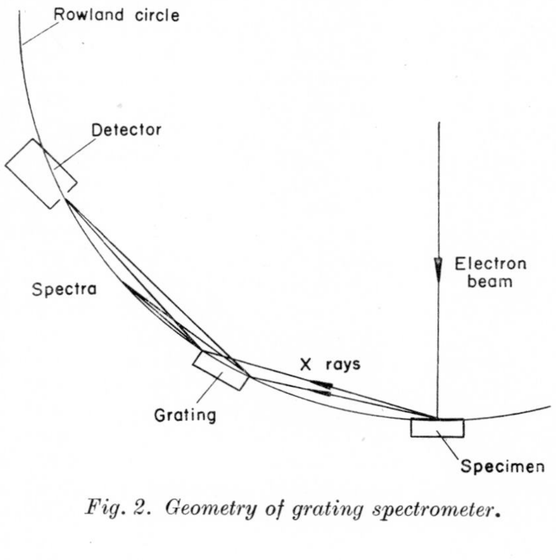

Fig. 2. Geometry of grating spectrometer.

(*ii*) *Resolving power.* The resolving power of the non-dispersive technique is so poor as to impose a severe limitation on its usefulness. High resolving power is important in cases where the L and M spectra of the heavier elements nearly coincide with the K spectra of the lighter elements and also for reasons referred to in the paragraph below on soft X-ray spectroscopy. The theoretical limit of the resolving power of a stearate crystal is, in the first order, equal to the number of effective layers of the crystals and usually lies between 50 and 100; the number of layers which can contribute to the diffracted beam being limited by absorption. The corresponding figure for the diffraction grating is typically two orders of magnitude greater than this.

(*iii*) *Some practical considerations.* The non-dispersive technique requires complex electronic equipment and elaborate precautions must be taken to maintain stability of the characteristics of the proportional counter. The stearate crystal can be used directly in a conventional crystal spectrometer, while the grating requires a separate spectrometer.

Soft X-ray detector. The most commonly used detector is the proportional counter, which differs from similar counters used for harder X rays in that the entrance window must be very thin to reduce the absorption of the X-ray beam to a minimum. Suitable counter windows are made of plastic materials and are usually between 0·1 μm and 1 μm thick. Since these thin windows are not vacuum-tight, the counter gas (e.g. 90 per cent argon, 10 per cent methane), supplied from a gas cylinder, is made to flow continuously through the counter. The output pulse from the counter is relatively small and the first stage of amplification requires the use of a good low-noise preamplifier.

Soft X-ray spectroscopy. The K spectra of the light elements and the L and M spectra of the heavier elements are relatively broad bands compared with the sharp lines found in the shorter wave-length region of the spectrum. The carbon K band, for example, has a half-width of about 1 Å; this is three orders of magnitude greater than that of a spectral line of the heavier elements. The wave-length and shape of the spectral lines of the light elements (in the solid state) is dependent on crystal structure and chemical combination. Light element microprobe analysis can therefore yield more information about the state of the element under examination than heavy element analysis, provided the resolving power of the spectrometer is sufficiently high.

Quantitative analysis. The same corrections must be made to the measured X-ray intensity as those described in the article on electron-probe microanalysis. The main difference is that the intensity in the light element case is also dependent on chemical bonding. The fluorescent correction for light elements is negligibly small but the atomic number and absorption corrections are by no means negligible. Relevant data such as back-scattering and absorption coefficients are not known with great precision and since the value of the latter are very high, accurate quantitative analysis is usually very difficult.

Bibliography

Advances in X-ray Analysis, Vol. 7 (1964) New York: Plenum Press.
Advances in X-ray Analysis, Vol. 8 (1965) New York: Plenum Press.
The Electron Microprobe (1964) New York: Wiley.
POOLE D. M. (1966) in *Encyclopaedic Dictionary of Physics* (J. Thewlis Ed.), Suppl. Vol. 1, 82, Oxford: Pergamon Press.
X-ray Optics and Microanalysis (1966) Paris: Hermann.
X-ray Optics and X-ray Microanalysis (1963) New York: Academic Press.

A. FRANKS

ELECTROSTATIC ACTUATOR. A device used to determine the relative sensitivity of laboratory condenser microphones, consisting of a flat slotted plate held parallel and in close proximity to the microphone diaphragm. A d.c. polarizing voltage E_0 and an a.c. drive voltage e applied between actuator and diaphragm create an electrostatic force on the diaphragm simulating a sound pressure p given by

$$p = \frac{\varepsilon_0}{d_0^2} E_0 e \sin \omega t,$$

where ε_0 is the free-space permittivity and d_0 the effective separation between actuator plate and diaphragm. The limitation on E_0 is set by the fact that the electrostatic force thus generated attracts the microphone diaphragm away from its back-plate and thus changes the response. In practice E_0 and d_0 are chosen so that the actuator force is less than 1 per cent of that generated by the normal polarizing voltage. The limitation on the amplitude of applied alternating voltage e is set by distortion, the second harmonic distortion being $e/4E_0$. Due to difficulties in measuring or deducing the effective d_0 for the slotted back-plate this method cannot be used for absolute calibrations, but as the factor ε_0/d_0^2 is independent of frequency it provides a convenient method for determining the relative frequency response of microphones. If the true pressure response of a microphone is required it is necessary to reduce the acoustic load on the diaphragm to zero using a $\lambda/4$ termination; otherwise the radiation impedance loading the diaphragm will introduce errors which may exceed 1 dB near the natural resonance of the diaphragm system where the radiation impedance becomes commensurate with the diaphragm impedance.

M. E. DELANY

ELEMENTARY PARTICLE PHYSICS, RECENT ADVANCES IN. With the commissioning of the giant proton synchrotrons in 1960 at Geneva and Brookhaven, U.S.A., the number of known elementary particles showed a dramatic increase; now the total is 150–200. This prolific abundance of particles raises doubts on the meaning of "elementary". Fortunately, a classification scheme (SU_3 symmetry) was found which created some kind of order in the apparent chaos. The present state of elementary particle physics may be compared with that of inorganic chemistry at the end of last century. Mendeleef had observed the regular pattern in the chemical properties of the elements and had devised his Periodic Table but we had to wait until the 1920's for the understanding of the scheme in terms of the electronic structure of the nuclei. Today we have a means of classifying the elementary particles but lack the full understanding of this classification.

The properties which may distinguish one particle from another are mass, lifetime, spin, isotopic, spin, strangeness and parity. The particles are subject to the forces of the *strong*, electromagnetic, *weak* and gravitational *interactions*. The relative strengths of these interactions and the present state of their theoretical understanding is given in Table 1.

A simple classification of the particles can be given in terms of their interaction properties. Those particles which interact with themselves and other particles only through the electromagnetic or weak interactions are called *leptons* and those which have, in addition, strong interactions properties are called *hadrons*. The hadrons themselves can be subdivided into *mesons* and *baryons* for which the baryon number $B = 0$ and 1 respectively. The various mesons are the quantum particles of the strong interaction between the baryons just as the photon is the quantum particle of the electromagnetic field. A list of the long-lived particles and some of their properties is given in Table 2.

Table 1

Interaction	Relative strength	State of theory
Strong	1	Only selection rules and classification scheme
Electro-magnetic	10^{-2}	Well understood—Quantum electro-dynamics
Weak	10^{-13}	Phenomenological theory based on analogy with electro-magnetic theory
Gravitational	10^{-39}	No significant effect on a single particle

Table 2. Long-lived particles

	Particle		S	$I(J^P)$	Mass (MeV)	Mean life (sec)	Principal decays
baryons	Nucleon, N	P n	0	$\frac{1}{2}(\frac{1}{2}^+)$	939	stable 1013	 $pe\nu_e$
	lambda	Λ^0	−1	$0(\frac{1}{2}^+)$	1115	$2{\cdot}6 \times 10^{-10}$	$p\pi$, $n\pi$
	sigma	Σ^+ Σ^0 Σ^-	−1	$1(\frac{1}{2}^+)$	1192	$0{\cdot}8 \times 10^{-10}$ $<10^{-11}$ $1{\cdot}6 \times 10^{-10}$	$p\pi$, $n\pi$ $\Lambda\gamma$ $n\pi$
	xi	Ξ^- Ξ^0	−2	$\frac{1}{2}(\frac{1}{2}^+)$	1317	$1{\cdot}8 \times 10^{-10}$ $2{\cdot}8 \times 10^{-10}$	$\Lambda\pi$ $\Lambda\pi$
	omega	Ω^-	−3	$0(\frac{3}{2}^+)$	1675	$1{\cdot}3 \times 10^{-10}$	$\Xi\pi$, $\Lambda\pi$
mesons	pion	$\pi^\pm$ π^0	0	$1(0^-)$	138	$2{\cdot}56 \times 10^{-8}$ 10^{-16}	$\mu\nu_\mu$ $\gamma\gamma$
	kaon	$K^\pm$	±1	$\frac{1}{2}(0^-)$	496	$1{\cdot}23 \times 10^{-8}$	$\mu\nu_\mu$
		K^0, $\overline{K^0}$	+1, −1			K_1^0 $0{\cdot}9 \times 10^{-10}$	$\pi\pi$
						K_2^0 6×10^{-8}	$\pi\pi\pi$ $\pi\mu\nu_\mu$, $\pi e\nu_e$
leptons	muon	$\mu^\pm$		$J = \frac{1}{2}$	106	$2{\cdot}2 \times 10^{-6}$	$e\nu_e\nu_\mu$
	electron	$e^\pm$		$J = \frac{1}{2}$	0·51	stable	
	μ-neutrino e-neutrino	ν_μ ν_e		$J = \frac{1}{2}$	0	stable	
	photon	γ		$J^P = 1^-$	0	stable	

1. S = strangeness $I(J^P)$ = isotopic spin (spin, parity).
2. Antiparticles have strangeness, charge and magnetic moment opposite to the corresponding particle, e.g. the anti xi-minus, $\overline{\Xi^-}$, has strangeness +2 and is positively charged.
3. By convention, the negatively charged leptons and positively charged mesons are defined as "particles".

The short-lived mesons and baryons, the resonances, have lifetimes typically of 10^{-23} sec. For short lifetimes, τ, it is customary to consider the mass width of the resonance, Γ, which is related to τ by $\Gamma = \boldsymbol{h}/\tau$ where $\boldsymbol{h} = 6{\cdot}6 \times 10^{-22}$ MeV sec. The resonances have short lifetimes because they have sufficient rest mass to make it kinematically possible for them to decay to other hadrons without violating the laws of the strong interaction. Otherwise, the decay would proceed slowly through the weak interaction which does not have the same rigid conservation laws. Some of the well-studied meson and baryon resonances are listed in Tables 3 and 4.

For interactions between the particles there exist conservation laws that are rigidly obeyed regardless of the type of interaction. These are, the conservation of mass-energy, linear momentum, angular momentum, electric charge, baryon number, lepton number and recently muon-lepton number and electron-lepton number. There are other quantum num. bers, isotopic spin I, and strangeness S, for which the final sum equals the initial sum only in some types of reactions. For example, the strong interaction conserves isotopic spin and strangeness the electromagnetic interaction conserves strangeness but not isotopic spin and the weak interaction conserves neither. Further, there is a set of rules which certain reactions processes obey, either exactly or approximately. For example, the rules concerning strangeness appear to be:

$\triangle S = 0$ for strong and electromagnetic processes.
and $\triangle S = 0$, 1 for weak interaction processes.

The principles of particle–anti-particle (or charge conjugation) symmetry, C, left-right (or spatial) symmetry, P, and time reversal, T, symmetry have been applied recently to many physical processes. By the invariance of a reaction under charge con-

Table 3. Meson resonances

Particle	$I(J^P)$	Mass (MeV)	Γ (MeV)	Principal decays
η	$0(0^-)$	549	< 10	$\pi\pi\pi$, $\gamma\gamma$
ω	$0(1^-)$	783	12	$\pi\pi\pi$
η'	$0(0^-)$	959	< 4	$\eta\pi\pi$, $\pi\pi\gamma$
Φ	$0(1^-)$	1020	3	$K\overline{K}$, $\pi\pi\pi$
f	$0(2^+)$	1253	118	$\pi\pi$
D	$0(1^+)$	1286	40	$K\overline{K}\pi$
E?	0(?)	1420	60	$K^*\overline{K}$
f'	$0(2^+)$	1500	80	$K\overline{K}$
ϱ	$1(1^-)$	765	124	$\pi\pi$
A_1?	$1(1^+)$	1072	125	$\varrho\pi$
B?	1(?)	1220	125	$\omega\pi$
A_2	$1(2^+)$	1324	90	$\varrho\pi$
$\varkappa$?	$\frac{1}{2}(0^+)$	725	< 12	$K\pi$
K^*	$\frac{1}{2}(1^-)$	891	49	$K\pi$
C?	$\frac{1}{2}$(?)	1215	60	$K\varrho$
K^*	$\frac{1}{2}(2^+)$	1405	95	$K\pi$

jugation we mean that the reaction R and its charge conjugate reaction $C(R)$, in which each particle is replaced by its anti particle, are identical. C is known to be conserved in the strong interaction but violated in the weak. The experimental situation concerning the electromagnetic interaction is confusing, some experiments in favour of invariance, some against. The experiments consist of measuring the energy distribution of π^+ and π^- in the electromagnetic decay $\eta \to \pi^+ + \pi^- + \pi^0$. Since η and π^0 are their own antiparticles the charge conjugate reaction is $\eta \to \pi^- + \pi^+ + \pi^0$ so that if C invariance holds the π^- and π^+ should behave identically in the decay. Certainly, any violation of C that may occur is of the order of a few per cent.

The effect of P operating on a reaction R is to change the signs of all the spatial coordinates concerned in the reaction, producing the mirror imaged reaction, $P(R)$. Conservation of parity would imply that the laws of physics apply equally to the reaction and its mirror image. P is conserved only in the strong and electromagnetic interaction. The classic example of the *violation of parity* in weak interactions was the observation of the violation of right-left symmetry in the β decay of polarized ^{60}Co. Another example of the "handedness" of the weak interaction is the decay $\Lambda^0 \to p + \pi^-$ in which the mesons are preferentially emitted to one side of the plane in which the Λ was produced.

Intuitive ideas of symmetry can be restored by replacing the mirror image by its charge conjugate mirror image, i.e. P by CP. CP invariance would suggest that the true reflection is a mirror image in which all particles are replaced by their antiparticles.

Table 4. Baryon resonances

Particle (mass)	S	$I(J^P)$	Γ (MeV)	Principal decays
$N^*_{1/2}(1518)$	0	$\frac{1}{2}\left(\frac{3}{2}^-\right)$	120	$N\pi$, $N\pi\pi$
$N^*_{1/2}(1688)$	0	$\frac{1}{2}\left(\frac{5}{2}^+\right)$	100	$N\pi$
$N^*_{1/2}(2190)$	0	$\frac{1}{2}\left(\frac{7}{2}^-\right)$	200	$N\pi$, ΛK
$N^*_{1/2}(2650)$	0	$\frac{1}{2}\left(\frac{9}{2}^+\right)$	200	$N\pi$, $N\eta$
$N^*_{3/2}(1236)$ or $\Delta(1236)$	0	$\frac{3}{2}\left(\frac{3}{2}^+\right)$	120	$N\pi$
$N^*_{3/2}(1924)$	0	$\frac{3}{2}\left(\frac{7}{2}^+\right)$	170	$N\pi$, ΣK
$N^*_{3/2}(2420)$	0	$\frac{3}{2}\left(\frac{11}{2}^+\right)$	200	$N\pi$
$N^*_{3/2}(2825)$	0	$\frac{3}{2}$ (?)	260	$N\pi$
$Y^*_0(1405)$	-1	$0\left(\frac{1}{2}^-\right)$	35	$\Sigma\pi$
$Y^*_0(1520)$	-1	$0\left(\frac{3}{2}^-\right)$	16	$\Sigma\pi$, $\overline{K}N$
$Y^*_0(1815)$	-1	$0\left(\frac{5}{2}^+\right)$	50	$\overline{K}N$, $\Lambda\pi\pi$
$Y^*_1(1385)$	-1	$1\left(\frac{3}{2}^+\right)$	44	$\Lambda\pi$
$Y^*_1(1660)$	-1	$1\left(\frac{3}{2}^-\right)$	44	$\Sigma\pi$, $\Sigma\pi\pi$
$Y^*_1(1765)$	-1	$1\left(\frac{5}{2}^-\right)$	75	$\overline{K}N$, $\Lambda\pi$
$Y^*_1(2065)$	-1	$1\left(\frac{7}{2}^+\right)$	160	$\overline{K}N$
$\Xi^*(1530)$	-2	$\frac{1}{2}\left(\frac{3}{2}^+\right)$	7·5	$\Xi\pi$
$\Xi^*(1816)$	-2	$\frac{1}{2}$ (?)	16	$\Lambda\overline{K}$, $\Xi^*\pi$
$\Xi^*(1933)$	-2	$\frac{1}{2}$ (?)	140	$\Xi\pi$

The combined operation of CPT on a reaction has been proved theoretically to produce a reaction which takes place with probability equal to that of the original process. The *CPT theorem* ensures the equality of particle and antiparticle lifetimes. From this theorem, *CP invariance* would imply T invariance, i.e. that the laws of nature do not depend upon the direction of the flow of time. It was a surprise, therefore, that CP invariance was observed to be violated. The neutral K-meson is produced in two forms, K^0 and $\overline{K}^0$ of strangeness $+1$ and -1 respectively. K^0 and $\overline{K}^0$ are superpositions of states K_1^0 and K_2^0 given by

$$\left.\begin{aligned} K_1^0 &= K_0 + \overline{K}_0 \\ K_2^0 &= K_0 - \overline{K}_0 \end{aligned}\right\} \quad \text{or} \quad \left.\begin{aligned} K_0 &= K_1^0 + K_2^0 \\ \overline{K}_0 &= K_1^0 - K_2^0 \end{aligned}\right\}.$$

Since

$$CP(K_1^0) = CP(K_0 + \overline{K}_0) = (\overline{K}_0 + K_0) = K_1^0$$

and

$$CP(K_2^0) = CP(K_0 - \bar{K}_0) = \bar{K}_0 - K_0 = -K_2^0,$$

then K_1^0 and K_2^0 are unique states of $CP = +1$ and $CP = -1$ respectively. The lifetimes of K_1^0 and K_2^0 are 10^{-10} and 6×10^{-8} sec, respectively, and the principal decay modes are $\pi^+\pi^-$ and $\pi^0\pi^0$ (states of $CP = +1$) and $CP = -1$ states of $\pi\pi\pi$, respectively. In 1964 it was observed that 1 in 500 of all decays of the long-lived neutral K were in the $K_2^0 \to \pi^+ + \pi^-$ modes, a decay forbidden by CP invariance. Consequently, it was concluded that CP is violated in weak interactions. The observed neutral K decays were renamed K_S^0 and K_L^0 to correspond to the states of short and long lifetimes (but no longer of unique CP). The small amount of the CP violation is still a mystery. From the CPT theorem, a violation of CP implies also a violation of T but no direct violation of T or, indeed, violation of CP in any other type of process has been observed in spite of numerous experimental investigations. The $K^0\bar{K}^0$ complex is a system unique in particle physics and thus may have properties that enhance the CP violation to an observable level. Attempts have also been made to associate CP violation with violations (if any) of the $\Delta S = \Delta Q$ rule. The experimental situation is, to say the least, confused. The $\Delta S = \Delta Q$ rule implies that in a weak decay involving leptons the change in strangeness of the strongly interacting particle is equal to the change in charge; thus the decay $\Sigma^- \to n + e^- + \nu$ is allowed but $\Sigma^+ \to n + e^+ + \nu$ is forbidden, as observed. The rules and conservation laws that apply to reaction processes are summarized in Table 5.

Table 5. Selection rules

Conserved quantity	Interaction		
	Strong	Weak	Electro-magnetic
S	Yes	No	Yes
I	Yes	No	No
C	Yes	No	?
P	Yes	No	Yes
$T \equiv CP$	Yes	No	?
CPT	Yes	Yes	Yes

Invariance principles play an important role in the strong interaction. The observation of the charge independence of the strong interaction has led to the grouping of particles into isotopic multiplets in which charge Q (or I_3, the third component of isotopic spin) serves only to label the individual members. The total isotopic spin, I, of the multiplet members is given by the number of members, N, through the relation $N = 2I + 1$. I_3 and Q are related by $Q = (I_3 + B/2 + S/2)\,\mathbf{e}$. Isotopic multiplets occur as singlets ($I = 0$; Λ, η, ω, Y_0^*), doublets ($I = \frac{1}{2}$; N, Ξ, K), triplets ($I = 1$; π, Σ, Y_1^*) and quadruplets ($I = 3/2$; $N_{3/2}^*$). The hadrons can thus be grouped into multiplets. We can say that the strong interactions are invariant under some group of operations which transform members of the multiplet into one another. In group theory, it can be shown that invariance under a group of transformations leads to a conservation law. Conversely, it is assumed that the conservation of a particular quantum number is due to the invariance of the interaction under a group of transformations in an abstract space. The concept of charge independence can be reformulated in a group theoretical manner as an invariance of the strong interaction to the *SU*(2) *group of rotations* in isotopic spin space. SU(2) is a set of unitary, unimodular 2×2 matrices with one independent generator which commutes with all the other generators and leads to the conservation of this generator. By a suitable choice of matrices, e.g. the Pauli matrices, we can transform one member of an isotopic multiplet into another. We associate I_3 with the independent generator of the group. From group theory, the allowed dimensionalities of the representations of SU(2) are all the integers 1, 2, 3, ... and in each representation the states are distinguished uniquely by their I_3 eigenvalue. The multiplet symmetry is broken by the electromagnetic interaction and any differences (e.g. in mass) between members are attributed to this source. The mechanical properties of each member are required to be identical e.g. spin, parity, etc.

The strong interaction is invariant with respect to both isotopic spin and hypercharge, Y, and this led Gell-Mann and Ne'eman to consider the *SU*(3) *group* of 3×3 matrices, associating the two independent generators with I_3 and Y. Hypercharge is related to strangeness, S, through the relation $Y = S + B$, where baryon number B has the values 1 for baryons and 0 for mesons, respectively. Another definition of hypercharge is that it is the average charge of the multiplet. The states of the basic 3-dimensional representation of SU(3) do not correspond to any observed physical particles. If they exist then they would have fractional charge, hypercharge and baryon number. We can identify the states of the octet representation with known elementary particles. Since SU(3) contains SU(2) then it specifies the constitution of the isotopic multiplets contained in the octets. Two examples of identified octets are the long-lived baryons with $J^P = \frac{1}{2}^+$ and long-lived mesons with $J^P = 0^-$. These are shown in Figs. 1 and 2, plotted in $Y - I_3$ space.

The big success of SU(3) was the identification of the decuplet representation comprising the $J^P = {}^3/_2{}^+$ particles as shown in Fig. 3. Just as charge was responsible for mass splitting within the multiplets, so hypercharge causes splitting between multiplets. In the decuplet of Fig. 3, the observed mass splitting is 145 MeV. At the time of the first derivation of SU(3),

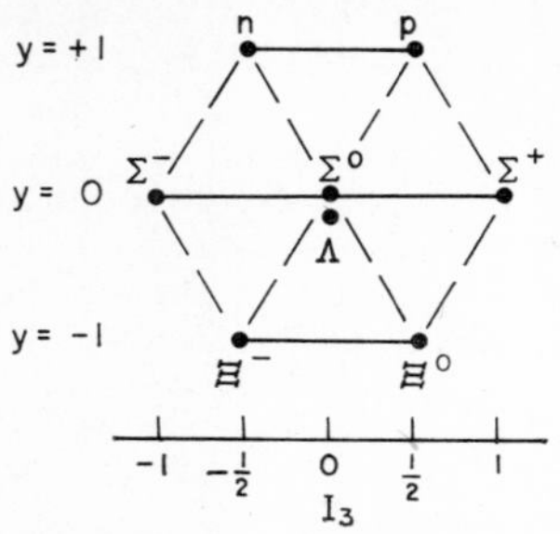

Fig. 1. The octet of long-lived baryons ($J^P = \frac{1}{2}^+$).

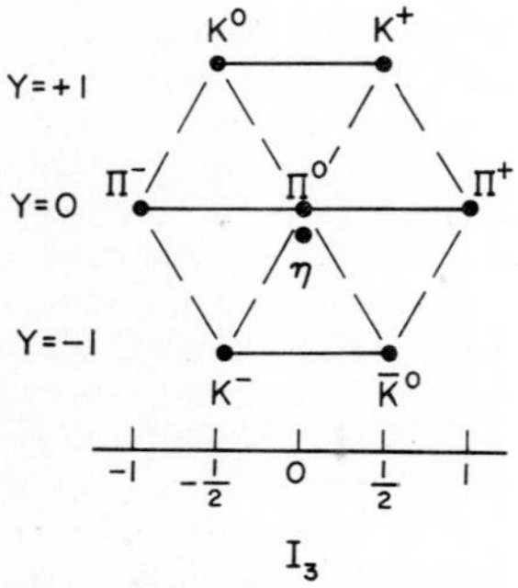

Fig. 2. The octet of long-lived mesons ($J^P = 0^-$).

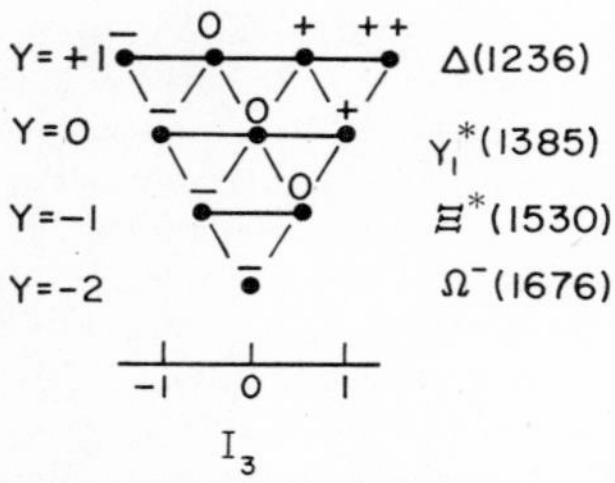

Fig. 3. The $J^P = {}^3/_2{}^+$ *baryons.*

the Ω^- had not been observed but using this value of mass splitting the properties of the missing particle were predicted to be $I = 0$, $Q = -1$, $J^P = {}^3/_2{}^+$, $Y = -2$ (or $S = -3$) and mass $= 1676$ MeV. The Ω^- was subsequently observed in early 1964 after the careful examination of several million bubble chamber photographs.

Another octet of particles can be formed from the vector mesons of $J^P = 1^-$. Nine such mesons are known but two of them have identical quantum numbers, the ω and Φ mesons with $Y = 0$, $I = 0$. These nine mesons could be accommodated into an octet and an SU(3) invariant singlet. The SU(3) particle of $Y = 0$ and $I = 0$ required in the octet to satisfy the mass relations was found to be a mixed state of ω and Φ. The vector meson octet is shown in Fig. 4. The mass relation for octet states may be written as

$$M(I, Y) = m_0 + m_1 Y + m_2(I(I+1) - \tfrac{1}{4}Y^2).$$

This equation holds for the decuplet with similar values for the coefficients m_1 and m_2. This similarity and the $\omega - \Phi$ mixing are not explained by SU(3) symmetry.

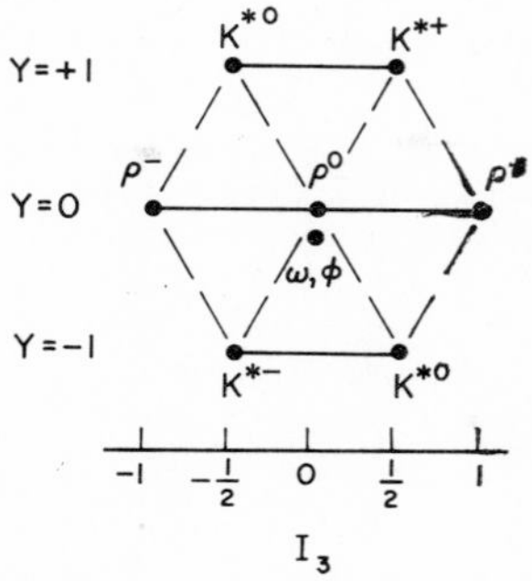

Fig. 4. The vector meson octet ($J^P = 1^-$).

The existence of singlets and octets for the mesons, and octets and decuplets for the baryons can be explained in terms of hypothetical particles, *quarks* and *antiquarks*, which are the names given to the members of the basic 3-component representation (3) of SU_3 and its complex conjugate $(\bar{3})$. The quantum numbers for the quarks are listed in Table 6. The baryon number for quarks, q, is $B = {}^1/_3$ and for antiquarks, $\bar{q}$, is $B = -{}^1/_3$. The meson states ($B = 0$) may be constructed from a $q\bar{q}$ pairing which yields a singlet and octet of mesons thus, $(3) \times (\bar{3}) = (1) + (8)$, while the baryon states ($B = 1$) require a qqq combination for which

$$(3) \times (3) \times (3) = (1) + (8) + (8) + (10).$$

This principle of quark combinations is used in the higher symmetries, SU(6) etc.

Table 6. Quantum numbers of the quarks

	B	I_3	Y	Q
q_1	$\frac{1}{3}$	$\frac{1}{2}$	$\frac{1}{3}$	$\frac{2}{3}$
q_2	$\frac{1}{3}$	$-\frac{1}{2}$	$\frac{1}{3}$	$-\frac{1}{3}$
q_3	$\frac{1}{3}$	0	$-\frac{2}{3}$	$-\frac{1}{3}$

The application of SU(3) to describe the strong interaction complications which arise in many weak interaction processes has led to the *Cabibbo theory*. The major discrepancies between the basic $(V - A)$ weak interaction theory and experiment have been removed. For example, the rates for the leptonic decays of the strange baryons had previously been a factor of 10 lower than expected. A striking prediction of the theory is that the coupling in the beta decay of the Σ^- is $(V + 0{\cdot}3A)$ and not $(V - A)$.

Spin is introduced into a higher symmetry group, *SU*(6), which includes SU(3) and SU(2) as subgroups. In SU(6) consideration of the product $q\bar{q}$ leads to a singlet of SU6 (X^0, which has $J = 0$, $I = 0$, $Y = 0$), a nonet of spin 1 particles comprising a singlet of SU(3) and an octet, and an octet of spin 0 particles, as observed experimentally. For the baryon states, a $\overline{qqq}$ combination leads to a set of states composed of the octet of spin $\frac{1}{2}$ and the decuplet of spin $^3/_2$ particles of figures 1 and 3. The presence of these baryon states in the same SU(6) group explains the fact that the same mass formula holds for these groups in SU(3). Further predictions of SU(6) are the amount of $(\omega - \Phi)$ mixing, and the ratio of the magnetic moment of the proton and neutron (experiment gives $-1{\cdot}49$ while SU(6) predicts $-^3/_2$). The difficulty with SU(6) is that it is not relativistically invariant and cannot be expected, for example, to give good relations concerning scattering processes. The many attempts to combine SU(6) with relativity have met with mixed success.

Further correlations between particles arise from the *Regge-pole theory*. The starting point of the theory is the idea of allowing the angular momentum which appears as a parameter in the radial Schrödinger equation to take not only continuous but even complex values. In the limit of high energies, Regge's method of analysis allows one to consider the scattering of the strongly interacting particles in terms of Regge trajectories. The ϱ meson and the nucleon trajectories are shown in Fig. 5. Physical particles occur every two units of angular momentum along the trajectory. The extension of the meson trajectories to the negative mass axis has made it possible to explain sharp dips in some scattering processes occurring at momentum transfers corresponding to the intercept of the trajectory with $J = 0$. The range of validity of the mass-spin relationship is being experimentally investigated. The higher mass particles are known as Regge recurrences and from SU_3 we

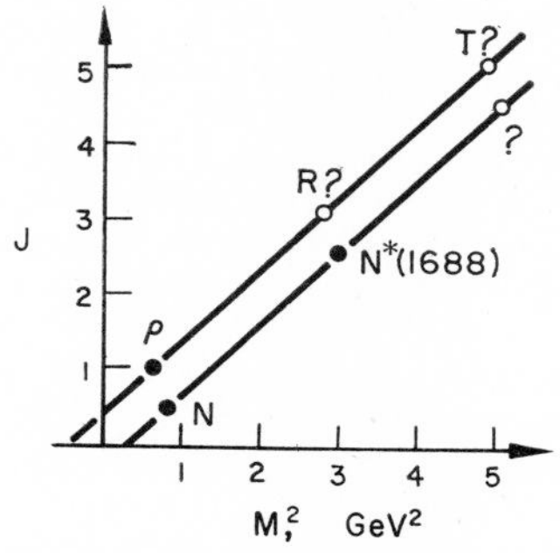

Fig. 5. Regge trajectories for ϱ-meson and nucleon.

expect to observe an octet or decuplet of particles corresponding to each of the recurrences. The ground states of possible trajectories have $J^P = 0^+, 0^-, 1^+, 1^-$ for mesons and $J^P = \frac{1}{2}^+, \frac{1}{2}^-, {^3/_2}^+, {^3/_2}^-$ for baryons. The origin of these ground states may well lie with the elusive quarks. The masses of the quarks may be so high that they cannot be produced by the present accelerators.

Bibliography

Kernan A. (1961) in *Encyclopaedic Dictionary of Physics* (J. Thewlis Ed.), **3**, 325, Oxford: Pergamon Press.

Roman P. (1961) in *Encyclopaedic Dictionary of Physics* (J. Thewlis Ed.), **3**, 123, Oxford: Pergamon Press.

T. G. Walker

EMITTANCE. Emittance is a property of a beam of particles which refers to the area occupied by the beam in a certain phase space.

The motion of a beam of particles can be considered as the motion of representative points in a six-dimensional phase space, defined by three position co-ordinates x, y, z and three momentum co-ordinates $p_x\, p_y\, p_z$. If interactions between individual particles are neglected and their equations of motion can be derived from a Hamiltonian, then Liouville's Theorem can be applied. This states that the density of points in phase space near any particular point is constant as the motion proceeds. Thus the hypervolume in six-dimensional phase space, occupied by the beam, is constant.

It often occurs in practice that the motions of particles in the three spatial co-ordinate planes are not coupled. In this case the six-dimensional hypervolume can be split up into three, two-dimensional phase planes with co-ordinates x, p_x, y, p_y and z, p_z.

The points representing the beam will then occupy a certain area in each of the two-dimensional phase planes. These areas are constants of the motion. A further simplification is possible if the motion along one axis, say the z axis, is uniform, as is the case with a beam of particles in a drift space. Now the angle a particle makes with the z axis, in the xz plane, θ_x is a measure of the ratio p_x/p_z, in the non-relativistic case. If p_z is constant then θ_x is a measure of p_x and the co-ordinates of the transverse phase planes x, p_x and y, p_y can be replaced by x, θ_x and y, θ_y which are all observable quantities.

In many practical situations encountered in the use of particle beams, e.g. particle accelerator devices, the area representing the beam in x, θ_x phase space is elliptical in shape. The area of the ellipse is πab, where a and b are the semi-axes.

The emittance, E, is defined by

$$E = ab.$$

The plot of the limiting values of θ_x, measured in milliradians, against x, in cm, is called an emittance diagram and the units of E are cm-mrad. The definition of E is usually retained even when the shape of the emittance diagram is not elliptical, i.e.

$$E = \frac{\text{area of emittance diagram}}{\pi} \text{ cm-mrad.}$$

The numerical value of E clearly depends on the particular value of p_z at which the measurement was carried out. To have a number which is independent of the axial momentum, a "normalized" emittance, E_n, is often used:

$$E_n = E\beta\gamma, \quad \text{where} \quad \beta = \frac{V_z}{c}, \quad \gamma = \frac{1}{\sqrt{(1-\beta^2)}}$$

and **c** is the velocity of light. An alternative definition, which is sometimes used for non-relativistic beams, is to multiply E by the square root of the beam energy in MeV, so that the units of E become cm-mrad-MeV$^{1/2}$.

Walsh has defined a "normal" beam as one with elliptical phase space contours. In the case of cylindrical symmetry the volume in four-dimensional phase space, V, is given by

$$V = \frac{\pi^2 E^2}{2}.$$

Van Steenbergen has used these results to define a quantity B, called the beam brightness or brilliance

$$B = \frac{I}{V} = 2I/\pi^2 E^2$$

where I is the beam current. The units which are most often used for B are mA/cm^{2}rad^2. A normalized brightness B_n can also be defined:

$$B_n = 2I/\pi^2 E_n^2.$$

The various methods of measuring emittance have been reviewed briefly by Banford.

Bibliography

Banford A. P. (1966) *The Transport of Charged Particle Beams*, London: E. and F. Spon.

Judd D. L. (1958) *Ann. Rev. Nuclear Sci.* **8**, 181.

H. Wroe

ENGEL–BREWER THEORIES OF METALS AND ALLOYS. The accepted electron theories of metals have not succeeded in predicting the types of crystal structure which the atoms of a given element will assume, and only in a few cases such as the α/β brasses is there any theory of phase-boundary limits in alloys. The Engel–Brewer theories are attempts to correlate the known crystal structures of the metallic elements with definite electron configurations of free atoms, which are either those of the ground states, or of excited states of relatively low promotion energies. These ideas were first advanced by Engel (1945) who noted that, in the Second Short Period of Mendeleev's Table, the crystal structures of the first three elements were:

Element	*Valency*	*Crystal Structure*
Na	1	body-centred cubic
Mg	2	close-packed hexagonal
Al	3	face-centred cubic

In these elements, the outer or valency electrons are in $3s$ or $3p$ states, and Engel therefore postulates that the b.c. cubic, c.p. hexagonal, and f.c. cubic metallic structures are characterized by 1, 2, and 3 (sp) electrons respectively. For Na, the ground state is ($3s$) and the electrons in the metal are regarded as being in states derived from this configuration. For Mg, the ground state is $(3s)^2$, and this configuration is unsuitable for metallic bonding because it involves electrons with paired spins. The atom is, therefore, regarded as undergoing excitation to the state ($3s$) ($3p$) with promotion energy 63 kcal/mole, and the electron states in the metal are derived from this configuration.

In the case of Al, the ground state $(3s)^2(3p)^1$ would have one unpaired electron available for metallic bonding, but excitation to the configuration $(3s)(3p)^2$ with promotion energy 83 kcal/mole makes 3 unpaired electrons per atom available for bonding, and the correspondingly increased bonding energy is regarded as more than compensating for the rather high promotion energy. The b.c. cubic, c.p. hexagonal, and f.c. cubic structures are, therefore, regarded as based on the configurations $(ns)^1$, $(ns)(np)$, and $(ns)^1(np)^2$ respectively, and the general process of crystallization from metallic vapour to solid may be looked on as involving two stages:

(1) Excitation from the ground state to the state characteristic of the crystal structure, involving promotion energy P.
(2) The condensation from excited state → solid, involving bonding energy B.

Looked on in this way, the process is favoured by a low value of P, and a high value of B.

In the transition metals, a group of 8 electrons expands into one of 18 by the building up of a sub-group of ten d electrons. It is known that the cohesive forces in the crystals of these elements involve the d-electrons. Engel assumes that all unpaired d electrons take part in the bonding process, but that they play no part in determining the type of crystal structure which is controlled solely by the numbers of s and p electrons. It is assumed that d electrons and $(s + p)$ electrons can be considered separately, and that hybridization does not take place. This is in contrast to the usual views, particularly for transition metals of the early Groups, and the idea that the *d* electrons play no share in determining the crystal structure has been strongly criticized. Brewer now considers *sd* or *spd* hybridization as taking place, but regards the factors which determine the crystal structure as relatively insensitive to the exact proportion of *d* function on which the hybrid is based, so that the structure is determined primarily by the proportions of *s* and *p* functions. No reason is given why this should be so.

Engel and Brewer accept the view that the elements of the First Transition Series must be considered by themselves, and differ from those of the Second and Third Transition Series which are closely similar. Referring to the latter, Mo in Group VIA has ground state $(4d)^5(5s)^1$. The single *s* electron, therefore, indicates a b.c. cubic structure and, since all the *d*-electrons are unpaired, there are 6 bonding electrons/atom, and a structure of great stability results. On passing to Tc, the ground state is $(4d)^5(5s)^2$, but by assuming the configuration $(4d)^5(5s)(5p)$, seven electrons/atom are made available for bonding, and the change to a c.p. hexagonal structure is understood, since a b.c. cubic structure would require a configuration $(5d)^6(5s)$ with only 5 bonding electrons per atom (two of the *d* electrons are paired because there are only five *d* states). The c.p. hexagonal structure persists on passing from Tc → Ru, and then Rh and Pd have f.c. cubic structures which, according to the Engel and Brewer interpretations, are favoured because, after Group VI a configuration $(d^x)(s)(p)^2$ has more bonding electrons than a configuration $(d^{x+1})(s)(p)$. In this way the crystal structures of the later Transition Metals of the Second and Third Series are interpreted satisfactorily. On passing back from Group VI A, Brewer shows that the interplay of promotion and bonding energies make the body-centred cubic structure stable in Groups II A to VA. In this way the persistence of the b.c. cubic structure in Nb and Ta, and at high temperatures in Zr, Hf, Y, La, Sr and Ba, is explained (in Ba, the b.c. cubic structure is the only one found). The existence of a f.c. cubic modification of Sr, stable over a wide range of temperature, and of f.c. cubic Ca in the First Long Period is a serious objection to the acceptance of the Engel postulate, because the supposed characteristic 3 valency electrons per atom could be obtained only by breaking into the very stable octets of the preceding inert gases, and this is most improbable. (It is to be noted that there is a very satisfactory electron theory of metallic Ca in terms of two valency electrons/atom.)

In the First Transition Series, the Engel–Brewer interpretation of the structures of Ca, Sc, Ti, V, and Cr is the same as that for the corresponding elements of the Second Transition Series. Difficulties and confusion arise over the elements Mn, Fe, Co, and Ni. Engel attempted to interpret the ferromagnetic properties of Fe, Co, and Ni by assuming that the $(3s)^2(3p)^6$ octet (the argon octet) was broken into, and that α-Fe had most of its atoms in a $(3p)^3(3d)^{10}(4s)^1$ configuration, and some in a $(3p)^6(3d)^7(4s)^1$ configuration, the ferromagnetism resulting from the unpaired spins of the $(3p)^3$ sub-group. Analogous structures were postulated for Co and Ni, but there are many difficulties in accepting this interpretation because, apart from the great stability of the $(3s)^2(3p)^6$ octet, it is the radial factor of the 3*d*, and not of the 3*s* function which is appropriate for ferromagnetism. In contrast to this, Brewer regards the ferromagnetism of iron as resulting from unpaired spins of 3*d* electrons—this is the conventional view—and he regards α-Fe and γ-Fe as based on the configuration $(3d)^7(4s)$ (b.c. cubic), and $(3d)^5(4s)^1(4p)^2$ (f.c. cubic) respectively. In later developments of the theory, non-integral configurations are introduced.

Both Engel and Brewer regard the sequence of f.c. cubic structures of Fe, Co, Ni, and Cu as forming a series involving the $(4s)^1(4p)^2$ configuration of outer electrons, with the number of 3*d* electrons increasing from 5 in Fe to 8 in Cu. This means that the f.c. cubic structure of Cu is based on the configuration $(3d)^8(4s)^1(4p)^2$ with 5 bonding electrons per atom (6 of the *d* electrons are paired). Engel (1945) also suggested that Cu was a superlattice of two kinds of atom with configurations $(3d)^{10}(4s)^1$ and $(3d)^8(4s)^1(4p)^2$, the arrangement being such that the bonding *d* electrons were all of paired spins. This last views is, however, not generally accepted. Ag and Au were regarded as having structures similar to copper, and based on $(ns)^1(np)^2$ configurations.

The above interpretation of the crystal structures of the metallic elements involves the assumption that the three main types of crystal structure are associated with definite numbers of $(s + p)$ electrons per atom, so that the structures may be regarded as electron compounds. The question then arises as to what variation may be tolerated on either side of the characteristic electron concentration. The quantitative developments are due mainly to Brewer (1966). In alloys of the Cu—Zn type it is known that the b.c. cubic or β-phases tend to occur at electron concentrations of about 1·5. Brewer, therefore, regards this as the upper electron concentration limit of $(s + p)$ electrons for the b.c. cubic phase, although this choice has been criticized. (In the system Li-Mg, the b.c. cubic structure extends up to a composition corresponding to an electron concentration of 1·72, as was pointed out by Engel.) In Cu-Zn and similar alloys the well-known c.p. hexagonal ε-phases occur at electron concentrations of about 1·85, and extend down to about e.c. = 1·7, and Brewer takes this as the lower electron concentration limit of $(s + p)$ electrons for the c.p. hexagonal structure, although objections to this choice have been raised. (In systems such as Cu-Ge and Ag-Sn, there are c.p. hexagonal phases whose compositions extend to electron concentrations lower than 1·5.) For the upper limit, the value 2·12 is chosen from the solubility of Al (3-valent) in Mg (2-valent, c.p.

hex.). For the f.c. cubic structure, the solubilities of In, Mg, and Zn in Al (f.c.c.) indicate that this structure can tolerate an electron concentration of $(s + p)$ electrons as low as 2·5, but the restricted solubilities of Si, Ge, and Sn in Al suggest that the electron concentrations of $(s + p)$ electrons cannot rise much above 3.

Having, thus, associated each crystal structure with a definite range of electron concentrations of $(s + p)$ electrons, it is then possible to predict the extents of solid solutions on the assumption that these are determined by electron concentration alone. If we consider the solution of Tc with configuration $(d)^5sp$ in Mo$(d)^5s$, we introduce one more electron for each atomic substitution and, since the d^5 cores are the same, we expect Mo to dissolve up to 50 atomic per cent Tc in order to reach the limiting $d^5sp^{0\cdot 5}$ configuration which corresponds to 1·5 $(s + p)$ electrons. If we consider the solution of Pt d^7sp^2 in Mo d^5s, 12 atomic per cent Pt will have to dissolve in order to reach the limiting value of $d^5sp^{0\cdot 5}$. (The reader may be confused by the fact that in some of the developments of the theory, it is assumed that the c.p. hexagonal structure of the pure metals is characterized by 1·5 electrons per atoms, instead of the original 2·0. These non-integral configurations are to be regardet as a more accurate refinement of the theory.) In this treatment the Pt atoms are regarded as giving up d-electrons until they attain the d^5 configuration. Actually it appears to be difficult to reduce the d electron concentration as low as d^5 when the solute is as far to the right in the Periodic Table as Pt.

The Engel–Brewer correlation is in agreement with the fact that intermediate phases with the c.p. hexagonal structure are found in alloys of metals of Groups VIA and VIII (e.g. Mo-Ru, Mo-Rh) because the c.p. hexagonal structure is associated with an electron concentration of $2(s + p)$ electrons which can result from a mixture of atoms with one (b.c. cube) and three (f.c. cube) $(s + p)$ electrons per atom. The values given above enable the composition limits of these intermediate c.p. hexagonal phases to be predicted on the assumption that electron concentration alone is the controlling factor.

The solubility limits predicted as described above are not always confirmed, and have usually to be corrected to allow for the effects of the different sizes of the atoms, and the different internal pressures of solvent and solute. The critical mixing temperature T_c of two metals 1 and 2, whose molal volumes do not differ too greatly, is given by the *Hildebrand* (1950, 1962) *equation*:

$$T_c = \left(\frac{V}{2R}\right)\left[\left(\frac{\Delta E_1}{V_1}\right)^{\frac{1}{2}} - \left(\frac{\Delta E_2}{V_2}\right)^{\frac{1}{2}}\right]^2$$

where $\left(\frac{\Delta E_1}{V_1}\right)^{\frac{1}{2}}$ and $\left(\frac{\Delta E_2}{V_2}\right)^{\frac{1}{2}}$ are the solubility parameters, the ΔE and V terms referring to the heat of vaporization and molal volumes respectively. Corresponding equations allow calculations of the solubilities in one another below the critical temperature. Brewer used these equations to calculate the solubility limits of body centred cubic transition metals from the left-hand side of the Periodic Table in one another, and obtained satisfactory results. When transition metals from the left-hand side of the Periodic Table are mixed with those from the right-hand side of the Periodic Table, an attractive term is added proportional to the number of d electrons which are made available for bonding by transfer from the internally paired condition of the metal to the right because of transfer to the vacant orbitals of the metal of the left. A repulsive term is added for phases with atoms of different sizes.

In alloys of transition metals from the right–hand side of the Periodic Table, with those from the middle Groups, intermediate phases with the σ-structure, the χ (α-Mn) and μ structures, and the Laves structures are sometimes found. The compositions at which these occur have been discussed in the light of the Engel-Brewer views, and electron configurations have been suggested. Brewer discusses the composition limits of the different phases in alloys of transition elements from the left-hand and middle Groups of the Periodic Table with those from the right-hand Groups. He examines the binary equilibrium diagrams where these are known, and uses the general concepts of Engel and of Brewer outlined above to predict the maximum solid solubilities of the different phases in systems where data are lacking. The results are presented in a series of diagrams from which the maximum solid solubilities can be read off both for binary and for some multi-component systems. Thus in the sequence of elements Ta-W-Re-Os-Ir a diagram may contain a line which represents compositions in the binary systems Ta-Re. The same line represents alloys of Ta with varying amounts of equiatomic mixtures of W and Os, since the electron concentration of an equiatomic mixture of W and Os is the same as that of Re. For elements of the Second and Third Transition Series to the right hand of the Periodic Table, some of the lines in the Brewer diagrams are nearly lines of constant electron concentration, and where this is not so the difference is due both to calculations of the kind referred to above, and also to comparisons with phase diagrams which have been established. Brewer's diagrams are, thus, partly the result of interpolation between known facts, and partly due to calculations which involve both the Engel concepts and the other factors which have been referred to. These diagrams represent a considerable achievement. They summarize much data and are of undoubted value to those who wish to predict maximum solubility limits in unknown alloys of the metals concerned. They are more satisfactory for the transition metals of the Second and Third Long Periods than for those of the First, and further experimental work is needed to show whether they are really satisfactory. In view of the strong criticisms of the principles underlying the supposed Engel correlation, it is right to point out that the success of the Brewer calculations of maximum solubility limits is not necessarily a proof of the Engel theories. The Brewer calculations assume that the solubility limits depend primarily on electron concentrations, and are modified by the various complicating factors referred to above. Other workers have recognized that the c.p. hexagonal, σ, χ, μ etc. phases depend on electron concentration, and have used other valency or Group Number schemes to generalize the results. It is quite probable that Brewer's conclusions would have been substantially the same if he had used other valency schemes, provided that these gave a regular increase in the number of electrons per atom on passing along a sequence such as Ta $\rightarrow$ W $\rightarrow$ Re ...

Bibliography

BREWER L. *Prediction of High Temperature Metallic Phase Diagrams*, UCRL 10701, Research Report.

BREWER L. *High Strength Materials* (Ed. V. F. Zackay), *Proceedings of the Second Berkeley International Materials Conference.*

BREWER L. (1966) in *Phase Stability in Metals and Alloys* (Eds. P. S. Rudman, J. Stringer and R. I. Jaffee). Battelle Institute Materials Science Colloquia, Geneva and Villars. Mc Graw-Hill, 1967.

ENGEL N. (1945) Kem. Maanedsblad, No. 5, 6, 8, 9, and 10 (in Danish). A brief summary is given in Powder Metallurgy Bull. (1954) **7**, 8.

HILDEBRAND J. H. and SCOTT R. L. (1950) *Solubility of Non Electrolytes*, New York: Reinhold.

HILDEBRAND J. H. and SCOTT R. L. (1962) *Regular Solutions*, New York: Prentice Hall.

HUME-ROTHERY W. (1968) *Progress in Materials Science* **13**, No. 5, Pergamon Press.

LOMER W. M. (1962) in *Encyclopaedic Dictionary of Physics* (J. Thewlis Ed.), **4**, 602, Oxford: Pergamon Press.

W. HUME-ROTHERY

ENTROPY AND INFORMATION. Information, in the theory of communication, has a definition of the same mathematical form as that of entropy in statistical mechanics. They are both logarithmic measures concerned respectively with a number of possible messages and with a number of possible states in a given system. They are not concerned with the nature of the system measured: information is not concerned with the meaning of messages, and entropy is not concerned with the physical nature of the system except insofar as that affects the number of possible states of the system. Since information is a property of messages or signals and entropy a property of physical systems, the two measures are not always comparable. In certain circumstances, however, they may be applied to the same system, e.g. when information is stored in a small system. The question of their interrelation then becomes important.

Consider first the characteristics of entropy in classical thermodynamics. The entropy of the thermodynamic system is a function of the state of the system, as defined by the observed macroscopic properties (pressure, volume, magnetization, temperature, composition, etc. etc.). Since, by the second law of thermodynamics, the entropy of a system must increase in any natural process, and since this implies one direction of the process rather than another, entropy provides an indication of the direction of natural processes: a process will go in that direction for which the entropy of the system increases, and come to equilibrium when the entropy reaches a maximum. In consequence, entropy plays a fundamental part in thermodynamics.

In classical thermodynamics no detailed mechanisms are postulated, the theory providing only the relations between observable properties, with entropy playing a part in the formulation of those relations. Statistical mechanics provides an explanatory framework in which entropy is given a probabilistic meaning. In so doing it makes possible the comparison between entropy and information.

The statistical definition of entropy is derived from the recognition that systems which have the same observable properties may be different in their microscopic complexions. In the simplest form of definition, entropy is given by $S = \boldsymbol{k} \ln P$, where P is the number of (equally probable) microscopic complexions of the system, and $\boldsymbol{k}$ is Boltzmann's constant. If the complexions are not equally probable, the definition is generalized to the form $S = -\boldsymbol{k} \Sigma p_i \ln p_i;\ \Sigma_i p_i = 1$, where p_i is the probability that the system is in the complexion i, for the given observable properties. By these formulae, the statistical entropy of the system is associated with the number of different microscopic complexions that are possible in that they agree with the observed or postulated macroscopic properties of the system.

In communication theory, information is defined by equations of the same form as those given above for entropy, but the interpretation is at first rather different. The essential problem of communication, according to Shannon, is that of reproducing at one point a message selected at another point. For the designer of a communication system, it is the degree of selection which is important. The meaning of the message, or whether indeed it has any meaning, is of no relevance to its communication, meaning being relevant only to the sender and the recipient. Thus the amount of information is defined so as to take no account of meaning, only of degree of selection. A suitable measure of information would therefore be a monotonic function of the number of messages in the set from which the particular message was selected. The logarithmic function is used because it is mathematically simple and because it makes of information a linear quantity, proportional to such features as duration of message, number of storage elements, number of communication channels, etc.

For these reasons, the information conveyed by a

message is defined as $I = K \ln P$, where P is the number of messages in the set of (equally probable) messages from which that one has been selected for transmission, and K is a positive constant. If $K = 1$, then I is given in natural units. If also the logarithm is to the base 2, then the units are binary digits, or bits.

If a sequence of m symbols be transmitted, then the number of possible messages is given by the number of permutations which can be made of the symbols. With an alphabet of n symbols and with the ith symbol occurring m_i times in the sequence, the number of permutations is $P = m!/(m_1!m_2! \dots m_i! \dots m_n!)$. The amount of information transmitted by a message selected from this set would then be $I = K \ln P = -K \sum_i m_i \ln (m_i/n)$. For long messages from a source with stationary statistical properties, we may put $p_i = m_i/n$. The average information per symbol for messages from that source is then given by

$$I_{\text{per symbol}} = -K \sum_i p_i \ln p_i .$$

Shannon remarked that this expression for information had the same form as that given for entropy in statistical mechanics. He therefore regarded I per symbol (H in his notation) as the entropy of the set of probabilities $p_1 \dots p_i \dots p_n$. However, he also regarded I as a measure of the amount of uncertainty which would be removed by reception of the message. It is now not unusual to find entropy and uncertainty used as synonyms, and this double usage can lead to confusion about the relation between information and entropy. This confusion is greater in those works in which the uncertainty is not restricted to the eventual recipient of the message but is allowed to apply to the situation as a whole.

Ideas of the relation between entropy and information have diverged, depending on whether more emphasis is placed on the set of probabilities or on the uncertainty.

Consider the first of these, in which I is regarded as a measure of the set of probabilities $p_1 \dots p_i \dots p_n$. Shannon defines the "entropy per symbol" of an information source, but "entropy per symbol", being a property of messages, cannot be directly compared with the entropy of a physical system. If we consider instead a system in which information is to be stored, then the comparison becomes possible. If there are P possible states of the system (with equal probability of being selected), then the amount of information stored is $I = K \ln P$. When the states are not equally probable, the formula becomes $I = -K \sum_i p_i \ln p_i$, as before. Since in these formulae I is the amount of information stored in a physical system, it is directly connected with the physical properties of the system. Brillouin made the connexion explicit by introducing the idea of bound information.

Bound information occurs when the possible stored messages can be interpreted as complexions of a physical system. Then bound information is given by $I_b = K \ln P$ and entropy by $S = \boldsymbol{k} \ln P$. P has the same meaning in the two equations, and K can be put equal to Boltzmann's k, so that bound information is given in the same units as entropy. These units involve both energy and temperature, in the dimensions of $\boldsymbol{k}$, and this would mean that information is given unlikely dimensions. However, if temperature is measured in entropy units, then both entropy and bound information will be dimensionless. Bound information is thus closely connected with entropy.

These ideas of information bound into a physical system have been applied particularly to the organized structures found in biology, e.g. to the information content of D.N.A., and, less directly physical, to the information content of species distributions in living communities.

The negentropy theory of information. The idea of bound information and its application to physical problems has been most developed by Brillouin. He has gone further than this, however, by putting forward the negentropy theory of information, in which information and entropy are opposed.

Entropy is commonly regarded as a measure of the amount of disorder in a physical system. It is also said that entropy measures the lack of information about the microscopic structure of the system. For these reasons, Brillouin and other negentropy theorists suppose that information actually represents not entropy but negative entropy (negentropy). He therefore reverses the sign of information, in those problems which involve both information and entropy. Bound information then appears as a negative term in the total entropy of the physical system.

The idea of negentropy leads to a "generalization" of the second law of thermodynamics, allowing for interchange between information and entropy. The second law (Carnot's Principle) can be written as $\Delta S \geqq 0$, for any natural process in a closed system. The "generalized Carnot principle", then, states that in any such process $\Delta(S - I_b) \geqq 0$. This principle would allow the entropy to be reduced when information is obtained, without contravening the second law (which requires the total entropy to be increased) by postulating that the increase may be in $-I_b$ as well as in S.

The "generalized Carnot principle" has been applied to a number of physical problems in which information is involved, leading to interesting and important results. For example, it has been applied to Maxwell's demon, and to the efficiency of an observation. The application to Maxwell's demon leads to the conclusion that the demon cannot violate the second law. An observer, whether physicist or demon, requires sources of negentropy, since every operation is made at the expense of the negentropy of the surroundings; and the price paid in negentropy must always be at least equal to the amount of information received.

Brillouin goes on to define the efficiency η of an experimental method of observation as the ratio of the information obtained to the cost in negentropy

$|\Delta N|$, which is the entropy increase ΔS accompanying the observation. Thus:

$$\eta = \Delta I/|\Delta N| = \Delta I/\Delta S, \quad \Delta N = -\Delta S.$$

These results are not yet subject to experimental test. Their validity depends on the validity of the supposed opposition between information and entropy, and the consequent reversal of sign when comparing them. Here it should be noted that only in the neg-entropy theory are information and entropy given opposite signs. It remains questionable whether the reversal is either necessary or justified, in view of the identity between the definitions of bound information and (positive) entropy. It seems that the only reason for Brillouin's reversal of sign is the supposition that information must be connected with order and entropy with disorder.

Entropy and information as uncertainty. If we focus attention on the subjective aspect of probability (i.e. in which the probability of an event is that ascribed to it by an observer) then the receipt of a message will make it possible for the recipient to adjust his estimate of the probability of various events. The message carries information which reduces the uncertainty of the recipient with regard to the events. This use of information theory has been developed as a statistical method, comparable to analysis of variance, by Garner and McGill; in relation to thermodynamics it has been developed by Jaynes and Tribus into a form of statistical mechanics based on the idea of information as a means of reducing uncertainty.

In this approach, the methods of statistical mechanics are regarded as a special case of a general technique for guarding against bias. If we measure the properties of a physical system, the results of the measurement may be taken as a message concerning the system. This message will not, however, give us complete information about the state of the system, and when we interpret the message we must do it in such a way as to avoid accidentally including information which we do not have. Now, the central problem of statistical mechanics is to find the probability distribution for the possible complexions of the system. To find this distribution without accidentally including information based on unsupported supposition, we should assign the probabilities so as to maximize our uncertainty. This assignment of probabilities would be the one required by statistical mechanics.

The average amount of information in the message is given by $S = -K \sum_i p_i \ln p_i$. The least prejudiced or biased assignment of the probabilities is that which maximizes S subject to the given information. (S is defined as amount of information; it is also regarded as entropy, and as uncertainty.) By maximizing the entropy, therefore, we assume least about the physical system.

Entropy is maximized by the usual method of Lagrangian multipliers:

Let $\langle g_r \rangle$ be a known average of a function $g_r(X_i)$, where X is a property that serves to identify a state and X_i is the value of X associated with complexion i. Then express the known averages in terms of the averages derived from probabilities thus:

$$\langle g_r \rangle = \sum_i p_i g_r(X_i).$$

With these r equations and the additional equation $\Sigma p_i = 1$, maximize the entropy. The resulting probability distribution will be given by

$$p_i = \exp(-\lambda_0 - \lambda_1 g_1(X_i) - \lambda_2 g_2(X_i) - \ldots)$$

with

$$\lambda_0 = \ln \sum_i \exp(-\lambda_1 g_1(X_i) - \lambda_2 g_2(X_i) - \ldots)$$

The expectation values are given by

$$\langle g_r \rangle = -\partial\lambda_0/\partial\lambda_r$$

and the entropy by

$$S_{\max} = K\lambda_0 + K \sum_r \lambda_r \langle g_r \rangle.$$

When these results are applied, they lead to the usual results of statistical mechanics. For a simple example, if two systems are allowed to interact we will know only their common energy, not how the energy is shared between them; thus the two systems must be considered together because we have not sufficient information to consider them separately. The Jaynes' formalism will then give probability distributions for the two systems in which they have a common Lagrangian multiplier. Thus we arrive at the idea of thermal equilibrium, with temperature defined by the common multiplier.

The Jaynes-Tribus approach to statistical mechanics which is sketched above is in effect a detailed application of the suggestion made by Boltzmann that entropy is a measure of missing information. By taking the mathematical ideas of information as more fundamental than those of thermodynamics, a unified treatment is given to "thermostatics" and "thermodynamics". In addition to the use of the idea of information, however, the method involves a subjective notion of probability which may or may not be essential to the interrelation of information and entropy.

Bibliography

Brillouin L. (1956) *Science and Information Theory*, New York: Academic Press.

Chambadal P. (1963) *Evolution et applications du concept d'entropie*, Paris: Dunod.

Messer C. E. (1961) in *Encyclopaedic Dictionary of Physics* (J. Thewlis Ed.), **2**, 876, Oxford: Pergamon Press.

Quastler H. (1961) in *Encyclopaedic Dictionary of Physics* (J. Thewlis Ed.), **3**, 835, Oxford: Pergamon Press.

Shannon C. E. and Weaver W. (1949) *The Mathematical Theory of Communication, Urbana:* University of Illinois Press.

Stephenson H. P. (1962) in *Encyclopaedic Dictionary of Physics* (J. Thewlis Ed.), **7**, 290, Oxford: Pergamon Press.

TRIBUS M. (1961) *Thermostatics and Thermodynamics: an introduction to energy, information and states of matter with engineering applications*, Princeton: Van Nostrand.

YOCKEY H. P. (Ed.) (1958) *Symposium on Information Theory in Biology*, Oxford: Pergamon Press.

J. A. WILSON

ENTROPY AND LOW TEMPERATURES. A low temperature, T, is characterized by the fact that at this temperature the typical thermal energy $\boldsymbol{k}T$ is small compared to the characteristic energy difference, ΔE, under consideration. But another important fact is that as the absolute zero is approached, those aspects of a system that are in internal thermodynamic equilibrium tend to a state of complete order associated with a vanishing of their contributions to the entropy of the system. This follows from the third law of thermodynamics.

For this reason the absolute zero is of outstanding importance as a reference point for measurements of entropy. Although it can never be reached, it is possible in many cases to make reliable extrapolations of the entropy from the lowest accessible temperatures to the absolute zero.

Many of the remarkable phenomena observed at low temperatures such as superfluidity, superconductivity or the onset of magnetic ordering, may be thought of as manifestations of this tendency towards the ordered state. As illustrations we shall consider the entropy of (1) a material having electronic or nuclear magnetic moments; (2) solid and liquid ^{4}He; (3) solid and liquid ^{3}He; (4) some ^{3}He—^{4}He mixtures.

1. A paramagnetic solid. As the simplest example, take a paramagnetic solid in which each ion carries a spin $\frac{1}{2}$; thus at high temperatures each spin can take up one of two equally probable orientations. With N such spins the total number of possibilities is 2^N and so the entropy associated with the spin system at high temperatures is $k \ln 2^N = \boldsymbol{R} \ln 2$ per mole. If the spin were S, the corresponding entropy would be $\boldsymbol{R} \ln (2S + 1)$ per mole.

At low temperatures, according to the third law of thermodynamics, this entropy has to tend to zero provided that the spin system remains in thermodynamic equilibrium. The interactions between the spins (suppose this is characterized by an energy ε) will begin to be manifest when $\boldsymbol{k}T \sim \varepsilon$. Below this temperature some ordering process is to be expected (e.g. a ferromagnetic or antiferromagnetic transition) by which the spin contribution to the entropy is ultimately reduced to zero. Figure 1 illustrates how the entropy of a particular paramagnetic salt (iron ammonium alum) depends on temperature.

Similar considerations apply to the contribution to the entropy from the nuclear spins.

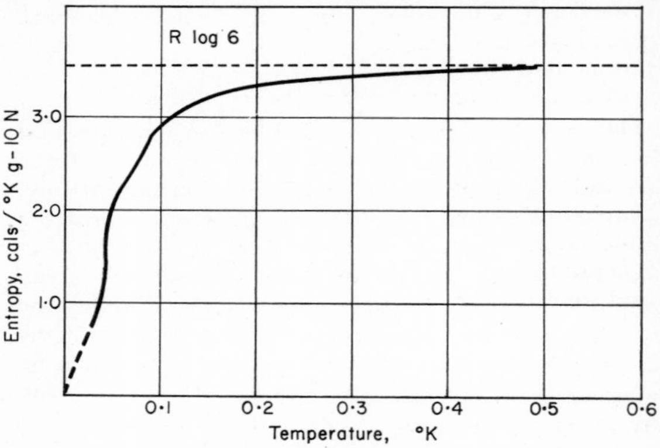

Fig. 1. The entropy of iron ammonium alum (spin $^5/_2$) as a function of temperature at low temperatures.

2. ^{4}He. In normal substances there is a triple point at which solid, liquid and gas are all in equilibrium and below which the liquid no longer exists. Both ^{4}He and ^{3}He are unusual in that the liquid form can exist in equilibrium with the solid right down to the lowest temperatures.

Because of this it is possible in the helium isotopes to observe how the solid and liquid take up ordered configurations which allow their entropy to vanish as $T \to 0$. Solid ^{4}He is more or less normal; as the melting temperature and pressure go down, the characteristic temperature of the lattice also goes down but the entropy at the melting point associated with the lattice vibrations nevertheless falls off towards zero (see Fig. 2).

In liquid ^{4}He, on the other hand, there occurs the λ-phenomenon in which at 2·17°K (under the equilibrium vapour pressure) the specific heat of the liquid undergoes a λ-shaped anomaly. ^{4}He atoms obey Bose–Einstein statistics. An ideal gas obeying these statistics has an anomaly in its specific heat (C_v) at a temperature T_c given by: $T_c = \boldsymbol{h}^2(N/2{\cdot}612V)^{\frac{2}{3}}/2\pi m\boldsymbol{k}$. (Here N is the number of particles of mass m in vo-

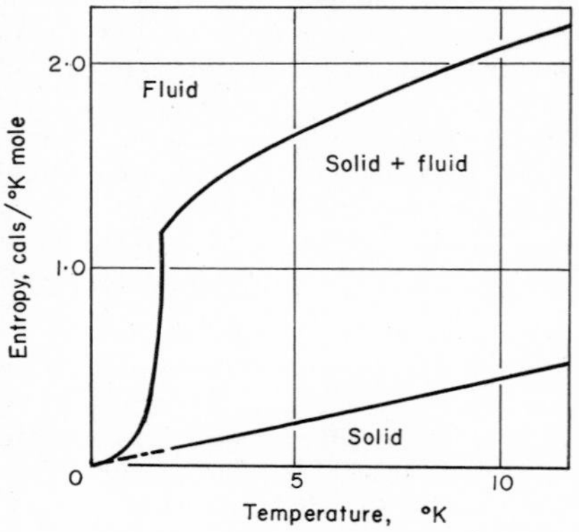

Fig. 2. The entropy of solid and liquid ^{4}He in equilibrium at low temperatures.

lume V; $\boldsymbol{h}$ and $\boldsymbol{k}$ are Planck's constant and Boltzmann's constant respectively.) The anomaly in liquid ^{4}He is associated with this effect (see Thewlis 1961).

Because of the large specific heat at the λ-point there is a sharp fall in entropy around this temperature so that the entropy of liquid ^{4}He falls off towards zero as indicated in Fig. 2. According to the Clausius–Clapeyron equation, the slope of the melting curve (melting pressure as a function of temperature) is related to the entropy difference ΔS between liquid and solid by the expression

$$\frac{dp}{dT} = \frac{\Delta S}{\Delta V}.$$

Here ΔV is the volume difference between liquid and solid which in this example remains fairly constant at the lowest temperatures. Thus, as ΔS tends towards zero, the slope of the melting curve must vanish, and indeed experiments show that in ^{4}He the melting curve does become very flat below the λ-point.

3. ^{3}He. In Fig. 3 is shown the entropy of solid and liquid ^{3}He in equilibrium at low temperatures. In ^{3}He the nuclear spin contribution (absent in ^{4}He) is all important. Liquid ^{3}He has an entropy which falls towards zero in a manner somewhat like that of an ideal Fermi–Dirac gas with a degeneracy temperature of about 0·5°K. In the solid on the other hand (because it is essentially a localized assembly) the entropy associated with the nuclear spins remains at the value $\boldsymbol{R} \ln 2$ corresponding to full randomness down to quite low temperatures ($\sim 10^{-3}$ °K). (At low enough temperatures, the mutual interaction between the atoms must, of course, produce some form of ordering in the spin system but this has not so far been directly observed.)

As can be seen in Fig. 3, the entropy of the liquid falls below that of the solid at temperature below 0·3°K. As with ^{4}He, ΔV the volume difference between liquid and solid, remains fairly constant at low temperatures so that according to equation 1, the slope of the melting curve must vanish when $\Delta S = 0$ and become negative below this temperature. This is confirmed experimentally: the melting curve has a minimum at 0·3°K and this has the interesting consequence that if the liquid is heated at a suitable constant pressure it can be made to solidify.

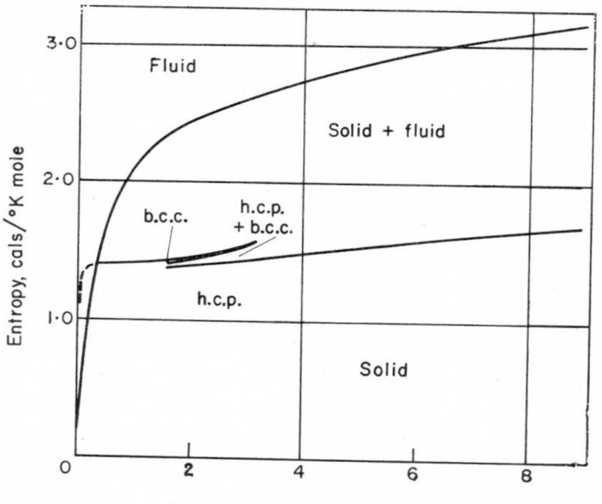

Fig. 3. The entropy of solid and liquid ^{3}He in equilibrium at low temperatures.

4. ^{3}He—^{4}He mixtures. To conform to the third law of thermodynamics, the entropy of liquid and solid mixtures of ^{3}He and ^{4}He must in some way tend to zero as T tends to zero. Both the solid and liquid mixtures show phase separation; in this, two separate phases are formed, one rich in ^{3}He and the other rich in ^{4}He. This occurs at a sufficiently low temperature, the precise temperature depending on the concentration of the mixture. One way of achieving zero entropy in these mixtures could thus be that this phase separation should become more and more complete as $T \to 0$ with each phase becoming richer in one of the isotopes; ultimately the two pure isotopes would separate out completely, each with zero entropy as in the separate, pure materials. In the solid phase, this seems to occur but in the liquid it is found that, when the concentration of liquid ^{3}He in ^{4}He falls below a few per cent, no further phase separation occurs and the ^{3}He molecules in the ^{4}He liquid behave very like a true ideal Fermi–Dirac gas of volume equal to that of the whole liquid mixture. The entropy of the ^{4}He tends to zero essentially as in the pure liquid ^{4}He while that of the ^{3}He goes to zero linearly in T like an ideal Fermi gas.

The production of low temperatures. The general principles of producing low temperatures are easily understood in terms of the behaviour of the entropy of systems at low temperatures. Figures 4, 5 and 6 il-

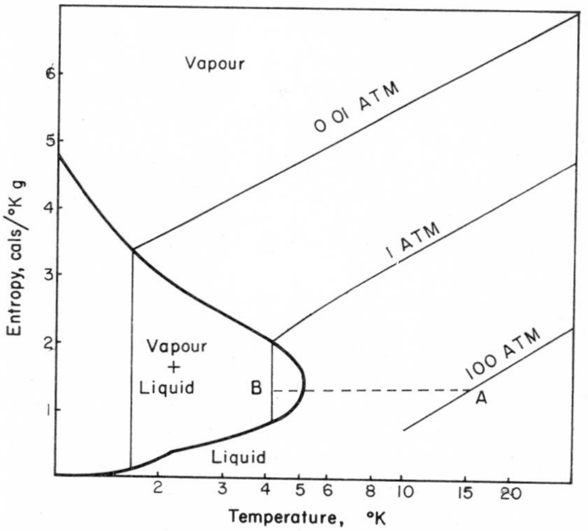

Fig. 4. The entropy of liquid helium and its vapour at low temperatures under different pressures. (The temperature scale here is logarithmic.)

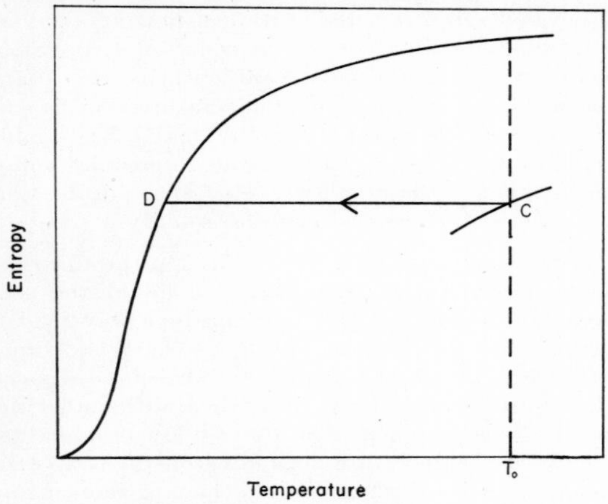

Fig. 5. The entropy of a paramagnetic salt (schematic) at two different magnetic fields. T_0 is the starting temperature. The path CD illustrates how the temperature of the salt changes under adiabatic, reversible demagnetization.

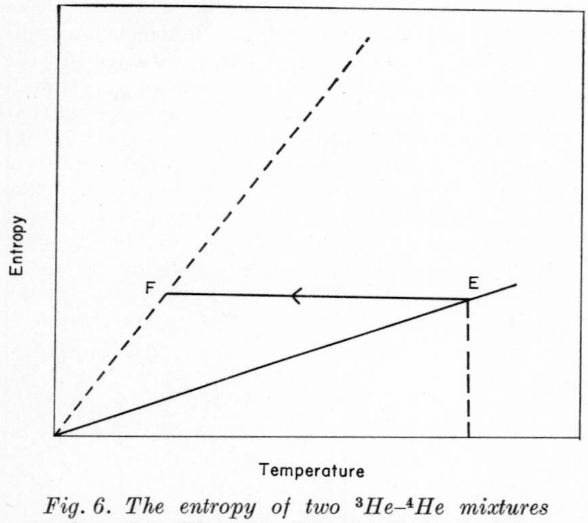

Fig. 6. The entropy of two ^{3}He–^{4}He mixtures (schematic). The state E corresponds to almost pure ^{3}He; the state F corresponds to a very dilute solution of ^{3}He in ^{4}He.

lustrate how (1) a gas–liquid system, (2) a paramagnetic salt and (3) a ^{3}He—^{4}He mixture may be used to produce low temperatures.

(*1*) *Gas-liquid system.* In this method the entr opy of the gas (e.g. helium at, say, 12°K) is reduced by pressure (here 100 atmospheres). The gas is then thermally isolated and allowed to expand back to atmospheric pressure. If this change is both adiabatic and reversible, the change is at constant entropy so that the representative point system (see Fig. 4) moves along the line AB. At B some of the gas is liquefied. This is the principle on which the Simon helium liquefier works.

(*2*) *Paramagnetic salt.* In this method, the entropy of a suitable paramagnetic salt (whose spins are still randomly orientated in zero field at the starting temperature, say 1°K) is reduced by applying a magnetic field of about 10,000 gauss. The salt is then thermally isolated and the field is reduced to zero. Again this is an adiabatic, reversible (and thus isentropic) change so that the representative point of the system (see Fig. 5) moves along the path CD back to the zero field curve. The temperature of the salt thus falls; the final temperature, which depends on the salt used, is typically about 10^{-2} °K.

(*3*) *^{3}He—^{4}He mixtures.* In this method, the starting temperature must be below that at which phase separation in the liquid mixtures occurs. ^{3}He is allowed to expand adiabatically and reversibly from the ^{3}He-rich phase in which the entropy is represented by the point E (see Fig. 6) into the ^{4}He-rich mixture where the ^{3}He behaves like an ideal gas occupying a volume equal to that of the ^{4}He. The ^{4}He makes a negligible contribution to the entropy at these temperatures. The entropy of the system is then represented by the point F. This forms the basis of the ^{3}He dilution refrigerator, although for practical reasons changes at constant entropy are not convenient.

See also: Thermodynamics, the Third Law; Entropy; Helium II; Paramagnetism at low temperatures.

Bibliography

MESSER C. E. (1961) in *Encyclopaedic Dictionary of Physics* (J. Thewlis Ed.), **2**, 876, Oxford: Pergamon Press.

PARK J. G. (1962) in *Encyclopaedic Dictionary of Physics* (J. Thewlis Ed.), **5**, 296, Oxford: Pergamon Press.

THEWLIS J. (Ed.) (1961) *Encyclopaedic Dictionary of Physics*, **3**, 674, Oxford: Pergamon Press.

WALKER P. A. (1962) in *Encyclopaedic Dictionary of Physics* (J. Thewlis Ed.), **7**, 293, Oxford: Pergamon Press.

WILKS J. (1961) in *Third Low of Thermodynamics*, Oxford: The University Press.

WILKS J. (1967) in *The Properties of Liquid and Solid Helium*, Oxford: Clarendon Press.

J. S. DUGDALE

ENVIRONMENTAL MONITORING. *Introduction.* Establishments in which radioactive materials are used are designed and operated in such a way that radioactivity is confined within their perimeters to the greatest extent practicable. A limited release of activity to the environment is, however, often unavoidable, especially in those cases where the work of the establishment results in the production of radioactive wastes. Wastes of very high specific activity, such as those produced in the first stages of chemical processing of irradiated reactor fuel elements, are stored indefinitely under carefully controlled conditions. It is generally accepted that long-term storage of low activity wastes, which may be produced in large volumes, is impracticable and that regulated disposal to the environment can be carried out with adequate standards of safety. Such releases of activity result in actual or potential irradiation of man, and measurements of radioactivity or radiation in the environment with the object of ascertaining the extent of human exposure to radiation is termed environmental monitoring.

Exposure pathways. Gaseous radioactive wastes may be discharged to atmosphere through ventilation extracts or through stacks. Liquid wastes, after chemical treatment when necessary, may be discharged into sewers, streams, rivers or direct into coastal waters. Solid wastes may be either buried on land or dumped in the oceans in containers specially designed to ensure that the wastes reach the ocean floor.

The ways in which the radioactivity so dispersed can cause human irradiation are numerous and complex, being dependent upon the environment into which the release is made, the use made of that environment by man and the identity and physico-chemical form of the radionuclides released. In Fig.1 and 2 some possible pathways are illustrated in simplified form for radioactivity released into ground and surface waters.

Pre-operational studies. Annual radiation dose limits for members of the public have been recommended by the International Commission on Radiological Protection, and before any release of activity to the en-

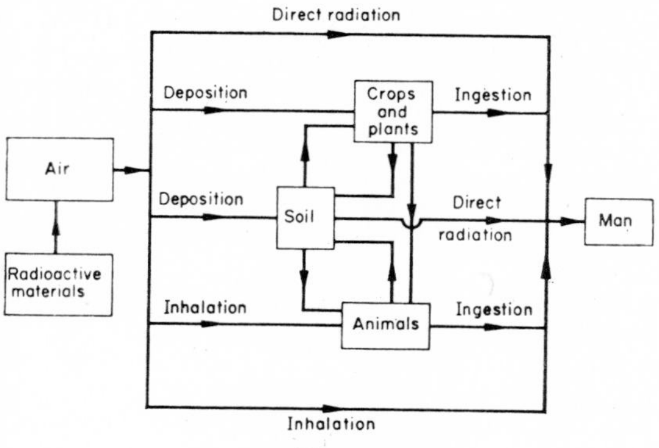

Fig. 1.

vironment can be considered acceptable it must be demonstrated that it will not result in these limits being exceeded. It is rarely possible to make direct measurements of the doses actually received by members of the public and they must be inferred from a study of the environment and a knowledge of the nature and quantity of the radioactivity released.

It is clearly desirable that such studies should, as far as possible, be undertaken before routine releases of activity begin, so that estimates can be made of

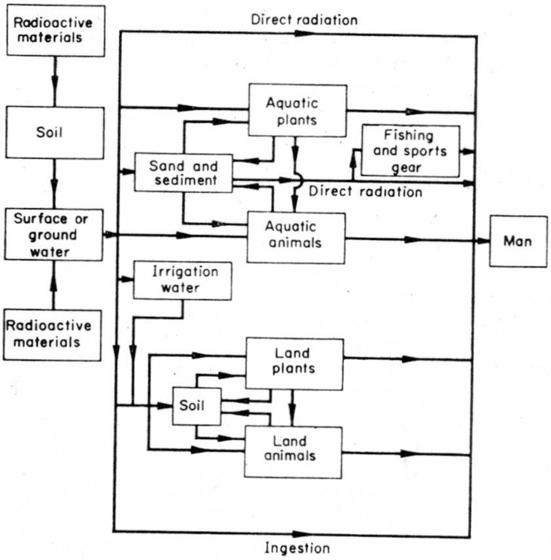

Fig. 2.

the relationship between release rate and the radiation exposure of persons affected. For any given location and mode of release, the first step is to identify the radionuclides which will be discharged. This information will be obtained from design data and proposed operating procedures. The next stage is the evaluation of the behaviour of these nuclides in the environment concerned and to identify the particular pathways which can lead to the return of this activity to man. Finally it is necessary to study the use made of the environment by man.

To reduce such studies to manageable proportions it is usually necessary to eliminate from the wide range of possible nuclides and exposure pathways those which experience and judgment show to have little significance. Although no general rules can be given it is often the case, for example, that nuclides of very short radioactive half-life, low radio-toxicity or small abundance in the wastes discharged can be dismissed after little consideration. This situation commonly arises in the case of nuclear reactors where a wide range of neutron activation products may be discharged in gaseous or liquid effluents and in reactor fuel re-processing establishments where mixed fission products may be similarly discharged. In the same way, some knowledge of the environment and the habits of the population often permits a substantial reduction in the number of exposure pathways requiring detailed study.

The remaining nuclides and pathways must all be carefully evaluated, laboratory and field studies being undertaken to obtain data not already available. If discharges of activity are to be made into coastal waters, for example, and it has been established that the consumption of locally caught fish may be an important exposure pathway, it will be necessary os determine first the dilution of activity as it disperste from the discharge point and the extent to which the diluted activity is re-concentrated by the various edible species of fish either directly from sea-water or as a result of feeding on contaminated lower organisms. An estimate must then be made of the amount of fish caught in the area concerned, the extent to which it is marketed along with fish imported from other areas, and the amount of fish consumed by the local population. In the case of activity released to the atmosphere, similar investigations may be necessary in relation to agricultural produce contaminated by deposited activity. These are examples only, but serve to illustrate the extensive study often required, embracing both the physical and biological environment and the occupational, dietary or other habits of populations.

When such investigations have been completed it will be possible to identify a very small number of pathways and nuclides which result in the largest contribution to the total radiation dose to man, and also to identify the group of persons who will receive the highest individual doses. These may then be referred to as the critical pathways, the critical nuclides and the critical group.

A wide range of critical nuclides and pathways have been identified in the study of individual nuclear establishments, marked differences sometimes being found between virtually identical establishments located in different environments. Iodine-131 in cow's milk has frequently been found critical when volatile fresh fission products are discharged into the atmosphere in an agricultural environment. The discharge of mixed neutron activation or fission products into estuaries or coastal waters can lead to a wide range of critical nuclides and exposure pathways; at three locations in the United Kingdom, for example, zinc-65 in oysters, ruthenium-106 in edible seaweed and zirconium-95/niobium-95 deposited on fishing nets have been identified as critical.

Design of environmental monitoring programmes. If the studies described above show that the rate at which activity is to be released will result in only trivial radiation doses to man, it may be concluded that no environmental monitoring is necessary, except perhaps for a few measurements to check the accuracy of the predictions. If the doses to be expected are not trivial then a programme will be required to assess the extent of human exposure. When the discharges begin, extensive measurements may be necessary to check that critical pathways and nuclides have been correctly evaluated, but the routine monitoring eventually established should be confined to critical pathways and nuclides so that the information obtained will be both positive and useful.

It is necessary to select from each critical pathway the material to be sampled and assayed. This should be as near the end-point as possible. In the case of food chains, for example, the measurement of activity in the commodity actually consumed by man will enable an estimate of radiation dose to be made which will be less liable to error than measurements on some other material appearing earlier in the chain.

Dose limits for members of the public are expressed in terms of the dose received over a full year without any restriction on the rate of dose accumulation over shorter time intervals. When the nuclides of interest have long radioactive half-lived it is therefore sensible to sample at whatever frequency is required to ensure representativeness, but to bulk the samples collected over a period for assay at relatively long intervals (e.g. quarterly or even annually). When the nuclides concerned are short-lived this method is not feasible and the frequency of both sampling and assay must be sufficient to ensure that short-term variations in the activity of critical materials will be detected.

The total number of samples of a given material collected at any one time must also be large enough to guarantee representativeness, but again the individual samples may be judiciously bulked prior to assay to obtain results which are typical for the critical group.

Interpretation of environmental monitoring results. The results obtained from environmental monitoring programmes are generally in terms of radioactivity per unit mass or volume of the material sampled, or the external radiation dose-rate associated with a particular item or location. The basic protection standards for persons, however, are in terms of annual radiation dose limits. If the monitoring results are to be meaningful, therefore, they must be related quantitatively to these dose limits. This can be achieved by the use of secondary protection standards which are termed *derived working limits* (D.W.L.s).

These D.W.L.s may be calculated for each critical material from the data accumulated during the pre-operational studies. In the case of a foodstuff containing a radionuclide x, for example, assume that the average daily consumption of the foodstuff by a member of the critical group is m (grammes/day). From the publications of I.C.R.P. it is possible to calculate the daily intake i_x(μCi/day) which, if maintained over a year, will result in a dose to the critical body organ equal to the appropriate recommended annual dose limit. The D.W.L. for this nuclide may then be expressed as i_x/m (μCi/g). Similarly, in the case of the external radiation dose-rate at a particular location, assume that a member of the critical group may be present at that location for h hours each year. Then if the appropriate annual dose limit is D rads, the D.W.L. for external radiation at the location concerned is D/h (rads/h).

Environmental monitoring results, when averaged over a year, may be expressed as a fraction of the D.W.L.s calculated in this way and a direct indication obtained of the extent to which environmental contamination is resulting in human irradiation.

A complicating factor in the interpretation of environmental monitoring results is that radioactivity detected in samples may originate from sources other than releases of activity from the establishment under consideration. Natural radioactivity due, for example, to potassium-40 or radium and its daughter products may be present. Difficulties due to natural radioactivity are best avoided by designing programmes so that assays are specific to the nuclides being released and not simply determinations of gross alpha or gross beta activity, which are in any case of little value in the interpretation of monitoring results in terms of radiation dose.

A more serious problem may be presented by the presence of radioactivity due to world-wide fall-out from nuclear weapons testing. The fission product nuclides involved are often those which feature prominently in the environmental monitoring programmes themselves. Allowance for fall-out activity can best be made by comparing the results of a monitoring programme with those of national fall-out surveys, although local variations in fall-out activity sometimes make it difficult to draw conclusions with much precision.

Bibliography

International Commission on Radiological Protection: ICRP Publication 2 (1959) Report of Committee II on Permissible Dose for Internal Radiation; ICRP Publication 7 (1966) Principles of Environmental Monitoring related to the Handling of Radioactive Materials; ICRP Publication 9 (1966) Recommendations of the International Commission on Radiological Protection, Oxford: Pergamon Press.

International Atomic Energy Agency (1966) *Manual of Environmental Monitoring in Normal Operation*, IAEA Safety Series No. 16.

International Atomic Energy Agency (1966) *Disposal of Radioactive Wastes into Seas, Oceans and Surface Waters.*

GODBOLD B. C. and JONES J. K. (1966) *Radiological Monitoring of the Environment*, Oxford: Pergamon Press.

SCOTT RUSSELL R. (1966) *Radioactivity and Human Diet*, Oxford: Pergamon Press.

F. MORLEY

F

FACSIMILE RECORDING OF PHYSICAL DATA.

Experimental data arising from the regular scanning of bodies having simple geometric form may be recorded and displayed by facsimile recorders.

In its simplest form a facsimile recording system consists of a sheet of suitable paper that can be traversed by a pen directly linked to an experimental scanning head so that every movement of the head is exactly reproduced by the pen. Signals from the scanning head can be arranged to activate the pen so that a pictorial representation of the property under examination is obtained on the paper sheet.

In a common form of recorder, the paper is electrosensitive and turns grey if an electric current is passed through it, the density of the grey being proportional to the current. The pen is replaced by an electrical contact which is connected via a suitable amplifier to the scanning head so that the pictorial record will show graded grey-tones corresponding to the amplitude of the signal from the scanning head. This form of recording has the advantage over others in that not only is a permanent record available very rapidly, but the record is in pictorial form which allows easy visual interpretation and also contains precise positional information.

A disadvantage of the system is that although the grey-tone on the record gives a qualitative measure of the signal amplitude, no real quantitative information is available, because the blackening of the paper by a given current depends on properties of the paper not easily measurable or controllable. This difficulty can be largely overcome by using a *quantizing system of signal processing*. In this system the signal from the scanning head is fed to a number of voltage discriminators which can be set to trigger at any desired level and the outputs of the discriminators are passed via a logic circuit to a grey-tone current generator which is in turn connected to the facsimile recorder.

In operation the voltage discriminators are set to known and predetermined levels, and the logic unit senses the most significant discriminator to have been triggered. The logic unit provides digital signals which control the output from the grey-tone current generator. In this way a smoothly varying input signal is split up into discrete steps so that the record will appear in a stepped or contoured form.

The contours are set by the discriminators and are therefore independent of recorder variations and the grey-tone reproduced at each contour level can be set by controls in the grey-tone current generator. The inherent flexibility in the input/output characteristic can be used to induce a degree of quantitative information in the record and at the same time produce a subjectively pleasing picture.

With present equipment the recorder resolution is limited by the 0·010″ diameter writing electrode and maximum speeds in the range 30″/sec to 300″/sec can be obtained.

Facsimile displays are widely used in meteorological and newspaper offices for the reception of maps and news-pictures and are now becoming popular in the non-destructive testing field, where they can be used to display and record test data.

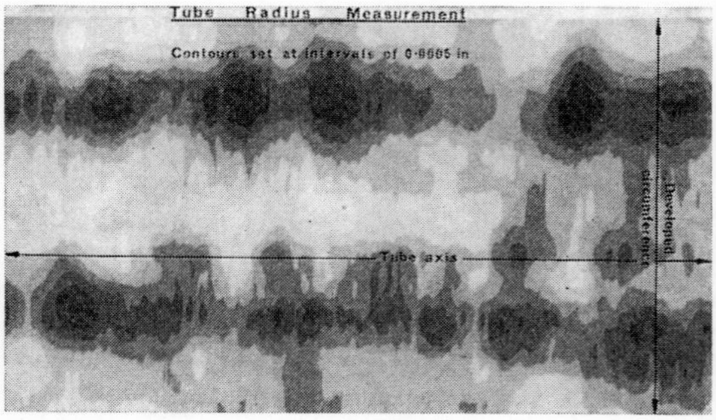

Fig. 1.

A typical application of facsimile recording in N.D.T. is found in a technique developed for the rapid measurement of the bore of metal tubing. In this particular technique a measuring probe which is connected to a facsimile recorder performs a helical scan down the inside of the tube under test.

The figure shows the record produced by such a scan down a $\frac{1}{2}''$ steel tube, the contours are set 0·0005 in. apart.

Bibliography

COLE H. A. G. (1966) *A quantizing half-tone generator for facsimile recording*, U.K.A.E.A. Report AERE-R 4878.

COTTERELL K. (1967) *Measurement of tube bore by capacitance gauging,* U.K.A.E.A. Report AERE-R 5382.

HODGKINSON W. L. (1964) *A quantized system of facsimile recording,* U.K.A.E.A. Report AERE-R 4688.

W. L. HODGKINSON

FISSION TRACK DATING. The recent discovery that insulating materials record the tracks of very heavy nuclear particles has led to the development of an extremely simple technique—the fission track method—for the dating of geological and archeological samples (Price and Walker 1963). At this writing, the basic validity of the new method has been demonstrated and a number of useful and unique results have already been obtained. The application of this technique is, however, in its infancy and much work remains to be done to determine the limits of its validity; and more important, to fulfil the promise of the method in making possible new types of dating measurements, particularly in determining numerous ages in a range hitherto difficult to study by other techniques.

The method depends on the fact that uranium nuclei undergo spontaneous fission. In the fission process, two heavy nuclear fragments are released, each possessing an energy of the order of 100 MeV and a consequent range in solids of $\sim$10 μ. These massive fragments produce a local disruption of the normal structure of an insulating solid for a distance of $\sim$15–25 Å around the particles' paths. This trail of radiation damaged material creates a distortion which can be seen directly, under favourable conditions, by examining a specimen at high magnification ($\times$ 100,000) in an electron microscope.

Fortunately, these damage trails can also be developed to optically visible size by a suitable etching treatment. The tracks so developed look very much like tracks of nuclear particles in a photographic emulsion. In effect, *natural minerals are particle detectors that can be developed and examined in much the same way as nuclear emulsion plates.* Figure 1 shows an example of developed particle tracks in a natural, unirradiated specimen of phlogopite mica (the reader who wishes to verify the phenomenon for himself can usually find tracks upon etching samples of granitic biotite for 10–20 sec in a 48 per cent solution of hydrofluoric acid).

Not all charged nuclear particles produce tracks in minerals. In fact, it has been shown that there exists for each mineral a threshold value for the density of ionizations caused by a charged particle below which no tracks are formed. This, in turn, implies a threshold in the mass of particles that will form tracks. In natural minerals, this mass threshold is generally between 30 and 40 atomic mass units. Thus, fission fragments will produce tracks, but α-particles will not.

A detailed consideration of the possible sources of heavy, energetic particles in nature (Price and Walker 1963) has shown that *only fission fragments from the spontaneous fission of ^{238}U give measurable track densities in terrestrial samples.* Strictly speaking, it may be possible to find cosmic-ray induced tracks in certain, special low uranium terrestrial minerals that are heavy element compounds. Such tracks have never been observed and form a very restricted exception to the stated rule. Extraterrestrial samples that have been exposed to the unshielded primary cosmic radiation contain tracks from other sources, but these are treated in a separate article (see *Nuclear particle tracks in meteorites*).

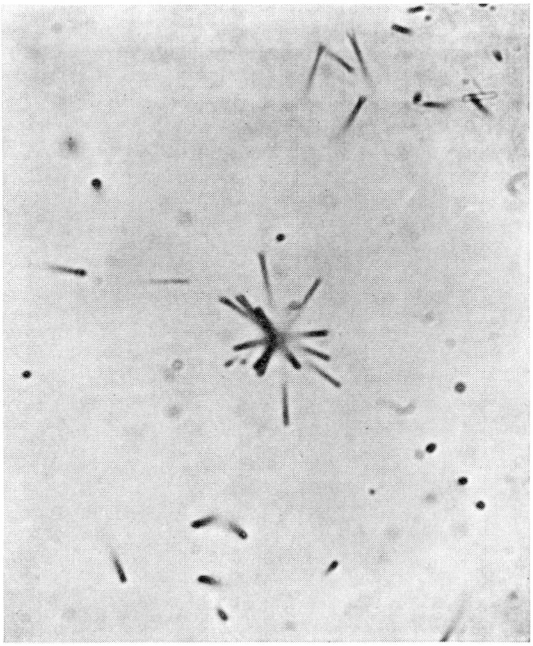

Fig. 1. Tracks from a segregated cluster of uranium: the star-shaped pattern in the centre is formed by a set of fossil tracks emmanating from a uranium rich region in a sample of phlogopite mica. The uranium content of this region was not sufficient to give a visible pleochroic halo.

Since essentially all tracks in natural samples of terrestrial material arise from the spontaneous fission of ^{238}U, a simple count of the number of stored tracks, coupled with a knowledge of the uranium concentration, can be used to determine the age of a specimen—provided, of course, that the tracks are retained in the sample for the full age and that the uranium concentration has also remained constant.

In practice, the uranium concentration is not generally determined explicitly. After a count of the fossil (stored) track density, S_s, the sample is sent to a nuclear reactor and bombarded with a known dose of thermal neutrons. The density of new tracks induced by the reactor irradiation, S_i, is then counted

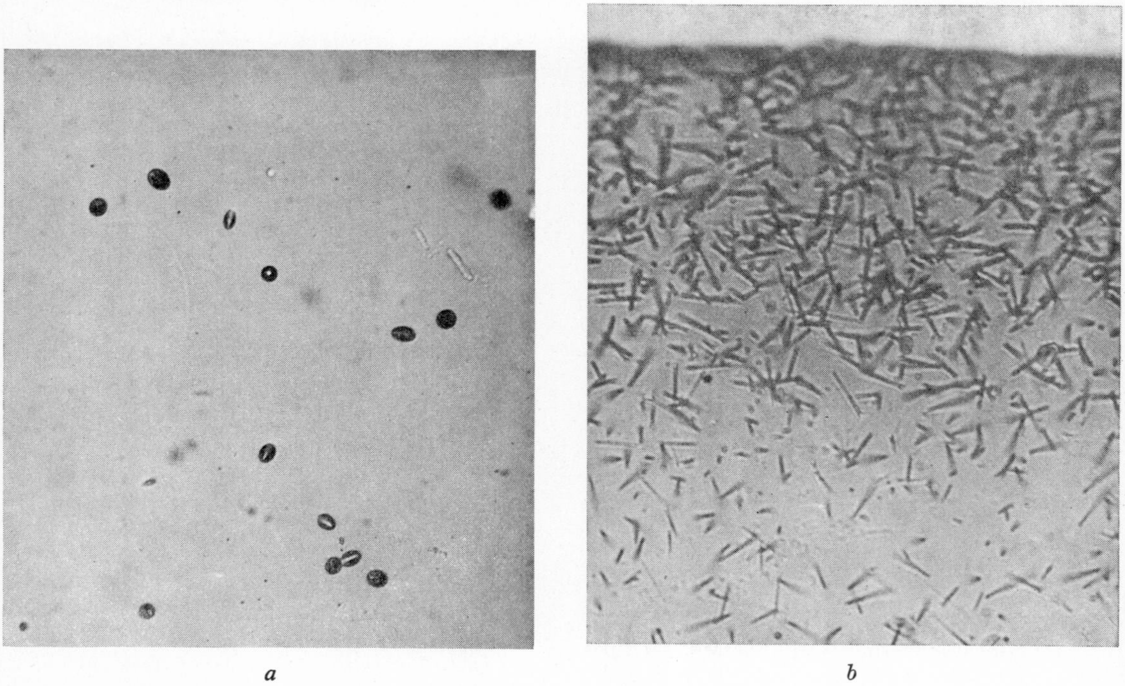

Fig. 2. Natural tracks in terrestrial materials: (a) Volcanic glass, Macusanite, from Peru (etched for 10 seconds in 48-percent hydrofluorid acid); (b) a zircon crystal from Australia (etchedfor 1 minute in boiling phosphoric acid at 500°C). The increase in track density as the free surface of the zircon is approached indicates that a nonuniform uranium distribution existed in the past.

and the age, T, determined from the following formula:

$$S_s/S_i = [\exp(\lambda \triangle T) - 1] (\lambda_F/\lambda_D f)$$

where λ_F and λ_D are the spontaneous fission and total decay constants of ^{238}U, and f is the fraction of the total uranium fissioned in the reactor irradiation.

The etching techniques for revealing tracks, as well as the appearance of the tracks after revelation, vary greatly from one material to the next. For example, hydrofluoric acid is usually used to develop tracks in mica and silica glass, while a boiling lye solution is found to be most effective on diopside. In most micas, the tracks are long and cylindrical (Fig. 1) while in natural glasses (Fig. 2a), they are revealed as characteristic, oval-shaped etch pits.

To apply the fission track dating method, it is necessary that the fission tracks, once formed, be stable against fading for the age interval of interest. Many natural minerals retain tracks in laboratory heating experiments from 400 to 1000°C and have extrapolated stability times $> 10^{10}$ years even at several hundred degrees centigrade (Fleischer *et al.* 1965a). In some minerals, however, autunite being a good example, the tracks fade in a relatively short time. Such minerals are obviously not practical to date.

The practical application of the method also requires that the sample contain enough uranium to give a measurable density of fossil tracks. An atomic concentration of ~ 1 per million, typical for many natural glasses, gives a track density of $\sim 200/\text{cm}^2$ in 10^5 years. The interesting region between 10^5 and 10^6 years is thus accessible for dating at this concentration. Some materials almost always contain so little uranium ($< 10^{-10}$) that they cannot be dated by the fission track method. Zircon crystals, on the other hand, frequently contain too much uranium to be useful—the track densities being too large to resolve. In general, uranium concentrations will vary markedly from one specimen to the next of the same material; thus, each specimen should be examined on its own merits regarding the possibility of using the fission track method.

The most useful materials that have been found for dating are silica glasses (obsidian), zircon, mica, hornblende, apatite, and sphene. Feldspars, quartz, and olivine generally contain too little uranium to be of use. Although the etching and annealing characteristics of a considerable number of minerals have been tabulated (Fleischer *et al.* 1965b) many minerals still remain to be examined.

A comparison between reported fission track ages and ages determined by other techniques is shown in Fig. 3. Below 2×10^8 years, the fission track ages

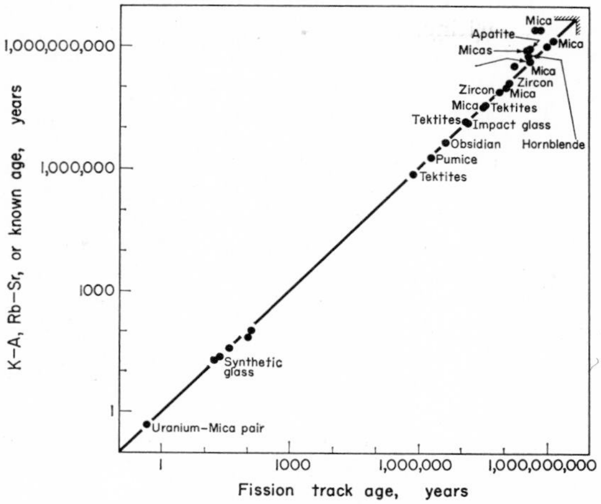

Fig. 3. Comparison of ages found by fission-track dating with those established by other means. The samples less that 200 years old are man-made; the older ones are geological.

are in concordance with the other measurements. Above 2×10^8 years, frequent discordancies are observed, even though correct ages are still obtained for T as great as $1{\cdot}4 \times 10^9$ years. The reasons for the discordant ages at $T > 2 \times 10^8$ years are not proved, though they are probably related to thermal or mechanical events in the history of the samples in question. The youngest age so far measured is 20 years in a man-made glass to which uranium had been deliberately added.

Among the most notable practical contributions of the fission track dating method to date have been the verification (Fleischer *et al.* 1965 c) of a previously reported K—A age ($T = 1{\cdot}8 \times 10^6$ years) for Bed I of Olduvai Gorge and the unequivocal demonstration of the close equivalence between the time of formation and the time of arrival on Earth of the Australian tektites (Fleischer and Price 1964). This latter result, which was obtained by comparing the age of the melted flange materials with the age of the unperturbed core material, illustrates one unique and promising aspect of the fission track method—the ability to make precise age determinations on a microscopic scale. In unpublished work, it has been possible to make simultaneous track measurements in several different types of crystals in a single thin section of rock.

Quite apart from its use as a dating tool, it should also be remarked that the fission track method is extremely useful for measuring the amount and distribution of small quantities of uranium in any sample (see Fig. 2). By irradiating specimens in easily available neutron fluxes, it is possible to measure concentrations as low as 10^{-14} atoms/atom. In the event that the sample to be studied does not itself register tracks, it can be placed next to a track detector such as mica or glass. Several plastics which register α-particle tracks have been found; and if these are used as detectors, the method also becomes a useful tool for studying trace distributions of boron and lithium. Irradiations with neutrons of different energies also make it possible to determine the $^{238}U/^{235}U$ ratio and the Th/U ratio.

Although much detailed work remains to be done to establish the ultimate utility of the fission track method, the unique capabilities of the method, coupled with its inherent simplicity, suggest that it should become a standard tool of geochronology. For a more detailed discussion of the techniques and results, the reader is referred to two review articles by the originators of the method (Fleischer *et al.* 1965 b, 1965 d).

Bibliography

Fleischer R. L. and Price P. B. (1964) *Geochimica et Cosmochimica Acta* **28**, 755.

Fleischer R. L., Price P. B. and Walker, R. M. (1965 a) *J. Geophys. Res.*, **70**, 1497.

Fleischer R. L., Price P. B. and Walker R. M. (1965 b) *Annual Reviews of Nuclear Science.*

Fleischer R. L., Price P. B. Walker R. M. and Leakey L. S. B. (1965 c) *Science*, **148**, 72.

Fleischer R. L., Price P. B. and Walker R. M. (1965 d) *Science*, **149**, 383.

R. L. Fleischer, P. B. Price and R. M. Walker

FLAMES, ELECTRICAL PROPERTIES OF. Although the discovery that flame gases would discharge an electroscope was made as early as 1600, the subject of ionization in flames became prominent only relatively recently, chiefly as a result of its various practical consequences and potential applications. Among more recent ones are the direct generation of electricity from rapidly flowing ionized flame gases, the control of some combustion processes by means of applied fields and the modification of carbon formation and deposition; the use of ionization probes in the detection and timing of flames and detonations and in gas chromatography is somewhat older. Among the less desirable consequences are the attenuation and refraction of radio waves by rocket exhausts.

Origin of ions. The concentration of free charge in flames is usually greatly in excess of what might be expected on the basis of thermodynamic equilibrium. Of the common product species, for flames in pure reactants, only NO has an ionization potential low enough (9·23 eV) to enter into calculations from this point of view. In one part per hundred, NO would cause between 10^7 and 10^{11} ions per cm^3 over the range of usual final flame temperatures. Ion concentrations in the reaction zones of flames of most pre-mixed reactants appear to be 10 to 10,000 times greater than this, depending on conditions. This is due to chemi-ionization, i.e. the process whereby the energetic molecular fragments, which propagate the main reaction, occasionally participate in a reaction step which releases an electron. CHO^+ and $C_3H_3{}^+$ have been proposed as the likely original chemi-ions. Charge exchange, attachment and clustering, however, occur in the flame gases and H_3O^+ is the most abundant molecular ion found in hydrocarbon flames. In sooting flames, particles of carbon black and their precursors carry an appreciable proportion of the charge.

Some flames, or rather some fuels, do not give rise to appreciable chemi-ionization; examples are flames burning H_2, H_2S and CO, in the absence of impurities. Notable among impurities which raise equilibrium product ionization very greatly are the alkali metals, by virtue of their low ionization potentials. As little as one part in 10^8 of potassium would produce more ions than is due to NO in normal flames. Correspondingly, stringent precautions must be taken to eliminate traces of these metals for measurements in which their ions must be avoided. For instance in pure hydrogen/oxygen flames, in the absence of NO and of chemi-ionization, minute traces of these metals may provide the sole source of flame ionization.

"Seeding" with metals of low ionization potential is thus an obvious method of increasing equilibrium product ionization and certain applications, or proposed applications, of flame ionization depend upon it. Thus the inclusion of a potassium seed is a necessary part of some proposals for MHD generation of electricity. Special rocket propellant compositions involving caesium compounds and giving rise to high final flame temperatures have been evolved with the ultimate aim, presumably, of "plugging holes in the ionosphere" for communication purposes, which may be necessary after high-altitude nuclear explosions.

An alternative, or additional, method of producing high equilibrium ionization levels in flame products is by raising the temperature beyond values attainable from the heat of reaction alone. Conventional preheating of the reactants is of some use but its range is limited by the melting point of any container walls. The adiabatic compression wave which leads a detonation produces such an effect. On burners, it is more convenient to associate the flame with an electric discharge. Such devices, which have become known as "augmented flames", have gained some prominence in recent years. They vary in design from those in which a high-voltage discharge is passed through hot products, to those in which the burner is associated with a plasma jet. In the latter, rapid rotation of the discharge by a superimposed magnetic field has been used to ensure uniform heating of the reactants and very stable "flames" with large throughput rates have been produced in this manner. The uses proposed for augmented flames are not confined to producing high ion concentrations but include increased rates of heat transfer (for welding, cutting, drilling, melting, smelting, sintering, spheroidizing, reducing ores, etc.) increased combustion intensities (for use in propulsive equipment), the synthesis of endothermic compounds, and many others.

Diagnostics. A large variety of methods has been used for the study of flame ionization. Generalizing for the sake of compactness of summary, they may roughly be subdivided into three groups. The first consists of methods in which probes or electrodes are inserted into the flame and are, more or less, in contact with the hot gases. It includes the measurement of conductivity and of the Hall effect and the use of Langmuir and other probes to deduce concentrations and mobilities of the charge-carriers. The main disadvantage of these methods lies in the interference of solid heat sinks with the delicate flame reaction processes and the limit to resolution set by the probe dimensions.

The second group is based on the use of electromagnetic radiation, usually microwaves or at radio frequencies, and includes measurement of attenuation, and the use of resonant cavities. Techniques in which the flame is surrounded by two coils and the Q of the circuit is measured at high frequencies also fall into this category. The use of electromagnetic interactions does not interfere with the flame processes but the spatial resolution is limited by the effective wave-length. This is usually very much larger than the thickness of reaction zones in flames (except at very low pressures) because ion concentrations are not high enough to permit the use of short wavelengths, e.g. the visible part of the electromagnetic spectrum.

The third group of methods consists of techniques in which the ions are withdrawn from the flame, by means of applied fields or flow, for examination in apparatus outside the flame. Thus the ion species have been investigated by sampling into a mass spectrometer and, with much less sophisticated systems, into devices which would measure ionic mobility under "cold" conditions. The rate of ion generation per unit area of flame front has been obtained by measuring saturation currents from plane flames between parallel plane electrodes. The distribution of such currents has been displayed by "ion photography"—a method in which the marking of photographic emulsions by minute amounts of electrolysis is utilized. In this group of methods interference with flame processes can also be avoided (except where sampling in the reaction zone by probes is involved); its generic disadvantage is that charge transfer, clustering and other interaction of ions with neutral species can occur not only in the flame gases as hitherto, but also with the cold gases in the outside apparatus.

Applications. Practical consequences and uses are all based, in some way, on the movement of free charges under the influence of fields. We can classify them roughly by the nature of the field applied which may be magnetic, electric, steady or oscillating, large or small. The application of a large magnetic field which has attracted most attention in recent years is the magnetohydrodynamic generation of electricity directly from moving combustion gases. In this scheme, the magnetic field is applied at right angles to the direction of movement of hot combustion products which travel at high velocities in a divergent duct. The free charges then travel at right angles to both the magnetic and the velocity vector, to be collected by electrodes which are embedded in the sides of the duct. In so doing, they slow down the gas, causing it to do work on them, analogously to conventional generation methods in which this work is done on turbine blades, for instance. The advantage lies in the absence of such an intermediate mechanical step with its inevitable energy losses. The disadvantages are mostly associated with the high gas temperature required for an appreciable conductivity to be maintained (even with seeding). This imposes severe restrictions on materials, aggravates the difficulties of introducing a large magnetic field into the gases and confines the usefulness of the scheme to the high-temperature end of a conventional power cycle.

The response of flame plasma to electromagnetic fields determines the consequences of, for instance, the interaction of radio waves with rocket exhausts. This is somewhat analogous to the diagnostic use of microwaves—it results in attenuation, refraction, diffraction and, at high plasma densities also reflection, of the waves. The applied interests in this subject arise from the difficulties these effects produce in radio communication with rocket-propelled vehicles (as well as from the previously mentioned concern about continuity of the ionosphere).

As regards uni-directional electric fields, applications can be divided according to the field strength. Methods of detection and measurement involve relatively small fields, usually between two closely spaced electrodes of an "ionization gap", to give a virtually instantaneous indication of the presence of the flame of detonation at a specified location. This allows flame phenomena to be processed by electronic circuitry, e.g. electronic timing devices in the measurement or detonation velocities, using several ionization gaps; electronic warning or safety devices in fire detection; current recording circuits in detectors for gas chromatography; and so forth.

In the case of high field intensities applied over appreciable distances, two further groups of applications can be distinguished; those based on the movement of ions and those based on the movement of gas induced by the ion drift. The movement of the charge carriers may be used for the purpose of transporting some species or particles to the electrodes as, e.g. in the "electrostatic" precipitation of solid flame products, using the flame-acquired charge. It has recently been shown that certain flame reactions can be greatly modified by moving the incipient particles with respect to the zone in which they are normally formed. Thus it is possible to control the formation of carbon in flames, as well as its subsequent deposition. A complementary example, in which it is the movement of charge that is the desired end-product is provided by the electrogasdynamic (EGD) generation of electricity. This is based on combustion products flowing at high velocities in a divergent duct, much as in the MHD schemes, except that the retarding force is applied here by an electric field between two electrodes in the duct, the field applied being such that it just does not prevent ions from reaching the downstream electrode.

The movement of neutral gas—sometimes called "the ionic wind"—arises because at normal pressures and field strengths ions do not accelerate but transfer the additional momentum acquired from the field to the molecules with which they collide. This causes an accelerating gas flow towards the electrodes, a flow which in order to accelerate, must entrain further gas. By confining the gas stream in particular ways it has been found possible to deploy the considerable entrainment velocities as a method of controlling certain combustion processes. Some examples are the aeration of diffusion flames, increasing combustion intensity by recirculating hot products, and controlling the rate of flame spread across solid propellant surfaces by applied fields. All these applications are limited, in a calculable manner, by the onset of secondary ionization and breakdown.

Bibliography

LAWTON, J. and WEINBERG F. J. (1968) *Electrical Aspects of Combustion*, Oxford: Clarendon Press.

Thewlis, J. (Ed.) (1961) *Encyclopaedic Dictionary of Physics*, articles beginning "Flame(s)", **3**, Oxford: Pergamon Press.

F. J. WEINBERG

FLIGHT MECHANICS. *1. Introduction.* The motion of a vehicle subjected to gravitational, aerodynamic, and propulsive forces is analysed by two complementary branches of aerospace engineering: (a) flight mechanics and (b) stability and control. While *flight mechanics* considers the purely translational motion of the vehicle regarded as a particle (the effect of the deflexion of the control surfaces on the aerodynamic forces is neglected), *stability and control* simultaneously considers both the translational and rotational motions of the vehicle regarded as a rigid body on which rotors and movable control surfaces are mounted (the effect of the deflexion of the control surfaces on the aerodynamic forces is accounted for).

In this article, the particle approach is taken, and the general equations governing the translational motion of an arbitrary vehicle (aircraft, missile, satellite, or spaceship) are presented with reference to non-steady flight over a spherical Earth. The main analytical simplifications of interest in performance analyses are indicated, together with the range of applicability of the resulting equations for different kinds of vehicles.

2. Vectorial Equations

Consider a variable-mass vehicle subjected to the following forces: the thrust **T**, the aerodynamic force **A**, and the weight **W** which is the product of the instantaneous mass m and the acceleration of gravity **g** (the vector **g** is due to the attraction of all the bodies of the Universe on the vehicle). The motion of this vehicle with respect to the Fixed Stars is governed by the differential equation

$$\mathbf{T} + \mathbf{A} + m\mathbf{g} - m\mathbf{a}_* = 0, \qquad (1)$$

in which $\mathbf{a}_*$ denotes the absolute acceleration of the vehicle, that is, the acceleration with respect to the Fixed Stars.

Now, assume that the vehicle moves in the immediate neighbourhood of the Earth for a time interval small with respect to the period of revolution of the Moon around the Earth. For this application, the Earth-vehicle system can be conceived as being isolated in space, which is the same as neglecting the differential effects of the Sun and the Moon on the motion of the Earth and the vehicle. Consequently, since the mass of the vehicle is negligible with respect to that of the Earth, the following idealization of the Earth's motion is possible: (a) the centre of the Earth moves with constant absolute velocity; and (b) the absolute angular velocity of the Earth is constant and has the same direction as the polar axis.

Because of these hypotheses and in the light of the theorem of composition of accelerations, the following relationship can be readily established between the motion of the vehicle with respect to the Fixed Stars and that relative to the Earth

$$\mathbf{a}_* = \mathbf{a} + 2\boldsymbol{\omega} \times \mathbf{V} + \boldsymbol{\omega} \times (\boldsymbol{\omega} \times \mathbf{r}). \qquad (2)$$

In this relation, **r** denotes the vector joining the centre of the Earth with the vehicle, **V** and **a** are the velocity and the acceleration of the vehicle with respect to the Earth, and $\boldsymbol{\omega}$ denotes the angular velocity of the Earth with respect to the Fixed Stars. As a consequence, equation 1 can be rewritten in the form

$$\mathbf{T} + \mathbf{A} + m\mathbf{g} - m[\mathbf{a} + 2\boldsymbol{\omega} \times \mathbf{V} + \boldsymbol{\omega} \times \times (\boldsymbol{\omega} \times \mathbf{r})] = 0, \qquad (3)$$

where **g** is the Earth's gravitational attraction per unit mass.

Notice that the transport acceleration $\boldsymbol{\omega} \times (\boldsymbol{\omega} \times \mathbf{r})$ depends on the instantaneous latitude of the vehicle, being zero at the Poles and a maximum at the Equator where its order of magnitude is $10^{-3}\, g_0$ (the symbol g_0 denotes the modulus of the acceleration of gravity at sea level). Furthermore, the Coriolis acceleration $2\boldsymbol{\omega} \times \mathbf{V}$ depends on the modulus and the direction of the velocity of the vehicle with respect to the Earth's axis, being zero when the flight path is parallel to the polar axis and a maximum when it is perpendicular to this axis; in the latter case, the order of magnitude of this acceleration is $10^{-3}g_0$ for present-day commercial aircraft but may increase to $10^{-1}g_0$ for vehicles travelling at either satellite speeds or velocities approaching the space velocity.

In the light of these considerations, equation 3 can be approximated by either

$$\mathbf{T} + \mathbf{A} + m\mathbf{g} - m(\mathbf{a} + 2\boldsymbol{\omega} \times \mathbf{V}) = 0 \qquad (4)$$

or

$$\mathbf{T} + \mathbf{A} + m\mathbf{g} - m\mathbf{a} = 0, \qquad (5)$$

depending on the degree of accuracy desired in engineering problems. In particular, equation 4 neglects the transport acceleration and is of considerable interest in many problems characteristic of ballistic missiles, satellite vehicles, and spaceships departing for interplanetary expeditions. On the other hand, equation 5 neglects both the transport acceleration and the Coriolis acceleration and is to be employed in those cases where the flight speed is small compared to the escape velocity or in problems where the emphasis is placed on preliminary design estimates or comparative performance analyses.

3. Scalar Equations

In the previous section, the vectorial equations governing the flight of a vehicle have been presented under several approximations. Regardless of the approximation employed, the derivation of the corresponding scalar equations is a problem of differential geometry which, for flight in a three-dimensional space, is quite complex (Miele 1962). This being the case, only one class of flight trajectories is discussed here, that of paths flown in a great-circle plane. For these trajectories, the projection of equation 4 on the tangent **t** and the normal **n** to the flight path yields the following relationships (Fig. 1):

$$T \cos \varepsilon - D - mg \sin \gamma - m\dot{V} = 0, \qquad (6)$$

$$T \sin \varepsilon + L - mg \cos \gamma - mV \times [\dot{\gamma} - V \cos \gamma/(r_0 + h) \pm 2\omega \cos \varphi] = 0,$$

in which T is the thrust, ε the inclination of the thrust with respect to the velocity, D the drag, L the lift, m the instantaneous mass, g the acceleration due to gravity, V the velocity of the vehicle with respect to the Earth, γ the inclination of the velocity with respect to the local horizon, ω the angular velocity of the Earth with respect to the Fixed Stars, and φ the smaller of the two angles which the polar axis forms with the perpendicular to the plane of motion. The dot sign denotes a derivative with respect to

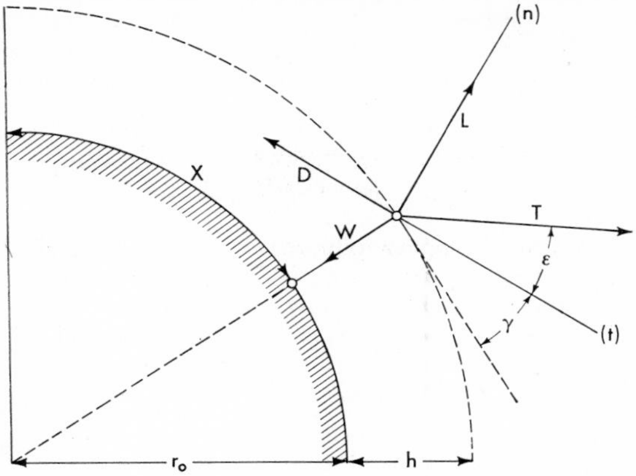

Fig. 1.

time. The sign preceding the Coriolis acceleration term in equation 6·2 is negative for motion in the same sense as that of the Earth's rotation and positive for motion in the opposite sense.

The dynamical equations (6) must be supplemented by the kinematic relationships on the horizontal and vertical directions

$$\dot{X} - V r_0 \cos \gamma/(r_0 + h) = 0,$$
$$\dot{h} - V \sin \gamma = 0, \tag{7}$$

where X denotes a curvilinear coordinate measured on the surface of the Earth, h the altitude above sea level, and r_0 the radius of the Earth. They must also be completed by the principle of conservation of mass

$$\dot{m} + \beta = 0, \tag{8}$$

in which β denotes the mass flow rate of fuel through the engine.

In order to understand the nature of the differential system composed of equations 6 through 8, a discussion of the functions g, D, L, T, β is needed. For the spherical Earth model, the acceleration of gravity varies with the altitude in accordance with the inverse square law

$$g = g_0[r_0/(r_0 + h)]^2 \tag{9}$$

in which g_0 denotes the acceleration of gravity at sea level. If the aerodynamic lag is neglected, the drag, the thrust, and the mass flow rate of fuel are functions of the following form:

$$D = D(h, V, \alpha), \quad L = L(h, V, \alpha) \tag{10}$$

and

$$T = T(h, V, \pi), \quad \beta = \beta(h, V, \pi), \tag{11}$$

where α denotes the angle of attack and π the engine control parameter. The latter is a variable controlling the engine performance (the rotor speed of a turbojet engine, the fuel-to-air ratio of a ramjet engine, the combustion chamber pressure of a rocket engine). With these considerations in mind, the set of five differential equations 6 through 8 involves one independent variable (the time t) and the eight dependent variables

$$X, h, V, \gamma, m, \alpha, \varepsilon, \pi. \tag{12}$$

Therefore, three degrees of freedom exist, as is logical in view of the possibility of controlling the time history of the angle of attack, the thrust direction, and the thrust modulus (by definition, the number of degrees of freedom of a differential system is the difference between the number of dependent variables and the number of constraining equations). This is the same as stating that, if the initial values of t, X, h, V, γ, m are specified, an infinite number of trajectories are physically possible, one trajectory for each arbitrarily prescribed set of functions

$$\alpha = \alpha(t), \quad \varepsilon = \varepsilon(t), \quad \pi = \pi(t). \tag{13}$$

It should be noted that the functions 13 need not be given explicitly but they can be implicitly defined through the equations of motion 6 through 8 and three supplementary constraints having the form

$$f_1(t, X, h, V, \gamma, m, \alpha, \varepsilon, \pi) = 0,$$
$$f_2(t, X, h, V, \gamma, m, \alpha, \varepsilon, \pi) = 0, \tag{14}$$
$$f_3(t, X, h, V, \gamma, m, \alpha, \varepsilon, \pi) = 0.$$

In this connexion, the following are examples of constraints of the type 14:

$$\varepsilon - \alpha = \text{const} \tag{15}$$

meaning that the engine is fixed with respect to the aircraft,

$$\varepsilon = 0, \tag{16}$$

meaning that the thrust is tangent to the flight path,

$$\varepsilon + \gamma = \text{const}, \tag{17}$$

meaning a constant inclination of the thrust with respect to the local horizon,

$$h = \text{const}, \tag{18}$$

meaning level flight,

$$X = \text{const}, \tag{19}$$

meaning vertical flight, and

$$\pi = \text{const}, \tag{20}$$

meaning flight with constant power setting.

4. Flat Earth Approximation

Assume, now, that the flight altitude above sea level is small when compared with the radius of the Earth and the flight velocity is small with respect to the escape velocity, that is,

$$h/r_0 \ll 1, \quad V^2/g_0 r_0 \ll 1. \tag{21}$$

In addition, assume that the Coriolis term appearing in equation 6·2 is small with respect to some of the other terms appearing in the same equation, for instance, the weight component on the normal to the flight path

$$2\omega V \cos\varphi/g_0 \cos\gamma \ll 1. \tag{22}$$

With these considerations in mind, equations 6 through 8 yield the following simplified set:

$$\begin{aligned} T\cos\varepsilon - D - mg_0\sin\gamma - m\dot{V} &= 0,\\ T\sin\varepsilon + L - mg_0\cos\gamma - mV\dot{\gamma} &= 0,\\ \dot{X} - V\cos\gamma &= 0,\\ \dot{h} - V\sin\gamma &= 0,\\ \dot{m} + \beta &= 0. \end{aligned} \tag{23}$$

which, formally, can also be derived by assuming a flat, non-rotating Earth (Fig. 2)

$$r_0 = \infty, \quad \omega = 0. \tag{24}$$

For this reason, whenever equations 23 are employed, it is customary to say that the *flat Earth approximation* is used. It should be emphasized, however, that this is merely a matter of terminology. In fact, as long as X

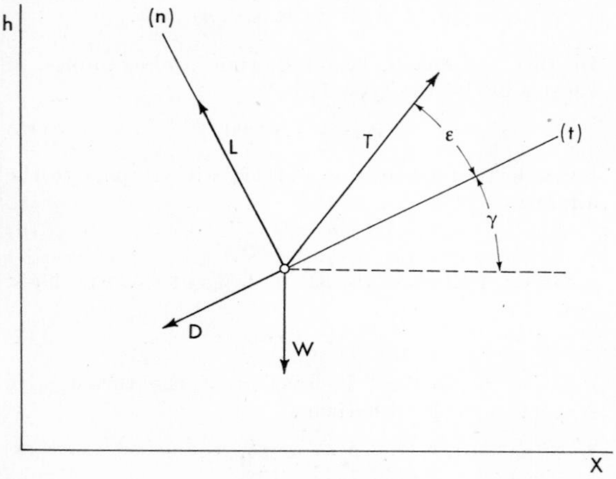

Fig. 2.

is interpreted as a curvilinear coordinate measured along the surface of the Earth and the gravitational field is considered to be quasi-uniform in modulus although centralin direction, equations 23 are applicable to extremely long range flight, the only requirement being that inequalities 21 and 22 are satisfied.

5. *Quasi-steady Approximation*

If the terms containing derivatives in the dynamical equations are small with respect to some of the remaining terms, further simplifications are possible. Under the flat Earth approximation, equations 23·1 and 23·2 reduce to

$$\begin{aligned} T\cos\varepsilon - D - mg_0\sin\gamma &= 0,\\ T\sin\varepsilon + L - mg_0\cos\gamma &= 0, \end{aligned} \tag{25}$$

which are no longer differential relationships, but relationships in finite terms. They enable one to study the so-called *point performance problem* within the framework of elementary algebra.

6. *Discussion*

In the previous sections, the main analytical relationships of interest for flight mechanics analyses have been presented for the non-steady flight over a spherical Earth, the non-steady flight over a flat Earth, and the quasi-steady flight over a flat Earth. The choice of the set of equations to be used depends on the particular vehicle as well as the flight condition being examined. Generally speaking, the analysis of the flight paths of rocket-powered aircraft, ballistic missiles, satellite vehicles, skip vehicles, and hypervelocity gliders requires the use of equations 6 through 8. On the other hand, the non-steady flight of aircraft propelled by air-breathing jet engines (e.g., climbing flight) can generally be studied using equations 23. Finally, the quasi-steady flight of aircraft propelled by air-breathing jet engines (e.g., cruising flight) can be studied by replacing equations 23·1 and 23·2 with equations 25·1 and 25·2. Depending on the particular application, further simplifications are possible. As an example, for the class of shallow flight paths defined by

$$\gamma^2 \ll 1, \tag{26}$$

the approximations

$$\cos\gamma \cong 1, \quad \sin\gamma \cong \gamma, \tag{27}$$

can be employed. The discussion of all of the analytical solutions obtained so far goes beyond the scope of this article, and the reader is referred to the specialized literature on the subject (Miele 1962, in press).

Bibliography

MIELE A. (1962) *Flight Mechanics, Vol. 1: Theory of Flight Paths,* Reading, Mass.: Addison-Wesley.

MIELE A. *Flight Mechanics, Vol. 2: Theory of Optimum Flight Paths,* Reading, Mass.: Addison-Wesley (in press).

A. MIELE

FUEL CELLS AND FUEL BATTERIES. Since 1964, when the earlier article (Liebhafsky 1966) was written, fuel-cell activity has continued at a high rate. The use of fuel batteries in actual applications continues to lag. As the research matures, more batteries are being built that approach practical devices, and the engineering difficulties of the step from cell to battery are becoming more widely known. It is still true that high unit capital cost (pounds or dollars per kilowatt) is a serious (perhaps the most serious) obstacle to the widespread commercial use of fuel batteries. The premium that the buyer is willing to pay for *convenience* dominates *low energy cost* in determining whether a reliable battery of adequate life can find a market today. The use of fuel batteries in space justifies the highest premium for the many characteristics that are grouped under convenience; these applications are followed at some distance by military uses and by uses in remote or inaccessible locations that demand reliable, unattended operation. These are the most promising markets now open to fuel batteries.

To *reactivity* and *invariance* as research requirements for the fuel cell must be added *uniformity* for the fuel battery. This is the lesson from recent experience. Only if conditions in a battery approach uniformity will the transport processes (for heat, electricity, momentum, and mass) proceed satisfactorily. Non-uniformity (local overheating, variations in pressure, and irregularities in the supply of reactants or in the removal of products) means that troubles are in the offing or at hand. The problems non-uniformity presents are often interrelated problems of electrochemical engineering.

Successful applications of fuel batteries. The many prototype fuel batteries that have been built since the earlier article cannot be described. The two that have proved themselves in actual service at this writing will be discussed.

The first, the General Electric hydrogen-oxygen fuel battery that served as the sole on-board power source for Gemini-5 attracted world-wide attention in 1965 and illustrates much of what was said in the introduction. A diagram of the system that included the two 1-kilowatt fuel batteries is Fig. 5 of the earlier article. A pictorial record of the development of the battery appears here in Fig. 1. Figure 2 shows two views of components for the complete cell. A comparison of this figure with the laboratory cell (Fig. 3 of the earlier article) gives an inkling of the role electrochemical engineering must play in making fuel batteries successful.

The unique feature of the fuel cells in this battery is the quasi-solid electrolyte—an *ion exchange membrane* that may be envisaged as a polymeric derivative of sulphuric acid. The polymer provides a skeleton that retains water through which the hydrogen ion is free to move while the anion (SO_3^-) remains firmly bound. As a consequence, the membrane can transfer hydrogen ion and (provided there has been no deterioration) can reject as *pure* liquid water the product of reaction in a hydrogen-oxygen cell. Water transport out of the cell is thus simplified, and the electrolyte can be as thin a sheet as can be made uniform and pore-free—important advantages in space applications.

The fuel cell proper (anode, electrolyte, cathode) is a thin sheet of ion exchange membrane coated on both sides with a thin layer of the electrocatalyst, platinum black, in which is imbedded a screen of titanium-palladium (0·1 per cent) alloy for the collection of current *within* the electrode (see the centre portion of Fig. 3 in the earlier article). A complete cell can be envisaged as a sandwich with the cell proper enclosed by the components A and B (Fig. 2) so that the wicks face the cathode, and View B is exposed. Hydrogen enters and leaves through the ports that appear as lateral projections in Fig. 2. The cell operates in an oxygen environment, and this gas has free access to the cathodes.

The measures taken to ensure easy transport and uniformity deserve attention. Each of the 32 cells in a stack (Fig. 1) is in parallel with the others as regards the removal of water, and the supply of hydrogen and oxygen (stored as fluids at high pressure and low

A *B*

Fig. 1. Development of the ion exchange membrane fuel battery by the General Electric Company. A, Physical Chemistry Section, Schenectady, N. Y., 1955–1960; Grubb-Niedrach cell, 0·02 watt; Cairns-Douglas battery, 15 watts; B, Direct Energy Conversion Operation, Lynn, Mass., 1959–1965; DECO stack, 300 watts; DECO Gemini battery, 1000 watts.

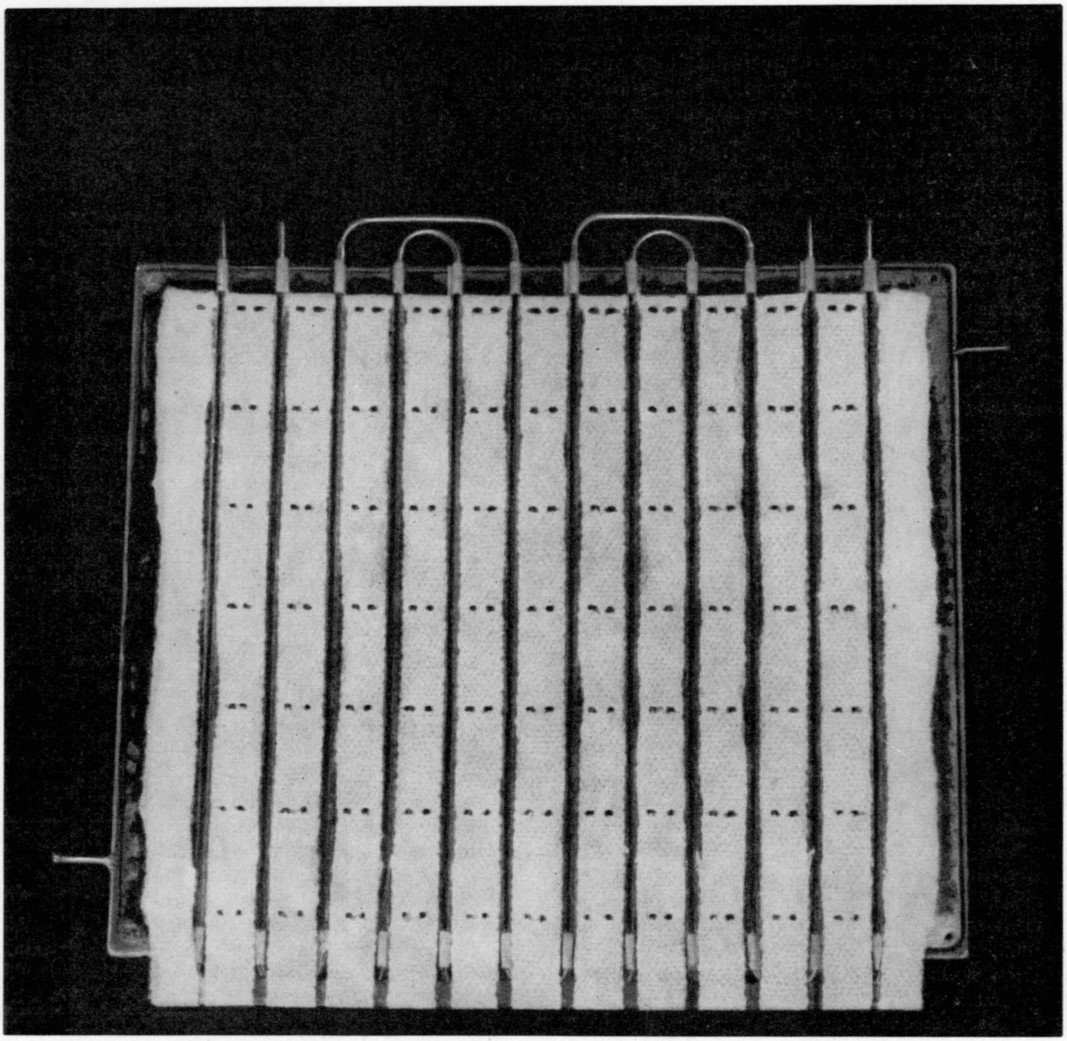

A. Fig. 2

temperature). Liquid water is transported independently of gravity by a wick system that terminates in a distributor wick on the face of a porous plate across which there is a differential pressure of oxygen. Heat is removed efficiently by a liquid coolant (temperature, 24°C) in good thermal contact with the cell. Abutment of the anode and cathode current collectors to give a *bipolar arrangement* internally connects the 32 cells in series so that about 25 volts is available at the stack terminals. In each battery (power, 1 kilowatt at 25 volts and 40 amperes), the three stacks are electrically in parallel, as are the two batteries on the spacecraft. To provide flexibility in sharing the load, each stack can be connected or disconnected individually. The fuel batteries were stored in an inaccessible chamber that was jettisoned during the return to Earth.

General Electric fuel batteries furnished power as needed on 7 Gemini missions during which they delivered 519 kilowatthours over 840 hours. Most of the difficulties encountered probably had their origin in the peripheral equipment. This, the most spectacular application yet recorded for fuel batteries, was highly successful. The weakest link in these batteries was probably the ion-exchange-membrane electrolyte. With greatly improved membranes assured, batteries with such electrolytes have a promising future in specialty applications—on Earth as well as in space.

The second application, though it attracted less

B

Fig. 2. Views of components for a complete General Electric H_2/O_2 cell for spacecraft. View A shows the wicks, sewn to the cathode current collector, for the transport of product water is liquid. The coolant ducts, projecting above, make possible the transport of heat. View B shows the anode current collector, which is also the wall of the anode (hydrogen) chamber.

notice, was equally successful and employed what is properly called a *fuel storage battery*. In this battery (Fig. 3), which owes much to W. Vielstich and was developed by Brown, Boveri and Co., the fuel (methyl alcohol) is stored in the electrolyte (potassium hydroxide) and is oxidized to carbonate at an anode with a low loading of a platinum-metal electrocatalyst. Because hydroxide ion is consumed, the solution must be replaced when the fuel is exhausted; the potassium hydroxide needed is costly. The carbon cathode operates on oxygen from air. This fuel storage battery is a simple, reliable, attractive, low-power device that should be capable eventually of producing electrical energy at costs at least competitive with those of other power sources for unattended service in remote or inaccessible locations.

The fuel storage battery exemplifies the interrelation, steadily growing closer, of fuel batteries and conventional storage batteries (in which neither a conventional fuel nor oxygen is consumed). A fuel battery, being a converter, requires a system to supply reactants; a storage battery does not. The Brown-Boveri fuel storage battery, operating as it does at low power and on air, needs no system. This hybrid device is closely related to metal-air batteries (e.g. zinc-air or magnesium-air batteries) in which the anode (rechargeable or not) stores energy in metallic form while the cathode operates on air. Such batteries are benefiting significantly from the progress made on air cathodes for fuel cells. Finally, fuel batteries and storage batteries complement each other, storage batteries being better suited to peak loads and fuel batteries to steady loads of long duration such as keeping a storage battery charged.

Fig. 3. Fuel-storage-battery-powered television relay station, Südwestfunk Baden-Baden, Germany. A similar station, at an altitude of 2300 metres on the Gebidem, is the power source for television for the Zermatt Valley, Switzerland. Both batteries were installed toward the end of 1965 and operated satisfactorily over the winter. The battery is rated at 24 watts, and 28 ± 2 volts is maintained by a special d.c./d.c. converter. The fuel is methanol dissolved in strong potassium hydroxide for ordinary service. For service at very low temperatures (to −25°C), a formate is added to the solution. The fuel-electrolyte solution is consumed during operation and must be replaced after 5000 to 6000 hours. The long period between replacements necessitates a large battery. (Photo courtesy Brown, Boveri and Co., Baden, Switzerland.)

Combining the two therefore makes a power source better than either for service that requires peak loads separated by intervals long enough so that the storage battery is never drained to the danger point. Examples: signalling devices; delivery trucks (vans); the "all-gas home".

Outlook. The outlook for fuel batteries, though still clouded, has become clearer than it was at the writing of the earlier article. It will be summarized first with respect to fuels and then with respect to applications.

The conventional fuels are hydrogen, compromise fuels (methyl alcohol, hydrazine, ammonia), and the fossil fuels.

Hydrogen is firmly entrenched as the prime fuel for space applications. Either in cylinders or generated by chemical reaction, it should find use in fuel batteries for other specialty applications. Hydrogen-oxygen batteries to serve any application can be built now, and it seems pointless to test this statement by the further building of costly prototypes for applications that such batteries can scarcely be expected to serve.

Methyl alcohol could be used successfully today in indirect systems (see below); as a fuel used directly, it is making slow but steady progress. Hydrazine, despite its toxicity and high cost, is gaining regard for specialty applications; being highly reactive at an anode, it may be looked on as hydrogen in convenient form; its cost is relatively unimportant in military applications, particularly in theatres for which transportation costs are overriding. The published literature on ammonia used directly as fuel is scanty; "cracked" ammonia provides hydrogen for the 200-kilowatt fuel battery in the Swedish submarine.

Of the fossil fuels, coal will never be used directly in fuel batteries. Present indications are that it is more likely to enter the picture as an eventual source of methane (when the growing supplies of natural gas have run low) than as a source of water gas (hydrogen and carbon monoxide).

The status of the direct hydrocarbon-air battery for low temperatures and ambient pressure, perhaps the most difficult assignment in the field, is as follows. On the basis of present results, it should be possible to build propane (or butane)-air batteries that will deliver between 100 and 200°C about 25 watts/ft^2 at a platinum loading of 10 g/ft^2 on the anode and 5 g/ft^2 on the cathode, which means 600 g Pt/kilowatt or about $ 2400/kilowatt for platinum alone. The size of these platinum loadings and the cost of the metal are both too high for *extensive* application of direct hydrocarbon/air batteries: the metal is limited in supply and badly needed for other uses. *Electrocatalysis* is thus the major research problem of the direct hydrocarbon/air cell today, and its future depends upon the finding of a satisfactory *replacement* for platinum. Indirect systems, in which carbonaceous fuels, anodically inert, are changed into gases more acceptable at the anode, promise soon to become useful in spite of complexity and high capital cost. In some systems, palladium transmits pure hydrogen to the anode. In others, hydrogen of low carbon monoxide content is used. In still others—those with molten carbonates (5–600°C) or doped zirconia (9–1100°C) as electrolytes—the temperatures are so high that the hydrocarbon and steam may be fed into (or close to) the anode chamber.

Methane commands special attention among fossil fuels as the principal constituent of natural gas, large new supplies of which are being discovered. Natural gas is easy to desulphurize (sulphur is unwelcome in any fuel battery); and methane, having the highest ratio of hydrogen to carbon, is of all hydrocarbons the best suited to the fuel battery. Its inertness in thermal reactions relative to other saturated hydrocarbons does not carry over to the fuel-cell anode. Methane-air batteries with either of the electrolytes in the previous paragraph would operate at temperatures high enough so that electrocatalysis could be accomplished without platinum, and such batteries could eventually be low in capital cost. Although the work on methane-air cells has not progressed far enough to warrant further description here, methane-air batteries are easily among the most rewarding objectives of fuel-cell research.

The outlook for various applications will now be given.

Space. The fuel battery has established itself for space missions. Future missions are scheduled to use fuel batteries, by Pratt and Whitney, based on the distinguished work begun over 30 years ago by F. T. Bacon (see earlier article).

Portable (carriable by one or two men). Successful application within three years seems certain, with power sources for military communication equipment in a preferred position.

Transportable (carriable by vehicle with batteries for propulsion excluded). Already installed by Brown, Bovery and Co. (Fig. 3). Successful military applications of other types at higher ratings seem likely.

Propulsion. Successful applications will come first on military vehicles. Golf carts with hydrazine batteries have been demonstrated by Allis-Chalmers. Military fork-lift trucks should operate on fuel batteries within ten years. The Swedish submarine effort was mentioned previously.

The passenger automobile seemingly affords the fuel battery a great opportunity, but unit capital cost is such a formidable hurdle now that other problems are scarcely worth discussing. For the present, effort should be concentrated on automobiles that use storage batteries, perhaps new types not yet in use, which might be replaced or kept charged by satisfactory fuel batteries. The locomotive or the fuel-battery-powered railroad car is a more promising application than the passenger automobile.

The home. Steady progress (Broers, TNO, Holland; Institute for Gas Technology, Chicago) being made on methane-air batteries with molten-carbonate electrolytes leads one to expect experimental home installations exceeding 20 per cent in comparative thermal efficiency within five years; such fuel batteries would be connected to banks of storage batteries as energy reserve for peak loads.

Central stations. The central station ranks with the passenger automobile in difficulty as an application for the fuel battery. It differs from the automobile in that unit capital cost is a less serious hurdle here than overall energy cost. The future of the fuel battery in the large-scale generation of electricity seems linked to the future of natural gas. The growth of atomic energy installations and the effect this will have on the coal industry both enter the picture because one must look perhaps a decade ahead for the earliest time that a central-station fuel battery might begin to be used. There seems no hope for fuel batteries in large central stations. But there is hope for the fuel battery in smaller central stations that serve a single community—stations in which the use of heat and of electric energy will be most efficiently combined and distribution costs will be reduced.

A logical position at present seems to be that fuel-cell research should continue so long as significant progress is made, and that the engineering development of H_2—O_2 and of H_2-air batteries for favourable applications should be emphasized. Fuel batteries will prove themselves indispensable in some applications and useful in many others.

Bibliography

Baker B. S. (Ed.) (1965) *Hydrocarbon Fuel Cell Technology,* New York: Academic Press.

Gould R. F. (Ed.) *Fuel Cell Systems, Advances in Chemistry Series No. 47,* Washington: American Chemical Society.

Liebhafsky H. A. (1966) in *Encyclopaedic Dictionary of Physics* (J. Thewlis Ed.), Suppl. Vol. 1, 110, Oxford: Pergamon Press.

Liebhafsky H. A. and Cairns E. J. (1968) *Fuel Cells and Fuel Batteries,* New York: Wiley.

Vielstich W. (1965) *Brennstoffelemente,* Weinheim: Verlag Chemie.

Note: The Proceedings of the Annual Power Sources Conference published each year by the P. S. C. Publications Committee, Red Bank, New Jersey usually contains late papers on fuel cells.

H. A. Liebhafsky

FUNDAMENTAL PARTICLES IN THE SERVICE OF MAN. At the present time researches into the structure of the atomic nucleus and of the forces that hold it together constitute a major activity of "big science" and employ a substantial proportion, in man-power and equipment, of the world's research effort in the whole field of physics. This is a far cry from the situation only a few decades ago when the efforts of individuals, often working in primitive conditions and on a shoe-string budget, pioneered the field with the establishment first of the existence of the electron as a constituent of all matter, and later of the fact that atoms have nuclei and that these nuclei consist of protons and neutrons. Now some of the world's most expensive research tools are being used in establishing order amid the apparent welter of so-called "fundamental" particles. Some of the ways in which a number of these particles have found practical applications in technology and medicine are described in this article.

Of all the fundamental particles so far discovered and classified only a very few have assumed importance in fields other than scientific research: these are the electron, the positron, the proton and the neutron. There is little doubt, however, that developments will arise in course of time which will make practical use of other particles besides these, and already two of the mesons—μ and negative-π—show promise of usefulness.

The electron. Easily the most important and versatile of the elementary particles, from the point of view of practical applications, is the electron. It is convenient to classify its role in practical affairs according to its origin—orbital or nuclear—and to the immediate physical environment in which it is used.

a) *Electrons in orbit.* In the conditions which obtain at the Earth's surface nearly all atoms carry sufficient orbital electrons of their own, or shared with other atoms, to balance the positive charge due to the protons on their nuclei. All chemical properties and changes depend upon the numbers and arrangement of electrons in their orbits, and the formation or breakage of the chemical links between atoms involve the redistribution of these orbital electrons. The formulation and development of the ionic theory and the electronic theory of valency, and concepts such as free radicals and electron density, have brought about a far clearer understanding of the nature of chemical structure and change than had previously been possible. With this understanding have come many of the major advances in chemical research and technology, ranging from commercial developments in the electrochemical and plastics industries, to the spectacular rise of the whole new science of molecular biology with its practical implications for medicine and agriculture. In addition the techniques of electron diffraction and electron spin resonance are valuable tools for the industrial scientist.

Electrons in conducting solids. Current electricity was already being used commercially on the grand scale well before Thompson's discovery of the electron; nevertheless its visualization as a shunting movement of loosely bound electrons in a conductor has helped considerably in the theoretical and practical development of current electricity for heat and light and of electromagnetism for motive-power and for communications, etc. The discovery and development of superconductivity is perhaps the most promising outcome in recent years of this concept of the electric current.

Electrons in semi-conducting solids. Certain substances or assemblies allow the passage of electrons in one direction more readily than in another. These include particularly some of the elements (and their compounds) whose orbital electron structures put them in the middle groups of the periodic table—between the obviously metallic and the obviously non-metallic elements. If the regular crystal structure of one of these materials is thrown out of step, as it were, by the introduction of "impurity" atoms, and if another material is placed in contact with it, electrons will flow across the boundary more readily in one direction than in the other. Such devices form the basis of solid-state diodes or rectifiers which are used both in low-current electronic devices and, increasingly, for handling large amounts of electricity in a.c./d.c. power rectifiers. The transistor is another solid-state semiconductor device in which an applied potential difference is used to control the size and direction of the current. It can act as a rectifier or an amplifier or as a switch without moving parts. These devices are the basic components upon which modern high speed computers are entirely dependent; they are used increasingly in all kinds of other electronic systems from domestic portable radios to advanced navigational guidance systems.

Free electrons in gases and plasmas. If sufficient energy is transferred to the atoms of a gas to ionize them the gas will become a conductor of electricity and in this state is called a "plasma". Gases, especially at low pressure, can be ionized by heat, by X rays, by electromagnetic or corpuscular radiation from radioactive decay, by high-frequency alternating current fields or by sufficiently high direct current potential differences. When a current flows in a plasma some of the atoms may be excited even further and their electrons pushed into abnormal orbits from which they revert with the emission of light or other electromagnetic radiation of characteristic frequency. This way of producing light is now very widely used in the form of fluorescent lighting, in mercury or sodium street lamps, in ultra-violet lamps and in electronic flash tubes and for other specialized applications. The earliest commercial form of electric light—the carbon arc lamp—relies upon a plasma initially created by the intense heat of an electrical short circuit and sustained by the heating effect of the current continuing to pass through the plasma. Arc lighting is still used in cinema projectors and searchlights, but the major use of the electric arc is now in welding, either in air or in an inert gas such as argon, usually at atmospheric pressure. The mercury arc operating in a near-vacuum is still the most important type of a.c./d.c. power rectifier.

The plasma which carries the current in an electric spark is produced by the breakdown of the insulating properties of the intervening gas, i.e. its ionization—by the potential difference across it. Intense local heating is produced in the path of the spark. The most familiar application is in the sparking plugs of internal combustion engines, and other uses include the firing of explosive charges and the spark erosion machining of complex shapes. Spark-gaps are used in the safeguarding of electrical equipment, and in the operation of some early types of radio transmitter.

Electrons in vacuo. Electrons "boiled off" from a hot cathode will carry a current to an anode held at a positive potential. This forms the basis of the diode valve which, by allowing a current to flow in one direction only, acts as a rectifier. With the addition between the cathode and the anode of a grid on which the potential can be varied at will to control the flow of electrons from cathode to anode, the valve can be used as an amplifier. These simple thermionic valves and developments from them—some extremely complex—form the basis of nearly all radio, radar and television transmission systems and of most receivers, as well as of many industrial instruments.

In the cathode-ray tube used in a television receiver a focused beam of electrons from a hot cathode is controlled in intensity by the incoming signal, while it is made to scan a fluorescent screen in step with the transmitter. Radar receivers and many kinds of industrial instrument use cathode-ray tubes to produce visible signals convenient to interpret or record. Scanning beams of electrons are also used in TV camera tubes; in one type the electron beam scans a light-sensitive plate on to the other side of which the picture is optically focused; electrons are re-emitted in accordance with the brightness of the optical image at each point of the scan, and these provide the signals which are amplified and transmitted to the receiving station.

X rays are produced when a beam of electrons from a hot cathode is suddenly decelerated or when it excites the orbital electrons in the atoms of the target anode; the energy of the X rays depends upon the atomic number of the target material and upon the energy of the electrons. For many practical purposes, including medical diagnosis and therapy, and industrial non-destructive testing, the target is a heavy metal such as tungsten. The electrons are given an energy ranging normally from about one thousand to one million electron volts, either by applying a simple potential difference between the cathode and the target anode or (for the highest energies) by employing some form of additional acceleration such as a linear accelerator or a betatron.

A beam of electrons *in vacuo* is used in the electron microscope; the beam passes through the specimen and is focused by means of a compound system of electromagnetic lenses (analogous to the optical lenses in a conventional microscope) upon a fluorescent screen; here it can be examined optically or recorded photographically. In a more recent version the electron beam scans the target in a manner similar to that in a television camera (see above) and the emitted electrons provide a current which is amplified and forms an image on a cathode-ray tube screen. Magnifications of up to several hundred thousand times are obtainable with the electron microscope. In electron microprobe analysis a beam of electrons excites characteristic X rays at a point on a specimen and measures their intensities at appropriate energies.

Beams of electrons *in vacuo* or emerging into air from a vacuum tube can deposit a great deal of energy in a short time. These are finding increasing applications in industry; for example, their heating effect is used for the welding of particularly difficult materials, while their ability to promote certain chemical reactions before causing a significant rise in temperature is used for cross-linking polyethylene to raise its melting point or to give it the property of heat-shrinkability. Electron beams are being used experimentally for the ultra-rapid drying of industrial paints, particularly on surface which cannot be heated without damage, and their biological effect enables them to be used for cold-sterilizing surgical materials or articles where no great degree of penetration is required, or where the dose rate has to be higher than that which can be achieved by γ-ray treatment.

b) Beta-particles—electrons from atomic nuclei. When a radioactive atom undergoes β-decay a neutron changes into a proton and an electron, the latter being flung out from the nucleus with a maximum energy characteristic of the particular radioisotope; in the case of many of these isotopes one or more γ-ray photons are also emitted. Beta particles have found many practical uses, particularly in industrial measurement and control devices where the extent of their absorption or scattering by the material under test is measured electronically; a signal is produced which can be made to indicate the thickness, density or mass per unit area of the material. Where low energy β-particles are involved the amount of scattering is also to some extent dependent upon the atomic number of the material. The signal can be fed into a metering or recording device, or back into the system which controls the characteristic being measured. Applications range from the laboratory measurement of the thickness of the gold conducting layer on a printed circuit to the automatic control of an entire sheet-metal rolling-mill.

Like cathodic electrons, beta-particles can be used to excite X rays from other atoms; these isotope-produced X rays are used in industrial devices, particularly for non-destructive analysis by X-ray fluorescence spectroscopy of alloys and minerals in factory or field, and for the continuous monitoring of the sulphur or metal content of hydrocarbon oils by the preferential absorption of X rays by elements of higher atomic number.

Because of the ease with which β-particles can be detected, β-emitting radioisotopes are used as tracers in a very wide range of studies, especially in biochemistry, medical diagnosis and industrial process investigations.

Because of their ionizing properties, beta-particles are used for medical treatment where limited penetration is required; for example, strontium-90 in specially shaped applicators is used for the treatment of the cornea of the eye.

Because of its low radio-toxicity, long half-life (12-years) and relative cheapness, tritium (hydrogen-3) is used as an activator in luminous signs, for which it has largely replaced radium and strontium-90. This last, however, being available in large quantities from spent nuclear fuel and having a half-life of 28 years, is finding application in the powering of devices such as flashing navigational beacons or other installations that are particularly difficult to service. Once installed, they can be expected to go on working for a decade or more without attention.

Positrons. Radioisotopes that are rich in protons, as opposed to the more plentiful and easily produced isotopes rich in neutrons, decay by emitting positrons (positive electrons). When a positron comes to rest in matter it reacts with an ordinary (negative) electron, both are annihilated and the whole rest mass of the two particles is converted into two photons each of 0·51 MeV energy, which can be very readily detected by scintillation counters. If two such counters, arranged to register only when both receive a signal at the same time, are set up on opposite sides of a source of positrons the annihilation photons will be detected unmistakably even against a background of other radiations. A positron emitter can in this way be located and its strength measured, even at very low concentrations and through a substantial bulk of surrounding material, more or less independently of the background radiation. A typical example is the use of cyclotron-produced oxygen-15 for the study of the heart-lung function; here the uptake of oxygen into the lungs, its progressive removal by the blood circulation and its final exhalation, can be studied in detail in different regions of each lung by external counting of the annihilation radiation.

Negative π-mesons. Although work is still at the early experimental stage it is possible that beams of negative π-mesons may one day prove to be valuable weapons in the fight against cancer. Unlike γ or X rays they can be made to deposit the bulk of their energy at a controlled depth and in a limited volume of tissue. The biological effect of a negative π-meson beam at the tumour depth will be produced mainly by alpha-particles that result from nuclear interactions of the π-mesons when they come to rest. Such alpha-particles have a high relative biological efficiency and a low oxygen enhancement ratio—that is to say, they do not need oxygen to be effective. Irradiation in all other regions is by lightly ionizing particles which have a low relative biological efficiency. At the present time potentially useful beams of π-mesons can only be produced in a few of the world's largest particle accelerators.

μ-mesons. A study is currently (1967) in progress on the suitability of cosmic-ray μ-mesons as probes for non-destructive examination of the Egyptian pyramids; preferential transmission of these radiations through any part of the structure might be taken to indicate the presence of internal cavities, leading to the location of hitherto unknown—and perhaps unentered—burial chambers.

Protons. Many chemical changes in which hydrogen takes part involve the existence—albeit fleeting—of free protons; in certain branches of organic chemistry the intra- or intermolecular migration of protons plays an important part.

Free protons are present in the partially ionized gas of the "atomic hydrogen" welding torch whose heating power derives largely from the catalytic reformation at the metal surface of H_2 molecules previously dissociated by passage through an electric arc.

Beams of protons produced by acceleration of hydrogen ions in a cyclotron are used in the production of some proton-rich radioisotopes which decay by positron emission.

Protons ejected from linear accelerators are used as plasma jets for the propulsion and manoeuvring of space vehicles.

Neutrons. Since the discovery in 1938 of nuclear fission, neutrons have played a role of rapidly increasing practical importance.

When a neutron causes fission of ^{235}U or ^{239}Pu, two or three fresh neutrons are produced in the process; this can lead in certain circumstances to a self-propagating chain reaction which can either be kept going at a steady rate in a nuclear reactor or made to build up explosively in a bomb. The energy produced is of the order of 10 million times that produced in chemical reactions involving the same mass of material.

In a nuclear reactor the chain reaction is sustained at a steady level by ensuring that on an average one neutron from each fission goes on to cause one further fission; the energy produced may be used to raise steam and generate electricity. In 1966 about 12 per cent of Britain's electricity was generated in nuclear power stations, and by the 1970's nuclear generation is expected to be the cheapest way of making electricity in most industrialized countries. In "breeder" reactors more neutrons are produced than are required to sustain an efficient reaction at the proper power level; some of these are absorbed in ^{238}U or ^{232}Th to produce fissionable ^{239}Pu or ^{233}U respectively. This new fissionable material can be used either *in situ* or extracted and fabricated into fresh fuel elements.

The hailstorm of neutrons present in an operating nuclear reactor can be used to make materials or objects radioactive by introducing further neutrons into their nuclei; such neutron-rich nuclei will decay by the emission of β particles with, in some instances, γ rays as well. Radioactive isotopes, both proton-rich and neutron-rich, are also produced as a direct result of the fission of uranium or plutonium; these fission products can be separated chemically from the bulk of the spent fuel and converted to the required chemical or physical form. Radioactive isotopes of almost every element are now available commercially as primary isotopes (elements or simple compounds), isotopically-labelled compounds, or sealed sources of radiation.

Specimens for analysis may be bombarded with neutrons and the radiations subsequently given off by the radioisotopes so formed may be examined; the nature of these radiations indicates which radioisotopes are present, and their intensities indicate the amount of each. Knowledge of the nuclear reactions leading to the production of these radioisotopes makes it possible to determine, quantitatively and qualitatively, many of the elements constituting the original specimens. This technique, known as activation analysis, is limited in its practical application to elements which readily absorb neutrons. Because of its extreme sensitivity, it is particularly valuable in trace analysis, for example in forensic work; it can also be adapted to certain types of routine non-destructive analysis in industry. The usual source of neutrons for activation analysis is a nuclear reactor, but in the case of oxygen in steel and other on-the-spot analyses where no reactor is available use is made of neutron generators; in these, 14 MeV neutrons are produced by reaction between electrically accelerated deuterium ions and a tritium target.

Portable industrial instruments for use on site, e.g. for borehole logging or for soil-moisture measurement usually depend upon neutrons produced on the spot by nuclear reactions such as that between beryllium and α-particles from radioactive decay; these instruments usually detect either prompt γ rays (e.g. from chlorine in the brine associated with oil-bearing formations) or thermalized (slow) neutrons, scattered back by the hydrogen nuclei of water molecules in, for example, the soil or other bulk material adjacent to the neutron source.

In the medical field neutron techniques under development include quantitative measurement by activation analysis of constituents of the living body (e.g. total sodium, chlorine or hydrogen), direct neutron-beam therapy and neutron activation *in situ* of implanted materials such as boron. Industrial neutron radiography can be used to reveal light materials that absorb or scatter neutrons in a mass of heavier material that is relatively transparent to neutrons but is too dense or massive to be examined satisfactorily by X rays or γ rays—for example, organic matter embedded in metal.

Bibliography

BRODA E. and SCHONFELD T. (1966) *Technical Applications of Radioactivity*, Vol. 1, Oxford: Pergamon Press.

COHEN J. B. (1966) *Diffraction Methods in Materials Science*, London: MacMillan.

FOWLER P. H. (1965) 1964 Rutherford Memorial Lecture, π -mesons versus cancer? *Proc. Phys. Soc.* **85**, June.

GAVIN M. R. and HOULDIN, J. E. (1959) *Principles of Electronics*, London: E.U.P.

GRAY A. and WALLACE G. A. (1962) (8th Edn.) *Principles and Practice of Electrical Engineering*, Princeton, N. J.: McGraw-Hill.

LENIHAN J. M. A. (1966) (2nd Edn.) *Atomic energy and its Applications*, London: Pitman.

LOFTNESS R. L. (1964) *Nuclear Power Plants*, Princeton, N. J.: Van Nostrand.

MOORE W. J. (1962) (3rd Edn.) *Physical Chemistry*, New York: Prentice-Hall.

POTTER E. C. (1956) *Electrochemistry: Principles and Applications*, New York: Cleaver Hume.

SMITH C. M. H. (1965) *A Textbook of Nuclear Physics*, Oxford: Pergamon Press.

WELCHER F. J. (Ed.) (1966) (6th Edn.) *Standard Methods of Chemical Analysis*, Vol. 3, Instrumental methods, Part A, Princeton, N. J.: Van Nostrand.

WHEATLEY P. J. (1959) *Determination of Molecular Structure*, Oxford: Clarendon Press.

R. M. LONGSTAFF

G

GAS LENS. A light beam can be focused as it passes through a volume of gas in which there are appropriate gradients of refractive index. A system of gases and hardware that maintains gas flow in a configuration suitable for focusing light is called a gas lens.

Invention and development of gas lenses have recently been stimulated by advances in laser technology. It has been suggested that modulated light beams from laser transmitters might be guided to receiving stations many kilometers away through a series of weak lenses with unusually low light loss. Gas lenses are weak because of the very small differences in refractive index among gases. However, such lenses are essentially free of the reflection and scattering that ordinarily occur at the surfaces of solid lenses. The extremely low reflection and scattering by gas lenses are due to the small refractive index differences and to the fact that the transition between regions of high and low refractive index is necessarily very gradual, because of gas diffusion, rather than abrupt, as at a solid surface.

Gas lenses may be composed either of a single type of gas, in which the refractive index is varied by maintaining temperature gradients in the gas, or of two or more types of gas that have different refractive indices at the same temperature and pressure. In either type of gas lens, the appropriate distribution of refractive index is maintained by maintaining uniform controlled flow of the gas or gases. Such flow must be non-turbulent if the lens is to be steady and free of scattering. When flow is non-turbulent, pressure differences are generally too small to have much effect on focusing.

It is possible to make a thermal gas lens with no moving parts, using thermal convection between warm and cool regions in solid surfaces surrounding the path of light through the gas lens. A simple form of convective thermal gas lens, and the first gas lens to be demonstrated successfully, is simply a warm wire helix, coaxial with a cooler tube surrounding the helix (Berreman 1964, 1965) (see Fig. 1). If the tube is approximately horizontal, the gas rises past the helix and falls in the outside region between the helix and the outer tube. The average temperature on a line along the axis of the tube and the helix is somewhat lower than the average temperature along parallel lines somewhat off axis and nearer the inner boundary

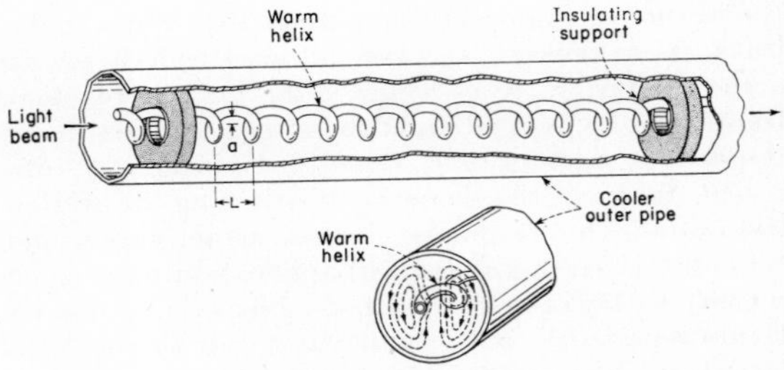

Fig. 1. A convective thermal gas lens (from Berreman 1965).

of the warm helix. The optical path along the axial line is greater than along an equal length of line off axis because the mean density of the gas in the cooler, axial region is higher than the mean density of the warmer, off-axis region. Consequently, a short length of such system acts as a positive lens. A warm ring or series of rings can also be used in place of the warm helix.

Thermal gas lenses of considerably greater strength, which are also less subject to aberrations, can be made by forcing gas to flow axially from a cool into a warm section of cylindrical tube (Marcuse and Miller 1964; Beck 1964). The cool gas near the axis moves a greater distance into the warm tube before it reaches approximately the temperature of the walls than does the gas nearer the walls. Consequently, such a gas lens is positive. A negative lens may be made by flowing warm gas into a cool tube.

Still stronger lenses can be made by causing a gas of high refractive index to flow into a gas of low

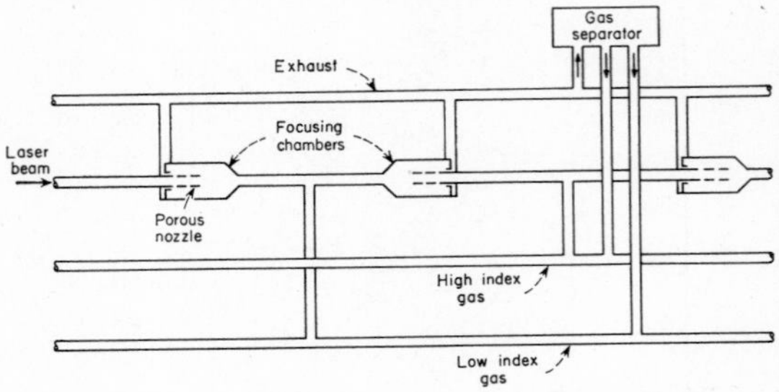

Fig. 2. A gas lens system using unlike gases (from Marcuse and Miller 1965).

refractive index in such a way that the gas near the axis flows farther along the axis, before it diffuses into a more or less homogeneous mixture, than that near the boundary of the lens (Berreman 1964; Beck 1964) (see Fig. 2). Such a lens can be made practically free of gravitational convection and consequent distortion, if the two types of gases used are of approximately equal density.

Bibliography

Beck A. C. (1964) Thermal Gas Lens Measurements, *Bell System Tech. J.* **43** (part 2), 1818, July.

Beck A. C. (1964) Gas Mixture Lens Measurements, *Bell System Tech. J.* **43** (part 2), 1821, July.

Berreman D. W. (1964) A Lens or Light Guide Using Convectively Distorted Thermal Gradients in Gases, *Bell System Tech. J.* **43** (part 1), 1469, July.

Berreman D. W. (1964) A Gas Lens Using Unlike, Counter-Flowing Gases, *Bell System Tech. J.* **43** (part 1), 1476, July.

Berreman D. W. (1965) Convective Gas Light Guides or Lens Trains for Optical Beam Transmission, *J. Opt. Soc. Amer.* **55**, 239, March.

Marcuse D. and Miller S. E. (1964) Analysis of a Tubular Gas Lens, *Bell System Tech. J.* **43** (part 2), 1759, July.

D. W. Berreman

GEOMAGNETIC INDICES. The degree of geophysical disturbance which affects the Earth during and after solar disturbances could be measured in various ways. However, the usual method makes use of the variations of the magnetic field at the surface of the Earth, which are almost certainly caused by the abnormal electric currents that flow in the ionosphere at such times. To provide a quantitative measure, it was decided in 1906 to assign to each magnetogram a "character figure": 0 for an undisturbed, or "quiet" record; 1 for a moderate disturbance; and 2 for a severe disturbance. The figures from all the reporting observatories were combined into an "international character figure" with a scale of 0·1, i.e. 0·0, 0·1, 0·2, etc. This was supposed to represent the disturbance for the whole Earth, rather than for any particular station, though in fact it was biased by the uneven distribution of the observatories.

The methods have been refined over the years, and modern practice, which is largely due to the work of Bartels, derives several related indices. These can be given in two forms: a "local" index representing the disturbance at a particular magnetic observatory; and a "*planetary*" *index* representing the disturbance for the Earth as a whole. The planetary indices, which are of the greater interest, are known as the *three-hour range index*, K_p; the preliminary international character figure, C^i; the *character figure*, C_p; and the *daily equivalent amplitude* A_p. They are derived as follows.

A magnetogram is first assigned a local *K index*, which is obtained by noting the greatest excursion

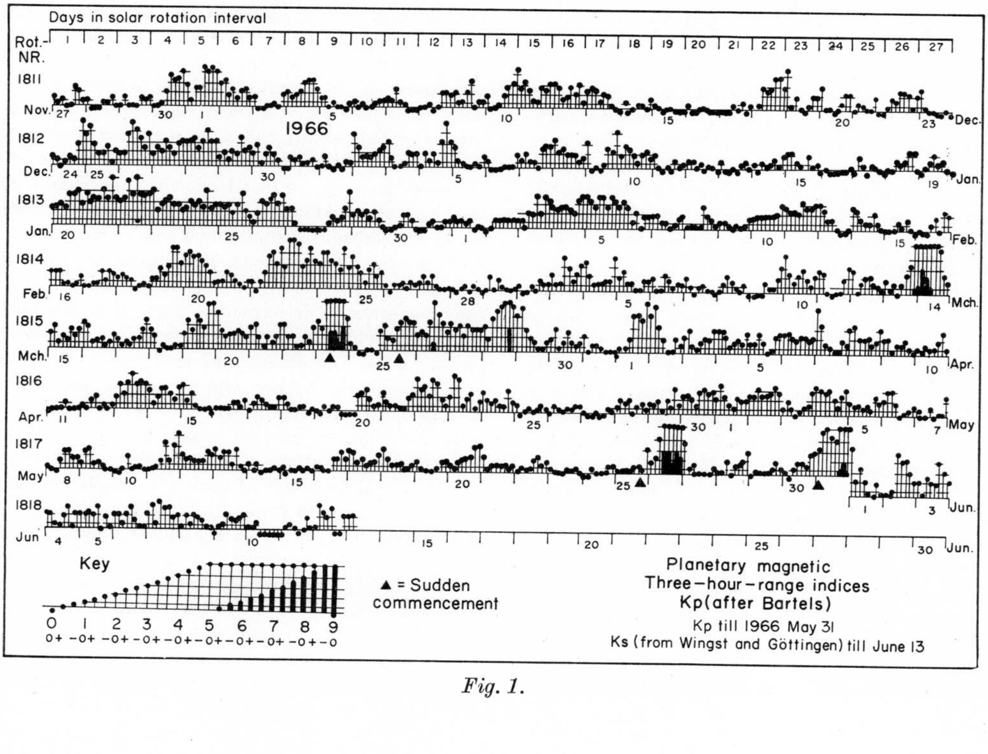

Fig. 1.

(i.e. the range) in any component, in a specified three-hour interval of the Greenwich day (00-03 UT, 03-06 UT, etc.) and converting it according to a logarithm scale into the range 0–9: 0 for "very quiet" and 9 for "very disturbed". A standardized K index, K_s, is then produced by application of a factor which depends on the geomagnetic latitude of the observatory. This takes account of the fact that some latitudes are more subject to disturbance than others. K_s is given to one-third of a unit (e.g. 5−, 5+, 6−). The K_s indices from 12 selected observatories in geomagnetic latitudes between 47° and 63° are combined to give the planetary index K_p, which is a good, if arbitrary, measure of disturbance for the whole planet and is virtually free of any diurnal or seasonal bias. From K_p, an intermediate three-hourly index ap is derived by converting to a linear scale in the range 0–400, so chosen that ap approximates to the actual deviation of a magnetogram trace measured in gamma. The daily average of the eight values of ap is A_p. The character figure C_p is now derived by converting the daily sum of a_p to the range 0·0–2·5. The preliminary character figure C^i is another version of the same index, but obtained as the average of the character figures which are also reported by the observatories. This maintains a certain continuity with the earlier procedures. The collation and analysis of the geomagnetic indices is carried out at the University of Gottingham and at the meteorological office at De Bilt, and the data are made generally available. A list of the quietest and most disturbed days in each month is also circulated. Kp is often displayed on "musical diagrams" like Fig. 1.

Other indices have been devised for more specific purposes and used to a limited extent. A *Q index* has been recommended for the highest latitudes, to be derived at 15-minute intervals at observatories between 58 and 90° geomagnetic latitude. The procedure is similar to that for the K index except that only the two horizontal components are inspected. More recently, an *R index*, giving hourly the range of the horizontal components in tens of gamma, has been proposed instead of Q for stations above latitude 65°. Various "*difference indices*" (u and D_{st}) have also been given. These require the comparison of records at the same time on successive days. D_{st} is of particular interest as it is intended to show the intensity of the ring current which may encircle the earth at a distance of several earth radii during magnetic storms. It is derived as the average decrease of horizontal force over all longitudes but restricted to low-latitude observatories.

Although arbitrary, in the sense that their physical significance was not always well defined, magnetic indices have long provided a common basis for geophysical studies, making it possible to intercompare different observations (for example, of the aurora) at a known level of activity. K_p and A_p are used most in such studies. More recently, space probes have found that the velocity of the solar wind (the ionized plasma of electrons and protons flowing from the Sun), and the strenght of the interplanetary magnetic field which it carries, are correlated with K_p. The reason for the correlation is not yet known, but it does seem that a satisfactory physical interpretation of geomagnetic indices may be at hand.

Bibliography

BARTELS J. (1957) *Annals of the International Geophysical Year* **4**, 215, 227, Oxford: Pergamon Press.

BARTELS J. (1962) *Collection of geomagnetic planetary indices Kp and derived indices A_p and C_p for the years 1932–1961*, I.A.C.A. Bulletin No. 18. Amsterdam: North-Holland.

Solar-Geophysical Data, U.S. Department of Commerce, Boulder, Colorado.

LINCOLN J. V. (1967) Geomagnetic Indices, in *Physics of Geomagnetic Phenomena* (eds. S. Matsushita and W. H. Campbell), Vol. I, p. 67. New York and London: Academic Press.

J. K. HARGREAVES

GRAVITATIONAL EFFECTS OF LUMINOSITY. The gravitational mass of the Sun is slowly dwindling due to the conversion of mass in its interior into nuclear energy and subsequent radiation of this energy from the surface of the Sun. If the nuclear processes were to cease, the Sun's mass would nevertheless continue to dwindle so long as it radiated energy from its surface, since the radiation energy trapped within the Sun constitutes an attractive mass that may be calculated using Einstein's famous equation $E = mc^2$. This reduction in the attractive mass of the Sun is so small as to be completely negligible in the calculation of planetary orbits or the internal hydrodynamics. Neutrinos are also produced in the centre of the Sun and, having enormously longer mean-free-paths than optical radiation, are lost directly from the centre out into space. The loss of mass (energy) due to neutrino radiation is comparable to that due to electromagnetic radiation. Recently, observational work on extragalactic radio sources has focused theoretical attention on the dynamical processes in massive objects. In such massive stars, it is expected that catastrophic events might well lead to gross losses of gravitational mass via neutrino radiation, and therefore it is of interest to contemplate what effects one might expect to find in such circumstances.

In Newtonian gravitational theory, as well as in Einstein's general theory of relativity, a uniform shell of mass (energy) will attract a particle provided the particle is outside the shell. When the particle is inside the shell, there is no longer any attraction to that shell. Consequently, in the case of a radiating body such as the Sun, the radiative energy emitted from the surface of the Sun continues to attract a particle until that radiation has gone past the particle. Afterwards the radiation no longer attracts, and therefore the Sun exerts a constantly decreasing attractive force, as is to be expected in view of its decreasing mass. If one considers the extreme example of the Sun or a star completely converted in an instant into neutrinos, an orbiting body would move on its course until the pulse of radiation reached it. Subsequently the body would continue with the same velocity but, no longer having a centrally attracting body, would no longer be bound. Thus in this extreme example objects formerly bound would have free velocities equal to that of their previous orbital velocities. In the case of the massive star, a catastrophic loss of mass from its centre would in the same way lead to a reduced gravitational binding of the outer shell of the star. In this way the internal energy obtained within that shell, which formerly served to resist the gravitational forces, would now cause it to expand outward to exceedingly high velocities, and the outer shell could be liberated from the star. It has been proposed that so-called run-away stars, namely stars that have unusually high velocities relative to their neighbours, are the result of the nearly complete dissipation by supernova activity of a companion in a closely bound binary star system. The reduction in mass of the companion thereby frees the other star which will now move with velocities comparable to its former orbital velocity. Whether such processes are actually of the astrophysical significance proposed has not been observationally verified, however they are interesting applications of gravitational theory.

The calculation of the above effects using the general theory of relativity differs in essential ways although the over-all result remains the same. In the general relativistic calculation, one obtains not only the same change in attractive acceleration due to the decreasing mass but also a new acceleration proportional to L/r (where L is the luminosity, not necessarily constant in time, measured at a test particle, and r is the distance of that particle from the centre of attraction). This term is exactly equal to the gravitational attraction of the radiation that occupies the space between the Sun and the test particle. The full expression for this acceleration is GL/rc^3 (G is the gravitational constant and c is the velocity of light) and, for the Sun, amounts to only $1{\cdot}1 \times 10^{-18}$ of the acceleration to the Sun itself. However, the general relativistic calculations automatically take into account the fact that one does not see the diminution in the mass until the radiation reaches the test particle. Thus the radiation between the gravitating object and the observer appears *twice* in the relativistic calculation, once as the not-yet-observed diminution of attraction and once in the new L/r term which has no Newtonian counterpart. A very brief outward pulse of radiation from the Sun would therefore give both a nearly discontinuous reduction in the attraction of the Sun for the test particle, and an inward impulse from the L/r term. This inward impulse, quite different from what one would expect on the basis of the Newtonian theory, is actually unobservable for the following reason. Consider an observer at some great distance from the system. He will see a gravitational red shift (reduction) in the frequency of any light from the test particle because it is bound in the gravitational field of the Sun and the frequency reduction simply represents the work that the radiative energy must do against the attraction of the star to escape from that star. Now, if the Sun loses mass, there is a lessening of that gravitational red shift. To the observer, this frequency change is indistinguishable from an outward velocity change which could also give the same change in frequency via the Doppler effect. Thus, the impulsive emission of energy from the star causes not only an apparent inward impulse due to the L/r term, but also gives an apparent outward impulse due to the discontinuous change in the gravitational red shift. In fact, the two effects exactly cancel, and the over-all effect of luminosity in the Einstein theory is essentially the same as in the Newtonian theory. The interesting facet here is not just that the two theories give the same result, but also the subtle interplay of the various effects found in the general theory of relativity.

Bibliography

BLAAUW A. (1961) *Bulletin of the Astronomical Institute of the Netherlands*, **15**, 265.

FOWLER W. A. and HOYLE F. (1963) *Nature*, **197**, 533.

HOYLE F. and FOWLER W. A. (1963) *Mon. Not. Roy. Astron. Soc.* **125**, 169.

LANDAU L. and LIFSHITS (1959) *Classical Theory of Fields*, New York: Wesley.

MICHEL F. C. (1963) *Astrophys. J.* **138**, 1097.

MICHEL F. C. (1965) *Phys. Rev.*, **140**, B514.

MISNER C. W. (1965) *Phys. Rev.*, **137**, B 1360.

THEWLIS J. (Ed.) (1961) *Encyclopaedic Dictionary of Physics*, articles beginning "Gravitation", "Gravitational", "Gravity", **3**, Oxford: Pergamon Press.

TOLMAN R. C. (1962) *Relativity, Thermodynamics, and Cosmology*, Oxford: Clarendon Press.

F. C. MICHEL

GRAVITATIONAL GEONS. The original term geon refers to a bundle of electromagnetic energy held together for a finite time by its own gravitational attraction. A gravitational geon would be a similar bundle of gravitational energy. It is possible to construct solutions to the Einstein gravitational field equations in the absence of both explicit mass sources and electromagnetic fields which, for a finite time, are singularity free and have the external characteristics of solutions with mass sources. Such fields would correspond to gravitational geons, but, as with the original, electromagnetic geons, have not been observed as such.

Bibliography

THEWLIS J. (Ed.) (1961) *Encyclopaedic Dictionary of Physics*, articles beginning "Gravitation", "Gravitational," "Gravity", **3**, Oxford: Pergamon Press.

C. H. BRANS

GRAVITATION, THEORIES OF (SURVEY). Although the primitive notion of gravitation is associated with the force felt by any body near the surface of the Earth, called its weight, the first scientific approach to the problem, Newtonian theory, grew out of a study of the motions of the planets in the solar system. This theory is based on the "Universal Law of Gravitation", which asserts that any two bodies are coupled by an attractive force along their line of centres which is proportional to the square of the distance between them. In retrospect, the development of this law is quite straightforward, once the planetary data is described relative to the proper coordinate system, that is, relative to an inertial coordinate system. The work of Brahe and Kepler indicated that the Copernican, or heliocentric, choice was the proper one, and Kepler's purely kinematic laws can be summarized by the statement that, relative to the heliocentric system, each planet accelerates centrally toward the Sun with a magnitude of acceleration proportional to the reciprocal of its distance to the Sun squared. The proportionality constant is independent of the planet. From this point an application of Newton's mechanical laws, $\mathbf{F} = m\mathbf{a}$, and the law of action-reaction, results in a force between the Sun and each planet of the form described above. The next step is to extend this result and apply it to terrestrial matter. Thus, Kepler's result that the accelerations of the planets are independent of their masses is made equivalent to the fact that all bodies fall with the same acceleration in vacuum near the surface of the Earth. This equivalence between terrestrial and celestial phenomena which appears quite natural by modern standards represents a significant departure from ancient astronomical thought.

The Cavendish experiment, which actually measures this attractive force between terrestrial objects, provides direct confirmation of the terrestrial validity of Newtonian theory to within limited accuracy. It also provides a measurement of the Newtonian gravitational constant which in turn makes possible a calculation of the mass of the Earth (and other objects in the solar system) from distance and gravitational acceleration measurements.

It is also possible to state Newton's theory as a conservative field theory rather than in terms of action-at-a-distance. Introducing a gravitational potential, φ, this theory can be summarized by the field equation $\nabla^2\varphi = 4\pi G\varrho$ and force equation $\mathbf{F} = m(-\nabla\varphi)$. In this equation, ϱ represents the density of inertial mass producing the field, φ, while the number m represents the inertial mass of the test particle experiencing the force due to this field. In this form Newton's theory bears a strong resemblance to the Coulomb electrostatic field theory but with one important difference. In Newton's theory the quantity which couples the particle to the field is the inertial mass of the particle rather than some other number (such as electric charge) unrelated to inertial mass. Hence, when the force equation is used in conjunction with the Newtonian mechanical force law, $\mathbf{F} = m\mathbf{a}$, the quantity which determines inertial reaction is the same quantity which determines how strongly the particle will react to the gravitational field. Hence, these quantities can be eliminated with the result that the gravitational field, in its effect on the motion of particles, appears to be purely an acceleration field. The first one to attempt a rigorous experimental verification of this fact was Eötvös. His experiment has been repeated more recently by Dicke with the demonstration that, to experimental accuracy, all bodies do experience the same acceleration at a given point in a gravitational field. In addition, further astronomical observations of planetary orbits gave increasing support to Newton's theory, at least a very good approximation. However, there did persist certain small inaccuracies, notably the fact that the motion of the planet Mercury, which experiences the strongest gravitational field, deviates very slightly from the path predicted by the Newtonian theory.

It is important to notice that the proper choice of inertial coordinate system is essential to the development of the theory. This choice was associated with the well-known philosophical debate between the geocentric and heliocentric astronomers. Newton himself believed that the choice of the inertial coordinate system was determined by an absolute structure of space itself. However, others did not. Berkeley, in rejecting the notion of an absolute structure for empty space, instead conjectured that the structure of matter itself was indispensable in understanding motion. Hence, extending Berkeley's ideas, it might be no mere coincidence that the coordinate system appropriate for describing planetary motions is the one in which the fixed stars are at rest. This could be considered a predecessor of Mach's principle which itself was of great importance in the development of the Einstein general relativistic theory of gravitation.

With the development of physics in the early twentieth century came special relativity and quantum theory. So far the interaction between quantum theory and gravitation has been negligible but this is not the case between gravitation and special relativity. Of course, the most natural approach might seem to be a straightforward adaptation of the Newtonian field theory to the frame-work of special relativity. For example, the three-dimensional scalar potential of Newton's theory could be replaced by a four-dimensional scalar or the time component of a four-vector (as with the electrostatic potential) or even a component of a tensor. However, it is also necessary to take into account the special relativistic equivalence between energy and mass. Hence, the energy carried by the gravitational field itself should contribute to the field equations as a source so that these equations must be altered to include non-linear terms associated with this self-interaction of gravitation. An analogous problem would be that of altering Maxwell's equations if photons were charged.

It is possible to carry out such a modification of

Newtonian theory and, if the gravitational field is chosen to be a symmetric tensor field, arrive at equations formally similar to Einstein's but expressed in the flat space time of special relativity. However, rather than viewing the Newtonian theory as one to be adapted to special relativity without essential change, Einstein took a different approach in dealing with the gravitational field.

In Einstein's approach fundamental significance is attached to the equality of gravitational and inertial mass, and this is formalized in his "Principle of Equivalence" which asserts that a gravitational field is locally indistinguishable from an acceleration of the coordinate system relative to an inertial system. In other words, in a sufficiently small elevator over a sufficiently small interval of time it is impossible to distinguish between an acceleration of the elevator relative to inertial coordinate systems in the absence of an external gravitational field and the presence of a Newtonian gravitational field in the elevator at rest in an inertial coordinate system. From the standpoint of Newtonian mechanics, the acceleration of objects to the floor of the elevator in the first case is due to "fictitious" inertial forces while in the second case the acceleration of these objects to the floor is due to the "real" force of gravity. However, since the principle of equivalence asserts that these are observationally indistinguishable (locally) a more satisfying theory would account for this equivalence. Further, this is indirectly related to the next logical step after special relativity; namely, a generalization of the theory to include not only coordinate systems separated by Lorentz, constant velocity, transformations but also accelerated or "curved" coordinate systems.

It is a standard problem in elementary mechanics to transform the description of the motion of a particle at rest in an inertial coordinate system to that observed in a rotating coordinate system and to calculate the acceleration measured in this coordinate system. If this rotating coordinate system is considered to be an inertial system then these acceleration terms (such as centrifugal and Coriolis) are associated with corresponding forces. These "inertial" forces are also proportional to the inertial mass of the particles experiencing them, just as in the case of gravitation. Thus, the gravitational field might be described by quantities of the same "geometric" type as these inertial forces. It is quite easy to generalize this procedure, using the notation of Riemmanian geometry on space-time. In this case an inertial coordinate system is a Minkowski coordinate system and the motion of a free particle is unaccelerated or straight line motion in space-time. An invariant description of this is to say that a free particle follows the path of a geodesic. In a more general coordinate system the equation of this geodesic would be expressed as follows:

$$\mathrm{d}^2x^i/\mathrm{d}\tau^2 + \Gamma^i_{jk}(\mathrm{d}x^j/\mathrm{d}\tau)\,(\mathrm{d}x^k/\mathrm{d}\tau) = 0.$$

In this equation the quantities Γ^i_{jk} are the Christoffel symbols and describe the way the components of a vector change when the vector is displaced parallel to itself. Hence in a Minkowski coordinate system all of these quantities are zero. However, in a non-inertial coordinate system these quantities produce accelerations or, when the equation is multiplied through by the mass, the negative of these quantities would be associated with the "fictitious" inertial forces experienced in the coordinate system due to the fact that the system is not inertial. There is, however, one important difference between gravitational forces and inertial forces. Inertial forces can be "transformed away", that is, in the absence of a gravitational field a coordinate system can be found in which the Christoffel symbols are zero over a region of space-time. However, for a gravitational field, it may happen that this field, although locally equivalent to an acceleration of a coordinate system relative to the inertial coordinate systems, cannot be exactly "transformed away" over a region of space and time. For example, in the case of the gravitational field of the Earth, the fact that the force lines are radially directed toward the centre of the Earth implies that the motion of particles accelerating under this force cannot simply be transformed away by acceleration of the coordinate system. In the language of Riemannian geometry this means that the space time is not flat in the presence of such a gravitational field.

Another way of arriving at the picture of a gravitational field as a curvature of space-time is to consider the motion of light through such a field, taking into account the principle of equivalence. The conclusion is that the frequency of light emitted from a standard oscillator at one point in a gravitational field will differ from the frequency of an identical standard oscillator when it arrives at another point in the field. Hence, a comparison of standard clocks at different points in the field by use of light rays leads to the result that the clocks do not give the same measure for the interval between events when the comparison is made by means of null geodesics, that is, light rays. In other words, parallel straight lines, or geodesics, in this space-time are not always the same "distance" apart so that the geometric structure of space-time associated with such measurements is curved. The full mathematical and quantitative treatment of a gravitational field in these terms is developed by Einstein in his general theory of relativity which directly relates the geometric properties of space-time to the presence of matter.

The predictions of the Einstein theory in the weak field limit coincide with those of Newton so that a necessary correspondence principle is satisfied. However, in stronger fields the discrepancies are significant. To date, the most direct experimental difference is in the prediction concerning the rotation of the perihelion of Mercury. The best available observations of this rotation indicate that Einstein's predicted value is correct within error, provided that there are no, as yet otherwise unobserved, perturbations to be considered from other sources. In addition the Einstein theory predicts the frequency shift of light in a gravitational field and the deflexion of light rays. The first of these two is a direct consequence of the principle of equivalence and thus to lowest order is not a decisive test of Einstein's theory. Recent observations have confirmed the Einstein value for this frequency shift even within the relatively weak gravitational field of the Earth. These experiments make use of the extremely sensitive frequency comparisons that can now be made by use of the Mössbauer effect. The experimental data concerning the deflexion of light has been somewhat inconclusive and cannot yet be regarded as providing incontrovertible support to Einstein's theory. Other techniques for gravitational experiments are being developed using stable and sensitive cryogenic phenomena such as superconductivity and superfluidity to detect small effects by allowing these to accumulate over long times.

From the beginning it was hoped that Einstein's theory would provide a unification, in some sense, of the two basic fields of classical physics: gravitational and electromagnetic. Clearly these fields do interact: the electromagnetic field carries energy that directly contributes to the gravitational field, while the latter affects the space-time structure containing the electromagnetic field. However, Einstein and others hoped that it might be possible to discover a deeper unification between the two, as for example had occurred for the electric and magnetic fields in special relativity and this would be the goal of a unified field theory. Of course, for the theory to possess any significance the unification would have to be non-trivial. One criterion is to require that the electromagnetic and gravitational fields transform into each other under certain transformations as is the case with the electric and magnetic fields in special relativity. Einstein's attempts at unified field theories were not successful in this regard. Other attempts were associated with the introduction of a fifth dimension, in which the components of the full five-dimensional metric tensor contain both the electromagnetic potential as well as the four-dimensional metric gravitational potential. In this case, however, there is no transformation between the two unless some physical significance is associated with the fifth dimension.

It has been shown that the combined Einstein and Maxwell equations by themselves constitute a unified field theory in another sense. Namely, it is possible to eliminate the electromagnetic field from the equations and express them entirely in terms of the gravitational field quantities. Thus the gravitational field, that is the metric tensor, carries the full information concerning both gravitational and electromagnetic interactions. In this sense the Einstein–Maxwell theory is "already unified".

The problem of the introduction of matter into the Einstein theory of gravitation has not been completely resolved. The original procedure of Einstein introduces mass as an explicit source term in the field equations using a stress-energy tensor modeled on fluid dynamics. However, if the explicit mass term is eliminated it is still possible to have solutions which "externally" resemble those corresponding to mass sources. These solutions fall into two classes. In the first place the solution corresponds to a topologically trivial space in which a central core is occupied by field energy of some sort, for example, electromagnetic or gravitational. These solutions correspond to "geons". In the second case, external solutions may be thought of as being produced by topological structures. Given a "$1/r$" solution in an "external" region, i.e. for large r, it may not be necessary to introduce an explicit source term in the differential equations to avoid the singularity at the origin that would occur if the space were topologically Euclidean. Rather, if the space is permitted to have a topology such that the region described by r less than or equal to a certain number is not a bound volume, so that $r = 0$ is not a point in the space itself but a limiting region such as $r = \infty$, then the singularity at $r = 0$ will not occur. In other words, the singularity associated with the convergence of gravitational lines of flux at the origin can be avoided by letting the space possess a "sink" or "wormhole" which allows the lines to pass through to another region where they emerge as diverging lines. The entire problem of the representation of mass in general relativity is still a very fruitful one.

A somewhat related matter is that of expressing the gravitational field energy itself. In special relativity energy is the fourth component of a four-vector, which for a field theory, can be represented as a spatial volume integral of the time-time component of a conserved energy momentum tensor. However, in general relativity where general coordinate transformations are allowed the concept of local energy must be associated with a pseudo-tensor. Further unless boundary conditions of asymptotic flatness are imposed even global energy questions may not be precisely resolved.

The question of energy transfer is critical in an analysis of gravitational waves. Early solutions, which were thought to represent gravitational waves were in fact only coordinate waves and could be shown to transfer no energy. More recently, analysis of the invariant structure of solutions to the Einstein equations has provided invariant and effective criteria for gravitational waves and solutions have been found in these forms. Of course, a relativistic generalization of the Newtonian theory could also have wave solutions so that this is not a result peculiar to Einstein's theory of gravitation. In either case, however, the weakness of the gravitational interaction insures that the energy carried by such waves, under normal circumstances, is extremely small and correspondingly difficult to detect.

Another direction in which research into Einstein's theory of gravitation has proceeded is that of cosmology and cosmological models. Some of the earliest exact solutions known to Einstein equations were those corresponding to cosmological models. Such models use various approximations to the real distribution of mass in the visible universe, usually in terms of fluid dynamics with continuous pressure and density. Some of these models give reasonably accurate accounts of the astronomical data such as the observed redshift and outward expansion of matter.

Other astronomical observations have caused renewed interest in the role of the matter tensor in Einstein's theory, namely the detection of extremely large and concentrated energy sources, indicating the possibility of matter in a state of gravitational collapse, i.e. in a state in which the gravitational forces have overcome all restraining pressures. Present models for material sources in Einstein theory indicate that once such a state has been initiated, it will continue until an intrinsic singularity occurs, corresponding to an infinite density as a final state. Unless some reasonable explanation can be given for the physical significance of such a catastrophe, it seems that the Einstein theory must be modified in some manner when such large concentrations of matter are considered.

Another question of cosmological significance concerns the role of Mach's principle in a gravitational theory, that is, the extent to which local inertial properties are influenced by the matter in the universe. There are many possible mathematical translations of this idea and, depending on their strength, Einstein's theory is, or is not, consistent with them. At any rate, it is clear that the principle of equivalence used by Einstein is a definite extension from the observed equality of gravitational and inertial mass to the conclusion that the *only* effect of external matter on the experiments in a laboratory is accelerative. Thus, in Einstein's theory, a sufficiently small freely falling laboratory is, in effect, completely isolated from the gravitational influence of the matter distribution in the universe, so that empty space-time is locally structured without any reference to matter. This is clearly a reversion to Newton's notion of absolute space, independent of matter. However, it is possible to weaken this result by the introduction of another field which can carry the influence of the matter distribution on local inertial properties into the freely falling laboratory. One choice for this field is the locally measured gravitational constant itself, and theories extending Einstein's to incorporate such a varying gravitational "constant" have been developed. However, all deviations of the experimental predictions of such theories from those of Einstein's are so small that as yet there is no decisive evidence.

Much attention has been given to the problem of fitting the Einstein theory of gravitation into the framework of quantum theory. In the weak field, linear limit, the Einstein equations are those to be expected for a quantum field theory of uncharged particles of zero mass and spin two, "gravitons". However, due to the extreme weakness of such fields the direct production and detection of quantum graviton effects seems outside the capability of contemporary experimental techniques. Of course, two of the distinctive characteristics of Einstein theory are the non-linearity of the equations and the generality of the coordinate transformations, so any satisfactory quantization must involve the full field equations and allow arbitrary transformations. However, many novel problems arise in attempts to apply standard quantum field theoretic techniques. For example, the true observables must be defined, taking into account the formidable gauge group of general coordinate transformations. In addition to the problems associated with the intrinsic complexity of the Einstein theory, there are also those associated with contemporary quantum field theory and particle physics. Perhaps of more significance, however, is the possibility that macroscopic metric and topological concepts of space-time, so fundamental to general relativity, may have little or no validity when extended to the microscopic quantum arena. Thus, the very notion of a quantized metric tensor may need profound alteration because of quantum observability problems. Further, alterations in the structure of space-time in the transition from the macroscopic to the microscopic level could likewise influence quantum theory itself, which presently uses a background space-time with characteristics derived only from macroscopic experience. It may well be in this area that the most significant interaction between gravitation and quantum theory will occur, rather than through predictions concerning super-weak gravitational interactions based on straightforward quantization of Einstein theory.

Bibliography

RINDLER W. (1962) in *Encyclopaedic Dictionary of Physics* (J. Thewlis Ed.), **6**, 268, Oxford: Pergamon Press.

THEWLIS J. (Ed.) (1961) *Encyclopaedic Dictionary of Physics*, **3**, 504, Oxford: Pergamon Press.

C. H. BRANS

GUNN EFFECT. In 1963 J. B. Gunn made the first observations of coherent microwave current oscillations in homogeneous, *n*-type, polar semiconductors subject to high electric fields. He discovered oscillations in gallium arsenide and indium phosphide. Similar oscillations have since been observed in cadmium telluride, zinc sellenide, gallium arsenide-phosphide alloys and indium arsenide under pressure. Most of the recent experimental and device work on the effect has been done with gallium arsenide for which the material technology is well advanced, and the numerical data quoted in what follows all refers to this material. The effect, which has become known as the Gunn effect, occurs in its purest form in samples of gallium arsenide over 100 μm long. At low fields the current is constant and merely falls slightly below the value to be expected from Ohm's law as the field increases. The current oscillations occur when the field exceeds a threshold value of about 3 kV/cm. There is then a sharp drop in the time-averaged value of the current and the current waveform consists of a periodic train of spikes separated by plateau regions as shown in Fig. 1. The maximum current is close to the constant current observed at threshold and is approximately twice the plateau current. The period τ of oscillation is close to the specimen length l divided by the electron drift velocity v corresponding to the plateau current, which is usually just under 10^7 cm/sec.

It now seems well established that the oscillations are due to the occurrence of a bulk negative conductivity in the homogeneous semiconductor at high fields, which makes the spatially uniform electron distribution unstable. To see how the negative conductivity arises let us consider the dependence of the electron energy E upon the electron wave number vector **k** in the conduction band of the semiconductor.

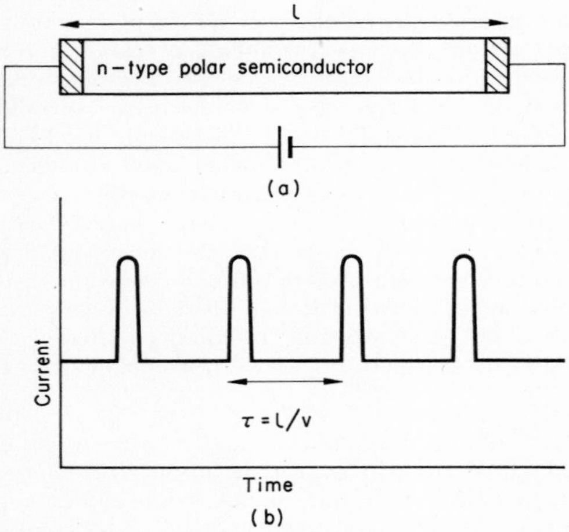

Fig. 1. The Gunn effect (a) schematic experimental arrangement; (b) current waveform above threshold.

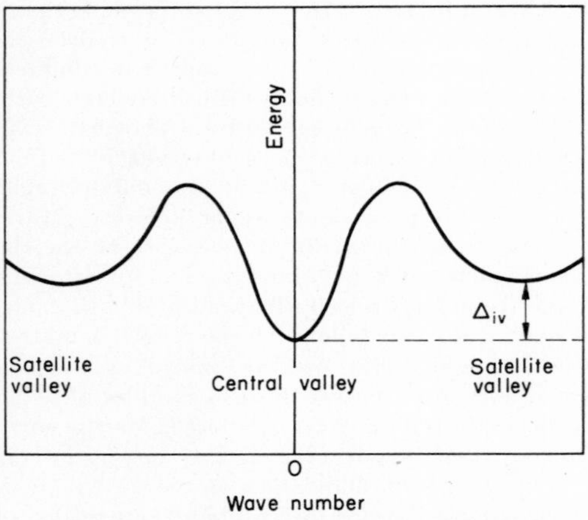

Fig. 2. Schematic plot of electron energy against wave number in the conduction band of a semiconductor showing the Gunn effect.

This dependence has the same qualitative features in all the materials which exhibit the Gunn effect. Figure 2 is a schematic plot of E against **k** for a particular direction in **k**-space. At $\mathbf{k} = 0$ there is a valley in the energy, known as the central valley, with a low effective mass (i.e. high curvature) while at finite values of **k** there are other valleys, which are symmetrically distributed and known as the satellite valleys, with higher energy and effective mass. The energy separation Δ_{iv} between the central and satellite minima is a few tenths of an electron-volt so that, at room temperature and zero field, the vast majority of the electrons are in the central valley. At low fields the satellite valleys play an insignificant role in determining the behaviour of the electrons. As the field strength is raised, however, more and more power is fed into the electrons from the field and their average energy (which we characterize by an electron temperature) must increase if the power dissipation to the thermal vibrations of the crystal lattice is to continue to balance the power input from the field. Eventually the electrons become sufficiently hot for a significant fraction of them to be transferred to the satellite valleys as a result of collisions with the short-wave-length lattice vibrations. At very high fields most of the electrons will be in the satellite valleys because the high satellite effective mass implies a correspondingly high density of states. The high satellite effective mass also implies that the mobility of electrons in the satellite valleys is much lower than that in the central valley. Consequently, as the field strength increases, there is a progressive transfer of electrons from the central valley to the satellite valleys which produces a progressive reduction of the average electron mobility. In semiconductors showing the Gunn effect, the average mobility decreases sufficiently rapidly over a particular range of fields for the average electron velocity to fall as well. The velocity-field characteristic therefore has the form shown schematically by the full curve in

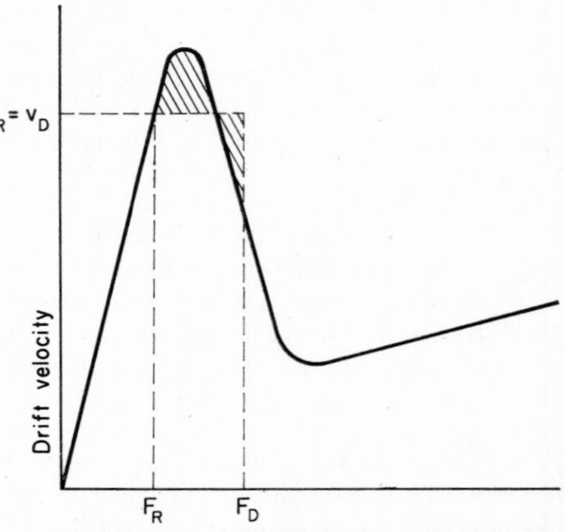

Fig. 3. Schematic velocity-field characteristic for a semiconductor showing the Gunn effect. The two cross-hatched regions have equal areas.

Fig. 3 and the material exhibits a bulk negative conductivity for fields between the peak and valley of this characteristic. The peak field is the threshold field for the Gunn effect in long samples.

When the semiconductor is biased into the negative conductivity range the homogeneous electron distribution becomes unstable. Any small fluctuation of the electron density will grow exponentially with a time constant equal to the dielectric relaxations time calculated from the slope conductivity at the bias point until it is limited by non-linearities. For a constant bias voltage, the ultimate form of the instability is a narrow domain of extremely high field, with equal low fields on either side, which moves uniformly through the specimen in the direction of electron drift. The velocity and shape of the domain may be determined by seeking the appropriate solution of Poisson's equation and the charge conservation equation when the drift velocity has the form shown in Fig. 3 and the diffusion current is taken into account. Numerical solution is generally necessary but semi-quantitative results may easily be obtained by ignoring the field-dependence of the diffusion coefficient. In that case, for a given value of the field F_D outside the domain, one finds that the domain velocity v_D is equal to the electron drift velocity v_R outside the domain and that the peak domain field F_D is determined by the simple geometrical condition that the two cross-hatched areas in Fig. 3 should be equal in magnitude.

The domain shape must be calculated numerically even when the diffusion coefficient is treated as constant. However, in the limit when the constant diffusion coefficients tends to zero the calculation is trivial. The domain has a right-angle triangular shape with a linear rise of field from F_R to F_D across a layer fully depleted of electrons at the leading edge and a vertical drop of field from F_D back to F_R across an infinitesimally wide and infinitely high electron accumulation layer at the trailing edge. The domain width is

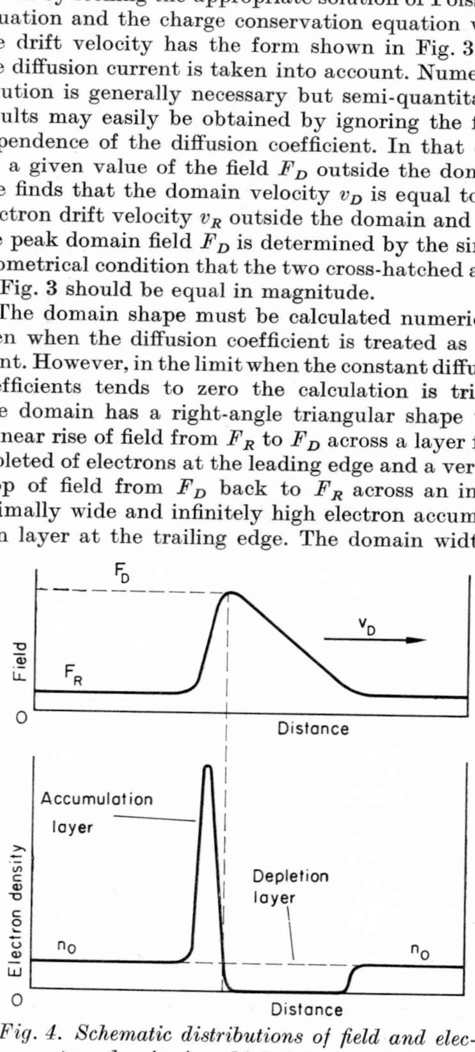

Fig. 4. Schematic distributions of field and electron density in a high-field domain.

equal to $\varepsilon_0(F_D - F_R)/4\pi e n_0$ where ε_0 is the static dielectric constant, e is the magnitude of the electronic charge, n_0 is the density of ionized donors and c.g.s. units have been employed. When the diffusion coefficient is finite, this triangular domain shape is rounded off, and the distribution of field and electron density have the forms shown schematically in Fig. 4.

When the field dependence of the diffusion coefficient is taken into account, the domain shape and the value of F_D are modified and the strict identity of v_R and v_D is no longer valid. It is still true, however, that for a given value of the field outside the domain, the domain velocity, peak domain field and domain shape are fixed. The value of the field outside the domain corresponding to a particular bias voltage and specimen length are determined by the condition that the area under the field curve between the ends of the specimen should equal the bias voltage.

The current waveform shown in Fig. 1 is due to the periodic recycling of high field domains through the specimen. The plateau regions correspond to the intervals when a stable domain is moving through the specimen; each current spike is associated with the passage of a domain into the anode and the formation of a new domain at the cathode.

For bias fields well above threshold the domain velocity is just under 10^7 cm/sec in gallium arsenide. Consequently the domain transit frequency, i.e. the current oscillation frequency, lies in the microwave region for samples a few tens of microns long. By inserting such samples into resonant cavities, microwave oscillators have been constructed delivering a few tens of milliwatts of power continuously and up to 100 W pulsed with efficiencies of a few per cent. The effects of the contacts, doping inhomogeneities and feedback from the cavity to the sample are all significant and the simple behaviour described above for long, homogeneous samples subject to constant voltages is not adhered to, although there is no change in the basic oscillation mechanism. The line width of the output spectrum is surprisingly narrow, less than 0·5 kc/sec, and Gunn oscillators promise to provide simple, low-voltage, solid-state sources of microwave power. Microwave amplification has also been achieved by reflecting the signal at a specimen of gallium arsenide in which the product of specimen length and carrier concentration is less than 10^{11} cm^{-2}. For such samples the domain build-up time is greater than the domain transit time and domains do not form; the negative conductivity remains, however, and is responsible for the amplification. The amplification mechanism is very noisy and is unlikely to be useful in practice.

At the time of writing (August, 1966) the evidence for the transferred electron mechanism as the origin of the Gunn effect is for the most part indirect but nevertheless seems overwhelming. The effect is found only in n-type, polar, semiconductors with the appropriate conduction band structure and the observed variation of the threshold field with the energy separation between the central and satellite minima in gallium arsenide under stress is as would be expected for this mechanism. The rapid transfer of electrons above the threshold field is due to the fact that the scattering of electrons by polar lattice vibrations, which is the dominant scattering mechanism in polar semiconductors at low fields, is unable to dissipate the Joule heating above a critical field. Consequently, above this field, the electrons in the central valley gain energy rapidly and are available for transfer to the satellite valleys by scattering of the short-wavelength lattice vibrations. Many polar semiconductors which have a conduction band structure of the right type suffer avalanche breakdown at high fields instead of exhibiting the Gunn effect. The threshold field for avalanche breakdown is determined by the energy gap between the conduction and valence bands and this process will occur first and mask the Gunn effect when the energy gap is less than the separation between the central and satellite valleys. In the case of indium arsenide, avalanche breakdown occurs at zero stress but when the energy separation between the central and satellite minima has been reduced sufficiently by unaxial stress the Gunn effect appears.

The velocity-field characteristic for gallium arsenide calculated by approximate solution of the Boltzmann equation for a multi-valley conduction band structure has a threshold field of 3250 V/cm, a slope mobility of magnitude 6860 cm^2/V-sec below the threshold field and 3000 cm^2/V-sec just above the threshold field. The characteristic then curves round to reach a valley at 20 kV/cm beyond which the velocity saturates at $0{\cdot}86 \times 10^7$ cm/sec. Low field mobilities in the neighbourhood of 7000 cm^2/volt-sec are found in good samples of gallium arsenide. The observed threshold fields vary from 1200 to 4000 V/cm in different samples. It seems likely that these wide variations are correlated with severe doping inhomogeneities in the material. Direct measurement of the velocity-field characteristic beyond threshold is difficult because domains form. Gunn has recently reported preliminary measurements using very short voltage pulses which indicate that the threshold field is about 3700 V/cm and that the initial slope mobility in the negative conductivity range has a magnitude of only 300 cm^2/V-sec. The latter observation is in poor agreement with the theoretical prediction and further work, both theoretical and experimental, is now under way to resolve the discrepancy.

The passage of narrow high field domains through long specimens has been observed by several workers using capacitive probe techniques. Of particular interest from the point of view of applications is the observation of triggered domains: once a domain has been initiated it will continue to propagate when the bias is reduced below threshold until the bias falls below the so-called minimum sustaining field which is close to half the threshold field. This behaviour is predicted by the theory outlined above; one finds that stable domains can be propagated, although not initiated, below threshold bias. Until recently the probe technique was too crude to resolve the domain shape, but it was possible to show that peak domain fields in excess of 75 kV/cm occur at high bias levels whereas the field outside the domain is only about 1·5 kV/cm. These observations indicate that either the velocity saturates at high fields, as the theoretical calculations imply, or that the valley field is extremely high as is suggested by the low slope beyond threshold reported by Gunn. Gunn has also recently reported the resolution of triangular domain shapes with a fully depleted layer at the leading edge in gallium arsenide samples with a resistivity of 10 ohm-cm.

The study of the Gunn effect has had a very promising start and the basic principles are already clear. Much work remains to be done towards obtaining a detailed understanding of the effect, resolving the present discrepancies between theory and experiment and exploiting the effect in active microwave devices.

Bibliography

Butcher P. N. (1967) *The Gunn Effect, Reports on Progress in Physics*, **30** (Part I), 97.

Engelbrecht R. S., (Ed.) (1966) Special Issue on Semiconductor Bulk-Effect and Transit-Time Devices, *Trans. Inst. Elect. Electron. Engrs., Electron Devices* **13,** 1.

Gunn J. B. (1964) Instabilities of current in III–V semiconductors, *IBM J. Res. and Dev.* **8**, 141.

P. N. Butcher

H

HARD SPHERE LATTICE GASES. A simple model system for theoretical study of the thermodynamic effects of molecular size and shape, the hard sphere lattice gas has two essential characteristics: the centres of the molecules may be located within a container only at certain regularly spaced ("lattice") positions; and the molecules are impenetrable, so that any particular molecule masks one site or more, depending on the lattice and the dimensions of the molecule. The primary desired information is an *a priori* calculation of the *equation of state* (or relationship between temperature, pressure and volume), particularly including phase changes.

Most simple, non-polar substances show a phase diagram such as Fig. 1(a) and an intermolecular potential energy such as Fig. 1(b). In a general way there is a relationship between the two diagrams:

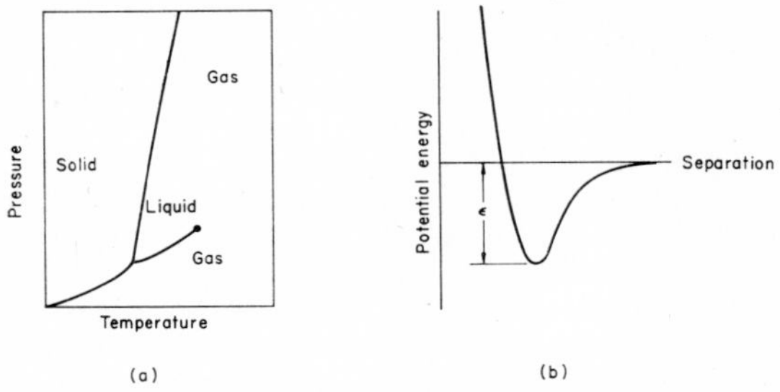

Fig. 1. Typical phase diagram (a) *and intermolecular potential* (b).

the liquid region occurs at temperatures T such that the average translational energy of a molecule ($3kT/2$) is roughly the same as the depth ε of the attractive (negative) part of the potential curve. Here k is Boltzmann's constant. $1{\cdot}38 \times 10^{-16}$ erg/deg. Implied is a definite involvement of this attraction in the behaviour of the liquid state. On the other hand, at temperatures considerably above the critical temperature, translational energy so greatly exceeds ε that the attractions between molecules must play a minor role; yet the fluid-solid phase transition still occurs and must be viewed as an effect of the (almost complete) impenetrability of the molecules. Physically then, hard sphere models of molecules, devoid of attractions, apply to the ultimate states of matter (solid and gas) under conditions of very high temperature and pressure. Such models, incidentally, have for decades been fruitful for studies of transport phenomena; more recently an increased interest has been placed on these systems for studies of equilibrium behaviour.

Classical thermodynamics reveals a significant simplification in the equation of state behaviour of these systems. Since any permitted arrangement of these hard molecules possesses zero potential energy, the total internal energy U is just the same kinetic energy possessed by an ideal gas, $3NkT/2$, where N is the number of molecules present. In other words, the internal energy depends only on temperature and not on volume, so that the heat capacity at constant volume, $C_v = (\partial U/\partial T)_V$ is actually the same as for the ideal gas, $3Nk/2$. Further, we find from the thermodynamic equation of state $P = T(\partial S/\partial V)_T - (\partial U/\partial V)_T$ and the Maxwell relationship $(\partial P/\partial T)_V = (\partial S/\partial V)_T$, that $(\partial P/\partial T)_V = P/T$. The pressure of the system is denoted by P, the volume by V and the entropy by S. Integrating this equation, we obtain $P/kT = f(\varrho)$, where $f(\varrho)$ is some function of the number density $\varrho = N/V$. The pressure then is rigorously proportional to the absolute temperature; the significant non-trivial problem is the relationship between the two variables P/T and ϱ. The function f has dimension of density (molecules/volume), and for the ideal gas is exactly equal to the density. This behaviour is shown in Fig. 2a; there is no maximum density since the molecules themselves are imagined to have zero volume.

If the molecules are more realistically postulated to have some non-zero volume, different behaviour is

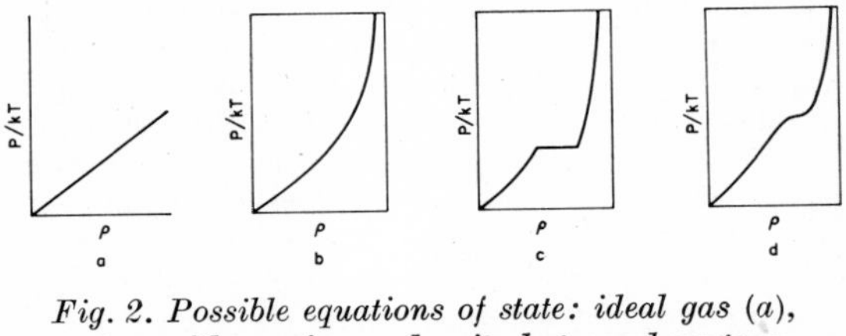

Fig. 2. Possible equations of state: ideal gas (a), *system with maximum density but no phase transition* (b), *discontinuous transition* (c), *continuous transition* (d).

expected. There must certainly be some maximum density attainable, as indicated in Fig. 2b. Real gases above the critical point upon compression, however, show behaviour as indicated in Fig. 2c. The horizontal portion represents the crystallization of the fluid attended by an abrupt increase in density. (Below the critical temperature, there is another horizontal portion corresponding to the condensation of the gas to the liquid.) Calculations of the behaviour of the *hard sphere gas* in two and three dimensions indicate that such systems probably behave in the manner of Fig. 2c, although the difference between the density of the crystal and that of the fluid is less than suggested by the figure.

With the lattice restriction on permitted locations, the calculations become considerably simpler; some examples of two-dimensional lattice gases are shown in Fig. 3. The examples shown involve at most nearest-neighbour exclusions and are the simplest computationally. Presumably finer grids would more closely approximate real systems and are currently under active investigation, as are lattice gases of asymmetric molecules and mixtures of molecules of different size.

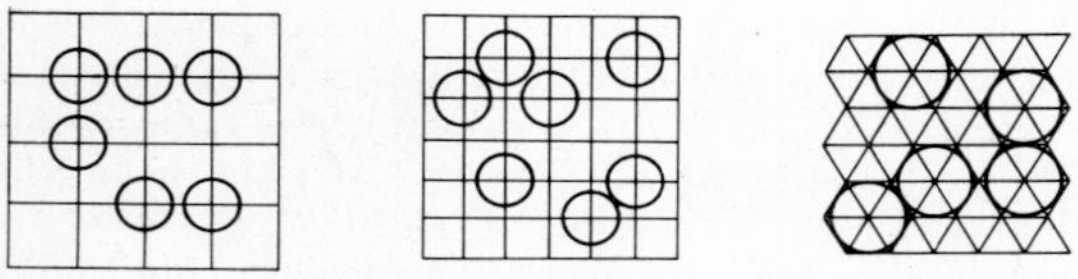

Fig. 3. Two-dimensional lattice gases: non-overlapping (a), square (b) and triangular (c) lattice gases with nearest-neighbour exclusion.

Formally the problem is the well defined one in statistical mechanics of calculating the entropy of the system as a function of density. This could be accomplished using *Boltzmann's formula* if the number of arrangements could be computed for an arbitrary number of molecules on a lattice of specified large size. This has been done exactly only for the non-overlapping case of Fig. 3a, which understandably shows no phase transition since the molecules are essentially independent and no cooperative behaviour is possible. The equation of state is of the type 2b. For more complicated and more interesting cases several techniques have been employed with varying degrees of success. Analytic approximations to the *partition function* in general give a faithful description in regions removed from a phase transition, but cannot be relied upon for accurate information about the transition. Some knowledge can be gained, however, about the probable existence of some sort of transition or singularity and about its approximate location. The same is true of other approaches based on the pair distribution function and Monte Carlo or random sampling techniques.

Two complementary methods have proved especially valuable in providing more detailed information about the transitions of these hard sphere lattice gases. A virial expansion is particularly successful for two and three dimensional grids due to the simplifications introduced by the lattice restrictions. A relatively large number of the expansion coefficients can be evaluated exactly; this expansion describes the gaseous region containing a low density of molecules. In addition a related high density expansion can be derived which describes the system at high densities. Using sophisticated extrapolation techniques, the probable behaviour in the intermediate density range can be inferred, including the likely nature of any transition or singularity present.

The other technique has been described as the exact finite method, or the strip method. Results have been reported only for two-dimensional systems, but in cases where it has been compared with the previous method the two approaches yield the same answer. The exact partition function is obtained for a two-dimensional lattice of infinite length but small finite width. By then letting the finite width become progressively larger, the behaviour of the thermodynamically significant lattice of very large width can be inferred. It is particularly fruitful to assign much of this work to an electronic computer due to the basic similarity of lattice gas configurations and the binary operations of electronic computers: the presence of a molecule corresponds to a 1 bit, while the absence of a molecule corresponds to a 0 bit. A primary advantage of the exact finite method is its equivalent treatment of all densities—high, low and intermediate.

All of the results obtained by these techniques for two and three dimensional lattices with nearest-neighbor exclusions are qualitatively as shown in Fig. 2d. A transition is observed, as expected for real systems, although the transition is "continuous": the two phases have the same density. However, at the transition point the "isotherm" appears to be horizontal, implying a very large or infinite compressibility. Limited information currently available indicates that with a finer grid the inflection point in Fig. 2d broadens into a horizontal region associated with an abrupt increase in density.

While the primary interest in lattice gases has been as an approximation scheme to real systems, it should be pointed out that certain adsorption problems correspond more directly to the lattice gas. In addition, with a simple change of notation the same mathematical treatment gives a description of other existing physical systems, such as magnetic materials, ferroelectric substances and binary alloys.

Bibliography

Brout R. (1965) *Phase Transitions*, New York: Benjamin.

FRIEDMAN A. S. (1961) in *Encyclopaedic Dictionary of Physics* (J. Thewlis Ed.), **3**, 2, Oxford: Pergamon Press.

GAUNT D. S. and FISHER M. E. (1965) Hard-Sphere Lattice Gases. I, Plane-Square Lattice, *J. Chem. Phys.* **43**, 2840.

GREEN H. S. and HURST C. A. (1964) *Order-Disorder Phenomena*, New York: Wiley.

RUNNELS L. K. and COMBS L. L. (1966) Exact Finite Method of Lattice Statistics. I, Square and Triangular Lattice Gases of Hard Molecules, *J. Chem. Phys.* **45**, 2482.

TOSI M. (1966) in *Encyclopaedic Dictionary of Physics* (J. Thewlis Ed.), Suppl. Vol. 1, 302, Oxford: Pergamon Press.

L. K. RUNNELS

HELICONS. P. Aigrain predicted in 1960 that it should be possible to propagate low-frequency electromagnetic waves in a metal in the presence of a constant external magnetic field. Soon after, Bowers, Legendy and Rose demonstrated such propagation experimentally in sodium. Subsequently these excitation ("helicons") have proven to be a powerful tool in the investigation of many properties of solids.

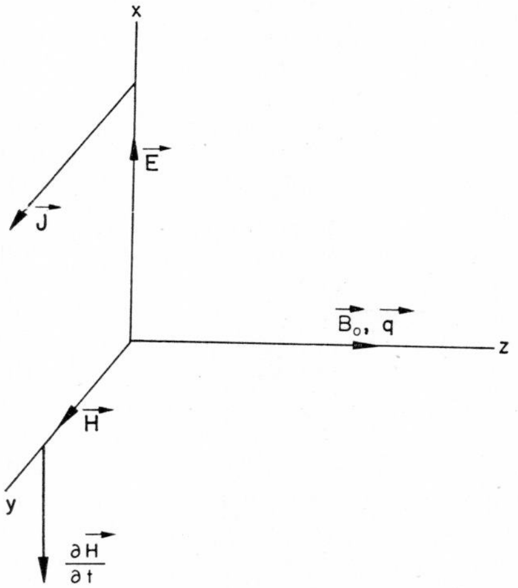

Fig. 1. Relation of fields, current and direction of propagation for helicons.

The propagation of helicon waves is made possible by the Hall current, which at sufficiently low temperatures and large external magnetic fields is almost at right angles to the field, and the fact that for *polarized* waves the oscillating magnetic field is in the direction of the current.

It is convenient in discussing these waves, to use a "polarization representation". If **A** is a vector field the rotation operator **inx** has the eigenvalues 1, —1 or 0. The first two eigenvectors are transverse to **n** and correspond to positive and negative angular moments respectively for the field about the direction of **n**. We shall refer to these as positively (right circularly) and negatively (left circularly) polarized fields (in optics, the opposite convention is common). For the eigenvalue zero, **A** is in the direction of **n** and will be called "longitudinally polarized".

If the field is that of a wave

$$\mathbf{A} = \mathbf{A}_0 \exp i(\mathbf{qr} - \omega t)$$

then we note that

$$\nabla \times \mathbf{A} = i\mathbf{q} \times \mathbf{A} = iq\mathbf{n} \times \mathbf{A} = q\lambda\mathbf{A}$$

where λ is the polarization eigenvalue.

The propagation of the helicon wave in the direction of a constant external magnetic field B_0 may now be understood as follows; the electrons of the solid process in the positive direction about the constant magnetic field. This produces an oscillating magnetic field given by

$$\nabla \times \mathbf{H} = (4\pi/\mathbf{c})\,\mathbf{J}$$

or

$$iq\boldsymbol{n} \times \mathbf{H} = \lambda q\mathbf{H} = (4\pi/\mathbf{c})\,\mathbf{J}\,.$$

For transverse fields

$$\mathbf{H} = \pm \frac{4\pi}{\mathbf{c}q}\,\mathbf{J}$$

for the case of right circular polarization (r.c.p.) or left circular polarization (l.c.p.) respectively (displacement current effects may be neglected at low frequency). An induced electric field is given by Faraday's law

$$\nabla \times \mathbf{E} = iq\mathbf{n} \times \mathbf{E} = \lambda q\mathbf{E} = -\frac{1}{\mathbf{c}}\,\frac{\partial\mathbf{B}}{\partial t}\,.$$

This leads (as may be seen from Fig. 1) to an electric field such that the Poynting vector $\mathbf{E} \times \mathbf{H}$ is in the direction of $\mathbf{q}$, and of magnitude $(\omega/cq)\,B = \dfrac{4\pi\omega\mu}{\mathbf{c}^2q^2}\,J$, μ being the permeability. The possibility of a self-sustaining wave now follows from the fact that, in the absence of relaxation, the electric field generates a current at right angles to itself—the Hall current. Its magnitude is

$$J = -E/RB_0,$$

where $R = -1/\boldsymbol{Nec}$ is the Hall coefficient.

This is precisely the original current provided

$$\omega = \frac{\mathbf{c}^2q^2RB_0}{4\pi}\,. \qquad (1)$$

(1) is the dispersion relation for helicons of low frequency (long wave-length). They are obviously polarized in the sense of rotation of the electrons, that is, in the convention used here, right circularly polarized.

Helicon waves have been observed in various ways: one of the simplest is to study the transmission through a slab of material (the direction of incidence and of magnetic field both being perpendicular to the slab). Transmission maxima should occur at "dimensional" resonances, that is, when the thickness of the slab is equal to an integral number of half-wave-lengths. For q's satisfying this condition the frequency which will be transmitted for a given field, or the field which will ensure transmission at a fixed frequency, are given by (1). Results from an experiment of Grimes are given in Fig. 2.

The analysis leading to (1) may be generalized; the dispersion relation is found to be

$$\frac{c^2q^2}{\omega^2} = \varepsilon_\pm \mu_\pm , \quad (2)$$

when the dielectric and permeability tensors may be expressed in terms of independent components for the various polarizations.

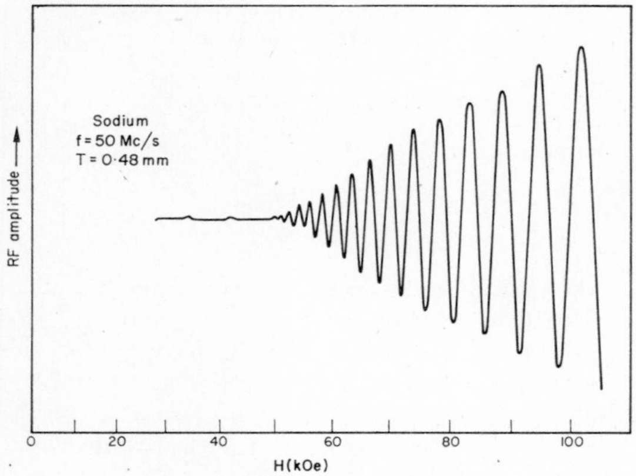

Fig. 2. Interference experiment due to Grimes involving helicons. A plot of voltage in the detector coil as a function of magnetic field, at a constant frequency of 50 Mc. The Figure shows the beat-pattern between the helicon wave and a reference signal.

The dielectric tensor is related to the conductivity tensor σ by

$$\varepsilon = \varepsilon_L + \frac{4\pi i}{\omega} \sigma \quad (3)$$

ε_L being, in a real solid, the dielectric constant of the lattice.

If the dielectric tensor has the form

$$\varepsilon_{ij} = \varepsilon_1 \delta_{ij} + (\varepsilon_3 - \varepsilon_1)\, n_i n_j + i\varepsilon_2 \eta_{ijk} n_k \quad (4)$$

where η_{ijk} is the permutation symbol and $\mathbf{n}$ is a unit vector in the direction of the magnetic field, it is diagonal in the polarization representation, and has transverse components

$$\varepsilon_\pm = \varepsilon_1 \mp \varepsilon_2$$

and a longitudinal one, ε_3.

For free electrons the transverse components are obtainable in terms of the Landau states of electrons in the constant field $\mathbf{B}_0$; these states are determined by the quantum numbers n, k_y and k_z, n being an integer and k_y and k_z quasi-continuous.

They have the form

$$\varepsilon_\pm = \varepsilon_L \left\{ 1 - \frac{\omega_p^2}{\omega(\omega + i/\tau)} \times \left[1 \mp \frac{\omega_c}{N} \sum_{n,k_y,k_z} \frac{(n+1) f(n, k_z + \frac{1}{2}q) - n f(n, k_z - \frac{1}{2}q)}{\pm\omega_c - \omega - i/\tau - (\hbar/m)k_z q} \right] \right\},$$

ω_c being the cyclotron frequency $\dfrac{eB_0}{mc}$ and ω_p the plasma frequency given by

$$\omega_p^2 = \frac{4\pi N e^2}{\varepsilon_L m} \quad (6)$$

$f(n, k_z)$ is the Fermi distribution function for the state of quantum numbers n, k_z. We shall assume in general that $\omega_p \gg \omega_c$. N is the total number of electrons in the conduction band. Relaxation has been taken into account through a relaxation time τ, here assumed constant.

For long wave-length ($q \to 0$) ε_+ may be divided into real and imaginary parts:

$$\varepsilon_+ = \varepsilon_r + i\varepsilon_i$$

where

$$\varepsilon_r = \varepsilon_L \left\{ 1 + \frac{\omega_p^2(\omega_c - \omega)}{\omega[(\omega_c - \omega)^2 + 1/\tau^2]} \right\} \quad (7)$$

and

$$\varepsilon_i = \varepsilon_L \frac{\omega_p^2 \tau}{\omega[(\omega_c - \omega)^2 + 1/\tau^2]} . \quad (8)$$

Damping of the helicon waves is small provided

$$\varepsilon_i/\varepsilon_r \ll 1 \quad \text{or} \quad \omega_c\tau \gg 1 .$$

Under these conditions and assuming $\mu = 1$, the dispersion relation becomes, for right circular polarization (r.c.p.)

$$\omega = c^2 q^2 \omega_c / \omega_p^2 . \quad (9)$$

The dissipative part becomes large beyond the frequency

$$\omega = \Omega = \omega_c - \frac{\hbar}{m} k_F q \quad (10)$$

where k_F is the wave number at the Fermi surface. This is known as the "Doppler shifted cyclotron frequency". Combining this with (2) and (7) we find that the "cutoff" comes at frequency ω given by

$$2\varepsilon_L \omega_p^2 \frac{E_F}{mc^2} \omega = (\omega_c - \omega)\,[(\omega_c - \omega)^2 + 1/\tau^2] . \quad (11)$$

If the magnetic field is sufficiently weak, the "cutoff" frequency is approximately at

$$\omega = \frac{mc^2}{2\varepsilon_L \omega_p^2 E_F} \omega_c^3,$$

a result which has been verified by M. T. Taylor.

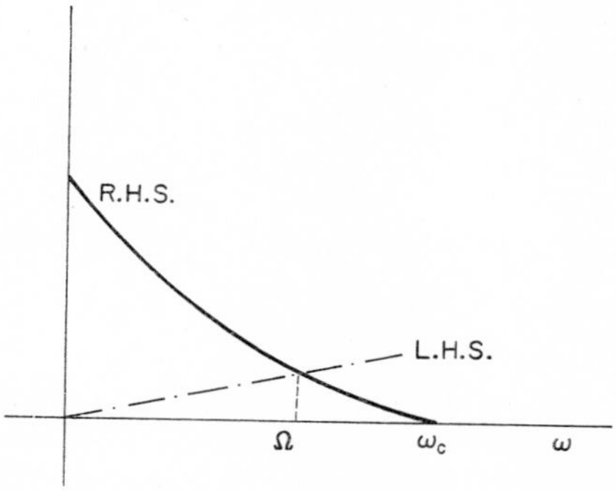

Fig. 3. Plot of right and left hand sides of equation 11 (R.H.S. and L.H.S. respectively) showing intersection at $\omega = \Omega$.

Use of helicons in determining band structures. 1. It has been suggested that the anisotropy of Fermi surfaces might be investigated by studying the variation of the absorption edge with orientation of crystals of a substance relative to the magnetic field and direction of propagation. Since the Doppler shifted cyclotron frequency depends on k_F, the variation of k_F with direction may be studied. For a given direction, the effective k_F is the largest value in the direction of the field for that orientation.

2. The dielectric constant may show oscillations of the Haas–Schubnikov type, which will be reflected in oscillations with the magnetic field of the helicon fringe pattern.

These oscillations will be superimposed on the interference pattern. The two oscillations may be separately identified by virtue of the fact that the basic pattern will shift when the frequency is varied, whereas the de Haas–Schubnikoff oscillations will not.

3. If we consider materials having equal numbers of positive and negative carriers, the low frequency dielectric constant will have the form

$$\varepsilon_{\pm} = \varepsilon_L \left\{ 1 + \frac{\omega + i/\tau}{\omega} \frac{\omega_p^2 + \omega_p'^2}{(\omega_c - \omega - i/\tau)(\omega_c' + \omega + i/\tau)} \right\}$$

where ω_c, ω_p are cyclotron and plasma frequencies respectively for the negative carriers and ω_c' and ω_p' those for the positive ones. In this case

$$\omega_c \omega_p'^2 = \omega_c' \omega_p^2.$$

The condition for negligible attenuation is now $\omega\tau \gg 1$ compared with $\omega_c\tau \gg 1$ for helicons. Since $\varepsilon_{\pm}$ go to a constant at low frequency, the dispersion relation in this case is *linear*, not quadratic. Such waves are *unpolarized*, and are known as "Alfvén waves". They have been investigated in bismuth and other "compensated" materials.

4. Helicon propagation is possible in doped semiconductors (InSb, Cd_3As_2, PbTe, Ge). Such substances may be better plasmas than metals. A substance will display plasma properties only if the density is high enough that there are substantial numbers of electrons inside the "screening length". The condition that this be so is

$$\frac{1}{2}\left(\frac{3\pi}{2}\right)^{3/2}\left(\frac{\varepsilon_s m}{m^*}\right)^{3/2}(a_0^3 N)^{1/2} > 1.$$

m^* is the electron effective mass, ε_s the static dielectric constant, a_0 the Bohr radius and N the electron density. Thus, the "effective" density is increased by the factor $\left(\frac{\varepsilon_s}{m^*}\right)^3$ due to dielectric and effective mass effects. This may be a very large factor for semiconductors.

In the case of Ge or PbTe, for propagation along symmetry directions, the dielectric tensor retains the form (4), giving rise to separation of longitudinal and transverse modes. For directions in which more than one cyclotron mass enters, a second pole will appear in the dielectric tensor, giving rise to a higher branch in the spectrum of right polarized helicons. *Left* circularly polarized waves may also be found, starting at a non-zero frequency and terminating at a Doppler shifted cyclotron frequency corresponding to energy surfaces "tilted" with respect to the magnetic field and direction of propagation. These result from the combination of several right *elliptically* polarized waves to form a left *circularly* polarized one.

For tilted surfaces also the coupling of cyclotron motion to the longitudinal motion of electrons associated with individual energy surfaces can give rise to a purely longitudinal mode in which the transverse fields associated with the various surfaces interfere destructively. This excitation is a density oscillation but differs from the usual plasma mode by having a frequency which is low and proportional to the magnetic field; it may thus couple to longitudinal phonons.

While it may easily be seen from (1) that the helicon dispersion relation will not depend on electron band parameters, those for the additional modes here discussed will depend on them.

5. Where there are open energy surfaces giving rise to open k-space orbits in planes perpendicular to the magnetic field, there is strong attenuation of the helicon or Alfvén waves. This follows from the fact that, when such orbits exist, the magnetic field cannot ensure a current perpendicular to the electric field of the wave. This field then loses energy in producing an electron current down the open trajectories.

Since these open orbits exist only for quite narrow ranges of crystal orientation in materials such as copper and silver, the attenuation is very sensitive to the orientation relative to the magnetic field and may be used to explore the geometry of the open surfaces. C. C. Grimes, G. Adams and P. H. Schmidt have in this way confirmed the existence in silver of surfaces open in the (100), (110) and (111) directions (see Fig. 4).

Interactions of helicons. The very low velocities of helicon waves makes possible the study of their interactions with other elementary excitations, e.g. with phonons and magnons.

Coupling to transverse phonons in potassium was studied by C. C. Grimes and S. J. Buchsbaum. Without such coupling, the phonon dispersion relation is linear, while that for helicons is quadratic. Near the intersection, coupling causes the curves to "repel" each other, and two branches of the coupled spectrum appear. This shows up in rather complicated interference effects in the helicon transmission at the frequencies and wave numbers in question.

Rosenman demonstrated helicon-phonon coupling more directly in Cd_3As_2. Using a quartz oscillator to excite transverse acoustic waves, he used a coil to detect the helicon waves generated.

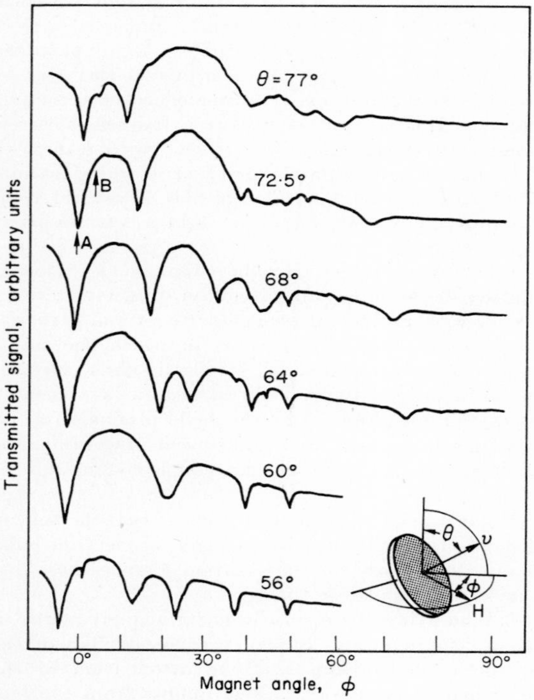

Fig. 4. Experimental traces showing the amplitude of the wave transmitted through a silver crystal as a function of the angle between the magnetic field and the crystal axes. Sharp minima in transmission occur for those magnetic-field directions where open orbits exist.

The interaction of helicons with spin waves (magnons) in ferromagnetics was predicted theoretically by Stern and Callen, and verified by Grimes in the case of nickel. Stern and Callen showed that strong coupling was to be expected even when the uncoupled dispersion curves were widely separated. This coupling is connected with the strong oscillating magnetic field of the helicons. For right circular polarization Faraday's law

$$\nabla \times \mathbf{E} = -\frac{1}{c}\frac{\partial \mathbf{B}}{\partial t}$$

leads to the ratio of magnetic to electric field in the wave:

$$|\mathbf{B}|/|\mathbf{E}| = \frac{cq}{\omega} = \frac{\omega_p^2}{cq\omega_c} = \frac{2c}{v}$$

where v is the velocity of the helicon wave.

It has been suggested that helicons might be used to do nuclear magnetic resonance experiments in large, oriented crystals of metal. No experiments of this nature have yet been reported.

See also: Hybrid resonances. Solid-state plasma.

Bibliography

CHYNOWETH A. G. and BUCHSBAUM S. J. (1965) Solid State Plasma, *Physics Today*, November.

LEHANE J. A. and THONEMANN P. C. (1965) *Proc. Phys. Soc.* **85**, 301.

ROSEMAN I. (1965) *Solid State Communications* **3**, 405.

WALLACE P. R. (1966) *Can. J. Phys.* **44**, 2495 and **44**, 1937.

7th International Conference on the *Physics of Semiconductors* (1964) Paris: Dunod.

P. R. WALLACE

HETEROGENEOUS CATALYSIS BY METALS (SURVEY). A catalyst is a substance which accelerates the rate at which a chemical reaction approaches equilibrium, without itself becoming permanently involved in the reaction. This definition is general, and applies not only to metal catalysis but also to catalysis by oxides, enzymes, and metal complexes in solution. This survey is concerned with systems in which the catalyst consists of solid metal and in which chemical reactions take place at the phase boundary between the metal and the liquid or gaseous reactants.

A large proportion of the known elements are metals, but the transition elements, which occupy Groups IIIA, IVA, VA, VIA, VIIA, VIII, and IB of the Periodic Table are the only metals which exhibit significant catalytic properties. Of these, the Group VIII metals iron, nickel, palladium, and platinum are the most widely employed.

Catalyst preparation. Catalytically active metal can be prepared in many forms. From the economic standpoint, the metal should be highly dispersed, so that as high a proportion of metal atoms as possible are present in surfaces and hence are available to catalyse reactions. However, the precise form of the catalyst is frequently dictated by the nature of the reaction system.

For liquid phase reactions supported metal or skeletal metal is frequently used. Supported catalysts consist of polycrystalline metal of high specific surface area supported on an inactive material which also has a high specific surface area (e.g. finely divided carbon is often used as a support). The catalyst is prepared by first impregnating the support with a suitable metal salt, and then reducing the salt to the metal. The Raney method of producing highly active skeletal nickel catalyst by dissolving aluminium from a 50:50 Ni:Al alloy in strongly alkaline solution is well known. This technique can also be used for making skeletal iron, cobalt, and copper catalysts. Other forms of catalyst used for liquid phase reactions include metal colloids and sponges produced by chemical reduction of metal salts or complexes using, for example, sodium citrate or alkaline formaldehyde as reductant. Colloidal metal may be very active indeed, representing as it does the ultimate in small particle size and hence also the ultimate in the economic use of metal. However, colloidal metal cannot be reclaimed easily whereas metal sponges, which have larger particle size and lower specific activity, can be filtered in the usual way. Adam's catalyst, a mild general purpose hydrogenation catalyst, is formed by the reduction of platinum oxide *in situ* to metallic platinum by hydrogen; the metal can be filtered off at the end of the reaction and is normally converted to oxide again by standard chemical methods. For details of experimental methods the reader should consult monographs such as Bond (1962) and Augustine (1965).

Supported metal or woven wire gauze may be used when reactants and products are gaseous. The use of foils, powders and films is restricted mostly to academic work. Supports commonly used are granular or pelleted alumina, silica, kieselguhr, etc. The carbon-supported catalysts used in liquid phase reactions would be unsuitable for gas phase reactions because the finely divided catalyst might be lost from the system by mechanical transfer.

Chemisorption. The necessary prerequisite for catalytic reaction is that one or more of the reactants must chemisorb on the catalyst surface. The chemisorbed species so formed then react together, or with gaseous molecules which collide with the surface, to give products which may or may not be chemisorbed. Chemisorbed products must later leave the surface, and this process is termed "desorption."

Physical methods that have been used successfully to examine the nature of chemisorbed species in non-reacting systems include infra-red spectroscopy, and measurements of work function, conductivity, and magnetic properties. However, these methods are of limited value in determining the nature of intermediates that participate in catalytic reactions because these intermediates are frequently present in very low concentrations. Information about such reaction intermediates is usually obtained by indirect methods, of which the two most important are (i) the interpretation of reaction kinetics (orders of reaction and activation energies), and (ii) the use of isotopic tracers. By the latter technique, the nature and behaviour of the intermediates can frequently be determined from an examination of the concentration and location of the tracer isotope in the products. Not surprisingly, the most used tracer is the hydrogen isotope of mass 2 (deuterium), but tritium, and the isotopes of carbon, nitrogen, oxygen and the halogens have all been used where appropriate. The widespread availability of mass spectrometers ensures that this technique will be used frequently in the future.

The energetics of chemisorption and of reaction have been discussed previously (see Malinowski 1961). Here we merely note that the strength of chemisorption has a complex effect on catalytic activity. Consider a catalysed unimolecular decomposition. If chemisorption of the reactant is weak, then surface coverage may be so low that catalysis is inefficient. On the other hand, if chemisorption is of such a strength that full surface coverage is just achieved, then maximum activity will be obtained. Between these conditions the catalytic activity will be approximately proportional to the strength of chemisorption. However, if chemisorption is very strong indeed, then the system will verge upon stable compound formation, and activity will be low in spite of full surface coverage having been achieved; that is, the adsorbate will have poisoned the catalyst. Thus, a graph of catalytic activity against strength of chemisorption exhibits a maximum. Frequently, it is found that reactants adsorb very weakly on the Group IB metals, and strongly on the early metals of the three transition series. In consequence, the metals which exhibit the highest catalytic activities for a given reaction are frequently those of Group VIII.

Classes of reaction catalysed by metals

General. The most important classes of reaction catalysed by metals are: hydrogenation of unsaturated hydrocarbons and carbon monoxide, ammonia synthesis and oxidation, dehydrogenation, and hydrogenolysis. These will be considered below, in this order.

The ability of transition metals to chemisorb hydrogen as atoms is the key to their behaviour as catalysts. This chemisorption can be demonstrated simply by admitting a mixture of gaseous hydrogen and deuterium to a reactor containing the catalyst.

The processes which take place are

$$\mathrm{H_2}(g) \underset{-1}{\overset{1}{\rightleftharpoons}} 2\underset{*}{\underset{|}{\mathrm{H}}}$$

$$\mathrm{D_2}(g) \underset{-2}{\overset{2}{\rightleftharpoons}} 2\underset{*}{\underset{|}{\mathrm{D}}}$$

$$\underset{*}{\underset{|}{\mathrm{H}}} + \underset{*}{\underset{|}{\mathrm{D}}} \underset{-3}{\overset{3}{\rightleftharpoons}} \mathrm{HD}(g)$$

where the asterisk denotes a site for chemisorption on the metal surface. The formation of gaseous HD can be monitored by mass spectrometry or by thermal conductivity measurements. Where HD is formed the reaction must necessarily involve first the dissociation of the reactants to give chemisorbed atoms (steps 1 and 2), and secondly the subsequent combination of chemisorbed hydrogen atoms with chemisorbed deuterium atoms (step 3). Processes (−1) and (−2) will also occur, of course, but will not be observable because the products will be indistinguishable from the reactants. All reactions to be described below are critically dependent upon the ability of the metal catalyst to form chemisorbed hydrogen atoms either by the dissociation of hydrogen molecules or by hydrogen atom abstraction from adsorbed hydrogen species, e.g.

$$\underset{*}{\underset{|}{\mathrm{CH_2}}}\mathrm{—CH_3} + (*) \longrightarrow \mathrm{CH_2} = \underset{*}{\underset{|}{\mathrm{CH_2}}} + \underset{*}{\underset{|}{\mathrm{H}}}$$

Hydrogenation. Metals are used extensively as catalysts for the reduction of organic compounds. Nickel, palladium or platinum are most usually employed—more from habit than for sound scientific reasons. The most important classes of hydrogenation are:

(i) the saturation of olefins, diolefins, and acetylenes,
(ii) the saturation of aromatic compounds,
(iii) the saturation of heterocyclic compounds,
(iv) the reduction of carboxyl groups which may be bonded either to aromatic, aliphatic, or heterocyclic molecules,
(v) the conversion of nitro-compounds and of cyanides to amines.

Where the reactant is multiple-unsaturated, the reduction may either take place in a step-wise manner to give a series of products in succession, or else complete saturation may occur during one residence of the reactant on the catalyst surface. Clearly, there is also the possibility that one of the intermediates may be formed selectively (the phenomenon of selectivity is briefly discussed in the last section of this survey).

It is generally found that even the most complex catalytic reactions proceed via a large number of elementary steps which are, themselves, formally simple. In hydrogenation these elementary steps are the successive addition of hydrogen atoms. For example, the simplest reaction of the classes listed above is ethylene hydrogenation, and this occurs as follows:

$$\mathrm{C_2H_4}(g) \longrightarrow \mathrm{CH_2} = \underset{*}{\underset{|}{\mathrm{CH_2}}} \underset{-\mathrm{H}}{\overset{+\mathrm{H}}{\rightleftharpoons}} \underset{*}{\underset{|}{\mathrm{CH_2}}}\mathrm{—CH_3} \xrightarrow[\;-\mathrm{H}]{+\mathrm{H}} \mathrm{C_2H_6}(g)$$

each step being reversible, in principle. Adsorbed ethyl is termed a "half-hydrogenated state". Clearly the hydrogenation of, say, nitrobenzene to cyclohexylamine will be a very complex process involving a large number of chemisorbed intermediates. For a thorough description of the mechanisms of hydrogenation reactions the reader should consult the bibliography.

Olefin isomerization. The interconversion of chemisorbed olefin and chemisorbed alkyl groups can lead to chemical change (isomerization) if the olefin is a normal butene or a higher olefin. The following reaction scheme demonstrates how cis-trans isomerization and double-bond migration occur when catalysed by Group VIII metals.

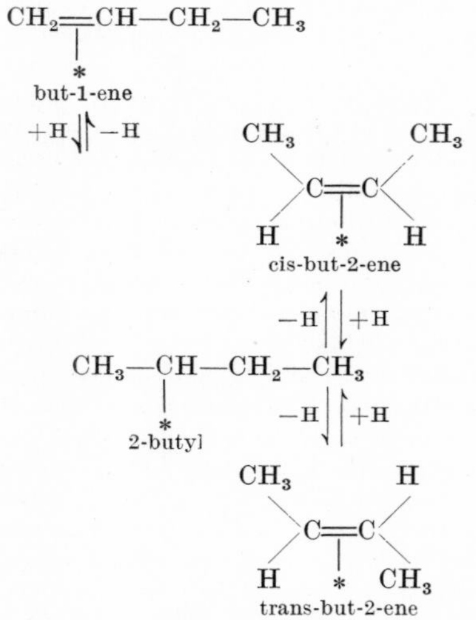

The geometrical configuration of the but-2-ene formed from but-1-ene depends upon which of the two hydrogen atoms of the —CH_2— group of adsorbed 2-butyl is lost. Since, to a first approximation, these hydrogen atoms are equivalent, the trans/cis ratio in the but-2-ene product is expected to be unity, and this is indeed observed in practice in the initial stages of reaction. This shows that the reaction is kinetically and not thermodynamically controlled under these conditions, since thermodynamic equilibrium strongly favours the trans-isomer. The isomerization of long-chain fatty acids and esters is of industrial importance, for example in the soap industry.

The Fischer Tropsch reaction. The reaction of hydrogen with carbon monoxide is catalysed by iron, cobalt, nickel, and ruthenium. The expected products—methane, methanol, and formaldehyde are obtained, together with large quantities of other substances which, at first sight, are unexpected, namely saturated and olefinic hydrocarbons, alcohols, aldehydes, ketones, acids, and esters. Carbon dioxide may also be produced. The most important reactions are

$$(2n + 1)H_2 + nCO = C_nH_{2n+2} + nH_2O \quad \text{(paraffin formation)}$$

$$2nH_2 + nCO = C_nH_{2n} + nH_2O \quad \text{(olefin formation)}$$

$$nH_2 + 2nCO = C_nH_{2n} + nCO_2 \quad \text{(olefin and carbon dioxide formation)}$$

$$2nH_2 + nCO = C_nH_{2n+1}OH + (n - 1)H_2O \quad \text{(alcohol formation)}$$

As expected from the stoichiometries of the above equations, the products obtained depend critically upon the carbon monoxide : hydrogen pressure ratio and upon the choice of catalyst, total pressure, and temperature. This process is not operated on a large scale except in South Africa, since it has been superseded by processes using oil as raw material. However, small-scale operation of the ruthenium-catalysed reaction in water at 130°C and under 1000 atmospheres pressure has been found to give high molecular weight paraffins (molecular weights of up to 100,000) with a high degree of selectivity, and this process may become of greater interest in the future.

The synthesis and oxidation of ammonia. Millions of tons of ammonia and nitric acid are consumed annually by the chemical industry in the manufacture of fertilizers, explosives, and dyestuffs. The first step in the process is the reduction of nitrogen to give ammonia:

$$N_2 + 3H_2 \rightarrow 2NH_3$$

which is achieved using a triply-promoted iron catalyst ($Fe—K_2O—CaO—Al_2O_3$). Although ammonia formation is an exothermic process the equilibrium under synthesis conditions favours the elements. Consequently, in practice, the mixture of nitrogen and hydrogen is passed over the catalyst at high speed, whereupon a few per cent of ammonia is formed. On leaving the reactor, the ammonia is washed out of the gas stream and the unconverted nitrogen and hydrogen are recycled.

The oxidation of ammonia to nitric oxide:

$$4NH_3 + 5O_2 = 4NO + 6H_2O$$

is achieved using a 90 : 10 Pt : Rh alloy wire gauze as catalyst, and a temperature of about 850°C. In contrast to ammonia production, this reaction occurs with extreme efficiency. Virtually all ammonia molecules which strike the catalyst react, and the reverse reaction is negligible. The nitric oxide is then converted to nitric acid.

Very different factors contribute to the choice of catalyst for these two processes. A large number of metals are active for ammonia synthesis (e.g. Fe, W, Mn, Ru, Os, U, Mo, and others) but in the pure state these metals are easily poisoned, and hence it is usual to incorporate small quantities of other substances which make the catalyst resistant to poisons. The addition of potassium, calcium, and aluminium oxides enhances the activity of iron as well as making it resistant to poisons, and so the oxides are termed "catalyst promoters." Many metals also catalyse ammonia oxidation, but the choice of the Pt—Rh alloy rests on metallurgical considerations. In the oxidation reaction the adsorbed species cause serious erosion of the catalyst surface; the alloy quoted above is the least susceptible to this erosion.

Dehydrogenation. Since catalysts do not modify the position of equilibrium of a system, the acceleration of a forward reaction rate must be accompanied by a corresponding increase in the rate of the back reaction. Thus, metals which are good hydrogenation catalysts are also expected to be good dehydrogenation catalysts. Metals will catalyse the dehydrogenation of primary alcohols to aldehydes and of secondary alcohols to ketones. Dehydrogenation of cyclic paraffins (e.g. the conversion of cyclohexane to benzene and hydrogen) occurs at moderate temperatures, whereas the dehydrogenation of aliphatic paraffins requires higher temperatures of about 500°C. At such a temperature, the catalytic cracking (hydrogenolysis) of carbon-carbon bonds occurs with the formation of lower molecular weight substances and carbon. The carbon frequently poisons the catalyst irreversibly by combining with the metal to form carbide. For this reason, commercial dehydrogenations are more usually carried out using oxide catalysts rather than metal catalysts, since the former give less cracking and they can be regenerated by oxidation when they become coked up.

Dehydrogenation is one of two important functions of the platinum reforming catalyst which is used on a vast scale in the petroleum industry for the production of aviation and motor fuel. Such fuels must contain a range of saturated and unsaturated hydrocarbons (C_5 to C_{11}). The catalyst consists of two components, each of which has a distinct catalytic function: platinum catalyses dehydrogenation, and acidified (often fluorided) alumina catalyses skeletal rearrangement. The concentration of platinum is normally only 0·05–0·10 per cent and it is present in an extremely high state of dispersion, so that almost every platinum atom is available for catalysis. The following is representative of the reaction sequences that hydrocarbons undergo; it should be appreciated that the molecules migrate back and forth between the platinum and the oxide, and for this reason this is termed a "dual function" catalyst.

n-hexane
↓ $-H_2$ Pt
hex-1-ene $\xrightarrow{\text{oxide}}$ 2-methyl pent-1-ene
↓ oxide
methyl cyclopentane
↓ $-H_2$ Pt
methyl cyclopentene
↓ oxide
cyclohexene $\xrightarrow[-2H_2]{\text{Pt}}$ benzene

Hydrogenolysis. This class of reaction includes all those processes in which single bonds are broken by reaction with hydrogen.

Typical examples are:

	bond undergoing fission
$C_2H_6 + H_2 \longrightarrow 2CH_4$	carbon-carbon
$C_2H_5NH_2 + H_2 \longrightarrow C_2H_6 + NH_3$	carbon-nitrogen
substituted ethers + $H_2 \longrightarrow$ alcohols	carbon-oxygen
aliphatic, acyl, aryl, heterocyclic } halides + $H_2 \longrightarrow$ HCl + appropriate residue	carbon-halogen

Conditions are known under which many supported noble Group VIII metals are active for these reactions; Raney nickel can also be used. These types of reaction are, on occasion, of value to the preparative chemist (Augustine 1965).

Selectivity. The sections above show that metals catalyse many types of reaction. Frequently, reactants may have several reactive groupings, and catalytic reaction with hydrogen may cause several reactions to occur simultaneously. Consider, for example, the reaction of hydrogen with bromobenzaldehyde. Three simple processes may occur, (i) the hydrogenation of the benzene ring, (ii) the reduction of the carbonyl group to give a primary aliphatic alcohol, or (iii) the hydrogenolysis of the carbon-bromine bond. Normally, only one product is required from a reaction, and in such a situation the catalyst is required to be selective in its action.

Many types of selective catalyst have been found, often by empirical methods. In some cases, unselective catalysts have been partially poisoned by nitrogen-, sulphur-, or oxygen-containing compounds; this treatment usually reduces specific activity, but this is often acceptable if the catalyst thereby becomes selective. (Poisoning of catalysts has been discussed in Mukherjee 1961.) A typical selective catalyst produced by this technique is the Lindlar catalyst which consists of palladium treated with quinoline (or pyridine) and a lead salt. This catalyst is selective for the hydrogenation of acetylene to ethylene, but will not catalyse the hydrogenation of ethylene on to ethane. It is probable that the poisons modify the electronic characteristics of the surface metal atoms and hence also the way in which molecules chemisorb on the catalyst. In the case of the Lindlar catalyst, the poisons may weaken the strength of ethylene chemisorption to the point where its surface coverage and hence its hydrogenation rate are negligible.

A second type of catalytic selectivity is purely mechanistic in origin. Consider a reactant A which reacts to give products, B, C, and D simultaneously, a typical example is the hydrogenation of buta-1,3-diene to give the three isometric n-butenes. This reaction is catalysed by all of the Group VIII metals and copper, although only Fe, Co, Ni, Cu, and Pd give butene free from butane. The three isomeric products are but-1-ene, cis-but-2-ene, and trans-but-2-ene. Copper supported on alumina at 100°C catalyses the preferential formation of but-1-ene by a 1 : 2-addition process (yields in excess of 90 per cent) whereas cobalt supported on alumina at the same temperature catalyses the formation of trans-but-2-ene preferentially by a 1 : 4-addition process (yield *ca.* 80 per cent). No metal catalyst has been found which provides cis-but-2-ene preferentially from buta-1,3-diene, but in the hydrogenation of the isomeric substance but-2-ene (dimethylacetylene) all of the Group VIII metals show considerable selectivity for cis-but-2-ene formation (yields 85–97 per cent) and copper is completely selective (yield 100 per cent). This type of selectivity is determined by the reaction mechanism which, in turn, is dependent upon the chemistry of the atoms in the metal surface.

Finally, catalysts are occasionally required for the selective removal of a component present in small concentration, such as a contaminant in a feedstock. For example, a poisoned palladium catalyst is used commercially to remove traces of acetylene from ethylene streams.

Bibliography

Advances in Catalysis, Volumes 1–19 dated 1948 to 1968 (series continuing), New York: Academic Press. Reviews contained therein.

ASHMORE P. G (1963) *Catalysis and Inhibition of Chemical Reactions*, London: Butterworths.

AUGUSTINE R. L. (1965) *Catalytic Hydrogenation*, London: Arnold; New York: Dekker.

BOND G. C. (1962) *Catalysis by Metals*, New York: Academic Press.

EMMETT P. H. (Ed.) *Catalysis*, Volumes 1–7 (series completed), New York: Reinhold. Reviews contained therein.

MALINOWSKI N. (1961) in *Encyclopaedic Dictionary of Physics* (J. Thewlis Ed.), **1**, 577, Oxford: Pergamon Press.

MUKHERJEE N. R. (1961) in *Encyclopaedic Dictionary of Physics* (J. Thewlis Ed.), **1**, 581, Oxford: Pergamon Press.

P. B. WELLS

HIGH DEFINITION RADIOGRAPHY. Industrial radiographic techniques generally make little use of image magnification. This is largely because image unsharpness resulting from the finite dimensions of the X-ray source, coupled with the large emulsion grains and double coatings on X-ray films, limit the advantage of magnification. The development of *projection microradiography*, however, has shown quite clearly that X-ray image resolution can be as high as 0·1 μm and growing interest in applications of this technique has fostered the associated development of ultra fine-focus X-ray microscopes with focal spot diameters between 10 and 0·1 μm. Whilst microradiography is usually confined to laboratory investigations, requiring minute specimens or specially prepared wafer thin samples to reduce image superposition, the availability of these "micron focus"

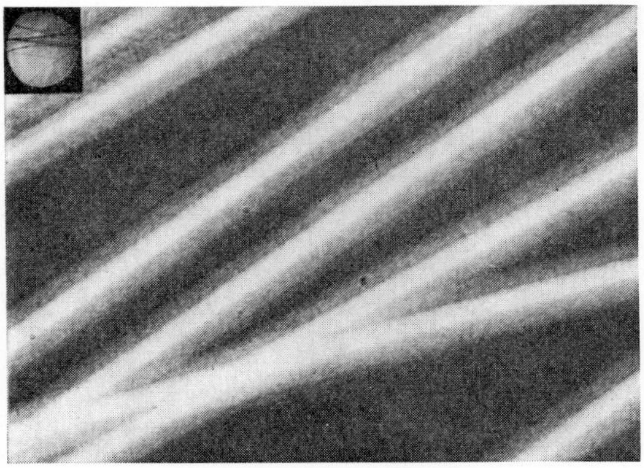

Fig. 1. High definition radiograph of 10 μm copper wire coated with a 10 μm layer of glass. (Inset is an unmagnified print of the original radiograph).

X-ray sources also opens up interesting possibilities in extending industrial radiographic techniques to resolve extremely fine detail and to make precise measurements on internal components or encapsulated systems. The absolute limit of resolution is set by the size of the focal spot.

The conditions chosen for any radiographic exposure are generally a compromise and in examining the interrelation between two of the relevant factors, exposure time and image definition, the potential advantage of the ultra-fine focus X-ray unit in radiography becomes apparent. In comparing the performance of X-ray units, the image unsharpness for equal exposure times is inversely proportional to the square root of the specific anode loading. Stated another way, the exposure times required to give a particular level of image unsharpness are inversely proportional to the specific loadings. From both points of view there is obviously an attractive gain to be achieved with a high specific loading. Although power dissipation at the small focal spot of an X-ray microscope is limited by problems of heat dissipation to only a few watts, specific anode loadings can rise as high as 10^4–10^6 W/mm². On the other hand for the

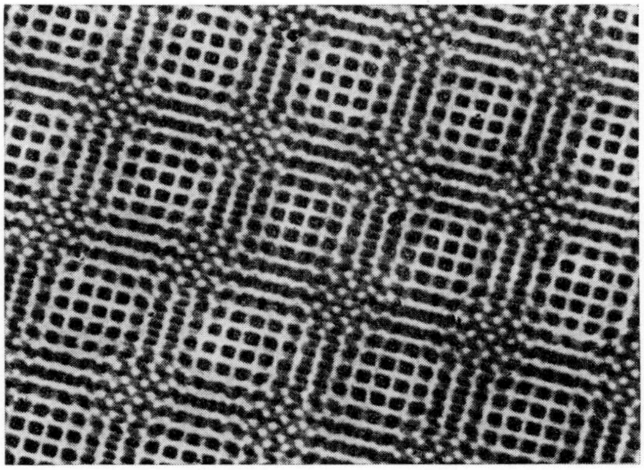

Fig. 2. A radiographic moiré fringe pattern produced by two adjacent 1500 t. p. i. meshes aligned with an angular displacement of 7°.

more conventional stationary target industrial X-ray tube power dissipations on the larger focal spots often rise into the kilowatt range, but the specific anode loadings are only in the 10^2–10^3 W/mm² range. In general, part of the decrease in exposure time required with these "micron focus" X-ray units can be offset by using fine grained photographic emulsions instead of X-ray film. In this way the radiograph, which will probably already have a primary projection enlargement of X2–X20, can be further magnified ×10 to reveal the fine detail. There is no clear demarcation between high definition radiography and microradiography, although the former might conveniently be reserved for non-destructive examinations that can be carried out without any modification or preparation of the sample. The present limitations of high definition radiography with "micron focus" X-ray tubes are that kilovoltages in excess of 50 are not yet commercially available, and the area of examination is restricted to a circle of about 20 cm² area. The extremely divergent beam of radiation can produce spurious contrast at curved surfaces by total reflection and in coarse grained samples by lattice diffraction. However, the trend towards miniaturization in microcircuitry, the increasing need to reveal microstructural deviations, and the continual lowering of dimensional tolerances collectively add importance to high definition radiography and indeed to any-non-destructive inspection technique that can be adapted to these special demands.

Bibliography

Halmshaw R. *et al.* (1966) *Physics of Industrial Radiology*, London: Heywood.

Nixon W. C. (1962) in *Encyclopaedic Dictionary of Physics* (J. Thewlis Ed.), **7**, 806, Oxford: Pergamon Press.

R. S. Sharpe

HIGH-INTENSITY HOLLOW CATHODE LAMPS. When two suitably spaced parallel plates are used for the cathode of a glow discharge the negative glows associated with each plate coalesce and this results in increased current density and a consequent increase in light output since each electron liberated from the cathode makes many more collisions with metal atoms before escaping to the positive column of the anode. A large number of parallel plate cathodes can conveniently be provided using a hollow cylinder. This type of discharge fulfils most of the requirements for the emission of sharp spectral lines of elements vaporized from the cathode surface by cathode sputtering since the temperature and pressure are low and there are no magnetic fields present. In particular the resonance lines of the sputtered metal are emitted with spectral widths of the order of 0·01 Å and the hollow cathode discharge lamp has thus proved extremely useful as a light source for atomic absorption measurements for which line widths of the above order are required.

Experience has shown that in the hollow cathode type discharge the onset of line broadening of resonance lines is due in most cases to self-absorption and self-reversal. To prevent this type of broadening the vapour pressure of the metal atoms liberated from the cathode must be kept low. Hence there is little possibility of achieving greatly increased intensity of the resonance lines whilst preserving the desired line width if a hollow-cathode lamp is used, since the one electrical discharge produces the atomic vapour and also excites it. Any attempt to increase the excitation by increasing the current will inevitably lead to a high pressure of metal vapour resulting in absorption broadening.

In the high-intensity hollow cathode lamps the two functions of the production of metallic vapour and its excitation are separated and instead of a single discharge two are used. The one produces the metallic vapour by the phenomenon of cathode sputtering and the other, electrically isolated from the former, causes the excitation of some of the atoms. In this manner the requisite vapour pressure of the metal atoms can be maintained, but the excitation can be increased by increasing the current in the second discharge which does not affect the metal vapour pressure and therefore does not produce the line broadening due to self-absorption or self-reversal. A simple form of high-intensity lamp is shown in Fig. 1.

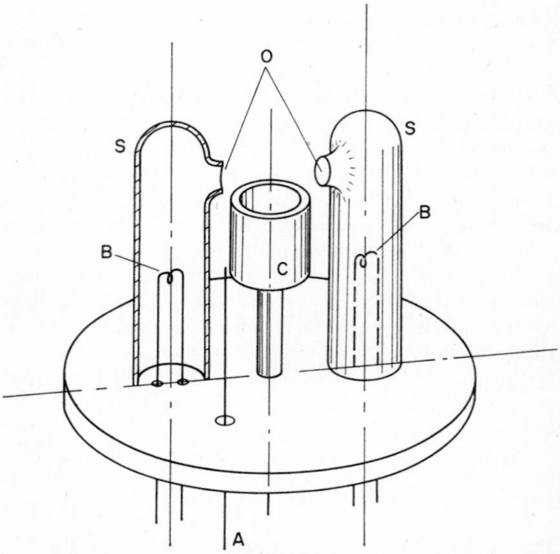

Fig. 1. Electrode arrangement in high-intensity hollow cathode lamp.

The electrode arrangement of the primary discharge is simply that of a conventional hollow cathode discharge whilst the secondary discharge occurs between two other electrodes positioned close to the top of the hollow cathode. The electrode assembly is housed in a glass envelope with an end viewing window. After standard outgassing procedures the envelope is filled with a rare gas at a pressure of approximately 3 torr. The positive column of the secondary discharge is located in the vicinity of the top of the hollow cathode and in order to restrict it to this region so that a high current density is achieved the auxiliary electrodes are shrouded as shown. This enables the atoms produced by sputtering in the hollow cathode discharge to be effectively excited by the secondary one. One electrode of the second discharge is usually in the form of an oxide-coated filament in order to lower the voltage across the discharge thereby avoiding excessive heating and difficulties due to cathodic sputtering of the secondary electrodes themselves. In operation such a type of lamp would operate at about 200 V and 20 mA for the primary discharge, depending on the metal of the cathode, and about 30 V and 500 mA for the secondary discharge. The auxiliary discharge is essentially one of low energy and favours the excitation of the metal spectra with respect to the rare gas which is used as the filler. Typical spectra obtained using such lamps are shown in Figs. 2, 3, 4, and 5. The increase in intensity of the resonance lines can be determined from the exposure times used in each case and their enhancement with respect to the argon filler gas is also shown.

Whilst no direct measurements have been made of the line widths of the enhanced resonance lines, the peak absorption of flames into which appropriate solutions of the various metals have been sprayed (see *Atomic absorption spectroscopy and its application to chemical analysis*) indicates that the secondary discharge has caused no measurable increase in line-width.

An improvement in the efficiency of high-intensity hollow cathode lamps can be effected if the geometrical configuration of the cathode is altered. Thus when the cathode is made in the form of a loop of the appropriate metal it is found that the auxiliary discharge will function equally well with argon or neon as the filler gas. The advantage of using neon, namely increased intensity of the resonance line for a given line width, is thus retained.

High intensity hollow cathode lamps are proving valuable in conventional atomic absorption spectrophotometers where their use for elements whose spectra is complex and contain many resonance lines results in better signal-to-noise ratio. The use of high-temperature flames to bring about the effective dissociation of the elements forming stable oxides leads to unwanted emission at the wave-length at which measurements are made and is a cause of noise. Acceptable signal-to-noise ratio can be obtained if the high intensity hollow-cathode lamp is used as the spectral light source. Furthermore in the case of nickel and cobalt the resonance lines at 2320·03 Å and 2407·25 Å each have neighbouring spark lines which are suppressed in the high-intensity hollow cathode lamp discharge as shown for the case of nickel in Fig. 6.

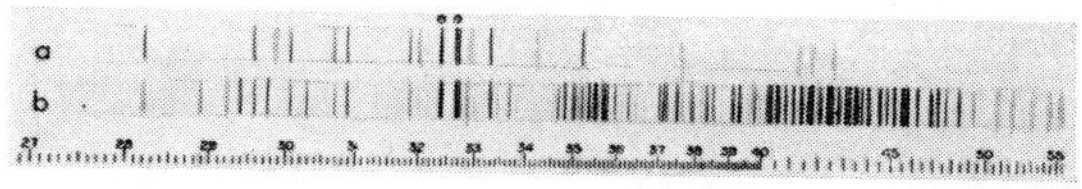

Fig. 2. (a) Spectrum of copper high-intensity hollow cathode lamp exposure 2 sec. (b) Spectrum of conventional copper hollow cathode lamp exposure 5 mins. The copper resonance lines are at 3247·54 Å and 3274·96 Å.

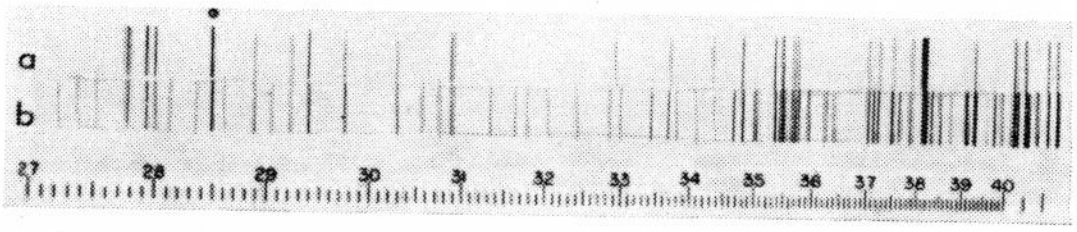

Fig. 3. (a) Spectrum of magnesium high-intensity hollow cathode lamp exposure 15 sec. (b) Spectrum of conventional magnesium hollow cathode lamp exposure 15 mins. The magnesium resonance line is at 2852·13 Å.

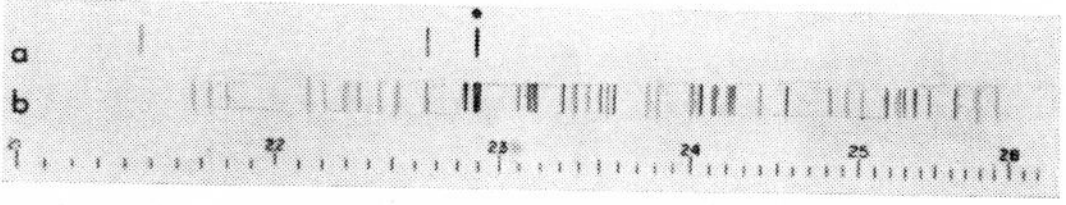

Fig. 4. (a) Spectrum of cadmium high-intensity hollow cathode lamp exposure 3 min. (b) Spectrum of conventional cadmium hollow cathode lamp exposure 5 hr. The cadmium resonance line is at 2288·02 Å.

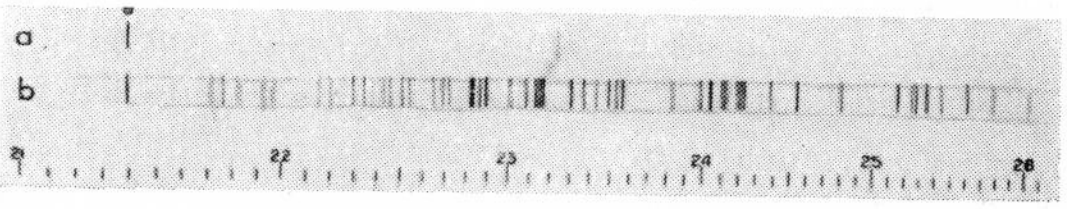

Fig. 5. (a) Spectrum of zinc high-intensity hollow cathode lamp exposure 3 min. (b) Spectrum of conventional zinc hollow cathode lamp exposure 5 hr. The zinc resonance line is at 2138·56 Å.

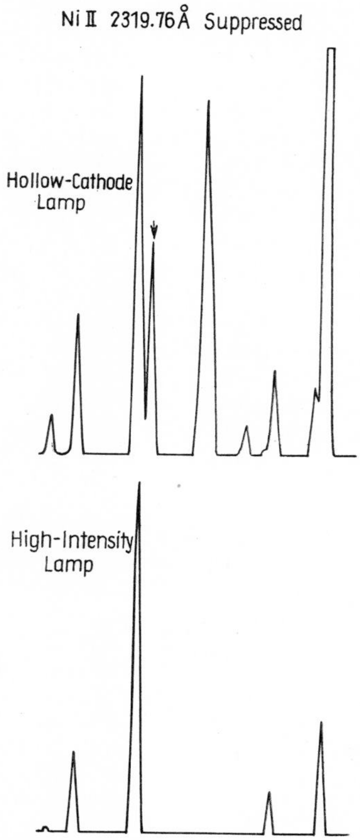

Fig. 6. Spectrum of conventional nickel hollow cathode lamp and a high-intensity hollow cathode lamp showing the suppression of the Ni II line at 2319·76 Å.

Apart from their use in atomic absorption spectrophotometric the high-intensity hollow cathode lamps have several other important applications.

When an atomic vapour is illuminated by light from a spectral lamp whose emitted spectrum includes the resonance lines of the vapour the absorbed quanta can be re-emitted as the atoms drop back to their ground state. The emitted radiation has the same frequency as the radiation originally absorbed and is emitted in all directions. Wood (1934) termed this phenomenon resonance radiation. In general, studies of resonance radiation have been largely restricted to elements having appreciable vapour pressure at relatively low temperatures. However, the atomic vapour of even high boiling point metals can be generated easily and without the need for high temperature furnaces by cathodic sputtering in an electrical discharge at low pressure. The light source used to produce the exciting light is required to be intense over the spectral width of approximately 0·01 Å which corresponds to the absorption width of the atomic vapour. High-intensity hollow cathode lamps possess this feature and are extremely useful in resonance radiation studies. An arrangement for isolating and detecting resonance lines by resonance radiation is shown in Fig. 7.

The region from which the resonance radiation originates will contain metal and filler gas atoms

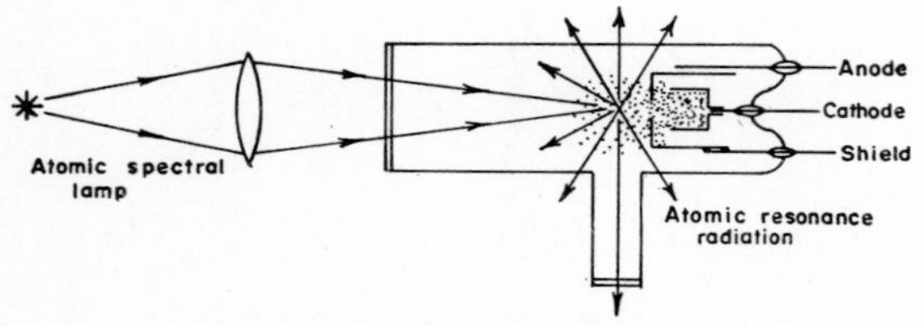

Fig. 7. Diagram of experimental arrangement for the production of resonance radiation using a high-intensity hollow cathode lamp and atomic vapour produced by cathode sputtering.

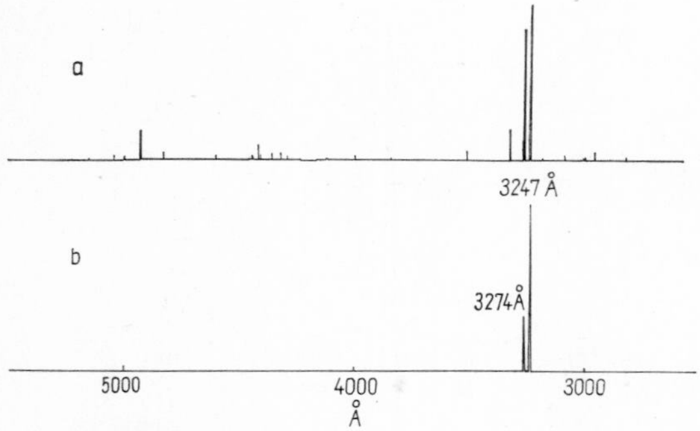

Fig. 8. Photoelectric recordings of (a) spectrum of copper high-intensity hollow cathode lamp; (b) resonance spectrum of copper isolated by resonance monochromator.

excited by the hollow cathode discharge and in order that the radiation from these atoms produce no output signal a modulation technique is used. The power supply to the high-intensity hollow cathode lamp is modulated whilst that to the resonance lamp is unmodulated and thus by the use of an a.c. detection system any radiation from the resonance lamp does not produce an output signal.

Photoelectric recordings of the resonance spectra of copper and magnesium together with the spectra of the respective high-intensity hollow cathode lamps are shown in Figs. 8 and 9.

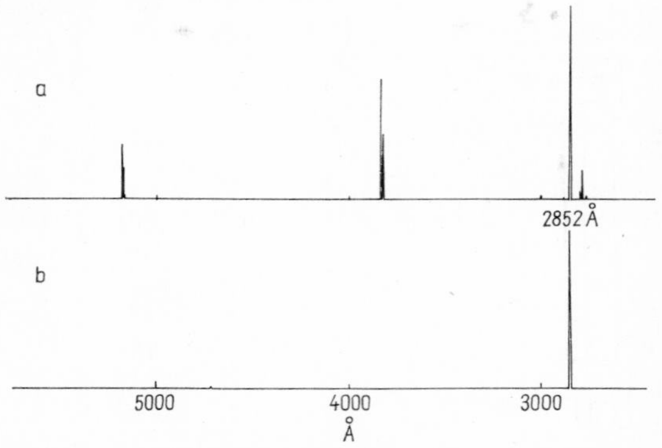

Fig. 9. Photoelectric recordings of (a) spectrum of magnesium high-intensity hollow cathode lamp; (b) resonance spectrum of magnesium isolated by resonance monochromator.

Bibliography

WOOD R. W. (1934) (3rd Edn.) *Physical Optics*, New York: Macmillan.

SULLIVAN J. V. and WALSH A. (1965) *Spectrochim. Acta*, **21**, 719.

SULLIVAN J. V. and WALSH A. (1965) *Spectrochim. Acta*, **21**, 727.

DAVIES D. K. (1967) *J. App. Phys.* **38**, 4713.

J. V. SULLIVAN

HIGH PRESSURE RESEARCH; RECENT ADVANCES. *Equipment for producing ultra-high pressures.* Most liquids freeze below 10 kb at room temperature (1 kb = 986·9 atm) and in general solids must be used for transmitting very high pressures. In spite of some inhomogeneity in pressure distribution the Bridgman anvil uniaxial press is the most convenient laboratory equipment for the study of small samples up to about 100 kb and 800°C. Two opposed anvils of carboloy (cemented tungsten carbide) compress a small disk-shaped sample retained by means of an annular pyrophillite gasket (Fig. 1). The heating is external by means of a split resistance-furnace, but higher temperatures have been reached by Bundy by internal heating. Higher pressures up to about 500 kb have been reached by Drickamer at room temperature by the use of smaller anvil faces and by increasing the lateral support of the anvils.

High temperatures and pressures are achieved industrially in the "belt" apparatus (Fig. 2), in which conical pyrophyllite gaskets serve as a seal and allow pressure transmission to a cylindrical sample, which may be compressed to 150 kb, and heated electrically to 2000°C by means of a condenser discharge (Bundy). An alternative to the "belt" apparatus is the tetrahedral press originated by Tracy Hall, in which piston thrust is transferred to the sides of a pyrohyllite tetrahedron in which the sample is embedded (Fig. 3). The piston thrust may also be applied to the faces of a cube. Other workers have used a supported piston-cylinder device.

X-ray diffraction may be studied on samples *in situ* by compressing them in a cylindrical hole drilled in a diamond or in a beryllium cylinder, through which an X-ray beam is passed, or by compressing them between diamond anvils. In the tetrahedral apparatus (Fig. 4) of Tracy Hall the X-ray beam is

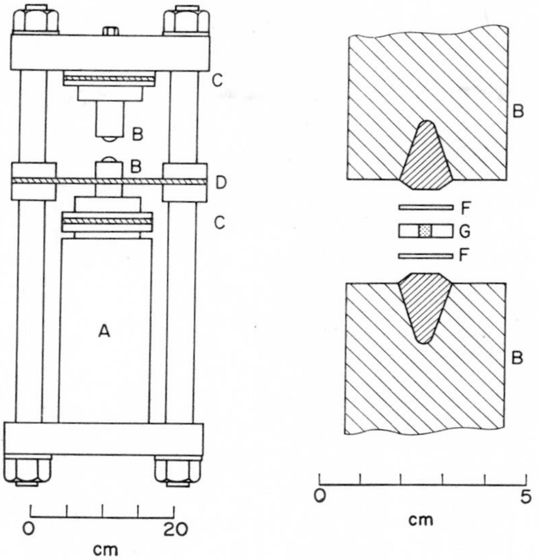

Fig. 1. Bridgman uniaxial anvil press. (Reproduced with permission from Fig. 3.9, p. 61 of Bradley and Munro, "High Pressure Chemistry", Pergamon Press, 1965. From Bradley, R. S., Grace, J. D. and Munro, D. C., "Trans. Faraday Soc.", **58**, *776 (1962); cf. Fig. 1 and Fig. 2, p. 777.)*

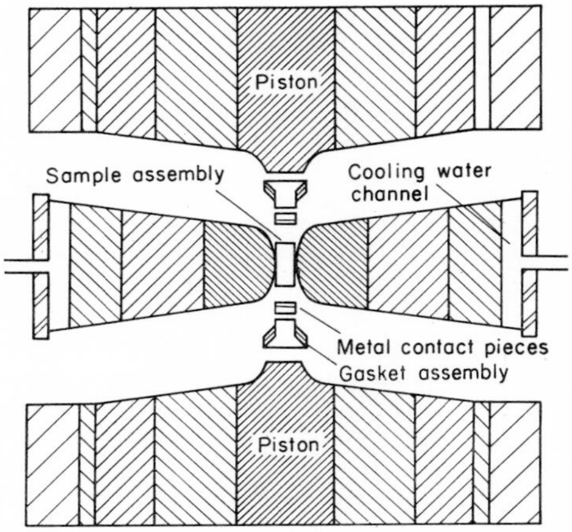

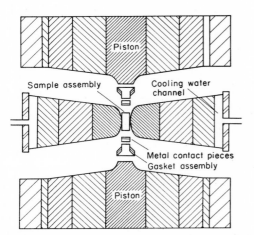

Fig. 2. Simplified cross-section of the "belt" apparatus. (Reproduced by permission from Fig. 2.7, p. 221 from Bradley and Munro, "High Pressure Chemistry", Pergamon Press, 1965. After H. T. Hall, 1960.)

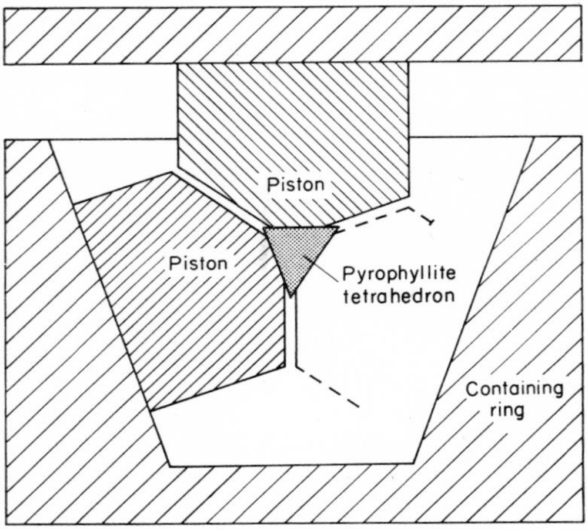

Fig. 3. Modified tetrahedal press. (Reproduced by permission Fig. 2.8, of p. 23 Bradley and Munro, "High Pressure Chemistry", Pergamon Press, 1965.)

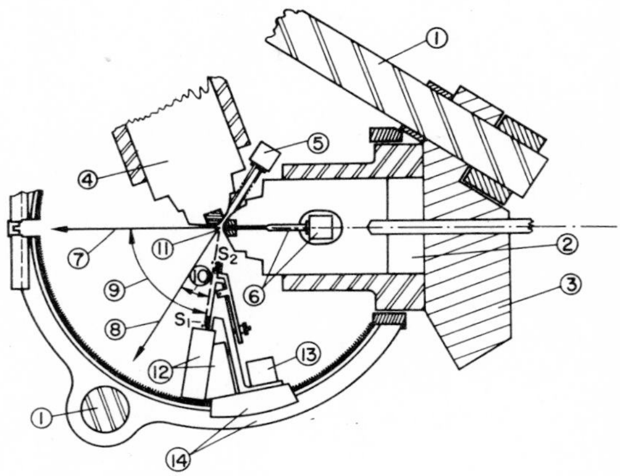

Fig. 4. High pressure high temperature X-ray diffraction apparatus. (Reproduced with permission from J. D. Barnett and H. T. Hall, "Rev. Sci. Instrum." **35**, *175 (1964).)*

led through the centre of the piston to the sample which is embedded in a tetrahedron of compressed lithium hydride, boron or boron-filled plastic. Pressures up to 75 kb and temperatures up to 1000°C have been achieved.

Magnetic studies at high pressure may be made by the use of Be-Cu alloy. Spectroscopic studies in the ultra-high pressure range may be made by the use of diamond anvils through which the beam passes axially, or by using NaCl to transmit the optical beam.

Measurement of ultra-high pressures. The dead-weight principle has been extended to 40 kb for solid samples by Kennedy and La Mori. Secondary calibration is usually made by observing phase transitions in metals, e.g. Bi, Cs, Tl, Ba, by means of studies on electrical resistance: absolute pressures are uncertain. Pressures arising from shock waves can be deduced by the application of the Rankine-Hugoniot theory, but the temperature is less certain.

Gases and liquids under very high pressures. At very high pressures gases behave similarly to liquids, e.g. they solidify (N_2 at 28·5 kb at 20°C), they dissolve solids, show partial miscibility, and the logarithm of the fluidity and diffusion coefficient are linear functions of $1/T$; at constant temperature, fluidity and diffusion coefficients decrease markedly with increasing pressure. For water at 0°C the viscosity at first decreases with increasing pressure and then increases and the anomaly in density at 4°C disappears with water under pressure. These results are most simply interpreted in terms of a two-fluid model of mixed open and close-packed liquids.

Phase changes under pressure. For Ga, Sb, Si, Ge, diamond, ice, etc. the melting point decreases on increasing the pressure; it is probable that this trend will be reversed at very high pressures by the intervention of solid-solid phase changes, e.g. for ice the highest melting point observed is 400°C at 200 kb (Pistorius). Maxima in the melting curve with $dT/dP = 0$ have been observed with Rb, and with Cs (two maxima) and these are not covered by the Simon equation $p/a + 1 = (T/T_0)^c$, where a and c are constant. The maxima are accommodated by the empirical equation of Kennedy (1966) $T/T_0 = 1 + b\Delta V/V_0$, where b is a constant and $-\Delta V/V_0$ is the compression at *room temperature* of the *solid* at the pressure corresponding to a temperature T.

Many alkali metal halides show phase transitions from the rock-salt to the caesium chloride structure under pressure. More covalently bonded solids such as CdTe show a transition: diamond or zinc blend type → rock salt type → white tin type. Semiconductors undergo a continuous alteration in electronic band structure under pressure and there is frequently an increase in electronic conductivity; there is some theoretical support for the belief that at sufficiently high pressures all solids will become electronic conductors. In addition to a continuous change in band structure some metals show an abrupt change, e.g. Ce, Cs, Yb at first-order transitions associated with rearrangement of electron bands.

Many molecular solids change irreversibly under pressure, e.g. CS_2 → black polymer, red P → black P. Similarly for covalent solids quartz → coesite → stishovite, BN → borazon, graphite → diamond. In the commercial process for diamond synthesis nickel is used as a catalyst in the "belt apparatus" at 60 kb and 1500°C. Diamond can be made by a direct process in the absence of catalyst by the use of a condenser discharge at 120 kb and 3000–4100°K. The triple point of carbon is estimated at 125 kb, 4100°K. Doped synthetic diamond crystals containing 0·1 per cent of boron are intensely blue and are good p-type semiconductors.

"Second-order" transitions have been studied under pressure with NH_4Cl and quartz.

Spectroscopic studies under pressure. Gases show broadening and shifts of spectroscopic lines owing to intermolecular perturbations; forbidden transitions may appear.

For the u.v. $\pi - \pi^*$ bands of aromatic hydrocarbons a shift to the red occurs under pressure. Infra-red spectra of stretching vibrations, e.g. of the CN and CO groups of organic molecules in solution, show a blue shift. With solids such as $CaCO_3$ shifts occur in the bands and forbidden transitions are allowed. Pressure increases the crystal field in inorganic complexes such as $Ni(NH_3)_6Cl_2$ and the effect on the Racah parameters has been interpreted in terms of an increase in the covalency of the binding with the ligand. Similar studies have been made on transitional metal ions in host lattices such as Al_2O_3, and these enable the local compressibility in the neighbourhood of the transitional ion to be determined. The effect of pressure on the internal electric field gradients in solids has been determined by studies on the pure quadrupole resonance, e.g. Cu_2O behaves as ionic and the vibration frequencies decrease with increasing pressure. Paramagnetic resonance of transitional metal ions in host lattice shows a splitting of the ground levels which is pressure dependent; for Cr^{3+} substituted in $K_3Co(CN)_6$ there is also a pressure dependence of the effective magnetic moment of the Cr^{3+} ion on pressure. A volume dependence of the Knight shift has been found for Li, Na, Rb and Cs. Nuclear resonance studies have been made on solid hydrogen. High pressure Mössbauer studies have been made on dilute solutions of ^{57}Fe in various metals; the quadrupole splitting with $CoCl_2$ under pressure agrees with calculations on the crystal field.

Shock waves. Very high transient pressures up to 9000 kb have been observed in shock waves. The pressure is given by the Rankine–Hugoniot equation $\Delta E = -\frac{1}{2}p\,\Delta V$, where ΔE is the increase in energy per unit mass and $-\Delta V$ the decrease in volume per unit mass. With gases the main effect is increase in temperature, and high-temperature chemical reactions have been studied. With solids shock waves generated by shaped explosive charges give very high pressures. A large increase in electrical conductivity is observed with P, S, Se and I_2, but this depends partly on an increase in temperature. Phase transitions have been observed in, e.g. Fe, Ge and the alkali metal halides. The self-ionization of water is increased by a factor of 10^6 owing to the combined effect of pressure and temperature.

Chemical applications of high pressure research. Increase in pressure may increase or decrease the velocity of a chemical reaction depending on whether there is a decrease or increase in partial molal volume ($\Delta \overline{V}^*$) in the conversion of the reactants to the transition state. $\Delta \overline{V}^*$ is dependent on the solvent since it partly depends on the change in volume of the solvent during the transition process. Recent work has emphasized steric effects, many reactions being possible under pressure which would normally be prohibited, e.g. the polymerization of tetramethyl ethylene. Pressure can also affect the relative yields of products, e.g. in the nitration of 1–3 xylene. Weak acids and bases become stronger under pressure since there is a volume contraction on dissociation arising from solvent electroconstriction. Glass electrodes have been used successfully under pressure by equalizing the pressure on both sides of the membrane and the increase in ionic product of water has been followed; much work has been done on the physical chemistry of sea water under pressure.

Metallurgical applications. The most important work appears to be the extrusion of normally brittle metals under hydrostatic pressure.

Bibliography

Benedek G. B. (1963) *Magnetic Resonance at High Pressures*, New York: Interscience.

Bradley J. N. (1962) *Shock Waves in Physics and Chemistry*, London: Methuen.

Bradley R. S. (Ed.) (1963) *High Pressure Physics and Chemistry*, New York: Academic Press.

Bradley R. S. (Ed.) (1966) *Advances in High Pressure Research*, Vol. 1, New York: Academic Press.

Bradley R. S. and Munro D. C. (Eds.) (1965). *High Pressure Chemistry*, Oxford: Pergamon Press.

Bridgman P. W. (1964) *Collected Experimental Papers*, Harvard: The University Press.

Bundy F. P., Hibbard W. R. and Strong H. M. (Ed.) (1961) *Progress in Very High Pressure Research*, New York: Wiley.

Giardini, A. A. and Lloyd E. C. (Eds.) (1963) *High Pressure Measurement*, London: Butterworths.

Gschneider K. A., Hepworth M. T. and Parlee N. A. D. (Eds.) (1964) *Metallurgy at High Pressures and Temperatures*, New York: Gordon and Breach.

Hamann S. D. (1957) *Physico-Chemical Effects of Pressure*, London: Butterworths.

Van Itterbeek, A. (Ed.) (1965) *Physics of High Pressures and the Condensed Phase*, Amsterdam: North Holland.

Paul W. and Warschauer D. M. (Eds.) (1963) *Solids under Pressure*, New York: McGraw-Hill.

Ryabinin Yu N. (1960) *Gases at High Densities and Temperatures*, Oxford: Pergamon Press.

Thewlis J. (Ed.) (1961) *Encyclopaedic Dictionary of Physics*, articles beginning "High pressure", **3**, Oxford: Pergamon Press.

Tomizuka C. T. and Emrick R. M. (1965) *Physics of Solids at High Pressures*, New York: Academic Press.

Tongue H. (1959) *The Design and Construction of High Pressure Chemical Plant*, London: Chapman and Hall.

Zeitlin A. (1964) *Annoted Bibliography on High Pressure Technology*, London: Butterworths.

R. S. Bradley

HIGH-SPEED COLLISIONS. The scientific study of collisions between two bodies began with the birth of mechanics in the latter part of the sixteenth century. Galileo did considerable experimental work on the subject but the first published statement concerning this phase of mechanics appeared in 1639 in a treatise called *De Proportione Motus* by Marcus Marci, a professor at Prague. No theories of impact were formulated until 1668, then, within a period of less than three months the Royal Society of London received papers on this subject from Wallis, Wren, and Huygens. Wallis' work dealt with the action of inelastic bodies only; Huygens approached the problem from the kinetic energy standpoint; and Wren's work included the results of his experiments. These experiments apparently had considerable effect on the formulation of Newton's laws, as he spoke highly of Wren's work and referred to his experiments in his *Principia*. These experiments had clarified the behaviour of elastic and inelastic bodies and Newton's theory of impact divided bodies into these two classes which he called "perfectly elastic" and "imperfectly elastic". In the case of the former, there is no loss of kinetic energy and in the other case, energy is dissipated on impact. Other early investigators who contributed to a knowledge of the subject include Bernoulli, Poisson, Rayleigh, and Hertz.

From the practical side, the study of high-speed collisions has been motivated by warfare. For millennias the contest between faster projectiles and stronger armour has continued; however, the incentive to study this phenomenon has changed in recent years. The projectiles which are now of primary interest are meteoroids travelling through space at speeds estimated to average about 20 km/s, more than an order of magnitude greater than the fastest conventional bullet, and with maximum speeds reaching 75 km/s. The targets, instead of being massive fortifications or heavy armour, are now the relatively delicate structures of space vehicles.

The current interest in the problem of high-speed collisions or impact is indicated by the enormous amount of literature being published on the subject. During the past decade many hundreds of papers have appeared, presenting scores of theories of impact and an even greater number of empirical relations. Workers at numerous laboratories are attempting to simulate meteoroid impact by firing projectiles into targets of various materials. The principal method employed is the use of a two-stage, light-gas gun in which an explosively driven piston is used to compress a light gas propellant (He or H_2), which in turn accelerates the projectile. Velocities up to about 12 km/s have been attained. Guns utilizing exploding foils, electromagnet fields, and high-voltage accelerators are capable of driving micron-sized particles to much higher velocities. It has been impossible, however, to reach average meteoroid velocity with projectiles whose size and properties on impact are fully known.

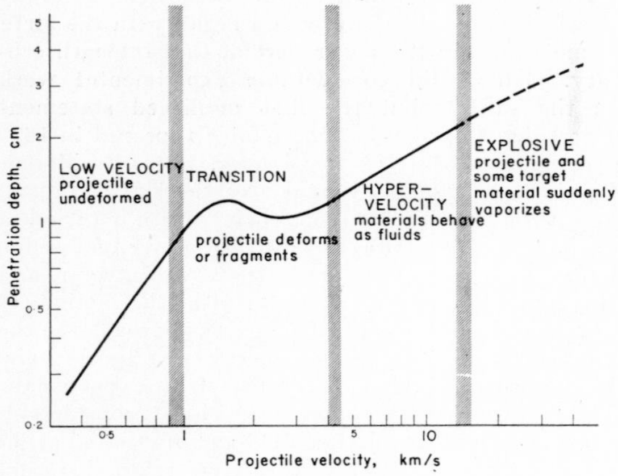

Fig. 1.

Regimes of impact phenomena. The impact of thick targets may be classified into several categories or regimes which depend primarily on the projectile velocity. The rather idealized diagram shows the effect of velocity on the projectile's penetrating power.

In the low-velocity region, the projectile remains intact and experiences little deformation. It produces a deep cylindrical cavity, the diameter of which is only slightly greater than that of the projectile. In this range, the motion of the projectile is strongly influenced by the strength of the two materials. The penetration, or cavity depth, generally varies as the four-thirds power of the projectile velocity.

As the velocity increases, a critical value is reached at which the projectile is unable to resist the forces of impact and begins to deform, or if brittle, to break into fragments. This marks the end of the low-velocity region and the beginning of the transition range of impact. As the velocity increases still further, a ductile projectile will mushroom and a brittle projectile will fragment into smaller and smaller pieces. The rate of increase in crater depth decreases as an increasing amount of the projectile's kinetic energy is used in widening the crater. At still higher velocities the crater depth actually decreases. It was at this point on the velocity scale that it seemed that the ballisticians had lost the contest with the armour designers. But eventually when they were able to increase the projectile velocities still further, the penetration depth again began to increase. The craters have now become almost hemispherical. At still higher speeds the penetration is roughly proportional to the two-thirds power of the velocity.

It is evident that a transition region has been crossed. This new region is commonly called the hypervelocity stage. Just where this regime begins depends upon the definition of the term hypervelocity. Several criteria for the beginning of hypervelocity have been suggested. One is that the craters are hemispherical in shape. Another is that the cratering efficiency (the volume of crater per unit of projectile energy) is a constant. A third, and most common, criterion is that the ratio of projectile velocity to stress wave velocity in the target exceeds unity. It should be realized that the regime boundaries are not precise and whatever criterion is used to define the boundary is one of convenience and not one of precision. One characteristic of this region is that the stresses developed by the projectile's deceleration and induced stress waves are much greater than the mechanical strength of either projectile or target. For example, when a copper ball strikes a steel target at 10 km/s, the interface pressure is approximately 60 million pounds per square inch. This is so much greater than the ultimate strength of either material that they act as if they had no strength at all, at least during the initial period of impact. A material with little or no

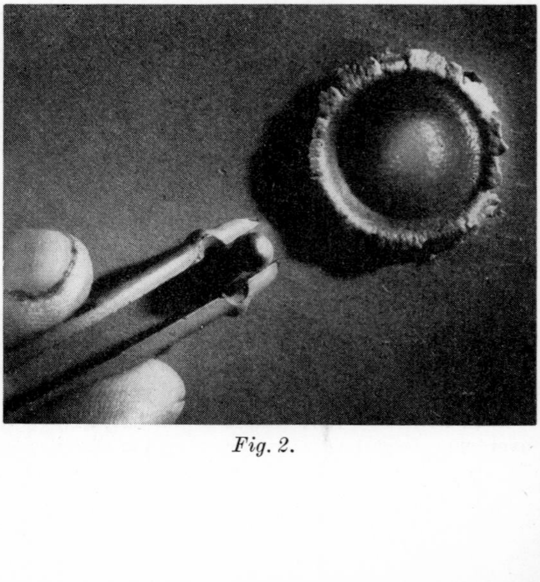

Fig. 2.

strength is fluid, so this is sometimes called the fluid impact or hydrodynamic phase. Resistance to the initial penetration is due largely to inertia forces.

Continuing up the velocity scale, an impact region is reached that is, as yet, virtually unexplored. It is called the regime of impact explosions, for the enormous amount of energy suddenly made available is sufficient to vaporize the projectile and a part of the target. It has been suggested that the threshold of this regime is where the ratio of projectile velocity to stress wave velocity is at least three. In an aluminium, for example, the stress wave velocity is about 7 km/s, which means that the onset of the explosive region may be greater than 20 km/s while the upper limit of present ballistic facilities is little more than one-half this value.

Description of the cratering process. The crater formation resulting from hypervelocity impact is extremely complex, involving many different types of material behaviour. The physical model given here is based upon the experiments of many investigators using high-speed photography, flash radiographs, and the observations of fracture and shock waves in transparent targets, together with post-mortem examinations of craters and the adjacent material.

The projectile approaches the target at a velocity considerably in excess of the stress wave velocity in either material. Impact is announced by an intense flash of light. Pressures exerted are in excess of millions of atmospheres. Shock waves radiate from the point of contact into both the projectile and the target. As the speed of the projectile is greater than that of the shock wave, the shock receding into the projectile is actually carried below the original target surface and the release of pressure at the projectiles boundary creates lateral flow in both materials. The wave being propagated into the target compresses and accelerates the target material and the crater now begins to form. The intense heat generated usually causes some melting. Material flows along the wall of the crater and is ejected at very high velocities, usually in excess of the projectile velocity. The rate of crater expansion is greater than the shock wave velocity so the region of highly compressed material is confined to a thin hemispherical shell adjacent to the crater surface. This phase of the cratering process can be described by hydrodynamic principles. This fluid impact phase persists for only a fraction of a microsecond.

Finally, as the velocity of the crater surface decreases, the front of the shock wave becomes detached and between the two is a shell of hot, compressed material. As this shell becomes thicker, the energy density behind the shock wave decreases. Stresses rapidly decay because of geometrical divergence and energy dissipation, to the point where the material strength becomes the important factor.

Ultimately, the stresses become too small to overcome the mechanical strength of the target. Some of the material being ejected from the crater will remain to form a raised lip around the cavity. Permanent deformations cease and the crater surface rebounds or shrinks somewhat as the elastic stresses in the target relax.

In the meantime, the initial shock wave decays into an elastic wave and continues to dissipate energy throughout the target. If this wave encounters a free surface it will be reflected as a release wave. The tensile stresses resulting from this reflected wave

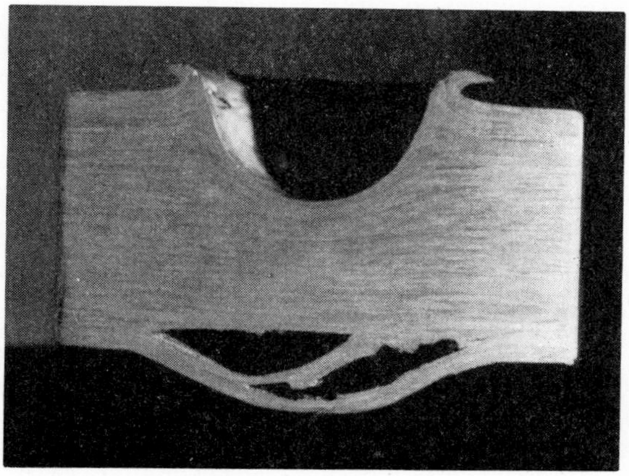

Fig. 3.

may exceed the strength of the material, producing secondary fractures. These fractures may occur as internal cracks, as rear surface bulges, or as complete detachments of target material. If the target is a spacecraft hull, internal cracks would weaken the structure; bulges could impede flow in pipes or jam mechanisms; and detachment of material would produce a shrapnel effect, creating a hazard to personnel and equipment.

Experimental and theoretical studies. This physical model of cratering is generally accepted by those who are engaged in the study of high-speed impact. Opinions differ widely, however, concerning the quantitative relations between the impact parameters and the cratering process. It is generally agreed that the depth of penetration is directly proportional to the projectile diameter and is in some way dependent upon its density and velocity. There is little agreement as to the significance of the target density, strength, hardness, brittleness, compressibility, or sonic velocity. The value of two-thirds for the velocity exponent in the hypervelocity region of the depth-velocity curve discussed earlier is based upon the assumption that the crater size is a function of the projectile's kinetic energy. This exponent will be different if some other function of the velocity, such as momentum, is the controlling factor.

Many data and empirical relations for penetration have been published. Each of these describes the results obtained for certain velocity ranges and projectile-target combinations. A correlation equation developed by Herrmann and Jones (1961) and widely used to estimate meteoroid damage is

$$p/d = K_1 \ln[1 + (\varrho_t V^2/K_2 H_t)],$$

where p is the penetration depth, d is the projectile diameter, P_t is the target material density, H_t is the target hardness (Brinell), v is the impact velocity, and the constants K_1 and K_2 have values relating to certain material combinations which were determined empirically over the experimental range up to 10 km/s. Those who developed this relation state that there is no justification in extrapolating this relation to other materials or to higher velocities. To extrapolate an empirical relation would imply that the partitioning of energy such as heating, energy transmitted throughout the target by the stress waves, kinetic energy of the material thrown out of the crater, and energy dissipated in plastic deformation, remains constant. Such an assumption has not been substantiated either from experimental evidence, or from theoretical considerations. Extrapolations can only be carried out when a penetration law is based on sound theoretical grounds and when there is assurance that phenomena neglected do not become operative at higher velocities.

Because of the difficulties and limitations of the experimental approach, it seems to many that an understanding of the problem of high-speed collisions must come primarily from theoretically considerations. Many theories have been presented over the past few years, the majority based upon rather arbitrary assumptions. As yet, there has not emerged a unified theory of impact.

Most of the theoretical analyses can be divided into the following categories: (1) rigid projectile, (2) thermal penetration, (3) blast wave, (4) hydrodynamic, and (5) viscoplastic.

The *rigid projectile theory* assumes that the projectile and target are both rigid but are separated by a thin layer of fluid material. This fluid is believed to move radially as the projectile advances, eroding the target to form the crater. The *thermal penetration theory* assumes that the energy released is sufficient to melt or vaporize both the projectile and a portion of the target during a very short interval of time. Much of the present theoretical work centres around the *blast wave theory* which has been developed to describe various high energy gas flows. This theory treats the impact problem as one-half of a spherically symmetric disturbance. The target is considered to be compressible and its equation of state that of a perfact gas. The *hydrodynamic theory* is based upon the fact that the impact pressure is so far above the shear strength of the materials that they should be treated as fluids. Application of the hydrodynamic model utilizes high-speed digital computers to integrate the equations of motion of compressible inviscid fluid. The *visco-plastic* model differs from the hydrodynamic model in that it provides for the inclusion of viscosity and dynamic yield strength.

Most of these theories have enlarged our understanding of the impact phenomena and each is probably applicable to some extent to some phase of the cratering process. There is little agreement, however, as to when the various theories might apply. In the early stages, the material does probably act as a perfect gas and then as an incompressible fluid. In the later stages, the material strength evidently becomes the controlling factor.

Perhaps the most promising approach is the quasi-theoretical one which involves the formulating of mathematical models based upon analyses of experimental data. At the present, however, such models must be based upon available data for relatively low impact velocities and the validity of extrapolation to higher velocities is hazardous.

Most of the studies, both experimental and theoretical, have dealt with orthogonal impact; however, most of the impacts of meteoroids on space vehicles will be at oblique angles to the surface. Oblique-angle impact has proved too complex to be analysed theoretically. Experimentally, it has been shown that craters are essentially hemispherical if the normal component of the projectile velocity is in the hypervelocity range.

Bibliography

CHRISTMAN D. R., GEHRING J. W., MAIDEN C. J. and WENZEL A. B. (1963) *Summary Report on the Phenomena of Hypervelocity Impact*, GM DRL TR 63–216, June.

COSBY W. A. and LYLE R. G. (1965) *The Meteoroid Environment and Its Effects on Materials and Equipment*, NASA SP-78.

DEARDEN E. (1961) in *Encyclopaedic Dictionary of Physics* (J. Thewlis Ed.), **1**, 361, Oxford: Pergamon Press.

GOLDSMITH W. (1963) Impact: The Collision of Solids, *Appl. Mech. Rev.*, **16**, No. 1963, 855.

HERRMANN W. and JONES A. H. (1961) *Survey of Hypervelocity Impact Information*, Aeroelastic & Structure Research Lab. MIT Report 99–1, Sept.

HOPKINS H. G. (1960) *Progress in Solid Mechanics*, Vol. 1, Ch. III, *Dynamic Expansion of Spherical Cavities in Metals*, Amsterdam: North-Holland.

JONES O. H., POLHEMUS J. F. and HERRMANN W. (1963) *Survey of Hypervelocity Impact Information II*, Aeroelastic & Structures Research Lab. MIT Report No. 99–2.

KINSLOW R. (1965) *Collisions at High Velocity*, International Science and Technology, April.

MSFC Meteoroid Damage Working Group (1963) *A Bibliography Concerning Aspects of the Meteoroid Hazard*, NASA, MDWG-63-2, April.

Proceedings of the Sixth Symposium on Hypervelocity Impact, Sponsored by: U.S. Army, Air Force, and Navy, August 1963.

Proceedings of the Seventh Symposium on Hypervelocity Impact, Sponsored by: U.S. Army, Air Force, and Navy, February 1965.

R. KINSLOW

HIGH-SPEED SHIPS. *The problem.* Since the advent of steam power, the speeds of most naval and merchant ship types have been steadily but slowly increasing—with the notable exception of bulk carriers that have grown in size at a phenomenal rate with very little change in speed. At the same time that ships have advanced moderately in speed, aircraft have shown tremendous increases. It is the purpose of this article to discuss developments in the direction of distinctly higher speeds for ships.

The principal obstacle to high speed on the surface of the ocean has been wavemaking resistance, the added resistance or drag associated with the generation of waves. This resistance can be minimized but never eliminated entirely. The establishment and maintenance of a train of waves requires an expenditure of energy that reveals itself in the added power required to drive the ship. Physically, this means that the integration of the longitudinal components of pressure over the surface of the forebody of the ship exceeds that on the afterbody. One can observe the buildup of a bow wave on any fast ship, but no corresponding buildup at the stern is to be seen.

The most significant parameter in the study of wavemaking resistance is the "Froude Number," $V/\sqrt{(gL)}$, where V is the ship speed, L is length, and g is the acceleration due to gravity, all in consistent units. William Froude, the famous British pioneer in ship model testing, showed that when this Froude Number is constant geometrically similar ships or models of different sizes generate similar wave patterns. Furthermore, the wavemaking resistance was found to be proportional to displacement (or total weight) under such conditions.

The historical increase in ship speeds has mostly been associated with an increase in ship size, with very little change in Froude number. In the case of naval vessels and passenger liners, there has been a trend toward some increase in length and slenderness, which have a beneficial effect on attainable speed. But above a certain Froude number the wavemaking resistance, and hence the required power to drive a ship, increases at a faster and faster rate, imposing an insurmountable barrier to increasing speed by conventional means.

Another obstacle to high speed on the surface of the sea is the existence of wind-generated surface waves, particularly on stormy routes such as the North Atlantic Ocean. Fortunately, trends to larger, as well as longer and more slender, ships has been favourable to good rough water speeds. But even the largest and fastest ships are often delayed for days in a rough winter crossing of the North Atlantic.

There are two possible ways to overcome the wavemaking barrier: one is to go beneath the surface, as a submarine does, and the other is to lift the main hull off the surface, as a hydrofoil craft can do.

Submarines. A submarine running a short distance below the surface makes waves and experiences wavemaking resistance. But as it goes deeper than about five diameters, wavemaking virtually disappears. In fact, if a well-formed body moves through an ideal fluid at great depth there is no resistance to motion; the integrated pressures on the afterbody balance those on the forebody. But in a real fluid with viscosity there is frictional drag, which may be expressed:

$$R_f = C_f \varrho S V^2$$

where C_f is a coefficient of friction, ϱ is density, S is the wetted area, and V is the velocity. If the body is not well-formed or has unfair appendages, the resistance will be increased by separation of flow or eddying. Hence, the frictional resistance clearly represents a minimum for practical vehicles moving in a fluid.

Theoretically a sphere would be the form for minimum frictional resistance, because it has the smallest area for a given volume. But in a non-ideal fluid such as water, it would experience large form drag or eddymaking. Optimum submarine hulls are "streamlined", tear-drop shaped forms, much stubbier than surface craft.

Early submersibles were surface craft that were able to travel short distances below the surface at slow speed for brief periods of time. The advent of nuclear power led directly to the true submarine capable of deep submergence and high speed. Its military value was immediately evident and has been exploited in several ways. The submarine also appears potentially valuable for the transport of military cargoes where secrecy is an important factor.

For commercial application the submarine does not show great promise. First, the ability to travel below the surface must be paid for at a terrific price in construction and operating cost. Second, the submarine suffers from serious limitation in internal volume. The hull of a surface ship can easily be made to enclose double (or more) of its submerged volume, so that it can accommodate cargoes of the usual density, much lighter than water. For dense cargoes such as petroleum products, which would be suitable for submarine transport, it is impossible for a submarine to compete with mammoth surface tankers in low cost transportation.

General limits on speed. It is of interest to consider the limits of attainable speed for bodies of different size, whether moving in water, air or at the interface—using the ideal submarine as a basis of comparison. Hence, in the figure is plotted the non-dimensional power-speed ratio, P/WV, where P is installed power, V is speed, and W is the weight of the body, as a function of non-dimensional speed coefficient, V^2/gl. Here $l = \nabla^{\frac{1}{3}}$, is an indicator of vehicle size, since ∇ is the submerged volume $= W/\varrho g$ for a floating body. The lines represent the resistance

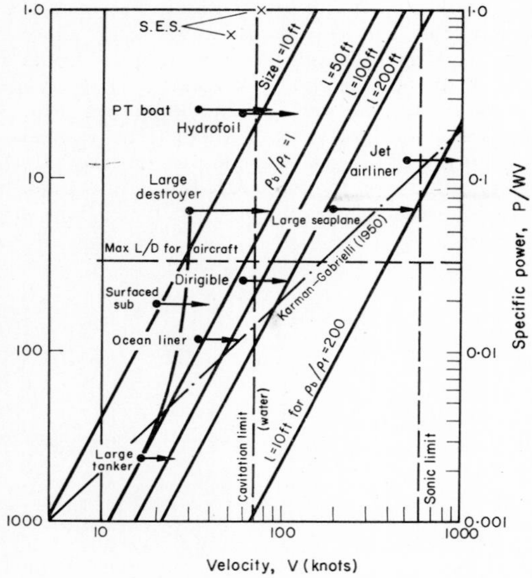

Fig. 1.

of ideal submerged bodies having only frictional resistance, and having surface areas which are minimum in relation to volume, i.e. spheres. The coefficient P/WV can also be interpreted in terms of resistance, since installed power, $P = RV/\eta$, where η is overall propulsive efficiency. Hence, $P/WV = R/W\eta$. Furthermore, if we wish to use aeronautical terminology we may introduce a scale of lift drag $= L/D$, which is the reciprocal of R/W. Hence, $P/WV = \dfrac{1}{(L/D)\eta}$. It is assumed in the figure that viscous effects are not affected by size, which, of course, is not strictly true.

The parameter identifying the parallel lines is the ratio of weight to displacement or buoyancy, $W/\Delta = W/\varrho\nabla$. For a floating or neutrally buoyant body, the ratio is 1·0. For aircraft it is generally 100 to 200, since the weight is supported by dynamic lift instead of buoyancy. Shown in relation to the lines are various typical submarines, ships, airplanes, and an airship at their design conditions of operation. The destroyer curve shows how as speed increases the ship draws away from the reference line as the result of wavemaking. This shows clearly the advantage of going below the surface of high speed is required. It also shows that increasing length by making the craft more slender is a partial alternative, however.

Although the speed ratio v^2/gl is significant for any moving body, the Froude number $V\sqrt{(gL)}$, involving length L, is customarily used for surface ships, as previously noted. Hence, a Froude number scale has been added to the figure for a typical case of $L = 5{\cdot}7 \times l$. Froude established that for geometrically similar ships specific resistance or P/WV is constant for any given Froude number, or v^2/gl, if viscosity changes with size are ignored.

The figure shows that attainable speed for all vehicles is limited both by available power and by size. Increasing either power or size (l) generally permits increases in speed. Going to high altitude where W/Δ is greater is also advantageous for airplanes. Unless some revolutionary new discovery appears, it is impossible for any craft to go to the right of the applicable line.

There is another limit of speed in water introduced by the necessity of avoiding cavitation. This limit varies with pressure and hence with depth. Here we are considering cavitation on the hull itself, not on the propulsion device, which may become a problem at even lower speeds. Similarly there is a sonic barrier in air that can be passed only by a large additional expenditure of power. If one goes sufficiently deep in water there is no problem.

Surface ships. As power plants improve in efficiency and new types of prime movers—such as nuclear power plants—appear, technical interest grows in the possibility of much higher speed displacement type ships. Model studies have been carried out at the David Taylor Model Basin (U.S.A.) and at the National Laboratory. Published results of the latter study have shown that a fine-hulled ship of 400-ft length would require over 350,000 horsepower to attain a speed of 50 knots. The deadweight or payload would be small because of the great weight of machinery and fuel required by present-day technology. Hence, the possibility of such high-speed surface ships for commercial use seems dim indeed.

So we may now raise the question of whether or not new concepts in surface ships might change this picture. The answer is in the affirmative if we can develop hull forms specifically to reduce wavemaking resistance and minimize the effect of storm seas. Considering the problem of wave resistance first, there are three possible approaches which can be drawn from the previous discussions: (1) to modify the form in such a way as to cancel the original waves by other waves, (2) to make the hull longer and more slender in order to reduce the wave disturbance, or (3) to put a large part of the hull far enough below the surface to reduce wavemaking. The first method is typified by the so-called bulbous bow, which is found on all large U.S. Naval vessels and passenger liners, as well as on many cargo ships and tankers. A fairly small enlargement of the forefoot has been found to give a significant alteration of the bow wave and hence reduction in resistance. Recently consideration has been given to much larger bulbs in the hope of having a more drastic effect on wavemaking. The theory is that a bulb near the surface will create its own train of waves; and if it is properly located somewhat forward of the bow of the ship, its wave train can partially cancel the ship's bow wave train. For example, the wave train generated by a sphere can be readily computed by wave theory. In actual practice no single shape can exactly counteract the somewhat complex wave configuration created by a hull. But significant reductions in resistance can be obtained.

Mention should be made of the large, protruding bulbs being installed in many large, slow tankers. Here the beneficial effect of a bulb on resistance appears to result from its improvement in the flow over the forebody, with less separation or eddy resistance. Any favourable effect on wave resistance is a minor factor, since wavemaking resistance is a very small part of the total in these low-speed ships.

Making a ship longer and more slender is an approach which has been followed generally in the past. The reason is, of course, that the wavemaking resistance is directly related to the Froude number $V\sqrt{(gL)}$. Hence, if the length can be increased, the speed can also be increased in proportion to $(L)^{\frac{1}{2}}$ without significant change in wavemaking resistance. Recent destroyer and destroyer leader types have shown significant increase in slenderness compared to earlier vessels. But there are practical limits to proceeding too far in this direction.

The problem in reducing wavemaking by the third method of putting much of the hull beneath the surface is rendered difficult by the requirement of going quite far below to have a significant effect, combined with the necessity of having only a minimum amount of hull projecting through the surface for navigation, and for air supply and exhaust for the power plant. The resistance of any appreciable projection through the surface may easily cancel the advantage of the submerged hull. However, taking advantage of the light machinery weight possible with advanced steam or gas turbine plants, this approach offers the possibility of maximum speed greater than modern submarines, combined with ample range at a reduced cruising speed. Furthermore, with sufficient power semi-submerged ships could be driven at high enough speeds to be in a "supercritical" condition in relation to the ocean waves that cause pitching motions—that is, the frequency of encounter with the longest waves would be well above the ship's natural frequency of pitch. High cost, as well as technical difficulties, have so far prevented the construction of semi-submerged ships. They remain an interesting possible development for military purposes.

Dynamic lift craft. The first type of craft to lift the hull partly out of the water to reduce wavemaking was the planing boat or hydroplane. By making use of dynamic lift it was able to obtain considerably higher speed than a displacement hull. A more recent extension of this idea is the hydrofoil boat, in which the entire hull is lifted out of the water supported on hydrofoils or water wings.

A number of such craft have been in successful operation for several years in high-speed ferry service in Europe, and to a lesser extent in the United States and Japan. Most of these boats have operated in sheltered waters, and they encounter serious limitations when open ocean operation is considered. An obvious problem in rough water is the need to raise the hull much higher above the surface than is required in smooth water. This causes serious structural and propulsion problems. Furthermore, if a larger size is considered in order to obtain greater range and payload, structural problems become critical. The "law of the squares and the cubes" causes a serious obstacle: hull weights increase as the cube of linear dimensions, while the strength of critical structural members such as the supporting struts increase only as the square of dimensions. This situation makes it very difficult to design and build really large ocean-going hydrofoil craft.

The largest hydrofoil craft at present is the U.S. Navy patrol boat, "Plainview," now nearing completion in Seattle. It is 220 ft in length, is powered by two 15,000 h.p. gas turbines, and is expected to attain a speed of better than 40 knots. Meanwhile, it has been announced that a larger craft will be built in Norway to carry 250 passengers, or 150 passengers and eight automobiles, at a speed of 40 knots between Denmark and Sweden.

Most of the hydrofoil boats now in operation have fixed foils, arranged in a V so that the boat can rise out of the water until the correct area of foil remains submerged to provide the needed lift at any particular speed. This is a simple and fairly inexpensive design, but it is not well-suited to rough water operation. Open-ocean hydrofoil craft, like the pioneering "Denison" and the "Plainview," are designed with fully submerged foils, so that an automatic control system is required to adjust the lift to suit the speed of the boat and the changing surface elevation of the sea. The system is designed to allow the boat to follow the contour of the long waves and swells, while the foils cut through the shorter waves. Unless the waves are unusually steep, the hull is high enough to allow the wave crests to pass by beneath it.

Another kind of craft that rides above the surface of the sea is a type variously known as a Surface Effect Ship, a Hovercraft, a Ground Effect Machine, or an Air Cushion Vehicle. Since they rely on aerodynamic rather than hydrodynamic principles, these craft are related to aircraft as well as to ships. Small ones are now in ferry service in England—e.g. across the Solent—and in San Francisco Bay. Much larger ones are in the construction and planning stages.

The principle of the surface effect ship is to utilize large fans to force the air under the hull at sufficient pressure to lift it off the surface. As the air escapes more air must be continuously supplied to maintain the air cushion; and therefore, a continuous expenditure of power is required to keep the craft in the hovering condition. And the higher the craft is lifted from the surface, the greater the power required. A revolutionary development was the recent introduction of a flexible skirt around two or more sides of the hull, which greatly reduces air losses and therefore permits drastic power reductions—or alternatively an increase in the surface clearance with the same power. This innovation has permitted the newer craft to operate successfully in moderately choppy seas.

Forward motion of air cushion vehicles is usually obtained by a separate propulsion system providing thrust either in the air or in the water. Directional control is provided by the necessary fins and surfaces operating mainly in the air.

The interesting feature of the surface effect ship is that it is not subject to an inherent limitation in size, such as that restricting the hydrofoil boat. The problem of increasing the size of surface effect ships can be visualized by a crude dimensional analysis. Considering a series of craft as rectangular platforms of different size, hovering at the same height above the surface, the air flow requirement is proportional to the perimeter, which is proportional to a linear dimension, L. Hence, the power requirement increases with L. But for the same air pressure the cushion can support the same load per square foot of the platform, regardless of size. In other words, the load on the platform can increase in proportion to L^2. Thus, there is an advantage in increasing the size of the craft, and this advantage can be used to increase the payload, to reduce the power, or to increase the hovering height.

Hence, a 160-ton cross-channel ferry is now under construction that is expected to carry 250 seated passengers and 32 automobiles across the English Channel at a speed close to 70 knots. And serious planning is under way for a 4000-ton freighter 480 ft in length, 144 ft in breadth, and powered by eight 18,500 h.p. gas turbines driving water propellers. In calm water it could operate at 50 knots for a period of 10 hours with a payload of 2000 tons, probably in the form of cargo containers. Such a vehicle might fill a gap that now exists between low-speed, economical surface ships and high-speed, expensive aircraft. Therefore, it may eventually serve a useful commercial purpose in carrying high-value express cargoes over moderate distances.

In summary, it may be stated that although the possibilities of significant increases in ship speeds on the surface of the ocean are not great, there are a number of new concepts for obtaining higher speeds below or above the surface. Meanwhile, the place of the surface ship—whether faster or not—in world transportation is assured.

E. V. LEWIS

HYBRID RESONANCES. If we consider propagation of r.f. radiation through a thin sample of a conducting solid in the presence of a sufficiently strong constant external magnetic field parallel to its surface, and if the solid contains more than one sort of carrier (either electrons and holes or electrons whose effective masses are different due to the different orientations of its energy surfaces relative to the magnetic field) it is found that there is a resonant absorption at a frequency corresponding to a coupled motion of the different carriers.

Consider, for example, a compensated material (i.e. one containing equal numbers of electrons and holes). The dispersion relation for an electromagnetic wave in the geometry in question is

$$c^2q^2 = \omega^2 \frac{\varepsilon_+\varepsilon_-}{\frac{1}{2}[\varepsilon_+ + \varepsilon_-]} \tag{1}$$

where q is the wave number, and $\varepsilon_+, \varepsilon_-$ are the dielectric constants for waves with right and left circular polarization respectively about the direction of the constant magnetic field. Neglecting relaxation,

$$\varepsilon_+ = \varepsilon_L \left[1 - \frac{\Omega_p^2}{(\omega - \omega_c)(\omega + \omega_c')}\right] \tag{2}$$

$$\varepsilon_- = \varepsilon_L \left[1 - \frac{\Omega_p^2}{(\omega + \omega_c)(\omega - \omega_c')}\right]. \tag{3}$$

Here, ε_L is the dielectric constant of the lattice, Ω_p is the plasma frequency, and ω_c, ω_c' the cyclotron frequencies of electrons and holes respectively. It follows that

$$\varepsilon_1 = \tfrac{1}{2}[\varepsilon_+ + \varepsilon_-] = \varepsilon_L \left[1 - \frac{\Omega_p^2(\omega^2 - \omega_c\omega_c')}{(\omega^2 - \omega_c^2)(\omega^2 - \omega_c'^2)}\right].$$

Propagation of waves through the material is possible so long as the right-hand side of (1) is positive. If the plasma frequency Ω_p is large compared with the other frequencies in the problem this will be so up to the frequency

$$\omega = \sqrt{(\omega_c\omega_c')} \tag{4}$$

at which point the right-hand side of (1) has a pole. Thus, taking account of relaxation, the imaginary or absorptive part should peak there. This absorptive resonance is hybrid in character, involving coupled excitation of the electrons and holes.

The condition $\varepsilon_+ + \varepsilon_- = 0$ implies that the field is along the direction of propagation and therefore corresponds to a density oscillation. The field of each type of carrier acts collectively upon the carriers of opposite sign, providing a coupled mode.

The existence of this sort of "hybrid resonance" was first noted by Smith, Hebel and Buchsbaum in Bismuth.

A similar hybrid resonance has been predicted in n-type germanium and PbTe by Wallace. Here it is not a question of electrons and holes but of electrons with different effective masses in the field because they belong to differently oriented energy surfaces. For external magnetic field in the (111) direction and propagation perpendicular to that direction, the frequency of the hybrid resonance is

$$\omega = \frac{eB}{m_T c} \sqrt{\frac{5m_T + 4m_L}{3(m_T + 2m_L)}} \tag{5}$$

where B is the external magnetic field, m_T and m_L the effective masses along the short and long axes of the energy ellipsoids respectively.

See also: Helicons. Solid-state plasma.

Bibliography

SMITH F. E., HEBEL L. C. and BUCHSBAUM S. J. (1963) *Phys. Rev.* **129**.

WALLACE P. R. (1966) *Canad. J. Phys.* **44**, 2495.

P. R. WALLACE

I

INDUCED VALENCE DEFECT STRUCTURES. When a foreign atom is introduced into a crystal, it is common to observe marked changes in a number of its properties. These changes occur even at very low concentrations of the foreign atoms. In this context, it will be assumed that these concentrations are small (1 at. per cent $= 10^{21}$ atoms per cm^3, or less). As the concentration of foreign atoms becomes larger, limiting solubility and the separation of new phases have to be considered.

When the foreign atom in solid solution has a valence different from that of the host atoms, these changed properties of the doped crystal are called *induced valence* or *controlled valence* effects. These terms were first introduced by Verwey and his co-workers in their classic studies of the electronic properties of such systems as Li_2O added to NiO. Among the properties of crystals which are subject to induced valence effects are electronic and ionic conductivity, diffusion, dielectric loss, optical and magnetic properties and chemisorption. Since such properties underlie many important processes such as sintering, tarnishing, luminescence, rectification and catalytic activity, there has been a great deal of study of these effects. Nevertheless, only comparatively few systems have been studied thoroughly as single crystals, and very often, interpretation of data obtained on powders alone is ambiguous.

The incorporation of the foreign atom into the host lattice can be considered in two ways. Firstly, the foreign atoms are themselves *atomic* imperfections, and secondly, they give rise to *electronic* imperfections or defects. The effects that these atoms have on the crystal properties arise directly from the interaction between these deliberate imperfections and the inherent atomic and electronic imperfections of the crystal. The interactions may be such that equilibrium conditions hold, but often the history of the incorporation may mean that true equilibrium has not been achieved. This is so because the solid diffusion processes that lead to solid solution are very slow and can involve careful heating and cooling over extended times. Non-equilibrium conditions may also apply to the defect state of the host crystal. The degree of control of the various properties by induced valence effects depends to a large extent on an adequate understanding of these facts, and the literature of this subject is full of conflicting reports, which can, in many cases, be traced to non-equilibrium.

It should be noted that the phenomenon of non-stoichiometry is also closely related to these induced valence effects. Thus, NiO_{1+x} can be regarded as crystals of $Ni^{2+}O$ in which Ni^{3+} ions are incorporated along with vacant cation sites, $\square^+$; and the properties, or some of them, are similar to those achieved by incorporating Li^+ ions in NiO crystals.

The importance of induced valence effects has been that the defect structures of crystals and their properties can be controlled more easily and more widely by specifically adding impurity atoms. The position the altervalent atoms occupy in the crystal may be normal lattice sites ($\square$) or interstitial sites ($\triangle$). In any particular case, this will be determined by the energetics of the various possibilities. When lattice positions have been substituted, say in a MX crystal, the altervalent atoms tend to occupy the sites of the atoms to which they most closely approximate in electronegativity. However, size is also very important, and it has been shown by electron spin resonance that Gd^{3+} ions may occupy Pb^{2+} sites in PbSe whereas the smaller Fe^{3+} ions are more probably incorporated interstitially.

For a pure crystal at equilibrium, the atomic defect structure involves the exchange between vapour and solid, and the free energy balance between the energy to form lattice defects and the gain in entropy in having them present. Likewise, the electronic energy levels at higher temperatures have an equilibrium disorder in which levels above the ground state are occupied. Some of these higher levels (excitons) are not mobile under electric fields, but other levels of the excited electrons are. Two models are used to describe this electronic mobility. One is a band model in which the excited electrons move, by quantum mechanical tunnelling, in energy levels which lie in bands (*conduction*). The band (*valence*) of energy levels from which the excitation occurred now contains empty levels, and mobility is possible in this band also. It is, in effect, the motion of a *positive* electronic charge (*hole*). The forbidden energy levels between the bands is referred to as the energy gap, and is shown in Fig. 1(a). The other model is proposed for the cases like transition metal oxides where such tunnelling is negligible between the atoms in their equilibrium positions. In these cases, electrons and holes can still move as the nuclei approach each other during atomic vibrations. This activated process for electronic mobility is known as the jump model.

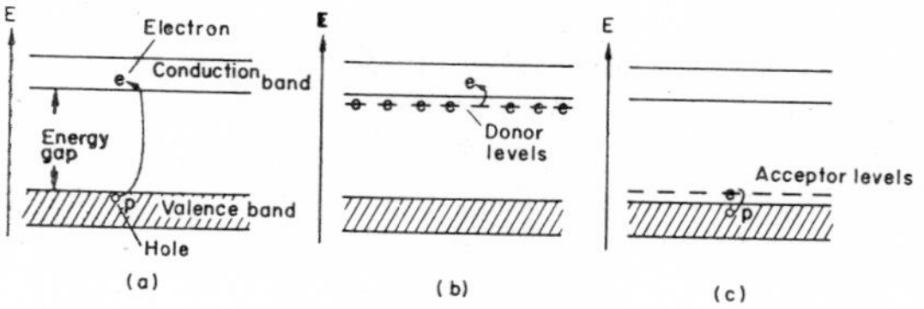

Fig. 1.

In addition to this continuous view of the electronic energy levels, the presence of the lattice defects can introduce energy levels which are localized in the vicinity of the imperfection and may either be occupied (*donor*) or empty (*acceptor*). If these levels are suitably located in relation to the bands, or the levels on the vibrating atoms, additional possibilities arise for mobility, and this is shown in Fig. 1(b) and (c) for the band model. Foreign atoms in the crystal may influence the equilibrium by combining with the atomic imperfections and by acting as donor or acceptor sites for electrons. In working out the equilibrium defect structure of doped crystals, these interactions with the foreign atoms must be included together with the contribution of their charge to overall electroneutrality. In principle, the equilibrium structures for different temperatures and vapour pressures can be calculated if the energies involved are known. Kroger and Vink have indicated the result of such calculations for a number of systems. The general result shows various regions of temperature and pressure at the extremes of which the effects of non-stoichiometry outweigh the influence of the foreign atoms, with intermediate regions in which the foreign atoms dominate the determination of the structure and its properties. These last regions which are often the only practical ones are the important ones for induced valence effects.

Ionic conductivity and diffusion. $CdCl_2$—NaCl and $CdCl_2$—AgCl provide examples of induced valence structures in which the number of vacant lattice sites is increased. The Cd^{2+} ion has one more positive charge than the host cations and incorporation leads to one vacant cation site ($\square^+$) or one interstitial anion ($Cl^- | \triangle$) per Cd^{2+} ion. At temperatures between −50°C and about 250°C, atomic transport processes such as ionic conductivity and diffusion reflect this increase in lattice defects. For example, at 210°C the addition of 1 at. per cent Cd^{2+}, increases the conductivity of AgCl more than 100 times. Density measurements also support the increase in vacancies. At lower temperatures there is evidence from dielectric loss measurements that the Cd^{2+} and the $\square^+$ are associated in the alkali halides. In other similar systems such as Ag_2S—AgBr, the extra Ag^+ ions have been shown from the conductivity data to be present interstitially, as have Mg^{2+} ions in ZrO_2.

Electronic conductivity. Induced valence structures have found their greatest usefulness because of the effects on electronic conductivity (and related phenomena like thermal e.m.f.). The position of acceptor and donor levels in the presence of altervalent atoms can only be determined experimentally. However, it is possible to give some general rules. Interstitial atoms which are more electropositive than the host atoms tend to act as donors, while those that are more electronegative tend to be acceptors. If the associated levels are sufficiently close to the bands, excitation at quite low temperatures is possible and conduction by either electrons (*n type*) or holes (*p type*) occurs. Substitutional atoms with more electrons than the host will tend to be donors and those with less electrons will be acceptors. The energies for these excitation processes can be approximated to in the following manner. The dielectric constant (ε) of the crystal is usually high and the electron that is donated or accepted will be located far enough from the nucleus of the foreign atom for the ionization of a hydrogen atom in a dielectric ($13{\cdot}5/\varepsilon$ eV), to be a rough approximation. As the above rules indicate, the position of the altervalent ion will influence its effect. Interstitial neutral copper atoms (Cu°) in ZnS behave as donors, raising the electronic conductivity, but univalent copper ions (Cu^{I}) occupying Zn sites are acceptors, enhancing hole conduction. On the other hand, aluminium which is both more electropositive and has more valence electrons than zinc produces donor levels in both positions. The effect of altervalent atoms in lattice positions is clearly seen in the case of a crystal like germanium or silicon, with quadrivalent atoms in a diamond structure, and is illustrated schematically in Fig. 2. If a germanium atom

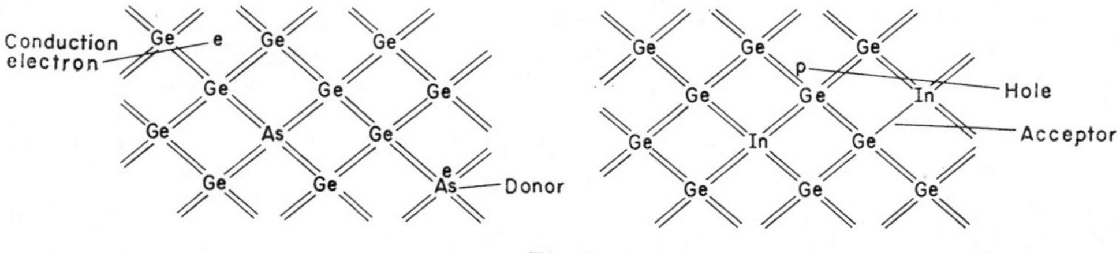

Fig. 2.

is replaced by an arsenic one with five valence electrons, four of these are used in bonding and the fifth will be weakly bound to the arsenic nucleus and can easily be excited into the conduction band of the crystal. The replacement of germanium by indium with only three valence electrons means a deficiency in bonding or an acceptor level which is weakly bound to the indium. If an electron from another bond (i.e. a valence band electron) is excited to this acceptor, a hole is formed in the valence band. Room temperature thermal energies provide excitation in these *doped* germanium crystals. The small lithium atom with a single valence electron occupies an interstitial position in Ge and acts as a donor.

Some MX compounds like germanium can be doped to give either n-type or p-type conduction. PbS is an example in which Ag produces holes and Bi gives electrons. Other systems can only be doped to be more or less conducting of the same type. An example is NiO in which Li^+ enhances p-type conduction while Cr^{3+} makes the oxide less conducting, and no case of induced n-type conduction has been established. If the altervalent atom in the crystal binds an electron or hole in a small orbital, then solid solution will affect its own properties, and the effect on the crystal properties will depend on the specific foreign atom even when valence and size are kept constant. However, if the bound electrons or holes are in orbitals which are contained on the neighbouring host atoms, the change in crystal properties will be non-specific. Which of these situations will occur depends on the nature of the electronic structures involved. If a completely filled shell would have to be disrupted, this situation is unlikely to occur. Thus, the solution of Li^+ atoms in NiO may be represented, for ionic bonding, as either

(a) $Ni^{2+}O^{2-}Ni^{2+}O^{2-}Li^{2+}O^{2-}Ni^{2+}O^{2-}$

or

(b) $Ni^{2+}O^{2-}Ni^{2+}O^{2-}Li^{+}pO^{2-}Ni^{2+}O^{2-}$ (where p represents a positive hole).

Since (a) disrupts a filled shell it can be rejected, and (b) can be expected to hold, with the hole, p, weakly bound to the Li^+ so that it resides on the adjacent Ni^{2+} ions. (b) can then be equally well represented by

(c) $Ni^{2+}O^{2-}Ni^{2+}O^{2-}Li^{+}O^{2-}Ni^{3+}O^{2-}$.

The conductivity of a crystal is equal to $n|-e|\,\mu$, where n is the number of *current carriers*, $-e$ is the electronic charge and μ is the mobility of the carriers. In regions of foreign atom control, n will be proportional to the concentration of altervalent atoms, but except when very small and at high temperatures, this concentration will also affect μ, reducing it as the impurity atoms act as scattering centres. Kroger and Vink (1956) have listed a large number of induced conduction systems.

Optical phenomena. Absorption and the subsequent emission of light may occur when crystals are doped with altervalent ions. For example, illumination of crystals of Ag_2S—CdS gives rise to excited electrons and holes (photoconductivity). The holes are in due course trapped by the altervalent Ag^+ ions, which are less positive than the host Cd^{2+} ions (luminescent emission). The electron is finally trapped by the Ag^+ ⅋p centre giving off infra-red or lattice phonons. In other systems the foreign atom may simply be involved by excitation from or to its own characteristic levels; and these processes may be accompanied by *luminescence* or *photoconduction*.

Magnetic properties. In a number of systems of transition metal ions in a host crystal, the cations of which have closed shells, the transition ion is found to take up the valence of the host. This requires that oxygen must also be incorporated. Mn^{2+} and Ni^{2+} in Al_2O_3 become Mn^{3+} and Ni^{3+}, and Cr^{3+} in TiO_2 becomes Cr^{4+}. Selwood has described this phenomenon as *induced valence*. However, if the foreign atom has a closed shell the atoms in the host may change their valence and a controlled valence effect is obtained, for example, $SrCoO_3$ in $LaCoO_3$ is accompanied by a suitable change, $Co^{3+} \rightarrow Co^{4+}$, in the Co ions of the host.

Mechanism of the incorporation. Because of the valence changes which are induced in these altervalent systems, it is found that the conditions of the solution process are very important. In an oxidizing atmosphere Li^+ dissolves in NiO as described above, but in the absence of oxygen metallic nickel separates, although in the oxidizing case some Li^+ does appear to dissolve by direct filling of existing vacant sites without involving the gaseous oxygen. The mechanism for the converse doping by Cr^{3+} is not well understood. The change in properties suggests that doping only occurs by reaction with the inherent Ni^{3+} and evolving oxygen. The natural p-type conduction is reduced and the doped oxide sinters more slowly than the pure oxide which suggests a decrease in vacancies. Detailed mechanisms of this type are only available for a few systems.

Metal oxidation. Oxidation of metals involves the formation of an oxide (sulphide, chloride, etc.) layer between the metal and the oxidant. If the process is to continue it is necessary to transport both ions and electrons across the oxide phase. Since the defect structures of the oxide layer are influenced, both electronically and atomically, by the presence of altervalent ions, it is not surprising to find marked effects on oxidation rates in some cases. Hauffe and his coworkers have studied a number of such systems including the oxidation of zinc, with and without small traces of lithium. In this case, a quantitative relation, $k/k' = [Zn^+|\triangle]/[Li^+|\square^+]$, was established which gives the ratio of the rate constants for the pure metal and the alloy in terms of the concentrations of interstitial $Zn^+|\triangle$ and of lithium in the doped oxide. Conversely, Al–Zn alloys do oxidize less rapidly than pure Zn. The effect the altervalent ion will have will depend on whether the conditions are such that electron transfer or ion transport is rate controlling.

Chemisorption and catalysis. Both chemisorption and catalytic activity in transition metal oxides are known to be sensitive to the defect structure of the oxide. Again it would appear that altervalent ions in the oxide should enable these phenomena to be controlled in a precise way. Much investigation of this possibility has been made in the last 15 years. Not only oxides but other semiconducting catalysts such as germanium have been studied. The results overall have been disappointing. A large number of effects on both the activation energy and the specific rate have been reported, and ions of higher and lower charge again have opposite effects. Nevertheless, in the systems which have been studied the most extensively, Li^+ and Cr^{3+} in NiO, conflicting and paradoxical findings are still reported. Only broad conclusions can be suggested and certainly nothing quantitative has been established. Thus, for the oxidation of CO, H_2 and C_2H_4 over NiO, doping with Cr^{3+} usually increases the activation energy, and with Li^+ there is usually a decrease. The complexity of catalytic processes with the requirement of chemisorption without strong bonding, and the fact that the surface of the solid only is involved, no doubt accounts for the difficulties that have been encountered in using altervalent effects to predict catalytic activity. For H_2—D_2 exchange on ZnO, doping with Ga^{3+} which increases n-type conduction leads to enhanced catalytic activity but reduces the coverage of chemisorption.

Bibliography

Kroger F. A. and Vink H. J. (1956) *Solid State Physics*, **3**, 373, New York: Academic Press.

Kubaschewski O. and Hopkins B. E. (1962) *Oxidation of Metals and Alloys*, London: Butterworths.

Rees A. L. G. (1954) *Chemistry of the Defect Solid State*, New York: Wiley.

Stone F. S. (1962) *Advances in Catalysis*, **13**, 35, New York: Academic Press.

Suchet J. P. (1965) *Chemical Physics of Semiconductors*, New York: van Nostrand.

P. J. Fensham

INDUSTRIAL FLUOROSCOPY, INCLUDING IMAGE INTENSIFIERS. In fluoroscopy the X-radiation which is transmitted through the specimen falls on to a screen which fluoresces, i.e. emits light within the visible part of the spectrum: thus, because of differential absorption in the different thicknesses of the specimen, a visible image is produced on the fluorescent screen. The thinner, less absorbent parts of the specimen are seen as brighter areas on the screen, so that the tonal range is reversed, compared with a film radiograph seen on an illuminated film-viewing screen.

Fluoroscopy has two major advantages over film radiography. First, an image is obtained without the use of consumable recording material. Secondly, the screen image can be viewed while the specimen is moving. The major disadvantage of fluoroscopy is that, compared with a film radiograph, the detail and sensitivity observable are usually very much poorer.

There are three main reasons for this poorer sensitivity:

1. A typical fluoroscopic screen image is very dim (typical brightnesses are (0·3–0·003 cd/m²) compared with the brightness of a film on a viewer (3–30 cd/m²)). At these low brightnesses the human eye, even when fully dark-adapted, is incapable of perceiving such small contrast or fine detail as can be discerned on a film radiograph.

2. Fluoroscopic screens are generally constructed to give the brightest possible image, and compared with film, are very coarse-grained and incapable of resolving fine detail.

3. Fluoroscopic screens have a contrast gradient of unity, whereas film has a contrast gradient at the densities normally used, of 4–6, so that a small difference in X-ray intensity (across, say, the image of a flaw) is enhanced by this factor in the film recording process, and so the sensitivity obtained is proportionately better.

Two techniques are used in industrial fluoroscopy. In the first, the equipment is designed to obtain the highest possible screen brightness. A high-output X-ray tube is used, which usually means using a relatively large tube focus; a short tube-screen distance is employed together with the most sensitive fluorescent screen. The image obtained is necessarily unsharp, but may be bright enough to view without much dark-adaptation of the observer.

In the second system, known as "enlarged image fluoroscopy", a very fine-focus X-ray tube is used (0·1–0·5 mm focus), and the specimen is placed a distance away from the screen so that the image is projected two to six times natural size. Because of its fine focus, the X-ray output of this type of tube is usually low and a lower screen brightness has to be tolerated, which almost always requires dark-adaptation of the observer. The reduced ocular performance at the lower screen brightness is, however, more than compensated by the increased size of the image details due to the projective magnification. Because of this magnification also, the effect of fluorescent screen graininess is reduced. Special high-output fine-focus rotating anode X-ray tubes have been built for this technique.

In both methods, the observer must be protected from the X-radiation which is transmitted through the specimen and through the screen. Either a mirror viewing-system is employed, or the screen is viewed through a sheet of lead glass of suitable thickness. Because of the protection problems X-ray energies greater than 150–200 keV have rarely been used for fluoroscopy.

Although image-quality indicator (I.Q.I.) sensitivities of $1\frac{1}{2}$–2 per cent have been claimed on specimens ranging in thickness from 5 to 100 mm of aluminium alloy, fluoroscopic techniques are not extensively used for flaw-detection in metals, except when only a relatively poor sensitivity is required. Commoner applications are for the detection of metal particles in ceramics, in tinned foodstuffs and in packaged goods; the detection of broken wires in insulated cables or in metal-jacketed valves; the checking of spacings of metal components in plastics, the correct filling of food cans, the cladding thickness of fuel elements.

The operation of a fluoroscopic set does not require a photographic darkroom and a set be made easily portable, by using a hood over the fluorescent screen to exclude extraneous light and suitable protection round the X-ray tube. Such small sets have considerable application to package inspection.

If the image on the fluorescent screen is photographed and this "photograph" used for interpretation, the process is known as '*fluorography*' or *photofluoroscopy*. The method, though not much used industrially, has three potential advantages:

1. A permanent record is obtained.
2. The photograph of the screen can be on small-size film, e.g. 70 mm strip, and so cost much less than full-size X-ray film.
3. Because there is no haza-d to an observer, high energy X rays can be used without difficulty.

Image Intensifier Systems

As already described, the prime limitation of a simple fluorescent screen system is the low screen brightness, which usually requires the observer to be dark-adapted and to work in a darkened room. There are several methods, all basically electronic, which enable the image brightness to be raised to a level where no dark adaptation is necessary. These are:

1. Electronic tube intensifiers. Essentially these consist of a fluorescent screen which converts the X-ray image into visible light: this screen is backed with a photoelectric layer which converts the light image into electrons; the electrons are accelerated and focused by electrostatic or other means on to a smaller viewing screen which produces a light image. A gain in brightness is obtained partly by energy given to the electrons on acceleration and partly by the reduction in area of the screen image. All the screens are in a vacuum envelope. Such tubes were first produced by Teves and Tol, and by Coltman.

2. Fluorescent screen, television systems. The image on a normal fluorescent screen is viewed by a television camera, either directly, or through a special optical system, or through a supplementary light amplifier tube. A vidicon or an image orthicon may be used, and the main problem in design is to minimize light losses between the primary screen and the much smaller face of the camera tube. Intensification is obtained through the usual electronic circuitry and the signal can be processed in a variety of ways, e.g. contrast enhancement.

3. Direct television systems. The television camera tube is directly sensitive to X rays and the X-ray image is formed on the pick-up surface of the tube. Obviously, the useful X-ray field cannot have an area greater than the pick-up surface.

4. Solid-state intensifiers. These are panels consisting of multiple layers of photoconductive or electroluminescent material and are used as substitutes for conventional fluorescent screens. There has been a variety of designs of such panels, but none have yet come into widespread commercial use with X rays.

In one typical panel, there is an electroluminescent layer of ZnS(Cu, Al) and a photoconductive layer of CdS(Cu, Ga). The layer thicknesses are 0·05 mm and 0·25 mm and an a.c. potential of 800 V is put across the layers. The layers are separated by a thin opaque layer to prevent optical feedback. The mechanism of operation is that the X rays absorbed in the photoconductive layer cause a high energy electron to be emitted which in turn causes a shower of conduction electrons through the photoconductive layer. These locally change the conductivity of this layer, so that the proportion of the potential which is across the electroluminescent layer is increased and the light output of this layer in this area therefore increases. Brightness gains of $\times 100$ over a conventional fluorescent screen have been obtained. The light build-up time is slow, varies with the incident X-ray intensity, and is non-linear. The phosphor has a long decay-time but this can be overcome by using a grooved photoconductive layer on which an auxiliary electrical supply can be applied to erase the image (Kazan).

Electronic tube X-ray intensifiers, although extensively used in medical radiography, particularly with a television camera to pick up the image from the viewing screen, have not yet found extensive industrial use. They are limited by the sizes of primary screen which can be built into a vacuum tube, and have a limited life. They have been superseded by indirect television systems which have several inherent advantages.

A typical indirect system is that developed by the Marconi Company (England), which uses a large diameter image orthicon television tube. The image on the primary fluorescent screen, which is 12″ diameter and of conventional type, is reduced in size and transferred optically by means of a large aperture Schmidt mirror system on to the pick-up surface of the $4\frac{1}{2}$″ image orthicon tube. A 1000-line scan television circuit is used and electronically it is possible to control image contrast, to use automatic brightness control, and to have a limited amount of electronic magnification. The amount of electronic amplification available is such that viewing brightness is no longer a problem and the performance is limited either by television tube noise or quantum fluctuation limitations (see below). Other workers have used a light amplifier tube between the mirror system and the television tube, or an X-ray image intensifier tube in place of the fluorescent screen.

The image is viewed on a conventional television monitor which can be remote from the X-ray equipment, and recording, if required, can be by means of a suitably synchronized cine camera, or a video tape recorder. Cine-fluoroscopy is possible with these equipments.

Although several direct X-ray pick-up television tubes have been demonstrated, including some with large diameter screens, only one type has, so far, been used industrially. This is a small-diameter Vidicon having a photoconductive sensitive surface 12 × 6 mm, developed by Machlett and Ohio State University (U.S.A.). Very high resolution in the images is possible and the image is presented on a 17″ television monitor, but comparatively high X-ray dose-rates are required on the Vidicon tube face (30–100 r/min is desirable). Applications are limited to specimens whose image can be encompassed within the very small pick-up area of the Vidicon tube face, unless a scanning system is used, but there have been successful applications to the inspection of small electronic components, to thin welds, and to the measurement of small orifices, generally using low energy X rays.

Performance of image intensifier systems. Brightness gain and contrast can be disregarded as factors of comparison in television type intensifiers, since both are under the control of the operator. In performance in terms of resolution of detail, the primary fluorescent screen is usually the limiting factor, with perhaps the limitations of the television raster if too low a line-standard is used.

The limitations in performance can also be considered from another standpoint—the statistical fluctuations in the number of quanta utilized through the system; this was first discussed in detail by Morgan and Sturm.

The information in the image is carried through the system by either X-ray quanta, light quanta, or electrons. A certain number of X-ray quanta are transmitted through unit area of the specimen per second and these must carry all the information about that area of the specimen irrespective of any subsequent conversions to light quanta or electrons.

The emission of X-ray quanta from the target of the tube is a random phenomenon and the absorption processes in the specimen and screen are similarly random. Thus while it is legitimate to give a value to the average number of quanta involved at any stage, there is a natural fluctuation in the number of quanta. Statistical analysis shows that if we are concerned with N quanta, the average fluctuation is $N^{1/2}$. Thus if

$N = 100$ the average fluctuation is 10, i.e. 10 per cent of N

$N = 10{,}000$ the average fluctuation is 100, i.e. 1 per cent of N

$N = 10^6$ the average fluctuation is 1,000, i.e. 0·1 per cent of N.

If therefore one considers a small area of the screen having a slightly different brightness to the background (i.e. a slightly different number of quanta emitted per unit area), and this area is formed as the result of the absorption of say 100 quanta, there will be a 10 per cent fluctuation in the brightness of this area. If the area is *on average* only 5 per cent brighter than the background it will be difficult to detect because of the fluctuations. If, however, the area is formed as the result of absorption of 10^6 quanta, the fluctuation will be only 0·1 per cent and the average 5 per cent difference should be easily detectable. Thus the minimum discernible contrast depends on the number of quanta utilized. It has been found that the threshold value for detection—the contrast-to-fluctuation ratio, K—is about three times the average fluctuation.

Minimum observable contrast, $C_{\min} = k/N^{1/2}$.

It has been shown that in any fluoroscopic process, the effective value of the average fluctuations in the number of quanta utilized at the end of the process, is dominated by the fluctuations at the stage where the number of quanta is smallest. In simple fluoroscopy, this stage is the absorption of light quanta on the retina of the eye: in other words, the limiting factor in sensitivity is the performance of the human eye. In television image intensifier systems, however, the stage at which the least number of quanta are utilized is usually the absorption of X-ray quanta in the primary fluorescent screen.

If the fluorescent screen absorbs only a few quanta per unit area, one can reach the stage where increase in amplification through the electronic circuiting produces a gain in viewing screen brightness, but no improvement in discernible detail. Using television image intensifier systems with high-energy X rays, this condition is often approached; the quantum fluctuations are seen as an unsteadiness in local image brightness on the viewing screen; to obtain further improvement in performance therefore, fluorescent screens which absorb and convert a greater proportion of the X rays incident on this screen, are necessary.

Bibliography

BEWLEY D. K. (1961) in *Encyclopaedic Dictionary of Physics* (J. Thewlis Ed.), **3**, 795, Oxford: Pergamon Press.

COPE A. D. (1960) in *Encyclopaedic Dictionary of Physics* (J. Thewlis Ed.), **3**, 796, Oxford: Pergamon Press.

HALMSHAW. R. *et al.* (1966) *Physics of Industrial Radiology*, London: Heywoods.

R. HALMSHAW

INFRA-RED RADIATION IN MODERN TECHNOLOGY. Mans' eyes are sensitive to electromagnetic radiations of wave-lengths from 0·4 to 0·7 microns—the "visible" region. Objects are seen mainly by the light they reflect, and one object is discriminated from another by colour, size, and shape. Only when objects have high temperatures above about 800°K, such as the Sun, the stars, and incandescent lamps, can the radiation emitted by the object itself be seen. In the infra-red portion of the spectrum which lies between 0·7 and 1000 microns, artificial detectors, in place of eyes, are used effectively to extend our sensing capability. In this region of the spectrum the radiation from objects is mostly due to self-emission rather than reflection.

Infra-red is generally known for its heating properties and for its importance in spectroscopy. However, it is not as well known that giant strides have been made in the development of infra-red system components for military purposes. The component which has undergone the most dramatic improvement is the infra-red detector—the heart of an infra-red system. During World War II when infra-red systems were being used successfully by the military, the detectors were sensitive from 0·7 to 3·0 microns. In the twenty years since that time, a variety of infra-red detectors have been developed with wide ranges in characteristics such as spectral response, detectivity and speed of response. Detectors are available for operation in any part of the infra-red spectrum which one might choose. Detectors can be obtained with sensitive areas as small as 10^{-6} cm^2. Response times are as short as 20 nanoseconds. Also, the detectivity of many of these detectors is limited by fluctuations in the arrival rate of photons at the detectors—an ideal situation. In fact, today one can choose a detector that is just about tailor-made for the particular system in mind. Image converters and intensifiers have also been developed. These are, however, generally limited to the 0·7–1·2 micron region for the very sensitive devices and 2 to 14 micron region for the less sensitive, more sluggish instruments.

Infra-red systems also contain optical elements, namely, mirrors, lenses, spectral and spatial filters. Their function is to collect the radiation from some region of space, process it optically, and focus it on the detector. Focusing optics for infra-red use are commonly telescopes with angular resolutions of about a milliradian or less. The collecting mirrors or lenses range in size from one inch diameter laboratory components, to three foot diameter mirrors used in long-range detection systems and astronomical devices. Various optical materials are available for lenses or windows; the choice depends upon the spectral region of interest. Some of the materials used for making lenses and/or windows in the 8 to 20 micron region include germanium and silicon in crystalline or cast forms, and zinc sulphide or zinc selenide in hot-pressed compacts, certain mixed crystals, and arsenic-sulphur glasses. In the 3 to 5 micron region calcium aluminate glass, hot-pressed magnesium fluoride, crystalline magnesium oxide, and some other materials can be used. While in the 0·7–2·7 micron region many different materials are available.

Spectral and spatial filters are used to process the incoming radiation. Spectral filters are used to reject unwanted portions of the spectrum. Many types are commercially available today. The most commonly used are the simple absorption filter and the dielectric interference filter. The simple absorption filter makes use of the selective absorption properties of various semiconductors, where these properties are determined mainly by the size of the forbidden gap. For the dielectric interference filters, films of materials selected for their refractive index are deposited on a substrate in certain thicknesses so that the waves reflected at the interface between the substrate and the films interfere destructively with the waves reflected from the film surface. By combining a series or stack of these films, one on top of the other, a great variety of spectral transmittance characteristics can be obtained. For example, narrow spike pass filters throughout the infra-red spectrum are commercially available today, and myriads of broader pass band filters can also be fabricated upon request.

Spatial filters are basically shaped field stops or apertures. Their purpose is to transmit radiation from targets with particular shapes. For example, if one wishes to detect point targets only, and reject extended objects, an aperture is used which consists of alternately translucent and opaque bars. The bar spacing is made to fit the image of an unresolved target. As the aperture is scanned across the scene where a target is present in a homogeneous extended background, the output from the detector will be d.c. until the target enters the field of view. At this point the output from the detector will contain an a.c. signal impressed on the d.c. Thus the point target may be detected. A great deal of analysis effort has been expended in determining configurations of the aperture for detecting targets of various shapes against a variety of backgrounds.

Two important factors involved in determining the spectral region of operation for a particular application are: the spectral distribution of radiation from the target and its contiguous background, and the properties of the intervening gaseous medium such as the atmosphere. The spectral distribution of radiation emanating from an object depends upon its temperature and the spectral emittance (i.e. the ratio of radiation intensity from the object to the radiation intensity from a blackbody at some wave-length, λ). Spectral emittance, in turn, depends upon the refractive index and extinction coefficient of the constituent material as well as the shape of the object and the microstructure of the surface.

The atmosphere, which is the second influential factor in determining the radiation arriving at a remote sensor from some object of interest, is a spectrally selective absorber. Portions of the spectrum where the atmosphere is transparent are called "windows". Windows of various spectral widths can be found throughout the infra-red region. The three windows most commonly used are, nominally, the 0·7 to 2·5 micron region, the 3 to 5 micron region, and the 8 to 14 micron region.

Most infra-red systems are passive devices (i.e. they sense objects by their self-emission and/or sometimes by reflected natural light). It is possible, however, to use an infra-red "spotlight" to illuminate objects for improved remote sensing in some cases. The advent of lasers has made this concept a real possibility. Such systems are currently being studied for ranging and navigation. The uses of infra-red are many and varied. The subsequent text touches on some of them which range from military applications, where infra-red has been the most exploited, to the gathering of agricultural data from space.

Possibly the best known military application of infra-red has been its use in heat seeking missiles for air-to-air and ground-to-air devices. An example of the former is the "Sidewinder", and of the latter, the "Redeye". In both these examples the missile locks onto the intense emission produced by the hot exhaust from jet engines.

In the area of airborne reconnaissance systems, infra-red has been employed to extend the range of standard photography. By having a wider spectral range in which to work, more information may be made available and operations are possible at night when photographic systems cannot function. However, extrapolation of imagery-interpretation from the visible into the infra-red is not straightforward, and is the subject of considerable study today.

Because of the secrecy inherent in pencil-beams of invisible radiation, infra-red has been applied by the military to various ranging systems. In addition to the use of passive systems for such ranging jobs as triangulation, active systems have also been developed for the same purpose. One of the earliest successful applications of the active system was the Sniperscope. The Sniperscope illuminates a target with infra-red radiation, which after reflection is received and presented as an image to the observer. Although invisible to the target, the reflected radiation is made visible to the observer.

Related to, and sometimes overlapping the military applications are the uses of infra-red in the space sciences. So far, such applications involve viewing either the Earth or the other planets. Tiros and Nimbus, which are meteorological satellites, have been instrumented with infra-red devices. For the satellite Tiros, the output data from the infra-red instrumentation has permitted the association of temperature distribution with cloud and storm formation, leading to more accurate weather forecasting. Further information on atmospheric temperature distribution is being obtained from the Nimbus satellite which is equipped with an infra-red spectrometer. By means of this spectrometer, the atmospheric absorption bands are examined to provide information of a thermal nature.

Determination of the temperature distribution of planets required knowledge of the spectral radiance from these planets, in the near and far infra-red. Such radiance measurements must be made from outside the Earth's atmosphere because of the many and variable infra-red absorption bands in the terrestrial atmosphere. For this purpose, a number of satellites and high-flying balloons have been equipped with radiometers. The average temperature of Venus was determined in this manner with surprising results.

Currently, studies are being made of the possibilities of using lasers for communications in space as evidenced by experiments in the Gemini manned space efforts. Here, the astronauts attempted to communicate with ground stations as their vehicle passed over the Earth. Whether the task is long-distance communication or tracking, the main difficulty in space application is the acquisition problem. If the position of the target is not known accurately enough, the time required for searching may exceed the tolerable limit for all present systems.

Some of the most promising applications of the new infra-red systems are in the Earth sciences: oceanography, meteorology, geophysics, geology, and agriculture. In oceanography, infra-red techniques have exhibited some success in mapping sea currents and in the detection, counting and tracking of icebergs during the polar night. For mapping sea currents, use is made of the fact that there is a temperature difference between the currents and the surrounding sea. Because, in the far infra-red, water acts as an excellent absorber (i.e. a black body), its emitted radiation is almost completely determined by its temperature. The infra-red systems are able to distinguish this difference and therefore map the currents. Similarly, ice behaves as a blackbody at long wave-lengths. Its temperature is lower than the ocean water temperature. As a result, icebergs can be detected and tracked.

As for the meteorological and geophysical applications, these have already been touched upon under space sciences, where the studies of weather conditions and planetary atmospheres were discussed.

Heat produced by processes taking place underneath the Earth's surface can be detected by means of temperature differences manifested at the surface. This introduces the possibility of finding mineral deposits or detecting underground fires and nuclear explosions. These techniques are also being employed in predicting volcanic and hydrothermal activity. Particularly encouraging is the successful use of infra-red systems in detecting crevasses hidden under arctic snow. The air trapped in the crevasses by the overlying snow is usually warmer or cooler than the surrounding snow, thus producing a temperature difference at the surface between the crevasse cover and the surrounding snow. Although invisible to the eye, the crevasses are easily detected by infra-red, radiometric methods.

To fight forest fires, it is necessary to delineate the

perimeter of the burning zone even when obscured by heavy smoke. Infra-red scanners are tailor-made for this purpose. Smoke particles, because of their size, obscure the fire in the visible part of the spectrum. But, there is more than a magnitude of difference between the wave-length of visible radiation and infrared radiation; therefore the smoke particles are not as effective in scattering infra-red radiation.

Presently under study are methods of using the infra-red characteristics of plants to detect diseased crops or to determine, from great altitudes (or space), the type and extent of crops present in large areas.

Although industry had an early start in using infrared for spectroscopic and heating purposes, it has just begun to apply this region of the spectrum to sensing and control. An early, widely used application of sensing instrumentation was in the railroad industry. A broadband radiometer measuring the radiation emitted by a journal box, was able to determine the temperature of that box. Since a rise in temperature is an early indicator of faulty lubrication, the journal box could be repaired or replaced before failure, leading to a considerable saving to the rail road industry. A more recent application of temperature sensing involves the use of a thermograph to find flaws in honeycomb structures. Knowing that the thermal conductivity of flaws is different to that of acceptable bonds, flaws are easily found when one side of the structure is uniformly heated and the other side mapped with a thermograph. Systems have also been adapted to controlling the sizes of hot steel bars and sheets during their manufacture.

Particularly exciting are the medical uses of infrared instrumentation. One technique uses infra-red absorption spectroscopy to determine quickly and accurately the concentration of carbon dioxide in the blood; the percentage of carbon dioxide present is a useful diagnostic tool. The possibilities of detecting cancer and other disorders are being investigated by trying to associate with these diseases certain anomalous temperature variations of the body as observed with a thermograph. Some success has recently been obtained in diagnosing dying or dead teeth with a radiometer capable of detecting the temperature difference between them and healthy teeth.

In short, infra-red technology has been growing rapidly over the past twenty years. This mushrooming technology has been applied to problems associated with space, geology, forestry, oceanography, medicine and industry. The uses of infrared in these areas have just begun to be exploited.

Bibliography

Holter M. R. *et al.* (1962) *Fundamentals of Infrared Technology*, New York: Macmillan.

Jamieson J. A. *et al.* (1963) *Infra-red Physics and Engineering*, New York: McGraw-Hill.

Kruse Paul W., McGlauchlin Laurence D. and McQuistan B. (1962) *Elements of Infra-red Technology*, New York: Wiley.

Lee P. A. (1961) in *Encyclopaedic Dictionary of Physics* (J. Thewlis Ed.), **3**, 839, Oxford: Pergamon Press.

Smith R. A., Jones F. E. and Chasmar R. P. (1957) *The Detection and Measurement of Infra-red Radiation*, Oxford: Clarendon Press.

I. W. Ginsberg and T. Limperis

INFRA-RED SCANNER. An infra-red scanner is a system used to generate an infra-red map of some scene. It usually contains a rotating mirror, reflective imaging optics, detector, preamplifier, amplifier, and a display device. Basically, the detector area and the optical train define the size and shape of the field of view. The field of view can be made to move across the object plane in any geometrical pattern desired by placing a mirror in the entrance optics of the system and mechanically moving the mirror. Most infra-red scanners today are mounted in airplanes and are used to map the radiation emanating from the Earth. To provide good quality images the field of view is typically small, on the order of milliradians. In airborne applications, the field of view scans at right angles to the aircraft path. The motion of the vehicle carries the scanner forward so that successive scans cover different strips on the ground. The scanning speed is adjusted so that these strips are contiguous. Thus an image is produced.

INSULATING-CORE TRANSFORMER. *Introduction.* The need for a relatively compact source of high-voltage d.c. power in the range from a few hundred kilovolts to several megavolts has been met by a novel approach to the transformation of alternating current to filtered high-voltage direct current. The insulating-core transformer principle minimizes many of the high-voltage limitations of conventional transformers without the complex voltage-multiplying circuits of the Cockcroft-Walton method, at output power levels of tens of kilowatts even in the megavolt region.

Principles of operation. The insulating-core transformer (ICT) is essentially a power transformer with multiple secondary windings, each of which is insulated from the other. The alternating current in each secondary is rectified and filtered, and the individual d.c. outputs are connected in series. The secondary insulation is achieved by segmenting the magnetic core and inserting a thin plastic layer between each segment. Thus, the high voltage is supported along the magnetic core instead of being insulated away from it, as in conventional transformer designs. The electric and magnetic fields can therefore exist simultaneously in the same space.

Figures 1 and 2 are diagrams of one embodiment of the ICT principle, that operates from 3-phase, 50/60-cycle mains at conventional voltages, with a power-conversion efficiency of more than 70 per cent. Units of this general design, with 7-inch diameter magnetic cores, supply 60 mA at 300 kV, or 20 mA at 500 kV,

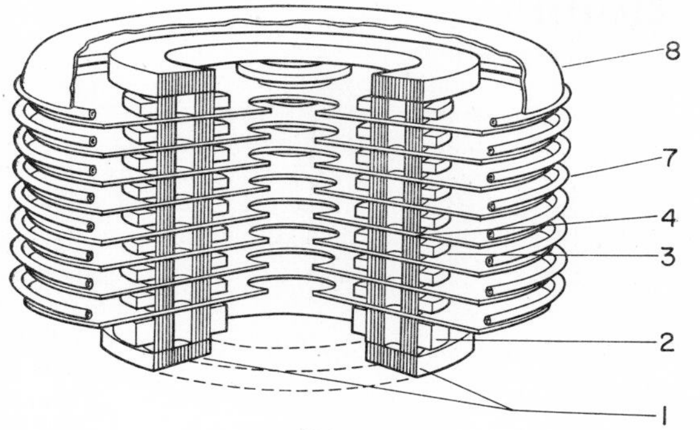

Fig. 1.

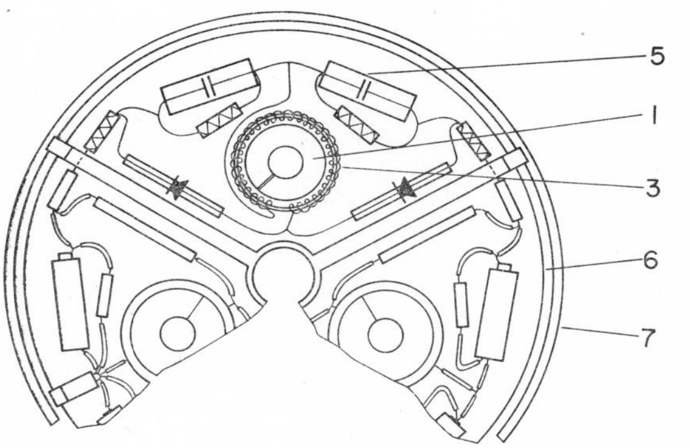

Fig. 2.

depending on the number of secondary rectifying circuits. A laminated magnetic-core structure (1) has a segmented leg for each of the three phases and toroidal flux-return paths at the ground and high-voltage ends. Three primary windings (2) induce the flux.

An alternating current at a voltage of about 10 kV is induced in each of the secondary windings (3), separated from each other by insulating layers (4). This current is rectified and filtered by a voltage-doubling circuit (5). Selenium or silicon rectifiers are used in this design. The d.c. outputs of each section of a single deck (6) are connected in series, and the last section of one deck is connected to the first section of the next deck. A high-voltage terminal spinning (8) and equipotential rings (7) keep corona leakage at a minimum. The entire stack is enclosed in a grounded pressure vessel, filled with sulphur hexafluoride (SF_6) gas at one atmosphere gauge pressure. The ICT output is obtained through a gas-to-air or gas-to-cable bushing.

Magnetic-flux leakage limits the number of decks that can be stacked with a particular magnetic-core diameter, thus limiting the maximum voltage attainable. With a proper selection of magnetic-core diameter and secondary windings, however, a broad variety ICT designs have already been developed, ranging from very powerful units at modest potentials, to 3-MV sources that provide 20 mA direct current.

Applications. Insulating-core transformers have been used as sources of high-voltage d.c. power for developing and testing electrical equipment and cable, not only in the laboratory but also in the field. Perhaps the greatest single application for the ICT has been the acceleration of electron beams for industrial radiation processing. A typical ICT electron accelerator for this purpose is shown in Fig. 3.

Other embodiments of the insulating-core principle are currently being studied for use in high-voltage power generation and transmission.

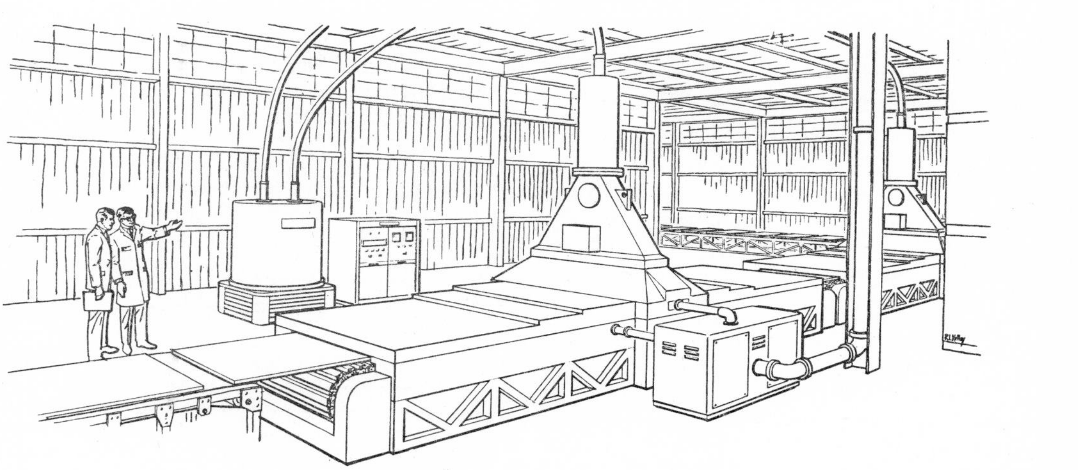

Fig. 3.

Historical background. The insulating-core transformer was invented by Dr. R. Van de Graaff, chief scientist of High Voltage Engineering Corporation, as a means of extending the technology of his electrostatic accelerator principles to higher power levels. The first working model of the ICT has been in continual service since 1958, as the power source of an accelerator delivering 25 mA of 1-MeV electrons.

This prototype ICT was designed as a single-phase transformer with the acceleration tube axially mounted in opposition to the ICT stack. Operating with primary power at a frequency of 10kc/s, the magnetic-core material is ferrite and the rectifiers are silicon diodes. The return path of the flux includes a large-area gap at the high-voltage terminal to the grounded tank liner, also of ferrite.

The 60-cycle version of the ICT was introduced in 1960, and to date more than 60 of these units, operating at 300–500 kV, have been produced. Three ICT power sources with 15-inch diameter magnetic cores have been built for operation in the 1·5–4 MV range.

Bibliography

Burrill E. A. (1963) Recent advances in d.c. methods of particle acceleration, *I.E.E.E. Trans. on Nucl. Sci.* NS-10, No. 3, 69, July.

Butler O. I. (1961) in *Encyclopaedic Dictionary of Physics* (J. Thewlis Ed.), **3**, 849, Oxford: Pergamon Press.

Van de Graaff R. J. (1965) U. S. Patent No. 3, 187, 208, June 1.

E. A. Burrill

IONIZING RADIATIONS IN INDUSTRY.

Introduction

Ionizing radiations from radioactive sources or electrical machines have a limited number of applications in industry. The present article deals with their uses in bringing about chemical or biological changes of commercial or social value. Their other major use, for the dispersal of static electrical charges, is dealt with elsewhere in this dictionary.

As the term implies, ionizing radiations interact with matter by dislodging electrons from their orbits, leaving positively charged ions or excited atoms or molecules. This can effect the physics, chemistry or biology of a system in rays which can be commercially useful. Exploitation has only become possible on a commercial scale with the availability of large radiation sources; two classes of source have been developed as a direct outcome of nuclear energy research—electron accelerators for nuclear physics studies, and radioisotopes as by-products of the operation of nuclear reactors. Both are now available as commercial products. In the 1950's, when large-scale radioisotope production from power reactors first became a practical possibility, many applications of radiation were expected to be commercially exploited, especially where alternative processes were technically or economically unsuitable; however, more detailed studies, especially of irradiation costs and the efficiencies of the proposed processes, eliminated most of these, while others were overtaken by improvements in conventional processing techniques. A small number of chemical reactions, having the nature of chain-reactions that can be initiated by radiation and will then proceed unaided, have proved to be worth exploiting, and others show promise but need further work. Similarly in the biological field many commercial uses were proposed for ionizing radiations, but only a few have survived close economic scrutiny and have found or look like finding practical application on an industrial scale.

Chemical Applications

1. Polyethylene. Irradiation of polyethylene at doses of about 5 Mrad induces cross-linking. This alters the physical properties of the plastic, which will no longer melt on heating, but assumes a rubbery nature at temperatures above 113°C. The process, usually carried out by electron-beam irradiation, raises the useful working temperature of the plastic, enabling it to be used as an insulating material for heavy-duty cables, under-floor heating elements and electronic circuits liable to heat up in operation; further, it enables soldering to be carried out with a minimum of damage to the insulation. If cross-linked polyethylene is stretched while hot it will retain its stretched shape on cooling, returning rapidly to its original dimensions on momentarily re-heating; polyethylene sheet or sleeving treated in this way is widely used as a wrapping material for food and for covering the exposed ends of cables, etc. A combination of radiation and heat treatment, with the incorporation of a foaming agent or a dissolved gas, is used to produce a foamed polyethylene.

2. Plastic-impregnated timber. Many monomers can be polymerized by radiation, and if penetrating γ-rays are used the process can be applied to wood that has been vacuum impregnated in depth with a liquid monomer such as methyl metacrylate; this can provide a range of composite materials having improved appearance, hardness and water-resistance and which are useful for special applications such as high-quality cutlery handles. The range of use is limited by the present high cost of suitable monomers.

3. Rapid curing of paints. Electron-beam irradiation of specially formulated paints will produce a mar-proof surface in a few seconds' curing time, without significant heating or the need for an added catalyst. Electron accelerators may thus replace ovens on small-produktion lines, with a saving of space, capital and running-costs, and cleaning-down time. The process can, moreover, be used in heat-sensitive materials.

4. Ethyl bromide production. In a process perfected by a chemical company in the United States pure ethyl bromide is produced directly from ethylene and hydrogen bromide in a reaction catalysed by γ-radiation.

5. Bio-degradable detergents. Synthetic detergents in common use are not fully broken down in bacterial sewage-treatment processes, with the result that effluent from the plants can be harmful to fish and can cause foaming on rivers. A process has been developed for the radiation synthesis of a fully degradable straight-chain alkane sulphonate detergent that is competitive in price with the present (non-degradable) branched-chain materials in current use.

6. Vulcanization of rubbers. Ionizing radiation can be used to vulcanize rubbers, alone or cross-linked with other materials, giving a range of products with some advantages over sulphur-vulcanized rubbers, particularly in resistance to heat and wear. However, the cost of the irradiation is at present too high for the process to be exploited commercially, and more research is needed on methods of reducing the necessary radiation dose.

7. Co-polymerization. Considerable research has been carried out on the radiation-induced co-polymerization of different monomers, and on grafting monomers onto different polymers, in order to combine the best properties of both in a single textile or plastic material. Textile fabric for which improved crease resistance is claimed is already on the market, and research continues in other promising fields.

Biological Applications

Because of the very great complexity of living matter, significant biological effects are produced by doses of ionizing radiation that are usually quite low by comparison with those needed to produce useful chemical effects. Partly for this reason the biological uses of ionizing radiations are at present being exploited on a larger scale than the chemical uses.

1. Sterilization of medical products. Ionizing radiations in sufficient quantity are lethal to all living things, and after extensive tests a dose of 2·5 Mrad has been found to be effective against bacterial spores and other infective micro-organisms, provided that the initial contamination level is not excessive. Gamma-ray or electron-beam irradiation is now widely used for sterilizing medical products, especially those that might be damaged by heat or steam, or which cannot with certainty be reached in all parts by a cold sterilant material such as ethylene oxide gas. In particular, γ-sterilization has made possible, and economically very attractive, the widespread use of pre-packed sterile hypodermic syringes mass-produced from inexpensive thermoplastic materials; these remain sterile until unpacked and used, after which they are destroyed. In this way one possible channel for cross-infection between patients is eliminated, in addition to which there is a valuable saving in time and trouble in emergency situations. Most hypodermic syringes now used in the U.K. are of this type, and because of its undoubted success the commercial exploitation of the process is spreading rapidly. The most usual source of radiation is cobalt-60, where the penetrating power of the γ rays enables bulk packages to be sterilized with complete reliability; electron accelerators are sometimes used where deep penetration is not required.

2. Sterilization against anthrax. The sterilization of animal products against anthrax is vitally important, especially where the disease is endemic in the country of origin; the spores are unusually resistant to heat and other conventional sterilants, and normal treatment is by a multistage chemical process which necessitates unbaling the material first. A 2·5 Mrad γ-ray dose is sufficient to inactivate the bacillus at all stages of its life cycle, and the radiation can be applied to the unopened bales at the place of importation. The world's first commercial γ-sterilization plant has been in successful operation in Australia since 1960 on goat-hair imported from S.E. Asia for carpet manufacture. Installations are planned in other countries for the treatment of wool and of hides.

3. Food treatment. Although earlier hopes for the widespread application of radiation to problems of food conservation have proved disappointing, a number of specialized fields of use are beginning to become established. These can be grouped under the headings of long-term preservation, short-term preservation, inactivation of pathogenic organisms and control of insect pests.

(i) Long-term preservation. All decay organisms can be destroyed by radiation, but the amount required for complete sterilization of food is liable to give rise to unpleasant odours and flavours; these become quite unacceptable if the food is exposed to further radiation sufficient to inactivate the enzymes which would otherwise cause autolytic decay even in the absence of micro-organisms. Research continues in the hope that adjustment of irradiation conditions, or a combination of radiation and heat-treatment, will result in a useful long-term preservation process for a limited range of foods.

(ii) Short-term preservation. Much lower doses of radiation can usefully prolong the storage life of many foods without giving rise to any significant loss of palatability. The sprouting of potatoes and onions in storage can be delayed until the new season's crop reaches the market—a process that is already in commercial operation on a pilot scale in some countries. The onset of mould growth in soft fruits can be inhibited for 3 to 4 weeks, which may result in strawberries, for example, being marketed fresh in areas where this is not normally possible. It has been shown

that white fish, if irradiated soon after being caught, can be stored on melting ice for 2 to 3 weeks without deteriorating or giving rise to "fishy" odours.

(*iii*) *Inactivation of food-poisoning organisms.* Salmonellae (and the parasite Cysticercus bovis) can be inactivated by a radiation dose low enough not to spoil flavour. The process is of considerable potential interest in the field of public health, and is particularly applicable to products such as frozen horse-meat, which is imported into the U.K. in large quantities as pet food; the frozen blocks could be treated effectively and quite cheaply without thawing, at a central installation at the landing port.

(*iv*) *Control of insect pests.* Stored grain can be freed from infestation by beetles and weevils by irradiating it with a dose of about 20,000 rads; unlike most chemical treatments, this dose will kill or sterilize all insects at all stages of their life cycle. It is only feasible economically, however, where a large plant with ample storage and handling facilities can be kept operating more or less continuously. To evaluate the process, a plant has been built in Turkey under the auspices of the UN Special found Programme for Developing Countries. Certain specific insect pests in the field have been tackled very successfully by the sterile male release technique, where very large numbers of artificially reared male insects are sterilized by a non-lethal dose of radiation and liberated in the infested area; here they find themselves in competition with the much smaller number of indigenous males, with the result that few fertile matings take place and the next generation is greatly reduced in numbers. The technique is at present limited to isolated communities of insects having suitable life cycles and breeding habits. Its most spectacular success has been the extermination of the screw-worm fly from the south-eastern states of the U.S.A.; more recently, the olive-fly has been successfully tackled in some Mediterranean areas.

Radiation sources. The ionizing effect of high energy electron beams does not differ significantly from that of γ rays from radioactive sources such as cobalt-60 or caesium-137. The main practical distinction between the two lies in the penetrating powers of the radiation and the rates of deposition of energy: electrons are much less penetrating but can be focused to give a very much higher dose-rate. Electron accelerators are complex machines requiring skilled supervision and maintenance and a reliable source of electric power; cobalt or caesium sources will continue to emit γ rays in all circumstances with 100 per cent reliability and at an accurately known energy, intensity and rate of decay. The choice between accelerator and radioactive source will depend partly on the technical requirements of penetration and dose-rate and partly on the relative costs and convenience of installation, maintenance and operation. Both types of installation are in increasing use; cobalt-60 radiation sources are at present being exploited more successfully for the sterilization of medical products, and electrical machines for processing films and coatings. The choice between cobalt-60 and caesium-137 is almost entirely an economic one—cobalt-60 is very conveniently produced by neutron activation of the inactive metal in a nuclear power reactor, and it emits two γ-ray photons per disintegration. Caesium-137 has to be separated chemically from spent nuclear fuel which is of course becoming more plentifully available—it is, however, much less easy to handle and emits only one γ ray of a rather lower energy than cobalt-60; on the other hand, caesium-137 has a much longer half life—28 years as against $5\frac{1}{4}$ years for cobalt-60. At present cobalt-60 is the less expensive of the two and is almost universally preferred. Unseparated fission products in spent nuclear reactor fuel in temporary storage in reactor "cooling ponds" are used for experimental and non-routine commercial irradiations. Aged fission product waste stored in vitrified form may possibly prove useful in the future as a cheap bulk radiation source having a low specific activity but a half-life that is essentially that of caesium-137.

Design of irradiation plants. The design of a commercial irradiation plant will of course depend very largely on the specific application, e.g. whether for packages or for bulk materials, for continuous or batch operation and whether or not deep penetration is required; it will also depend on the requirements of total dosage, uniformity of dose, dose rate and throughput. In all cases attention must be paid to the efficient use of the available radiation, by presenting the product in both sufficient area to intercept the maximum of radiation emitted from the source and in sufficient depth to absorb as much of this radiation as is practicable. Usually some kind of conveyor system is employed which moves the material to be irradiated around the source so as to ensure an equal deposition of energy throughout the bulk of the material irradiated. Safety considerations are of overriding importance, and interlocks are provided to ensure that there is no risk of accidental exposure of personnel to the unshielded source. The irradiation plant cell walls are usually constructed of concrete about 140 cm thick, or its equivalent.

By the end of 1966 details had been published of about 20 commercially operational irradiation plants using radioisotope sources (nearly all cobalt-60) and a similar number of plants using electron accelerators.

Bibliography

Black R. M. (1965) *Reports on the Progress of Applied Chemistry*, I, 144.

British Nuclear Energy Society (1967) *Proceedings of the Symposium on an Application of Ionizing Radiations in the Chemical and Allied Industries*, London: BNES.

Charlesby A. (1964) *Radiation Sources*, Oxford: Pergamon Press.

Cornwell P. B. (1966) *The Entomology of Radiation Disinfestation of Grain*, Oxford: Pergamon Press.

International Atomic Energy Agency (1963) *Industrial leger of large Radiation Sources,* Vienna: IAEA.

International Atomic Energy Agency (1966) Food irradiation, *Proceedings of an International Symposium held in Karlsruhe,* Vienna: IAEA.

Jefferson S. (1964) *Massive Radiation Techniques,* London: Newnes.

Ministry of Health (1964) *Report of the Working Party on Irradiation of Food,* London: H.M.S.O.

Putman J. L. (1962) in *Encyclopaedic Dictionary of Physics* (J. Thewlis Ed.), **6**, 843, Oxford: Pergamon Press.

Swallow A. J. (1960) *Radiation Chemistry of Organic Compounds,* Oxford: Pergamon Press.

R. M. Longstaff

ION SOURCES: RECENT DEVELOPMENTS.

1. Introduction

The main developments in ion sources in recent years have been in the direction of greatly increased beam intensities and, in the case of particle accelerator applications, a much greater concern with beam quality. Several sources of polarized protons have been brought into service and development of these continues vigorously. In the field of controlled thermonuclear reactions, continuous hydrogen ion beams of up to 1 A have been successfully produced. The continued development of tandem Van-de-Graaff accelerators has created a demand for more intense beams of negative ions and for more reliable and versatile sources.

Specialized fields of ion source work have arisen out of the requirements for ion propulsion of space vehicles and for ion implantation in semiconductors.

2. Sources for Particle Accelerators and Thermonuclear Devices

2.1. R.F. source. The radio frequency source, though not changed in principle, has been developed for particle accelerators to deliver pulsed proton beams of hundreds of mA.

2.2. Duoplasmatron source. For the most intense beams the duoplasmatron is now almost universally used. The general arrangement of a typical duoplasmatron is shown in Fig. 1. The ions are created in a low pressure arc discharge which operates between thermionic cathode and an anode containing a small orifice, through which the ions pass and are to be formed into a beam by a highly negative probe electrode. The extraction process is discussed more fully below. Between the anode and cathode is an intermediate electrode, sometimes called the capillary, which contains a constriction. The latter is usually between 2 and 5 mm in diameter and 5–15 mm long and creates an increase in the arc current density and therefore in the ion density. The intermediate electrode is also part of a magnetic circuit which produces a strong, non-uniform field increasing in magnitude towards the anode, as shown in Fig. 1. This has the effect of increasing the plasma density in the space between the intermediate electrode and anode, so that the ion density is increased still further, reaching a high value near the anode orifice. The use of an intermediate electrode in this way is the unique feature of the duoplasmatron and enables large ion currents to pass through a small orifice in the anode. Thus the flow of neutral gas from the arc chamber is reduced to a minimum.

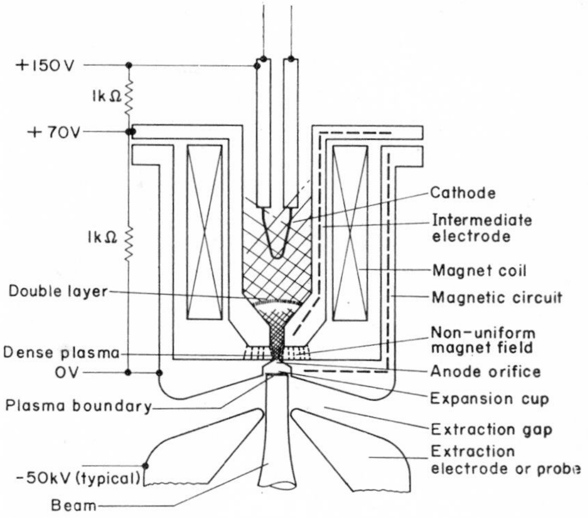

Fig. 1. Main features of a duoplasmatron.

The detailed processes occurring in the arc are complex and difficult to study because of the small physical dimensions of the critical regions. Von Ardenne, and more recently Demirkhanov, have studied the variation of plasma potential along the axis of the arc. A sheath or double layer forms in the plasma where the bore of the intermediate electrode decreases towards the construction. The potential difference across this layer, which is convex towards the cathode (see Fig. 1), can be a few tens of volts so that electrons are accelerated across it and focused into the capillary. Intense ionization of the neutral gas in the arc chamber by electron collision then builds up the plasma density so that the arc current can be carried through the narrow capillary. In the space between the intermediate electrode and anode a positive space charge develops which produces a maximum in the plasma potential. This can be a few tens of volts more positive than the anode, so that ions formed on the anode side of the "potential" hill are accelerated down it towards the anode orifice, emerging with appreciable directed velocities. Von Ardenne gives 10–20 eV as the mean energy of the ions (at a density of 10^{14} ions/cm^3) in the vicinity of the anode orifice for arc currents of a few A. Ion current densities in excess of 100 A/cm^2 are possible.

The mechanical design of the arc chamber and the materials used in its construction depend on the particular application. High energy particle accelerators require pulsed beams of protons with intensities from tens to hundreds of mA. This means arc currents of 10–50 A with arc voltages between 100 and 200 V, so that the power consumed by the arc can be several kW. However, the duty cycle is small (10^{-3}–10^{-4}) so that the mean thermal power to be dissipated is low and cooling is no problem. Also, the errosion of the anode by the arc is not troublesome.

Peak axial magnetic fields of 1–3 kG are typical. The gas pressure in the arc chamber varies between 5×10^{-2} Torr and 1 Torr, the higher pressure being required for output currents of about 1 A.

The size of the anode orifice depends on the maximum output current required. A diameter of 0·5 mm is quite adequate for proton currents of about 100 mA, but sources capable of an output of 1 A have an orifice diameter of 1–2 mm. For pulsed sources, the flow of neutral gas between pulses becomes too high if larger orifices are used. The gas consumption varies of course with the orifice size and source pressure, but a typical value for a 100 mA proton source is about 100 cm^3/hour at atmospheric pressure.

Sources for experimental thermonuclear devices have to satisfy different requirements. The emphasis is on the maximum possible output current of protons or molecular hydrogen ions with continuous operation. The problems of cooling the source and minimizing the erosion of the anode are severe. One of the most successful designs is that of Kelly, who used a well cooled copper anode with continuous arc currents of 50 A.

A variety of thermionic emitters has been used in duoplasmatron sources. Allison *et al.* used a simple "hair pin" of 1 mm diameter tantalum wire. The most widely used cathode at the present time is the directly heated oxide coated type.

Ion Extraction. The extraction of ions from sources in general has been reviewed by Gabovich. In the case of the duoplasmatron the current density at the anode aperture is so high that it is impossible to extract the ions directly because the electric fields required are far too high and result in breakdown of the extraction gap. The solution is to allow the plasma passing through the anode orifice to expand until the area of the plasma boundary is much larger than the area of the orifice. In this case the current density at the plasma boundary is greatly reduced and lower, more practicable electric fields are adequate to accelerate the desired beam across the extraction gap. The plasma expands into a recess, called the expansion cup, which is made in the source anode immediately behind the orifice (see Fig. 2). The plasma boundary forms somewhere inside the expansion cup, its actual shape and position depending on the shape of the electrodes forming the extraction gap, the applied extraction potential and the distribution of ion density in the plasma. Kirstein and Hornsby have produced a computer program which will calculate the profile, perveance and current density in a space charge limited beam from an emitter of arbitrary (but fixed) shape and any set of accelerating electrodes. The use of this program can give a valuable insight into the probable performance of a proposed extraction geometry, but practical design is still empirical.

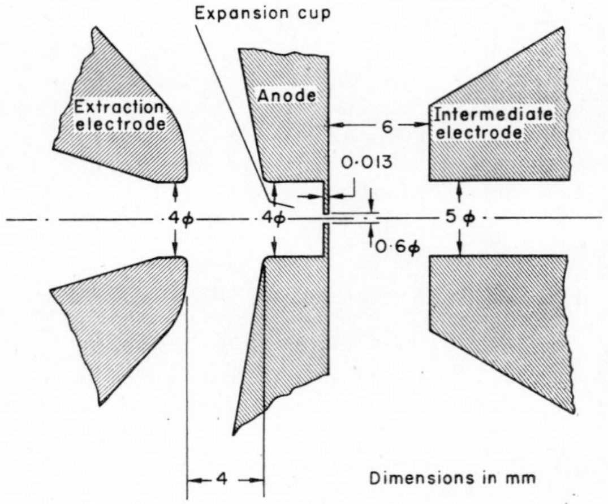

Fig. 2. Critical dimensions of a typical duoplasmatron designed for outputs of about 100 mA.

All duoplasmatrons use an expansion cup of some sort but the details vary widely depending on the application. Three examples, out of many, are now described and the operating conditions of several sources are given in the table.

Figure 2 shows the important dimensions of a duoplasmatron developed at Brookhaven National Laboratory, which is typical of proton sources with output currents of about 100 mA.

Gabovich, working with 10 cm diameter expansion cups, showed that the presence of a magnetic field at the plasma boundary can distort it and lead to poor focusing. This was confirmed by Vosiki *et al.*, Kelley *et al.* and Fauré. However, an axial magnetic field inside the expansion cup reduces the loss of plasma to the walls. Figure 3 shows a source developed by Vosiki *et al.* which uses a small coil and the expansion cup, the field from which falls rapidly towards the plasma boundary. This design uses a plane gridded extraction electrode and is capable of pulsed output currents of up to 1 A of hydrogen ions with a pulse length of 10 μs.

Kelley *et al.* have used a 2·5 cm diameter expansion cup with an axial magnetic field which was cancelled at the plasma boundary by the field of a powerful solenoid used to focus the beam. The whole source is made in copper, thus being well cooled and is capable of producing continuous beams of hydrogen ions at up to 1 A.

2.3. Modified duoplasmatron. Demirkhanov has modified the duoplasmatron by the addition of an anti-

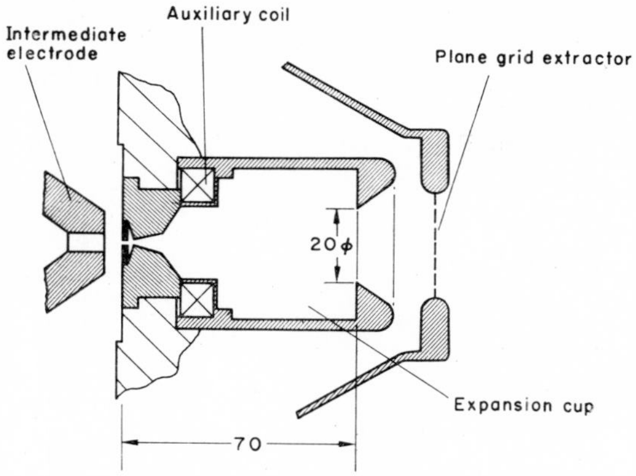

Fig. 3. Duoplasmatron with small ancillary coil and plane grid extraction electrode.

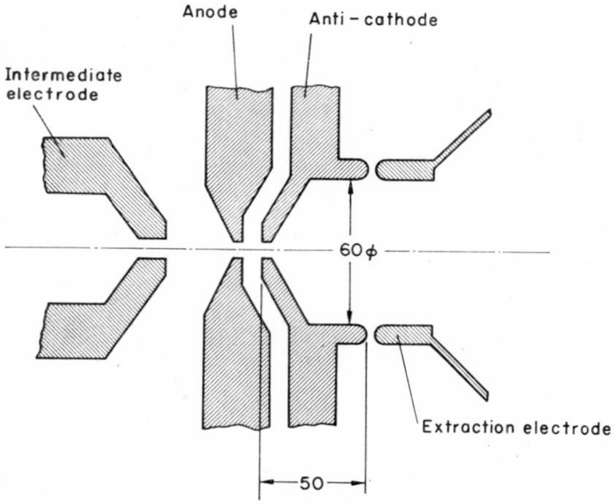

Fig. 4. Modified duoplasmatron with anti-cathode.

cathode behind the anode as shown in Fig. 4. The arc then operates like a P.I.G. discharge and a lower gas pressure (typically 5×10^{-3} Torr) can be used which permits larger apertures for ion extraction without excessive gas flow. This type of source is capable of producing 1·5 A of hydrogen ions with 85 per cent protons, using a 20 A arc current and 6 cm diameter expansion cup.

Kelley *et al.* have used the same principle in the development of a source of molecular hydrogen ions delivering continuous beams of 200–300 mA with 65 per cent H^+.

2.4. Negative ion sources. The production of negative ion beams has been fully reviewed by Rose and Gelejs.

2.5. Polarized proton sources. Sources of polarized protons are now in operation in a number of laboratories. The subject has been comprehensively reviewed by Dickson.

2.6. Beam quality. Methods of measuring the optical quality, the "emittance", of an ion beam have been reviewed briefly by Banford.

The emittance of the beams produced by the sources described above are given in the table. It must be emphasized that these values should be treated with caution. The results of emittance measurements on the output beams of sources are complex and difficult to interpret.

Wroe has described a duoplasmatron with a 1·5 mm diameter, 1 mm long expansion cup which produces a 60 mA beam at 40 kV with an invariant emittance of 6×10^{-3} cm-mrad. The emittance diagram is not distorted and the emittance value is roughly what would be expected from simple theoretical consideration.

3. Sources for Propulsion of Space Vehicles

The use of ion beams for propulsion of space vehicles has been reviewed by Stuhlinger. Though the final particle beam must always be electrically neutral, some systems make use of an initial beam of ions from a source. The source requirements are quite different to any other application. It must have a reliable operating life of about two years, an output beam with a current density of 20 mA/cm², an energy consumption of not more than a few hundred eV per ion pair produced, a total mass of a few grammes per mA of beam and an ion to neutral ratio of about 100 to 1.

Most effort has been devoted to sources of the surface ionization type.

4. Sources for Ion Implantation and Studies of Atomic Reactions

In this field, ion beams of a wide variety of elements are required with easy means of changing the ion species. Hill and Nelson have developed a source which uses the sputtering mechanism to produce metallic ions of Fe, Au, Ag, Al and several other elements. A simple low voltage arc discharge in a "support gas" produces ions which bombard a biased metallic specimen. Metal ions are sputtered off, some of which are ionized in the arc and can be extracted. In the case of Ag and Au, about 50 per cent of the beam is in the form of metal ions and currents of up to 10 A are readily obtained.

5. Sources for Electromagnetic Separators

Freeman has obtained improved resolution in an electromagnetic separator with an ion source using an arc discharge of novel geometry. A thermionic emitter in the form of a tantalam bar is used, inside a graphite arc chamber with a slit, parallel to the emitter, for ion extraction. The whole source can operate over a temperature range of 450–1000°C and output currents of up to 10 mA are obtained. An important property of the source is that the output beam is free from high frequency modulation. This phenomena, often known as "hash", is a feature of most

Some ion source data

Author	Type of source	Duty cycle		Plasma source parameters	Magnetic field	Gas pressure (Torr)	Extraction voltage	Output current	Beam composition	Invariant emittance (Area of diagram × π) cm-mrad	Application
		Pulse length	Rep. rate								
Tallgren	RF—Schneider design.	10 μs		Up to 10 kW peak RF power at 140 Mc/s	50 G transverse field	2×10^{-3}	30 kV	up to 300 mA	over 90 per cent H_1^+	0·08	Proton synchrotron source
Wroe	RF—Thonemann type extraction gap	1 ms	1 pps	10–20 kW peak RF power 125 Mc/s	50 G transverse field	0·01–0·02	20 kV	100 mA	over 80 per cent H_1^+	0·27	Proton synchrotron source
Wroe	Duoplasmatron with 15 mm ∅ × 11 mm long expansion cup	50 μs	1 pps	30 A arc at 150 V	1–2 kG peak axial field	0·1	40 kV	60 mA	about 80 per cent H_1^+	6×10^{-3}	Proton synchrotron source
van Steenbergen	Duoplasmatron—expansion cup 4 mm ∅ × 3 mm long. Fig. 2	50 μs	1 pps	30 A arc at 150 V	1–2 kG peak axial field	0·1	40 kV	140 mA	about 80 per cent H_1^+	0·40	Proton synchrotron source
Vosicki *et al.*	Duoplasmatron—expansion cup 30 mm ∅ × 60 mm long with auxiliary coil Fig. 3	10 μs	1 pps	70 A arc	5 kG peak axial field	1	70 kV	up to 1 A	over 85 per cent H_1^+	0·4	Proton synchrotron source
Kelley *et al.*	Copper "duo-plasmatron"—expansion cup 20 mm ∅ × 40 mm long with auxiliary coil	Continuous		50 A arc	No data available	No data	100 kV	up to 1 A	over 85 per cent H_1^+	not measured	Injector for DCX thermonuclear device
Kelley *et al.*	Duoplasmatron with anticathode and 20 mm ∅ × 40 mm long cup	Continuous		50 A arc	No data available	0·05–0·07	100 kV	200 to 300 mA	over 65 per cent H_2^+	not measured	Injector for DCX thermonuclear device

Demirkhanov *et al.*	Duoplasmatron with anticathode and 60 mm ∅ × 50 mm long cup Fig. 4	100 μs	15 kc/s	20 A arc	1·5 kG peak axial field	$5–7 \times 10^{-3}$	30 kV	1·5 A	85 per cent H_1^+	not measured	experimental source
Lawrence	Duoplasmatron with offset intermediate electrode	Continuous		Arc voltage 120–150 V	No data available	$4–7 \times 10^{-2}$	No data available	80 to 100 μA	H_1^-	not measured	Negative ion source for tandem van-de-Graaff
Brooks	Collisional type negative ion source. Duoplasmatron with conical expansion cup and separate donor canal	Continuous		No data	No data available	No data	70 kV	830 μA	H_1^-	not measured	Negative ion source for tandem van-de-Graaff
Hall and Nelson	Sputtering source. Low voltage arc in axial magnetic field	Continuous		Arc voltage 100 V arc current up to 3 A	About 100 G	No data	7 kV	1 mA 10 mA	A^+, Xe^+, Cu^+, Ag, Au^+ and other elements	not measured	Studies of atomic interactions in metals and ion implantation
Freeman	Coaxial filament arc with axial magnetic field	Continuous		Up to 80 V arc with up to 3 A arc current	Up to 150 G	Min. Press 10^{-4}	40 kV	Up to 10 mA	B, N_2, O_2, Me, A, Kr, Sr, Cr, Sm, Gd, Eu, Pu	not measured	Electromagnetic separator

sources which use a thermoionic arc discharge in a magnetic field and, if severe, it can destroy the focusing of the separator ion beam.

Bibliography

BANFORD A. P. (1966) *The Transport of Charged Particle Beams*, London: Spon.

DICKSON J. M. Polarized Proton Sources and the Acceleration of Polarized Beams, *Progress in Nuclear Techniques and Instrumentation*, **6** (F. J. M. Farley Ed.).

GABOVICH M. D. (1963) *Pribory i Technika Eksperimenta* No. 2, March-April (1963) (English translation).

HOBBIS L. C. W. (1962) in *Encyclopaedic Dictionary of Physics* (J. Thewlis Ed.), **4**, 75, Oxford: Pergamon Press.

Proceedings of the International Conference on Electromagnetic Separators (1965) *Nuclear Instruments and Methods*, **38**, 1.

Proceedings of the Los Alamos Linac Conference (1966).

ROSE P. H. and GALEJS A. Production and Acceleration of Ion Beams in the Tandem Accelerator, *Progress in Nuclear Techniques and Instrumentation*, **2** (F. J. M. Farley Ed.).

STUHLINGER E. (1964) *Ion Propulsion for Space Flight*, New York: McGraw-Hill.

H. WROE

ISOCHRONOUS CYCLOTRON. During the decade 1956–66 some tens of isochronous cyclotrons have been constructed, many of them capable of accelerating several different types of particle over a continuous range of energy. Technological advances in the design of magnet and high frequency systems have made it possible both to exploit the "azimuthally varying field" (or "*sector-focusing*") principle to remove the fundamental energy limitation of the earlier machines, and to design machines in which both the shape and strength of the magnetic field and frequency of the accelerating electric field can be varied over the wide limits required for variable energy and multi-particle operation.

In this article the principles of the sector focused isochronous cyclotron are outlined, and a brief description of a typical hypothetical machine is given. It is assumed that the reader is familiar with the principles of the original classical cyclotron, described by Broadbent (1961). In the earlier article the reasons for the energy limitation in the classical cyclotron are discussed. The argument is repeated here as an introduction to the sector focusing principle.

In order to ensure stability of particle motion about the median plane ("vertical stability") it is necessary to have a magnetic field which decreases with radius. In a true isochronous machine on the other hand an increase of field is required. This increase is to compensate for the relativistic increase of mass as the particles gain energy; the ratio of field at some radius r to its value at the centre is just $\gamma(r)$, the ratio of total energy to rest energy of the particles at the radius; for 20 MeV protons for example γ is 1·02.

It was soon recognized that these contradictory requirements impose a limitation on the energy obtainable in a classical machine. The angular velocity of the particles decreases as they are accelerated to larger radii and this leads to an increasing delay between the time at which the field at the accelerating gap is a maximum and the time at which the particles cross it. If this phase difference becomes $\pi/2$ then the particles experience no accelerating field; if it exceeds $\pi/2$ the particles experience a field of opposite sign and are decelerated back to the centre. The precise energy at which this limitation occurs depends on the degree of vertical focusing required and the voltage applied to the dees: the highest energy reported for protons (with the very high dee to dee voltage of 400 kV) is 22 MeV.

A method of avoiding this energy limitation was suggested by L. H. Thomas in 1938. He proved that if a sinusoidal azimuthal variation of periodicity $\pi/2$ is introduced in the magnetic field, then it is possible to achieve both stable vertical motion and a field which increases with radius, so that, within limits, the isochronous condition can be satisfied. Such a field shape can be provided by modulating the pole surface, or more simply by fitting four radial wedge shaped ridges to the pole. It was later shown that any number of ridges greater than two may be used, and that the harmonics in $B(\theta)$ which are multiples of the ridge periodicity are unimportant. The fitting of ridges has two effects; first, it distorts the particle orbit from a circle to a more nearly polygonal shape (triangular for three ridges) and second, it introduces circumferential components of magnetic field (see Fig. 1). Now the radial components of velocity caused by the orbit distortion interact with the circumferential component of field to produce a varying force which is always towards the median plane. If the field modulation is great enough, this force overcomes the defocusing force arising from the interaction of the radial component of field (associated with the gradient of the azimuthally averaged field) and the circumferential component of velocity.

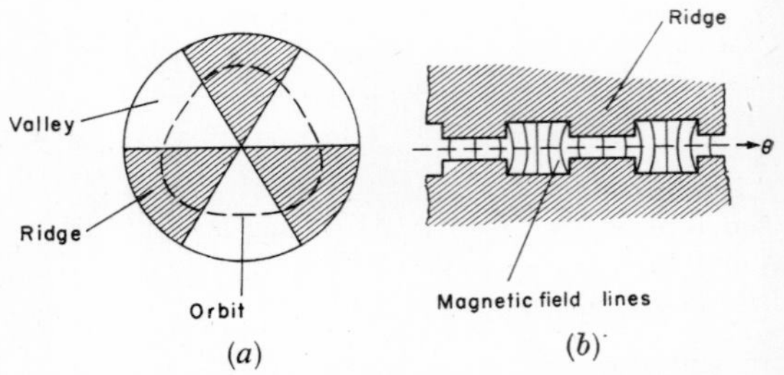

Fig. 1. (a) Plan view of pole with three radial ridges, showing distorted equilibrium orbit; (b) developed section of magnet gap, showing circumferential components of magnetic field.

As will be shown quantitatively below, the applicability of this remedy is limited. At high energies the radial gradient of field required to ensure isochronism is large, and the field modulation required becomes impracticably high. Fortunately, however, as discovered in 1955 by D. W. Kerst and K. R. Symon, the focusing due to the ridges may be increased if they are spiralled, as shown in Fig. 2. If this is done, the focusing when crossing from a valley to a ridge is greatly increased, and decreased by a corresponding amount when crossing from ridge to valley. This gives a rapidly alternating focusing and defocusing force, which yields a net focusing effect (Courant 1962).

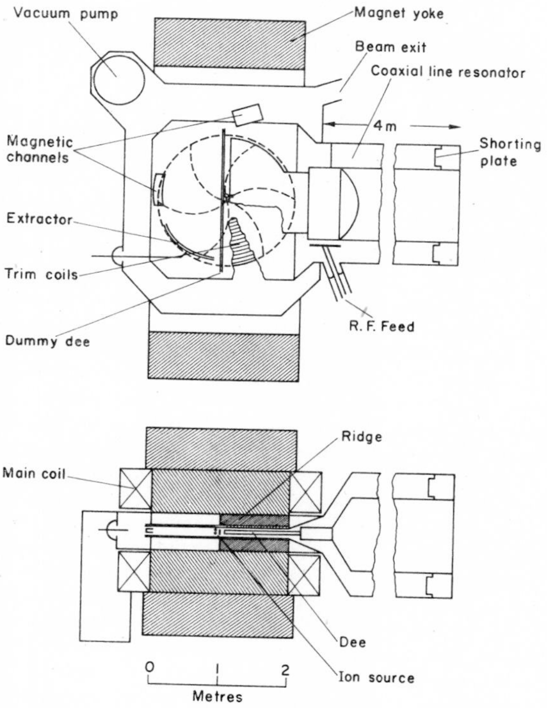

Fig. 2. Semi-schematic view of a typical variable energy cyclotron for an energy of up to 70 MeV for protons and 100 MeV for alpha particles. (The design is compounded from several existing machines, and there are many possible variations of detail, especially in the r.f. system.)

Quantitatively the requirement for focusing and isochronism can be expressed in terms of the "flutter" δ and the "spiral angle" ψ. For a general magnetic field in a machine with N ridges of form

$$B(r) = \bar{B}(r)\left\{1 + \sum_s \delta_s \cos\left(sN\theta + \Phi_s(r)\right)\right\} \qquad (1)$$

these may be defined as

$$\delta = \left(\sum_s \delta_s^2\right)^{1/2}, \qquad \psi = \tan^{-1}\frac{1}{r}\,\frac{dr}{d\Phi_1}.$$

The angle ψ is very roughly the angle that the centre line of the ridge makes with the circumference, as may be seen in Fig. 2. In terms of this parameter, the energy γ and velocity β normalized respectively to the particle rest mass and the velocity of light, and the radial and vertical oscillation frequencies Q_R and Q_V normalized to the rotation frequency, the essential features of an isochronous cyclotron may be summed up in the following approximate relations. These are not accurate enough for design purposes, but they illustrate well the physical factors involved:

$$\bar{B}(r) \approx B_0\gamma(\bar{r}) \qquad (2)$$

$$Q_R \approx \gamma \qquad (3)$$

$$Q_V \approx \left\{\frac{\delta^2}{2}\left(1 + 2\cot^2\psi\right) - \left(\gamma^2 - 1\right)^{1/2}\right\}^{1/2} \qquad (4)$$

Equation 2 shows that the field increases with radius parabolically at first, rising rapidly as γ increases, becoming infinite at the limiting radius c/ω where the particle velocity equals the velocity of light. For vertical stability Q_V^2 must be positive, and Eq. 4 shows that to achieve this for large values of γ, the flutter must be large and the spiral angle small. The former can be increased to be of order unity by splitting up the magnet into a ring of separate sector shaped magnets, but a "tight" spiral implied by small ψ makes the design very awkward.

A further limitation of a fundamental nature sets in at $\gamma = 2$, corresponding to a particle kinetic energy equal to the rest energy. From Eq. 3 it is seen that $Q_R = 2$ at this energy. When Q is integral, resonant oscillation amplitudes can be set up by small harmonic errors of periodicity Q in the guiding field. It is thought to be impracticable to reduce them to such a level that substantial (if not total) loss of beam is prevented. Other resonances (for example $Q = 1{\cdot}5$) occur in these machines, but their harmful effects can be made negligible by good design. Indeed, introduced in a controlled fashion they can be used to aid beam extraction.

At the time of writing (end of 1966) there are over twenty sector-focused cyclotrons in operation, and a larger number planned or under construction. Acceleration up to $\gamma = 2$ has been demonstrated in an electron model (corresponding to an energy of 0·51 MeV); for machines accelerating heavier particles the maximum energy to date is 65 MeV ($\gamma = 1{\cdot}07$) though considerably higher values are planned.

A typical design for a general purpose isochronous cyclotron to be used for nuclear physics, radio chemistry, and radiation chemistry might be as follows.

The magnetic field, with variable excitation so that B_0 can be varied between about 6 and 16 kg is provided by an electromagnet of weight about 250 tons and pole diameter 80″. Three spiral ridges shape the field in such a way that δ and $\cot\psi$ rise from zero to 0·35 and 1 respectively at maximum radius. The minimum gap of 7″ is reduced to $5\frac{1}{2}$″ by two copper plates on which are mounted 12 independent concentric trim coils which are used to adjust the radial field configuration for different particles and energies. In the valleys between the ridges are weaker coils to provide adjustment of the first harmonic field.

The accelerating field is produced between a single dee and "dummy dee" consisting of a pair of bars at earth potential, rather than with two dees as in a classical machine, since a high energy gain per turn is not so important. A voltage of 30–60 kV between dee and dummy dee with frequency variable from 7–22 Mc/s is provided by a 150 kW power amplifier which feeds a resonator, which is typically a large concentric line with the dee supported at the open end and a movable shorting plate at the shorted end. (Sometimes the resonator forms part of a self-oscillatory system rather than being driven.)

Particles are produced in a "hooded arc" ion source at the centre of the machine, and extracted by a puller electrode mounted on the dee, through suitable defining slits to give a good quality beam. On reaching the edge of the machine (after some hundreds of revolutions) they enter the deflector, consisting of a channel formed by two electrodes between which is an electric field variable between 20 and 100 kV/cm. After emerging from the deflector they pass through a magnetic channel which reduces the local magnetic field and allows the beam to leave the cyclotron and be collected and transported to the experimental area by a suitable array of magnetic quadrupole lenses and bending magnets.

The acceleration takes place in a vacuum of order 2×10^{-6} Torr, provided by a large oil diffusion pump. Adequate stabilization of magnet, trim coil supplies, accelerating frequency and voltage is provided to ensure stable acceleration and extraction.

Such a machine, based on a number of actual designs, is shown in Fig. 2. There are many variations in detail, particularly of accelerating and extraction system. Different applications require a different emphasis in the design.

Probably the largest use of these machines is for research in nuclear structure, where they compete with modern Van de Graaff accelerators. For such work good energy resolution and control is essential. They are also used for studies in radio-chemistry, metallurgy (for example studies of radiation damage effects), radiation chemistry, biological irradiation, and isotope production. For some applications secondary neutrons produced in a target are used; for time of flight studies, time structure of the beam may be exploited.

At higher energies the type of machine outlined above becomes rather cumbersome, and some modifications in concept, such as separate sectors (and even completely separated orbits) become necessary. Several design studies have been made, and machines are planned, though none is yet built. Some technical problems associated with high beam power and induced activity remain to be solved, and the ultimate potentiality of isochronous cyclotrons in the 100–1000 MeV range remains to be seen.

Bibliography

BROADBENT D. (1961) in *Encyclopaedic Dictionary of Physics* (J. Thewlis Ed.), **2**, 252, Oxford: Pergamon Press.

COURANT E. D. (1962) in *Encyclopaedic Dictionary of Physics* (J. Thewlis Ed.), **7**, 70, Oxford: Pergamon Press.

LIVINGOOD J. J. (1961) *Principles of Cyclic Particle Accelerators*, Princeton: Van Nostrand.

Proceedings of the International Conference on Isochronous Cyclotrons, Gatlinburg, May 1966. Published as *I.E.E.E. Transactions on Nuclear Science*, Vol. NS-13, No. 4.

RICHARDSON, J. R. (1965) *Progress in Nuclear Techniques and Instrumentation*, Vol. 1 (Ed. F. J. M. Farley), Amsterdam: North-Holland. Publishing Company.

J. D. LAWSON

ISOTOPIC SPIN. Elementary particles are observed to occur in groups of almost similar mass and properties but differing in electric charge. Such groups of particles are known as isotopic multiplets (e.g. the neutron, n, and the proton, p, are an isotopic doublet, the pions, π^+, π^0, π^- form a triplet, etc.). Apart from effects resulting from the charge, particles in a multiplet interact with one another independently of their pairing (e.g. nn, pp or np). The neutron and the proton can be considered as two charge states of the same particle, the nucleon. Multiplets are characterized by a multiplet number I, isotopic spin, where $(2I + 1)$ is the number of members of the multiplet. For the nucleons $I = \frac{1}{2}$ and for the pions $I = 1$. Only multiplets with 1, 2, 3 or 4 members ($I = 0, \frac{1}{2}, 1$ or $^3/_2$ respectively) have been observed. All members of the multiplet have the same values of spin, parity and strangeness. In atomic physics, an atom of spin J in a magnetic field can take up $(2J + 1)$ orientations relative to the field direction, each state differing from the next by one unit of angular momentum. By analogy, isotopic spin, I, is regarded as a vector in isotopic spin space, each of the $(2I + 1)$ orientations designated by the value of the third component of isotopic spin, I_3, represents a physical particle of the multiplet. The nucleon doublet can take values of $I_3 = +\frac{1}{2}$ and $I_3 = -\frac{1}{2}$ corresponding to the proton and neutron, respectively. The pion takes values of $I = 1, I_3 = +1, 0, -1$ corresponding to the π^+, π^0 and π^-. In strong (nuclear) interactions the total I and I_3 do not change. For example,

$$\pi^- + p \rightarrow \pi^0 + n$$

$$I = 1 \quad I = \tfrac{1}{2} \quad I = 1 \quad I = \tfrac{1}{2}, \quad \text{i.e. } \triangle I = 0$$

$$I_3 = -1 \quad I_3 = +\tfrac{1}{2} \quad I_3 = 0 \quad I_3 = -\tfrac{1}{2}, \quad \text{i.e. } \triangle I_3 = 0.$$

In electromagnetic processes only I_3 is conserved (e.g. $\pi^0 \rightarrow \gamma + \gamma$) and in weak processes (e.g. $\underset{(I=0)}{\Lambda^0} \rightarrow p + \pi^-$) neither I nor I_3 are conserved.

See also: Elementary particle physics, recent advances in.

T. G. WALKER

L

LASER CASCADES. A laser cascade can be defined as a cascade of laser lines. Although the laser effect has also been obtained in solids, liquids and gases; up to now, laser cascades have only been observed in gases. In the usual fluorescent spectrum of atoms or molecules, a cascade emission occurs when a unique electron produces several photons related to one another by common energy levels as represented in Fig. 1. On this diagram, we can see that an electron excited to the level 1 goes down in several successive steps of quantified energy by the emission of photons of wave-length λ_{ij} with

$$\lambda_{ij} = \frac{hc}{W_i - W_j}$$

where

h is Planck is constant ($6{\cdot}55 \times 10^{-34}$ J × s);
c is velocity of light (3×10^8 m/s in vacuum);
W_i, W_j are energy of levels i and j;
$i = 1, 2, 3$ $j = i + 1$.

To obtain a laser cascade, it is necessary to invert the populations of these excited states. In the case of a cascade with 3 optical transitions as in Fig. 1, this condition can be expressed by the relation

$$N_1 > N_2 > N_3 > N_4$$

N_k is the population density of the state k — N_k is usually expressed in atoms/cm³.

However, in equilibrium conditions the Boltzmann law applies. In the case of 2 levels 1 and 2 with $W_1 > W_2$ we can write

$$N_1 = N_2 \exp\left[-(W_1 - W_2)/kT\right]$$

with k = Boltzmanns constant ($1{\cdot}38 \times 10^{-23}$ J/°K);
T = thermodynamic temperature in °K,
so that the population of lower level exceeds that of upper level. As a consequence, the medium is absorbent for electromagnetic waves of wave-length λ_{12}.

To experience gain and construct laser amplifiers or oscillators, it is necessary to use selective excitation techniques. This excitation is optical in solid or liquid lasers; then it is called "optical pumping" (see this expression). In gas systems, optical excitation is rarely used because it requires strong monochromatic lines in precise coincidence with absorption lines of material a condition rarely satisfied in practice. Then excitation of gases for laser action is mainly obtained from an electrical discharge where various excitation processes act. Among these processes, the collisions of electrons with atoms or molecules are frequently the most efficient in attaining noticeable population inversions.

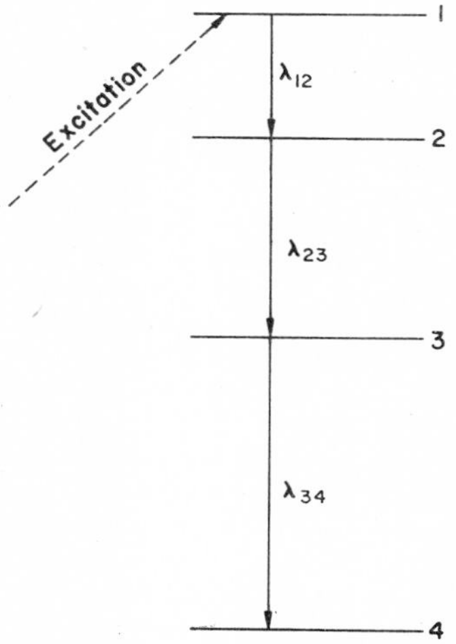

Fig. 1. Schematic representation of the cascade emission of an atom.

However, in certain cases other selective processes are particularly suitable and have been successfully used. In the helium-neon laser, selective excitation of neon atoms is attained from collisions of neon atoms in the fundamental state with metastable helium atoms. Helium possesses 2 metastable states, 2^3S and 2^1S, represented on the partial energy level diagram of Fig. 2, where we have also drawn the principal Ne levels of interest here. The metastable levels of helium have a very long lifetime (1 msec as an order of magnitude which is much longer than the lifetime of normal excited states, typically 10^{-6}–10^{-8} sec) because radiative transitions are forbidden from these levels. In pure helium discharges, metastable levels are principally depopulated by collision with the walls, but as they are connected to upper levels by radiative transitions, a very high density of atoms exists at these levels. As shown in Fig. 2, the energy difference between the 2^3S level of helium and $2s_2$ level of neon

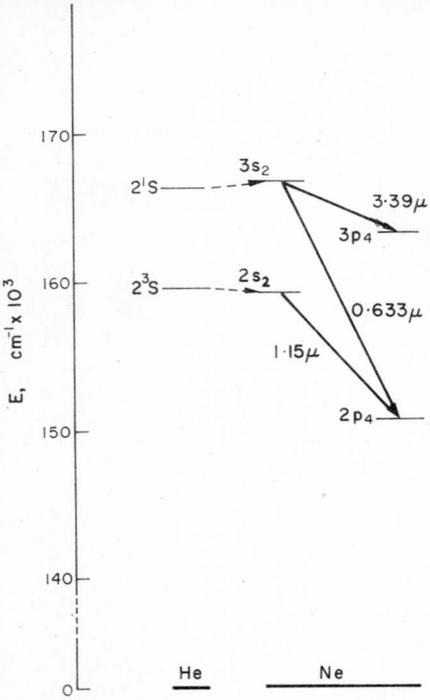

Fig. 2. Partial energy level diagram of helium and neon. Three characteristic laser lines are represented.

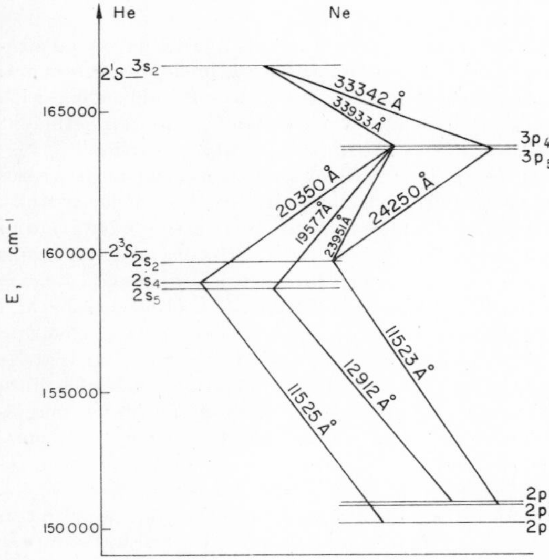

Fig. 3. Energy level diagram showing the different cascades of neon produced from collision of neon atoms with the 2^1 S helium atoms.

and between the 2^1S level of helium and $3s_2$ level of neon is very low; quantitatively, it is 2 to 3 times the thermal energy of atoms at ambient temperature so that, a very efficient transfer of excitation occurs leading to laser action from these neon levels. In Fig. 2, 3 transitions of neon are shown, selected among several tens using this pumping scheme. The $3p_4 - 3s_2$ transition has certainly the highest gain among the neon laser transitions (40 dB per metre have been effectively measured). In a laser working at 3·39 μ, the $3p_4$ level is strongly populated by stimulated emission whereas direct excitation of this level by electron impact from the ground state of neon is particularly difficult because these levels have the same parity. As a consequence, inversion of $3p_4$ relative to $2s_2$ and $2s_5$ levels is produced or enhanced by stimulated emission which acts here as a new pumping mechanism specific to lasers. Let us mention particularly the $3p_4 - 2s_2$ transition ($\lambda = 2{\cdot}395\ \mu$) which exists as a laser line only if the 3·39 μ line oscillates. The oscillation of these 2s — 3p transitions transfers the excitation from 3p levels to 2s levels and similarly as in the upper level product or enhances stimulated emission of several 2p — 2s lines. A characteristic example is the $2p_7 - 2s_4$ transition ($\lambda = 1{\cdot}1525\ \mu$), the parent of the well known $2p_4 - 2s_2$ transition (Fig. 2). This line is observable in helium-neon mixture only when the $3p_4 - 2s_4$ transition oscillates. Figure **3** shows several sequences obtained experimentally in a helium-neon laser.

Other cascades have been observed in pure gas systems such as pure neon and also in carbon monoxide for molecular transitions oscillating in pulsed operation.

The kinetic of laser cascades has been studied theoretically and also experimentally for several of them.

Bibliography

DER AGOBIAN R., OTTO J. L. CAGNARD R. and ECHARD R. (1964) *J. de Physique* **25**, 887.

HAKEN H. DER AGOBIAN R. and PAUTHIER M. (1965) *Phys. Rev.* **140**, A 437.

MAIMAN T. H. (1962) in *Encyclopaedic Dictionary of Physics* (J. Thewlis Ed.), **5**, 223, Oxford: Pergamon Press.

MCFARLANE R. A., FAUST W. L., PATEL C. K. N. and GARRETT C. G. B. (1964) *Quantum Electronics* III, (P. Grivet and N. Bloembergen Eds.), Paris: Dunod.

TROUP G. J. (1967) in *Encyclopaedic Dictionary of Physics* (J. Thewlis Ed.), Suppl. Vol. 2, 141, 144, Oxford: Pergamon–Press.

R. DER AGOBIAN

LASER SAFETY. *Introduction.* With the increasing use of lasers the question of laser safety is assuming greater and greater importance and the present article, which is an edited version of one appearing in *Scientific Research* (July 1966), and is published by kind permission of the editors of that journal, is designed to make laser users aware of the problems involved. The article contains, in the main, the conclusions of a seminar on laser safety held in May 1966 in the U.S.A. The importance of such a seminar is clear when it is realized that lasers are now operating in many countries at powers ranging from milliwatts to gigawatts, and producing radiation that extends right across the electromagnetic spectrum.

Of particular concern are the effects of laser beams on the eyes because these have caused permanent blindness in a number of cases. Other effects—such as burns—are also significant. Although no inadvertent exposures have caused burns more serious than second-degree so far, the hazards are very real—particularly with the extreme high-energy pulses produced by Q-switched and CO_2 lasers.

The effects on the eyes. Laser beams can harm the eyes by burning them, in much the same way as any other bright source might, e.g. the Sun or an electric arc. W. T. Ham has pointed out that the hazard of a laser beam is enhanced because the beam is emitted as parallel radiation. Such a beam produces images on the retina that vary in diameter from 200 microns (without optical equipment) to 10 or 20. According to him "small spot sizes come to equilibrium very quickly and at a much lower temperature than larger spot sizes. If we compare the effect of 20-micron spots with 800-micron ones in rabbit eyes, there is roughly a factor of 10 between them, which means that, all other factors being constant, the 20-micron spot size can take 20 times as much irradiance for the same temperature. We believe that it is the rise in temperature of the biological materials that causes the thermal damage and denaturation which may be irreversible and which leads to burn or a lesion".

Exposure time too has a great deal to do with the extent and type of damage caused. In studies of mild lesions of the retina—lesions that are just visible ophthalmoscopically within 5 min after exposure—produced by radiation pulses of relatively short time duration (0·2–2·0 msec), Geeraets noticed that histological damage occurred only in the pigment epithelium. Longer exposures (more than 10 msec), on the other hand, caused damage to the choroid as well. But both the pigment epithelium and the choroid have about the same light-absorption values. For this reason, the difference in the effect on them is explained by their difference in thickness. Since the choroid is about 10 times thicker than the pigment epithelium, it should be able to absorb about 10 times as much energy per unit volume without heating to destructive temperatures.

Because of this dependence on exposure time, CW (continuous-wave) radiation with its longer exposures will almost certainly affect the choroid, according to Ham. With Q-switching and multiple-spiked operation, which usually involves less than 1-msec exposure times, only the pigment epithelium will be involved.

The theory of thermal injury developed by J. J. Vos and others indicates that—for pulsed-laser exposure times of less than 1 msec—the choroid temperature will not add any significant heat flow to the pigment epithelium. "Because most laser pulses rarely exceed 1–2 msec", Ham concludes, "it follows that only the pigment epithelium will be involved for mild lesions. But for exposure times above 10 msec, there is time for heat flow from the choroid to make appreciable contributions to the pigment epithelium. Histological evidence confirms the involvement of the choroid as well as the pigment epithelium for mild lesions produced by long exposure times".

The power density of the beam is the third major cause of irreversible damage to the retina. For long exposure times Ham has shown that the power density needed to produce mild lesions becomes almost constant, defining a limiting retinal irradiance for these lesions. This implies that if one maintains above-ambient temperature in retinal tissue for long periods of time, one will not produce visible tissue coagulation. "We produced a mild lesion on an 800-micron-dia. spot size in 3 min with an irradiance at the retina of 6 W/cm^2," Ham said. "Perhaps somewhere below this it might be permissible to expose the eye continuously".

In the case of the Q-switched lasers which involve very high power density (MW/cm^2) and short exposure times (5–50 msec), Ham maintains that a simple thermal concept of retinal injury is inadequate to explain the biological effects that are induced. Under these conditions, he says, physical effects play a more important role. For example, the pigment granules or absorbing sites in the pigment epithelium would reach high temperatures during the early stage of a giant pulse, producing ionization and possibly a plasma. A number of processes including two-quanta excitation, self-focusing Raman and Brillouin scattering, shock waves, frequency doubling and re-radiation by black-body emission must be considered as possible modes of energy dissipation that can occur before heat conduction has time to play a prominent role.

A further complication, according to Ham, involves applying the classical temperature concept to time intervals of 10^{-8} sec and macromolecules (proteins and nucleic acids). Thermal "after effects" may account for the observed phenomena at relatively low power densities—but not at power densities above about 20 MW/cm^2.

Histological evidence indicates that for mild lesions induced in the rabbit retina by Q-switched ruby pulses with an image diameter of 800 microns (3–5 MW/cm^2), damage is confined to the pigment epithelium and the inner layers of the retina; subretinal haemorrhaging occurs in the region between the pigment epithelium and the choroid. When power densities are increased still further (20–50 MW/cm^2), material is extruded into the inner retina and the vitreous humour. At these power densities, haemorrhaging into the vitreous humour occurs—which may result in loss of the eye. Ham points out that these observations suggest that shock waves are a possible mode of biological damage to the retina.

Just what the actual effects on vision of these minimal lesions are has not been completely evaluated yet because the number of cases is still small. Zaret feels that if they occur in the periphery, they are relatively insignificant. However, in the macular region of the eye there might be some significance. And if a lesion is located in the neural pathways from the macular to the optic disk, results could be severe.

Because of the importance of these minimal lesions to sight, the question of "thresholds"—just what is the lowest level of biological change to the retina one should use as the borderline between normal retina and damage—is a very real one; and it is one on which there is no general agreement.

For example, Geeraets points out that the minimal ophthalmoscopic visible lesions are not true "threshold" lesions because more sensitive measuring devices have detected smaller biological changes following laser exposure. "If we use, for instance, histochemistry." He said, "in which we are not looking for ophthalmological changes, but for enzyme inactivations, we shall find that our threshold can be lowered by about 10 or 15 per cent."

Geeraets also mentions *electroretinagram* (*ERG*) tests in which the electrical response of the retina is measured under certain light stimulation conditions. "Many people take the amplitude of ERG to be equal to visual function," he says. "That has never been proved and depends on many factors— and the animal cannot say if he has a visual loss or not. So we can only make the statement that permanent ERG changes can be produced with energies that are 50 per cent below the threshold of ophthalmologic visible lesion in our laboratory."

On defining a safety level—a maximum tolerance—that it would be reasonable to adopt, Geeraets feels that his criterion would be based 50 per cent below permanent ERG changes which is about 75 per cent below the data points given by Ham. Zaret agrees with this thinking—and would put the minimum safety level an order of magnitude below the level of radiation needed to produce an injury "of any sort, no matter how subtle the techniques must be to elicit it."

Ham, on the other hand, is a "little more radical." "I think our curves for minimal lesions should be divided by a factor or two for military situations," he said, "and for those people who want to plan industrial and laboratory safety, I'd use a factor of 10."

The effects on other tissues. (*a*) *Acu teeffects.* Whatever the consensus for "minimum" levels of laser energy required to damage the eye, some of the "macro" effects of the laser—including their lethal effects—are clear from the work with mice and large animals of Edmund Klein and Samuel Fine.

"Most of our observations deal with direct and gross effects of laser radiation," Klein says, "and most of them are confined to the events subsequent to a single exposure. Furthermore, most of what we know deals with the immediate effects; we just have hints as to what might happen as delayed phenomena or as a chronic sequelae, and we don't even have hints—only perhaps hunches—about what may happen as a result of cumulative exposures." The effect of lasers on mice are largely non-linear, Klein points out, and an extrapolation from relatively low levels of energy and power in the mouse to higher energy and power levels pertaining to larger mammals such as man is not entirely appropriate.

While thermal effects—as Ham pointed out—are important in producing damage, Klein feels that secondary effects are equally so; these he summarizes as "pressure and possibly shock-wave generation at the time of impact, photosensitizing effects, photoactivating effects and a motley assortment of phenomena known as photobiological effects." He bases his feeling that secondary effects are important on the fact that when mice are exposed to intense laser beams, particulate matter, which travels at initial velocities as high as 20,000 ft/sec is ejected. Some of the particles may be representative or associated with free radicals, Klein indicates, and certain non-linear effects may occur, including frequency multiplication, scattering and "possibly some other phenomena."

When the surface of the mouse was irradiated by 30–50 J/cm^2 the mouse surface was extended into a hemispherical balloon from which emanated, primarily in the direction of the incident radiation, a plume containing the backscattered radiation and particulate matter. "Clearly this phenomenon, while it contains a thermal component, is going to have effects other than those which could be produced by the application of heat *per se*," says Klein.

This observation is backed up by the case of a mouse which was hit in the head with 50 J/cm^2. Only a little discoloration showed on the animal's forehead, but as the skin was peeled back, considerable haemorrhaging was found throughout the brain itself. However, the bones, skin and muscles for the most part remained intact. "Obviously, this is not a direct thermal effect," says Klein, "or else you would have a hole burned through them."

The non-thermal components are not confined to the head alone. Klein finds that mice hit in various parts of the body also suffer massive internal haemorrhage although on the exterior there is only a bluish discoloration of the skin.

Surprisingly, even with extensive damage to their bodies, mice that were irradiated other than in the head (where neurological defects occurred) often did not die and could produce normal progeny. For example, some anatomical sites where important structures such as nerves, blood vessels or important viscera come close to the surface are more vulnerable to any kind of trauma. Yet even in these areas, rather extensive laser injury seemed surprisingly to be compatible with life according to Klein. For example, when an anterior aspect of the neck was rather extensively injured, it did not seem to interfere with the animal's well-being or survival, despite some obvious source of discomfort, and within 3–7 days it had regressed completely; the lesion simply healed over and the animal was fine. "Needless, to say," Klein continues "if the energy levels are higher—about 200 J/cm^2—then the effects will be proportionately more severe, and it will either result in permanent paralysis or death."

(*b*) *Chronic effects.* In contrast to these acute effects, Klein also hints at some of the chronic, or delayed, effects that one might experience from laser irradiation. For example, Klein and his group irradiated a melanoma and collected some of the black material that came off the pigmented tumour in the plume. They found that this material grew when it was transplanted into another mouse, indicating that it is viable. "Certainly," says Klein, "the consideration arises that just as it is ejected by the laser radiation, it might be injected, thus leading both to local dissemination and—if it happens to enter the circulatory or lymphatic system—distant dissemination and metastasizing of the tumour."

Other interesting observations about the transplanted black tumour material, according to Klein, include the fact that "it is no longer black; it may be non-pigment forming." Furthermore, he says, unlike the melanoma from which it was derived (which kept growing until it killed the mouse), "the derived tumour following laser radiation was much more slowly growing, and over a period of 6 months of observation in about 10 animals has not yet killed a single one of them. It nevertheless is a tumour. Now this indicates that not only is viable material retained in the product of pulse radiation, but that the material may undergo a genetic change.

Klein's group has irradiated about 3000 mice at various levels of energy and power, ranging from millijoules to megajoules and from kilowatts to 500-megawatts. Over a two-year period, Klein says, they have seen no higher increase of cutaneous neoplasm—benign or malignant—than they would see in the ordinary population. However, about a 10 per cent higher incidence of spuamosal carcinoma was found if the animals were treated with an inflammatory agent such as Croton oil. "Of course," says Klein, "we can't extrapolate from the mouse to man as far as carcinogenesis is concerned."

(*c*) *The question of a threshold.* Klein also discusses thresholds, which he defines as the amount of irradiation producing the minimal biological change (grossly or microscopically observable) that can be observed in the living animal. The smallest such change he has been able to find, he says, is temporary pigment arrestment.

For example, a white spot was produced on a mouse by 0·5 J/cm^2 at a wave length of 6·943 Å. This showed that two biochemical processes were differently affected (pigment formation and cellular multiplication) since hair growth at the site was perfectly normal, indicating that the exquisitely sensitive process of cellular multiplication was retained. However, pigment formation had only been inhibited, not destroyed, because when the new hair cycle started pigment formation was normal; thus this is a reversible process. "To our knowledge," says Klein, "this represents the most sensitive indicator of laser radiation."

Some corresponding thresholds at other wavelengths are 0·12 J/cm^2 at 5300 Å (*Q*-switched at a power level of 2 megawatts) and 0·9 J/cm^2 at 10,600 Å (non *Q*-switched).

To produce histological changes it has been found that the energy density required is greater than that for threshold effects by about an order of magnitude. Gross changes are produced by energy densities about twice as great as those that will produce histological changes.

Bibliography

Ham W. T. Effects of laser radiation on the mammalian eye, *Proceedings of the Seminar on Laser Safety, May, 1966*, published by Martin Co., 1967, for the U.S. Army Office of the Surgeon General.

Klein E. Injurious effects of laser radiations on mammals, *Proceedings of the Seminar on Laser Safety, May, 1966*, published by Martin Co., 1967, for the U.S. Army Office of the Surgeon General.

Vos J. J. (1962) A theory of retinal burns, *Bull. Math. Biophys.* **24**, 115.

J. Thewlis

LIGAND FIELD THEORY. Ligand field theory is the theory of the origins of and the consequences of the splitting of inner orbitals of ions by their environment in chemical compounds. As such, it is of major importance in the theoretical treatment of transition metal complexes. The foundations of the theory were laid in 1929 by Bethe, in a paper concerned with *crystal field* theory; this latter theory differs from ligand field theory in that it only credits ligands in complexes with the ability to produce an electrostatic field about the metal ion, whereas in ligand field theory, molecular orbital considerations are included in the treatment.

Bethe indicated that in a chemical environment the degeneracy of atomic orbitals may be removed, the degree of removal being a function of the symmetry of the environment. In particular, if an octahedral complex is formed by bringing six identical ligands along the *x*, *y*, and *z* axes to form a regular octahedron about a metal ion, the five d-orbitals on the latter do not remain degenerate (i.e. of equal energy), but split into a set of three, conventionally labelled the d_{xz}, d_{yz}, d_{xy} orbitals, or the d_ε (Van Vleck), t_{2g} (Mulliken) or γ_5 (Bethe) orbital sets, and a set of two, conventionally labelled the $d_{x^2-y^2}$ and d_{z^2} orbitals, or the d (Van Vleck), e_g (Mulliken) or γ_3 (Bethe) orbital sets. In octahedral complexes, the d_ε set lies lower in energy, but in tetrahedral complexes the reverse is true. This conclusion is suggested by the electrostatic model for an octahedral complex; in the d_ε set of orbitals, the lobes of electron density point between the axes, whereas in the d_γ set the lobes point along the axes. In this model, an electron in an orbital whose lobes point at the ligands (represented as point negative charges) will be destabilized with respect to one in an orbital whose lobes point between the ligands.

However, on both theoretical and experimental grounds the electrostatic model is known to be inadequate. In particular many lines of experimental evidence indicate that electrons which should be entirely in d-orbitals according to crystal field theory, actually spend part of their time in orbitals associated with the ligand atoms, i.e. in molecular orbitals encompassing both the metal and the donor atoms.

Moreover the differentiation of the d-orbital energy levels by the crystal field is a comparatively small effect superimposed on the much larger repulsive effect arising from the spherical components to the field. These repulsions are related to the lattice energy in the case of crystals or to the ligation energy in the case of complex ions. For example, the $d_\varepsilon - d_\gamma$ energy difference (measured in terms of the equivalent parameters Δ, 10 Dq or $E_1 - E_2$) is 57 kcal/mol (20,000 cm^{-1}) in the ion $Ti(H_2O)_6^{3+}$, whereas the hydration energy of the ion is ~1000 kcal/mol. It may be shown that the stabilization of the d_ε set is 4 Dq whereas the destabilization of the d_γ-set is 6 Dq (see figure).

Electrons tend to occupy the more stable d-orbital set preferentially, subject to the restrictions of the Pauli exclusion principle and interelectronic repulsions. The first principle holds in the same way as it does for free atoms, whereas the latter may be understood as follows. If a second electron is placed in an orbital which is already occupied instead of in an unoccupied orbital, then a certain interelectronic repulsion energy, known as the *pairing energy* (P) must be overcome. For octahedral complexes containing 1, 2, or 3 d-electrons, the electronic configuration of the ion is said to be t_{2g}^1, t_{2g}^2 or t_{2g}^3. If the octahedral complex ion has from 4 to 7 d-electrons, then two possible electron configurations arise in each case, a spin-free and a spin-paired one, expressed as $t_{2g}^3e_g^1$ or t_{2g}^4, $t_{2g}^3e_g^2$ or t_{2g}^5, $t_{2g}^4e_g^2$ or t_{2g}^6, $t_{2g}^5e_g^2$ or $t_{2g}^6e_g^1$ (see Table). The configuration actually adopted in any case is determined by whether Δ is less than or greater than P; e.g. if $\Delta < P$ for a d^4 complex ion, then the spin-free configuration $t_{2g}^3e_g^1$ is adopted, as in salts of chromium (11).

By similar arguments, the electron configuration of tetrahedral complexes can be expressed, although it is important to note (a) that the d_γ orbitals lie lower than the d_ε ones in this case, and (b) the magnitude of Δ_{tet} is much less than that of Δ_{oct} for a given metal ion and set of ligands. A consequence of (b) is that no spin-paired tetrahedral complex ions are known.

The magnetic properties of $d^4 - d^7$ complex ions are obviously dependent on whether $\Delta < P$ or $\Delta > P$, i.e. if the complexes are high spin, then there are 4, 5, 4 or 3 unpaired electrons respectively, whereas if they are low spin, then there are only 2, 1, 0 or 1 unpaired electrons respectively.

The fact that for most transition metal complexes, the d-electrons do not occupy all five orbitals with equal probability but instead tend to occupy the more

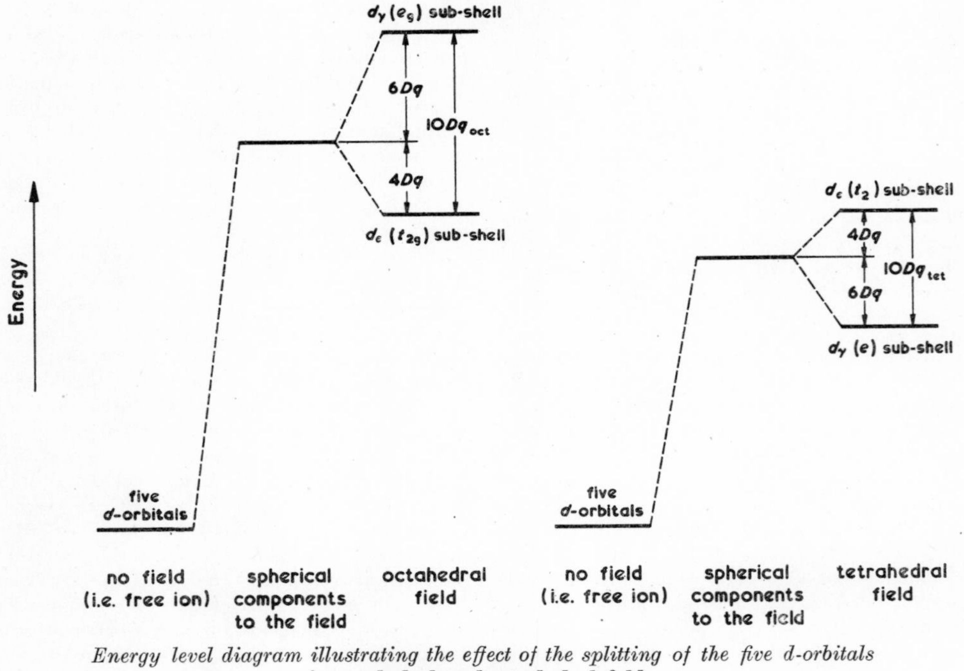

Energy level diagram illustrating the effect of the splitting of the five d-orbitals in octahedral and tetrahedral fields.

Octahedral ligand field stabilization energies

Electronic configuration	Examples	Electronic distribution					
		Weak field spin free			Strong field spin paired		
		t_{2g}	e_g	L.F.S.E.*	t_{2g}	e_g	L.F.S.E.*
d^0	Ca^{2+}, Sc^{3+}	0	0	0	0	0	0
d^1	Ti^{3+}, V^{4+}	1	0	0·4	1	0	0·4
d^2	V^{3+}, Cr^{4+}	2	0	0·8	2	0	0·8
d^3	V^{2+}, Cr^{3+}	3	0	1·2	3	0	1·2
d^4	Cr^{2+}, Mn^{3+}	3	1	0·6	4	0	1·6
d^5	Mn^{2+}, Fe^{3+}	3	2	0	5	0	2·0
d^6	Fe^{2+}, Co^{3+}	4	2	0·4	6	0	2·4
d^7	Co^{2+}, Ni^{3+}	5	2	0·8	6	1	1·8
d^8	Ni^{2+}	6	2	1·2	6	2	1·2
d^9	Cu^{2+}	6	3	0·6	6	3	0·6
d^{10}	Cu^+, Zn^{2+}	6	4	0	6	4	0

* In units of Δ_{oct}.

stable d_ε set in preference to the d_γ set (in the octahedral case) has important energetic and structural consequences. In the absence of ligand field effects, it would be expected for a given series of metal complexes (e.g. for Ca^{2+} ... Zn^{2+}) that there would be a fairly smooth variation of bond energies from the d^0 to the d^{10} ion. This is because both the energy and the size of the orbitals could be expected to be a monotonic function of the nuclear charge. However, while the number of d-electrons is increasing steadily from Ca^{2+} ... Zn^{2+}, their distribution within the t_{2g} and e_g subshells is not uniform. The increase in stability of the high-spin complex ions arising from the preferential filling of the more stable t_{2g} sub-shell, known as the *ligand field stabilization energy* (*L.F.S.E.*), is given in the Table. Different L.F.S.E. apply to low- and high-spin octahedral complexes, and for complex ions of different symmetries, e.g. in the spin-free (i.e. weak field) octahedral cases, maximum L.F.S.E. arises for the d^3 and d^8 configurations, whereas in the spin-paired octahedral cases, maximum L.F.S.E. arises for the d^6 configuration. In general it is found that metal-ligand bond energies do not vary smoothly from d^0 to d^{10} complexes unless a correction for L.F.S.E. is made.

The magnitude of Δ may be estimated from electronic spectral data and in the case of divalent transition metal ions, Δ amounts to some 10,000 cm^{-1} or 30 kcal/m. Subtraction of the correct L.F.S.E. for each ion from say the hydration energies of the bivalent ions of the first transition series (which amount to some 500–800 kcal/m) leads to a series of corrected hydration energies which then vary smoothly as expected from Ca^{2+} through to Zn^{2+}. Similar results are obtained for lattice energies and heats of ligation of complex ions.

The most obvious stereochemical consequences of L.F.S.E. lie in the irregular variations of the ionic radii of bivalent and tervalent ions of the first transition series with increasing atomic number. The radii would be expected to decrease monotonically from Ca^{2+} to Zn^{2+} owing to the steady increase in nuclear charge. However, as indicated above, there is appreciable L.F.S.E. for certain ions and in particular for the d^3 and d^8 spin-free octahedral species. As a consequence, the ionic radii of the d^1, d^2, d^3, d^4, d^6, d^7, d^8 and d^9 ions are all less than those expected from the smooth curve joining the radii for the d^0, d^5 and d^{10} ions (for which there is no L.F.S.E.). In molecular orbital language, the t_{2g} electrons are *non-bonding* to a first approximation, whereas the e_g electrons are *anti-bonding*, i.e. they tend to weaken and thus lengthen metal-ligand bonds.

One of the most important properties of transition metal ions is their colour, i.e. the intensities and frequencies at which electronic transitions occur from the *ground electronic state* (which is related to the electron configuration of the ion) to various *excited states* (which may be related to different electron configurations for the ion). The simplest case is that of the reddish-purple $Ti(H_2O)_6^{3+}$ ion, for which the absorption band at $\sim$20,000 cm^{-1} ($\varepsilon = 4$) has been identified with the $e_g^1 \leftarrow t_{2g}^1$ electronic transition. The electronic spectra of the $d^2 - d^9$ ions are more complicated, but the absorption bands can still be related to the parameter Δ. Analyses of a large number of spectra of various ions have permitted the following conclusions to be made regarding the magnitude of Δ:

(i) For a given set of ligands, Δ_{oct} is $\sim$45 per cent larger for second-row transition metal ions and $\sim$75 per cent larger for third-row transition metal ions than for the first-row transition metal ions of the same periodic group and oxidation state, e.g. for the ions

$Co(en)_3^{3+}$, $Rh(en)_3^{3+}$ and $Ir(en)_3^{3+}$, $\Delta = 23{,}200$, $34{,}600$ and $41{,}400$ cm^{-1} respectively (en = ethylenediamine, a bidentate nitrogen-donor ligand). As a consequence, second- and third-row transition metal complexes are normally low-spin.

(ii) Δ_{oct} is 40–80 per cent larger for tervalent than for bivalent ions of the same metal, and ~100 per cent larger for quadrivalent ions—the same set of ligands being assumed in each case.

(iii) Δ_{tet} for four tetrahedrally-disposed ligands coordinated to a given metal ion is ~40 per cent of Δ_{oct} for the same ligands octahedrally coordinated to the same metal ion. Δ_{tet} varies with the oxidation state of the metal in the same way as Δ_{oct}.

It is found that ligands may be arranged in order of the Δ_{oct} values for their complexes with almost any given ion. This series of the ligands is known as the spectrochemical series, and the order of increasing Δ_{oct} values is as follows for the commonly occurring ligands: $I^- < Br^- < Cl^- < F^- < H_2O \sim$ other oxygen-donor ligands $< NH_3 \sim$ other nitrogen-donor ligands < ethylenediamine $< CN^-$ (the value for I^- being a little less than half that for cyanide). Thus the first ligand field band of the $CrCl_6^{3-}$ ion (which is a measure of Δ_{oct} for Cl^- coordinated to Cr(111)) occurs at 13,200 cm^{-1}, whereas for the $Cr(H_2O)_6^{3+}$ ion the corresponding band occurs at 17,400 cm^{-1}. Chloride is therefore said to lie lower in the spectrochemical series than water.

Essentially, the position of a ligand in the spectrochemical series parallels the order of decreasing radii of the donor atom, i.e. $I < Br < Cl < S < F < O < N < C$. However, other factors such as the π-electron donor or acceptor properties of the ligand affect the position of a ligand in this series, particularly in the case of sulphur-donor ligands. Moreover the position of a ligand in the spectrochemical series is not determined by whether it is anionic or neutral. It must be emphasized that Δ is principally used as an empirical parameter, because even the most sophisticated attempts at *a priori* calculation of its magnitude, using *Hartree-Fock orbitals* for the metal and the ligand, have met with only limited success.

Another property of a ligand of interest in ligand field theory is its position in the *nephelauxetic series*. The nephelauxetic ratio β is the ratio of the value of a representative parameter of interelectronic repulsion in a complex to its value in the free gaseous ion. The nephelauxetic series of ligands is quite different from the spectrochemical series of ligands and is as follows, in order of decreasing values for β: $F^- > H_2O > NH_3 > NCS^- > Cl^- > CN^- > Br^- > I^-$. Broadly speaking, the order of ligands in the series reflects the higher deformability (or polarizability) of the ligands to the right in the series, and hence the greater degree of covalency of the metal ligand bond, i.e. fluoride is recognized as forming the least covalent bonds to a metal ion. With respect to the metal ion in a given oxidation state within a periodic group, the parameter β for a given set of ligands increases in the order first-row transition metal < second-row transition metal < third-row transition metal.

Recent electron spin resonance (ESR) and nuclear magnetic resonance (NMR) studies have also shown that appreciable delocalization of the central metal's d-electrons into the ligand orbitals occurs.

Distortions of regular octahedral or tetrahedral complexes lead to important changes in the energy levels of ions. This may affect the number of unpaired electrons and hence the magnetic moment of a complex. Most obviously, however, distortion of an ion from cubic symmetry increases the number of absorption bands in the electronic spectrum of the complex through a lifting of the degeneracies of certain of its excited states. The most common distortion of octahedral and tetrahedral complexes is tetragonal, i.e. the complex ion is distorted along a four-fold symmetry axis. The limiting case of a tetragonal distortion of an octahedral complex is the complete removal of the ligands along the four-fold axis, i.e. the complex is now square planar. Extensive magnetic and spectral work has been carried out on such complexes, and also on complexes of other symmetries such as trigonal bipyramidal, square pyramidal, square antiprismatic and dodecahedral.

Ligand field spectra are usually determined in solution, but may also be carried out on solid materials by the technique of diffuse reflectance spectroscopy, or by polarized single crystal spectroscopy. The diffuse reflectance from a very fine powder consists of radiation which has penetrated the very fine crystallites and therefore lost intensity according to the Lambert law, $\log_{10} I_0/I = \varepsilon cd$, and reappeared at the surface after multiple scattering. Thus the electronic spectra of materials which are either insoluble in all solvents or soluble only with decomposition, may also be determined. The ligand field also removes the degeneracy of f-orbitals in lanthanide and more especially in actinide ions. The magnitudes of f-orbital splittings are not as great as those of d-orbital splittings, because electrons in f-orbitals are better shielded from the ligands.

Ligand field theory has therefore important consequences in magnetism and in the electronic, infrared and ESR spectra of complexes. It also has a bearing on the rates of chemical reactions, on redox reactions and in the photochemical and catalytic properties of ions.

Bibliography

BALLHAUSEN C. J. (1962) *Introduction to Ligand Field Theory*, New York: McGraw-Hill.

BROWN D. A. (1961) in *Encyclopaedic Dictionary of Physics* (J. Thewlis Ed.), **1**, 321, Oxford: Pergamon Press.

CARLIN R. L. (1963) *J. Chem. Ed.*, **40**, 135.

CLARK R. J. H. (1964) *J. Chem. Ed.*, **41**, 488.

COTTON F. A. (1963) *Chemical Applications of Group Theory*, New York: Wiley.
COTTON F. A. (1964) *J. Chem. Ed.*, **41**, 466.
DUNN T. M., MCCLURE D. S. and PEARSON R. G. (1965) *Some Aspects of Crystal Field Theory*, New York: Harper and Row.
FIGGIS B. N. (1966) *Introduction to Ligand Fields*, New York: Wiley.
GRAY H. B. (1964) *J. Chem. Ed.*, **41**, 2.
JØRGENSEN C. K. (1962) *Absorption Spectra and Chemical Bonding in Complexes*, Oxford: Pergamon Press.
SUTCLIFFE B. T. (1967) in *Encyclopaedic Dictionary of Physics* (J. Thewlis Ed.), Suppl. Vol. 2, 171, Oxford: Pergamon Press.

R. J. H. CLARK

LINEAR MOTORS. A linear motor may be loosely defined as an electromagnetic machine designed to produce force and/or motion in a straight line. An easy way of describing such a motor is to say that it is like a conventional rotary machine which has been cut along one side and unrolled as shown in Fig. 1.

Just as the most widely used conventional motor is the induction machine, so in linear form, the induction type has found most applications, and for basically the same reason, namely that the secondary member is fed with current by induction so that neither electrical nor mechanical contact with it is essential. The secondary member can consist in its simplest form of a solid piece of metal in which form it is extremely robust. In the conventional motor the secondary member or *rotor* is usually of "cage" construction, the main body consisting of axially-slotted, laminated steel with conducting bars filling the slots, as shown in Fig. 1(a). All the conducting bars are connected together at each end of the machine by thick conducting rings. The conductor is usually copper or aluminium and it is common practice to cast the bars and end rings into the rotor at a single operation, often incorporating fan blades at one end for ventilation purposes.

Examination of Fig. 1(b) shows that the "unrolling" process has produced essential differences, which must influence the design of linear machines to a great extent. For example, any relative movement between primary and secondary members will result in portions of each becoming magnetically open-circuited and if the motion is continued the stator and 'rotor' will soon part company entirely. (The word 'rotor' is retained for the secondary member in the linear machine since no ambiguity is thereby introduced and a new word therefore seems unnecessary.)

For linear motors designed to produce motion it is therefore necessary to extend either the rotor or the stator member, and such extension enables linear machines to be classified into two groups depending on which of the members is extended. Figure 2 illustrates the difference between the two. Both types are

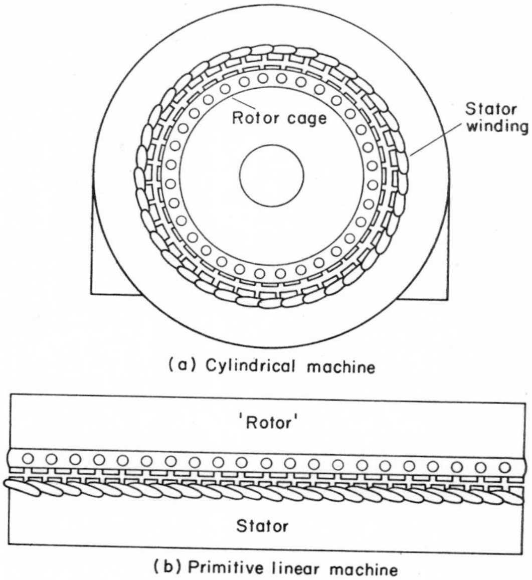

Fig. 1.

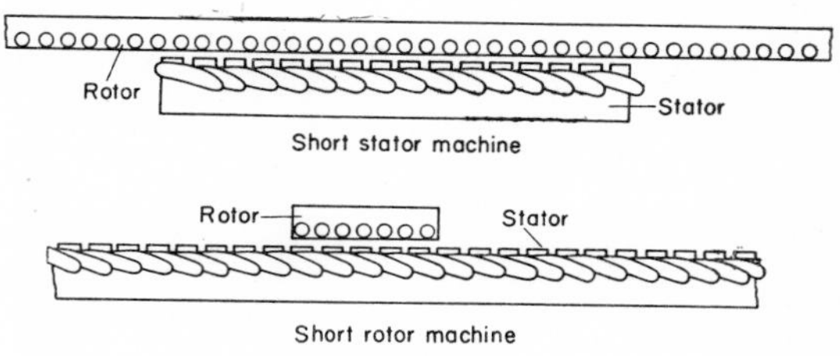

Fig. 2.

clearly wasteful of material. The active zone of each machine is confined to those regions in which both members face each other so that all other parts represent extra capital cost above that of the corresponding conventional machine. The short-rotor machine is more expensive than the short-stator motor since the inactive parts of the former consist of insulated windings in slots, whereas the simple cage construction of an elongated rotor is cheaper to produce. In addition, the inactive parts of a short-rotor machine must be fed with current unless some form of switching is incorporated which can be arranged to cut off the feed to most of the inactive zone. In the latter case the switching arrangements will also add to the cost of the system.

Either type can be used with the roles of rotor and stator interchanged, i.e. with the stator as the moving member. A disadvantage of this method of operation in induction machines is that the power must then be fed to the moving member requiring brushes and current feeders as are required with all d.c. systems. However, the inverted rotor/stator arrangement can be advantageous with induction machines in which the motion is required over very long distances, since the rotor is cheaper to produce.

A second difference between rotary and linear machines arises as the result of purely *magnetic* forces between primary and secondary members. Electromagnetic (induction) forces can be likened to shearing stresses in mechanical systems, the forces acting in a direction perpendicular to the principal direction of the magnetic field. Purely magnetic forces, within this analogy, represent compressive stresses, acting *along* the magnetic field, and in the case of linear motors, tending to pull the two halves of the motor together. In rotary machines the magnetic pull is largely balanced out around the periphery by the cylindrical symmetry.

Sheet rotor machines. Substantial modifications can be made to the design of linear motors for specific purposes in order to eliminate or at least minimize the disadvantages arising for both of the reasons given above. In particular a type which has become known as the "sheet rotor" motor is free from magnetic pull and uses a very cheap form of rotor, making the short stator machine attractive even for systems with very long rotors. The several stages of development which lead to this type are illustrated in Fig. 3. The induction machine is again used as an example.

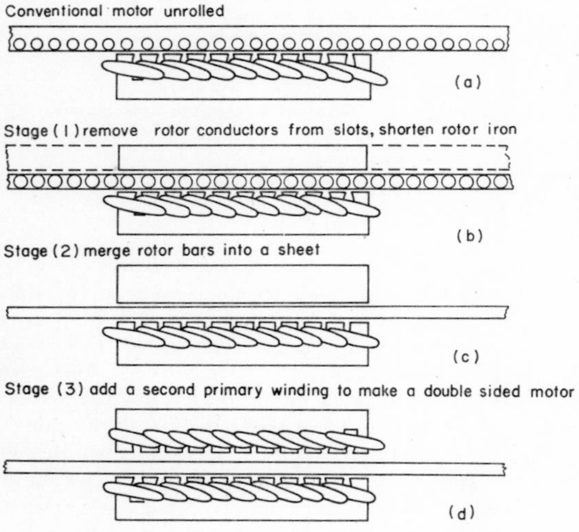

Fig. 3.

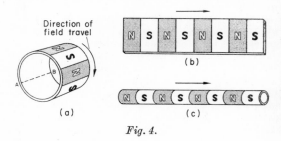

In the counterpart of the conventional machine shown at (a), the force is produced on the steel teeth of the rotor. Removal of the conductors from the slots as shown at (b) transfers the force from the steel to the conductors themselves. The purpose of the steel is now only the closure of the magnetic circuit and it is therefore no longer necessary to extend this steel to the full rotor length, since it is not required to move with the conductor. When the rotor conductor is entirely contained within the airgap it is not necessary to divide it into bars and end-rings. The construction can be simplified to a single sheet as shown at (c). Since the steel closing the magnetic circuit is a part of the stator it can be fitted with a second primary winding, as shown at (d). The benefits from the use of a second winding may be used in a variety of ways, as follows:

(i) The second winding may be housed in the same size of slots as those of the first, thereby enabling the use of *twice* the current loading from virtually the same size and weight of motor. The gain in this case is in *power/weight ratio.*

(ii) Each winding may be run at one half the current density of that of a single winding alone. The output remains the same but the stator I^2R loss is halved. Here the gain is in *efficiency.*

(iii) Addition of a second winding enables the slot depth of each winding to be reduced to one half of that of a single winding for the same output. This reduces stator leakage flux and results in improved *power factor.*

The designer may choose to use a compromise design to obtain reduced benefits but to obtain them in more than one aspect.

Tubular motors. Another useful form of linear machine is the tubular or "axial flux" motor. The development of this type is illustrated in Fig. 4. The rotating magnetic field pattern of a conventional 4-pole machine is illustrated by the paper cylinder at (a). When cut along the line AB and unrolled, as at (b) it represents the field of a flat linear machine. If re-rolled about a perpendicular axis, as shown at (c) it represents the field of a tubular motor in which the field pattern travels along the axis of the cylinder. In the double-sided flat machine of Fig. 3(d)

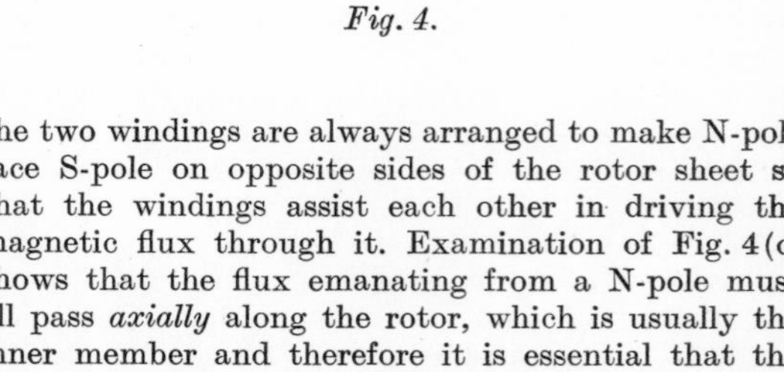

Fig. 4.

the two windings are always arranged to make N-pole face S-pole on opposite sides of the rotor sheet so that the windings assist each other in driving the magnetic flux through it. Examination of Fig. 4(c) shows that the flux emanating from a N-pole must all pass *axially* along the rotor, which is usually the inner member and therefore it is essential that the rotor member contains a ferromagnetic core, otherwise the constriction of the magnetic field in the bore of the stator leads to an unacceptably high magnetizing current.

One of the principal advantages of the tubular motor is that constructionally it can be very simple. The stator windings of flat linear motors are similar to those shown in developed diagrams of conventional motors. An example is shown in Fig. 5. When the system is rolled up so that P falls on R and Q on S, the active currents need no longer be connected together by the wasteful end windings but can circulate in simple coils. The stator winding of a tubular motor therefore consists of a single row of identical circular coils on a common axis. The rotor may, in its simplest form, consist of a solid mild steel rod inside a copper or aluminium sleeve.

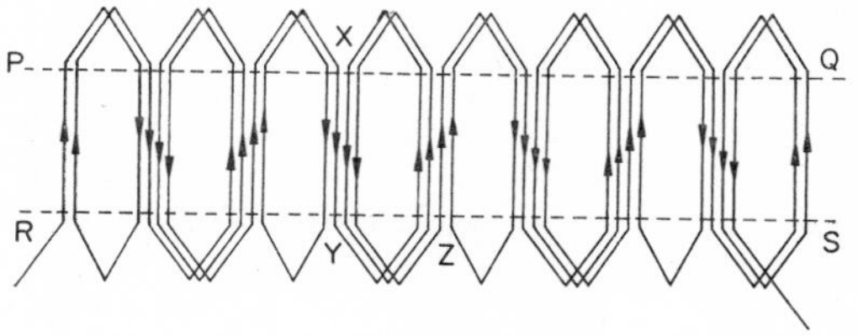

Fig. 5.

Characteristics of linear motors. The characteristics of any type of linear machine can generally be deduced from those of the corresponding rotary machine. The following important differences should, however, be noted.

Whereas the speed of rotation of a conventional induction motor field is usually expressed in rev/sec as the supply frequency divided by the number of pairs of poles, the synchronous speed of a linear motor is most conveniently expressed in ft/sec or m/sec as twice the pole pitch (in feet or metres) multiplied by the frequency. Since the periphery of a rotary machine is given by twice the pole pitch multiplied by the number of pole pairs, the above two results are consistent.

The technique of removing secondary conductors from slots and placing them in the airgap as described with reference to Fig. 3 incurs the penalty of a larger airgap and therefore a larger magnetizing current than is required in the conventional machine. The extent to which this is important depends on the factors which determine the ratio of magnetizing current to load current. Generally, the most potent of these factors is the pole-pitch and one machine having twice the pole-pitch but four times the airgap of another will have the same ratio of magnetizing current to secondary current if all other dimensions of the two machines are identical.

In linear machines in which magnetizing current is not dominant the running characteristics are also modified from those of conventional machines by the transient secondary currents which are introduced at the edges of the rotor or stator, whichever is the shorter member. The effects of these transients in an induction motor are dependent on whether the coils of each phase of the primary winding are connected in series or in parallel. Generally it is preferable to connect a short rotor machine in parallel in which case the effect of the edge transients is to give the rotor conductor an apparently higher value of resistivity than its physical properties indicate. Such increase depends on rotor length but for rotors greater than 2 pole-pitches long, the effect is negligible.

Short stator induction machines should generally be series-connected, in which case the effect of the edge transients is to produce excess losses for speeds greater than $n/(n + 1{\cdot}5)$ times the synchronous speed where n is the number of poles along the length of the stator. This implies that the efficiency of a short stator machine is always less than $n/(n + 1{\cdot}5)$ and that highly efficient machines must contain large numbers of poles.

Applications of linear motors. Linear motors may be grouped so far as application is concerned into three classes as follows:

(*a*) *Power machines.* This class is required to have high efficiency and power factor and to handle large power. They are necessarily high speed machines and are unlikely to be designed for operation at frequencies less than 50 c/s.

(*b*) *Energy machines.* These machines are primarily *accelerators.* They may produce large forces at standstill and attain high velocities but energy efficiency and energy/weight ratio are more important than efficiency and power/weight ratio. Their main advantage is in their compactness and relatively low cost compared with gravity-fed or pneumatically-driven impact machines for forging, blanking, testing etc. The energy efficiency of an induction accelerator may be improved by increasing the field speed as the missile accelerates. This can be done either by a graded pole-pitch scheme or by using a variable frequency programme.

(*c*) *Force machines.* For low-speed or standstill operation efficiency is of no importance. The design criteria include such quantities as force/input, force/weight and force/cost. Electromagnetic machines may produce about 6 lbf/in^2. of pole face for continuous rating and up to 50 lbf/in^2. for a few seconds. These values are very small compared with those obtainable by hydraulic or pneumatic rams. Induction-type actuators produce force for about 30–40 W/lb.f when designed for 50 c/s supply but this figure can be reduced below one half these values by using much lower frequencies, of the order of 5 c/s.

The linear motor actuator is therefore mainly applicable where only small forces but long strokes are required and where the advantage of requiring no additional apparatus such as pump, sump, etc, and the ease with which power can be fed to the actuator, along flexible leads if required make it economically superior to the hydraulic system. It has been used for such purposes as parcel sorting, sliding door and valve operation and as a travelling crane drive. Tubular motors are popular as force machines.

Special applications. In some cases a single flat stator unit may be used without closing the magnetic circuit. Such units may be designed with sheet rotors to produce, in addition to the tangential thrust, either a lifting force or an attractive force on the rotor sheet and this type of motor has been used for unloading conveyor belts with the motor entirely beneath the belt.

Double-sided motors can be used to drive large conducting disks at relatively low speeds. Absence of gearing and therefore of noise and backlash may be attractive features. The motors can be designed for 2-phase operation for position control.

Both tubular and flat machines have been used as liquid metal pumps, for linear machines will operate with liquid rotors. Both a.c. and d.c. types have been used to pump liquid sodium and potassium in connexion with nuclear reactor projects. An open-sided machine has been used in the U.S.S.R. for pumping liquid steel.

Self-oscillating systems may be set up by arranging two linear induction motors back-to-back so that their travelling fields are directed towards the centre of the track. A short rotor will build up oscillations and reach a stable amplitude depending on its inertia and the electrical parameters of the system. The efficiency of the system is low but can be improved by fitting mechanical end-springs to limit the amplitude.

The history of linear motors. The earliest patent on linear machines appears to have been granted to the Mayor of Pittsburgh in 1890 for a linear induction motor. Between that date and 1940 there are numerous patents on the application of linear induction motors to shuttle propulsion in weaving looms. The first proposal to use a linear motor to propel trains was made in 1905 although it was then proposed to fit the stator to the track, a scheme which could only hope to be economically successful in very high traffic densities. In 1914 M. Bachelet proposed a combined induction propulsion and levitation scheme both for shuttle propulsion and public transport. 1918 is recorded as the date of the earliest tubular motor, Birkeland's cannon, which was a d.c. reluctance motor which never emerged beyond the model stage.

During the 1930's the liquid metal pump was produced and in 1945 came the first large scale linear machine, an aircraft accelerator made by Westinghouse and known as the "Electropult". It was also the first example of an inverted rotor/stator combination with the stator mounted on the moving carriage. Developing 10,000 h.p. it had a synchronous speed of 225 m.p.h. Two tracks were built, one 5/8 mile long, the other over a mile. D.c. braking was incorporated at the end of the track for bringing the carriage to rest. The track was very expensive, being of the type shown in Fig. 3(a).

During the 1950's new records were set up for both maximum speed and acceleration. A team at the National Aeronautics and Space Administration in the U.S. produced an acceleration of 1200 g with a d.c. system. At the Royal Aircraft Establishment, Farnborough a d.c. linear motor rotor reached a speed of 1500 ft/sec. The self-oscillating back-to-back motor was invented in 1954. Tubular actuators first appeared on the market about that time. The first linear motor travelling crane was manufactured commercially in 1965.

Bibliography

LAITHWAITE E. R. (1966) *Propulsion Without Wheels*, London: English Universities Press.

LAITHWAITE E. R. (1966) *Induction Machines for Special Purposes*, London: Newnes.

E. R. LAITHWAITE

LOW ENERGY ELECTRON DIFFRACTION (LEED). Although the phenomenon of electron diffraction was first discovered by Davisson and Germer using low energy (50 eV) electrons, subsequent work has been carried out mainly by the use of high energy (40–100 keV) electrons. This is because the diffraction of low energy electrons is believed to be due entirely to one or two layers of atoms on the surface of the crystal. Consequently, the diffraction patterns obtained will be very sensitive to the presence of adsorbed gas. Since at a pressure of 10^{-6} torr an adsorbed monolayer of gas is formed in about one second, it is only under ultra high vacuum conditions (pressures of 10^{-9}–10^{-10} torr) that meaningful results can be obtained. It is therefore only during the past decade, when reliable apparatus for producing such high vacuum became readily available, that low energy electron diffraction (LEED) began to be used for the study of surface structure.

Two types of apparatus are commonly used. Farnsworth (1932) and his school at Brown University have used a system in which the diffracted electron beam is collected in a Faraday cage and measured (Fig. 1). In the latest development of this apparatus (Park and Farnsworth 1964), the electron beam is pulsed at 1000 pps and the output from the Faraday cylinder is passed through an a.c. amplifier before being rectified and applied to control the intensity of the

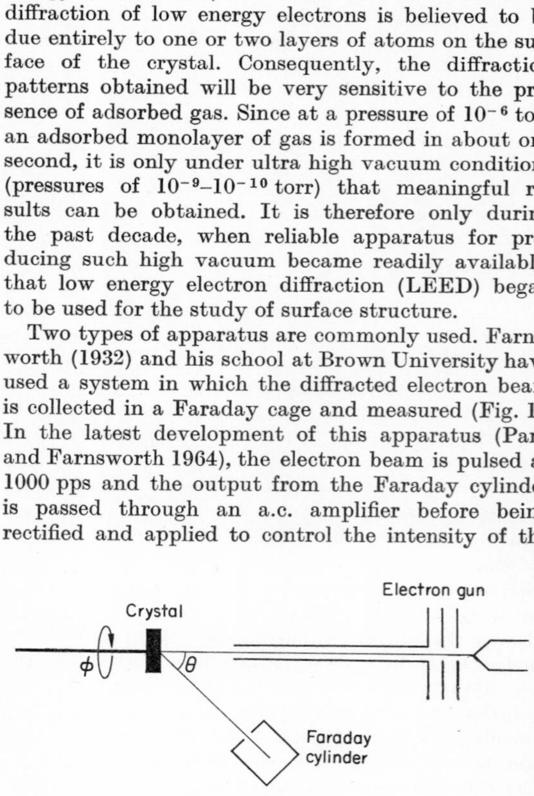

Fig. 1.

beam in a cathode-raydisplay tube of the storage type. The crystal is rotated at mains frequency (3600 r.p.m.) about an axis coinciding with the incident electron beam, thus varying the azimuth φ of the diffraction. At the same time the Faraday cylinder is slowly moved so as to vary the angle θ between the incident and the diffracted beam. The spot of the display cathode-ray tube moves in synchronism with the variation in φ in a circular path of radius proportional to $\sin \theta$. The result is that the diffraction pattern of the entire hemispherical solid angle, except for a small angle excluded by the gun, is displayed on the cathode-ray tube, a complete scan occupying about 90 seconds. When it is desired to measure the intensity of a diffracted beam, the crystal is rotated until this beam lies in the plane in which the collector moves. The gun voltage is then swept through a small range in synchronism with the X deflexion of a cathode-ray tube, and the collector current provides a signal to the Y deflector plates. In this way an intensity profile of the beam is displayed on the cathode-ray tube.

Germer and his associates at Bell Telephone Laboratories and Cornell University have developed a different form of apparatus (Fig. 2). In this, the diffraction pattern is displayed directly on a hemispherical fluorescent screen, whose centre of curvature coincides with the diffracting crystal. Immediately in front of the screen are mounted two fine mesh hemispherical grids. The first of these is maintained at the same electrostatic potential as the crystal; this ensures that the space round the crystal is free of electric fields, and that the diffracted electrons travel in straight lines from the crystal. The second grid is maintained at a potential slightly above that of the gun filament, and serves to repel all electrons which have been inelastically scattered. The metal backing of the fluorescent screen is maintained at a potential of a few kilovolts positive to the gun filament. The electrons passing through the grids therefore reach the screen with sufficient energy to excite fluorescence. This apparatus is obviously very convenient for visual observation of the diffraction pattern. The pattern can be photographed by a camera placed outside the tube. The intensity of the spot in the diffraction pattern can be measured by viewing the spot through a telephotometer.

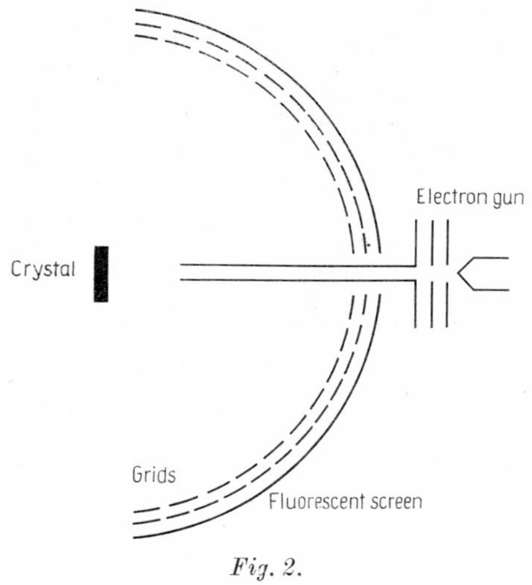

Fig. 2.

In a modified form of the direct display type of apparatus (Fujiwara *et al.* 1966), the fluorescent screen and grids are of cylindrical, instead of hemispherical, shape. This has the advantage that the screen almost completely surrounds the crystal. By arranging the crystal so that the incident beam strikes it at glancing incidence, diffracted beams can be observed making angles less than $\frac{1}{2}\pi$ with the incident beam. On the other hand, the cylindrical geometry changes the general appearance of the diffraction pattern.

LEED results. The main features of a diffraction pattern from clean metal surface can be interpreted by supposing the diffraction to occur mainly at a single surface layer of metal atoms (MacRae 1963). If the energy of the incident electron beam is varied over a range of about 300–1000 eV, it is found that the intensity of any particular diffracted beam is a maximum for certain critical values of the energy, and therefore wave-length, of the electrons. Qualitatively, this can be explained by supposing that there is a small penetration of electrons into the crystal, so that interference occurs between the electron waves scattered by successive layers of atoms. In order to explain quantitatively the variation of the intensity of a diffracted beam with electron energy, it is usually necessary to make supplementary assumptions, e.g. that the separation between the first two or three surface layers of atoms is different from that of the bulk material, or that the amplitude of the thermal vibrations of the surface atoms is different from that of the bulk material. For energies of the incident beam less than about 300 eV, the variation of the intensity of the diffracted beam with electron energy is much more complicated, and it is usually impossible to devise a satisfactory explanation of all the features of the pattern.

Miyake and Hayakawa (1966) have suggested that even the lowest energy electrons may in fact penetrate a considerable distance into a crystal, and the fact that the general features of the diffraction pattern are explicable as diffraction by a two-dimensional grating formed by a single surface layer of atoms is due to dynamical effects (Whelan 1961). They suggest that if dynamical effects are fully allowed for, it may not be necessary to invoke special hypotheses, such as abnormal spacing between surface layers of atoms, in order to explain the observed diffraction patterns. Although there is little doubt that LEED can give a reliable indication of the general arrangement of atoms in a clean metal surface, little reliance can be at present be placed in more detailed deductions from the patterns.

If gas is admitted to the system at a pressure greater than about 10^{-6} torr, the metal surface rapidly becomes contaminated by an adsorbed layer of gas, and extra spots commonly appear in the pattern. These can be interpreted by assuming that the adsorbed gas atoms are arranged in a regular two-dimensional array on the metal surface. The principal application of LEED at the present time is for the study of such gas adsorption effects, which are of particular importance for the understanding of catalysis.

Bibliography

FARNSWORTH H. E. (1932) *Phys. Rev.* **40**, 684.

FUJIWARA K., HAYAKAWA K. and MIYAKE S. (1966) *Japanese J. Appl. Phys.* **5**, 295.

MACRAE A. U. (1963) *Science* **139**, 379.

MIYAKE S. and HAYAKAWA K. (1966) *J. Phys. Soc. Japan* **21**, 363.

PARK R. L. and FARNSWORTH H. E. (1964) *Rev. Sci. Instrum.* **36**, 1592.

THEWLIS J. (Ed.) (1961) *Encyclopaedic Dictionary of Physics*, articles beginning "Electron diffraction", **2**, Oxford: Pergamon Press.

WHELAN M. J. (1961) in *Encyclopaedic Dictionary of Physics* (J. Thewlis Ed.) **2**, 540, Oxford: Pergamon Press.

T. B. RYMER

1. Introduction

LUNAR EXPLORATION, TECHNOLOGY OF. Recent advances in technology have resulted in a sudden increase of knowledge about the Moon. Ancient observations of lunar motions, followed in the last century by telescopic mapping, photometry and polarimetry, yielded clues to the Moon's peculiar nature. Modern investigations using radars, microwave radiometers, and infra-red scanners confirmed and extended the evidence showing that the crater-riddled material of the lunar surface must be in a strange physical state, being of a sooty darkness, rather smooth on an average macroscopic scale, and highly vesicular or rough and underdense at small-to-microscopic scale. Then space flight became a reality, and the effective resolving power of our visual instruments soon improved by a factor of a thousand, and then a million (Fig. 1). Spacecraft revealed the detailed structure of the remarkable lunar soil, and added the vital fact that it can support substantial loads. Still in the future are the elucidation of its chemistry and mineralogy, the determination of its variability over the lunar plains and highlands, and the discovery of what may lie beneath it. Thus the exploration of the Moon, though spectacularly started,

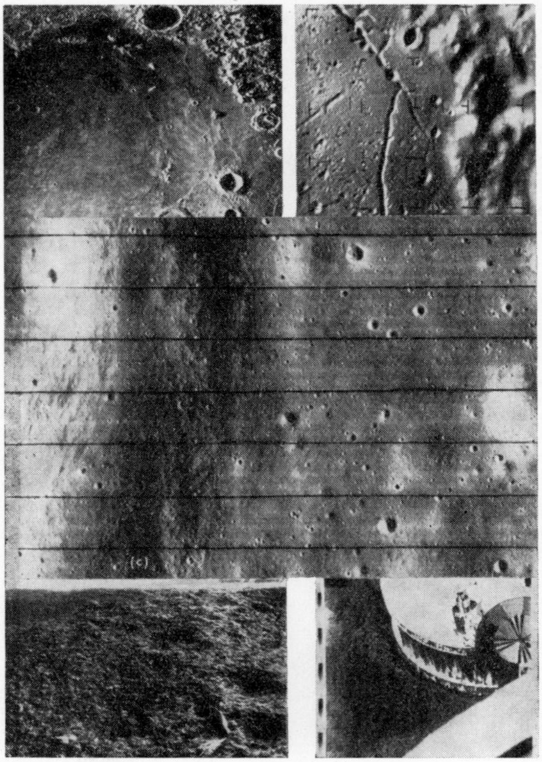

Fig. 1. Lunar images. (a) Photo from Earth. (b) Ranger 9 view of rilles and dark halo craters on floor of Alphonsus. (c) Lunar Orbiter photo of typical mare surface. (d) Portion of Luna 9 panorama. (e) Surveyor TV picture showing footpad of spacecraft and lunar soil disturbed by landing impact.

has still a long way to go, and a continuing elaboration of lunar technology can be expected.

The history of lunar probes to January 1968 is outlined in Table 1. For this review we will concentrate on recent and near-future efforts, since these incorporate and extend the technology of their precursors.

2. Energy and Mass

The first requirement for going to the Moon is a speed of nearly 11 km/sec above the Earth's atmosphere. With present rocket propulsion techniques, the mass accelerated to the lunar transfer speed can be only about one per cent of the initial mass. For orbiting or landing on the Moon, about one-third or two-thirds, respectively, of the remaining mass must be expended; for starting back toward the Earth, a further mass expenditure is of course required. These mass ratios give rise to two evident aspects of lunar

Table 1. Lunar flights, 1959–1968

Date	Name	Result
Jan. 1959	Mechta, or Luna 1*	First escape from Earth
March 1959	Pioneer 4	Distant lunar fly-by
Sept. 1959	Luna 2	First lunar impact; no magnetic field found on sunlit side of Moon
Oct. 1959	Luna 3	First photographs of far side
Jan. 1962	Ranger 3	Lunar fly-by
April 1962	Ranger 4	Lunar impact; no data
Oct. 1962	Ranger 5	Lunar fly-by; tracking data only
April 1963	Luna 4	Lunar fly-by
Jan. 1964	Ranger 6	Impact on target; television failed
July 1964	Ranger 7	First high-resolution images of Moon; 4316 pictures, Mare Cognitum
Feb. 1965	Ranger 8	7137 pictures, Mare Tranquillitatis
March 1965	Ranger 9	5814 pictures, Alphonsus; public TV pictures in real time, best resolution of Ranger series (about 1/3 meter)
May 1965	Luna 5	Impact in Mare Nubium
June 1965	Luna 6	Midcourse failure, missed Moon
July 1965	Zond 3*	Fly-by; pictures of far side area not covered by Luna 3
Oct. 1965	Luna 7	Impact in Oceanus Procellarum
Dec. 1965	Luna 8	Impact in Oceanus Procellarum
Feb. 1966	Luna 9	First lunar landing; panorama pictures, radiation measurement on surface of Oceanus Procellarum
April 1966	Luna 10	First lunar orbit; radiation survey near Moon
June 1966	Surveyor 1	Soft landing in Oceanus Procellarum; more than 11,000 TV pictures taken; spacecraft survived several lunations
Aug. 1966	Lunar Orbiter 1	200 Apollo site survey photographs taken, Earth photographed from near Moon, some photographs made of far side
Aug. 1966	Luna 11	Lunar orbit
Sept. 1966	Surveyor 2	Impact on Moon; during midcourse manoeuvre one vernier engine failed to ignite; spacecraft tumbled
Oct. 1966	Luna 12	Lunar orbit; photographs taken in S. W. Mare Imbrium
Nov. 1966	Lunar Orbiter 2	Additional Apollo site and far side photographs; high-resolution survey, oblique views of Crater Copernicus
Dec. 1966	Luna 13	Lunar landing; panorama pictures, surface experiments
Feb. 1967	Lunar Orbiter 3	Lunar orbit; 307 pictures including photo of Surveyor 1.
April 1967	Surveyor 3	Soft landing; TV and soil mechanics experiments.
May 1967	Lunar Orbiter 4	Lunar polar orbit; 326 pictures covered all of Moon's near side and much of far side.
July 1967	Surveyor 4	Lunar impact; spacecraft failed during retrofire.
July 1967	Explorer 35	Lunar orbit; field and particle experiments.
Aug. 1967	Lunar Orbiter 5	Lunar polar orbit; 424 pictures including numerous sites of scientific interest.
Sept. 1967	Surveyor 5	Soft landing; TV and chemical analysis experiments.
Nov. 1967	Surveyor 6	Soft landing; TV and chemical analysis experiments.
Jan. 1968	Surveyor 7	Soft landing in lunar highlands; TV, soil mechanics, and chemical analysis experiments.

* Russian series.

exploration technology today: (1) launch vehicles are large (Fig. 2), and (2) rendezvous is a required technique. In the Apollo project, the rendezvous manoeuvres are to be made in lunar orbit. Only a minor fraction of the 43,000 kg spacecraft mass is to be taken all the way down to the lunar surface and back up again. Without this technique, the Apollo rockets would have to be even bigger than they are, or else the total mass would have to be increased through multiple launches with rendezvous and propellant

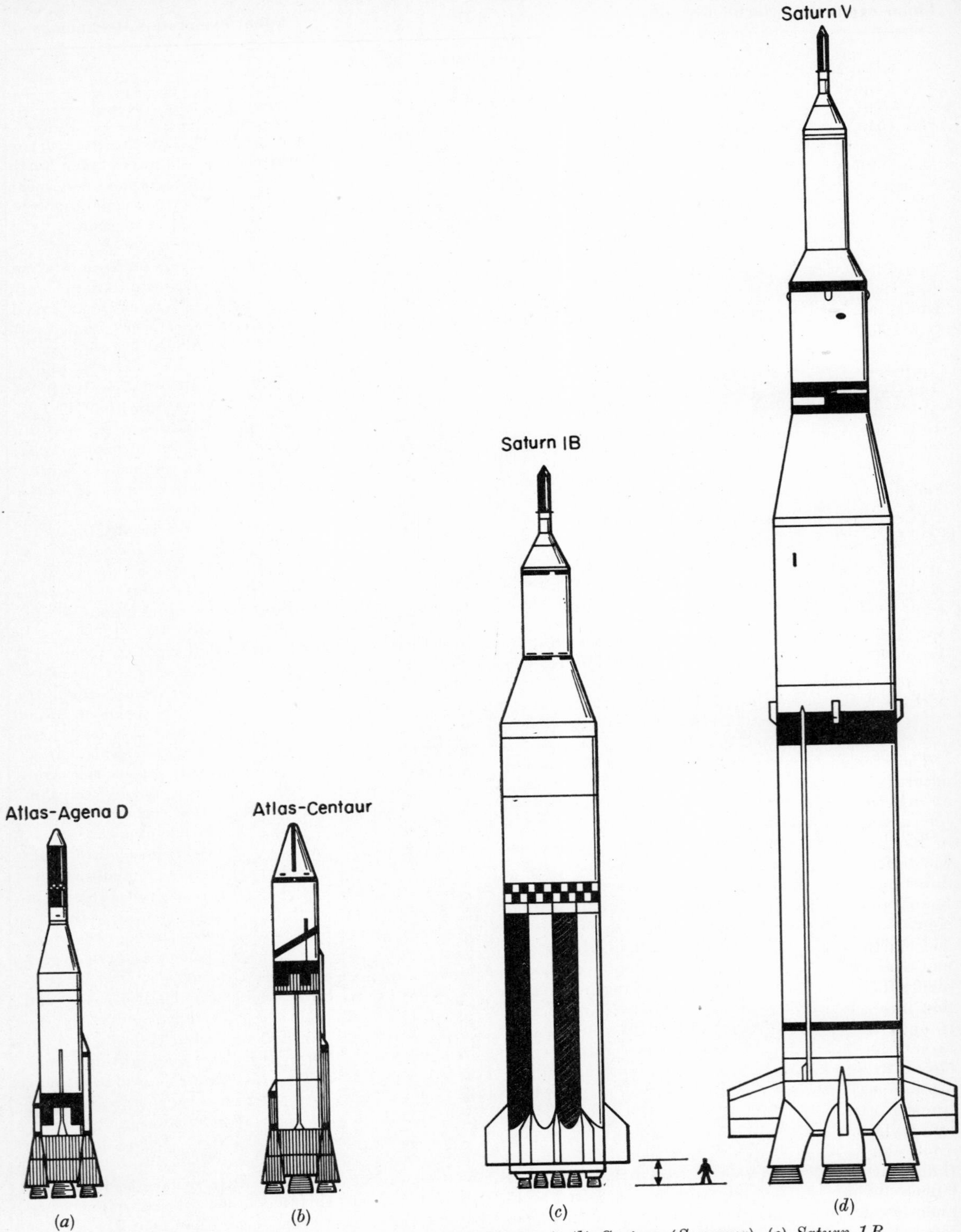

Fig. 2. U.S. Launch Vehicles. (a) Atlas Agena (Ranger). (b) Centaur (Surveyor). (c) Saturn 1B (Apollo test). (d) Saturn V (Apollo lunar flight).

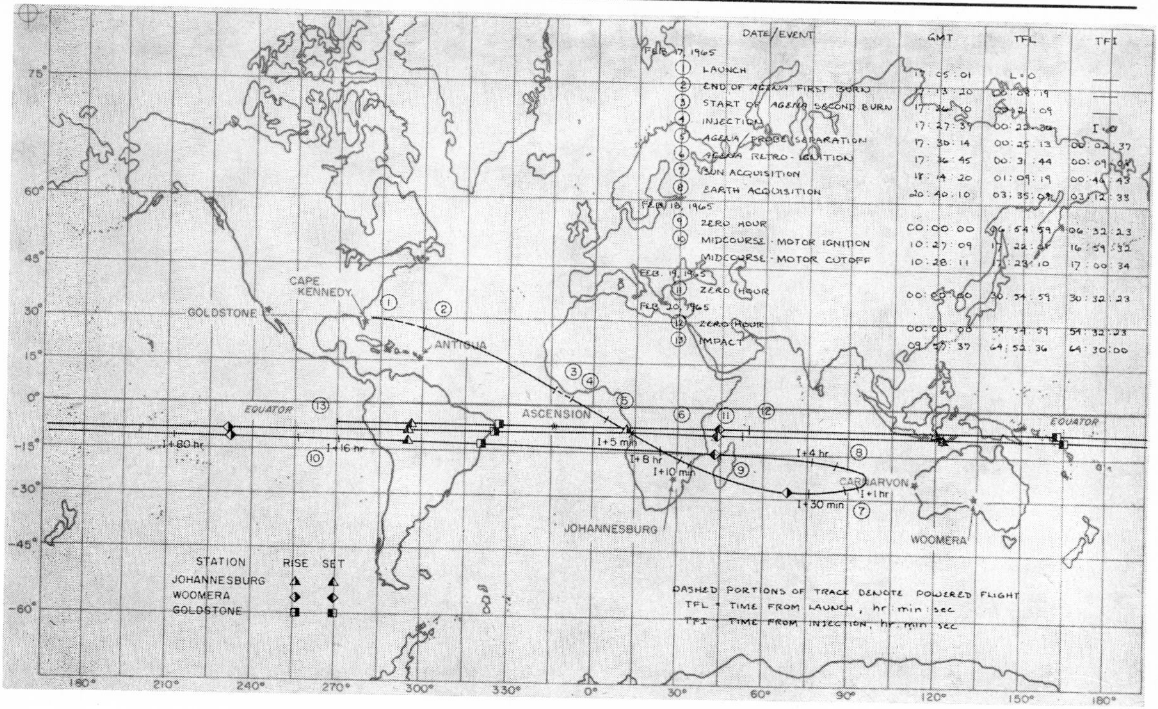

Fig. 3. Earth trace of Ranger VII trajectory.

or crew transfers, either in Earth orbit before departure or after landing on the Moon.

3. Injection Guidance

The second requirement for going to the Moon is accurate position and velocity control of spacecraft at injection into the transfer orbit, with a capacity for varying the injection vector to match the motions of Earth and Moon. Allowable injection errors are of the order of kilometres in position, milliradians in direction, and metres per second in speed; these can be obtained with extensions of the guidance and control technology developed for ballistic weapons and Earth satellites. The compensation for celestial motions, however, has required a further development: the parking orbit. By launching the spacecraft first into a low Earth orbit and then later propelling it onto the desired transfer trajectory toward the target, launch-site geographical constraints are relieved. By varying the time of coasting in the parking orbit, injection can be placed always near perigee of the transfer orbit (most efficient on an energy basis). Also, by varying the azimuth of launch, acceptable transfer orbits can be reached despite Earth rotation during a time interval (typically an hour or two) called the launch "window". The azimuth limits, and consequently the window duration, are set either by range-safety boundaries or by the coverage limits of down-range instrumentation sites. The Earth trace of a typical parking-orbit departure path is shown in Fig. 3. The loop in the trace occurs when Earth rotation overtakes the diminishing eastward component of the spacecraft's motion; by the time the spacecraft reaches the Moon, its apparent motion is almost sidereal.

4. Telecommunications, Orbit Determination, and Midcourse Guidance

The third requirement for going to the Moon is midcourse guidance. In principle, given extreme component accuracies, guidance could be terminated at injection, but in practice it is better to accept errors such as those mentioned earlier and then to make one or more en-route corrections to the flight path, using the very powerful methods afforded by radio tracking, computation, and command. The demand for communication to and from spacecraft at interplanetary ranges has called forth radio developments so advanced that lunar communications are now relatively straightforward for information rates typical of command, telemetry, voice, and still-picture transmission. Using a very stable oscillator on the ground and phase-coherent systems in both ground and fligth equipment, the radio link that carries the command and telemetry modulations can serve also, via the two-way Doppler principle, as an extremely precise range-rate indicator. By taking

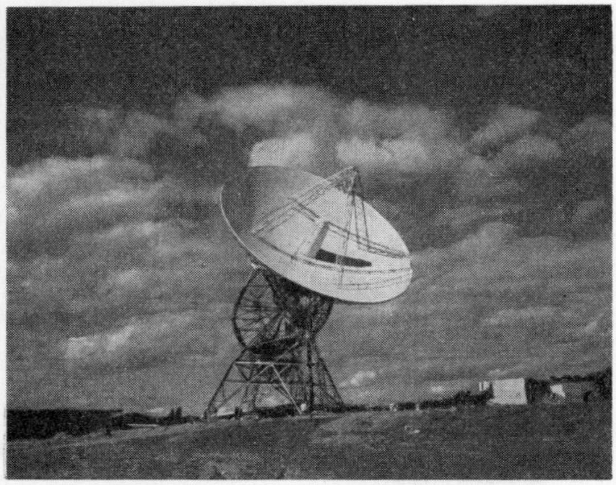

Fig. 4. 85-foot antenna of Deep Space Network Station, Robledo, Spain.

many hours of two-way Doppler data together with some angle tracking data, removing the biases and random errors, and fitting the results to a gravitational path, a highly accurate estimate of the orbit can be obtained. Absolute range data, obtained by measuring the time delay of a coded modulation sent to and from the spacecraft, can provide a further refinement. By these means, we now measure the range and radial velocity of spacecraft en route to the Moon with residual errors of the order of metres and cm/sec, respectively.

In order to maintain continuous contact with lunar and planetary spacecraft, the United States and cooperating nations have established a network of stations in North America, Spain, South Africa, and Australia. The Robledo station in Spain is shown in Fig. 4. Each such station has an 85-foot paraboloid antenna with S-band (2200-MHz) transmitting and receiving equipment, data handling equipment, and communications to and from the mission control point in California.

In order to make the orbit correction, the spacecraft must have a propulsion system whose total impulse can be varied by command. Also, it must have an attitude control system and some orientation reference so that the direction of the manoeuvre can be specified. Optical references are commonly used; the Sun is the obvious target for controlling one axis, and the Earth, a star, or the Moon can be used for the other. The manoeuvre commands tell the spacecraft to turn through two angles, to fire its rocket engine either for a given time or for a given speed change as measured by an on-board integrating accelerometer, and then to reacquire the references. A nominal manoeuvre to correct the injection errors listed earlier would involve a speed change of only 20–50 m/sec; thus the propulsion requirement is small. The mid-course-manoeuvre strategy, however, can be quite complex. Not only does the manoeuvre influence the location and speed at arrival on or near the Moon; it also can provide a correction to the time of flight, and this is important when it is desired to place a critical operation within view of a particular ground station. For U.S. lunar missions launched from Cape Kennedy, outbound flight times of $1\frac{1}{2}$, $2\frac{1}{2}$ or $3\frac{1}{2}$ days are used so that spacecraft arrival at the Moon will occur over California. The slower transits give more payload but require more accurate injection guidance.

On a typical flight to the Moon, the midcourse manoeuvre is executed 16 to 40 hours after injection. One of the major new technological developments called forth by the lunar programme is the complex operating system, involving intercontinental communications, large high-speed computers, and, above all, hundreds of highly trained people, that makes it possible to determine the orbit, evaluate the condition of the ground and flight equipment, and send the requisite midcourse commands in what has come to be called "real time".

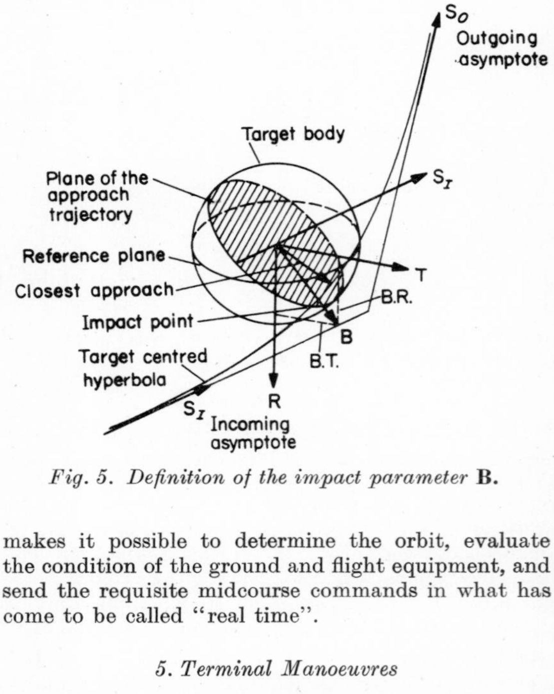

Fig. 5. Definition of the impact parameter **B**.

5. Terminal Manoeuvres

The fourth and last requirement for going to the Moon is the manoeuvre at arrival. For the Ranger television missions, the terminal manoeuvre was just a reorientation to aim the spacecraft's cameras along the line of flight. For lunar orbiters, a velocity reduction of about 1 km/sec is needed; for landing, about 2·6 km/sec. The geometry of the arrival is illustrated in Fig. 5.

The resultant of the Moon's orbital velocity and the spacecraft's motion, nearly radially away from Earth, is such that the arrival path for vertical impact falls typically in the western equatorial region of the Moon, site of the Oceanus Procellarum. For describing paths other than the vertical one, the concept of the

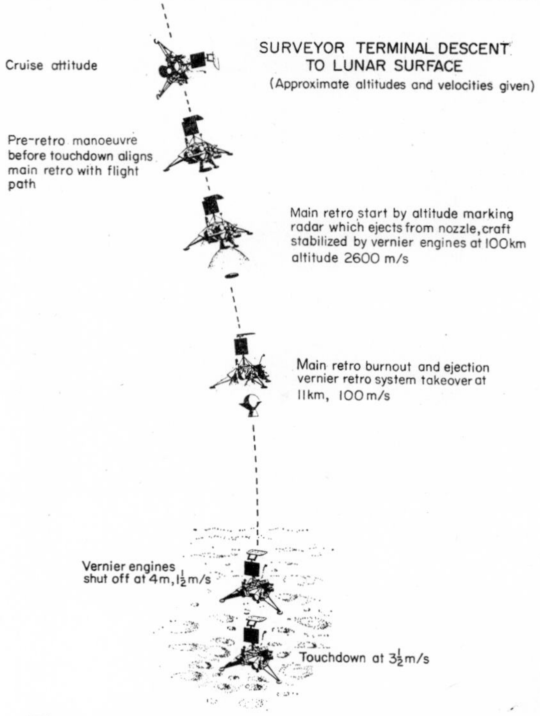

Fig. 6. Surveyor descent and landing sequence.

impact parameter **B** has proved useful; this is a vector defining the location of the approach asymptote relative to the vertical path, as shown in Fig. 5. Since the picture is symmetric about the vertical path, all impacting trajectories with a given magnitude of **B** will have the same inclination to the local vertical at impact, and all fly-by trajectories with a given magnitude of **B** will have the same periselene altitude. The components of **B** are controlled by the midcourse manoeuvre. A typical aiming strategy involves selecting the value of **B** to place the impact in a region on the Moon where (a) the angle of sunlight gives good shadows for imaging the surface, (b) the approach is close enough to vertical to be within the capacity of the retro-manoeuvre control system, and (c) the landing will occur in a desirable area from the stand-point of safety or scientific interest. When entry into orbit is intended, or orbit followed by landing as in Apollo, the parameters controlled are the altitude at retro-fire and the inclination of the orbit.

The landing sequence of the Surveyor spacecraft is illustrated in Fig. 6. About one-half hour before landing, the spacecraft is turned so as to point its radars and rockets along the flight path. About 100 km above the surface, the altitude-marking radar triggers the retro-fire sequence. Throttleable vernier engines ignite. A few seconds later the main retro-rocket starts, the altitude-marking radar is ejected, and control signals begin to come from the ranging and Doppler radar system which senses both vertical and lateral velocity components. At an altitude near 10 km, the main retro-rocket burns out and is jettisoned, leaving the spacecraft descending at about 100 m/sec toward the Moon's surface. Thrust of the vernier engines is controlled to maintain attitude, speed, and altitude close to a preprogrammed profile, until the final condition (speed about 1 m/sec downward, altitude 4 m) is reached; the engines are then shut off, and the spacecraft falls to the surface.

6. Spacecraft Designs

To meet all of the requirements described above, and to execute experiments on and near the Moon, unique unmanned machines have been devised incorporating subsystems for thermal and attitude control, power, telecommunications, instrumentation, data storage and processing, propulsion, guidance, and structural integrity. To these functions in the near future will be added the additional ones required for manned flight: life support, on-board navigation, rendezvous, re-entry and return to Earth. In each of the subsystems, the technology has roots in the disciplines

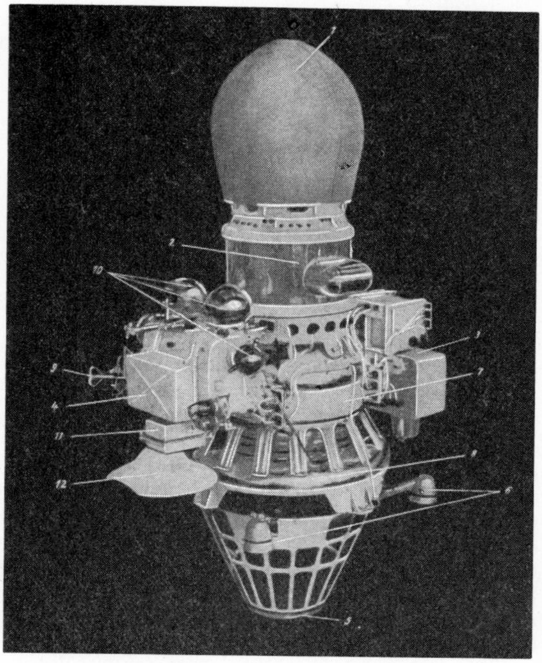

Fig. 7. Luna 9 spacecraft. (1) Automatic station (landing capsule). (2) Control system compartment. (3, 4) Control components. (5) Liquid-propellant rocket engine. (6) Control engines. (7) Spherical oxidizer tank. (8) Toroidal fuel tank. (9) Attitude-control jets. (10) Attitude-control gas supply spheres. (11) Radio altimeter, with (12) Parabolic antenna.

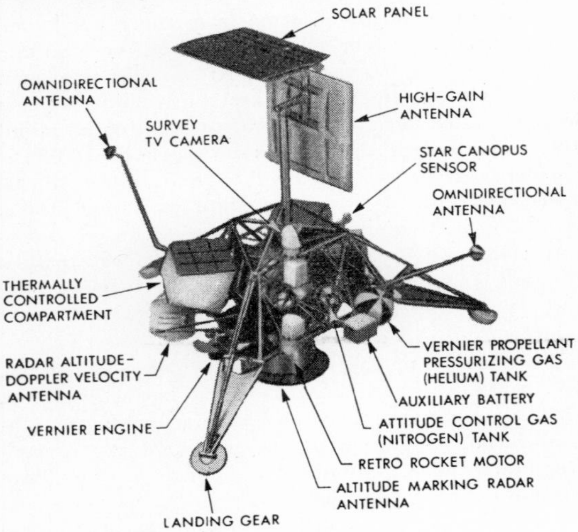

Fig. 8. Surveyor spacecraft.

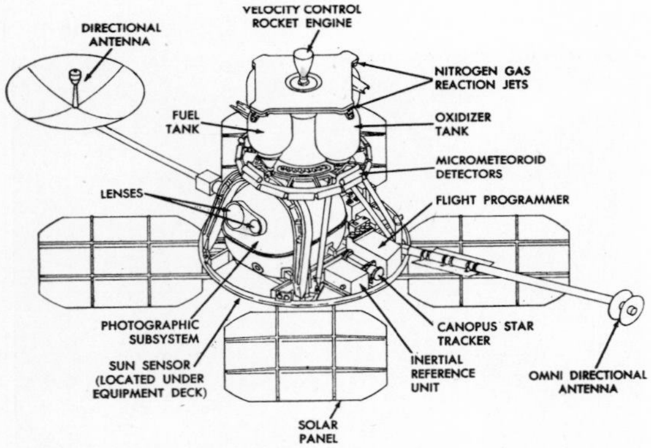

Fig. 9. Lunar Orbiter spacecraft.

of circuit theory, electronic parts and packaging design, solid-state physics, chemistry, optics, and so forth. In the ensemble, however, as in other modern complex systems, the whole has come out to be greater than the sum of the parts, and the new technology of spacecraft system integration is critical to the design. One of the chief technological products of the lunar programme has been the further elaboration of the management and engineering skills identified in past developments by words such as "interface control", "configuration management", and so on. Without attempting an explicit description of these skills, we can recognize their importance by noting advances toward successful completion of deep-space projects in accordance with pre-planned schedules, budgets, and performance criteria. Because of the competition between such projects and other national goals, this improvement in technology is essential to the progress of lunar exploration.

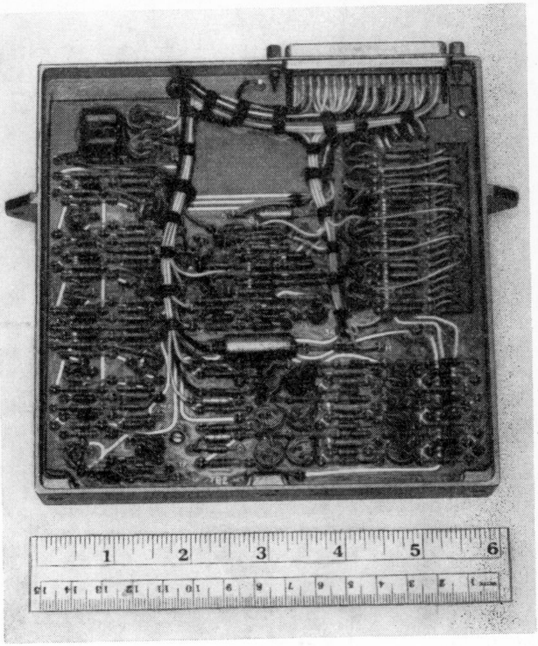

Fig. 10. Electronic module showing packaging concept used in Ranger spacecraft.

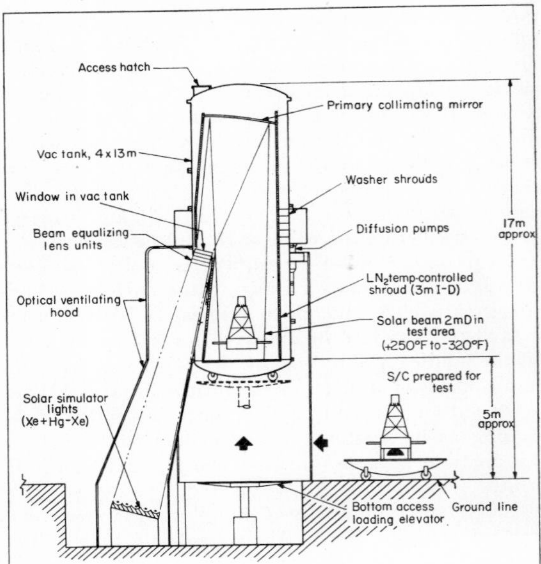

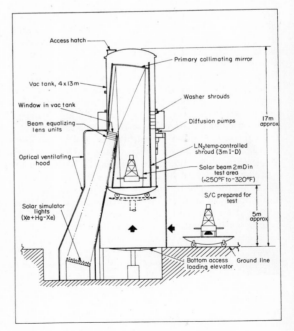

Fig. 11. Test spacecraft in solar-thermal-vacuum simulator.

Examples of present-day lunar spacecraft are shown in Figs. 7, 8, and 9. In each figure can be seen some of the subsystems whose required functions were outlined earlier. Not visible are the complex internal parts; a typical example of one of these is shown in Fig. 10. The Surveyor, whose weight at injection is about 1000 kg, operates on 100 to 200 watts of power and has 25,000 electronic parts.

7. *Reliability*

At the present state of technology, lunar missions are barely achievable. This is not because any single element is near the limit of electromechanical performance or human comprehension; it is because these complex systems are being operated very early in their evolution, with severely limited diagnostic instrumentation and only small performance margins. In such a situation, detailed understanding and control of the quality and operating regimes of the components is vital. But even with such control, elaborate simulation testing (Fig. 11) has proved to be necessary. A typical example of the sort of defect which could have caused a complete mission abort, had it not been discovered in preflight testing, is shown in Fig. 12. One of the major fronts of progress in lunar exploration technology today is the art of eliminating such defects, designing around the ones that cannot certainly be eliminated, and establishing by preflight analysis and tests that the expected risk has really been brought to an acceptable level. Only as these goals are met can we proceed with confidence toward the further exploration of the Moon.

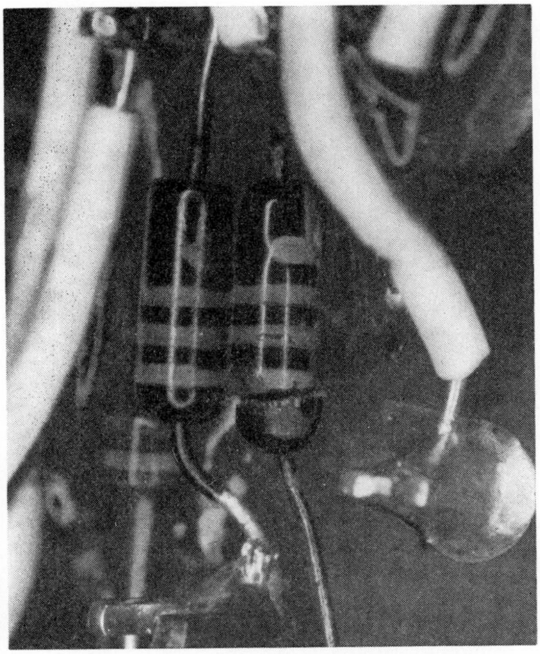

*Fig. 12. Fault discovered during preflight inspection (for scale, compare with **Fig. 10**).*

Bibliography

BARABASHOV N. P. *et al.* (1961) *Atlas of the Other Side of the Moon*, Moscow: Academy of Sciences of the USSR, and New York: Pergamon Press.

First Panoramic Views of the Lunar Surface (1966) Moscow: Academy of Sciences of the USSR, and Washington: NASA Technical Translation F-393.

HEY J. S. (1962) in *Encyclopaedic Dictionary of Physics* (J. Thewlis Ed.), **6**, **73**, Oxford: Pergamon Press.

JEFFREYS H. (1962) in *Encyclopaedic Dictionary of Physics* (J. Thewlis Ed.), **4**, 720, Oxford: Pergamon Press.

KOPAL Z. *et al.* (1965) *Photographic Atlas of the Moon*, New York: Academic Press.

The Lunar Orbiter (1966) Seattle: The Boeing Company for NASA Langley Research Center.

Manned Space Flight; Project Apollo (1965) NASA Facts, Vol. III. No. 1. Washington: U. S. Government Printing Office.

PARKS R. J. *et al.* (1966) *Surveyor 1 Mission* Report (Part 1, *Mission Description;* Part 2, *Scientific Data and Results*) TR No. 32-1023, Pasadena: Jet Propulsion Laboratory.

Ranger VII Photographs of the Moon (1964) NASA SP-61, 62, and 63, Washington: U. S. Government Printing Office.

RECHTIN E. (1963) *Lunar Communications*, TM No. 33-133, Pasadena: Jet Propulsion Laboratory.

SALISBURY J. W. (Ed.) (1964) *Bibliography of Lunar and Planetary Research*, 1960-1964, Bedford, Mass.: U. S. Air Force Cambridge Research Laboratories.

SCHURMEIER H. *et al.* (1966) *Ranger VIII and IX* (Vol. 1, *Mission Report;* Vol. 2, *Experimenters' Analysis and Conclusions*) TR No. 32-800, Pasadena: Jet Propulsion Laboratory.

SJOGREN W. *et al.* (1964) *The Ranger VI Flight Path and its Determination from Tracking Data*, TR No. 32-605, Pasadena: Jet Propulsion Laboratory.

U. S. Aeronautical and Space Activities (Annual) Reports to Congress from the President of the United States (1959-66) Washington: U. S. Government Printing Office.

The View From Ranger (1966) NASA EP-38, Washington: U. S. Government Printing Office.

WHITAKER E. *et al.* (1963) *Photographic Lunar Atlas with Supplements No. 1 and 2*, Tucson: University of Arizona Press.

J. D. BURKE

M

MEAN ELECTRON ENERGY IN NON-ATTACHING GASES. In most cases when free electrons are present in a gas the steady state mean energy of the electrons does not correspond to the ambient temperature of the gas. Usually, but not invariably, the source of energy which is responsible for raising the mean electron energy above the mean molecular energy is a steady or alternating electric field which supplies energy to the electrons as a result of the drift of the electrons in the direction of the field. This phenomenon has many important consequences and a large number of investigations have been carried out to examine it in detail.

In a large number of cases the electron number density n is very small compared to the gas number density N. As a consequence, electron-electron interaction can be neglected as can electron-ion interaction when ionization occurs. The discussion and data presented here apply to this situation.

The energy distribution of the electrons, which may, to some extent, be characterized by the mean energy, is of central importance. Together with the cross-sections for the various collision processes between the electrons and neutral molecules, the distribution determines the overall behaviour of the electron assembly or "swarm". This behaviour is conveniently described in terms of appropriate transport coefficients, for example the diffusion coefficient D, the drift velocity W, the ionization coefficient α and so on, and each coefficient can be calculated once this information is known. It is not surprising, therefore, that a great deal of theoretical and experimental work has been done to determine electron energy distributions in a large number of gases for a wide range of experimental parameters.

Parameters which determine the mean energy. The factors which determine the energy distribution can be discussed more easily by considering the special case of an electron swarm moving through a gas under conditions which ensure that only elastic collisions occur between the electrons and neutral particles. These conditions exist, for example, in monatomic gases provided the electrons have insufficient energy to cause electronic excitation of the atoms. The Boltzmann transport equation can then be integrated to give the following distribution function for the electron energies:

$$f(\varepsilon) = A \exp\left[-\int_0^{\varepsilon}\left(\frac{ME^2e^2}{6mN^2q_m^2(\varepsilon)\,\varepsilon} + kT\right)^{-1} \mathrm{d}\right] \quad (1)$$

where ε is the electron energy

m and M are the electronic and atomic masses respectively,

e is the electronic charge,

$q_m(\varepsilon)$ is the energy dependent momentum transfer cross-section,

k is Boltzmann's constant,

T is the gas temperature,

E is the electric field strength,

and

A is a constant whose value is such that $\int_0^\infty \varepsilon^{1/2} f(\varepsilon)\,\mathrm{d}\varepsilon = 1$.

The mean energy $\bar{\varepsilon}$ is given by the following relation:

$$\bar{\varepsilon} = \frac{\int_0^\infty \varepsilon^{3/2} f(\varepsilon)\,\mathrm{d}\varepsilon}{\int_0^\infty \varepsilon^{1/2} f(\varepsilon)\,\mathrm{d}\varepsilon}. \quad (2)$$

Two limiting cases serve to demonstrate the dependence of $\bar{\varepsilon}$ on the experimental parameters E, N and T:

1. *Zero electric field.* When $E = 0$, equation 2 becomes

$$\bar{\varepsilon} = \frac{\int_0^\infty \varepsilon^{3/2} \exp(-\varepsilon/kT)\,\mathrm{d}\varepsilon}{\int_0^\infty \varepsilon^{1/2} \exp(-\varepsilon/kT)\,\mathrm{d}\varepsilon} = \frac{3}{2}kT.$$

As would be expected, the mean electron energy equals that of the gas molecules and depends only on the temperature T.

2. *High electric field:* $\bar{\varepsilon} \gg 3kT/2$. When the ratio E/N is sufficiently large, kT becomes small compared

with the first term of the integrand in equation 1 and may be neglected. If, for the sake of this discussion, the further simplifying assumption is made that $q_m(\varepsilon)$ is constant, then equation 2 becomes

$$\bar{\varepsilon} = \frac{\int_0^\infty [\varepsilon^{3/2} \exp(-3mN^2q_m^2(\varepsilon)\,\varepsilon^2/ME^2e^2)]\,d\varepsilon}{\int_0^\infty [\varepsilon^{1/2} \exp(-3mN^2q_m^2(\varepsilon)\,\varepsilon^2/ME^2e^2)]\,d\varepsilon}$$

$$= 0{\cdot}43\left(\frac{Me^2}{m}\right)^{1/2}[(E/N)/q_m(\varepsilon)]. \qquad (3)$$

Thus, when the electron energy is controlled by the electric field, it is important to note that the mean energy of the electron swarm depends on the ratio of E/N rather than on E and N separately. (Since many experiments are done at room temperature the gas pressure p has been frequently used in the past as a measure of the gas number density N; as a consequence the parameter E/p appears much more frequently than E/N in the literature.)

Between the two limiting cases discussed above it can be shown, as would be expected, that the mean energy is a function of both E/N and T. Furthermore, although the discussion has, for simplicity, been limited to the special case of elastic collisions, the results obtained from this discussion are generally true, and it has been found by experiment that the mean energy depends on E/N and T for all gases over a wide range of these parameters. Arguments of a more general nature, which will not be reproduced here, can be used to explain this experimental observation.

Figure 1, in which $\frac{3}{2\mu} De$ (where $\mu = W/E$) is plotted as a function of E/N and T, shows a typical example of the dependence of the electron energy on these

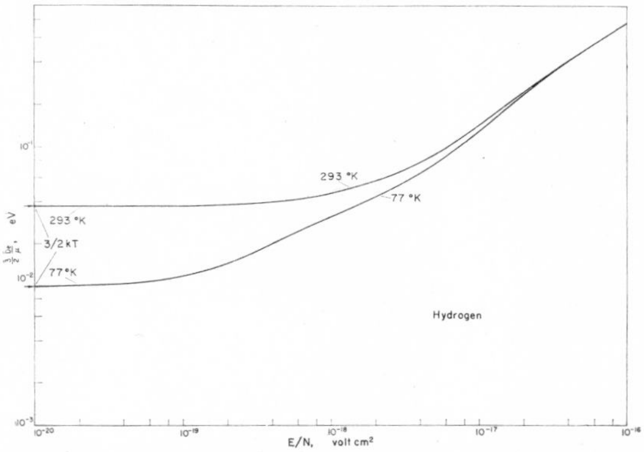

Fig. 1. Illustrating the dependence of the mean electron energy on the ratio E/N and the gas temperature T.

parameters. As will be described later the quantity $\frac{3}{2\mu} De$ is a measure of the mean electron energy. When E/N is sufficiently small it can be seen that the mean energy becomes almost independent of E/N and that the curves corresponding to each temperature approach asymptotically the value $\frac{3}{2}kT$ appropriate to that temperature. On the other hand, when E/N is sufficiently large the two curves converge showing that the gas temperature is of little importance in determining the energy distribution when the ratio of the power received from the field to the rate of energy loss in collisions is sufficiently high. Furthermore, since the curves shown in the figure were constructed from measurements made over a large range of pressure at each temperature, these results confirm that the energy distribution is a function of the ratio E/N rather of E and N separately.

Methods of determining the mean energy. As yet there appears to be no method of directly measuring the mean electron energy as a function of E/N but several methods have been used which enable the mean energy to be estimated when information about the form of the energy distribution function is either known or postulated. Probe and microwave methods have been used with success in some investigations although the repeatability and hence accuracy of the results has not always been good. References to these techniques and the results from them will be found in the literature.

The largest and perhaps most reliable source of data comes from steady state experiments in which the ratio of the electron drift velocity to diffusion coefficient is measured in a single experiment. Experiments of this type, although restricted to situations in which the electron number density is low, enable data to be obtained for mean electron energies from the thermal value to energies such that primary and secondary ionization are significant processes.

Details of the experimental method are given elsewhere in this Dictionary (see Dutton 1961). In the most precise measurements care must be taken to ensure a high degree of uniformity in the electric field within the diffusion chamber since the measurements are particularly sensitive to the presence of small radial components in the field. A section through a typical apparatus which has been developed to ensure adequate uniformity is shown in Fig. 2. Electrons enter the diffusion chamber through a hole 1 mm in diameter in the cathode and from there drift and diffuse towards the anode in the uniform field established by the guard electrode structure. The distribution of current over the anode is determined by measuring the percentage of the total current received by the central disk, the remainder of the current being received by the surrounding annular section of the electrode. A wide range of values of E/N can be covered by choosing suitable combinations of

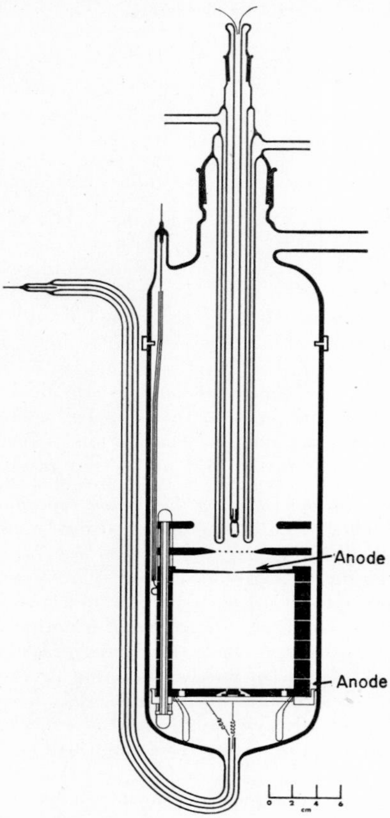

Fig. 2. Schematic section of a typical apparatus for measuring the ratio W/D by the Townsend-Huxley method.

the two parameters. Electric field strengths commonly used are in the range 1– 100 V cm^{-1} and gas pressures in the range 1–700 torr.

In the absence of attachment and ionization a simple expression relates the ratio W/D to the fraction of the total current received by the central disk of the anode, the diameter of the disk and separation of the cathode and anode. Moreover it can be shown (see *Townsend energy factor: Townsend energy ratio*) that the quantity $\varepsilon_K = De/\mu = De/(W/E)$ is related to the mean energy $\bar{\varepsilon}$ through the equation

$$\bar{\varepsilon} = F \cdot \frac{3}{2} \varepsilon_K .$$

where the factor F is within 20 per cent of unity in many cases. ($F = 1$ when the electron energy distribution is Maxwellian.) Thus although the method does not lead to accurate values of the mean energy unless the value of F can be determined it does enable a close estimate of the mean energy to be made in very many important cases.

It should be noted that this method is not restricted to those situations in which there is no attachment or ionization. Although the interpretation of the experimental results is more complex, an extension of the method enables both the ratio W/D and the attachment coefficient to be determined when electron attachment takes place. When ionization occurs it is still possible to determine W/D although it is necessary to know the value of the ionization coefficient before the ratio can be calculated from the ratio of the currents measured in the lateral diffusion experiment. Very recent work has shown that the method can be further extended to enable measurements to be made even when secondary processes make a significant contribution to the total current.

Where data are presented in the next section unmodified experimental results are given rather than more accurate estimates of the mean energy which can be obtained from them. The reader is referred to original papers for such information. Thus the quantity $\frac{3}{2}\varepsilon_K$, which is directly determined by experiment, is quoted as a function of E/N and T for each gas rather than a derived mean energy $\bar{\varepsilon} = F \cdot \frac{3}{2}\varepsilon_K$. In many cases the differences are small, and, for many applications, unimportant.

Experimental data for electrons in some typical gases. Figure 3 shows a few examples of the variation of the mean electron energy with E/N. Some comments can be made about the behaviour in each gas in explanation of the very great differences which have been observed experimentally.

(*i*) *Helium and argon.* These two gases are chosen to typify the monatomic gases. This choice was made not only because there is a large difference in the atomic weights of the gases but also because it is known that the elastic scattering cross-section for electrons in helium is reasonably constant whereas

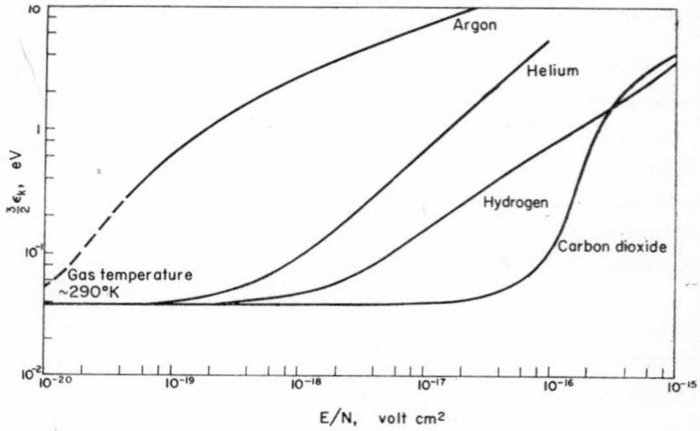

Fig. 3. Representative examples of the variation of the electron mean energy with E/N. The dotted section of the curve for argon has been obtained by calculation.

the cross-section in argon changes rapidly with electron energy at low energies (Townsend–Ramsauer effect).

At low values of E/N, before the onset of electronic excitation, only elastic collisions are possible between the electrons and atoms and the energy exchange per collision between electrons and gas atoms is very small as a result of the large mass discrepancy between the electron and atom. It is to be expected, therefore, that relatively large values of the electron mean energy will occur for small values of E/N and that higher energies will be observed in the heavier monatomic gases at the same value of E/N. The experimental results for helium and argon shown in Fig. 3 certainly confirm this prediction, although the ratio $\varepsilon_{K_{\text{argon}}} : \varepsilon_{K_{\text{helium}}}$ is in fact much larger than would be expected from this simple picture. As a result of the difference in atomic weight of the two gases it can be seen from equation 3 that the mean energies would be expected to be in the ratio $(M_{\text{argon}}/M_{\text{helium}})^{1/2} : 1$, that is a little more than 3 : 1. At $E/N \cong 5 \times 10^{-19}$ V/cm² the observed ratio is, however, about 30 : 1. The explanation lies in the large differences which exist in the elastic scattering cross-section. For electron energies near 0·5 eV the cross-section in argon exhibits a pronounced minimum and becomes less than one twentieth of the cross-section for helium which remains reasonably constant below about 3 eV. Equation 3 shows that this should result in the ratio of the energies being greater than 20 : 1 so that the overall ratio should be greater than 60 : 1. The observed ratio is not as high as this at the value of E/N chosen since the mean energy in argon is above the Townsend–Ramsauer minimum. The fraction of electrons with energies near the minimum is therefore small.

The results for helium and argon shown in Fig. 3 indicate that the initial very rapid rise of $\bar{\varepsilon}$ with E/N is curtailed at higher values of E/N. This is because an appreciable number of the more energetic electrons now have sufficient energy to cause electronic excitation of the atoms with which they collide; finally at still higher values of E/N the process of ionization adds still further to the large energy losses caused by inelastic collisions.

(*ii*) *Hydrogen and carbon dioxide.* There is a considerable contrast between the situations which have just been described and those for electrons in molecular gases. Even for the lowest energy electron swarms, some electrons have sufficient energy to cause rotational excitation of the molecules; thus inelastic collisions help to control the energy distribution even at the lowest values of E/N. As E/N is increased, the swarm becomes sufficiently energetic for some of the electrons to excite molecular vibration as well as rotation until finally electronic excitation and ionization are added to the other two inelastic energy exchange processes.

The importance of inelastic collisions in determining the mean electron energy can be demonstrated by examining the results for hydrogen. Using equation 3 and assuming a constant cross-section of about 10^{-15} cm² the mean energy predicted on the basis of elastic collisions only is found to be just less than 1 eV for $E/N = 3 \times 10^{-17}$ V cm². This is to be compared with the value of 0·35 eV for $\frac{3}{2}\varepsilon_K$ obtained experimentally (cf. 1·86 eV for the same value of E/N in helium). A detailed examination of the way in which the energy gained by the electrons from the field is distributed among the competing energy exchange processes shows that, for hydrogen at this value of E/N, very approximately 50 per cent of the energy is absorbed in rotational excitation, 25 per cent in vibrational excitation and 25 per cent in the energy exchange in elastic collisions. Thus, although the cross-sections for the inelastic processes are very much smaller than the elastic scattering cross-section (particularly the cross-sections for rotational excitation), this factor is outweighed by the very much larger fraction of the electron energy which is transferred at each inelastic collision.

Finally electron energies in carbon dioxide are given to contrast with the data in hydrogen. Strictly speaking carbon dioxide lies outside the scope of this article since there is appreciable electron attachment for E/N greater than about 2×10^{-16} V cm². However, the behaviour below this value of E/N is sufficiently interesting to warrant its inclusion and the remainder of the data, obtained by the extension of the method referred to earlier, is included for completeness.

In this case it will be noticed that the electrons remain practically in thermal equilibrium with the gas until E/N has reached a comparatively high value. This result is, apparently, a consequence of the very high momentum transfer cross-section at low energies (where it is approximately 50 times larger than the hydrogen cross-section) and the existence of vibrational levels which can be excited by electrons of very low energy.

The examples listed here have been selected simply to illustrate the very great diversity which exists in the behaviour of low-energy electrons in gases. The data for many other gases for a wide range of values of E/N and gas temperature are available in the literature.

Bibliography

ALLIS W. P. (1956) *Handbuch der Physik,* Vol. 21, Berlin: Springer-Verlag.

CHAPMAN S. and COWLING T. G. (1939) *The Mathematical Theory of Non-uniform Gases,* Cambridge: The University Press.

DUTTON J. (1961) in *Encyclopaedic Dictionary of Physics* (J. Thewlis Ed.), **2**, 403, Oxford: Pergamon Press.

HARRISON J. A. (1967) in *Encyclopaedic Dictionary of Physics* (J. Thewlis Ed.), Suppl. Vol. 2, 407, Oxford: Pergamon Press.

HUXLEY L. G. H. and CROMPTON R. W. (1962) in *Atomic and Molecular Processes* (Ed. D. R. Bates) Chapter 10, New York: Academic Press.

LOEB L. B. (1955) *Basic Processes of Gaseous Electronics*, California: The University Press.
MCDANIEL E. W. (1964) *Collision Phenomena in Ionized Gases*, New York: Wiley.

R. W. CROMPTON

MEMBRANE TECHNOLOGY. *Introduction.* When the transport of different species of matter across an interface takes place at different rates, we call that interface a *membrane*. The existence of solutions with differences in chemical potential in the components across the membrane develops various coupled phenomena or effects which can be utilized in nature or by man to carry out various transport processes. These topics plus an outline of some of the routes to making synthetic selective membranes are the subject of this article. The emphasis will be on liquid solutions, although in many respects the subject matter also pertains to gases.

The easiest concept of a membrane is that of a thin, flexible film, but one can think of more exotic situations which might be considered to be selective membranes. If a mixture of rare gases were brought in contact with a zone of hydroquinone molten on both sides but crystallized in the centre, differential transport of those gases which do and do not form a clathrate complex with this organic crystal might be achieved. A vacuum gap is another example of a less obvious membrane, as is a waxed thread across the air-water interface on a film balance. Recently the concept of a thin supported layer of particles swollen by the solvent has come into favour as a selective, high flux membrane. A liquid oil membrane is easily prepared. In any case, what may be termed the *membrane situation* is shown in Fig. 1. Membrane M...M is the imperfect barrier which differentiates between the transport of the components A and B. For the various membrane phenomena to be present there must be some transport and there must be distance from equilibrium. That is C_{A1}, C_{B1}, T_1, P_1, and E_1 can not all be equal to C_{A2}, C_{B2}, T_2, P_2, and T_2. Given some driving force across the membrane due to this difference in composition, temperature, pressure, or electrical potential then various transport phenomena or effects can develop at the membrane. These membrane phenomena are coupled so that transport of one species leads to some extent to transport of the other in practical cases, and one aspect of disequilibrium engenders others. It is extremely difficult or impossible in an actual experimental situation to have only one of these phenomena occur in isolation. For some of the membrane phenomena of Table 1 to be present one component must form ions in the solution. It is convenient to think of water as a major component, but this is not necessary.

The membrane situation

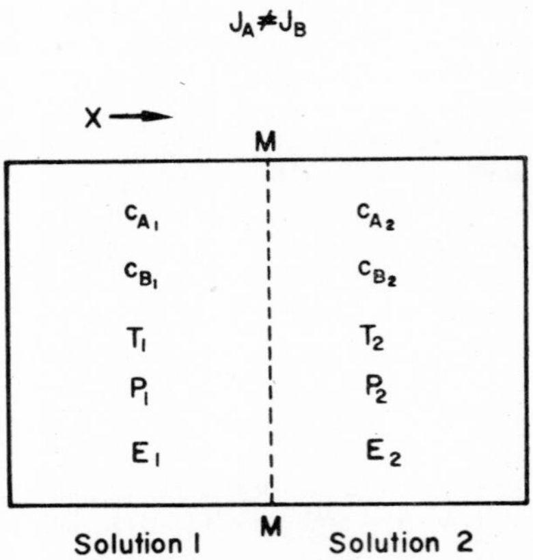

Fig. 1.

Membrane transport phenomena

Phenomenon	*Driving force*	*Primary flux*	*Effect generated*
streaming potential	pressure	solvent	electrical potential
molecular ultrafiltration (hyperfiltration, reverse osmosis)	pressure	solvent	selective transport of solvent
hydraulic permeation	pressure	solvent	transport of solvent
piezodialysis	pressure	solute	selective transport of solute
Soret effect	temperature	solute	(selective) transport of solute
thermoösmosis	temperature	solvent	(selective) transport of solvent
thermal potential	temperature	ionic solute	electrical potential
electrodialysis (ionophoresis)	electrical potential	ionic solute	(selective) transport of solute
electroösmosis	electrical potential	solvent	(selective) transport of solvent
dialysis, diasolysis	concentration	solute	(selective) transport of solute
osmionosis	(triple) concentration	ionic solute	selective transport of solute
concentration potential	concentration	ionic solute	electrical potential
osmosis	concentration	solvent	selective transport of solvent
Dufour effect	concentration	heat	thermal disequilibrium

Streaming potential, hydraulic permeation, thermo-ösmosis, and electro-ösmosis are phenomena which can take place with a single component. Therefore a selective membrane is not necessary. It is possible to envisage a crude electrodialysis without selective membranes but only from a compartment of higher to one of lower concentration. Modern electrodialysis "uphill" into a compartment of higher concentration demands a selective membrane. Depending on what mixture is being purified by dialysis a selective membrane may or may not be required. The other phenomena generated by differences in concentration obviously require a selective membrane. Although what we have termed molecular ultrafiltration, that is separation at the molecular level by pressured permeation, is often called "reverse osmosis", this name is confusing for several reasons. One being that at low concentrations osmotic pressure is not a factor, another being that at high concentrations the selectivity is diminished and again osmotic pressure is lowered. Thirdly, piezodialysis may be considered the opposite of "reverse osmosis" and this certainly is confusing. For a mixture of salt and water, for example, it is preferable to think of molecular ultrafiltration as the preferential solution (hence permeation) of the solvent in the membrane, and piezodialysis as the preferential solution of the salt in the membrane without trying to relate these two effects to osmosis. In a mathematical expression for the flux of matter and heat across a membrane it is possible to delineate these membrane phenomena in separate terms of the equation, hence these terms of Table 1 have some significance. In the practice of laboratory or plant processes to use only one name for what is taking place is an oversimplification.

Membrane transport. In a uniform continuous medium the flux of one component can be expressed as a sum of products of a conductivity with a driving force. The terms are typically of the following form:

Concentration effect $\dfrac{Dc}{\boldsymbol{R}T} \dfrac{-\mathrm{dln}\, c}{\mathrm{d}x}$

Pressure effect $\varkappa \overline{V} \dfrac{-\mathrm{d}P}{\mathrm{d}x}$

Electrical effect $\dfrac{uc}{z\boldsymbol{F}} -z\boldsymbol{F} \dfrac{\mathrm{d}E}{\mathrm{d}x}$

Thermal effect $\dfrac{\mathfrak{L} \cdot Q^*}{T} \dfrac{-\mathrm{d}T}{\mathrm{d}x}$

$$J = -\frac{Dc}{\boldsymbol{R}T}\frac{\mathrm{dln}\, c}{\mathrm{d}x} - \varkappa \overline{V}\frac{\mathrm{d}P}{\mathrm{d}x} - uc\frac{\mathrm{d}E}{\mathrm{d}x} - \frac{\mathfrak{L}Q^*}{T}\frac{\mathrm{d}T}{\mathrm{d}x} \quad (1)$$

Nomenclature

a = activity
A = a component of the mixture
B = a component of a mixture
c = concentration of a component or ion
c_1 = upstream concentration
c_2 = downstream concentration
c_{coion} = concentration of a coion in an ion-exchange membrane
c_{ext} = concentration of a solution external to an ion-exchange membrane
c_{gegen} = concentration of counterions in an ion-exchange membrane
c_{ionex} = ion-exchange capacity of a membrane
c_{intr} = concentration of intruded solute in an ionexchange membrane from the external solution
D = diffusion coefficient
E = an electromotive force
E_{meas} = a measured electrical potential
E_{theory} = a theoretical membrane potential
f = diffuse reflection factor
J = the flux of a component through a membrane per unit time per unit area
K_1 = the Hagen-Poiseuille permeability constant
K_2 = the Knudsen permeability constant
$\varkappa$ = permeability of a continuum
l = length of a pore
L = a generalized coefficient
$\mathscr{L}$ = thermal diffusivity
M = molecular weight
p = vapour pressure
P = pressure
Q^* = heat of transport
r = radius of a pore
$\mathscr{R}$ = rejection figure of merit for a membrane
t = thickness of a membrane
t^+, t^- = transport number of an ion
T = temperature
u = electrical mobility
$\overline{V}$ = partial molar volume
x = direction of membrane transport
z = valence of an ion
α = number of pores per unit area
η = viscosity of a mixture
Λ = mean free path
Π = osmotic pressure
ϱ = density of a mixture

Since we are perforce dealing with a non-equilibrium situation, we must remember that the thermodynamic expressions describe only the driving forces and not the dynamic system.

The notation and concepts of irreversible thermodynamics are well suited to describe the coupled forces and flows, although they do not treat the causes or handle the change in coefficients at the membrane interfaces. Using the notation of Fig. 1:

$$\begin{aligned} J_A = & -\boldsymbol{R}T\, L_{AA}\, \mathrm{dln}\frac{a_A}{\mathrm{d}x} - \boldsymbol{R}T\, L_{AB}\, \mathrm{dln}\frac{a_B}{\mathrm{d}x} \\ & - (L_{AA}\overline{V}_A + L_{AB}\overline{V}_B)\frac{\mathrm{d}P}{\mathrm{d}x} \\ & - (L_{AA}Z_A + L_{AB}Z_B)\, F\frac{\mathrm{d}E}{\mathrm{d}x} \\ & - \frac{1}{T}(L_{AA}\mathfrak{L}_A Q_A^* + L_{AB}\mathfrak{L}_B Q_B^*)\frac{\mathrm{d}T}{\mathrm{d}x} \end{aligned} \quad (2)$$

$$J_B = -\boldsymbol{RT}\, L_{BA}\, \mathrm{dln}\, \frac{a_A}{\mathrm{d}x} - \boldsymbol{RT}\, L_{BB}\, \mathrm{dln}\, \frac{a_B}{\mathrm{d}x}$$

$$-(L_{BA}\overline{V}_A L_{BB}\overline{V}_B)\frac{\mathrm{d}P}{\mathrm{d}x} - (L_{BA}Z_A + L_{BB}Z_B)\, F\frac{\mathrm{d}E}{\mathrm{d}x}$$

$$-\frac{1}{T}(L_{DA}\pounds_A Q_A^* + L_{BB} + \pounds_B Q_B^*)\frac{\mathrm{d}T}{\mathrm{d}x} \qquad (3)$$

Some simplification of this formalism results from the equality of cross coefficients (Onsager 1931):

$$L_{AB} = L_{BA} \qquad (4)$$

One consequence of equations 2 and 3 is that setting a given driving force at zero does not eliminate any fluxes. Furthermore, the coupling of fluxes implies that it would be almost impossible to carry out an isothermal experiment.

There are three facets of membrane structure which must be analysed in any particular case to assign their respective contribution to membrane transport. Normally each of the three makes some contribution no matter how small. These contributions by the structure of the membrane itself are:

(a) sieving effects,
(b) diffusive effects, and
(c) electrostatic effects.

If a membrane has a rigid, microporous, crystalline structure then the steric contribution to semipermeability can be appreciable or even paramount depending on the relative size of the transported components and the interstices. It is assumed that for rigid, microporous membranes the transported fluid does not dissolve in the material of the membrane. This sieving action is the basis of separation in the particulate or colloidal range using membrane ultrafilters which discriminate at from 100 ängstroms to several microns. Recent advances in the control of polymeric microstructure have enabled the manufacturers of this sieving type of thermoplastic membrane to make in the laboratory materials of narrow size distribution down to 50 and perhaps even 25 ängstroms. Although synthetic aluminosilicates (molecular sieves) are routinely produced which have pore sizes of about three to twenty ängstroms, the use of this latter type of material is not considered here because it is available only as particles not sheets. When the radius of the pores in a membrane is at least several times larger than the mean free path of the molecules in the feed stream, then the Hagen-Poiseuille equation for tubular flow describes the flux:

$$J_A = K_1 C_A (P_1 - P_2) \qquad (5)$$

$$K_1 = \frac{\alpha \pi r^4 \varrho}{8 l \eta M}. \qquad (6)$$

If the radius of the pore is small enough to be of the same magnitude as the mean free path of the fluid being transported, then there is interaction between the walls and the fluid some of which is reflected diffusely:

$$J_A = K_2 C_A (P_1 - P_2) \qquad (7)$$

$$K_2 = \alpha \frac{8\pi r^3}{3l} \frac{1}{\sqrt{(2\pi MRT)}} \frac{2 - f}{f}. \qquad (8)$$

This equation, due to Knudsen, relates the number and the dimensions of the pores with the fraction of molecules emitted from the walls. There may be a transitional region of effective pore size which is intermediate between Hagen-Poiseuille flow and Knudsen flow where the radius of the pore is only slightly larger than the mean free path of the transported fluid. In this transitional region both the properties of the fluid and the pores are related thus:

$$J_A = \frac{\alpha C_A}{8\eta l}\left(1 + \frac{16(2-f)}{f}\frac{\Lambda}{r}\right)\frac{\pi r^4}{512\boldsymbol{R}T}(P_1^2 - P_2^2). \qquad (9)$$

For rigid membranes such as porous ceramics, leached inorganic glasses of specific structure, metal membranes made by removing only one component of an alloy (e.g. distilling the zinc from brass to make a porous copper), or any stable material with effective pore size greater than about 40 ängstroms, this sieving action is the main mode of separation.

When the membrane is an amorphous solid such as an elastomer or glassy polymer, then no configuration of the polymer exists for a sufficient duration to consider the transport in terms of fluid flow in a pore. Diffusion on a molecular basis of the permeant becomes the chief mechanism of transport. Eq. 2 and 3. Solubility, not molecular size, becomes the criterion of transport, although molecular cross-section does determine the diffusion constant. Pure Fickian diffusion with a diffusion constant insensitive to concentration is rare in liquids although experimentally observed for gases:

$$J = -D\frac{\partial C}{\partial \chi} \qquad (10)$$

$$\frac{\partial C}{\partial t} = D\frac{\partial^2 C}{\partial \chi^2} \qquad (11)$$

In many practical cases D, the diffusion "constant", is not constant but is a function of concentration and also direction (anisotropy). When a mixture of two or more permeants are present to interact with the membrane both the solubility and the diffusivity become multibody problems of the interacting, cross-coefficient type shown in equations 2 and 3.

When one or more components being transported dissociate to form ions, then the presence of ion-exchange groups in the membrane becomes the dominant consideration. So important are ion-exchange membranes to the study of membrane transport that perhaps three-quarters of the literature is devoted to experiments involving ionic selectivity. The degree of control of transport number by the presence of fixed ionic charges in the membrane is described by the Donnan relation. When the concentration of fully ionized fixed charges of the same sign is sufficiently high (>1 meq/dry g), then the intrusion of external ions is so low that the concentration of counterions in the membrane is approximately the same as the concentration of ion-exchange material. This means that in an electric field only the counterions are transported and the transport number for them approaches one, while that of the coion approaches zero. This electrolytic selectivity of an ion-exchange membrane for ions of the same charge as the counterions is often called permselectivity, while semipermeability is used to describe selective transport by sieving or strictly diffusive mechanisms. In a typical case the effective concentration of the fixed-charge ion-exchange material in the membrane is several moles. If the external salt solutions is a typical value such as 0·1 N, then by Donnan exclusion only about 0·002 N of salt will be intruded for transport:

$$c_{\text{intr}} \approx c_{\text{coion}} \approx \frac{c^{\text{a}}_{\text{ext}}}{\bar{c}_{\text{ionex}}} = \frac{(0{\cdot}1)^2}{5} = 0{\cdot}002\ \text{N}. \qquad (12)$$

In an electric field the flux of counterions (gegenions) will be high, with a transport number approaching one:

$$J_{\text{gegen}} = \mu c_{\text{gegen}} \frac{\mathrm{d}E}{\mathrm{d}x} \quad \text{high} \qquad (13\text{a})$$

$$J_{\text{intr}} = \mu c_{\text{intr}} \frac{\mathrm{d}E}{\mathrm{d}x} \quad \text{low}. \qquad (13\text{b})$$

A low area-resistance for an ion-exchange membrane (<20 ohm-cm²), a value independent of thickness, is an imperfect criterion for ion-exchange membranes because it could be achieved by an amphoteric or mosaic membrane of both types of fixed-charge (positive and negative). The usual criterion is to measure the membrane concentration potential for two known solutions of the same salt, classically potassium chloride, and compare this value with the theoretical one:

$$\text{Permselectivity} = \frac{E_{\text{meas}}}{E_{\text{theory}}} = \frac{\dfrac{t^+ - t^-}{t^+ + t^-} - \dfrac{\boldsymbol{RT}}{\boldsymbol{F}} \ln \dfrac{a_2}{a_1}}{\dfrac{\boldsymbol{RT}}{\boldsymbol{F}} \ln \dfrac{a_2}{a_1}}$$

$$= 2t^+ - 1 \text{ for cation-exchange membranes.} \qquad (14)$$

The permselectivity value is dependent on the concentration (activity) of the external solution, but should be greater than 90 per cent in solutions of 1 N or less for an effective membrane. The area-resistance is less dependent on the concentration of external solution but increases markedly for uni-univalent salt solutions of less than 0·001 N indicating that a minute amount of coion may be necessary to facilitate the transport of gegenions from one fixed-charge site to another.

Membrane processes. We shall now briefly describe some of the processes employed in a practical way to separate mixtures by means of membranes. The ones emphasized in current research are pressure permeation, pervaporation, dialysis, electrodialysis, and electrogravitation. In the last section some routes to making the membranes employed in these processes will be outlined.

Pressure permeation (Fig. 2) is the separation of a mixture by applying high pressure to a stirred solution on the upstream side of a selective membrane

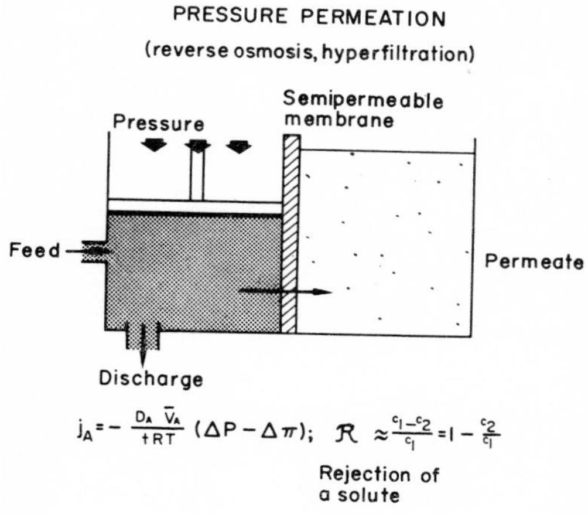

Fig. 2.

which transports only one component. Since the applied pressure must overcome at least the osmotic pressure of the upstream mixture the process is often termed reverse osmosis particularly in the desalination industry. Other names are hyperfiltration and molecular ultrafiltration. The process is used on an experimental scale for the renovation of sewage, the purification of raw sugar juice, the dehydration of fruit juices, and the purification of pulping wastes in the paper industry. Typical operating pressures are 30 to 125 atmospheres. Transport of a single component such as water is described by the Poynting equation:

$$J_A = \frac{-D_A V_A C_A}{t\,\boldsymbol{RT}}(\Delta P - \Delta\pi). \qquad (15)$$

Differentiation between components is achieved by using a membrane which gives different values of C, aided sometimes by different values of the diffusion constant. By far the most widespread membrane for aqueous separations is cellulose diacetate, cast in such a way as to provide a thin layer about one-third of a micron thick (low t) on a porous mass of the polymer which serves to give this asymmetric membrane physical strength. The largest known installations have several thousand square feet of membrane area at this writing.

Since a higher gradient of concentration can be provided in a membrane by withdrawing the transported material on the downstream side as a gas under vacuum than by pressuring the upstream side, there is merit to the scheme of providing a differential in pressure in this fashion. It has been less studied than pressure permeation, however, because it cannot tolerate solid components such as salt and sugar nor can high boiling permeants be used. If the vacuum pump is not effective enough to vaporize the material on the downstream side (actually part of the membrane is desiccated), then this process just becomes pressure permeation with one atmosphere of pressure. Most of the research work has been done with petroleum fractions using compositions containing ethyl cellulose as the membrane. One interesting area of research is the separation of the isomers of xylene. In the U.S.A. the words permeation and evaporation have been combined to form the term *pervaporation* for this process. No installations greater than a few square feet in area are known to the writers (Fig. 3).

For many years particulate matter or dissolved polymeric material has been washed free of small molecules by placing it across a membrane from a stream of fresh water. Under the concentration gradient the impurities "dialysed" out while the membrane, usually a form of regenerated cellulose, held back the protein, clay, or other material. With the advent of polyvinyl chloride membranes of good chemical resistance dialysis has been used to recover copper and nickel sulphate from their solutions in sulphuric acid, and the latter from steel "pickling liquors". At the laboratory level finer separations such as bacitracin-A from gramicidin-S have been performed using swollen cellophane as the membrane.

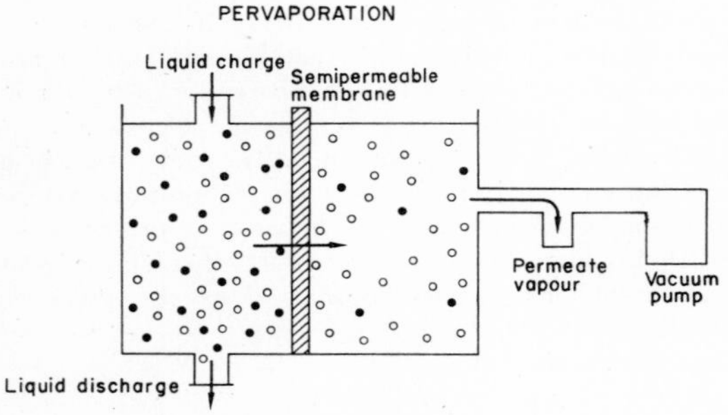

Separation factor $= \frac{c_{A_2}/c_{B_2}}{c_{A_1}/c_{B_1}}$

For non-interacting permeants $J_A = \frac{D_A}{t}(X_A P_A - P_A)$ $\quad P_A \rightarrow 0$

For interacting permeants $J_A = \frac{1}{t}\int_{c_2}^{c_1} D_A\,dc$

Flux = (diffusivity)(solubility)

Fig. 3.

When the downstream solvent is a different material from all of the upstream components, the term *diasolysis* should be used for this variant of dialysis.

Electrolytic processes. The word *electrodialysis* refers to transport of ions across a membrane. Although inefficient deionizing processes can be carried out without ion-selective membranes, modern, conventional electrodialysis, employs ion exchange membranes in alternate array as shown in Fig. 4, which illustrates desalination. By placing the membranes in this alternate fashion every other compartment becomes a diluting compartment and every other compartment becomes a concentrating one, since the ions can move through only the type of ion-

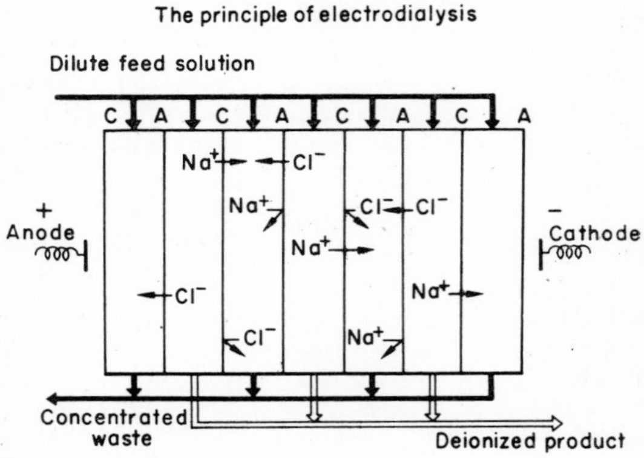

Fig. 4.

exchange membrane for which they are the counterion. Several hundred small to medium size electrodialysis plants (up to 10,000 square feet of membrane surface) are operating in arid zones of the world today desalinating brackish water by means of this type of equipment. In Japan there are several plants run under different operating conditions to produce a 20 per cent saline brine from sea water (3·5 per cent saline) in order to make salt. The largest of these has 400,000 square feet of membrane surface; the diluted brine, not being potable under these conditions, is discarded. In addition to the concentration of brine or desalination, other processes using electrodialysis are the removal of salts from molasses, the partial dimineralization of milk and whey, recovery of paper pulping wastes, lowering of citric acid levels in citrus juice, recovery of spent acids, the purification of colloidal preparations, and metathetical inorganic reactions. If carried out for long times, electrodialysis can be used to produce very high grade conductivity water (25 million ohm-cm).

The proper design of an electrodialysis plant is a complex engineering task since there are so many variables. The so-called "stacks" of several hundred membrane pairs may be operated in series or parallel under continuous or batch conditions. The internal flow paths of small stacks may be internally staged. Some design parameters in addition to general layout of the plant are power level, type of waveform, current density, current reversal, direction of hydraulic flow, electrode rinse streams, cell spacing, cell dimensions, flow patterns, flow velocities, residence time, type of gasketing, method of turbulence promotion, as well as types of membranes. Many groups attack this type of problem with a computer. Since temperature plays an important role in determining conductance of ionic solutions, it should be noted that modern practice calls for high temperature operation to increase efficiency of this electrolytic process. The more recent Japanese plants are in the southern part of that country and American experience is more favourable in the southwest rather than the north-central states. The Israelis also prefer higher temperature operation in the Negev desert. Desalinating plants cannot be operated at still higher temperatures because of chemical and mechanical degradation of the ion-exchange membranes, most of which are limited to 60–70°C.

Since the economic optimization of an electrolytic membrane process involves the weighing of capital costs, power costs, and membrane area in relation to the amount of product, it may be advantageous to decrease capital costs at the expense of higher membrane area. One alternative is the use of one-type of ion-exchange membrane only in the electrogravitational process shown in Fig. 5. Here cation-exchange membranes, which are cheaper to make and have longer life than anion-exchange membranes, are simply hung in a tank with no supports, gasketing.

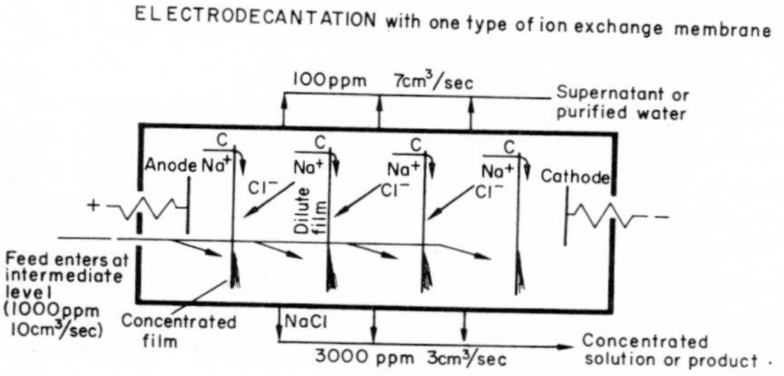

Fig. 5.

or intrastack flow paths. When the current is turned on, by transport depletion a layer of less dense liquid rises on the anodic side of each membrane and a layer of more dense liquid falls by gravity on the cathodic side. This is true whether the electrolytic transport is by dissolved ions or colloidal particles such as rubber latex. In the case of ionic solutes the difference in density based on concentration is augmented by the fact that the layer of lower concentration is heated more and the layer of higher concentration heated less by electrical power losses further reinforcing the difference in density. The moving films are about one-sixteenth of an inch thick. The figures in Fig. 5 are for 50 sheets of membrane eight inches square separated by compartments one-quarter inch thick. At a current density of two milliamperes per square centimetre the product water coming to the top in the case of desalination is 5°C hotter and the brine is 1°C hotter than the feed. For a latex any membrane can be used which discriminates between water and a colloid. In this case it is particularly necessary to reverse the current periodically to prevent the latex from coagulating on the membrane. Reversal of current is simpler in electrondecantation than conventional electrodialysis since the purified water is always drawn from the top and the brine from the bottom. This process has been used commercially in Malaya in the natural rubber industry. Although extensive laboratory work has been done with electrodecantation, no commercial desalination plants are known to be in existence at this writing. Labour costs and capital costs are low for this process, but power costs are necessarily higher.

The preparation of membrane surfaces. One thinks first of synthetic, organic polymeric films as the raw materials for making membranes. Many other materials may be considered for the actual working surface that makes the separation at the molecular level, or for the support or matrix in which the working surface is found. Some examples are clay, shale, glass, copper, silver, ceramics, resin-impregnated paper, gelatin, and woven material of any type. Figure 6 shows alternative operations in the handling of organic polymers. There are analogous steps for handling glass or other inorganics.

Different processes require different types of membranes, but in any event some definitions may be helpful at this point. A *homogeneous membrane* is one that visually is all of one phase and has uniform characteristics from one side to the other. It is usually transparent or at least translucent. Crystalline membranes may be homogeneous from a phase point of view, but normally they are specifically named, e.g. gold, clay, palladium, shale, graphite, copper membranes. A membrane is termed homogeneous if it, visually, has one phase even if it obviously has more than one phase on a submicroscopic level. For example, a hydrophilic membrane is called homogeneous even if water is a necessary component. A *heterogeneous membrane* is one that, visually, has one component dispersed in another. Usually one phase is amorphous. Common examples include rubber containing an inorganic filler, or ion-exchange resin ground to a fine powder and milled into a plastic. In the case of the rubber, the filler plays only a mechanical role. In the second example, the powdered ion-exchange resin markedly affects the transport properties. Precipitate membranes are heterogeneous membranes made *in situ* by chemical or electrochemical reaction. A *reinforced membrane* is one in which a woven material, or a non-woven material such as paper, is used to add mechanical

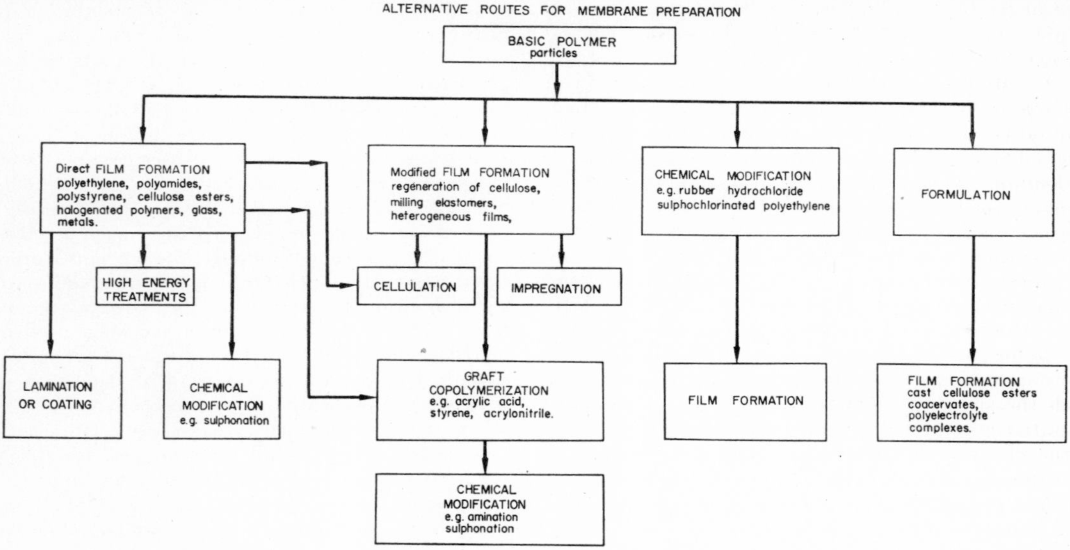

Fig. 6.

support. An *asymmetric membrane* is one having a graduated or laminated structure that influence its transport properties. Coated cellophane, "skinned" molecular ultrafiltration membranes, or graft copolymer membranes having a gradient in the graft are examples.

An *ion-exchange membrane* is one having sufficient ion-exchange capacity to render it electrolytically conductive. Arbitrarily this is stated as having an area resistance of less than 20 ohm-cm^2 in a uni-univalent electrolyte. Cation-exchange membranes have a cation as the electrolytically mobile counterion; anion-exchange membranes, correspondingly, have a mobile anion. *Amphoteric ion-exchange membranes* have cationite and anionite regions so intermingled that they are physically indistinguishable. A *mosaic* membrane is a single ion-exchange membrane whose cationite and anionite areas are distinguishable and in parallel. Apparently many natural membranes have a mosaic (and asymmetric) structure. A *bipolar membrane* is an integral, laminar combination of anionite and cationite regions; it is the electrolytic analogue of an electronic p-n junction. A bipolar array is a laminar conjunction of individual cationite and anionite membranes either physically touching or separated by a porous spacer. Ion-exchange membranes may also be made asymmetric, usually to discriminate between ions of the same sign but different charge.

Originally, investigators used naturally occurring materials for membranes, such as apple skin, gelatin, natural rubber, pig and sheep entrails, and parchment. A second stage of development was the use of regenerated cellulose and the esters of cellulose. Even today, these synthetic cellulosic esters are among the most widely used membrane-forming materials. Within the last 30 years, polyesters, alkyds, polyethylene, sulphonated polystyrene, polytetrafluoroethylene, and many other pure synthetics have become important in the preparation of membranes for laboratory and commercial use.

It is the flexibility in processing, as shown in Fig. 6, along with comparatively good uniformity, and availability at a reasonable price, that leads one to consider first organic polymers. In the laboratory, it may be possible to make essentially the same product by different routes: for example, chemical modification followed by film formation, or film formation followed by chemical modification. When various sequences are attempted on a large scale, one route generally shows itself to be superior to the others from a practical point of view. One step that has been omitted from Fig. 6 may be termed physical modification or conditioning. Normally, specific physical treatment such as polishing, orienting, drawing, relaxing, or soaking are interspersed throughout a process. This is especially true at the end, when one or more treatments may be used so as to have the membrane conform to the specifications generated by its intended use. By far the most common way to make a membrane is to form a film directly. Billions of square feet of plastics and metals are formed into protective sheaths and diffusional barriers by this process. Chemical modification of a film already formed is one route to the preparation of ion-exchange membranes. Sulphonation of a film (or thin sheet) containing polystyrene is a good example. Another reason for carrying out a chemical modification is that the prepolymer is thermoplastic or soluble, but the final product is either thermoset or intractable. An example of this is the cyclization by dehydration of a film of polyhydrazide to make polyoxadiazole. Graft copolymerization is a process that either covalently links additional polymer to the original film (grafting) or entangles the new polymeric addendum to the substrate so intimately that, functionally, the modified film appears to be grafted. This process can be initiated chemically, thermally, mechanically, or by high-energy radiation. The asymmetric diffusion membrane and several ion-exchange membranes are synthesized using a grafting step. Although grafting of one polymer onto another is a technique at least 35 years old, it was reborn with the development of radiation methods. Concomitant with the study of high-energy radiation as a means of initiating graft copolymerization has been a study of the effects on the substrate polymer alone. Radiation cross-linking is a potential tool in the control of the physical properties of membranes for gaseous diffusion and liquid permeation. A novel means for making isoporous membranes is the "nucleopore" process. In this process, a film is exposed to massive, high-energy particles such as fission fragments. Defects are created that, upon chemical etching, lead to uniform pores of controllable size from 25 Å up to hundreds of Ångstroms in mica, minerals, and amorphous plastics. The term *cellulation* in Fig. 6 refers to such processes as the formation of a void by leaching out a component, or the blowing of a foam. A porous battery-separator can be made by dispersing fine starch particles in polyvinyl chloride and then leaching out the starch. A thin section of conventional closed-cell foam may be considered a membrane for gaseous diffusion.

Modified film formation means a procedure more complex than direct extrusion. Of prime importance is the regeneration of cellulose to make a plasticized film, cellophane. Cellulose membranes have been extensively used by laboratory workers for decades and are still the membrane of choice for biological separations, haemodialysis, and other uses. The incorporation of powders in elastomers to make heterogeneous membranes has already been mentioned. In the early period of making ion-exchange membranes, various procedures were published for impregnating or adsorbing dyes onto cellulose and its derivatives. Examples are the adhering of basic dyes to cellulose acetate and the adsorbing of dyes to collodion. These methods lead naturally to the incorporation of additives in casting solutions before film formation. The casting of polymeric membranes goes back almost a century and is intertwined with the history of regenerated cellulose. Much of the pioneering work by Michaelis, Meyer, Sollner, Grabar, Elford, Bechhold, Zsigmondy and their students on the preparation of membranes for ultrafiltration, electrodialysis, electro-osmosis, thermo-osmosis, dialysis, and electrolytic cells was devoted to the formulation and casting of cellulosic membranes and mixtures therewith.

For the various electrolytic processes, the goal is to make a chemically and physically stable film containing enough ion-exchange capacity to result in a low electrolytic resistance. For dialysis and ultrafiltration, a precise, reproducible, isoporous microstructure with the proper balance of discrimination and solvent flux is sought. In the period between 1915–1950, one might say that homogeneous ion-exchange membranes were built up by starting with a castable film-forming polymer and adding ionogenic material, dyes, proteins, synthetic polyelectrolytes, and the like. The latter cannot be cast directly themselves because, being polar, they are brittle. On the other hand, graft copolymerization of an ion-exchange group directly or indirectly to a film may be looked upon as the controlled destruction of its crystallinity. This is stopped at the optimum balance of mechanical strength and electrolytic conductivity. Analogously, for non-electrolytic membrane separations, one builds up the maximum amount of plasticizer (water), usually by the incorporation of hydrophilic salts such as zinc chloride and magnesium perchlorate, in order to get a high, isoporous void volume and sufficient mechanical strength. Superimposed upon this search for hydrophilic-hydrophobic balance and controlled isoporosity is the desirability of increased mechanical strength by cross-linking, polycondensation, entanglement, radiative post-treatment, or polymerization in a mould.

To conclude, mention will be made of two contemporary developments of particular importance in the field of pressure permeation. Both are successful efforts to provide a thin membrane interface in order to achieve high flux without sacrificing selectivity. In both cases desalination was the immediate goal although the membranes will remove other compounds from water as well. An asymmetric membrane of cellulose diacetate comprising a thin selective layer about one-third of a micron thick which supported on a more porous layer of the same material for mechanical support is made by casting a solution of the polymer dissolved in acetone containing formamide as a plasticizer. Immediately after casting a layer about one-tenth of a millimetre thick on a glass plate, one plunges the glass plate in water at about 75°C. The tough skin forms on the top side; both acetone and formamide are washed out leaving the asymmetric, hydrophilic membrane on the glass, ready for use. Another thin, highly selective membrane is made by adding a small amount (10^{-3} to 10^{-4} molar) of boiled zirconyl chloride to the saline water. In the heating a hydrous zirconia colloid is formed. By filtering this suspension, through a porous supporting plate, an active membrane of zirconia gel is formed on the surface of the plate. Both of these membranes are employed at several hundred pounds of external pressure to remove dissolved salts from water at high efficiencies.

Bibliography

JOHNSON J. S., DRESNER L., and KRAUS K. A. and SHAFFER L. H. and MINTZ M. S. (1966) in *Principles of Desalination* (K. S. Spiegler, Ed.) New York: Academic Press.

KLEINZELLER A. and KOTYK A. (Ed.); Symposium *Membrane Transport and Metabolism* at Prague 1960, New York: Academic Press.

LAKSHIMINARAYANAIAH N. (1965) Transport Phenomena in Artificial Membranes, *Chem. Rev.* **65**, 491.

Membrane Phenomena (1956) *Disc. Faraday Soc.* **21**.

SCHÖGL R. (1964) *Stofftransport durch Membranen*, Darmstadt: Steinkopff.

WILSON J. R. (Ed.) (1960) *Demineralization by Electrodialysis*, London: Butterworths.

H. Z. FRIEDLANDER and R. A. GRAFF

MICROWAVE FREQUENCY STABILIZERS. It is necessary to consider in detail the parameters required to describe the output of an oscillator in order to understand the action of any microwave frequency stabilizer. The power spectrum of an oscillator is shown in Fig. 1. The useful output power at any instant of time is contained in a bandwidth $\Delta\nu_i$ with a mean frequency ν. The background noise pedestal is assigned a width $\Delta\nu_n$. It is often more convenient to use a standard deviation σ_i to describe the width of the power spectrum, since frequency measurements usually involve the statistical analysis of a series of numerical values. The effective Q of the power spectrum, defined as the Q of an equivalent tuned circuit oscillator, i.e. $Q \simeq \nu/\Delta\nu_i$, is another alternative parameter. In general, the mean frequency ν is not constant, but varies with time. This is indicated by the dotted power spectra in Fig. 1. The range over which the mean value can fluctuate is a function of time and is denoted by $\Delta\nu(t)$. The range of mean values depends on the time required to make one frequency measurement and the period over which a series of measurements is made. For example, a single measurement may take t seconds to make, and subsequent analysis made of a series of measurements over a period of T seconds, minutes or days, so that a complete description requires all this information. The stability may be described as x parts in 10^n per t seconds for T seconds, minutes or days. For the purpose of comparison in this article, typical values for x and n are given assuming that $t = 1$ second and T a few minutes or a few hours, usually one day. These values refer specifically to the stability of continuously operating 10 GHz reflex klystrons. The stabilizers described herein may be used with any continuously operating microwave oscillator which is voltage tunable, such as a klystron, a backward wave oscillator or a solid state microwave power source.

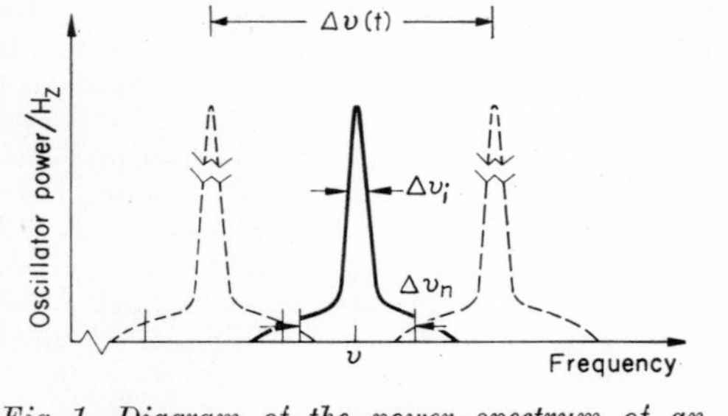

Fig. 1. Diagram of the power spectrum of an oscillator and its fluctuation illustrating the symbols ν, $\Delta\nu_i$, $\Delta\nu_n$ and $\Delta\nu(t)$ defined in the text.

High Q cavity stabilizer (*Stalo*). The frequency fluctuations of an oscillator with a low Q cavity may be reduced considerably by presenting a high Q transmission cavity as a load impedance. This cavity behaves as a bandpass filter with the additional property that the phase of the input coupling impedance provides negative feedback to stabilize the frequency of the oscillator. This is called a stalo (*sta*bilized *l*ocal *o*scillator) stabilizer. The Q of the external cavity load replaces the oscillator Q as the factor determining the frequency fluctuations. The changes caused by ripple or other instabilities in the oscillator power supply are reduced by the ratio of the oscillator cavity Q to that of the external cavity. The stability of the stabilized system is therefore dependent on the quality of the unstabilized system. The stabilization factor S is defined as the modulation sensitivity of the unstabilized oscillator divided by that for a cavity stabilized oscillator.

The advantages of a passive device such as a stalo stabilizer are that it requires no external power supply and that there is no limit to the frequency of the disturbance which can be reduced. The electronic stabilizers to be described in later sections must have some form of bandwidth limiting to avoid oscillations in the control circuits. The main disadvantages of a stalo system are two-fold. Firstly, there is a practical limit of about 100,000 to the maximum Q of a cavity, so that the effective Q of the power source cannot be greater than this value. Secondly, the coupling to the external high Q cavity results in a loss of about 10 dB in the available microwave power. The coupling may be reduced to provide more available power if reduced stabilization is acceptable. The long-term stability is poor because the stalo cavity is susceptible to mechanical strains and external temperature fluctuations even if it is constructed from invar.

Typical values of the parameters for an oscillator stabilized with a stalo are

$$\Delta\nu_i = 5 \times 10^5 \text{ Hz}$$

$$\Delta\nu(t) = 1 : 10^5/\text{minute}$$

$$\Delta\nu(t) = 1 : 10^4/\text{day}$$

$$S = 20.$$

D.C. electronic frequency stabilizers. The first d.c. electronic frequency stabilizer for a reflex klystron was described in detail by R. V. Pound in 1946. The effective Q of a microwave oscillator was increased by employing a high Q reference cavity as a frequency discriminator. An error signal, dependent in phase and amplitude on the difference between the reference cavity resonant frequency and the oscillator frequency was produced and used to reduce frequency fluctuations in the output. The system is shown schematically in Fig. 2 (a). A description of the operation of a discriminator of this type requires knowledge of the properties of a magic-T waveguide junction and the change in impedance of a resonant cavity near resonance. When the microwave frequency is far from the

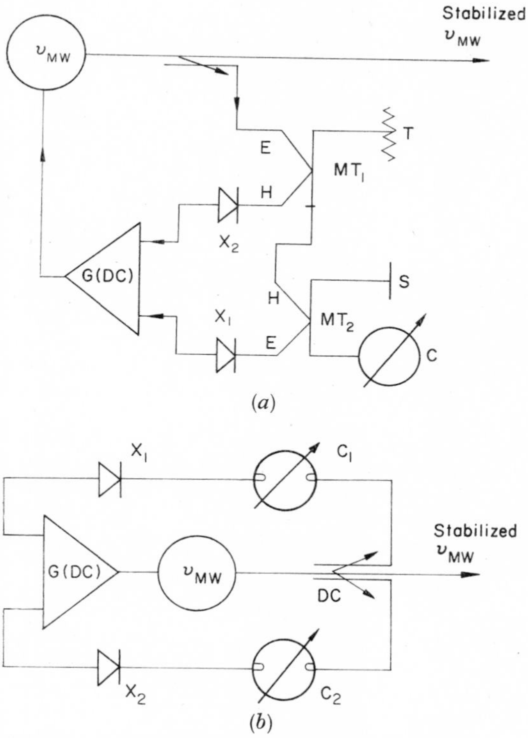

Fig. 2. Block diagram of d.c. electronic, balanced bridge frequency stabilizer using (a) a reference cavity and (b) two reference cavities as a microwave discriminator.

cavity resonance, or at its resonant frequency its impedance is real. When a small difference exists between the microwave frequency and the cavity resonant frequency, the impedance has a reactive component. The path length of the coupling from magic-T MT_2 to the cavity C and the short S are chosen so that waves reflected from these loads returning to MT_2 are $\pi/2$ radians out of phase. When the cavity resonant frequency corresponds to the oscillator frequency, a property of a magic-T is that when incoming waves in the 1-arm and 2-arm are $\pi/2$ radians out of phase the incoming power divides equally between the asymmetric E- and H-arms. The microwave crystal detector X_1 rectifies the power leaving the E-arm. The microwave crystal detector X_2 rectifies half the power leaving the H-arm of MT_2, because of the power division in MT_1. When a small difference exists between the resonant frequency of the cavity and the oscillator frequency, the imaginary part of the reflection coefficient of the cavity gives rise to an increase in the power at X_1 and a decrease in the power at X_2 for a difference of one sign and conversely for the other sign. A frequency discriminator is established by adjusting the power levels incident on X_1 and X_2 so that each crystal detector has the same d.c. output level when the oscillator frequency corresponds to the resonant frequency of the cavity. Frequency changes in the oscillator give rise to voltage changes which can be applied to the frequency controlling electrode of the microwave generator after amplification and phase correction. This minimizes the difference between the frequency of the oscillator and the cavity.

This stabilizer is difficult to use because it depends on the stability of the balance of two microwave bridges and on the stability of the difference in the microwave path lengths in the symmetric arms of MT_2. All these factors must be readjusted for satisfactory operation at different frequencies. A simplified d.c. stabilizer with approximately the same stabilization factor as the Pound system was described by M. J. A. Smith in 1962. Figure 2b shows the circuit diagram for this stabilizer which uses no balanced circuits and has no critical path lengths. Two identical high Q-resonant cavities are employed, each fed with an equal sample of the power output of the oscillator. The resonant frequencies of the two cavities, C_1 and C_2, are equally displaced above and below the desired operating frequency by an amount less than the half-width of the cavity resonance. The oscillator is initially tuned to the operating frequency. When the frequency increases C_1 transmits more power to crystal X_1 and C_2 reduces the power to X_2. The combination acts as a frequency discriminator. The differential d.c. output voltage between X_1 and X_2, after amplification, provides the correction signal controlling the oscillator. The operating frequency may be changed by simultaneously tuning the two resonant cavities maintaining the same frequency difference.

The properties of these two d.c. electronic frequency stabilizers are very similar. The latter is more convenient to use. The spectral purity of the output will be governed by the Q of the reference cavities and the large low frequency noise power characteristic of the microwave crystal detectors.

Typical values of the parameters for an oscillator stabilized by a d.c. electronic frequency stabilizer are

$$\Delta\nu_i = 5 \times 10^5 \text{ Hz}$$

$$\Delta\nu(t) = 1 : 10^6/\text{minute}$$

$$\Delta\nu(t) = 1 : 10^4/\text{day}.$$

A.C. electronic frequency stabilizers using a reference cavity. The first widely used a.c. frequency stabilizer was also described in detail by R. V. Pound and, consequently, is known as the "Pound" stabilizer. It avoids the gain instabilities common to all d.c. amplifiers by introducing an intermediate frequency (i.f.) which acts as the information carrier. The crystal demodulators are operated in a frequency band where their noise figure is minimized. The control loop employs an inherently more stable high gain i.f. amplifier.

Figure 3 is the circuit diagram for this stabilizer. A sample of the microwave power is extracted from the output by a directional coupler and fed into a magic-T bridge MT_1. The power division in the bridge results in one half of the input power reaching the reference cavity C and the other half arriving at the microwave crystal demodulator X_1. This crystal is

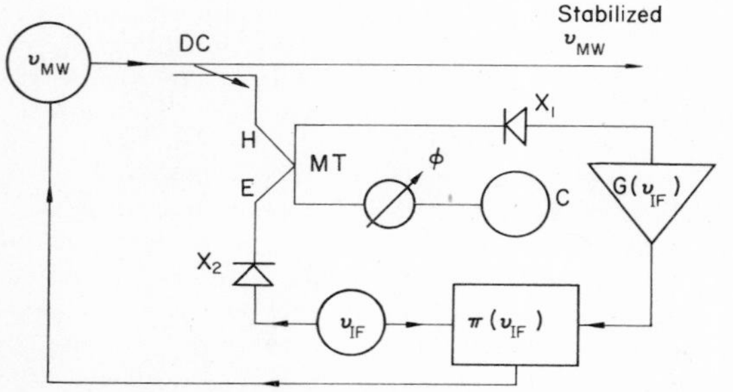

Fig. 3. Block diagram of "Pound" a.c. electronic frequency stabilizer using a reference cavity.

adjusted so that it appears as a matched termination to the waveguide at the operating frequency. The frequency dependent reflection coefficient of the reference cavity is adjusted to be zero when the cavity is tuned to the desired output frequency. Power reflected from C returns to MT_1 where one half is coupled out of the stabilizer through the H-arm. The other half reaches the microwave crystal demodulator X_2. X_2 is initially matched; however, its reflection coefficient is modulated by the i.f. oscillator, producing sidebands in the reflected power, displaced above and below the oscillator frequency by the intermediate frequency. This is usually 30 MHz, however, satisfactory operation can be obtained at much lower frequencies. The reflected power containing the sidebands passes into MT_1 where it again divides. One half reaches C and plays no further role in the operation of the stabilizer because its power level is an order of magnitude below the incident power level. The other half returns to X_1. Microwave synchronous demodulation occurs using the coherent power at the oscillator output frequency as the reference signal. The i.f. voltage developed in X_1 is proportional in magnitude and phase angle to the microwave power reflected from C if the phase shifter Φ in the cavity arm of the balanced bridge is correctly adjusted. Synchronous i.f. demodulation produces a d.c. error signal dependent in sign on the phase of the i.f. input, which itself is dependent on the difference between the cavity resonant frequency and the oscillator output frequency. This d.c. error signal is applied to the tuning electrode of the oscillator to reduce its frequency fluctuations. The spectral purity of the output of a Pound stabilized system is approximately the same as it would be if the klystron cavity Q was equal to the reference cavity Q. The stability of the mean value of the output frequency will be the same as for a d.c. electronic system, over a longer period of time, i.e. $1:10^6$ or 10^4 Hz at X-band.

Numerous other circuits have been developed which operate in essentially the same manner as the Pound system. Microwave synchronous demodulation is used to produce an i.f. voltage which is dependent in phase and amplitude on the difference between the oscillator output frequency and the reference cavity resonant frequency. The major difference in these systems is the method of generating the microwave sideband frequencies which carry the required information.

Two of the more noteworthy alternatives are shown in Fig. 4. The system in Fig. 4(*a*) relies on frequency modulation of the oscillator to produce the

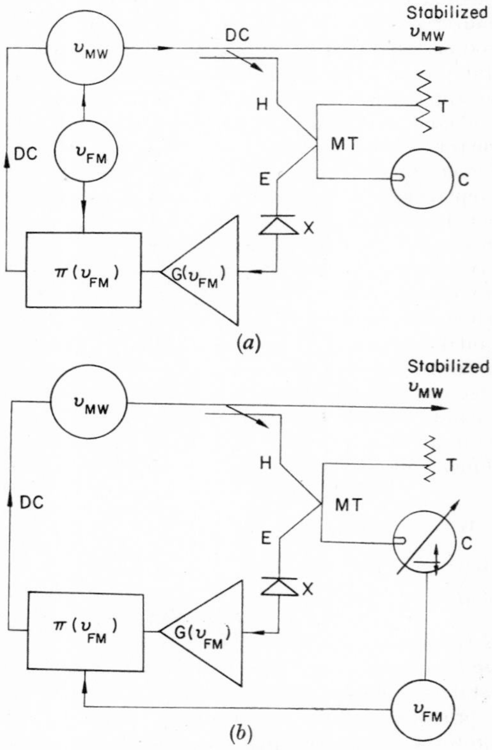

Fig. 4. Block diagram of modified "Pound" a.c. electronic frequency stabilizer using (a) frequency modulation of the oscillator and (b) frequency modulation of the reference cavity.

sideband power. The microwave power entering MT through the H-arm contains unmodulated power and sidebands displaced from the mean frequency by ν_{IF}. MT_1 divides the power; one half is dissipated in the termination T, the other half is incident on C. The i.f. output voltage is zero when the mean oscillator frequency coincides exactly with the cavity resonant frequency because the microwave power reflected from the matched cavity contains no frequency components separated by ν_{IF}. Unmodulated power is not reflected from a correctly tuned and matched cavity. The upper and lower sidebands are reflected, however, their frequency difference is $2\nu_{IF}$ and is rejected by the filtering action of the i.f. synchronous demodulator.

C does not present a matched load to the unmodulated power if a fluctuation occurs in the oscillator output frequency, so that the reflected power contains a component at the oscillator frequency. Microwave synchronous demodulation takes place at X producing an i.f. output voltage dependent in phase and amplitude on the difference between the mean microwave frequency and the reference cavity resonant frequency. The i.f. output voltage is also synchronously demodulated producing an error signal which tunes the microwave oscillator providing frequency stabilization. This stabilizer is very simple to operate; however, it does introduce microwave sideband frequencies into the useful power output. This may preclude its use in some applications. Figure 4 (*b*) shows the second modified Pound stabilizer which uses modulation of the resonant frequency of the reference cavity to generate the microwave sidebands. The modulation is obtained by replacing one wall of the cavity by a diaphragm which is moved physically by an electromechanical transducer. For high frequency operation a piezoelectric crystal may be used while at low frequencies a moving-coil mechanism similar to that employed in a telephone earpiece is convenient. The sideband demodulation system is identical to that described in the previous paragraph. The modulated cavity stabilizer has the advantage that it does not introduce sidebands into the microwave output power.

Typical values for the parameters of an oscillator stabilized by an a.c. electronic frequency stabilizer are

$\Delta\nu_i = 5 \times 10^5$Hz

$\Delta\nu(t) = 1 : 10^8$/minute

$\Delta\nu(t) = 1 : 10^4$/day.

A.C. electronic frequency stabilizer using a reference crystal oscillator. The use of a harmonic of a reference crystal controlled oscillator instead of a reference cavity provides improved frequency stabilization in a microwave power source. The functions of frequency discriminator and frequency standard are separated. The frequency deviation information is obtained by heterodyning the microwave oscillator against the crystal reference frequency. Frequency discrimination takes place at the intermediate frequency instead of the microwave frequency.

Figure 5 illustrates a typical control circuit. A sample of the output power is mixed with a harmonic of a high stability crystal oscillator. An i.f. voltage is produced with a frequency equal to the difference between the microwave source frequency and the crystal harmonic frequency. Amplification followed by frequency discrimination takes place at the intermediate frequency. The d.c. output voltage is fed back to the tuning electrode of the microwave power source to reduce fluctuations in its output frequency.

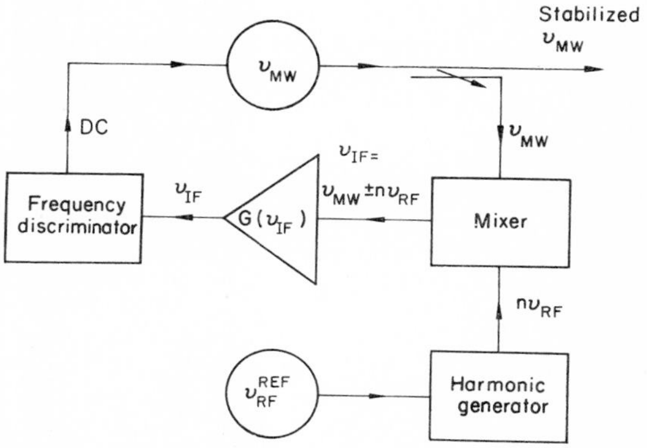

Fig. 5. Block diagram of a.c. electronic frequency stabilizer using frequency locking to a reference crystal oscillator.

The spectral purity of a microwave oscillator stabilized by this system is governed by the stability of the i.f. discriminator. It is always lower than the spectral purity of the crystal, however, it may be considerably better than that for a cavity-stabilized system. The long-term stability of the mean frequency is also improved in comparison with the cavity-stabilized system.

Typical values of the parameters for an oscillator stabilized with an a.c. electronic frequency stabilizer using a reference crystal oscillator are

$\Delta\nu_i = 1 \times 10^3$ Hz

$\Delta\nu(t) = 1 : 10^7$/minute

$\Delta\nu(t) = 1 : 10^6$/day.

A.C. electronic frequency stabilizer using phase-locking to a reference crystal oscillator. The various electronic frequency stabilizers described above have one feature in common, i.e. they require the existence of a frequency error before they can provide a correction signal. The residual frequency error may be minimized by increasing the gain of the feedback loop. There is no advantage in increasing the control loop gain beyond a certain value because the stability of the frequency discriminator then becomes the limiting factor in determining the ultimate stability of the microwave frequency. This difficulty can be overcome by replacing the frequency discriminator with a phase discriminator. The phase discriminator produces an error voltage in response to fluctuations in the relative phase of frequency of the microwave generator and that of the harmonic of the reference crystal oscillator. The correction signal minimizes the phase difference thus preventing a difference in the frequencies from appearing. This type of system was first suggested by M. Peter and M. W. P. Strandberg in 1955. A block diagram of a typical phase-lock oscillator synchronizer is shown in Fig. 6. This is identical with Fig. 5 except that the phase of the i.f. output is measured instead of its frequency. Phase comparison is obtained by synchronously demodulating the i.f. output with respect to the i.f. reference oscillator.

The various possible methods of operating a phase-locking frequency stabilizer may be listed according to whether one employs (i) a single high stability crystal oscillator of frequency ν_{RF} and a harmonic generator of order n which produces a fixed comb of microwave lock frequencies given by $n\nu_{RF} \pm \nu_{IF}$ where ν_{IF} is the tuned frequency of the phase-lock receiver, (ii) an individually selectable band of about 10 crystals which oscillate at frequencies $\nu_{RF} \pm n\nu$ where $n = 1, \ldots, 5$ and ν is about 10 kHz if ν_{RF} is about 10 MHz which produces a nearly continuous frequency resolution comb of frequencies in the range 1–10 GHz, or (iii) a crystal oscillator where ν_{RF} is about 10 MHz which can be voltage tuned by about 0·1 per cent to produce a continuous comb of frequencies in the range 1–10 GHz. Each of these systems can be further classified according to whether the phase-lock receiver is of fixed or variable frequency. The receiver bandwidth must be broader in the latter case but is generally not greater than 5 MHz if $\nu_{IF} = 30$ MHz. The stability of the oscillator which generates the radio frequency (r.f.) reference frequency determines the overall frequency stability and hence the effective Q of the microwave oscillator. It is typical that this stability is about $1:10^8$/sec and about $1:10^7$/day when the reference frequency crystal is temperature stabilized to $\pm 0{\cdot}1$°C. The stability of the i.f. reference oscillator must be such that it does not degrade the overall stability of the system. If $(\Delta\nu_{RF}/\nu_{RF}) \approx 10^{-8}$ then $(\Delta\nu_{IF}/\nu_{IF}) \approx (\Delta\nu_{RF}/\nu_{IF}) \approx 10^{-5}$ at 10 GHz. The short-term frequency stability of a tunable i.f. oscillator will satisfy this criterion, however, the long-term stability will not satisfy it by an order of magnitude.

The stability of the i.f. reference oscillator frequency and phase and the stability of the i.f. amplifier frequency response determine the microwave frequency stability when a primary frequency standard is used as the r.f. reference oscillator. The degree of stability can be increased by about three orders of magnitude if the i.f. reference frequency is also generated from the primary frequency standard. This increase in stability is gained at the expense of flexibility since there is only a fixed number of lock frequencies with no possibility of tuning about these frequencies.

The spectral purity of a typical microwave oscillator stabilized by a phase-lock synchronizer is comparable with a Pound stabilizer whose reference cavity $Q = 10^8$. Further details of this type of stabilizer are given by Buckmaster and Dering (1966). They give

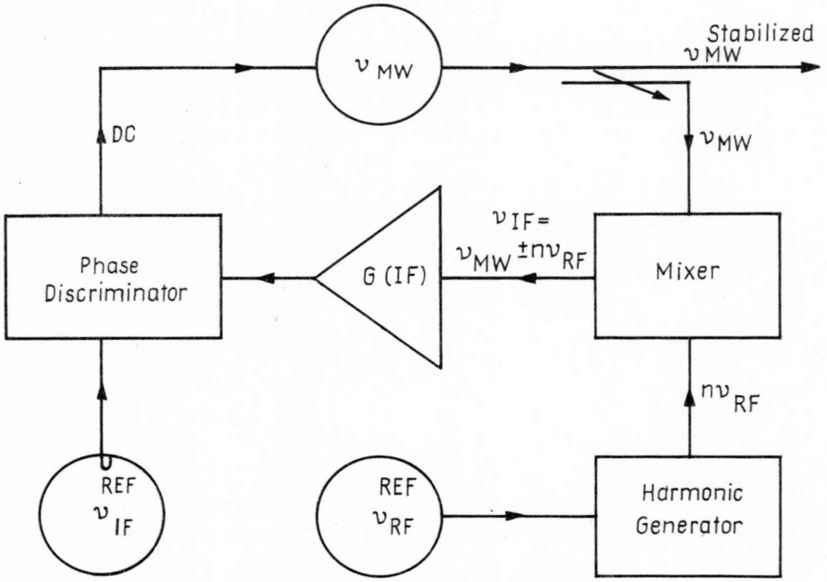

Fig. 6. Block diagram of a.c. electronic frequency stabilizer using phase locking to a reference crystal oscillator.

methods for simultaneously phase-locking a microwave oscillator to a reference crystal oscillator while tracking the resonant frequency of a resonant cavity.

Typical values of the parameters for an oscillator stabilized by an a.c. electronic frequency stabilizer using phase-locking to a reference crystal oscillator are

$$\Delta\nu_i = 10\text{Hz}$$

$$\Delta\nu(t) = :10^9/\text{minute}$$

$$\Delta\nu(t) = 10^8/\textbf{day.}$$

Bibliography

BUCKMASTER H. A. and DERING J. C. (1966) *J. Sci. Instrum.* **43**, 544.

ESSEN L. (1962) in *Encyclopaedic Dictionary of Physics* (J. Thewlis Ed.), 4, 675, Oxford: Pergamon Press.

PETER M. and STRANDBERG M. W. P. (1955) *Proc. Inst. Radio Engrs.* **43**, 869.

POUND R. V. (1946) *Rev. Sci. Instrum.* **17**, 490.

SMITH M. J. A. (1962) *J. Sci. Instrum.* **39**, 127.

H. A. BUCKMASTER and J. C. DERING

N

NEUTRON IMAGE INTENSIFIER. Just as electronic techniques for improving the detection of X-ray images have contributed to the usefulness of X-radiographic techniques, so too the development of a thermal neutron image intensifier has enhanced the usefulness of neutron radiographic techniques. The neutron image intensification technique which has been described in the literature and which is commercially available is an image intensifier tube similar to those used for the detection of X-radiation. The radiation beam is directed toward the evacuated intensifier tube, in which it is converted into light by a phosphor or scintillator layer. The emitted light causes photoelectron emission from an adjacent photoemissive layer. These electrons are voltage accelerated toward a phosphor output screen where a small, bright visible signal is obtained. The image is intensified in the tube by the acceleration of the electron image and by the "minification" of that electron signal. Figure 1 is a cross-sectional diagram of such a neutron image intensifier tube.

The neutron scintillator, as indicated in Fig. 1, is a mixture of ^{6}LiF and phosphor powders, in a resin binder. The LiF contains 95·72% ^{6}Li. The optimum mixture for light yield is one part ^{6}LiF to four parts phosphor, by weight. Two different phosphors have been used in the neutron scintillator. These are ZnCdS(Ag) and ZnS(Ag). The latter is preferred for two reasons. First, the emission spectrum from this phosphor provides a good match for the spectral response of the adjacent photoemitter surface, Sb—Cs—(0). Second, the lower density of the ZnS(Ag) yields less absorption for gamma radiation, thereby resulting in an improved neutron-gamma response ratio for the intensifier tube. The detection mechanism in either case is the same. The thermal neutrons are absorbed by the ^{6}Li; an alpha particle is promptly emitted, leaving a triton remaining. Each particle stimulates light emission from the phosphor. A typical scintillator target thickness of 0·4 mm absorbs about 30 per cent of an incident thermal neutron beam.

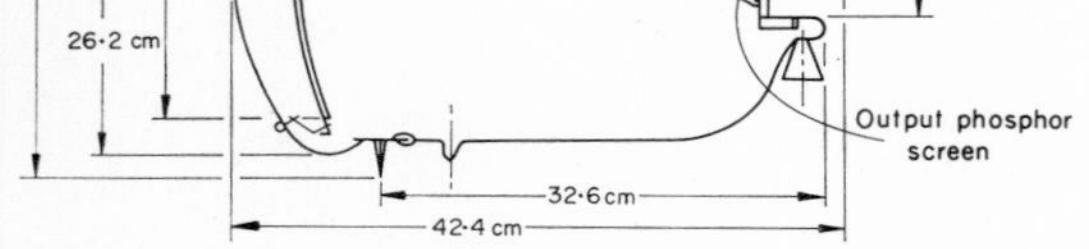

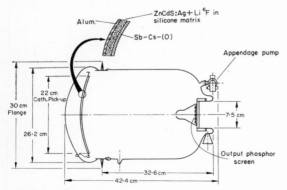

Fig. 1. A diagram of a neutron image intensifier tube. The expanded view shows the cross-section of the neutron scintillator-photoemissive target inside the tube.

Two sizes of neutron image intensifier tube have been made and used. Figure 1 shows the larger tube; a useful detection diameter of 22 cm can be obtained. A smaller tube, providing a neutron detection diameter of 15 cm has also been studied. The demagnification ratios for the two tubes are each about 9 : 1. In either case, a useful television signal can be obtained by optically coupling the small output phosphor light image to a television camera.

The light yield from a 15 cm diameter neutron intensifier tube [with ZnS(Ag) phosphor] is shown in Fig. 2. The tube provides linear response to thermal neutrons over at least four orders of neutron intensity. The excellent light yield characteristics of the intensifier tube have permitted the observation of gross brightening of the output phosphor for incident thermal neutron intensities as low as 100 neutrons/cm^2-sec for direct visual observation of the output phosphor screen by dark-adapted observers. For television response, a vidicon television camera has provided similar gross images for incident thermal neutron intensities as low as 2×10^4 neutrons/cm^2-sec; an image orthicon television system has provided improved low light level response by a factor of 100 times.

Of course, for such low neutron intensities, the detail and contrast which can be detected is limited by the statistical variation in the neutron imaging beam. The experimental results of tests to determine the smallest detail (at essentially 100 per cent contrast) one can detect by direct observation of the output phosphor correlate well with theory. This is shown in Fig. 3; neutron beam sizes which should be detectable lie above the lines, those which should not be detectable lie below them. Line B represents the theoretical prediction for the visible-not visible dividing line for a visual information accumulation time of 0·2 sec (the time normally used for the accumulation time of the human eye), a factor of 15 per cent of the incident neutrons contributing to the image (allowing 50 per cent absorption of the neutron beam in the intensifier tube window and 30 per cent absorption in the scintillator), and allowing a signal-to-noise ratio of three. The straight line A, for a beam size of 0·35 mm, is the limiting resolution of the intensifier tube. The experimental curve is in fair agreement with the lines A and B, the predicted limit for the conditions outlined above. The results would be poorer for lower contrast situations, and better for longer neutron image accumulation time, or better use of available neutrons. Line C represents the improvement one could obtain if the neutron absorption in the intensifier tube window could be eliminated; in this case 30 per cent of the incident neutrons would be used. This should be possible. Line D represents the ideal case in which all of the incident neutrons contribute to the image.

For radiographic problems, another property of interest is the contrast sensitivity which can be observed. For vidicon television system presentation of the neutron image with a collimated neutron intensity of 2×10^7 thermal neutrons/cm²-sec incident on the inspection sample, the contrast results for natural uranium, steel, and lead are shown in Fig. 4. The smallest thickness changes observed were four per cent for steel and uranium over a narrow thickness range.

The shape of the contrast curves is similar to those obtained for X-ray fluoroscopic systems, and can be explained similarly. For small sample thicknesses, a relatively large percentage change in material is required to yield a transmitted beam intensity change in the order of five per cent. A change in this order is about what one would expect to detect through the

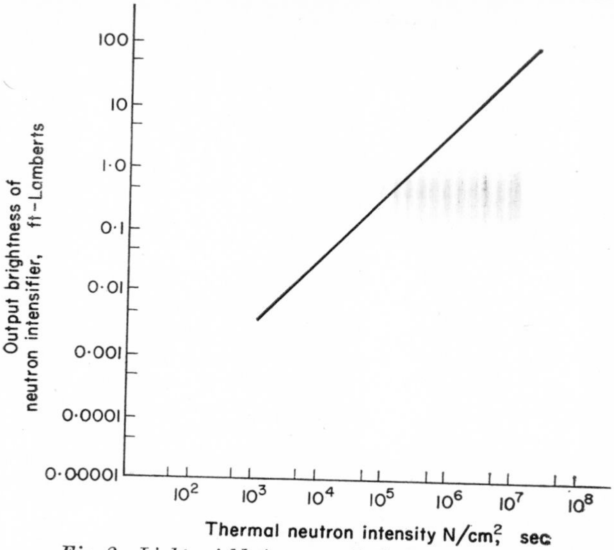

Fig. 2. Light yield from a ZnS(Ag) phosphor, neutron image intensifier as a function of incident thermal neutron intensity.

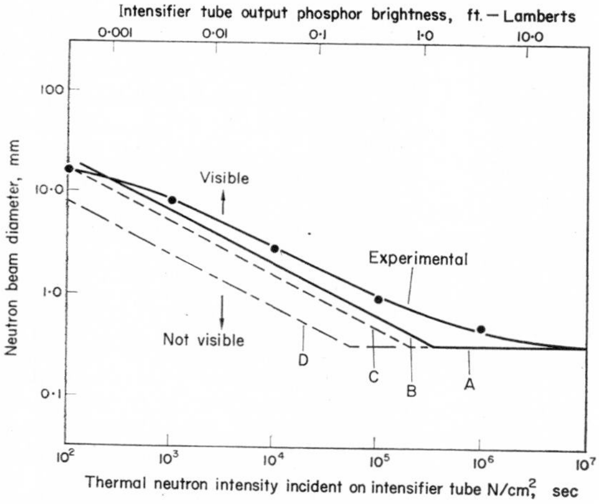

Fig. 3. Detail (at 100% contrast) detectable by direct observation of the output of a neutron image intensifier, as a function of incident thermal neutron intensity. The straight line A is the resolution limit of the evaluated intensifier tube (0·35 mm). *Points above the line should be detectable; those below the line should not be. Similarly, the lines B, C, and D represent the observation limits for the evaluated tube, for a similar tube in which all of the incident neutrons reached the target (instead of 50 per cent absorption in the window of the evaluated tube), and for the ideal case in which all of the incident neutrons contribute to the image, respectively. The experimental values lie very close to the prediction for the evaluated tube, lines A and B.*

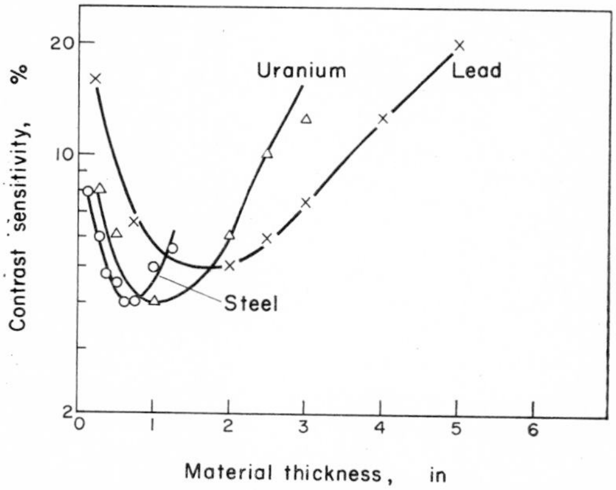

Fig. 4. Thickness changes visually detected in several metallic inspection materials by means of a neutron television system.

television system. As the sample thickness increases, a smaller percentage thickness change can yield the necessary change in transmitted radiation intensity, and the observable contrast optimizes. For larger sample thicknesses, scatter becomes important, also a smaller number of neutrons is involved in the production of the image, and the information becomes limited by the increasing deviation in that small number of image-forming neutrons. The data in Fig. 4 were for direct visual observation of the television screen; therefore the signal accumulation time would be in the order of 0·2 sec. A neutron image storage system, electronic or photographic, which would permit one to accumulate the signal for a longer time, would improve the contrast and detail one could observe for low neutron intensities reaching the intensifier.

The relative response of the intensifier tube to neutrons and to gamma radiation is of some interest in nuclear applications, since the inspection of radioactive materials has been one of the larger areas of application for neutron radiography. An intensifier tube containing ZnS(Ag) phosphor has yielded an equivalent output phosphor brightness when the intensifier tube was irradiated with either a thermal neutron intensity of 10^6 neutrons/cm^2-sec, or a ^{60}Co gamma radiation intensity of 600 R/hr. In tests at Argonne National Laboratory's Juggernaut Reactor neutron radiographic facility, it has been determined that useful neutron television images can be obtained even in the presence of ^{60}Co gamma radiation fields as high as 10,000 R/hr. The above facility provides a thermal neutron beam of 2×10^7 neutrons/cm^2-sec. The television neutron images obtained under these conditions yielded a high contrast resolution of about 0·5 mm, and displayed a contrast sensitivity of about 10 per cent.

Speed of response of the neutron imaging system is of interest since one of the primary advantages of the television technique over the neutron radiographic film technique is that motion can be observed. In the television technique (with a vidicon camera), high contrast motion as fast as 5 m/min has been observed (without appreciable blurr). The limitation on speed in this case is undoubtedly in the vidicon, rather than the image intensifier. An image orthicon camera has followed faster object motion, by factors of 2 to 3 times.

Applications of the neutron image intensifier include dynamic neutron image studies involving heated, irradiated reactor fuel materials, water flow and bubble pattern studies in heat transfer systems (such studies can be accomplished even through metal pipes or other filters), and the casting of heavy metals. In addition, of course, stationary images can be observed, with the advantages of excellent sensitivity (in terms of the number of neutrons required) and speed. The intensifier tube also appears to be potentially useful as a detector for neutron diffraction patterns.

Other techniques for neutron image intensification have been discussed in the literature, but these have not reached the stage of development of the intensifier tube described above. Other possibilities include arrays of small neutron detectors, spark counters, and other television techniques. The latter methods would include the use of neutron scintillator optically coupled to standard light image intensifier tubes and/or closed circuit television equipment. A neutron scintillator used with an image orthicon television system might provide useful neutron television images for relatively high neutron intensities; a light image intensifier in the optical system would improve the sensitivity of such a system. Neutron sensitive television camera tubes also appear to be possible. A neutron sensitive vidicon, for example, might be prepared by placing a neutron conversion layer such as cadmium directly behind an X-ray sensitive photoconductor (such as are used in X-ray sensitive vidicon camera tubes). The gamma rays emitted from the neutron bombarded cadmium would stimulate the photoconductor. Another vidicon-type tube approach might make use of an alpha emitting material, such as ^{6}Li or ^{10}B, in an insulating target; the charge pattern created in the insulator by the prompt neutron-alpha reaction could possibly provide a video signal.

If such television camera tubes could be made, and displayed reasonable neutron sensitivity, they would offer the advantage of improved resolution over the other techniques in which phosphor screens are involved.

Bibliography

BERGER H. (1965) *Neutron Radiography*, Amsterdam: Elsevier.

BERGER H. (1966) Characteristics of a Thermal Neutron Television Imaging System, *Materials Evaluation*, **24**, 475.

BERGER H., NIKLAS W. F. and SCHMIDT A. (1965) An Operational Thermal Neutron Image Intensifier Tube, *J. Appl. Physics*, **36**, 2093.

BEWLEY D. K. (1961) in *Encyclopaedic Dictionary of Physics* (J. Thewlis Ed.), **3**, 795, Oxford: Pergamon Press.

COPE A. D. (1961) in *Encyclopaedic Dictionary of Physics* (J. Thewlis Ed.), **3**, 794, 796, Oxford: Pergamon Press.

KALLMANN H. (1948) Neutron Radiography, *Research*, **1**, 254.

KALLMANN H. I. and KUHN E. (1942) *Neutron Image Converter*, U. S. Patent 2,270,373.

THEWLIS J. (1956) Neutron Radiography, *Brit. J. Appl. Physics*, **7**, 345.

THEWLIS J. (1963) *Encyclopaedic Dictionary of Physics*, **6**, 104, Oxford: Pergamon Press.

VINCENT B. J., Pollitt C. G. and Halmshaw R. (1966) Image-Intensifier Systems, in *Physics of Industrial Radiology* (Ed. R. Halmshaw), London: Heywood.

H. BERGER

NEUTRON IMAGE STORAGE. The purpose of image storage devices in imaging systems is basically to accumulate sufficient image information to present a useful picture. Taken in the broadest sense, the term neutron image storage can be applied to any neutron image detection device capable of integrating or storing signals. A photographic film technique would be one of the more common examples of such a detector. However, neutron image storage may take on some unusual aspects, as compared to storage systems for other radiations, if it is indeed desired that the final image be a true neutron image rather than one which is a combined neutron and gamma radiation result.

If we consider first the simple situation in which one merely wants to obtain a neutron image over some period of time, then there are a number of film techniques which can be used. For the most part, these include some conversion screen exposed with the film; the screen converts neutrons into radiation more readily detectable by the film. For example, cadmium emits prompt gamma radiation upon neutron bombardment, indium becomes radioactive, emitting beta and gamma radiation, and lithium and boron emit prompt alpha particles. In each case, the photographic detection of the neutrons is faster if the film emulsion itself contains some neutron conversion material, or if a conversion screen is used with normal film, than if the film is used alone.

Complications arise in those cases in which the desired neutron image may be masked such as by gamma radiation in the radiation beam, or from a radioactive inspection object. In such cases, useful photographic neutron images can still be obtained with proper detection methods. A detection technique which offers excellent discrimination against gamma radiation is the transfer technique. In this case a neutron image is detected by a screen of some potentially radioactive material, such as indium, silver, gold, or dysprosium. The image detection process is completed by taking an autoradiograph of the radioactive image on the exposed screen. In this situation, as in the case in which film was exposed directly to the neutron imaging beam, one can accumulate image information for some time. This time is not unlimited, however, because one gains very little for neutron beam exposures, or autoradiographic exposures, beyond 3 half-lives, because of saturation effects.

For very low neutron intensities and mixed radiation beams, therefore, one may not be able to use the transfer method to eliminate gamma effects from a neutron image. Problems of this type can be encountered in cases in which radioactive or small accelerator-type neutron sources are used as thermal neutron radiography sources. Other detectors may be applicable, depending upon the situation. If we return the film to the neutron imaging beam with some conversion screen, we can integrate information for an unlimited time. The problem, however, concerns the relative response of the detector to neutrons and to the interfering radiation. For methods in which film is exposed to the neutron beam with a metal conversion screen, it has been determined that thermal neutron exposures of about 10^5 neutrons/cm^2 produce a photographic effect equal to that of 1 mR of cobalt-60 gamma exposure. This is true for such metal conversion screens as cadmium, gadolinium, indium, and rhodium. One can improve this relative neutron-gamma response by using a neutron scintillator with the film, instead of a metal conversion screen. Scintillators containing boron-10 or lithium-6 enriched compounds in combination with phosphors, emit light upon neutron exposure. These detectors are more sensitive to thermal neutrons. Therefore, a thermal neutron exposure of only 10^4 neutrons/cm^2 yields a result equal to 1 mR of cobalt-60 gamma exposure.

Another integrating, image detection technique which offers some advantage in imaging thermal neutrons in the presence of gamma radiation is a thermoluminescent method. A thermoluminescent material such as lithium-6 enriched LiF will emit light if heated to temperatures of 210°C or higher, after having been irradiated with thermal neutrons. Such methods are used for dosimetry. Images can be prepared if the LiF is in the form of a sheet. This detector will also respond to gamma radiation but the sensitivities are such that a thermal neutron exposure of only 5×10^3 neutrons/cm^2 is required to produce an effect equal to that of an exposure of 1 mR of ^{60}Co gamma radiation.

Each of the above techniques can be used to provide neutron images and, in each case, the fact that the image information can be integrated for some period of time improves the quality of the images which can be obtained. The characteristics of the radiation properties of the imaging beam and of the individual detector used will determine whether or not the resultant image is a thermal neutron image, or one formed by combined effects of neutrons and other radiation.

From a different point of view, storage systems generally refer to electronic detection systems, and to some device which can provide storage times greater than those normally used for television frame rates. Until the recent development of television techniques for thermal neutron imaging, this definition had little practical meaning. Now that neutron image intensifier tubes are available for use in neutron television systems, a number of electronic storage techniques can be considered for improving the quality of the presented image.

One can store information within the television system itself. For example, with a vidicon camera tube one can turn off the scanning beam and collect image information for at least several seconds before turning on the electron beam to read the signal. This technique, combined with a high persistence phosphor in the kinescope, can be very useful.

A more versatile technique, of course, is to make use of an electronic storage tube system, in which information signals can be integrated. One can display the image signal after depositing 10 scans, or 1000 scans, or more, and by this means improve the quality of the signal. A simpler technique would be to combine a neutron image intensifier and a storage tube into one developer. The image-carrying electrons in the intensifier tube could be accelerated toward a storage target, instead of a phosphor screen. The storage target could then be scanned by an electron gun after a suitable period of image signal accumulation, to present a television image of the improved signal. The technology for such devices is known and could be applied to the neutron image storage problem.

See also: Neutron image intensifier.

Bibliography

BERGER H. (1965) *Neutron Radiography*, Amsterdam: Elsevier.

COPE A. D. (1962) in *Encyclopaedic Dictionary of Physics* (J. Thewlis Ed.), **7**, 44, Oxford: Pergamon Press.

GREATOREX C. A. (1962) Image Storage Techniques Applied to Diagnostic Radiology, *Advances in Electronics and Electron Physics*, Vol. 16 (J. D. McGee, W. L. Wilcock and L. Mandel Eds.), New York: Academic Press.

JACOBS J. E. (1961-2) Technical Solutions to the Problem of Reducing Patient Dosage, *Trans. A.I.E.E.*, **80** (Part I—Communications and Electronics), 721.

KASTNER J., BERGER H. and KRASKA I. R. (1966) LiF Thermoluminescence For Neutron Image Storage, *Nuclear Applications*, **2**, 252.

KNOLL M. and KAZAN B. (1952) *Storage Tubes and Their Basic Principles*, New York: Wiley.

ROSE A. (1948) The Sensitivity Performance of The Human Eye on an Absolute Scale, *J. Optical Soc. Am.*, **38**, 196.

STURM R. E. and MORGAN R. H. (1949) Screen Intensification Systems and Their Limitations, *Am. J. Roentgenology and Radium Therapy*, **62**, 617.

THEWLIS J. (Ed.) (1962) *Encyclopaedic Dictionary of Physics*, **6**, 109, Oxford: Pergamon Press.

H. BERGER

NEUTRON STANDARDS. The existence of the neutron was proposed in 1932 by Chadwick to explain the penetrating radiation which had been observed when the beryllium nucleus was bombarded with alpha particles. This process, which forms the basis of many present-day radioactive neutron sources, can be described by the reaction $^{9}Be(\alpha, n)^{12}C$, or by the equation $^{4}He + {}^{9}Be \rightarrow {}^{12}C + n$.

With the discovery of uranium fission in 1939 neutron physics developed at a tremendous pace. The first chain reactor was operated in 1942, and there are now hundreds in operation as research instruments, in power stations, and as a source of power for submarines and ships. In the postwar years many laboratories have installed neutron-producing particle accelerators.

Radioactive neutron sources. The necessity for standard neutron sources was soon realized. Emitting a constant number of neutrons per second, such sources produce directly a standard fast neutron flux density. Placed in a moderating medium they produce an easily calculated thermal neutron flux. Thus they serve as a means of measuring neutron reaction probabilities, of calibrating neutron detectors and dosemeters, and of inter-relating neutron measurements at different laboratories. They are relatively cheap to obtain, and can be made with reasonably good long-term stability, but they have certain disadvantages, notably the complexity of the neutron energy spectrum which they produce.

Being uncharged particles, neutrons cannot be detected directly by ionization methods. Broadly speaking there are two principal techniques which have been used to measure neutron source emission rates. In the associated particle technique a reaction is chosen in which a charged particle is emitted at the same time as a neutron. One can then count the neutrons by counting the associated charged particles. In the other technique the neutrons are detected by the effects which they produce when they take part in an observable nuclear reaction, the cross-section for which is already known. The reaction cross-section is a measure of the probability that a neutron will initiate the reaction.

An example of the associated particle method is the absolute counting of photoprotons produced in the $^{2}H(\gamma, n)^{1}H$ reaction.

In another example, alpha particles are counted when accelerated deuterons bombard a tritium target to produce the $^{3}H(^{2}H, n)^{4}He$ reaction. In this case the target of the particle accelerator is surrounded by a moderator, or slowing down medium, and the reaction is used to calibrate a slow neutron detector. The accelerator target is then replaced by the standard source, and the calibrated detector is used to determine the source strength. In a third example the $^{9}Be(\gamma, n)2^{4}He$ reaction is used to produce helium which is assumed to remain trapped in the beryllium metal. The helium is subsequently extracted and measured quantitatively. Needless to say, all these measurements require extreme care and there are many corrections to be made.

As an example of the other type of method, known briefly as the integration method, the source is placed at the centre of an effectively infinite mass of moderating material such as water or graphite and the capture rate of neutrons is measured throughout the medium. This is equal to the neutron emission rate. For example, thin foils of gold, indium or manganese, which capture neutrons, to form radioactive isotopes, may be used as detectors and the integrated capture rate throughout the medium may be derived numerically. Alternatively, in the case of liquid moderators, the integration may be carried out by attaching the foil to an arm which moves in such a way that the final activity is proportional to the required integral. In the well-known manganese sulphate bath technique the detecting element, manganese, is mixed homogeneously with the moderating material, water, and the integration is carried out physically by stirring the solution after removal of the source. In these examples a radioactive isotope of the detecting element is formed on neutron capture, and the capture rate is calculated from the amount of radioactivity produced. Another type of slow neutron detector which can be used is the boron trifluoride proportional counter, which detects the alpha particles produced in the $^{10}B(n, \alpha)^{7}Li$ reaction.

Some methods of neutron source calibration do not fall into the two main categories described above. In one method a nuclear reactor is used. The increase in reactor power observed when the source is introduced into the reactor is compared with the decrease in power produced by the introduction of an absorber of known capture cross-section (a negative source). In another method a proportional counter filled with hydrogen or methane is used to measure the neutron flux from the source. Recoil protons are counted and the neutron field is calculated from a knowledge of the neutron-proton scattering cross-section. These latter methods have not been developed to the high degree of accuracy achieved recently by the principal techniques. A detailed account of all the methods in use is beyond the scope of this article; for further information the reader is referred to the literature on the subject.

In two recent manganese bath experiments, the dependence of the method on predetermined reaction cross-section has been almost eliminated. Normally the neutrons are captured in roughly equal proportions by hydrogen and manganese nuclei in the solution, but if nearly all the neutrons were captured by manganese the manganese to hydrogen capture cross-section ratio would not need to be known so accurately. At the National Bureau of Standards in Washington this condition was achieved by replacing the hydrogen with deuterium which has a much lower capture cross-section. At the National Physical Laboratory in Teddington, the relative proportions of manganese and hydrogen nuclei was varied by changing the chemical concentration of the solution. In this way it was possible to extrapolate the data to zero hydrogen concentration. By these methods it has been possible to calibrate neutron source emission rates with an accuracy of better than one per cent.

International comparisons. It is clear that there is a strong interdependence of the measurements of neutron flux density and of neutron reaction cross-sections. A knowledge of the one will permit the measurement of the other. Since few reaction cross-sections were known very accurately it is not surprising that wide discrepancies occurred between the measurements of standard sources at different laboratories in the early intercomparisons. In 1951, after intercomparisons involving laboratories in USA, Britain, France, Italy and Switzerland, it was demonstrated that there was a spread of nearly twenty-five per cent in the neutron standards of the five countries, although all the standards were of the same type—the radium-beryllium alpha-neutron source.

Three years later, with laboratories in Belgium and Sweden added to the intercomparison network, the overall spread of values had been reduced to ten per cent, and now a different type of source was involved, the National Bureau of Standards radium beryllium photoneutron source, with quite a different neutron energy spectrum.

In 1961 all recent comparison results were analysed by a least squares method which showed that the overall spread of values for the neutron standards of ten laboratories was 3·8 per cent. Thus a steady improvement in the agreement achieved between various laboratories is observed as the techniques improve.

Hitherto, international comparisons had been arranged on the initiative of the individual scientists working in the field. However, in the late nineteen-fifties the importance of international standards for radiation measurement were discussed by the International Committee which supervises the activities of the International Bureau of Weights and Measures. As a result, in 1958 a new Consultative Committee was appointed to coordinate work throughout the world on the standards for the measurement of ionizing radiations. In 1960 working groups were formed to initiate radiation standards work at the International Bureau in the fields of X and gamma-radiation, radioactive nuclides, and neutron measurement. In 1963–1964 two new buildings were erected to provide laboratory space for the new work, since in addition to coordinating standards internationally, the Bureau also maintains its own standards. The working groups meet at regular intervals to discuss results of comparisons and to plan new activities.

Under the auspices of the International Bureau, the Canadian national standard radium beryllium neutron source has been circulated for measurement by eleven countries. A report on the results of this intercomparison is to be published by the International Bureau.

Need for more elaborate standards. As we saw above, the radioactive neutron source makes a very convenient, cheap, portable standard, but it is not the ultimate solution to the problem of neutron standards owing to the complexity of the neutron energy spectrum. Neutrons undergo many different types of reaction with nuclei, and the neutron reaction cross-section for the particular reaction, which has the dimensions of cm^2, is used as a measure of the probability that a neutron will take part in the reaction. Some radiative capture cross-sections vary fairly closely with the inverse of the neutron velocity. Many have strong resonances in the low energy regions and at high energies, closely spaced resonances which cannot be resolved with the resolution of existing neutron sources and detectors. Radiative capture events are (n, γ) reactions. (n, p), (n, α), (n, d), (n, 2n) are other types of reaction that can occur, as well as elastic scattering and fission. Hence the interaction of neutron radiation with matter is very much a function of the neutron energy and of the material.

The biological effect of neutron radiation has been shown to depend strongly on neutron energy as well as on the total amount of energy deposited in the tissue. The International Commission on Radiological Units and Measurements has published a table of neutron radiation dose in terms of neutron flux as a function of energy, and many practical dosemeters are designed to match the I.C.R.U. response curve. Hence for neutron dosimetry as well as for nuclear data measurement it is generally considered preferable to establish standards of flux density rather than, as in the X and gamma-ray case, to set up standards of exposure or of absorbed dose. Ideally what is required is the ability to provide accurately known monoenergetic neutron fluxes over as wide an energy range as possible.

Slow neutron flux standards. Neutrons undergoing collisions with nuclei in an effectively infinite moderating medium cease to lose any further energy when their average energy becomes comparable with that of the individual atoms in the moderator. They are then described as thermal neutrons, and their spectrum approximates to a Maxwellian distribution of velocities characterized by the temperature of the moderator. At 20°C the most probable neutron velocity is 2200 m/s; thermal neutron reaction cross-sections are normally tabulated for neutrons of this velocity which have an energy of 0·025 eV. Since thermal neutron fluxes are fairly readily available the low energy cross-sections are known more accurately than those for fast neutrons.

Many laboratories have established standard thermal neutron fluxes. These usually consist of a number of radioactive neutron sources placed in fixed geometry in a graphite moderator with a cavity at the centre for samples, where the flux gradient is very small if not zero. Usually a layer of paraffin wax or polyethylene is placed outside the graphite to reflect back some of the escaping neutrons thus raising the thermal neutron flux density at the sample position, typically a few thousand neutrons/cm^2 s.

For some years the National Physical Laboratory has used the Atomic Energy Research Establishment's reactor GLEEP as an interim thermal neutron flux standard. GLEEP is a natural uranium, graphite moderated reactor operated at very low power level so that changes due to fuel burn-up and fission product poisoning are very small. At the three kilowatt level the thermal neutron flux density at the sample position is 10^9 neutrons/cm^2 s. Calibrations carried out over a period of about five years indicate that the flux density when the reactor is operated at the standard power level has not changed by more than two per cent.

Another method of producing a standard thermal neutron flux density has been developed at the National Physical Laboratory. A deuteron beam from a 3 MV Van de Graaff positive ion accelerator is used to irradiate twin beryllium targets placed about one metre apart in a large mass of graphite. By means of feedback signals from suitably placed neutron detectors, the aim is to produce a zero-gradient, well moderated thermal neutron flux which is constant to within $\pm 0{\cdot}1$ per cent. The flux density is about 10^7 n/cm^2 s, which is a convenient intermediate level between those of the reactors and those of the conventional radioactive source assemblies.

Thermal neutron flux measurements are nearly always carried out by absolute determination of the radioactivity induced in detector foils and the flux density is calculated in terms of the known capture cross-section.

Few international comparisons have so far been carried out, but a review published in 1962 revealed discrepancies of up to six per cent in the earlier measurements. A comprehensive international intercomparison of thermal fluxes involving eleven countries has been arranged under the auspices of the International Bureau, and measurements by exchange of activated gold foils were in progress throughout 1966. The results will be examined by the appropriate working group and will be promulgated by the Bureau in 1967.

Fast neutron flux standards. In order to produce a field of fast neutrons which all have approximately the same energy it is necessary to be able to accelerate charged particles so that they can bombard a suitable target. By careful choice of bombarding particle, particle energy, target material and direction of emission of neutrons from the target, nearly monoenergetic neutrons can be produced over quite a large proportion of the energy range from a few keV up to about 19 MeV. Ideally the target should be mounted in a large room far from the walls, and scattering material in the vicinity of the target should be kept to a minimum, otherwise neutrons scattered back with changed energy would interfere with the monoenergetic field. In a given reaction energy and momentum must be conserved, and the energy equivalent of the differences in mass between the initial and final particles is added to (or subtracted from, in the case of endothermic reactions) the energy available for the kinetic energy of the final particles. 14 MeV neutrons are produced when a tritium target is bombarded by deuterons in the reaction

$$^3\mathrm{H} + \mathrm{H}^2 \rightarrow \mathrm{n} + \mathrm{H}^4\mathrm{e} + 18\ \mathrm{MeV}.$$

The cross-section for this neutron-producing charged particle reaction exhibits a maximum at a deuteron energy of 110 keV, so only a relatively low voltage accelerator is required. However, if the deuteron energy can be raised to 3 MeV the neutron energy range from 13·5 MeV to 19 MeV can be covered. The neutron flux can be estimated by counting the associated alpha particles or by observing neutron-proton scattering events in hydrogenous material and assuming the n-p scattering cross-section.

Neutrons with energy 2·5 MeV are obtained with the $^{2}H(^{2}H, n)^{3}He$ reaction which has an energy release of 3·3 MeV. As with the previous reaction, a useful yield can be obtained with a low voltage accelerator, but with 3 MeV of bombarding energy the neutron energy range from 2·5 MeV to 6 MeV can be covered. This reaction runs in competition with the $^{2}H(^{2}H, p)^{3}H$ reaction, and care is needed to separate the various charged particles, including scattered deuterons which trigger the charged particle counter.

The energy range from 6 MeV to 13 MeV is not covered. Neutrons in this energy range can be produced by the $^{9}Be(d, n)^{12}C$ reaction but they are not monoenergetic because several groups of neutrons with different energies are emitted according to the final energy state of the ^{12}C nucleus.

For production of neutrons in the intermediate, or keV, energy region a number of endothermic reactions are available. Here, a higher energy particle accelerator is essential because these reactions will only proceed with incident particle energies above a certain threshold value. Just above this value neutrons are produced with essentially zero energy in the centre-of-mass system of coordinates. In the laboratory system two neutron energy groups are projected at forward angles, corresponding to those emitted forwards and those emitted backwards in the centre-of-mass system. At bombarding energies above what is known as the backward threshold, neutrons are emitted in all directions and the energy in each direction becomes single-valued. Thus, in principle, neutrons with extremely low energy can be obtained at backward angles at bombarding energies just above the backward threshold. Examples of this type of reaction are $^{3}H(p, n)^{3}He$, $^{7}Li(p, n)^{7}Be$, $^{12}C(d, n)^{13}N$, and $^{51}V(p, n)^{51}Cr$. In some of these reactions the neutron flux can be calculated by a form of associated particle measurement, that of the radioactivity of the recoil particles retained in the target. The recoil particle energy would be too small for direct detection and the same would apply to the proton energies if the recoil proton techniques were used. These neutrons can, however, be standardized with activation detectors if the cross-sections have been predetermined, or by some variation of the manganese sulphate bath technique, which is a form of energy-independent, or flat-response detector.

The techniques described have been developed at a number of laboratories during the years since particle accelerators became available and in some cases accuracies as good as two or three per cent are claimed.

No comprehensive intercomparisons of these fast neutron flux density measurements have so far been arranged, but the International Bureau of Weights and Measures are planning such comparisons to take place as soon as a suitable comparison technique can be developed. The first comparisons will involve the higher neutron energies because many of the interested laboratories have only the lower voltage accelerators. On the other hand and for the same reason, flux measurements and cross-section measurements in the intermediate energy region have not received so much attention in the past and in many fields of work the need for improved accuracy in neutron standards at the lower energies is felt most keenly.

Bibliography

Physical Aspects of Radiation, Recommendations of the International Commission on Radiological Units and Measurements. Handbook 85, Washington, D. C.: National Bureau of Standards.

Measurement of Neutron Flux and Spectra for Physical and Biological Applications, Handbook 72, Washington, D. C.: National Bureau of Standards.

MARION J. B. and FOWLER J. L. *Fast Neutron Physics*, Parts I and II, Interscience monographs and texts in physics and astronomy, Vol. 4, New York: Wiley.

E. J. AXTON

NON-LINEAR OPTICS. When light waves pass through a material medium an electric polarization is induced which is a function of the optical electric field. It is convenient to express this function as a power series. Non-linear optics is concerned with the phenomena arising from the non-linear terms in the series. The subject has built up rapidly with the development of high power lasers with coherent output power densities in excess of 1 MW/cm². Incoherent optical sources are not usually powerful enough to produce significant non-linear effects.

At low field strengths only the linear term in the polarization is significant. It has the familiar effect of reducing the velocity of propagation below the value in vacuo and, in general, results in some absorption. At high field strengths the non-linear terms in the polarization become significant. In non-centrosymmetric media the first non-linear term is quadratic in the field. It consists of contributions oscillating at the sums and differences of all pairs of frequencies present in the incident field and consequently radiates light at these combination frequencies into the medium. Thus, the quadratic polarization produces mixing phenomena of which the most extensively studied is the generation of the second harmonic of an incident monochromatic light wave. In centrosymmetric media the polarization must reverse when the field is reversed and consequently the quadratic polarization is absent. The first non-linear term is then cubic in the field. It generates light waves at the sums and differences of all triples of frequencies in the incident light. In addition to these mixing phenomena, the cubic polarization allows one to control the velocity and attenuation of light at one frequency with an intense light wave at another frequency. The most interesting effects of this type are two-photon absorption in which the induced attenuation is positive and the stimulated Raman effect in which it is negative, i.e. one light wave is amplified in the presence of the other.

In what follows we establish the mathematical formalism used to describe the induced polarization and discuss the effects arising from the linear, quadratic and cubic terms in more detail. Non-linear phenomena can also arise from terms in the polarization involving higher powers of the electric field, gradients of the electric field and the magnetic field, but they are of little significance even with the light intensities available from lasers.

Let us consider a medium under the influence of an electric field which is the sum of a number of monchromatic contributions $E(\omega)\exp(-i\omega t)$ with time factors $\exp(-i\omega t)$ and space factors $E(\omega)$. Here, ω denotes the frequency and it is convenient to distinguish the contributions with frequencies $\pm\omega$. The linear polarization is the sum of all the monochromatic contributions $P^{(1)}(\omega)\exp(-i\omega t)$ with

$$\mathbf{P}^{(1)}(\omega) = \boldsymbol{\chi}^{(1)}(\omega)\cdot\mathbf{E}(\omega) \tag{1}$$

where $\boldsymbol{\chi}^{(1)}(\omega)$ is a tensor of the second rank called the linear susceptibility tensor. Similarly, the quadratic polarization is the sum of all the monochromatic contributions $\mathbf{P}^{(2)}(\omega', \omega'')\exp[-i(\omega'+\omega'')t]$ with

$$\mathbf{P}^{(2)}(\omega', \omega'') = \boldsymbol{\chi}^{(2)}(\omega', \omega'') : \mathbf{E}(\omega')\,\mathbf{E}(\omega'') \tag{2}$$

where $\boldsymbol{\chi}^{(2)}(\omega', \omega'')$ is a tensor of that third rank called the quadratic susceptibility tensor. Finally, the cubic polarization is the superposition of all the monchromatic contributions

$$\mathbf{P}^{(3)}(\omega', \omega'', \omega''')\exp[-i(\omega'+\omega''+\omega''')t]$$

with

$$\mathbf{P}^{(3)}(\omega', \omega'', \omega''') = \boldsymbol{\chi}^{(3)}(\omega', \omega'', \omega''') \,\vdots\, \mathbf{E}(\omega')\,\mathbf{E}(\omega'')\,\mathbf{E}(\omega''') \tag{3}$$

where $\boldsymbol{\chi}^{(3)}(\omega', \omega'', \omega''')$ is a tensor of the fourth rank called the cubic susceptibility tensor.

In equations 1, 2 and 3 the dots indicate tensorial contraction. All the optical properties of the medium, both linear and non-linear, may be derived by combining these "constitutive relations" between polarization and field with the relation of field to polarization imposed by Maxwell's equations:

$$\text{curl curl}\,\mathbf{E}(\omega) + \frac{\omega^2}{c^2}\,\boldsymbol{\varepsilon}(\omega)\cdot\mathbf{E}(\omega) = \frac{4\pi\omega^2}{c^2}\,[\mathbf{P}^{(2)}(\omega) + \mathbf{P}^{(3)}(\omega)] \tag{4}$$

where c is the velocity of light in vacuo and $\boldsymbol{\varepsilon}(\omega) = \mathbf{1} + 4\pi\boldsymbol{\chi}^{(1)}(\omega)$ is the dielectric tensor which takes account of the linear polarization. The vectors $\mathbf{P}^{(2)}(\omega)$ and $\mathbf{P}^{(3)}(\omega)$ are respectively the space factors of the quadratic and cubic polarizations at frequency ω. They are obtained by summing (2) and (3) over all sets of the frequencies involved which add up to ω. It should be noted in this connexion that different permutations of a particular set of frequencies make separate but equal contributions to these sums.

For completeness, and by way of introduction, we consider first the case when the field strengths are small enough for $\mathbf{P}^{(2)}(\omega)$ and $\mathbf{P}^{(3)}(\omega)$ to be neglected in (4). We are then usually concerned with plane light waves for which

$$\mathbf{E}(\omega) = \mathbf{a}\exp\left[i\,\frac{\omega}{c}\,\eta(\omega)\,\mathbf{n}\cdot\mathbf{r}\right] \tag{5}$$

where $\mathbf{r}$ is the position vector in the medium and $\mathbf{n}$ is a unit vector in the direction of propagation. The refractive index $\eta(\omega)$ and the constant vector $\mathbf{a}$ may be determined by substituting (5) into (4) and neglecting the right-hand side. Of primary importance, from the point of view of non-linear optics, are the birefringent properties of transparent uniaxial crystals belonging to the tetragonal, trigonal and hexagonal crystal classes. The symmetry of $\boldsymbol{\varepsilon}(\omega)$ in these crystals is such that they propagate both ordinary waves, with a refractive index $\eta_0(\omega)$ which is independent of the angle θ between the direction of propagation and the optic axis, and also extraordinary waves with a refractive index $\eta_e(\omega)$ which depends on θ. The two refractive indices are equal only when $\theta = 0$. Triclinic, monoclinic and orthorhombic crystals are also birefringent but the reduced symmetry makes their behaviour more complicated. Cubic crystals and liquids on the other hand are isotropic in their linear optical properties and propagate only ordinary waves.

We turn now to the effects arising from the quadratic polarization. The simplest of these is the *Pockels* or *linear electro-optic effect*. When light with frequency ω is propagated through a medium which is subjected to a time-independent (d.c.) electric field $\mathbf{E}(0)$ we see from (2) that $\mathbf{P}^{(2)}(\omega)$ is *linear* in $\mathbf{E}(\omega)$ and may be taken over to the left-hand side of (4). The effect of the d.c. field is therefore to modify the dielectric tensor by $8\pi\boldsymbol{\chi}^{(2)}(\omega, \mathrm{o})$. (o$\mathbf{E}$). The resultant change of the refractive indices is only of the order of 10^{-4} for fields of 10 kV/cm in materials like potassium dihydrogen phosphate for which the effect is relatively large. Nevertheless, a phase change of 2π can be produced in a path length of a few centimetres because the light wave-length is under 1 μm.

Next in order of complication is the phenomenon of optical rectification. We see from (2) that light with frequency ω produces a d.c. polarization $2\boldsymbol{\chi}^{(2)}(\omega, -\omega)$: $\mathbf{E}(\omega)\,\mathbf{E}(-\omega)$ which will generate a d.c. field in the medium. Optical rectification has been observed in potassium dihydrogen phosphate, ammonium dihydrogen phosphate, quartz and other materials using, primarily, focused ruby lasers with a wave-length of 6943 Å. Calculations based on the equations of motion of the charged particles in a medium have shown that the tensors $\boldsymbol{\chi}^{(2)}(\omega, -\omega)$

and $\chi^{(2)}(\omega, 0)$ differ only in the arrangement of their elements. This surprising relationship has been verified by comparing the results of experiments on optical rectification and the Pockels effect.

Much experimental work has been done on optical second harmonic generation. We see from (2) that light with frequency ω produces a second harmonic polarization $\mathbf{P}^{(2)}(2\omega) = \chi^{(2)}(\omega, \omega): \mathbf{E}(\omega)\,\mathbf{E}(\omega)$. To study the generation of the second harmonic field from $\mathbf{P}^{(2)}(2\omega)$ we have to replace ω by 2ω in (4). When $E(\omega)$ is given by (5) we see that the driving term on the right-hand side depends on position through the factor $\exp[2i(\omega/c)\,\eta(\omega)\,\mathbf{n}\cdot\mathbf{r}]$. Now, in the absence of the driving term, the second harmonic waves have the same form but with $\eta(\omega)$ replaced by $\eta(2\omega)$. A spatial resonance occurs when $\eta(2\omega) = \eta(\omega)$. Second harmonic generation is strong only when this "phase-matching" condition is satisfied so that the second harmonic field and polarization stay in a fixed phase relation as they propagate through the medium.

In transparent isotropic and uniaxial optical media it is usually the case that $\eta_0(2\omega) > \eta_0(\omega)$ and $\eta_e(2\omega) > \eta_e(\omega)$. However, in strongly birefringent uniaxial crystals it may be possible to find propagation directions for which $\eta_e(2\omega) = \eta_0(\omega)$ or $\eta_0(2\omega) = \eta_e(\omega)$. In potassium dihydrogen phosphate, for example, a fundamental ordinary wave is phase-matched to a second harmonic extraordinary wave when the direction of propagation is at 50° to the optic axis. Fundamental to second harmonic conversion efficiencies of several tenths have been obtained under phase-matched conditions with focused laser beams. In quartz, on the other hand, the birefringence is too small to allow phase matching and the conversion efficiency is in the order of 10^{-8}.

Perhaps the most interesting phenomenon arising from the quadratic polarization is parametric amplification. It occurs when we propagate a weak light wave at frequency ω_s, called the signal, in the presence of a powerful light wave at a higher frequency ω_p, called the pump. The two waves beat together to produce a third wave at the difference frequency $\omega_i = \omega_p - \omega_s$, called the *idler*. To study the behaviour of the signal and idler waves in the presence of the pump we have to set $\omega = \omega_s$ and ω_i in (4) and substitute on the right-hand side the signal and idler polarization, viz. $2\chi^{(2)}(\omega_p - \omega_s): \mathbf{E}(\omega_p)\,\mathbf{E}(-\omega_i)$ and $2\chi^{(2)}(\omega_p - \omega_s): \mathbf{E}(\omega_p)\,\mathbf{E}(-\omega_s)$ respectively. Phase matching occurs for collinear plane waves when $\omega_p\eta(\omega_p) = \omega_s\eta(\omega_s) + \omega_i\eta(\omega_i)$, in which case the signal and idler waves grow at the expense of the pump. The amplification is said to be parametric because the pump may be regarded as modulating the linear optical parameter of the medium, i.e. the dielectric tensor. When the pump is sufficiently intense and the ends of the crystal are highly reflecting, signal and idler waves at frequencies satisfying the phase matching condition build up from the noise background. Some tens of Watts of tunable pulsed output at frequencies close to 1 μm have been obtained from potassium dihydrogen phosphate and lithium niobate crystals pumped with the second harmonic of light from a neodymium doped calcium tungstate laser.

We turn now to the phenomena associated with the cubic polarization. In centro-symmetric media this is the first non-linear term in the polarization. It is the first non-linear term in most liquids, for example. Apart from the obvious mixing phenomena we also see from (3) that an analogue to the Pockels effect arises from the cubic polarization. A d.c. field $\mathbf{E}(o)$ changes the dielectric tensor by $12\pi\chi^{(3)}(\omega, o, o): \mathbf{E}(o)\,\mathbf{E}(o)$. This is the quadratic electro-optic or Kerr effect. The Pockels and Kerr effects both provide means of electronically controlling light waves.

An "optical" Kerr effect also occurs when the d.c. field is replaced by a powerful optical pump field with frequency ω_p. The change of dielectric tensor in this case is $\delta\varepsilon(\omega) = 24\pi\chi^{(3)}(\omega, \omega_p, -\omega_p): \mathbf{E}(\omega_p)\,\mathbf{E}(-\omega_p)$ and it produces more interesting results than might appear ar first sight. It will be recalled that the linear dielectric tensor has absorption peaks in its imaginary part when ω is equal to a transition frequency ω_t of the medium, where ω_t is equal to the energy of an excited state above the ground state divided by Planck's constant over 2π. The nonlinear suspectibility tensors also contain terms which are resonant when combinations of the frequencies involved are equal to a transition frequency. In particular, $\chi^{(3)}(\omega, \omega_p - \omega_p)$ is resonant when $\omega_p \pm \omega = \omega_t$ and the form of the resulting resonances in $\delta\varepsilon(\omega)$ leads to attenuation of light at frequency ω for the plus sign and to amplification for the minus sign. Attenuation occurs when $\omega_p + \omega = \omega_t$ because the system is being excited from the ground state to the excited state by the stimulated *absorption* of a pump photon and a signal photon. The process is therefore known as two-photon absorption. It has recently been used to obtain spectroscopic data about potassium iodide. Similarly, amplification occurs in the case $\omega_p - \omega = \omega_t$ because the system is then reaching the excited state by the stimulated absorption of a pump photon and *emission* of a signal photon. This amplification process is known as the stimulated Raman effect because of its close relation to the familiar Raman effect which takes place through the stimulated absorption of pump photons and the spontaneous emission of photons at the frequency $\omega_p - \omega_t$. The stimulated Raman effect is extremely powerful in organic liquids such as benzene, nitrobenzene and toluene and has also been observed in some solids. When the pump is applied, light at the frequency $\omega_p - \omega_t$ is built up by stimulated Raman amplification from the noise background. In the case of the organic liquids as much as a few tenths of the pump power is converted in this way. Moreover, light at the frequencies $\omega_p - 2\omega_t$, $\omega_p - 3\omega_t$ etc. frequently appears as a result of a cascade of stimulated Raman processes. Finally, light at the frequencies $\omega_p + \omega_t$, $\omega_p + 2\omega_t$ etc. is produced by mixing processes. The transition frequencies ω_t involved are either molecular vibration frequencies in the order of 10^{13} c/sec or sound wave frequencies in the order of 10^{10} c/sec. In the latter case the phenomenon is usually referred to as stimulated Brillouin scattering.

The reason for the surprising strength of the stimulated Raman effect remained unexplained until recently. The observed growth constants were up to two orders of magnitude larger than was expected from the power density of the pump. It now appears that intense laser beams have a self-focusing action. The dielectric constant at the laser frequency is raised by the optical Kerr effect and as a result the beam is still further concentrated. This self-focusing action is sufficiently powerful to create damage tracks in some materials.

Bibliography

BLOEMBERGEN N. (1965) *Nonlinear Optics*, New York: Benjamin.

BUTCHER P. N. (1965) *Nonlinear Optical Phenomena*, Bulletin 200; Ohio State University Engineering Publications, Columbus, Ohio.

KELLEY P. L. *et al.* (Ed.) (1965) *The Physics of Quantum Electronics*, New York: McGraw-Hill.

P. N. BUTCHER

NON-LINEAR WAVE PROPAGATION. It is customary to use the term *wave* when referring to a disturbance that propagates through a continuum with a finite speed. When the speed of propagation of this wave is independent of the form of the wave the propagation is said to be *linear* and the response of the continuum to a series of waves arriving simultaneously at a point may then be determined by superposition of the waves at that point. In the event that the speed of propagation depends on the waveform, the wave propagation is said to be *non-linear* and the principle of superposition is no longer applicable. The most common mathematical formulation of linear wave propagation is in terms of the familiar second-order linear wave equation

$$\frac{\partial^2 \Phi}{\partial t^2} = c^2 \nabla^2 \Phi,$$

describing the quantity Φ which propagates with velocity c. In mathematical terms, linearity of equations is described by the statement that if α and β are constants, and Ψ is another solution of the wave equation, then

$$\alpha\Phi, \beta\Psi \quad \text{and} \quad \alpha\Phi + \beta\Psi$$

are also solutions of this wave equation.

It is, however, a more usual situation that no such relationships exist between linear combinations of solutions. Simple examples of non-linear equations of different types describing wave phenomena are provided by the second-order non-linear wave equation

$$\frac{\partial^2 \Phi}{\partial t^2} = c^2 \left(1 + \varepsilon \frac{\partial \Phi}{\partial x}\right)^{\alpha} \frac{\partial^2 \Phi}{\partial x^2}$$

which can be used to describe waves in certain solids and the set of first-order equations describing conservation of mass, momentum and energy of gas dynamics:

$$\frac{\partial \varrho}{\partial t} + \operatorname{div}(\varrho \mathbf{v}) = 0 \qquad \text{(mass)}$$

$$\frac{\partial y}{\partial t} + (\mathbf{v} \cdot \operatorname{grad})\, \boldsymbol{v} + \frac{1}{\varrho} \operatorname{grad} p = 0 \qquad \text{(momentum)}$$

$$\frac{\partial S}{\partial t} + (\mathbf{v} \cdot \operatorname{grad})\, S = 0, \qquad \text{(energy)}$$

in which ϱ is the gas density, $\mathbf{v}$ is the gas velocity vector, p is the gas pressure and S is the entropy. In the second order equation the non-linearity enters through the product of a function of $\frac{\partial \Phi}{\partial x}$ and $\frac{\partial^2 \Phi}{\partial x^2}$ that appears on the right hand side of the equation. However, in the set of first-order equations describing gas dynamics the non-linearity enters in a number of different ways; in particular through the term $(\boldsymbol{v} \cdot \operatorname{grad})\, \boldsymbol{v}$ and through the term $\frac{1}{\varrho} \operatorname{grad} p$ due to the dependence of p on ϱ as determined by the equation of state of the gas.

A number of striking phenomena result from non-linearity in wave equations of which the occurrence of discontinuities in the solution (e.g. gas shocks) a finite time after the start of a continuous motion are perhaps the most remarkable examples. These are situations in which a continuous wave, or even a smooth wave, can modify its profile when propagating until an actual discontinuity occurs in the physical variable described by the wave. Striking examples of this are to be found in the formation of a bore wave, sometimes called a hydraulic jump, and in the breaking of waves on beaches.

The breaking of a water wave on a beach provides an interesting example of the effects of non-linearity and it also permits a simple physical explanation. In shallow water the speed of propagation of an element of the wave profile is proportional to the square root of the height of the element in question above the sea bed. Consequently, the crests of the waves propagate faster than the troughs with the result that the wave profile steepens with advancing time and ultimately, when the crest advances over the trough, the wave breaks. As already mentioned, actual jump discontinuities can also occur in gas flows where they are called *gas shocks*. Across a gas shock, density and pressure suffer a discontinuous change and, as with hydraulic jumps, the discontinuity when formed propagates at a distinctive velocity of its own, quite different from the velocity of propagation of an infinitesimal disturbance such as would form part of the continuous wave before the formation of the shock. Remarkable results also occur in magneto-hydrodynamics in which it is possible to distinguish three basically different types of wave, each of which has associated with it a characteristic form of discontinuity of the field variables involved.

Equations describing wave propagation are said to be of *hyperbolic type* on account of the algebraic form of the equation that describes the surfaces across which a normal derivative of the solution is discontinuous. Whereas for a linear wave equation these surfaces, which may be identified with wave-fronts, may be determined without reference to the solution, in the case of non-linear wave propagation their form can only be determined when the solution is known.

Bibliography

COURANT R. and FRIEDRICHS K. O. (1948) *Supersonic Flow and Shock Waves*, New York: Interscience.

COURANT R. and HILBERT D. (1962) *Methods of Mathematical Physics* II, New York: Interscience.

JEFFREY A. and TANIUTI T. (1964) *Nonlinear Wave Propagation*, New York: Academic Press.

JEFFREY A. (1966) *Magnetohydrodynamics*, Edinburgh: Oliver & Boyd.

STEWART R. W. (1962) in *Encyclopaedic Dictionary of Physics* (J. Thewlis Ed.), **7**, 732, Oxford: Pergamon Press.

STOKER J. J. (1957) *Water Waves*, New York: Interscience.

A. JEFFREY

NUCLEAR GEOPHYSICS. Techniques involving the measurements of nuclear radiations are applied to mineral exploration and exploitation, geochronology, meteorology, hydrology, oceanography and studies of sediment movement offshore or in rivers, under the broad heading of nuclear geophysics.

Borehole Logging and Mineral Exploitation

Nuclear techniques have been extensively developed for borehole exploration especially in oil-well investigations. Similar principles have been used for mineral exploration and exploitation. Although the various techniques are considered separately, it is common practice to use a combination of several to obtain the maximum information from new or existing boreholes.

Natural gamma radiation measurements. Most strata contain traces of radioactive uranium, thorium and potassium (^{40}K) which produce a count rate on a gamma-ray detector. In general, shales and volcanic ash show high gamma activity, sands are intermediate and limestones, salt, coal and anhydrites show low activity.

The technique is extensively used for identifying the boundaries of strata and for the correlation of horizons or sequences of strata in different boreholes. It is used for prospecting for uranium ores, both in boreholes and by overland or aerial surveys. The thickness of workable deposits of potassium salts is determined from boreholes. The use of a scintillation detector enables the spectrum of gamma-ray energies to be analysed from which some information can be obtained about the radioactive constituents of the strata. The interpretation is made difficult by the energy degradation due to the absorption and scatter of gamma-rays within the strata.

Gamma-gamma backscatter measurements. The gamma radiation from a source (often ^{137}Cs) penetrates the strata surrounding a borehole where it is scattered and absorbed. A detector is situated in the borehole some 40–60 cm from the source and is shielded from direct radiation by lead. The detector responds to the intensity of the radiation back-scattered from the strata by Compton scattering and this intensity reduces as the strata density increases.

When low energy gamma sources (less than 100 keV) such as ^{170}Tm or ^{75}Se are used, differences arise in the mass absorption coefficients of different elements. In general, the mass absorption coefficient of heavy elements is much higher than light elements at these energies and the intensity of backscattered radiation at a distance of 20–40 cm from the source is reduced by the presence of heavy elements. The method is used to locate and measure deposits of mercury, lead, barium, iron and manganese. Many of these applications are in short dry boreholes in mines.

Another application of this low energy technique to mining uses an ^{241}Am source to position an automatic coal cutting machine within the coal seam. The source and detector (which are about 20 cm apart) are mounted at the base of the machine and the thickness of coal left by the machine is displayed on a control panel. The proximity of shale or rock beneath the coal results in a reduced radiation intensity due to the higher mean atomic number of elements in the shale relative to those in coal.

Neutron-neutron measurements. A fast neutron source is used with a detector of slow neutrons in the borehole. The fast neutrons are slowed down in the strata, principally by collision with hydrogen nuclei. A high hydrogen content (high porosity) results in a high count rate in a slow neutron detector situated near the fast neutron source or a reduced count in a similar detector situated 50–60 cm away from the source.

The technique is widely used to determine strata porosity and can be operated in a cased borehole due to the ease with which neutrons penetrate steel or concrete.

The neutrons may be produced by the reaction of alpha particles on beryllium, using sources such as

^{241}Am or ^{238}Pu. The detector is either a $^{10}BF_3$ proportional gas counter or a scintillation detector using a lithium iodide crystal.

The method is somewhat sensitive to the presence in the strata of elements of high neutron absorption cross-section, such as chlorine. This problem is largely avoided by measuring epithermal neutrons using slow neutron detectors surrounded by paraffin wax and cadmium.

In ore exploration, the neutron capture can be used to locate areas containing high capture cross-section elements. Boron, manganese and mercury have been located by this technique.

Neutron-gamma borehole measurements. After the fast neutrons emitted by a neutron source are reduced to thermal energies by scattering, they are captured by the elements present in the strata or borehole. The reaction involves the instantaneous emission of a gamma ray (the n, γ reaction) and the measurement of the gamma radiation can provide information about porosity and elements in the strata.

Energy selection of the emitted gamma rays has been used in particular to identify the interface between oil and brine in strata. The high capture cross-section of chlorine in the brine results in an increase in the number of high energy gamma rays (5–7 MeV) in the brine saturated strata. The technique will operate in cased boreholes.

Neutron activation measurements. Irradiation of the strata by a neutron source in the borehole causes some elements to become radioactive and enables differences in strata composition to be inferred. The two common elements, silicon and aluminium both react to produce ^{28}Al (half-life 2·27 mins.).

High energy neutrons (14 MeV) from the d-t reaction in a neutron generator react with oxygen ($^{16}O(n, p)^{16}N$) to form ^{16}N which can be identified by its high energy gamma rays (6–8 MeV) or half-life of 7·4 secs.

This type of logging is both slow and expensive.

The application of neutron activation techniques using a reactor for the analysis of trace elements in rock and borehole core samples is dealt with in Helmholtz (1962). A much less refined version is of some value for field measurements using a neutron isotope source for rapid analysis of some ores. Applications include the determination of aluminium and silicon in alumino-silicate rocks and coals and the estimation of manganese, copper, vanadium and fluorine in ores.

Time-dependent neutron logging methods. The development of small ion-accelerator sources of neutrons capable of pulsed operation is opening up a new field of nuclear logging techniques.

When fast neutrons enter the strata, they are slowed down by elastic or inelastic collisions and are then captured rapidly but not instantaneously by the elements in the strata, usually with the simultaneous release of a gamma-ray. By measuring the number of slow neutrons or gamma rays in the borehole following a pulse of fast neutrons, the neutron or gamma-ray "die-away" time can be established. From this, the porosity or hydrogen content of the rocks can be inferred. The method also enables the position of water-oil and gas-liquid interfaces in strata to be established from within a cased borehole.

X-ray fluorescence applications. The development of X-ray isotope sources enables analytical measurements to be carried out in the field using X-ray fluorescence methods.

The technique uses an isotope source of X rays or low energy gamma rays (such as ^{55}Fe, 5·9 keV, ^{109}Cd, 22 keV, ^{57}Co, 122 keV) or bremsstrahlung radiation (such as $^{3}H/Zr$, 2 to 12 keV, $^{147}Pm/Al$, 10 to 100 keV) to excite the X rays of the element to be measured. These X rays are then detected either by a gas proportional counter or, more usually, by a scintillation detector. Energy selection is achieved by the use of a pair of selected balanced filters to isolate the required X rays. The intensity of the X rays is then obtained by a difference measurement. As an example, Zn K X rays (8·6 keV) are strongly attenuated by a nickel filter (absorption edge 8·33 keV) and transmitted by a copper filter (absorption edge 8·98 keV).

The use of this technique has led to field applications for measurement of ores *in situ* or as core or powdered samples. Tin has been measured *in situ* with a detectable limit of 0·1 per cent and copper and zinc (powder samples) with a limit of 0·25 per cent. Lead, molybdenum and titanium have also been measured.

A number of X-ray backscatter methods have been used for the analysis of the ash content of coal. These are limited either by small penetration into the coal sample or by errors arising from the variable iron content of the coal. A technique which measures the average atomic number of the elements in coal (and hence the ash content) by the backscatter of low-energy X rays from an $^{3}H/Zr$ bremsstrahlung source compensates for variations of the iron content by simultaneous measurement of the Fe K X rays excited in the sample.

Geochronology

A description of the methods of age determination by radioactivity is given in Poole (1961). The methods are capable of dating many igneous and metamorphic rocks but are less generally applicable to sedimentary deposits. The results are of considerable value in providing an approximate absolute time scale for geological events and for correlation with the fossil stratigraphic scale. Their use permits the interpretation of palaeomagnetic measurements relative to time and this has contributed to the theory of continental movement.

The use of ^{14}C dating enables recent geological deposits to be dated (back to 40,000 years or 70,000 years with ^{14}C enrichment). This covers some of the period of glacial activity and also provides calibration of the comparative chronology of pollen zones.

Meteorological Applications

The high sensitivity of detection of radioactive tracers and their extensive production by cosmic radiation, from terrestrial sources and from nuclear explosions enables them to be used for large-scale atmospheric investigations. The results supplement those obtained by measurements of ozone and water vapour.

Cosmic radiation produces a number of radioactive nuclides by neutron activation or by spallation reactions. These include ^{3}H, ^{7}Be, ^{14}C, ^{32}Si, ^{32}P, ^{33}P and ^{35}S. Together with RaD from terrestrial sources, they are of value in studies of the turn-over time (or mean lifetime) of elements or particles in the troposphere and in establishing the isolation of the tropospheric air mass from the stratosphere. The low specific activity and universal steady rate of production of cosmic-ray-produced tracers limits their value to meteorology but not to such subjects as dating in the biosphere (^{14}C, ^{3}H) and hydrology (^{3}H, ^{14}C).

The advent of man-made radioactive tracers from nuclear explosions provided tracers for large scale atmospheric movements. The isotopes principally of use include ^{3}H, ^{14}C, ^{89}Sr, ^{90}Sr, ^{137}Cs and ^{140}Ba. Studies in the troposphere show a circulation time round the earth of 12 to 25 days at mid-latitudes and a mean residence time of the tracers of one month.

More powerful fusion explosions inject tracers into the stratosphere and provide much information on stratospheric air circulation and interchange with the troposphere. Tracer information broadly supports the Dobson–Brewer stratospheric circulation pattern with air rising at the equator and spreading out to sink at the poles with occasional transfer to the troposphere via discontinuities in the tropopause.

In 1958 an injection of ^{185}W was made (by neutron activation of tungsten in the casing of a nuclear weapon) into the lower stratosphere near the equator. The bulk of the tracer remained at low latitudes and showed a stagnant zone in the equatorial stratosphere at about 70,000 feet. This showed the limited height of the Dobson–Brewer circulation and this was confirmed by the release in the Northern hemisphere of nuclear devices at 100,000 feet altitude. Debris labelled with ^{102}RL was distributed up to 300,000 feet. This tracer spread to the lower stratosphere over both poles and showed a high altitude zone of rapid meridianal transport.

Hydrological Applications

Radioactive tracers and instrument techniques have widespread application in hydrology. Since tritium is an isotopic constituent of water, it is of particular value as a tracer, especially under conditions where chemical tracers may be removed from the water, as in underground water seepage or where a change of phase takes place.

Tritium is formed in the atmosphere by the action of cosmic rays and enters the hydrological cycle via rainfall. The concentration is between 3 and 8 tritium units (one tritium unit, TU, is 1 tritium atom in 10^{18} hydrogen atoms and represents 7·2 disintegrations per minute per litre of water). The half-life of tritium is 12·25 years and this, in principle, enables underground waters to be "dated" over a period of 30 or 40 years using refined techniques of tritium reconcentration and measurement. Since the advent of thermonuclear explosions (about 1954) the tritium content of the atmosphere and rain has greatly increased. An annual variation occurs, showing the high contribution of the stratospheric tritium to the troposphere in the Spring and this is superimposed on a reduction of concentration from several thousand TU in Spring 1963 (in the United Kingdom) to some 500 TU in Spring 1966. This influx of tritium modifies the original "dating" method but enables studies of ground water movement, replenishment, storage volume and water mixing to be undertaken.

Carbon-14 is of some value as a tracer of ground water in which it occurs naturally as bicarbonate. This undergoes little exchange with the bound carbon even in aquifers such as limestone and chalk and allowance can be made for any exchange by measuring the $^{12}C/^{13}C$ ratio.

Stable isotope ratios of $^{1}H/^{2}H$ and $^{16}O/^{18}O$ are used to identify and differentiate between ground waters in adjacent aquifers. The origin of aquifer water can sometimes be inferred from the isotopic ratios since precipitation at higher altitude is isotopically lighter than that at lower altitude. Studies of the evaporation from lakes or other surface water can also be undertaken by stable isotope measurement and Antarctic snow deposits have been dated by annual variations of the isotopic ratios.

Radioactive tracers emitting gamma rays are useful for ground water studies since they can be detected *in situ*. In general, anions are less adsorbed in the ground than cations and ^{131}I and ^{82}Br have proved to be useful ground water tracers. Other complex ions such as potassium cobalticyanide (^{60}Co labelled) and chromium-EDTA (^{51}Cr labelled) are also satisfactory in many strata but none of these can be considered as reliable as tritium, although they may be more convenient.

Surface water studies include transit time measurements on the surface and in streams and in particular, quantitative flow measurements in rivers. In this application, the high sensitivity of detection of such tracers as ^{82}Br, ^{24}Na and ^{131}I has proved useful together with the advantage over chemical alternatives that they can be used in highly polluted or saline waters. Tests in laboratory flumes at flows of 150 l/sec (5 cusecs) have shown the radioactive methods to be accurate to ± 1 per cent. Measurements of flow rates of 200 m^3/sec (7000 cu ft/sec) have shown good agreement with current meter results. The problem of flow measurement using tracers is the establishment of the complete lateral mixing of the tracer across the width of the river, which becomes increasingly difficult in large rivers.

An instrument has been developed for measuring ground water velocities from a single borehole. A radioactive tracer is introduced into a vertically isolated section of a perforated borehole. In the absence of ground water flow, the concentration reduces very slowly by diffusion. The water velocity can be measured from the reduction of the tracer concentration by dilution, the lowest level of measurement being a few cm per day. The method is rapid but the results must be interpreted with care since they represent a point measurement of water velocity in an area of the strata disturbed by the presence of the borehole. Other tracer techniques are used for measuring the vertical flow velocity in perforated boreholes from which the rate at which water enters and leaves the borehole in different zones can be evaluated.

The neutron-neutron logging technique is applied in hydrology to water balance, irrigation and drainage problems. Portable instruments are used for measurements in the top few feet of soil and the sensitivity is such that a 1 per cent change of moisture content of the soil may result in several per cent change in the detector count rate.

Oceanography Applications

Natural radioactivity in the sea arises from river discharge and land run-off which introduces uranium and thorium decay products, from cosmic-ray-produced isotopes such as ^{14}C and from the ^{40}K content of the sea-water.

During the deposition of sediment, radioactive nuclides become incorporated in the material undergoing consolidation and can be used to "date" their formation. The ionium-thorium method provides a technique for recent deposits, using the decay rate of ionium-230 (half-life 80,000 years) in terms of which the sedimentation rate can be calculated from the relative concentrations of the two nuclides.

Carbon-14 measurements have been used to estimate the exchange time for the transfer of CO_2 between the atmosphere and surface ocean water as 3 to 10 years.

Radioisotopes also arise due to fall-out from nuclear explosions and these have been used to evaluate the process and time of mixing of water above and below the thermocline in the oceans.

The introduction of radioactive tracers in the sea has been used to study deep sea diffusion and dispersion.

Instrumental techniques for the measurement of very slow currents (down to 0·5 cm/sec) have long been a problem. An instrument (Deep-Water Isotopic Current Analyser) uses a circle of 16 scintillation detectors surrounding a central point from which a pulse of γ-active solution is released. The speed and direction of the movement of the tracer is determined relative to a compass system. A similar instrument is designed to work in depths of 1800 metres.

Borehole logging techniques have been adapted for ocean bed surveys. One instrument used for density measurement in sediments consists of a $3\frac{1}{2}$ in. dia. tube, 20 ft. long which penetrates the sediment. A ^{137}Cs source is then raised up the tube and the intensity of the backscattered gamma radiation provides a measurement of the sediment density profile.

Sediment Movement Studies

Radioactive tracers are widely used in studies of coastal silt, sand and pebble movement where the highly sensitive detection of the gamma radiation allows large-scale studies to be undertaken.

The tracers used include ^{46}Sc, ^{110}Ag, ^{60}Co, ^{140}La, ^{198}Au in amounts from tens of millicuries to over 100 curies. The tracer is either incorporated in an artificial medium such as glass which is ground and size matched to the sediment being studied or is attached by chemical deposition or adsorption directly to the surface of the particles of sediment. Tracing is carried out by submersing a scintillation counter detector on the sea or estuary bed to obtain either a local measurement or a series of measurements as the detector is towed slowly along the bed. A number of problems of considerable economic importance concerning the siltation of harbours or the movement of dredged material in an estuary have been solved.

The transport of suspended solids in rivers has been measured over the range 500–50,000 ppm by density gauges using the absorption of low energy gamma-rays from ^{109}Cd or ^{241}Am sources. The instruments are standardized with pure water to allow for density changes with temperature and in one instrument this is done automatically by means of a source oscillating between a pure water sample and the suspended sediment sample. The accuracy at 1000 ppm is about ± 30 per cent and this improves at higher sediment concentrations.

Bibliography

BERZIN A. K. *et al.* (1966) Present state and use of basic nuclear geophysical methods for investigating rocks and ores, *Atomic Energy Review*, **4** (2), 59, Vienna: I.A.E.A.

CURRAN S. C. (1961) in *Encylopaedic Dictionary of Physics* (J. Thewlis Ed.), **2**, 165, Oxford: Pergamon Press.

CURTISS L. F. (1962) in *Encyclopaedic Dictionary of Physics* (J. Thewlis Ed.), **5**, 8, Oxford: Pergamon Press.

HAMILTON E. I. (1965) *Applied Geochronology*, New York: Academic Press.

HELMHOLTZ H. R. (1962) in *Encyclopaedic Dictionary of Physics* (J. Thewlis Ed.), **6**, 36, Oxford: Pergamon Press.

MCCONNELL J. R. (1961) in *Encyclopaedic Dictionary of Physics* (J. Thewlis Ed.), **1**, 499, Oxford: Pergamon Press.

(1958) *Oceanographic Applications*, several papers, *Proc. Int. Conf. on the Peaceful uses of Atomic Energy*, Vol. 13, U. N., New York, (1956) and *Proc. Second U. N. Int. Conf. on the Peaceful uses of Atomic Energy*, Vol. 18, U. N., Geneva (1958).

PALEVSKY H. (1961) in *Encyclopaedic Dictionary of Physics* (J. Thewlis Ed.), **2**, 170; **4**, 817, Oxford: Pergamon Press.

PALMER G. H. (1961) in *Encyclopaedic Dictionary of Physics* (J. Thewlis Ed.), **4**, 508, Oxford: Pergamon Press.

PEDRICK R. A. and MAGIN G. B. (1966) Radioisotopes in oceanographic research, *Nucleonics*, **24**, 42, June.

POOLE J. H. J. (1961) in *Encyclopaedic Dictionary of Physics* (J. Thewlis Ed.), **1**, 95, Oxford: Pergamon Press.

Radioisotopes in Hydrology, (1963), Vienna: I.A.E.A.

RHODES J. R. (1966) *Radioisotope X-ray spectrometry: a review*, *Analyst*, **91**, 683, November.

ROGERS D. (1961) in *Encyclopaedic Dictionary of Physics* (J. Thewlis Ed.), **2**, 28, Oxford: Pergamon Press.

SHEPPARD P. A. (1963) Atmospheric tracers and the study of the general circulation of the atmosphere, *Prog. Phys.*, **26**, 213.

Isotopes in Hydrology, Proc. of a Symposium, Vienna, 1966. I.A.E.A. Vienna, 1967.

D. B. SMITH

NUCLEAR PARTICLE TRACKS IN METEORITES. It has recently been discovered that insulating materials register tracks of heavy nuclear particles (Price *et al.* 1965a, 1965b). The tracks are developed for observation in the optical microscope by immersing the samples in a suitable etching solution that preferentially attacks the radiation-damaged regions produced by the passage of the nuclear particles. Only particles which produce ionization at a rate greater than a critical threshold value produce enough damage to give etchable tracks. For the common silicate minerals, only particles with masses $\simeq 40$ atomic mass units—and these only near the end of their range—will produce tracks. Thus, *the silicate crystals in meteorites act as a form of insensitive nuclear emulsion that can be developed to reveal details of their radiation history.*

Although the study of particle tracks in meteorites is still in its infancy, over forty different meteorites, with representatives from each of the major classes (stony, stony-iron, and iron) have been found which exhibit large densities of tracks; it is thus certain that fossil tracks are a common meteoritic phenomenon.

In terrestrial samples, shielded from the primary cosmic radiation, it has been shown that the only significant sources of fossil tracks in minerals is the spontaneous fission of ^{238}U impurities. This has given rise to a new dating technique described elsewhere in this volume (see *Fission track dating*). Meteoritic samples have two additional classes of potential track sources: (a) those arising from the bombardment with primary cosmic rays, and (b) fission tracks from presently extinct transuranic elements that were incorporated in the meteoritic material early in the history of the solar system. (Strictly speaking, this is also a possible source of tracks in terrestrial samples; however, no terrestrial samples have ever been found that approach the antiquity of the meteorites. As a practical matter, therefore, extinct isotopes need only be considered in meteorites.)

Primary cosmic rays can either register directly or they may produce registerable particles as a result of nuclear collisions within the meteorite. A complete discussion of the various possibilities is beyond the

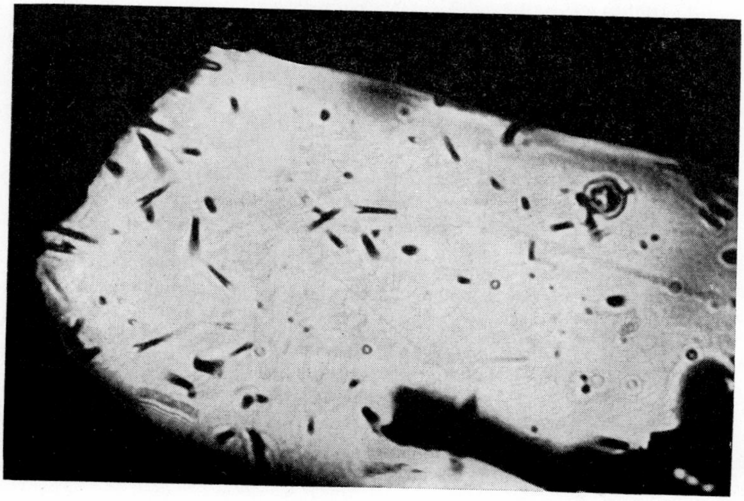

Fossil tracks of cosmic-ray primaries, with atomic number between ~20 and ~28, in a bronzite grain (40 by 60 μ) in a chondrule from the Clovis meteorite.

scope of the present article and, unfortunately, does not yet exist in the published literature. A detailed treatment is currently being prepared for publication in 1966.

The most obvious candidates for cosmic-ray tracks are very heavy particles in the primary beam itself. As a very heavy primary slows down after entering a meteorite, its rate of energy loss increases; and, if it is sufficiently heavy, it will produce a track towards the end of its range when the ionization density exceeds the critical value for track formation. The outstanding identifying characteristic of tracks from heavy primaries is the rapid attenuation of number with depth due to the large nuclear interaction cross-sections of heavy particles.

It has long been known that recoiling residual nuclei produced in high energy spallation reactions may also produce tracks. However, only recoils from heavy weight impurities will produce tracks and, to assess the importance of such tracks in a given sample, it is necessary to perform a calibration experiment using a beam of high energy particles (for example, 3 GeV protons) capable of simulating the reactions produced by incident cosmic rays.

Other possibilities for cosmic-ray tracks include induced fission of heavy element impurities and the existence of such bizarre (hypothetical) particles as magnetic monopoles.

Detailed studies of the tracks found in various meteorites have shown that most of the tracks are due to slowed down heavy primaries of nuclear charge >22. No evidence has been found for any other kind of track related to the incident cosmic-ray beam. It can be shown theoretically that the observable track length of a heavy primary increases rapidly with increasing mass above the critical value. This fact has led to the proposal that a study of long tracks in meteorites may be used to measure the presently unknown abundance of primary cosmic rays with masses greater than iron. Positive evidence for such particles has been reported (Fleischer *et al.* 1965a), but details have not been published.

The absence of cosmic-ray induced tracks in tektites has been used to set an upper limit of 300 years for the space exposure time of these objects (Fleischer *et al.* 1965b).

Evidence for the existence in certain meteorites of fossil tracks from the spontaneous fission of presently extinct trans-uranic elements, notably ^{244}Pu, has also been presented (Fleischer *et al.* 1965c). In principle, such tracks give a method for measuring the time interval between the end of nucleosynthesis and the formation of cooled down meteoritic material. If the present results are confirmed, the cool-down time of the early solar system would be of the order of several hundreds of millions of years.

Because of the generality of the phenomenon and the important information which it conveys, it is likely that the study of meteoritic particle tracks will increase in the future.

Bibliography

FLEISCHER R. L., PRICE P. B. and WALKER R. M. (1965a) *Proceedings of the IX International Conference on Cosmic Rays*, London: Institute of Physics and The Physical Society.

FLEISCHER R. L. NAESER C. W., PRICE P. B. WALKER R. M. and MAURETTE M. (1965b) *J. Geophys. Res.*, **70**, 1491.

FLEISCHER R. L., PRICE P. B. and WALKER R. M. (1965c) *J. Geophys. Res.*, **70**, 2703.

MAURETTE M. (1965) Doctoral Thesis, *Etudes Des Traces Des Particules Nucléaires Dans Lens Substances Naturaux*, University of Paris.

PRICE P. B., FLEISCHER R. L. and WALKER R. M. (1965a) *Annual Reviews of Nuclear Science.*

PRICE P. B., FLEISCHER R. L. and WALKER R. M. (1965b) *Science*, **149**, 383.

R. L. FLEISCHER, P. B. PRICE and R. M. WALKER

NUCLEAR REACTOR CLASSIFICATION. *Introduction.* In the *Encyclopaedic Dictionary of Physics* and its supplementary volumes descriptions of various types of nuclear reactor have appeared. The choice of which particular types to describe in detail has been based to some extent on the need for including, on the one hand, an account of the basic principles underlying reactor operation and construction, and on the other, the desirability of saying something about the newer or more important types of reactor. No attempt has been made to describe all the known types of reactor; and indeed a little thought will show that this would neither be possible nor desirable. A reactor may well be specified by a string of descriptive terms, and it would manifestly be impossible to list and define all the combinations of such terms in a Dictionary of this (or any other) type.

If, however, the various descriptive terms could be broken down into groups dealing with specific aspects of reactor design and performance, and if these terms could be systematized and defined, any known reactor type, specified by a combination of appropriate terms, could be recognized and its characteristics understood. If occasion arose, new terms could be added and defined.

This point has already been recognized by a Committee of the International Standards Organization dealing with terminology in nuclear energy, and most of the present article is, indeed, based on the work of that Committee.

System of classification. All the descriptive terms used in specifying a reactor type describe one of the following:

(a) the purpose of the reactor;
(b) the engineering design of the reactor;
(c) the nuclear design of the reactor.

These form three convenient classes which, when subdivided into appropriate groups, form a basis for the present system of reactor classification.

Table of descriptive terms classified into groups and classes

Purpose

Experimental 1
- Zero-power: zero-energy ... 2
- Critical experiment ... 3
- Engineering development . 4
- Reactor experiment ... 5
- Prototype ... 6
- Demonstration ... 7

Research 8
- Low-flux ... 9
- High-flux ... 10
- Pulsed ... 11
- Materials-testing ... 12
- Zero-power: zero-energy ... 13
- Beam ... 14
- Source ... 15

Training 16

Power 17
- Electric-power ... 18
- Propulsion ... 19
- Process-heat ... 20
- Magnox ... 21

Irradiation 22
- Food-irradiation ... 23
- Chemonuclear (chemical processing) ... 24
- Biomedical ... 25
- Materials-processing ... 26
- Materials-testing ... 27
- Isotope production ... 28

Production 29
- Irradiation ... 30
- Fissile-material production 31
- Plutonium-production ... 32
- Isotope production ... 33

Multipurpose 34

Engineering design

Nature of primary coolant circuit
- Pressurized ... 35
- Boiling ... 36
- Circulating-fuel ... 37
- Direct cycle ... 38
- Indirect-cycle ... 39
- Dual-cycle ... 40

Nature of primary coolant circuit
- Forced-circulation ... 41
- Natural-circulation ... 42
- Natural-convection ... 43
- High-temperature ... 44

Nature of coolant

Gas
- Gas-cooled ... 45
- Air-cooled ... 46
- Helium-cooled ... 47
- Carbon-dioxide-cooled .. 48
- Nitrogen-cooled ... 49

Liquid
- Liquid-cooled ... 50
- Water-cooled ... 51
- Light-water-cooled ... 52
- Heavy-water-cooled ... 53
- Organic-cooled ... 54
- Molten-salt-cooled ... 55
- Liquid-metal-cooled ... 56

Other
- Dust-cooled ... 57
- Fog-cooled ... 58

Nature of moderator

Liquid
- Water-moderated ... 59
- Light-water-moderated .. 60
- Heavy-water-moderated 61
- Organic-moderated ... 62

Solid
- Graphite-moderated ... 63
- Beryllium-moderated ... 64
- Beryllium-oxide-moderated ... 65
- Zirconium-hydride-moderated ... 66
- Plastic-moderated ... 67

Nature of fuel structure

Solid
- Pebble-bed ... 68
- Oxide-fuelled ... 69
- Carbide-fuelled ... 70
- Metal-fuelled ... 71
- Ceramic-fuelled ... 72

Liquid
- Aqueous ... 73
- Liquid-metal-fuelled ... 74
- Molten-salt-fuelled ... 75

Dispersion
- Slurry ... 76
- Paste ... 77
- Fluidized bed ... 78

Mobility of reactor
- Transportable ... 79
- Mobile ... 80
- Package ... 81

Overall Construction of reactor
- Pool, Swimming pool ... 82
- Tank ... 83
- Pressure-tube ... 84
- Integral ... 85

Nuclear design

Nature of neutron spectrum
- Fast ... 86
- Intermediate ... 87
- Epithermal ... 88
- Thermal ... 89
- Mixed-spectrum ... 90

Nature of fissionable material
- Natural-uranium-fuelled 91
- Enriched-uranium-fuelled 92
- Plutonium-fuelled ... 93

Nature of fertile material
- Thorium ... 94
- Natural-uranium ... 95

Nuclear configuration
- Homogeneous ... 96
- Heterogeneous ... 97
- Seed-core ... 98
- Diluted-core ... 99
- Circulating-fuel ... 100
- Bare ... 101

Nature of control system
- Absorption controlled ... 102
- Configuration controlled .. 103
- Poison controlled ... 104
- Spectral shift. ... 105
- Self-regulating ... 106

Conversion properties
- Burner ... 107
- Converter ... 108
- Breeder ... 109

In the table, therefore, a list of descriptive terms (most of which are adjectives or adjectival phrases) is given which are arranged in groups within the three main classes listed above. In some instances the names of the groups themselves also constitute descriptive terms and these are then numbered serially along with the rest of the descriptive terms. All these terms are listed alphabetically below with appropriate notes or definitions where necessary and with their numbers appended for ease of reference to and from the table.

Descriptive Terms: Notes and Definitions

Air-cooled (46)

Aqueous (73)
A homogeneous reactor in which the fuel is in aqueous solution.

Bare (101)
A reactor without a reflector.

Beam (14)
Specially designed to produce external beams of neutrons for research.

Beryllium-moderated (64)

Beryllium-oxide-moderated (65)

Biomedical (25)
Designed for biological and medical irradiation.

Boiling (36)
The primary coolant is allowed to boil.

Breeder (109)
A reactor which produces more fissile material than it consumes, i.e. which has a conversion ratio greater than unity.

Burner (107)
A reactor in which no significant conversion takes place.

Carbide-fuelled (70)

Carbon-dioxide-cooled (48)

Ceramic-fuelled (72)
The fuel contains non-metallic materials of high melting point.

Chemo-nuclear (*chemical processing*) (24)
Designed as a source of radiation for effecting chemical transformations on an industrial scale.

Circulating fuel (37, 100)
The operation of the reactor involves a circulation of fuel (fluid or fluidized) through the core.

Converter (108)
Significant conversion takes place.

Critical experiment (3)
Denotes an assembly of reactor materials which can be gradually brought to the critical state for the determination of the nuclear characteristics of a reactor.

Demonstration (7)
Designed to demonstrate the technical feasibility and explore the economic potential of a given reactor type.

Diluted-core (99)
A fast reactor whose core contains an excess of non-fissile material to improve a desired reactor characteristic, such as heat transfer or Doppler coefficient.

Direct-cycle (38)
The primary coolant is used directly to produce useful power.

Dual-cycle (40)
Useful power is produced by the utilization of heat from both the primary and secondary coolant circuits.

Dust-cooled (57)
The coolant consists of a suspension of solid particles in a gas (e.g. graphite in carbon dioxide).

Electric-power (18)

Engineering-development (4)
Designed and operated at a sufficient power level to yield data on the thermal and mechanical performance of a reactor type.

Enriched-uranium-fuelled (92)
Fuelled by enriched uranium in the form of metal, oxide, carbide, etc.

Epithermal (88)
Fission is induced predominantly by epithermal neutrons.

Experimental (1)
Operated primarily to obtain reactor physics or engineering data for the design or development of a reactor or reactor type.

Fast (86)
Fission is induced predominantly by fast neutrons.

Fissile-material production (31)

Fluidized-bed (78)
The fuel is in the form of a bed of fine particles which are maintained in a state of non-circulating suspension during reactor operation by the flow of the fluid coolant.

Fog-cooled (58)
The coolant consists of a suspension of water droplets in a gas.

Food-irradiation (23)
Used for the preservation of food.

Forced-circulation (41)
The circulation of the primary coolant is ensured by pumping.

Gas-cooled (45)

Graphite-moderated (63)

Heavy-water-cooled (53)

Heavy-water-moderated (61)

Helium-cooled (47)

Heterogeneous (97)
The core materials are segregated to such an extent that the neutron characteristics of the reactor cannot be accurately described by the assumption of a homogeneous distribution of the materials throughout the core.

High-flux (10)
Usually taken to mean capable of sustaining a maximum flux density greater than 10^{14} n/cm^2 sec.

High-temperature (44)
The reactor design and operation pose specific technological problems associated with high temperature.

Homogeneous (96)
The core materials are distributed in such a manner that the neutron characteristics of the reactor can be accurately described by the assumption of a homogeneous distribution of the materials throughout the core.

Indirect-cycle (39)
The primary coolant transfers its heat to a secondary coolant to produce useful power.

Intermediate (87)
Fission is induced predominantly by intermediate neutrons.

Integral (85)
The heat exchanger between the primary and secondary coolant circuits is contained within the reactor vessel.

Irradiation (22, 30)
Used primarily as a source of nuclear radiations for the irradiation of materials or for medical purposes.

Isotope-production (28, 33)
Used for the production of radioisotopes other than fissile isotopes.

Light-water-cooled (52)
Cooled by water of natural isotopic composition.

Light-water-moderated (60)
Moderated by water of natural isotopic composition.

Liquid-cooled (50)

Liquid-metal-cooled (56)

Liquid-metal-fuelled (74)
The fissile metal is dissolved in a liquid metal.

Low-flux (9)
Imprecise. Sometimes taken to mean having a maximum flux density less than 10^{12} n/cm^2 sec.

Magnox (21)
A reactor in the first generation of British-designed power reactors—from the use of "magnox" (aluminium-magnesium alloy) as the canning material.

Materials processing (26)
Used for modifying the properties of materials by irradiation.

Materials-testing (12, 27)
Used for testing materials and components in intense radiation fields.

Metal-fuelled (71)

Mixed-spectrum (90)
Having widely different neutron spectra in different parts of the reactor core.

Mobile (80)
Designed to be mobile during operation.

Molten-salt-cooled (55) }
Molten-salt-fuelled (75) }
A mixture of molten salt serves both as fuel and as primary coolant.

Multipurpose (34)
Designed for different purposes not normally combined in one reactor.

Natural circulation (42)
The circulation of the primary coolant is ensured by using the heat generated in the core to produce boiling or convection.

Natural convection (43)

Natural-uranium-fuelled (95)
Fuelled by natural uranium in the form of metal, oxide, carbide, etc.

Nitrogen-cooled (49)

Organic-cooled (54)

Organic-moderated (62)

Oxide-fuelled (69)

Package (81)
Denotes a compact power reactor specially designed to simplify shipping and assembly.

Paste (77)
Refers to a homogeneous reactor in which the fuel is a suspension of solid particles in a paste.

Pebble-bed (68)
Some or all of the essential nuclear material (e.g. fuel, fertile material, moderator) is in the form of a stationary bed of small balls ("pebbles") in contact with each other.

Plastic-moderated (67)

Plutonium-fuelled (93)
A significant amount of plutonium is present in the fresh fuel in the form of metal, oxide, carbide, etc.

Plutonium production (32)

Pool: Swimming pool (82)
The fuel elements are immersed in a pool of water which serves as moderator, coolant, and biological shield.

Power (17)
The primary purpose is to produce power.

Pressure-tube (84)
The fuel assemblies and coolant are confined in tubes that withstand the pressure of the coolant.

Pressurized (35)
The primary coolant is maintained under such a pressure that no bulk boiling occurs.

Process-heat (20)
Designed for the production of heat, conversion to other forms of energy being excluded.

Production (29)
The primary purpose is to produce fissile or other materials, or to perform irradiation on an industrial scale. Unless otherwise specified the term usually refers to a plutonium-production reactor.

Propulsion (19)

Prototype (6)
The first of a series of reactors of the same basic design. Sometimes used to denote a reactor having the same essential features but on a smaller scale than the final series.

Pulsed (11)
Designed to produce intense bursts of neutrons for short intervals of time.

Reactor experiment (5)
Designed to test the feasibility of a specific reactor design.

Research (8)
Used primarily as a research tool for basic or applied research. May be of any power level.

Seed-core (98)
Having a core containing local regions ("seeds") of enriched fuel distributed in a lattice of fuel of lower enrichment or of fertile material.

Self-regulating (106)
The reactor has an inherent tendency to operate at a constant power level owing to the compensating change in reactivity that follows a change in power level.

Slurry (76)
Refers to a homogeneous reactor in which the fuel takes the form of a circulating suspension of fine particles in a liquid.

Source (15)
Designed to supply a stable flux of neutrons having a well-determined energy spectrum, principally for conducting exponential or shielding experiments or for calibrating detectors.

Spectral shift (105)
The neutron spectrum may be adjusted by varying the properties or the amount of the moderator.

Swimming pool (82)
See under "Pool: Swimming pool".

Tank (83)
A variation on the pool-type reactor. The core is contained in a closed tank.

Thermal (89)
Fission is induced predominantly by thermal neutrons.

Thorium (94)

Training (16)
Operated primarily for training in reactor operation and instruction in reactor behaviour.

Transportable (79)
Capable of being moved, but only when not critical and possibly partly dismantled.

Water-cooled (51)
Cooled by water of natural isotopic composition (but sometimes used for cooling by water of any isotopic composition).

Water-moderated (59)
Moderated by water of natural isotopic composition (but sometimes used for moderation by water of any isotopic composition).

Zero power: zero energy (2, 13)
Designed to be used at such a low power that no cooling system is needed.

Zirconium-hydride-moderated (66)

Bibliography
Various articles with headings beginning "Nuclear reactor" in the *Encyclopaedic Dictionary of Physics* and Supplementary Volumes.

J. Thewlis

NUCLEAR REACTORS, HIGH TEMPERATURE (HTR). 1. In one sense all nuclear reactors are high temperature reactors since the effect of fission in uranium and plutonium is to release energy, which appears as heat and, hence, high temperature in the fissile material. In any nuclear reactor, this heat is eventually transferred to the coolant of the reactor and transported to the heat exchangers or boilers where it is removed and either rejected or used to generate electricity.

2. In practice, the name "high temperature reactor" is now widely used for a particular type of reactor system with a chemically inert gas as the coolant and a reactor core made of refractory materials. There are, at the time of writing, several reactors of this type under construction or in operation. Each one uses helium as the coolant, has a graphite moderator, and has fuel assemblies containing coated particles with an oxide or carbide kernel and a pyrocarbon coating. The use of beryllia as a moderator has been considered but this work

has not yet reached the same advanced stage as that on the use of graphite.

3. When high temperature reactors were first proposed, it was not expected that it would be possible to make fuel assemblies which could retain most of the fission products produced at the expected operating temperatures of around 1500°C. The first ideas for this type of reactor, therefore, were based on the use of fuel assemblies which would emit most of the volatile fission products, many of which have high capture cross-sections for neutrons. Although this would mean a highly radioactive coolant circuit, it was hoped that by removing most of these fission products in a coolant purification plant, the neutron economy of the reactor would be improved because of the removal from the fuel of many of the fission products which would otherwise absorb neutrons. In practice, the early irradiation tests of typical fuels in an experimental reactor showed that only the gaseous fission products were emitted to any great extent and that the fraction of the total fission product absorption which could be removed was quite small. This meant that there would be very little improvement in the neutron economy but considerable complication in the design and operation of the reactor because of the highly radioactive coolant circuit.

4. In the meantime, further research work on fuel assemblies showed that it was possible to coat small spherical particles of uranium with pyrocarbon and that these "coated particles" retained most of the fission products produced in the uranium. A considerable amount of development work has now been carried out on these types of fuel and, as a result, it is now possible to fabricate coated particle fuels which will release less than one part in ten thousand of the fission products produced, even after long irradiation times. The fabrication of these fuels typically consists of a number of processes:

(i) the making of a spherical kernel of uranium oxide or carbide of about 500 microns diameter;

(ii) the deposition of a low density pyrocarbon layer of about 30 microns thickness by decomposing a gaseous hydrocarbon in a fluidized bed furnace at 1200°C;

(iii) the deposition of a silicon carbide layer of about 30 microns thickness by decomposing a gaseous silicon compound at about 1400°C;

(iv) the deposition of a final coating of high density pyrocarbon of about 60 microns thickness at a temperature of about 2000°C;

(v) the introduction of these coated particles into the graphite structure of the fuel element.

5. The successful development of these coated particle fuels with their good fission product retaining properties meant that it was then possible to consider a reactor which was similar in general design to other gas cooled reactors with metal clad fuel elements but with the difference that the fuel element could be operated at higher surface temperatures in a helium atmosphere, since it was made of refractory materials. The two essential differences between high temperature gas cooled reactors and other gas cooled reactors are, therefore, the use of helium as the coolant gas and the use of ceramic or refractory fuel elements. Although all power reactors at present use a conventional steam cycle power plant for the final stage of conversion of heat into electricity, in principle it is possible with a high temperature reactor to achieve high enough coolant outlet temperatures of 800°C and above to make the use of a closed cycle gas turbine power plant an attractive alternative, on the grounds of both thermal efficiency and capital cost.

6. At present, two types of high temperature reactor are under active development. One type, exemplified by the Dragon Reactor Experiment at Winfrith, Dorset, England, and the Peach Bottom Reactor at Peach Bottom, Pennsylvania, U.S.A., uses prismatic type fuel elements. The fuel assemblies are, essentially, graphite tubes containing fuel compacts which are composed of coated particles in a graphite matrix. The kernels of the coated particles contain a mixture of uranium and thorium oxides or carbides. The helium coolant passes over the outside of the graphite tubes. The fuel assemblies are in the form of rods the whole length of the reactor core and are changed either individually or as clusters of rods. The inlet temperature of the helium coolant to the reactor core is about 350°C and the outlet temperature from the reactor core is about 750°C. In the Dragon Reactor Experiment, which has a heat output of 20,000 kW, the heat is rejected to the atmosphere through an auxiliary water cooling system. The Peach Bottom Reactor has a heat output of 116,000 kw and will generate 40,000 kw of electricity using a conventional steam cycle.

7. The other type of high temperature reactor, exemplified by the AVR reactor at Jülich, West Germany, uses spherical fuel elements. This reactor also uses helium as the coolant and has similar operating temperatures. The fuel is in the form of spheres of graphite which contain coated particles of similar composition to those in the other two reactors. The spherical fuel elements can be added to or withdrawn from the reactor one by one and either discharged or recirculated through the reactor with new spherical fuel elements being added as required. The AVR reactor has a heat output of 56,000 kW, and will generate 20,000 kW of electricity.

8. The other development which was needed to make helium cooled high temperature reactors a practical possibility was the development of the helium technology itself. Helium is a relatively expensive gas so that problems of leak tightness are more important than in gas cooled reactors using carbon dioxide, which is a relatively cheap gas, and ways of constructing demountable seals or flanges with acceptable leakage rates have now been demonstrated. Helium also needs to be kept chemically pure in order to avoid chemical reaction between the impurities and hot graphite. Methods of keeping the total impurity level down to a few parts per million

by volume, as well as methods for removing fission products from helium, have now been successfully demonstrated. The experience gained in these and other aspects of helium technology in the several reactor projects under construction or in operation provide a sound basis for exploiting the high temperature reactor commercially.

9. Because the high temperature reactor need have no structural or cladding materials in the core, which can be made entirely of moderator and fuel, it has a very good neutron economy. This is reflected in the low fuel cycle costs which can be achieved with this type of reactor, either with the highly enriched uranium/thorium fuel cycle or with the slightly enriched uranium fuel cycle. The good neutron economy also leads to good utilization of the fuel needed for these reactors. The use of helium as the coolant leads to higher power densities than in other gas cooled reactors and also to higher gas outlet temperatures. These factors together can lead to lower capital costs for the reactor itself and, combined with the lower fuel cycle costs, can lead to lower generating costs for electricity.

10. At present the high temperature reactor is at the stage when the basic principles have been successfully demonstrated and the next stage of commercial exploitation can begin. The incentives for carrying on to this next stage are summarized in the following ten reasons:

(i) a conveniently high power density in the core and heat exchangers giving a compact reactor with low capital cost;

(ii) high reactor outlet gas temperatures, permitting the use of the most modern low cost steam cycle power plant, and giving good thermal efficiency;

(iii) high specific ratings of about one electrical megawatt per kilogram of fissile material, leading to low fuel inventory costs;

(iv) high burnup, leading to low fuel cycle costs in a once-through cycle;

(v) good conversion factor, leading to lower fuel cycle costs with recycled fuel;

(vi) the ability to use thorium to conserve supplies of natural uranium;

(vii) ceramic coated particle fuel which is relatively easy to fabricate and can withstand very high temperatures and high burnups;

(viii) flexible core geometry, giving scope for using various fuel cycles;

(ix) good development potential for higher temperature applications, such as closed cycle gas turbines;

(x) relatively cheap development costs.

See also: Coated-particle nuclear fuel.

Bibliography

HILL J. F. (1962) in *Encyclopaedic Dictionary of Physics* (J. Thewlis Ed.), **5**, 97, Oxford: Pergamon Press.

C. A. RENNIE

NUMERICALLY-CONTROLLED MACHINE TOOLS, MEASURING AND CONTROL SYSTEMS FOR. *Introduction.* Numerically-controlled machine tools are so called because information about the required position of the cutting tool at each stage of the machining operation is fed to the machine in numerical form. This is usually achieved in the same way that numbers are fed into a digital computer, that is as patterns of holes punched in paper tape or patterns of magnetic signals on magnetic tape, but the information can also be fed in manually by, for example, setting up a series of decade switches. The usual measuring scales on a machine tool are replaced by transducers which produce a signal (usually, but not always, electrical) which represents the position of the cutting tool with respect to some known datum on the machine. The moving parts of the machine tool are driven via control systems which ensure that, at each stage of the machining operation, the cutting tool is placed in the position indicated by the numerical information which has been fed to the machine tool. Usually, but not always, the motors driving the moving parts of the machine tool are actuated by feedback control systems which continually seek to reduce any difference that there may be between the actual position of the cutting tool and the position which it should occupy at that stage of the machining operation as demanded by the input data.

The essential features of the control system of a numerically-controlled machine tool are:

(i) Measuring systems or transducers which produce signals (usually electrical) representing the positions of moving parts of the machine tool. The movements so measured may be linear movements of slides or rotary movements of a positioning table or spindle slide.

Simple numerically-controlled machine tools may be controlled in only two axes whereas more elaborate machines may be controlled in as many as six axes, translational and rotational.

(ii) Drive mechanisms for effecting the necessary linear or rotary movements in response to the information supplied. Linear movements are effected by electric or hydraulic motors driving leadscrews or pinions engaging with a rack, or by hydraulic rams. Rotary movements are usually effected by worm and wheel drives. (For numerical control of screw-cutting operations on a lathe, continuous control of the spindle rotation is also necessary.)

Servomechanisms, actuated by the difference between the actual and demanded positions, are generally employed, but drive systems without feedback—open-loop systems which merely move the slide or table a specified amount—are sometimes used.

(iii) A control system which accepts instructions, usually in the form of coded signals on punched cards, punched paper tape or magnetic tape, but sometimes also by manual setting of decade switches, and which controls not only the relative positions of the different parts of the machine tool but also the tools selected, spindle and feed speeds, the flow of coolant, hydraulic clamps, etc.

Measuring Systems

Ideally the quantity measured and controlled should be the position of the machined surface but, as yet, it has been possible to achieve this in only a very few instances. For example, a numerically-controlled cylindrical grinding machine (the "Multi-finitron") has been built in which the diameter of the workpiece close to the grinding wheel is measured continuously, and the machine is controlled on the basis of this measurement and the information about the desired workpiece size. In general, however, in-process measurement—the continuous measurement of workpiece size—presents considerable difficulties and less-direct methods have to be used.

Failing a measure of workpiece size, the next best thing is a measure of the position of the cutting edge with respect to some datum on the machine, and many numerically-controlled machine tools use measuring systems of this kind. It should be noted, however, that factors such as wear of the cutting tool and deformation of the workpiece or machine tool under the cutting forces can lead to errors and care must be taken to minimize these effects.

On numerically-controlled machine tools in which the slides are moved by leadscrews or by rack-and-pinion drives, even-less-direct measuring systems are often used. With these systems the rotation of the leadscrew or pinion is measured and the position or movement of the cutting tool is then inferred from this measurement and the known pitch of the lead-screw or rack. In addition to the errors mentioned above, such systems are also liable to error as a result of deformation, and of wear or backlash between leadscrew and nut or between rack and pinion. The problems involved have, however, been successfully solved by many manufacturers.

In addition to distinctions based on the position of the measuring equipment, a number of different basic types of measuring transducer can be distinguished. A transducer may be digital in which case a given distance is represented by a train of pulses according to some recognized code, or analogue in which the dimension is represented by a proportional voltage, current or frequency; a system may also be absolute if the quantity measured is the actual distance from a fixed datum, cyclic if the range is limited and repetitive over the range of the machine, or incremental if only changes in position are measured. Analogue incremental systems have not, as yet, been used for machine tool control but each of the other combinations is used. (It should be noted that it is often possible to use a transducer in several different ways. Thus the same transducer may be used as either a digital incremental system or as an analogue cyclic system. Similarly a number of analogue cyclic systems can be used to form an analogue absolute system.)

The output of some cyclic transducers is approximately proportional to the size of the displacement and cannot be measured accurately enough. Such transducers are usually used at a null position, the position of the null being varied by controlling the inputs to the transducer, and are not, therefore, truly analogue.

Brief descriptions of the measuring systems used on numerically-controlled machine tools are given below.

1. Generally-used leadscrew systems. In order to obtain the required accuracy—typically 0·001–0·0001in.—it is necessary for the measuring system to have a resolution of the order of 0·1–1·0° and in order to achieve this it is usually necessary for the signal to repeat after one revolution of the leadscrew. Thus, in order to be able to measure displacements or distances corresponding to more than one revolution of the leadscrew, it is necessary to be able to count revolutions. This is conveniently done from an auxiliary shaft or shafts geared to the leadscrew and, in practice, a combination of coarse and fine, or coarse, medium and fine measuring systems is usually employed when absolute measurements are required (see Fig. 1). In this way the accuracy with which measurements can be made is made independent of the gearing used but depends, of course, on the accuracy of the measuring system itself.

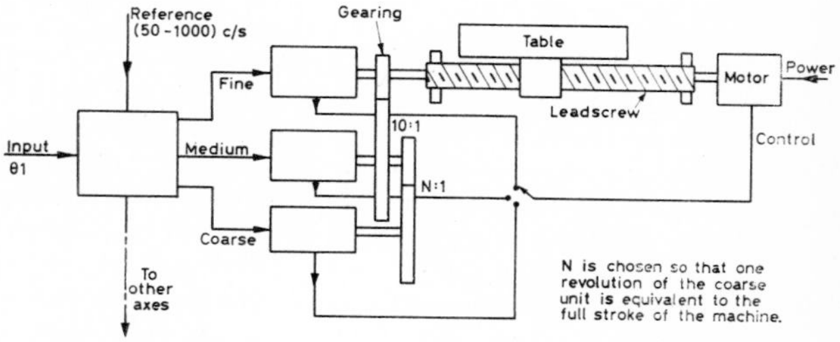

Fig. 1. Machine tool control system showing coarse, medium and fine controls of table position.

(*a*) *Digitizers* (*incremental digital*). A radial grating, or a commutator or series of electrical contacts, attached to the leadscrew, together with a fixed photo-electric head or a wiping contact can be used to produce one pulse for each unit of angular rotation.

The effective accuracy of a digitizer depends upon the number of digits per revolution and on the pitch of the leadscrew. Optical digitizers of reasonable diameter and about 1000 lines are capable of an accuracy of about 0·0002 in. when used with a leadscrew of pitch 0·25 in.

(Radial gratings can also be used as cyclic analogue transducers as described below under "Optical gratings").

(*b*) *Synchro-resolvers* (*cyclic analogue*). A synchro-resolver is similar in construction to an electric motor, having a fixed outer casing carrying a stator winding consisting essentially of two coils at right angles, and a rotor carrying a single winding. In the usual mode of operation the stator windings are fed with alternating voltages proportional to $V \sin \theta$ and $V \cos \theta$, where θ is the required position of the rotor relative to the stator. The magnetic field produced by the currents in the stator windings will induce an alternating voltage in the rotor unless the rotor is at such an angle that the axis of its winding is perpendicular to the field. The rotor thus acts as an error detector, giving an output signal which can be used, via a servo motor, to drive the leadscrew until the rotor is in the position demanded. (Zero error voltage is also obtained when the rotor makes an angle of 180° with the stator field, and the error voltage must be fed to a phase-sensitive rectifier to produce a d.c. voltage the polarity of which will indicate the direction in which the leadscrew must be rotated to bring the rotor to the correct zero position.)

In an alternative mode of operation the stator is energized from a two-phase supply and the phase of the voltage induced in the rotor windings is then a measure of the angular difference between the demanded and actual positions of the rotor.

The over-all accuracy of a synchro-resolver and associated electrical circuits is about ±20 minutes of arc.

Inductosyns (see below) can also be made in a form suitable for the measurement of leadscrew rotation.

2. *Generally-used linear systems.* (*a*) *Resolvers* (*analogue cyclic*). Several transducers operating on the same principle as synchro-resolvers have been developed for linear measurement.

In the *Inductosyn* (Harrison *et al.* 1957) (see Fig. 2(a)) rotor and stator windings are inductively coupled. A single-phase "rotor" winding of rectangular form is printed on to a flat glass or steel-backed plastic scale of the required length. A short two-phase "stator" or cursor is attached to the moving part of the machine so that it moves over the "rotor" with a working gap of about 0·005 in. The stator windings are displaced by one-quarter of a pitch and are fed with alternating currents proportional to $\cos \theta$ and $\sin \theta$ where the desired linear position is $\dfrac{\theta}{360} \times$ pitch.

In the *Helixyn* (Buckerfield 1960, 1962) (see Fig. 2(b)) the windings are helical on concentric cylinders and are capacitatively coupled so that d.c. voltages proportional to $\sin \theta$ and $\cos \theta$ will produce a null in the "rotor" output at the defined position. In practice, the "stator" windings are energized by trains of pulses so that transformer coupling of input and output circuits is possible. The windings have a pitch of 0·1024 in. which is chosen since $1024 = 2^{10}$, and digital computation of the input voltages is simplified when a digital unit of 0·0001″ is used. However, the error signal is truly analogue and a resolution of better than 0·0001 in. is possible. The helical construction has

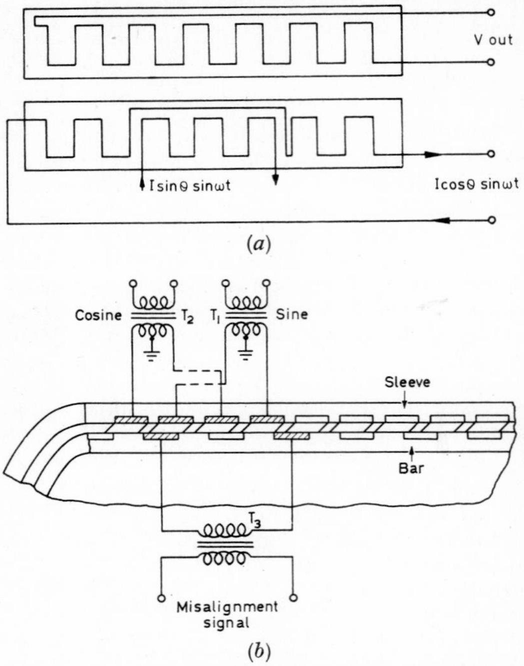

Fig. 2. Linear resolvers: (*a*) *Inductosyn;* (*b*) *Helixyn.*

the advantage that relative rotation of the two cylinders is equivalent to relative linear displacement, and advantage is taken of this in setting zero positions and in the use of a correction cam to reduce errors caused by errors in the spacing of the windings.

In practice variations in the voltages (or currents) $V \sin \theta$ and $V \cos \theta$ are not important so long as their ratio remains proportional to $\tan \theta$. Linear resolvers are capable of an accuracy of 0·0001″.

(*b*) *Optical gratings* (*incremental digital or cyclic analogue*). A uniformly-ruled grating, with suitable illumination and a photocell as detector, can provide a linear measuring scale.

In its simplest form a measuring system based on an optical grating involves a grating of the desired length and a short section of an identical grating which moves over it, both sets of rulings being parallel. Light transmitted through the two gratings is received on a photocell and the output signal goes from a maximum through zero to a maximum again as the short grating is moved a distance of one ruled pitch in a direction at right angles to the rulings. In practice such a system can be used only by detecting the peaks in the output of the photocell and converting these into pulses which can be counted; no information about the direction of motion is available.

A more-useful system is produced by inclining the rulings on the two gratings at a small angle to each other so that Moiré fringes are produced at the right angles to the rulings (see Fig. 3). Movement of the reading grating through one ruled pitch causes the Moiré fringe pattern to move one fringe pitch, and if the direction of relative motion of the gratings reversed the direction of movement of the Moiré fringes will also reverse. At the same time a high degree of magnification of the fringe pattern can be obtained since the actual fringe pitch can easily be as great as 0·5 in. even with gratings with 1000 lines

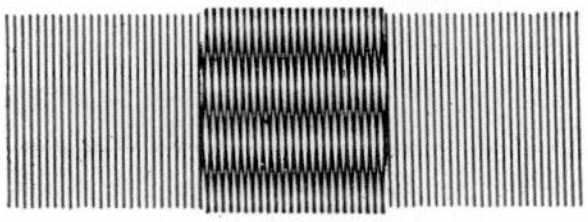

Fig. 3. Moiré fringes formed between gratings inclined at a small angle to each other. Relative movement of the gratings by one grating pitch causes the Moiré fringe system to move through one fringe pitch.

per inch. It is, therefore, possible to use two photocells spaced one-quarter of a Moiré fringe apart, the output signals from which are the same as if they were spaced apart one-quarter of a grating pitch. The output signals from these photocells contain information about both the magnitude and the direction of motion of the gratings. In practice, four photocells are used (Williamson 1958), spaced across the Moiré fringe to give an averaging effect and so as to produce four pulses for a relative movement of one grating pitch. Thus with a 2500-line per inch grating, increments of 0·0001 in. can be detected.

However, most of the optical gratings now used on numerically-controlled machine tools are of coarse pitch—20–100 lines per inch—and can, therefore, easily be of the reflection type, ruled on stainless steel. In order to maintain the required resolution each Moiré fringe must now be sub-divided into 100 parts and this is done by effectively modulating the system so that, when the gratings are stationary or moving with constant relative velocity, the output from the photocell is a signal of constant frequency. The phase difference between the modulating signal and the output from the photocell is proportional to the relative displacement of the gratings, the difference in frequency between these two signals at any instant being proportional to the relative velocity of the gratings. Two methods of effecting the modulation are used:

(i) A spiral-ruled disc is rotated at constant speed between the gratings and a photocell (Walker 1963).

(ii) The outputs from suitably-spaced photocells are sampled cyclically (Davies *et al.*, 1960).

Accuracies of 0·0001 in. correspond to controlling the phase to 3·6°.

(*c*) *Ferromagnetic scales* (*cyclic analogue*). A row of accurately-positioned steel pins mounted on a non-magnetic base provides a scale which can be "read" electromagnetically. Two types of scale of this kind have been used on numerically-controlled machine tools.

The *Accupin* (Evans and Kelling 1963) comprises precision-ground steel pins 0·098 in. diameter positioned on 0·1 in. centres. The reading head contains a coil on a C-core and the scale acts as a variable-reluctance path to the head so that the coil has maximum inductance when the core is positioned across a pin, and minimum inductance half a pitch later. To give increased accuracy, each single reading head contains four C-cores spaced one pitch apart, and two reading heads with a half pitch difference between them are used. Thus when the inductance of one head is a maximum the inductance of the other is a minimum and the two can be used in an inductance bridge. The output varies sinusoidally with displacement and can be used with that from a second pair of reading heads spaced a quarter of a pitch away from the first (in practice, a whole number of pitches plus a quarter) to obtain signals proportional to the sine and cosine of displacement. In practice the two pairs of reading heads are fed from a two-phase supply like a resolver and when the bridge outputs are added electrically the phase of the resultant signal depends on the position of the slide just as with the resolver. Accuracies of 0·0003 in. can be achieved.

The AEI inch bar is used as a line standard of 1-in. pitch. A coarse-positioning element defines the required section and fine positioning is achieved by using synchro-resolvers to move the reading head

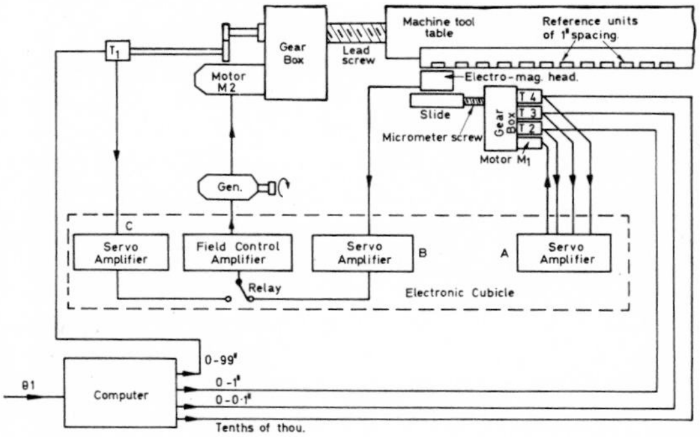

a

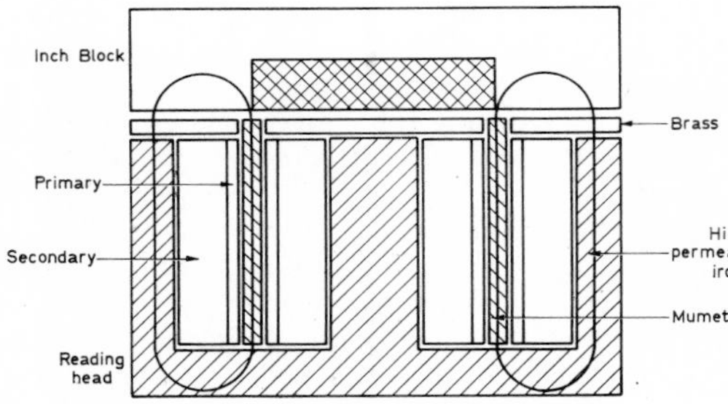

b

Fig. 4. A.E.I. inch-bar measuring system.

with respect to the moving part of the machine tool. Since its stroke is limited to 1 in. and this servo has only to position the head and not the table, high accuracy can be achieved. Final positioning is achieved by driving the table until a differential transformer in the reading head is accurately centred on the appropriate magnetic pin (see Fig. 4). Accuracies better than 0·0001 in. can be achieved.

(*d*) *Hydraulic actuator* (*absolute digital*). In theHydrapoint system the linear measuring system forms an integral part of the hydraulic ram which is used to move the machine tool table (see Fig. 5). The wall of the actuating cylinder has accurately-placed holes at 0·1 in. intervals and a pneumatically-operateds witch

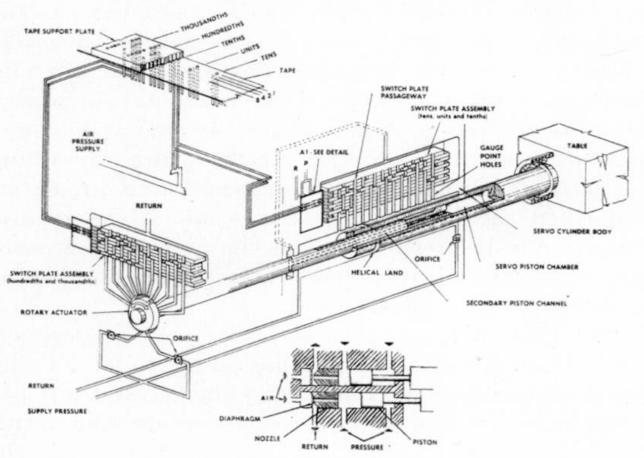

(*a*)

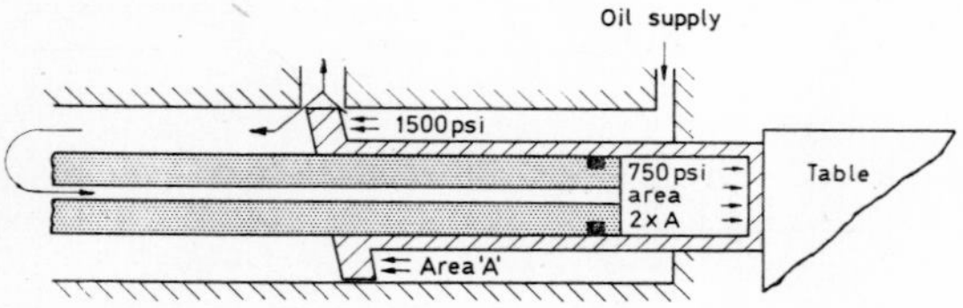

(*b*)

Fig. 5. Moog Hydrapoint control system: (a) Complete system; (b) Detail showing how piston is located at the chosen position.

plate assembly allows only one hole to be open at a time. There are two coaxial pistons in the cylinder, the inner piston being fixed axially whilst the outer piston, which carries a helical land at one end and is attached to the machine table at the other, is free to slide between the inner piston and the cylinder wall. The pistons are, however, free to rotate about the axis of the cylinder and are keyed to rotate together. The cross-sectional area of the inner piston is twice that of the helical land on the outer piston. A hole along the axis of the inner piston connects the space between the end of this piston and the outer piston to the switch plate assembly and thus to one of the holes in the cylinder wall. Oil under pressure is fed to the space between the outer piston and the cylinder, and then flows across the land to a sump at the end of the cylinder.

Coarse positioning is achieved by opening the desired hole in the cylinder wall. When the helical land is central across the open hole the pressure in the oil between the inner and outer piston is just one-half of the supply pressure but, since this pressure acts on an area twice that of the helical land, the outer piston is in equilibrium in this position. Opening any other hole on either side of the helical land will cause the outer piston to move until the land is central across that hole.

Fine positioning is achieved by rotating the pistons. A rotary-vane actuator divides the effective lead of 0·1 in. into 100 equal divisions, thus permitting table positioning in increments of 0·001 in. There are 100 holes in the vane housing and again a pneumatically-operated switch-plate assembly allows only one hole at a time to be open. The vane positions itself over the open hole in a manner similar to that described above.

A pneumatic tape reader actuates the switch-plate assemblies in accordance with the required position data in the form of holes punched in papertape.

The Hydra-point is an absolute digital system and, as the smallest increment is 0·001 in., it is not suitable for continuous-path control. The repeatability is about 0·0003 in.

3. *Other rotary and linear measuring systems.* The measuring systems described above include most of those in current use. There is, however, a number of other systems which have been proposed and these include:

(a) Resistive potentiometers (Absolute analogue)
(b) Inductive potentiometers (Absolute analogue)
(c) Capacitive potentiometers (Absolute analogue)
(d) Differential transformers (Cyclic analogue)
(e) Encoders (Coded discs) (Absolute digital)
(f) Line standards (Cyclic analogue)

Feed Drives

Most of the control systems used on numerically-controlled machine tools are of the feedback type; the desired and actual positions are compared in a comparator and any error in the position of the moving part causes a force to be applied to it in a direction which tends to reduce the error. However, some of the simpler positioning systems use open-loop control in which there is no feedback. The measuring system is used merely to disconnect the drive once the desired position has been reached. Usually as et of commutators is driven from a leadscrew and only one particular set of segments, corresponding to the desired position, is energized. When the desired position is reached the signal produced stops the motor but there is normally some overshoot. The same signal, therefore, reverses the motor and reduces the traverse rate. This time, because of the slow speed of approach, there is no overshoot on reaching the desired position. (In practice the system is modified slightly so that the desired position is always finally reached from the same direction so as to minimize errors due to backlash in the transmission.)

Electric or electro-hydraulic stepping motors also form the basis for open-loop control systems. These are driven by a train of pulses, each pulse rotating the motor through a given angle. The required displacement is thus produced by supplying the appropriate number of pulses to the motor. Obviously, open-loop stepping motor drives can be used only when the resistance to the motion of the table is much less than the force available to move it since if the motor were to stall there would be an undetected error. They would appear, therefore, to be more suitable for positioning systems than for continuous-path control but have been used (Opitz 1967) for systems of the latter type.

Stepping motors can also be used, in conjunction with any type of measuring system, as servo motors which can be driven by a digital signal; all the other servo motors and other types of feed drive require the control signal, whether originally analogue or digital, to be converted to analogue form.

In general, however, the slides or tables of numerically-controlled machine tools are driven by hydraulic rams, or by rack-and-pinion drives or, more usually, leadscrews rotated by electric or hydraulic servo motors. The servo motor or the valve controlling the flow of oil to the ram is usually energized by an error signal, which is generated by comparing the desired position with the actual measured position of the table. Table speeds range from 0 to 500 inches per minute, depending on the accuracy required and on whether or not machining is taking place whilst the table is moving.

Important factors which must be considered when designing feed-drive systems are:

1. Stiffness. With positioning control systems the slide or table can be clamped when the desired position has been reached but with a continuous-path system the table must follow the desired path despite the cutting forces which may tend to disturb its motion. Any error must, therefore, generate a large restoring force, i.e. the drive must be stiff. Stiffness is achieved essentially by making the gain in the control loop large; the measuring system and comparator must be capable of detecting small errors, and a small error signal must generate a large force.

Gearing down the leadscrew or pinion from the servo motor (with a hydraulic ram the same effect is achieved by increasing the ratio of the cross-sectional area of the ram to the maximum rate of flow of oil through the valve) increases the stiffness but at the expense of reducing the rate at which the system can respond to changes in demanded position. Gearing also introduces the possibility of backlash.

2. Natural frequency. The rate at which a numerically-controlled machine tool can respond to a change in the demanded relative position of tool and workpiece depends, in the first place, on the bandwidth of the servo system. This includes the measuring system, the comparator and the means for energizing the servo motor or hydraulic ram. A large bandwidth means that the table can be moved with large accelerations and thus that the system can respond rapidly to demanded changes of position. In practice this means that high rates of traverse—whether actually machining or merely positioning—can be used. (If a large bandwidth is to be used it is also necessary to ensure that information can be supplied to the comparator at the necessary rate.)

A large bandwidth is, of course, a desirable feature of most servo systems and the usual techniques for ensuring adequate stability are employed also on the servo systems used with numerically-controlled machine tools. But more is involved than just the design of a control system; the structure of the machine tool must be such that the large bandwidth can be utilized. In particular, this means that mechanical resonances likely to be excited during operation of the machine tool must be of frequency higher than the bandwith of the servo system since otherwise the change of phase between force and displacement that is associated with resonance can give rise to instability.

The mechanical structures of numerically-controlled machine tools should, therefore, have high natural frequencies. This comment applies to the basic structure of the machine tool itself (high natural frequencies are also helpful in avoiding unwanted machining vibrations—chatter), to attachments such as counter-balance weights, to the system formed by the machine tool and its foundation but particularly to the system formed by the mass of the moving table and workpiece vibrating on a spring comprising the leadscrew or hydraulic ram. In general, as the product of the mass of the moving portion and the maximum distance through which it has to be moved—the mass $\times$ stroke product—increases, the natural frequency decreases and it is on the larger machines that the problem of ensuring that the natural frequency of the feed-drive system is high enough becomes most acute (Ogden 1964).

A hydraulic ram forms a relatively-cheap means of driving a machine tool table but the column of oil in a long ram makes a very weak spring and, at all but the smallest values of the mass $\times$ strokes product, high natural frequencies can be obtained only by using rams of large cross-sectional area. This means that large oil-flow rates are needed to move the table rapidly, and, in practice, the rate of response of the system then becomes dependent more on the rate at which oil can be supplied than on the bandwidth of the servo system. The practicable upper limit of mass $\times$ stroke product for hydraulic ram drives is about 10^5 lb in. At this value the natural frequency of a 5-in. diameter ram is about 40 c/s.

Leadscrew drives can have higher natural frequencies than ram drives but increasing the natural frequency means increasing the diameter of the leadscrew and, again, a point is reached when the rate of response becomes limited by the inertia of the leadscrew rather than by the bandwidth of the servo system. A 1-in. diameter leadscrew gives a natural frequency of about 40 c/s at a mass $\times$ stroke product of 5×10^5 lb in.

Rack-and-pinion drives can maintain high natural frequencies up to even greater values of mass $\times$ stroke product. They are, however, difficult to design because of the problems of backlash and friction, and are not much used.

Several ingenious feed-drive systems (De Barr 1964–65) have been devised in attempts to increase the natural frequency of the drive but none of these is yet used.

3. Friction. In general, the greater the power of a motor or other drive, the more difficult it is to make it respond rapidly to a change in demand. It is desirable, therefore, to keep as small as possible the power required from feed drives and this means that all practicable steps must be taken to reduce the power required to overcome frictional forces. The non- linear character of frictional forces also complicates the design of servo systems and introduces the possibility of "stick-slip" behaviour instead of the smooth movements required for good surface finish and rapid and accurate positioning.

The weight of the moving table and workpiece is usually supported on slideways, and on conventional machine tools these are often ground or scraped surfaces on which a similar ground or scraped surface slides, lubrication being obtained by means of oil ways in the lower of the two surfaces. The surfaces are sometimes hardened to resist wear. In efforts to still further reduce wear and friction, one of the sliding surfaces is today often of plastic material. However, in all these systems sliding friction is still present and in many numerically-controlled machine tools sliding friction is eliminated by supporting the weight of the table and workpiece on rolling-element bearings or on a film of oil under pressure—hydrostatic slideways (Geary 1962).

The elimination of friction between leadscrew and nut is also important and one solution to this problem has been found in the recirculating-ball leadscrew (Green 1965) which is now used on most numerically-controlled machine tools with leadscrew drive. Hydrostatic nuts, in which there is a film of oil under pressure between leadscrew and nut, have also been developed and offer even less frictional resistance. They are not, however, much used as yet, mainly because of manufacturing difficulties.

4. Backlash. Backlash in gearing or other transmission elements also represents an undesirable nonlinearity in the control system. In machine tools on which the position of the moving slide or table is not measured directly but only inferred from the rotation of a leadscrew, backlash also introduces the possibility of an uncertainty in the actual position of the table at any instant. Special arrangements are usually made to minimize backlash in the feed gearing of numerically-controlled machine tools and in the leadscrew/nut combination and, in addition, wherever possible arrangements are made for the final position to be approached always from the same direction.

It will be clear from this short discussion of some of the problems attending the design of feed drives for numerically-controlled machine tools that the development of a machine tool of this kind is not just a matter of adding a control system to an existing machine tool. The design of a numerically-controlled machine tool calls for careful integration of structure and control systems.

Many mechanical features that have been developed to meet the needs of numerical control are now used on conventional machine tools also. These include hydrostatic slideways and recirculating-ball leadscrews.

Control Systems

A programme, containing all the information necessary for the machining of the component, must be prepared for each component to be machined. This programme, which will contain all the necessary dimensional information as well as information about the cutters to be used, spindle and feed speeds, clamping actions and coolants, is usually prepared as a series of patterns of holes on paper tape. One-inch wide paper tape with up to eight holes in a row is becoming standard, and various codes are used to enable the data to be represented in this way; the American E.I.A. and the I.S.O. machine tool codes are the ones most used (Leslie 1966).

With some systems of machine tool control the programme may be prepared as a pattern of holes punched in a card, or even by manual setting of banks of switches.

Punched tape or cards are usually fed to electrical or photoelectric tape or card readers, although pneumatic tape readers are also used. But, whatever the form of the programme or the method of reading, some arrangement must be made for converting the essentially-digital information in the programme into the analogue or digital information required for the control of the machine tool. For this purpose a control system or controller is required.

Although some of the operations of the machine tool are of an on-off or digital character, e.g. coolant control, selection of correct spindle speed by energizing the appropriate combination of clutches, control of clamps, etc.—control of the relative position of cutting tool and workpiece involves smooth movement of a table or slide so that the output of the servo system must usually be in analogue form. (When stepping-motor drives are used, a digital output from the control system is acceptable.) Control systems may be either digital or analogue, depending on whether the digital input information is processed in digital form and then changed to analogue form for activation of the servo system, or is changed to the analogue form before it is processed. Both digital and analogue control systems can be either incremental or positional.

In a digital incremental system the command signals consist of trains of pulses, each pulse representing a specified unit of distance. In digital positional systems the input and feedback signals represent position and their difference is fed to a digital/analogue converter to provide an analogue control signal. Analogue incremental systems are not used, but with an analogue positional system the input data is converted to analogue form—voltage, frequency, phase, etc.—for comparison with the analogue feedback signal.

In addition to controlling the relative position of tool and workpiece the control system must also control clamps on machine slides, select spindle and feed speeds, control the flow of coolant, and, when tool-changing facilities are provided, select the correct tool.

1. Positioning systems. When the only dimensional information involved is that concerning the relative position of tool and workpiece prior to starting a machining operation, e.g. the drilling of a hole, a relatively-simple control system will suffice (see Fig. 6). Facilities can also be provided for:

(a) Straight-line milling, parallel to any linear axis or between any two points.

(b) Undirectional approach to provide final approach to the desired position at slow speed and always in the same direction so as to minimize the effects of backlash.

(c) Tape preparation whilst machining the first component of a batch. For this component the machine tool is controlled manually and at the same time a tape is prepared to enable the remainder of the batch to be machined under numerical control.

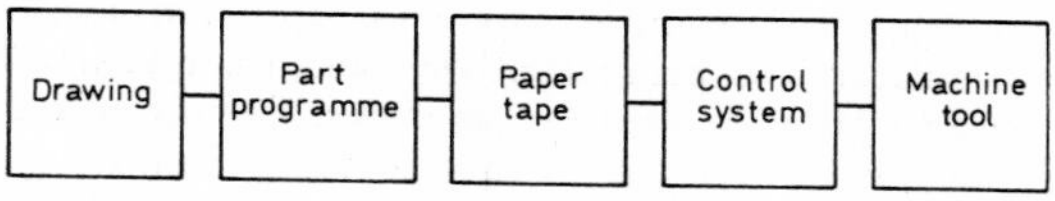

Fig. 6. Schematic representation of a positioning or point-to-point control system.

2. Contouring systems. For continuous-path machining the control system must arrange for the movement of the cutting tool relative to the workpiece to follow the desired path. For any but the simplest (straight line) machining operations, therefore, the amount of information to be handled and the rate at which it must be handled are very much greater than for positioning control systems, and contouring control systems are correspondingly more elaborate and expensive than positioning systems (see Fig. 7). It would be impracticable to put all the necessary dimensional information for the machining of, say, a die on to punched tape; instead, only sufficient information is included to specify the path to be followed and the information on the tape is processed in a computer which generates the large amount of information required for the control of the machine tool by interpolation between the specified points.

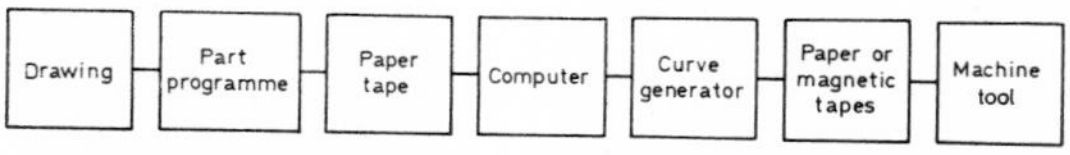

Fig. 7. Schematic representation of a continuous-path or contouring control system.

With some systems a separate general-purpose or special-purpose computer is used to process the input data and produce a magnetic tape for the control of the machine tool. A special-purpose computer for this operation is often called a director. Alternatively the computer may be an integral part of the control system. The magnetic tape produced by the director contains, in a form suitable for feeding to the control system on the machine tool, the instructions for machining the component. Separate channels on the magnetic tape are used for dimensional information related to each machine axis, and also for machine instructions such as spindle speed, tool changing, etc. Note that the director converts information in one code—depending on the way in which the programme has been prepared—to information in another code—suitable for a particular machine tool control system. Thus, in general, a different director is required for each combination of programming system and machine tool control system.

A curve generator which forms part of the director generates the data about the path to be followed by the cutting tool by interpolating between the points at which changes occur in the form or the gradient of the path to be followed (these points are included in the programme and specify the path to be followed.) Linear, circular or parabolic interpolation, or combinations of these, can be used.

The magnetic tape contains, for each axis, a signal which is an analogue of the desired position. Depending on the system, amplitude or frequency of an alternating signal may represent position or velocity along the axis. The tape thus contains information about the desired position of the cutting tool with respect to the workpiece at any instant and is played at constant speed in the machine tool control system. The effective accuracy of machining (assuming no tool wear or deflexion of the workpiece or cutting tool) depends not only upon the feed rate and the speed with which changes of amplitude, frequency or phase can be recognized, but also upon the time taken for the servo system to respond to changes in input.

Contouring systems may allow control of the cutting tool or workpiece along or about one or more axes and may also include facilities for:

(*a*) *Cutter-diameter compensation.* The path followed by the centre of a milling cutter is not the same as the contour produced and allowance has to be made, when planning to machine a given contour, for the radius of the cutter to be used (see Fig. 8). Cutter-diameter compensation allows the same programme

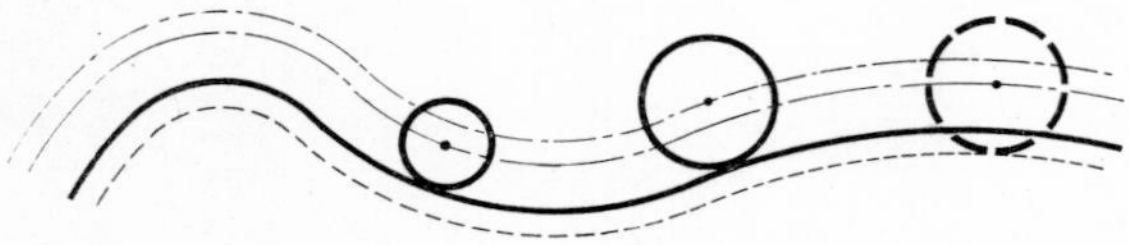

Fig. 8. Cutter diameter compensation can be used to machine a given path with cutters of varying diameter, or to enable roughing and finishing cuts to be taken with the same programme.

to be used with a range of cutter diameters. It can also be used to allow two cuts—a roughing and a finishing cut—to be taken.

Cutter-diameter compensation requires a signal component normal to the contour, the amplitude of which can be set up manually on the control system. This is easily arranged with analogue systems but has not yet been applied to digital systems.

(*b*) *Zero shift.* This enables the zero point on each axis to be changed at will by manual operation on the control system.

(*c*) *Mirror image arrangement.* This can be used in two different forms:

i. The programming of components with one or two planes of symmetry can be simplified by programming one or two quadrants only and using a mirror-image arrangement which reverses the direction of motion along one or two axes to enable this programme to be used for machining the other quadrants also (see Fig. 9a).

ii. Right- and left-handed components can be produced from the same programme by using the mirror-image facility (see Fig. 9b).

(*d*) *Manual override.* Although the programme contains information about the rate at which the cutting tool is to be fed through the work at each stage of machining, it is sometimes desirable to allow the operator to override the programme and increase or decrease the feed rate by manual control. Provision is sometimes made also to allow the operator to change or override the numerical information on the tape.

3. Lathe systems. The control system required for a numerically-controlled lathe is essentially a two-dimensional contouring system with some special facilities (Tipton 1965). These may include:

(a) Provision for automatic change of spindle speed so as to maintain a constant or approximately-constant cutting speed as the diameter of the component changes.

(b) Provision for accurate synchronization of spindle rotation with movement of the tool slide so as to permit accurate screw cutting.

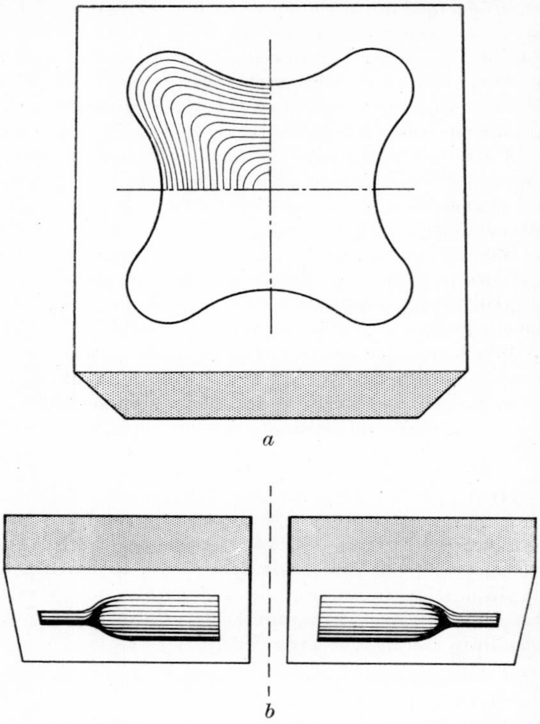

Fig. 9. Mirror image facility: (*a*) *A symmetrical component can be machined by programming only part of the path—in this case one quadrant;* (*b*) *Right- and left-handed components can be manufactured using the same programme.*

(c) Provision for using several cutting tools, perhaps simultaneously, on one workpiece. It may be necessary to take special precautions to avoid interference as the turret rotates to bring another tool into use, or between tools on separate cross slides. It may also be necessary to provide means for compensating for any errors in the setting of tools in the toolholders.

Bibliography

BUCKERFIELD P. S. T. (1960) *Control*, **3**, June, 90; July, 88; *Industrial Electronics*, November 1962.

DAVIES B. J., ROBBINS R. C., WALLIS C. and WILDE R. W. (1960) *Proc. I.E.E.* **107** B, No. 36, November.

DE BARR A. E. (1964-5) *Proc. I. Mech. E.* **179**, Part 3 H.

EVANS J. J. and KELLING L. V. C. (1963) *Control Engng.*, **10**, May, 112.

GEARY P. J. (1962) *Fluid-film bearings*, British Scientific Instrument Research Association.

GREEN R. E. (1965) *Machinery* **107**, 27 October, 908.

HARRISON L. H. R., HORLOCK B. A. and HUNT F. D. (1957) *Electronic Engng.* **29**, 254.

LESLIE W. H. P. (1966) *Machinery*, **108**, 20 April, 877.

OGDEN H. (1964) *Proc. 4th Int. Mach. Tool Des. Res. Conf.*, Oxford: Pergamon Press, 37.
OPITZ H. (1967) *Proc. 7th Int. Mach. Tool Des. Res. Conf.*, Oxford: Pergamon Press.
TIPTON H. (1965) *New Scientist*, 12 May, 362.
WALKER D. F. (1963) *Int. J. Mach. Tool Des. Res.*, **3**, 61.
WILLIAMSON D. T. N. (1958) *Control*, **1**, July 17; August, 70.
(Hydrostatic units) (1964) *Metalworking Production*, **108**, March, 62.
(Hydro-point system) (1966) *Metalworking Production*, **110**, 23 February 57.
Multi-finitron, Fortuna-Werke Spezialmaschinenfabrik, A.G. Stuttgart: Bad Cannstatt.

A. E. DE BARR

NUMERICALLY-CONTROLLED MACHINE TOOLS, PRINCIPLES AND APPLICATIONS OF.

Introduction

Machine tools are power-driven machines for changing the shape or size of a workpiece by either displacement or removal of material. Thus they include, for example, metal-cutting machine tools such as lathes and milling machines, and also metal-forming machine tools such as forging machines, punch presses and bending machines. As yet, numerical control has been applied mainly to metal-cutting machine tools and will be described here in relation to machine tools of this kind but it is equally applicable to machines for machining wood or plastics, and to many types of machine for shaping metal by forming rather than by cutting.

A machine tool is usually required to produce components of a specified shape and size. The shapes of simple parts—cylinders, plates, etc.—are usually determined by the characteristics of the machine tool, but their size is dependent on the position of one or more cutting edges. To take a simple example, suppose a right circular cylinder is to be manufactured by turning a piece of bar in a lathe (Fig. 1(a)); the construction of the lathe will ensure that the cylinder is truly circular in section and that its diameter is exactly the same from end to end, but the actual diameter and length of the cylinder depend upon the distance of the cutting edge from the axis of the spindle, and upon the distance through which the cutting edge is moved. Or, to take another example, suppose a planing machine is to be used to machine a flat surface (Fig. 1(b)); the construction of the planing machine will ensure that the surface is flat but the actual height of the flat surface above the bed of the machine tool will depend upon the position in which the cutting edge is placed. With other types of workpiece, for example a mould or die of the type shown in Fig. 1(c), both the shape and size depend upon the positioning of the cutting tools.

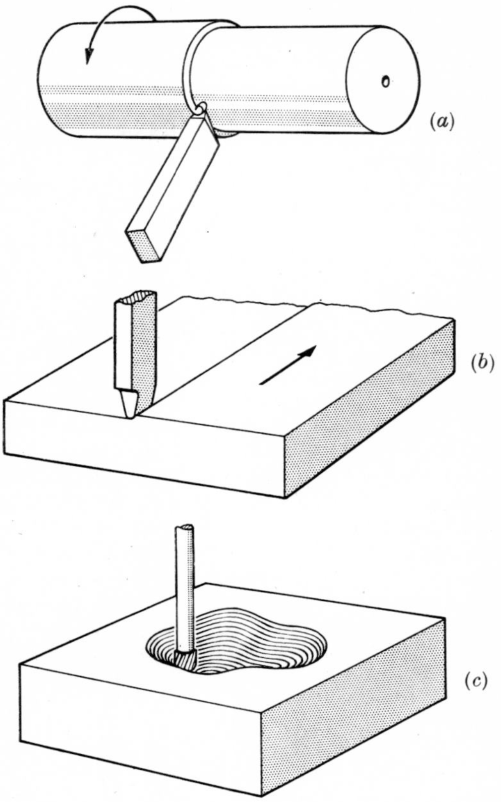

Fig. 1. Metal cutting (a) Turning; (b) Planing; (c) Die sinking.

Thus the accuracy of a metal part manufactured on a machine tool depends to a large extent upon the accuracy with which cutting tools have been positioned with respect to the workpiece. All machine tools include some means by which the position of the cutting tools can be varied, and the job of the operator of a manually-operated machine tool consists largely of varying the position of the cutting tools so that the machine tool produces a component of the desired shape and size. In this task he is assisted by scales of various kinds fixed to the machine tool which enable him to measure just how far he is moving the cutting tool in any particular direction. The operator is also required to load and unload the workpiece, to select the correct spindle and feed speeds, and to control the flow of coolant or cutting fluid.

If large numbers of identical parts are to be manufactured the operator can be relieved of much of the burden of setting the cutting tool in the correct position by providing him with equipment which automatically guides the cutting tools into the required positions for each operation. This equipment often takes the form of special jigs and fixtures but is sometimes built into the machine tool itself. The most-used machining operation is turning and turned parts are required in such large numbers that special kinds of lathe have been developed to produce them. "Automatic" lathes allow completely automatic production of a wide variety of turned parts, all the operations of the machine—feeding the stock, advancing the tools, selecting spindle speeds, etc.—being controlled by cams which actuate levers, switches or valves to perform the required movements.

Obviously, however, components have to be required in large numbers—the batch size (the number manufactured at one time) must be large—to justify the cost of manufacturing all the cams and of setting up the machine. When the batch sizes do not warrant this expenditure turret or capstan lathes can be used. Up to eight tools can be used on one set up, and two or more tools can be cutting simultaneously. All the operations are controlled manually but the use of pre-set stops to determine the starting and finishing positions of cuts allows semi-skilled labour to be used. Turret and capstan lathes can be converted for automatic operation by adding what have come to be called plugboard control systems. Pre-setting of the stops controlling dimensions is still required but the particular speeds and movements required for each stage of the operation are pre-selected by setting a switch or inserting a plug into the appropriate hole to select the required electrical circuit (see Fig. 2). In use, completion of one stage of the machining operation initiates the start of the next, the appropriate hydraulic valves or clutches to produce the required movements being energized as determined from the settings of the plugboard or switch panel.

More-complicated parts can be produced on a copying lathe, the cutting tool being made to follow the path of a stylus moving over a template. In a similar way complex three-dimensional shapes—dies, for example—can be produced by copy milling, a small milling cutter being constrained to follow the movements of a stylus moving over a wooden or plastic replica of the part required. Again, however, the manufacture of replicas or templates is expensive and can be justified only if appreciable numbers of parts are to be made.

Fig. 2. Plugboard for control of a capstan lathe.

When very large numbers of a particular component are required, as in the motor industry for example, machine tools or sets of machine tools—transfer lines—are specially designed and made to manufacture just that one component. Usually several machining operations are required and these are carried out in sequence, the component being moved automatically from one machining stage to the next. However, in most parts of the engineering industry components are manufactured in small-to-medium sized batches of, say, 5–500 parts and batches of this size certainly do not justify the construction of special machine tools and often do not justify the cost of manufacturing models, templates or jigs, or of setting up an accurately-placed system of stops. In these circumstances the over-all cost of a part is less when it is made by a skilled operator than when it is made by relatively-unskilled labour with the aid of jigs, etc. (see Fig. 3). When larger quantities are involved the position is reversed but, until recently, the manufacture of most machine parts has been wholly dependent upon the skill of an operator.

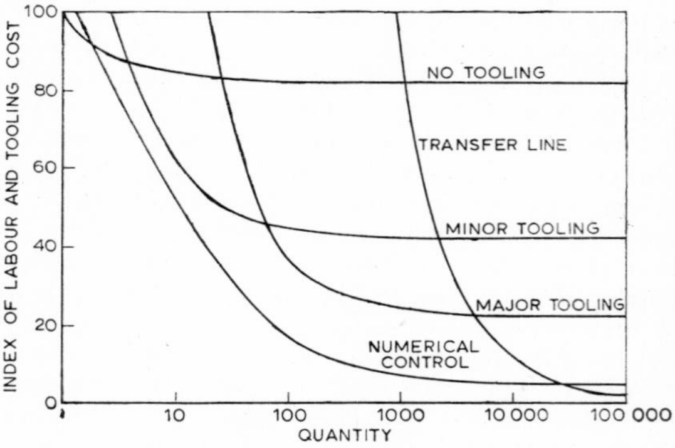

Fig. 3. Comparative costs of different methods of manufacture.

In recent years, however, largely as a result of developments in control system technology and electronics, numerically-controlled machine tools have been developed which enable components of the desired shape and size to be produced automatically without large expenditure on jigs or the need for expensive setting up of stops. These machine tools are relatively expensive but control of the size of a part as well as of its shape now rests with the machine tool and, as a result, skilled operators can be released for other work and the total time and cost of manufacture of a part greatly reduced.

The situation is summarized in general terms in Fig.3 (see also Brewer (1963)). If only one part of a given kind is required then the cheapest method of manufacture is to use conventional manually-operated machine tools and a skilled operator; if very large numbers of a given part are required then the total cost per part is least when conventional machine tools are used with extensive tooling and relatively unskilled operators. For a wide range of intermediate batch sizes, however, the cheapest method of manufacture often involves the relatively-expensive numerically-controlled machine tool.

Numerically-controlled Machine Tools

Numerically-controlled machine tools are so called because information about the required position of the cutting tool at each stage of the machining operation is fed to the machine in numerical form. This is usually achieved in the same way that numbers are fed into a digital computer, that is as patterns of holes punched in paper tape or patterns of magnetic signals on magnetic tape, but the information can also be fed in manually by, for example, setting up a series of decade switches. The usual measuring scales on a machine tool are replaced by transducers which produce a signal (usually, but not always, electrical) which represents the position of the cutting tool with respect to some known datum on the machine. The moving parts of the machine tool are driven via control systems which ensure that, at each stage of the machining operation, the cutting tool is placed in the position indicated by the numerical information which has been fed to the machine tool. Usually, but not always, the motors driving the moving parts of the machine tool are actuated by feedback control systems which continually seek to reduce any difference that there may be between the actual position of the cutting tool and the position which it should occupy at that stage of the machining operation as demanded by the input data.

Several different types of numerically-controlled machine tool can be distinguished, the differences being essentially in the control systems rather than in the basic machine tool. In general, the control system is designed to suit the work that the machine tool does although the design of the machine tool itself is often varied to suit the type of control system to be used.

Positioning or point-to-point systems. Most machining operations—parallel turning, boring, drilling, reaming, tapping, planing, cylindrical grinding, surface grinding, face milling, slab milling, etc.—involve the initial setting of a cutting tool in the correct position with respect to the workpiece, and then relative motion of the cutting tool and workpiece parallel to one of the axes of the machine tool (rotation of the cutting tool and/or workpiece about its own axis may also be involved). The cutting tool must be accurately positioned prior to the start of the machining operation but thereafter it is necessary usually only to control the speed of the relative movement of the cutting tool and the workpiece. The simplest type of numerically-controlled machine tool is, therefore, one incorporating a positioning or point-to-point control system which positions the workpiece relative to the tool prior to the start of the actual machining operation but which does not control the path of the cutting tool during machining.

An obvious application of this type of system is to a drilling machine. The control system places the workpiece under the drill in the correct position for each hole, usually by moving the table of the machine tool parallel to orthogonal axes X and Y (sometimes by one translational movement and one rotational movement) (see Fig. 4). The speed with which the drill moves along the Z axis—rapid approach to the surface of the workpiece, controlled feed for drilling,

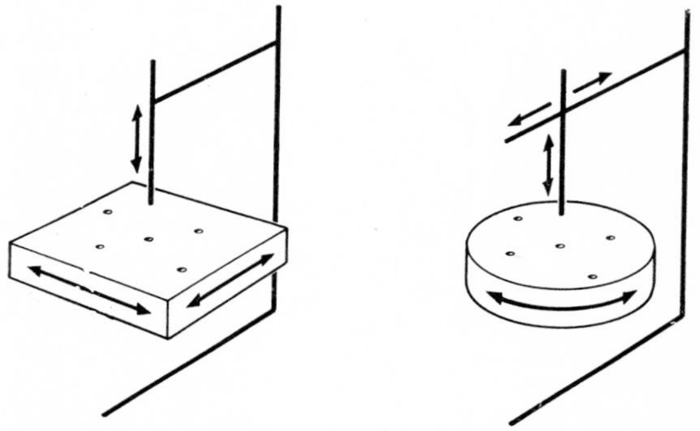

Fig. 4. Positioning (or point-to-point) control system.

stop at the correct depth and withdrawal—is usually controlled by pre-set switches or valves on the Z axis. The size of the hole is, of course, determined by the size of the drill used. Thus the control system has no control over the diameter or depth of the hole but only over its position with respect to the workpiece.

Usually the two movements necessary to position the workpiece are carried out simultaneously but this is clearly not essential and sometimes, in order to reduce the cost of the machine tool, they are carried out sequentially. In this case only one control system is needed, this being used to control each axis in turn. Sometimes also, only one controlled feed motor is used, the drive being transferred from one axis to the other by means of clutches.

Positioning control systems of this kind are widely used on drilling and boring machines. Another application for positioning control systems is on punch presses, the sheet metal workpiece being correctly positioned and the appropriate punch selected automatically.

Straight-line systems. A relatively simple development of a positioning control system enables it to be used to control the path of the cutting tool with respect to the workpiece along a line parallel to one of the axes of the machine tool. Control of the depth of penetration of a drill would be one example, but straight-line control systems are more often used on milling machines. A simple straight-line control system enables lines or surfaces to be milled parallel to any of the controlled axes of the machine tool.

A more-complex type of straight-line control system enables a milling cutter to follow a straight-line path between any two given points on the workpiece, whether parallel to a machine axis or not.

Contouring or continuous-path systems. More-complicated machining operations such as non-parallel turning, die sinking, the milling of cams, etc. require the cutting tool to follow a series of prescribed, not necessarily straight, paths across or through the workpiece. Thus the position of the cutting tool must be controlled whilst it is actually cutting. Three types of control system for this purpose can be distinguished, according to the number of axes controlled. (Continuous control on one axis is denoted by 1 A; on two axes by 2 A etc. Discontinuous control on one axis is denoted by $\frac{1}{2}$ A. At one time a $2\frac{1}{2}$ A system was indicated as $2\frac{1}{2}$ D. A recently-introduced nomenclature, due to Leslie (1966), uses C for continuous control and L for discontinuous control; a $2\frac{1}{2}$ A system is thus denoted by 2 C L.)

(a) $2\frac{1}{2}$ A. Continuous control is possible in two dimensions but movement parallel to the third axis can be effected only discontinuously. Cams can be shaped by milling around the desired outline, and three-dimensional shapes—dies, for example, can be generated by machining around a series of contour lines, the tool being moved along the third axis prior to each cut (see Fig. 5(a)).

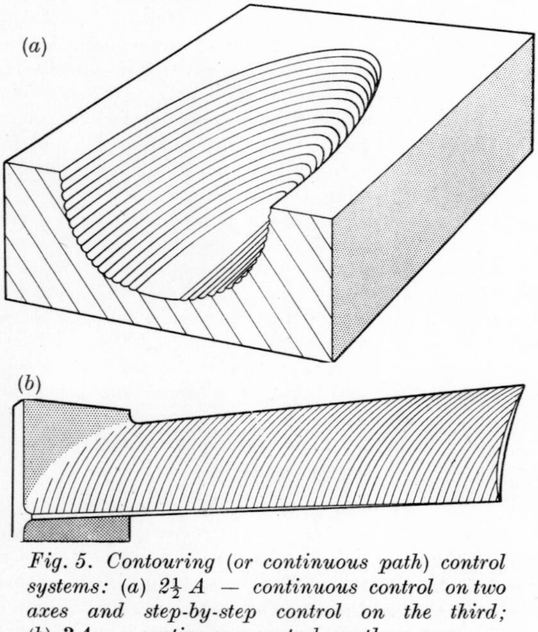

Fig. 5. Contouring (or continuous path) control systems: (a) $2\frac{1}{2}$ A — continuous control on two axes and step-by-step control on the third; (b) 3A — continuous control on three axes.

(b) 3 A. Full three-dimensional control is possible and the tool can be made to follow any given path through the workpiece (see Fig. 5(b)).

(c) Multi-axis. Although most shapes can be machined on a machine tool which can be controlled continuously in three axes, the preparation of the numerical data is often facilitated if the machine can be controlled in four, five or even six axes (see Fig. 6). In addition to translation along three principal direc-

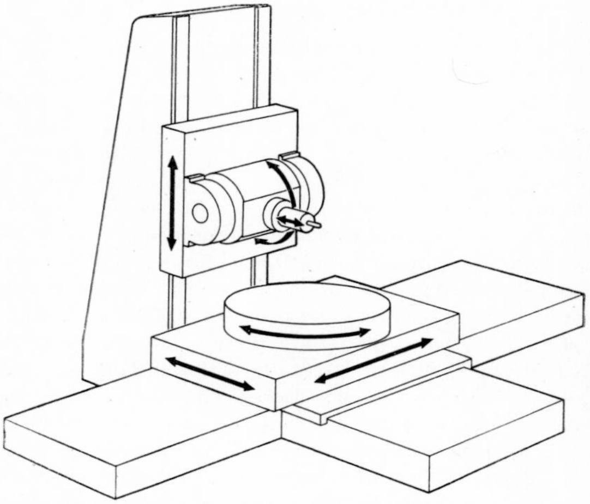

Fig. 6. A machine tool with multi-axis control. The workpiece can be moved along the X and Y axes, and rotated about the Z axis. The spindle can be moved along the Z axis and also tilted about the Y axis.

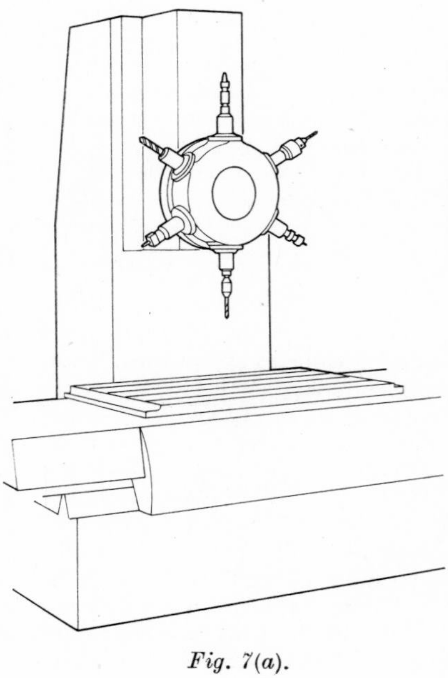

Fig. 7(a).

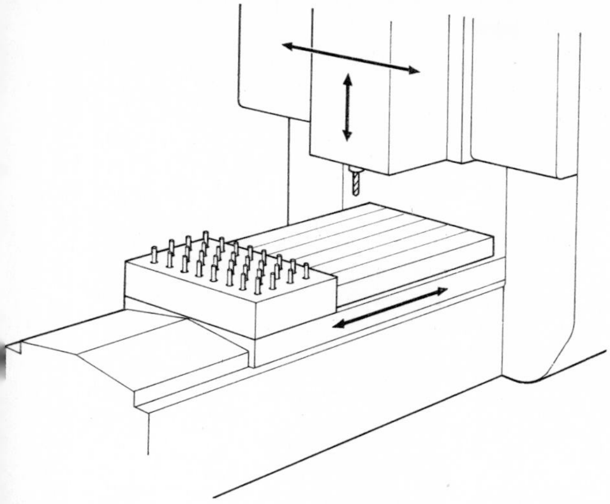

(b)

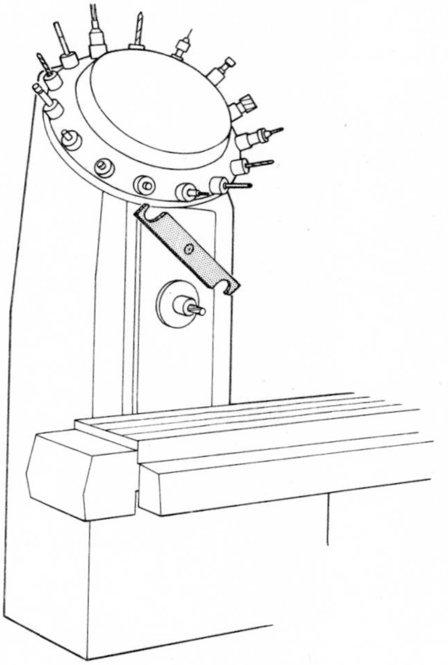

(c)

Fig. 7. Methods of automatic tool changing (a) Turret; (b) Egg-box magazine; (c) Drum magazine.

tions, rotation of the workpiece about two or more axes, and of the machine spindle about one of the axes of the machine may also be involved.

Many machining operations require the use, in sequence, of several different cutting tools, e.g. drills of different sizes, taps, reamers, etc. in the same spindle. On a conventional machine tool the tools are changed by the operator, either by removing one and replacing it by another, or by rotating a turret which carries several tools, any one of which can be connected to the spindle. This process of tool changing can, however, also be carried out under numerical control, information about the tool required for each phase of the machining operation being fed to the machine along with the dimensional data. The indexing of a turret or the removal of one tool from the spindle and its replacement by another chosen from a magazine are then carried out automatically at the appropriate stage in the operation (see Fig. 7).

A numerically-controlled machine tool with tool-changing facilities, control of two, three or more linear axes and perhaps control of the rotation or indexing of rotary table can perform a large variety of operations on a workpiece with the minimum of handling and re-setting. Such a machine tool is often referred to as a machining centre and is particularly useful when milling, drilling, boring, tapping and reaming operations have to be carried out on one workpiece (see Fig. 8).

Part Programming

The preparation of instructions for a numerically-controlled machine tool is called part programming (Leslie 1966): it involves the specification of the geometrical features of the part to be manufactured and also the specification of cutting speeds, spindle speeds, control of coolants, etc.

Manual programming. The necessary instructions for most numerically-controlled machine tools fitted with positioning systems can be written out entirely by hand. The programme specifies the coordinates of the centre of each hole, say, and also the size of drill, the spindle speed, etc. This programme is then translated into punched tape in the appropriate machine code.

The labour of programming can often be reduced by using ready-prepared sub-routines. For example, a sub-routine might specify all the operations required for the drilling, tapping and spot-facing of a hole. All the programmer then has to do is to specify the coordinates of the hole with respect to a given datum, specify the size of the hole and call for the appropriate sub-routine.

Computer programming. The preparation even of programme for positioning systems is greatly facilitated by the use of a computer. For example, the computer can be instructed to introduce the appropriate sub-routine and will then produce a machine-control tape containing all the necessary instructions. Some of the geometrical aspects of programming can also be simplified by the use of a computer. For example, if a number of holes is to be drilled on the circumference of a circle, it is necessary only for the programmer to specify the diameter of the circle,

Fig. 8. A machining centre.

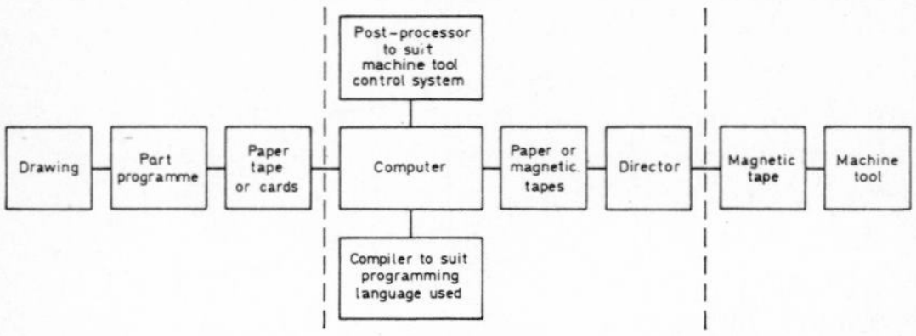

Fig. 9. Schematic representation of machine tool control system. The operations shown on the left are always carried out off-line, i.e. not necessarily at the same rate as the machine tool operates. The operations on the right are carried out on-line, i.e. in time with the actual machining operation. Those operations shown in the centre of the diagram may be carried out on-line or off-line. With some systems two or more of the stages shown may be integrated.

the number of holes and the position of one hole. The computer will then calculate and prepare a machine-control tape containing the coordinates of the centres of all the other holes.

But when contouring control systems are involved a computer is essential, if only because it is obviously impossible for the programmer to write down the coordinates of every point on the path to be followed by the cutting tool. The basic programme must be written in a language which can be understood by the computer: in general a special programme called a compiler is necessary to interpret each programming language to each computer. The output of the computer has to be in a form which is acceptable to the machine tool control system and, in general, a special programme called a post processor is necessary for each combination of computer and control system. A special-purpose computer for machine tool control will usually incorporate a curve generator for interpolating between the programmed points. It is supplied with the part programme and the appropriate compiler and post processor and produces a tape which will control the machine tool (see Fig. 9).

Many programming languages have been devised to facilitate the preparation of the part programme but, as the N.E.L. points out (1965), there is an obvious need for a universal language acceptable to all users. The main feature of a programming language is the way in which geometrical features—points, lines, centres of circle, curves, etc.—are specified. There are, however, many other requirements of an ideal system. The nearest approach so far to a universal programming language is APT and its derivatives developed in the U.S.A. A universal programming language is expensive to develop, however, and it is for this reason that a large number of special-purpose languages has been developed to suit limited requirements.

The machine tool control tape can be checked by feeding it to what is, in effect, a numerically-controlled drawing machine which will trace the outline of the path to be followed by the cutting tool (Johnson, 1963). Conversely, a drawing machine with a curve-following attachment can be used to convey the necessary information about the geometry of the part to be machined by allowing the curve follower to trace the outline of the part: a tape containing all the information about the line is then prepared. Drawing-measuring machines can also be manually operated (Blackie, 1966). The operator positions a stylus over a particular part of the drawing and, by pressing a key, the coordinates of this point can be read into the computer. As some dimensions must be known more accurately than they can be obtained from a drawing, provision is included for dialling in accurate values for any dimension.

Dimensional information can also be read from a model or prototype by allowing a tracing head to follow the outline of the part. Experiments are reported in *Steel*, 1964 in which photogrammetry is used as a way of conveying information about a three-dimensional shape to the machine tool control system.

The Economic Advantages of Numerically-controlled Machine Tools

A numerically-controlled machine tool may be from 50 to 250 per cent more expensive than a conventional machine tool of the same type. But the total cost of manufacture of a component on a numerically-controlled machine tool, particularly when small batches only are involved, can often be appreciably less than if a conventional machine tool is used (Van Raalte 1965). The principal savings achieved by the use of numerically-controlled machine tools, and which can more than offset the increased capital cost, are:

(i) Only simple work-holding fixtures are required and most of the cost of designing, manufacturing and storing jigs and fixtures is saved. No marking out is required.

(ii) As there is no need for marking out or for elaborate jigs and fixtures, little machine time is wasted in putting a new workpiece into position. The machine tool can, therefore, be cutting metal for a much larger proportion of the total time than is usual so that, in a given time, capital and overhead costs are spread over a larger number of workpieces. It is obviously important, however, that the machine tool is kept fully loaded; capital costs of numerically-controlled machine tools are too high for them to be left idle for any appreciable time.

The above are the principal sources of saving but there are many others also, including

(iii) Reduction of lead time. As there is no need to make jigs and fixtures, manufacture of a component need not be delayed whilst these are designed and made.

(iv) Increased accuracy. Errors in setting up or in reading scales are eliminated and the cost of inspection can often be greatly reduced.

(v) Increased repeatability. Each component is exactly like the next and there is no wastage. Moreover, the increased reproducibility can lead to savings at a later stage, assembly for instance.

(vi) Ease with which modifications can be introduced. There is no need to interrupt production in order to alter jigs or fixtures when modifications have to be introduced.

(vii) Ease of scheduling. The time required to machine a component is known in advance and work planning is greatly facilitated.

(viii) Reduction in the value of work in progress. A single numerically-controlled machine tool will often replace several conventional machine tools and thus machine tools can be kept fully occupied with less work in progress.

(ix) Reduced inventory costs. There is no lower limit to the number of parts of a particular type that can be produced economically and so stocks of parts can be reduced.

It is, however, important to realize that the introduction of numerical control is not just a matter of replacing one or more machine tools by another. If the full economic benefits of numerical control are to be realized, careful consideration must be given to the whole process of design and manufacture. Decisions that were previously made on the shop floor have now to be taken at the planning stage and efficient planning is essential. Moreover, the full advantages of numerical control are obtained only if designs take fully into account the possibilities and limitations of the machine tools. Reference has already been made to the need to keep machine tools fully loaded, and two- or even three-shift working is essential with the larger and more expensive machines.

Future Developments

In addition to the improvements in servo system design, transducer design and machine tool design that, taken together, may be expected to bring about improvements in performance and/or reductions in the cost of numerically-controlled machine tools, several more far-reaching developments can be foreseen. These include:

In-process measurement. At present most numerically-controlled machine tools are controlled on the basis of measurements of the position of the cutting tool with respect to some datum on the machine tool. Wear of the cutting tool or deflexion of any part of the system under the imposed cutting forces can, therefore, lead to dimensional errors. Already attempts are being made (Tipton 1966) to develop methods of measuring the dimensions of the workpiece as it is actually being machined, and eventually numerically-controlled machine tools may be controlled from measurements of actual workpiece size.

Adaptive control. Although numerically-controlled machine tools can carry out, automatically, most of the operations previously performed by the machine tool operator, one feature at least is at present lacking. The operator of a manually-controlled machine tool can adjust cutting conditions continuously to deal with variability of workpiece material, size of blank or tool wear, and to ensure best surface finish or maximum production. With a numerically-controlled machine tool, however, all or most of the cutting conditions have to be fixed in advance by the programmer. Although the programmer can probably choose cutting conditions which ensure the desired workpiece quality, this will usually be achieved at the expense of productivity, i.e. he will play safe and choose slower speeds, say, than could perhaps actually be used. The result is that the part takes longer and costs more to machine.

Adaptive control (De Barr 1966) offers the possibility of altering this situation by allowing the machine tool itself to vary cutting conditions continuously so as to ensure the desired end result. If, for example, tool wear and surface finish of the workpiece could be monitored continuously, the machine tool could always choose the feeds and speeds that would ensure the desired workpiece quality at least cost. Only a relatively simple control system is required for this purpose but there are considerable difficulties in the continuous measurement of tool wear, surface finish, workpiece size, etc.

Computer control. The shape of machine parts is already controlled by computers when they are made on numerically-controlled machine tools; adaptive control would allow a computer to choose the machining conditions also. Some of the obvious possible extensions of computer control include:

Planning. Computer programmes can be devised to control the order in which machining operations are carried out on a given workpiece so as to relieve the programmer of the necessity of writing this into his program (I.C.T. 1965). They could also be used to allocate jobs to particular machine tools and to determine the order in which jobs are carried out.

Economics. As, when numerically-controlled machine tools are used, the exact state of each job is known at any instant, a computer could exercise over-all control over a whole workshop, optimizing cutting conditions not only on each machine but also for the workshop as a whole.

Application to other manufacturing processes. Although numerically-controlled machine tools have so far been used mainly for metal cutting, they can be envisaged for many other kinds of metalworking operation—forming, spinning, bending, etc. The principles of numerical control can also be applied to other manufacturing processes—assembly, soldering, winding, armature winding, etc.

Bibliography

Blackie I. T. B. (1966) *Machinery*, **108**, April 27, 929.

Brewer R. C. (1963) *Engineer's Digest*, **24**, No. 9, 109.

De Barr A. E. (1966) *Machinery*, **108**, April 20, 868.

I.C.T. Milmap Program, I.C.T. Technical Publication 3317, International Computers and Tabulators, September 1965.

Johnson W. B. (1963) *Tool Mfg. Eng.* (July), **51**, 47.

Leslie W. H. P. (1966) *Machinery*, **108**, April 20, 877.

N.E.L. Report No. 187 (1965) National Engineering Laboratory, East Kilbride, Glasgow.

Tipton H. (1966) *Metalworking Production*, **110,** December, 7, 77

Van Raalte H. A. (1965) *Numerically-controlled machine Tools*, London: The Machine Tool Trades Association.

A. E. De Barr

O

OPHTHALMIC LASER. *Definition.* An ophthalmic laser is a combination of an ophthalmoscope (the normal instrument used for visual examination of the retina) and a laser which can produce a tissue alteration suitable for the treatment of certain ocular conditions.

Structure and function of the human eye relevant to detachment surgery. To understand the need for such an instrument and appreciate its use we must know something of the structure and function of the human eye. This (Fig. 1) is a roughly spherical structure with a thick outer wall, the sclera. About one-fifth of the anterior wall, the cornea, is transparent and behind this lies the anterior chamber filled with aqueous humour and containing the iris. Separating

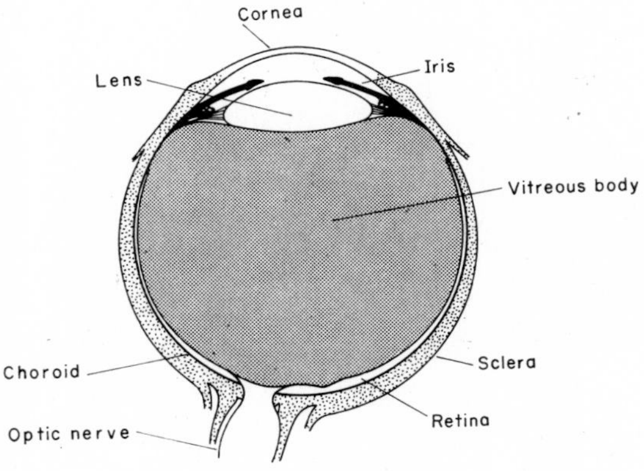

Fig. 1. Section of the human eye.

the anterior and posterior chambers of the eye is the lens which is encapsulated and attached to the ciliary body. The posterior chamber is filled with a gel-like substance, the vitreous humour, and as far forward as the ora serrata, the wall of the posterior chamber supports the choroid and retina.

The retina (Fig. 2) is a layered structure and light entering the eye passes down through the transparent layers of the retina as far as the pigmented epithelium of the choroid. The rods and cones in contact with pigment epithelium are stimulated and the nervous impulses they produce travel back through the various retinal layers to the uppermost one. This layer consists of nerve fibres which are connected to the optic nerve where it enters the eye.

It is evident with this structure that any splitting of the retinal layers or detachment from pigment

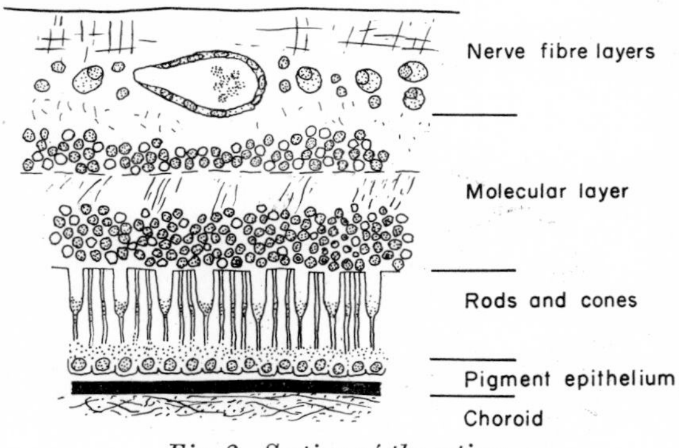

Fig. 2. Section of the retina.

epithelium interrupts the pathway for nervous impulses to the brain and the area involved becomes blind. Retinal detachments may result from injuries, degenerative changes or diseases and their early diagnosis and treatment is important to avoid blindness.

Detachment procedures. The aim of any detachment surgery is to effect a firm bond between retina and choroid. In the simplest case, that is prophylactic treatment, the retina is still opposed to the choroid and any procedure aims at a firm bond which will resist detachment. In the case of an actual detachment surgical procedures are aimed at reducing the detachment so that the retina is firmly opposed to the choroid and then effecting the bond as before. If this is impossible the detachment must be limited to prevent a progression towards total detachment (Figs. 3 and 4) and the borders of the detachment only are treated.

A number of procedures exist for producing a bond between retina and choroid. These include diathermy, the use of liquid nitrogen cooled probes, the use of silicone implants to bring the separated layers together by pressure, or the light from a powerful xenon arc lamp focused on to the retina. Essentially

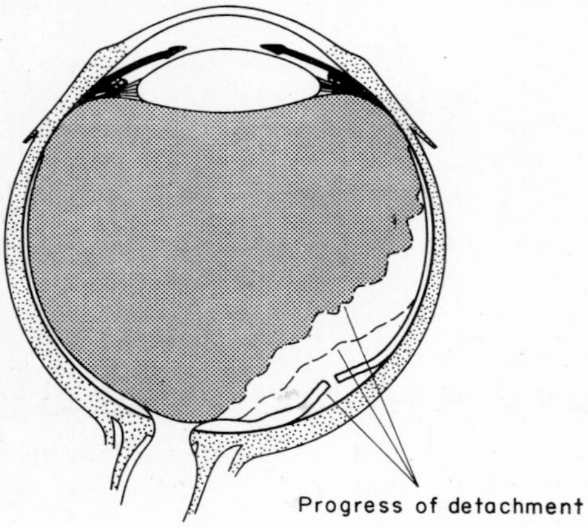

Fig. 3. Progression of retinal detachment from small initial lesion.

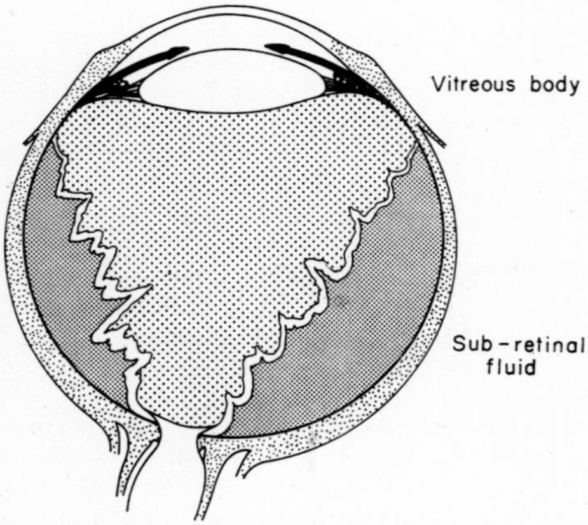

Fig. 4. General appearance of total detachment.

all these procedures induce a coagulation of some retinal and choroidal tissue. An aseptic choroiditis, or inflamation, is produced and the scar, which results in ten to fourteen days, forms a firm fibrous bond between the choroid and retina. From an aseptic point of view the xenon arc lamp is the most desirable procedure as the light from the arc lamp is focused by the optical system of the eye on to the retina but the operation requires anaesthesia and often mechanical fixation of the eye as the immediate reaction of the patient to the bright light is to move his eye. This can cause difficulties as the application may last a second or so. In this procedure also the infra-red and ultra-violet portion of the lamp spectrum must be filtered out to avoid damage to the cornea and lens.

Selection of the type of laser. An examination of the transmission curve of the refractive media of the human eye (Fig. 5) shows that, of the pulsed lasers presently available, the ruby laser is the most suitable. Emitting light at **6943** Å some **98** per cent of the radiation traversing the cornea reaches the retina. Experiments showed that energies of a few joules could be transmitted through the clear media without damage to them provided that the peak power was not raised by Q-switching to above 100 kW.

Animal experiments have been carried out to determine the most suitable dose range for patients and one of the adhesions produced between choroid and retina using a ruby laser is shown (Fig. 6). In this section the adhesion between the retina (upper portion) and choroid (lower portion) is well developed at seven days. The separation of the retinal layers on either side of the adhesion is an artefact produced during sectioning (the microtome knife was cutting upwards during sectioning and this procedure invariably detaches the retina from the choroid) emphasizing the strength of the adhesion shown in the picture.

A number of experiments have been carried out using a 2-in. by $\frac{1}{4}$ in. ruby laser to establish the design criteria for a clinical instrument as up to this stage ordinary laboratory units had been used in the animal experiments. The clinical dose range lay between 0·1 and 0·5 joules delivered in approximately one

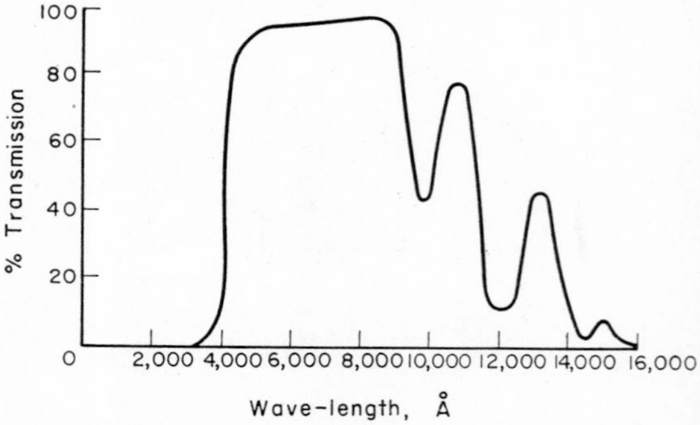

Fig. 5. Spectral transmission curve of the refractive media of the human eye.

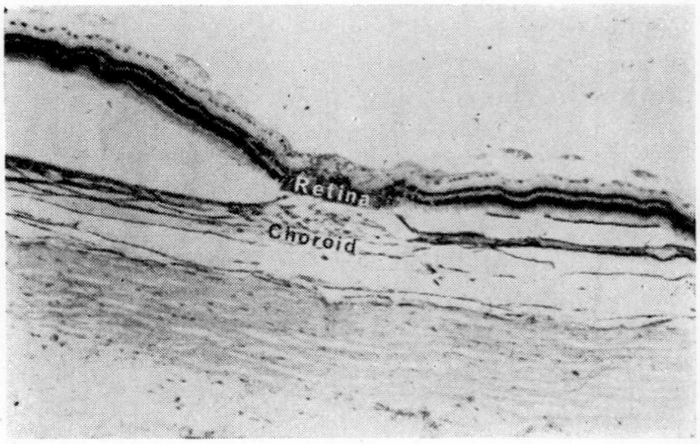

Fig. 6.

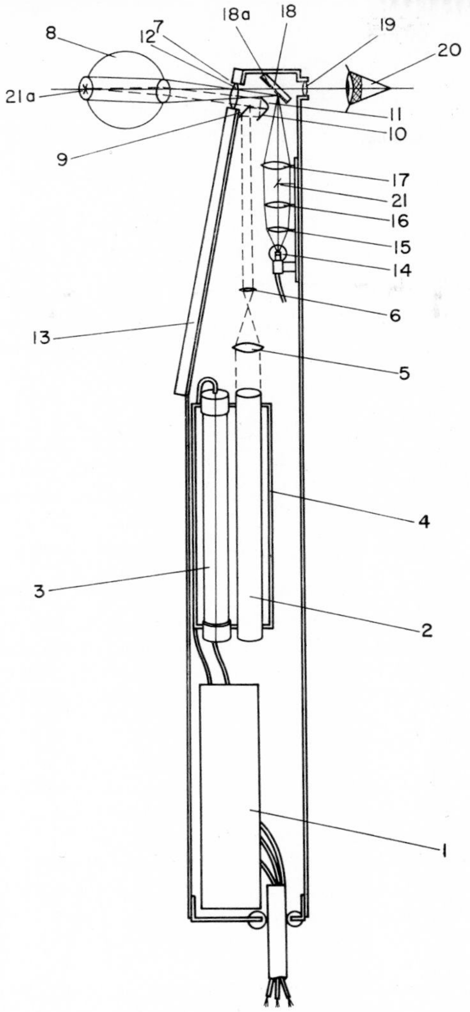

Fig. 7 (a).

millisecond with the peak power of the output "spikes" not exceeding 50 kW. The beam divergence was chosen to be about 1·5° for two reasons. Firstly, the beam had to be reduced from about 6 mm to 2·5 mm to ensure that it would clear the iris easily (otherwise the beam striking the iris produces a rapid contraction of the pupil rendering further treatment impossible, even when the drugs have previously been used to dilate the pupil). Secondly, with the normal beam divergence from the ruby laser (normally about 0·5°) there might be a tendency, especially when using the higher energies, to punch a hole through the retina (which is more likely to produce, than prevent a detachment). In general, the aim was to produce a focal zone in the retina with a flux of about 80 J/cm². In this region biochemical studies of changes in cell metabolism follow-

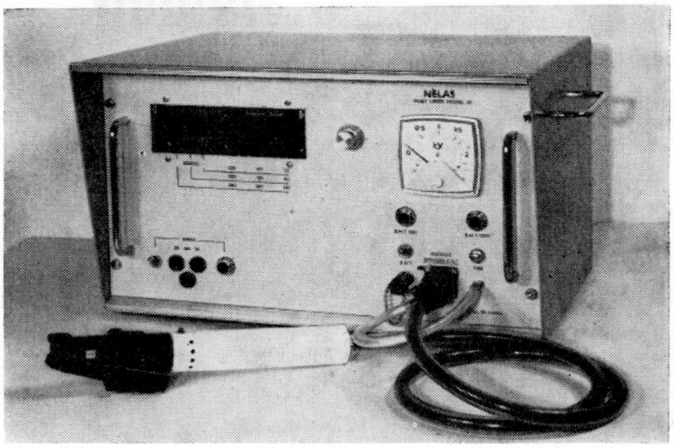

(*b*)

Fig. 7. Laser ophthalmoscope. (a) Diagrammatic. 1 Trigger transformer mechanism; 2 Ruber laser rod; 3 Xenon flash tube; 4 Elliptical cavity of polished aluminium; 5, 6 Telescope to reduce the laser beam from 6 to 2.5 mm dia.; 7 Aperture for emerging laser and illuminating beams; 8 patient's eye; 9, 10, 11 Three-mirror system approximating the laser beam to the illuminating beam as much as possible; 12 Ametropic correction for laser and illuminating beams; 13 Lens rack for ametropic corrections; 14 6-volt light source for viewing beam; 15, 16, 17 condensing system to produce a divergent viewing beam; 18 Mirror for viewing beam with observer's aperture; 19 sphero-cylindrical correction to correct observer's ametropia; 20 Observer's eye; 21 Projection graticule producing retinal target 21a; 22 An interposable defocusing lens is placed between the telescope (5 and 6) and the mirror system (9, 10, 11) to defocus the laser beam, so altering the size of the retinal image. (b) External view.

ing laser irradiation showed that a favourable response from the retina and choroid as regards collagen production and scarring could be expected.

Using these criteria a miniature ruby laser has been developed which would fit into the handle of a normal sized ophthalmoscope. This was combined with a modified illumination system containing a graticule which is projected on to the retina in the centre of the surgeon's field of view. This enables the surgeon to locate precisely the point at which the laser beam will impinge on the retina when the instrument is fired. A section (Fig. 7(a)) together with a photograph (Fig. 7(b)) of the instrument which was finally developed was shown.

Clinical experience with the laser ophthalmoscope. It was found during preliminary tests that no anaesthesia was necessary and because of the short duration of the laser flash (about one millisecond) the patient had no time to react to the bright light and move his eye. This meant that the patient could co-operate fully and look in the direction required by the surgeon and that the treatment was no longer necessarily an operating theatre procedure. Indeed, depending on the type and extent of the detachment, patients can be treated without being moved from their beds in the ward and some can even be treated as out-patients. The instrument is small and portable and can be used from any mains socket and in some circumstances has advantages over conventional therapy. In none of the cases treated and followed up during the past three years has any ocular or other pathological change been observed which could be attributed to the use of a laser. Experience with the instruments developed abroad seems to have been similar. A fundal photograph of a typical treated area is shown (Fig. 8).

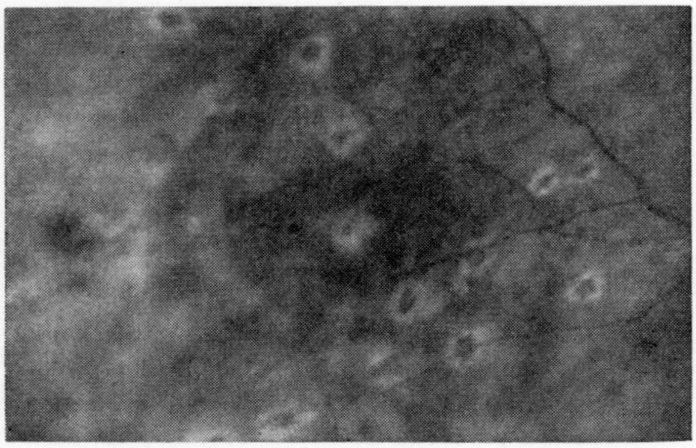

Fig. 8.

The mechanism of laser induced changes in living tissue. Early in the work on ophthalmic lasers the mechanism by which laser induced changes in living tissue might be produced had to be considered so that possible hazards might be evaluated before clinical work was undertaken.

The obvious hypothesis to examine was that the changes were produced by the absorption of light and its subsequent degradation to heat. This hypothesis becomes untenable if one compares the amount of energy required to produce changes in similar volumes of tissue by protein denaturation using coherent and incoherent light. With a xenon arc lamp coagulator between 4·0 and 8·0 J may be used to produce a given change in retinal tissues whereas 0·1 to 0·2 J from a ruby laser will produce a similar change. A second hypothesis that it is the peak power which is important does not fit in with the fact that as little as 0·5 mJ of laser light at 6943 Å will always produce a retinal change whereas 0·5 mJ of incoherent light at 6943 Å with the same peak power and the same exposure time will not produce a retinal lesion.

It can be shown that in the focal zone of a laser beam any material experiences mechanical forces acting on it and that ultrasonic shock waves are produced. The measured intensity of these, however, is too low to produce significant tissue changes.

Apart from these facts, laser light seems to be capable of inducing tissue changes which no amount of thermal or mechanical stimulation will. In certain cases a laser beam will stimulate some cell activities to almost twice their normal level (e.g. thermal burns in the retina scar in 10–14 days whereas, with increased collagen synthesis, laser lesions in the retina scar in 4–7 days). In other situations laser irradiation of part of a tumour sometimes results in a regression of the whole tumour beginning after about ten days. It seems to provoke an antibody reaction resulting in the destruction of the tumour.

On the evidence available at present it seems most likely that laser induced changes in living tissue will have to be interpreted in terms of the scattering of photons by the charge clouds surrounding various larger molecular species in the cells taking account of the probability that the coherence and intensity of the light will produce resonance phenomena.

D. Smart

OPTIC-ACOUSTIC EFFECT. When a radiation-absorbing gas or vapour is exposed to periodically-interrupted thermal radiation, a corresponding heating and cooling of the gas occurs thus producing periodic pressure fluctuations and sound emission. This phenomenon, known as the *Tyndall-Röntgen* or optic-acoustic effect, is used in the *Luft-type gas analyser*.

A further application is as a method of investigating the rate of energy transfer between the vibrational and the translational degrees of freedom of gas molecules. If the spectral distribution of an intermittent beam of radiation is such that it covers an absorption band of the gas under consideration, then absorption of amplitude-modulated radiation is, in general, made apparent as acoustic oscillations in the gas. However, if the modulation frequency is so great that the time of irradiation is much less than the time required for energy in those degrees of freedom which absorb the radiation directly to transfer to the translational degrees of freedom, then the amplitude of the corresponding pressure fluctuations will be negligible. Study of the amplitude and phase of the oscillatory pressure relative to the incident radiation, as a function of modulation frequency, therefore yields information on the rate of energy transfer. Apparatus employing this principle is often called a *spectrophone.*

Bibliography

Cottrell T. L. and McCoubrey J. C. (1961) *Molecular Energy Transfer in Gases,* London: Butterworths.

Delany M. E. (1959) *The optic-acoustic effect in gases, Sci. Prog.* **187**, 459.

M. E. Delany

OPTICAL FLATNESS, MEASUREMENT OF. A simple method of testing the flatness of an optical surface, that of using the proof-plane, is described in Dyson (1961). The technique, although widely used, has severe limitations since either or both surfaces can be distorted by the contact forces between them. In addition both surfaces can be damaged by trapped dust particles and, in the case of the proof-plane, the effect is cumulative. Precise measurements are best made with the Fizeau interferometer which avoids the objections to which the proof-plane method is subject. The essentials of the instrument are given in Dyson (1961). A more sophisticated version, illustrated in Fig. 1, embodies: (a) a "folding" mirror which makes for compactness and brings the viewing system to a more convenient position, (b) a twin

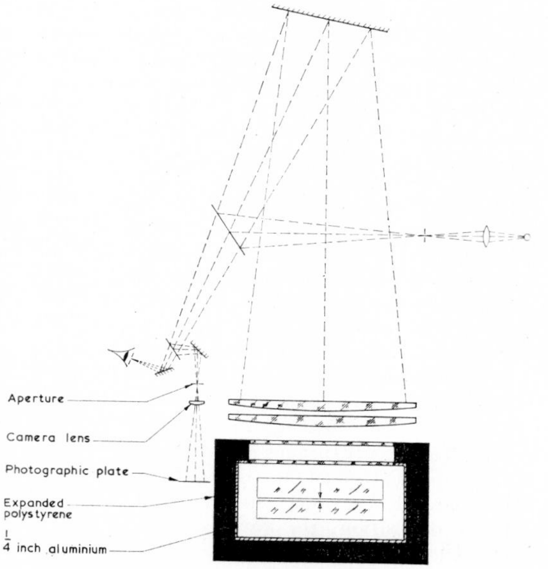

Fig. 1. The Fizeau interferometer.

lens collimator designed to reduce spherical aberration which would otherwise constitute a source of error (Taylor) and (c) beam-splitters for the light source, camera and viewing system so that all three lie on the axis of the collimator. The Fizeau interferometer is necessarily a comparator, that is the interference pattern produced can be interpreted in terms of variations of the thickness of the air-gap between the two surfaces under examination and this is equivalent to the sum of their errors of flatness. One surface must therefore be regarded as the master or reference surface and the determination of its topography is of fundamental importance.

Calibration of the reference surface. In the past the approach has been to take three flats which were identical except in respect of surface contour and make interferometric comparisons between each surface and each of the others in turn. A number of diameters and chords were evaluated in this way and the results assembled to give an overall contour map of each of the three surfaces. The main disadvantage of the method arises from the fact that the flats are conventionally, and often most conveniently supported horizontally on three points equally spaced round the periphery. The "sag" or deflexion due to gravitational influence which inevitably results from this method of support is virtually the same for both upper and lower flats. Mutual cancellation of the sag terms therefore occurs, and the results finally obtained are the "undeflected" figures of the flats, i. e. the figures they would assume if uniformly supported and experienced no deflexion due to gravity. Further, although the sag between the centre and the plane through the support points can be calculated to a first approximation (Timoshenko and Woinowsky-Krieger 1959), it is difficult to predict the overall form of the sag pattern.

This difficulty has been overcome at the National Physical Laboratory (Dew 1966) by employing, for the fundamental intercomparison, three glass "beams" or straight-edges of rectangular section. Supported at points near the ends, the sag of these beams can be precisely calculated or can equally well be determined experimentally. The sag term can then be applied to the "undeflected" figure obtained from the intercomparison of the three, and the figure of each beam supported at the ends either face up or face down is obtained. A beam calibrated in this way can then be employed to determine the figure of an extended

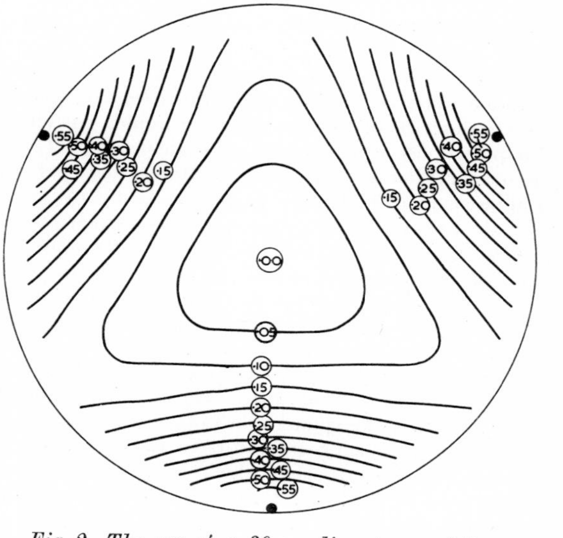

Fig. 2. The sag of a 30 cm diameter × 3·8 cm flat supported at three points equally spaced round the periphery. Contours at intervals of $^1/_{20}$ fringe ($\lambda/40$).

surface (the reference flat of the Fizeau) by evaluating a series of chords and diameters. The results derived from all these separate interferograms can then be assembled to provide an overall map of the surface, and the determination can be made with the worked face of the reference flat either uppermost or lowermost. This allows the Fizeau to be used subsequently to calibrate test flats mounted either way up according to their intended application. It also enables the sag of the master flat to be determined independently of the figure by taking half the difference between the two maps at corresponding points over the entire surface. The contour map shown in Fig. 2, which was obtained in this way, represents the sag of a 30 cm diameter $\times 3{\cdot}8$ cm glass flat supported at three points on the edge. The contours are at intervals of 1/20 fringe ($\lambda = 0{\cdot}5461\ \mu$: 1 fringe $= 273$ nm $= 1{\cdot}07 \times 10^{-5}$ in.). The sag between the centre and the plane through the support points is 0·69 fringes, slightly greater than the 0·63 fringes predicted by Timoshenko's formula.

Another method of calibrating the reference flat is to use a *liquid reference surface* (Lord Rayleigh 1893; Barrell and Marriner 1949; Bünnagel 1956; Dew 1964). If a liquid surface is to be flat (or more strictly to assume the curvature of the earth's surface), a number of conditions must be observed:

(1) The containing vessel must be large enough to avoid objectionable boundary effects.

(2) The viscosity must be high enough to damp out vibration.

(3) Surface effects caused by evaporation or atmospheric water absorption must be avoided.

(4) The liquid must be protected from settling dust particles, thermal disturbances, etc.

A silicone oil of 2 to 5 poises viscosity is the most suitable material. Additional difficulties are caused by the long relaxation time of the liquid. An oil of 2·8 poises viscosity in a 38 cm diameter vessel will take approximately 24 hours to attain a flat surface, and subsequently the effect of any slight movement of the vessel will take a similar time to disperse. It is therefore important that the supports for the flat and the liquid vessel should be exceptionally rigid and surrounded by thermal insulation so that disturbances of all kinds are avoided. Under good conditions, a surface which is flat to $\pm 1/100$ fringe over 30 cm diameter can be obtained (Dew 1966). The effect of the Earth's curvature over this diameter can be neglected. Two explanations for the long relaxation time have been suggested. Firstly that, at the extremely low rates of shear which are involved in the final stages of settling, the material behaves as a non-Newtonian liquid. Bünnagel, on the other hand, maintains that the phenomenon is due to static charges which are slow to disperse owing to the high electrical resistiv ty of the liquid. The use of a radioactive static eliminator has been suggested.

Interpretation of the interferogram. Gross errors of flatness of several fringes magnitude are most conveniently assessed by adjusting the surfaces to give as symmetrical a fringe pattern as possible. The resulting interferogram can then be regarded as a map in which the contours are at intervals of $\lambda/2$. Convexity and concavity can be distinguished by making small changes in the separation of the flats and observing the effect on the interference pattern. Most flatness measurements are concerned with errors of a fraction of a fringe, and it is then more convenient to employ a "*wedge*" *interferogram* obtained by setting the surfaces at a slight angle to each other. Two perfectly plane surfaces would, under these conditions, give a set of straight, parallel and equidistant fringes, and any departure from this ideal pattern will represent an error in one or other of the flats. A wedge interferogram can be evaluated in one of two ways: (a) by measuring the departure from straightness of a fringe in terms of the fringe spacing, or (b) by measuring the relative position of the fringes along an ordinate at right angles. Visual estimates are best made using the first technique. An accuracy of a tenth of a fringe interval is the best that even an experienced observer can hope to achieve and, if the reference surface has been evaluated in the same way, the accumulated errors may be several times greater. A slight improvement can be effected by making measurements on a photograph of the interferogram. The fringes are not, however, sharply defined but have a $\cos^2$ intensity distribution, and estimating the centre to better than 1/20th fringe is difficult.

Sharply defined fringes are obtainable by using the multiple-beam technique due to Tolansky, and estimates to 1/100th fringe can be made without difficulty. The disadvantages are (a) that the separation of the surfaces must be reduced to a few wavelengths and (b) that both surfaces must be silvered by vacuum deposition, the lower surface to give maximum reflectivity and the upper, a transmission of a few per cent. It must therefore be stipulated that neither the deposition of the metallic coatings nor their subsequent removal affects the contours of either surface.

The technique of equi-densitometry (Lau and Krug 1957) provides a photographic method of sharpening two-beam fringes until they are comparable with those obtained by the multiple-beam technique. A negative and a positive transparency are prepared under carefully defined and controlled photographic conditions. When the pair are superimposed, two sharply defined fringes are obtained for every fringe of the original interferogram.

A photoelectric method of evaluating a photographic negative of a two-beam interferogram, described by Dew (1964), employs the second method of interpretation mentioned earlier. A series of ordinates perpendicular to the fringes is selected to correspond with convenient unit separations on the flat and the points at which the fringes intercept each of these ordinates is measured. Interpolations from these measurements enable the relative optical paths at grid points over the entire surface to be obtained, and the effect of the wedge is then removed arithmetically. The measurements are made by projecting an enlarged image of the negative (Fig. 3) on to a pair of photocells which are separated by a distance equivalent to approximately half the fringe

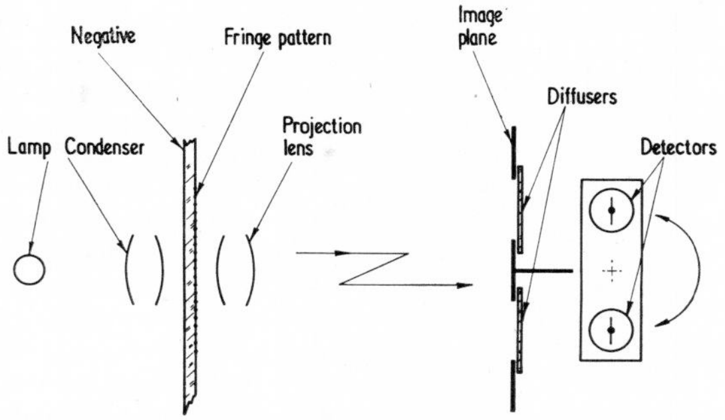

Fig. 3. Photoelectric equipment for the evaluation of a two-beam interferogram (Dew).

spacing. The lateral position of the negative, which is mounted on co-ordinate slides, is then adjusted until each fringe along the particular ordinate under examination is in turn brought on to the optical axis of the system. Each setting corresponds to a condition under which equal responses are obtained on the two photocells and is identified by a null-detection device. An accuracy of 1/100 fringe is readily obtainable, and only the effects of "plate-noise" and random blemishes on the negative prevent an accuracy five times better than this being realized.

A method due to Roesler is unusual in employing the "fluffed-out" fringe condition, obtained when the surfaces are most nearly parallel. Differences of optical path are indicated by intensity variations which are measured (without recourse to photographic recording) by means of a photoelectric scanning device. The greatest sensitivity is obtained when the mean optical path corresponds to a condition mid-way between completely constructive and destructive interference, and this is controlled by adjusting the air-pressure in the space between the two surfaces.

Transient thermal stresses. It is well known that a flat which has been handled, even for a brief period, may undergo gross distortions of figure induced by differential expansion of the material. Some time must elapse before equilibrium is regained, and it is standard practice to leave the flat to "soak" under the interferometer until observations of the fringe pattern indicate that a stable condition has been reached. Regrettably, economic reasons are often responsible for the soaking time being curtailed, with consequent errors in measurement.

If precise measurements on large flats are required, even more stringent precautions must be taken. A simple calculation shows that a temperature gradient of 0·001°C between the two faces of a glass flat measuring 30 cm diameter and 5 cm thick will result in a change of figure approaching 1/100th fringe. If, however, two similar flats are undergoing comparison and simultaneously experience the same gradient, then compensation will occur and the error of measurement will be zero. It is clear therefore that it is the resultant gradient which is significant.

The first step towards eliminating this source of error is to place the instrument in a temperature-controlled environment. The best control equipment will, however, rarely maintain the temperature to closer than $\pm 0{\cdot}1$°C, and measurements with thermocouples show that, under these conditions, the resultant gradient is likely to lie anywhere in the range $\pm 0{\cdot}02$°C. In the case of 30 cm diameter $\times 5$ cm glass flats this corresponds to a change in figure of $\pm 0{\cdot}15$ fringes. Considerable improvements can be effected by surrounding the flats with a good thermal insulator (e.g. 5 cm expanded polystyrene) and providing double-glazed windows where necessary (Fig. 1). Experiments with an enclosure of this type show that after an initial soak of 24 hours, the resultant gradient will not exceed $\pm 0{\cdot}003$°C if the ambient is maintained to $\pm 0{\cdot}1$°C. If this is still inadequate, as is the case if an accuracy of 1/100 fringe over 30 cm diameter is required, thermocouples must be used to enable the most favourable thermal conditions to be selected.

These tolerances can of course be relaxed for flats of smaller dimensions. The effect of a given gradient is proportional to the square of the diameter of the flat and inversely proportional to its thickness. The effect on a flat constructed of fused silica will also be less on account of its smaller coefficient of expansion, but careful consideration should be given to the case where flats of different materials are being compared.

Supports for flats. The effect of supporting a circular flat at three points equally spaced round the periphery has already been mentioned, and an experimental result for a flat of specific dimensions is given. The sag for flats of other dimensions may be derived from this result if it is assumed (Timoshenko) that the sag is proportional to the fourth power of the radius of the flat and inversely proportional to the square of its thickness.

If it is required to reduce the sag, other methods of support can be employed (Dew 1966). The best of these comprises six self-aligning supports equally spaced round a circle whose radius is seven-tenths that of the flat. A simpler but slightly less effective alternative consists of three points spaced round a circle whose radius is six-tenths that of the flat. Rectangular flats are best supported on four self-aligning supports situated at the corners of a rectangle whose sides are 0·55 the length and breadth of the flat. The use of elastic supports (e.g. sheet foam plastics) is not recommended on account of possible variations of thickness and elastic constants of the material.

Bibliography

BARRELL H. and MARRINER R. (1949) A liquid surface interferometer, *Brit. Sci. News.*, **2**, 130.

BÜNNAGEL R. (1956) Untersuchungen über die Eignung eines Flüssigkeitsspiegels als Ebenheitsnormal, *Z. Angew. Phys.*, **8**, 342.

DEW G. D. (1964) A method for the precise evaluation of interferograms, *J. Sci. Instrum.*, **41**, 160.

DEW G. D. (1966) The measurement of optical flatness, *J. Sci. Instrum.*, **43**, 409.

DEW G. D. (1966) Systems of minimum deflection supports for optical flats, *J. Sci. Instrum.*, **43**, 809.

DYSON J. (1961) in *Encyclopaedic Dictionary of Physics* (J. Thewlis Ed.), **3**, 880, Oxford: Pergamon Press.

LAU E. and KRUG W. (1957) *Äquidensitometrie — Grundlagen, Verfahren und Anwendungsgebiete*, Berlin.

RAYLEIGH, LORD (1893) Interference bands and their applications, *Nature, Lond.*, **48**, 212.

ROESLER F. L. and TRAUB W. (1966) Precision mapping of pairs of uncoated optical flats, *Appl. Optics*, **5**, 463.

TAYLOR W. G. A. (1957) Spherical aberration in the Fizeau interferometer, *J. Sci. Instrum.*, **34**, 399.

TIMOSHENKO S. and WOINOWSKY-KRIEGER S. (1959) (2nd Edn.) *Theory of Plates and Shells*, New York: McGraw-Hill.

TOLANSKY S. (1948) *Multiple Beam Interferometry*, Oxford: Clarendon Press.

G. D. DEW

OPTICAL TRANSFER FUNCTION. Subject to two simple conditions, which are detailed below, a grating-like object across which the intensity varies according to the function

$$I(x) = \alpha + \beta \cdot \cos(2\pi Rx)$$

may be shown to have at its image an intensity distribution of exactly the same form, namely

$$I'(x') = \alpha' + \beta' \cos(2\pi R'x' + \theta(R', \psi)).$$

The primes denote the image space of the optical system, R' is the spatial frequency, being the reciprocal of the grating period, and the definition of the remaining symbols is seen by reference to the first figure. The contrast of the object grating is measured by the ratio β/α and that of the image grating by β'/α'. The ratio of the two contrasts, known sometimes as the contrast transfer function, is the modulus of the optical transfer function and is denoted by $T(R', \psi)$ where the angle ψ is included to indicate the angle between the lines of the grating and the meridian plane of the optical system forming the image.

The coordinates of object and image space, x and x' respectively, are chosen to have units in the ratio of the magnification so that $x' = x$ gives the geometrical image of the object point x. The formula representing the image includes a factor to allow for a lateral displacement of the actual image by

$$\delta x' = -\theta(R' \cdot \psi)/2\pi R'$$

relative to the geometrical image. Figure 1 illustrates the ideal image, a curve identical to the distribution of intensity in the object, the perfect image influenced only by diffraction but having the contrast transfer

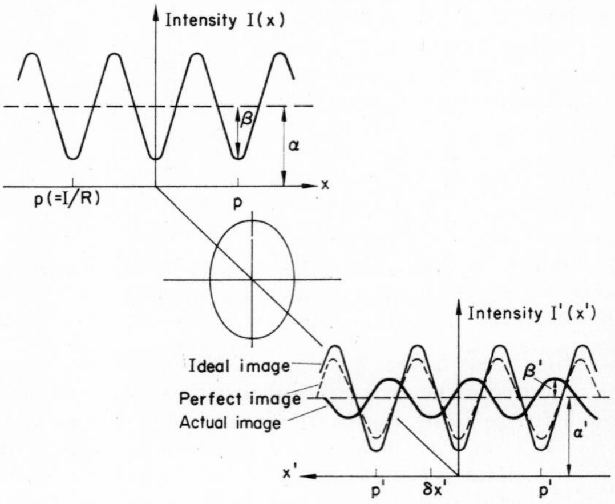

Fig. 1.

function less than unity for all spatial frequencies and thirdly an actual image as given by an optical system with aberration and showing a further reduction of contrast and a spatial phase shift. The contrast transfer function and the spatial phase shift are combined in a single complex number, the optical transfer function

$$D(R, \psi) = T(R, \psi) \cdot \exp i\,\theta(R, \psi).$$

The above statements relating to the images of cosine gratings require two conditions to be satisfied. Firstly the image intensity distribution should be obtained as the simple addition of the intensities produced by each elemental area of the object. This condition requires incoherence between neighbouring points of the object, a condition satisfied by self-luminous and most naturally illuminated objects. Secondly it is necessary for the aberrations of the optical system to change sufficiently slowly for the spread function, the intensity distribution in the image of a limitingly narrow slit, to remain constant in form over the region of the image plane in which its intensity is sensibly non-zero. These two conditions, superposition and isoplanatism respectively, if fulfilled bring optical systems into the category of linear systems.

Knowing the optical transfer function for a given region of the image plane as a function of spatial frequency, R', and of azimuth, ψ, completely and uniquely specifies the image forming properties for that region of the image. This follows from the knowledge that any spatial distribution of intensity may be represented, by means of Fourier analysis, as the superposition of a set of cosinusoidal intensity distributions of differing periods and azimuths. Each member of this set appears in the image with its contrast reduced by a factor $T(R', \psi)$ and its position laterally shifted by an amount $-\theta(R, \psi)/2\pi R'$. The image intensity distribution may then be found as the superposition of those modulated cosinusoidal components.

Further it should be noted that the spatial frequency variable R' is sometimes replaced by the reduced spatial frequency, s, defined as

$$s = (\lambda / N' \sin \alpha') \cdot R'$$

where $N' \sin \alpha'$ is the numerical aperture of the optical system in the image region. The reduced spatial frequency system is well suited to representing systems of all numerical apertures when it is required to illustrate their departure from the optimum state of optical correction or their relative degrees of correction; whereas the spatial frequency scale in lines per millimetre is more commonly used in practical applications when the optical system aberrations are large and the performance of the system is required in terms of actual spatial frequency.

Figure 2 illustrates in curve "a" the optical transfer function for an aberration free, and thus diffraction limited, optical system of circular aperture. It shows the image contrast to be less than the object contrast for all spatial frequencies and that the image contrast falls to zero for the reduced spatial frequency $s = 2$. Curve "b" illustrates the influence of the presence of a quarter of a wave-length of wavefront aberration, approximately the Rayleigh tolerance for permissible aberration, as obtained by defocusing the optical system. The optical transfer function is clearly sensitive to such small wavefront errors. Curve "c" shows the effect of increased defocusing, threequarters of a wave-length wavefront error, on the contrast and an example of a shift of the grating image by half a grating period as evidenced by negative contrast. The spatial frequency variable expressed in lines per millimetre is also included in the second figure, the values being for an optical system operating at $F/2$ with the illumination of wave-length 500 nm.

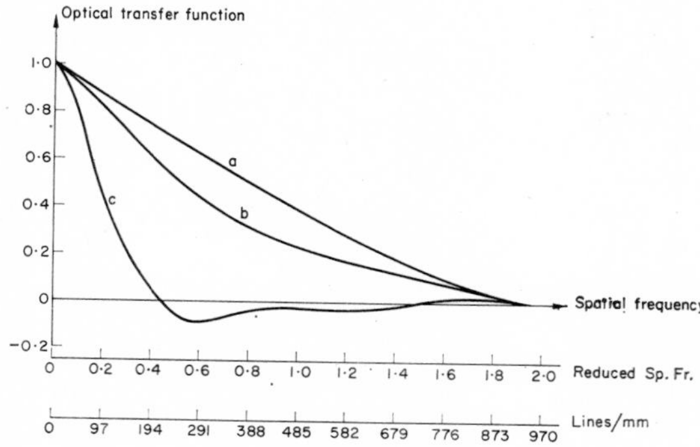

Fig. 2.

The earliest measurements of the optical transfer function were made using sets of gratings of sinusoidal transparency to establish the image to object contrast ratio at discrete spatial frequencies. Such grating may either be in the form of opaque plates with suitably profiled apertures such that when scanned by a slit the transmission is sinusoidal or they may be true transmission gratings as might be produced by photographing suitable two beam interference patterns.

An alternative is to measure the line spread function and to calculate from this, as its Fourier transform, the optical transfer function. Fourier analysis indicates that an infinitely narrow line object contains all spatial frequencies with equal amplitude while similar analysis of the line spread function indicates the relative amplitude of each spatial frequency as occurring in the image and therefore the desired function. The use of slits of finite width in the object plane and in scanning the image is necessary in a practical instrument and requires a correction factor to be applied to the measured optical transfer function. The presence, or otherwise, of the phase shift term, $\theta'(R', \psi)$, will be evident as it leads to asymmetry of the line spread function.

More recent instruments have been designed to evaluate the Fourier transform of the line spread function directly. They include a single grating of constant spatial frequency but cause the spatial frequency under examination to vary from zero to the maximum value of the grating by the use of a suitable masking aperture before the grating. A continuous range of spatial frequency is thus available and may be used as a scanning function to examine the line spread function and determine its Fourier transform. It may be noted that, due to the reversibility of optical systems, it is also possible to use the variable period grating as the object and to scan its image with a slit thus repeating the first methods described for the now complete spatial frequency range. The optical transfer function measured by any of these methods is, by convention, normalized to be unity for zero spatial frequency to remove the influence of the photometric transmission of the optical system.

Other methods of measuring the optical transfer function make use of interferometers and the analysis of the images of object structures other than sinusoidal targets. The interferometric instruments operate as analogue computators to evaluate the autocorrelation integral of the complex amplitude distribution in the exit pupil of the optical system, this function when suitably normalized being the optical transfer function. As an example of naturally occurring object structures that may be used in the evaluation of complete optical systems the sharp edge occurring between areas of uniform but different brightness may be considered as it is frequently to be found in practical photography. The distribution of intensity present in the image of such an edge may be measured by microdensitometer and numerical methods used to evaluate the image of a grating structure with sharp edges, that is a square-wave grating. The complete optical transfer function for square-wave object gratings may be found in this manner and a correction applied to convert this to the optical transfer function for sine-wave object gratings. The example quoted is a method of measuring the optical transfer function that operates for the complete optical system, whereas the previously described methods were applicable only to the lens and did not take into consideration other stages of the image forming system.

It should be noted that if the many stages of an image forming system are each separate linear systems it is possible to describe each stage by a transfer function and to combine these into a final system transfer function. A threshold function, denoting the contrast required for detection as a function of spatial frequency, for the viewing system may then be compared with the system transfer function to establish the limiting spatial frequency resolvable by the combination of the optical imaging system and the viewing system. The effect of changing one stage of the image forming system or the parameters of the viewing system may readily be introduced and the new value of the limiting spatial frequency established.

Bibliography

HOPKINS H. H. (1962) The application of frequency response techniques in optics, *Proc. Phys. Soc.*, **79**, 889.

The Practical Application of Modulation Transfer Functions (1963) Perkin-Elmer, Electro Optical Division, Norwalk, Connecticut, U.S.A.

LINFOOT E. H. (1964) *Fourier Methods in Optical Image Evaluation*, London: Focal Press.

K. G. BIRCH

OPTICS IN SPACE. Space optics is concerned with the use of image forming systems at visible wave-lengths and in adjacent regions, especially the ultraviolet, often combined with subsequent spectrometric analysis. The subject is extended to include special problems in design caused by launch requirements, stabilization in orbit and environmental effects peculiar to free space.

This article excludes systems operating at optical wave-lengths where image-forming is either rudimentary or non-existent. For example, monitoring of fluorescence in the atmosphere is more conveniently described under atmospheric physics. All ground-based optics are excluded, except those required for testing or observation of satellites.

The effect of the Earth's atmosphere on astronomical investigations imposes two limitations. Wave-lengths below 3000 Å are absorbed (Liller 1961) by ozone, oxygen and nitrogen, and many parts of the infra-red spectrum are absorbed by ozone, water vapour and carbon dioxide. This limitation restricts investigation of planetary atmospheres and particularly of stellar composition and evolution. The second limitation is that random refraction or "twinkle" imposed by the atmosphere prevents high resolution study of planetary features or of distant star clusters. The angular stability of an image lies in the region 10^{-4} to 10^{-6} radians, the latter figure being for exceptionally good seeing at selected sites. Hence, the resolution of a 60 cm diameter telescope is not (in general) exceeded by any ground-based instrument, however large, the advantage of the latter being in greater "light grasp" and speed of taking pictures rather than in better quality.

A space vehicle, in addition to being immune from these restrictions, sees a large area of the Earth and can (within limits) make astronomical observations whether or not the Sun is above its horizon. Its useful wave-length coverage is still restricted, because stellar radiation, other than solar, is probably substantially absorbed below the cut-off of the hydrogen spectrum at 912 Å (13·6 eV) by interstellar gas, and much of the infra-red is not used, either because it is not of interest or on account of its being too close to the black-body continuum of the object being viewed. The range of wave-length 0·1 μ to 5 μ encompasses most of the observations.

Missions in space. It is convenient from a descriptive point of view to recognize three classes of optical system, according to the purpose of the mission:

(1) Surveillance of the Earth itself
(2) Lunar and planetary probes
(3) Ultraviolet, (u.v.) or infra-red (i.r.) observations giving intensity versus wave-length information on astronomical objects.

The nature of the space vehicle required for class (2) is obvious; class (1) includes not only orbiting satellites, since high-flying aircraft perform a similar if more restricted service; balloons rising to 25 km are above the main infra-red absorbing regions for class (3) use, and rockets giving ~10 minute flights are useful in this class for solar and stellar observations.

Surveillance. Study of the Earth from space has applications in military reconnaissance, meteorology, geography and geophysics.

There is little publication of details or results of military satellites; surmises that a number exist are occasionally published (*Space/Aeronautics* 1964). Reports from astronauts show that man in space can recognize ground objects, and it is reported that Samos satellite pictures have a ground resolution of 50 cm taken at a height of 160 km. Low altitude and correspondingly short orbital life time are required for this application. Proposals for manned orbiting laboratories (McGraw-Hill 1966) are already in hand. Some details of surveillance optical systems are available (Millburn and Horton 1966).

Meteorological satellites are among the most successful vehicles in this class, including TIROS, NIMBUS and ESSA. Many hundreds of thousands of cloud cover pictures have been received, and the later satellites in the series have automatic picture transmission, whereby local cloud information can be picked up by any ground station, suitably instrumented. Vidicon television systems transmit pictures in daylight, and night-time coverage is given by an infra-red scanning device, essentially transmitting temperature information which corresponds closely with cloud conditions. Hurricanes Cleo and Dora in the U.S.A., 1964, were observed and tracked by three weather satellites.

Later Nimbus satellites extend the infra-red coverage to the CO_2 emission band at 15 μ (see NASA 1964, and McClain 1966).

Further applications of surveillance are in geology, geography, agriculture and forestry, etc. These applications are not well publicized. Striking pictures taken on Gemini missions by astronauts are widely published (e.g. A.W. & S.T., 1), see also Lowman (1965).

Mars probe. The only deep space pictures taken of a planet surface are those on the successful Mariner IV flight to Mars, which approached within 10,000 km of the planet on July 14th, 1965. (The successful Mariner II flight to Venus, December 14th 1962, carried infra-red scanning equipment at 8·4 μ and 10·4 μ; these showed no breaks in the planet's cloud cover.)

Details of the Mariner IV flight, together with 19 of the complete set of 21 pictures taken of the surface, have been published (Leighton 1966). These pictures are of 200 lines/frame, recorded on a Vidicon television camera tube, and have a surface resolution of about 2 kilometres, some 20 times improvement on the best visual Earth-based observations.

Since the velocity of the spacecraft relative to the Martian surface was 5 km/sec, and no image movement compensation was provided, the exposure time per frame was fixed at 0·2 sec. The optical system consisted of a Cassegrain telescope with an aperture of 1·5 in., working at f/8, the vidicon output being scanned onto magnetic tape and replayed to Earth. The areas photographed are distributed across the surface of the planet, the pictures being taken over a period of 26 min. Red and green filters were alternately interposed in front of the vidicon. The recording level was adjustable, and prior to launch the system was tested by photographing the moon to check the calculated settings. Each picture point was recorded as a 6-bit figure, affording 64 levels of brightness between black and white, and 8 hours were required to transmit each one of the 240,000 bit pictures.

Estimates of crater heights and slopes have been made by detailed examination of the pictures. Comments on the consequences for the various theories of formation of planetary surface features have appeared in several journals (McCall 1966; Sagan and Pollack 1966).

Infra-red emission from Mars was studied with the balloon-borne Stratoscope experiments, the second of which used a 90 cm diameter collecting mirror, stabilized to 0·1 arc sec, at a height of 25 km (Danielson 1964).

Lunar probes. Many tens of thousands of pictures of the lunar surface have been received from U.S.S.R. and U.S.A. lunar probes in the Luna, Ranger, Lunar Orbiter and Surveyor series. A major purpose of these was to examine possible landing sites for manned flights.

The far side of the Moon was first observed by the Russian Lunik 3 in 1959, and the later Zond 3, 1965 (Lipsky 1965). Both used on-board film processing, and subsequent scanning of the negatives. In general, results of the Luna missions, and particularly of their instrumentation, are less accessible than the details of the corresponding U.S.A. models.

The last three Ranger probes, VII, VIII, IX (July 1964–March 1965) were successful in sending back a total of 15,000 pictures of the Moon before they crashed on the surface (Schurmeier 1966). A direct vidicon read-out of pictures was used, since there was no time for other processing of 5000 pictures in the 15 min. of useful recording time available to each spacecraft. Six cameras were used in each probe, and the final part-picture of the Ranger VII mission was taken only a few hundred metres above the surface. (The final moments of Ranger IX were presented live on television in U.S.A.)

Difficulties in image movement are obvious in attempting to photograph the surface from a rapidly approaching spacecraft (Martz 1963). If a 600-line picture is required, centred on the point of impact, the magnification of an object occupying one-half the field of view must not change by more than $^1/_{300}$, and thus the ratio ($\mathrm{d}R/R$) where R is the distance to the surface, must not change by more than this value in an exposure time. A body impacting on the Moon with a velocity of 3 km/sec can photograph with this resolution with 0·1 sec exposure at 100 km height but only 0·001 sec, at 1 km height. The Ranger cameras in fact had exposure times of $^1/_{200}$ and $^1/_{500}$ sec.

Vehicles circling the Moon can give more generally useful data. At least five are now in orbit, these being

Luna 10, 11, 12 and Lunar Orbiters 1, 2 (see Fig.). Lunar Orbiter 1 (T.W. 1966; A.W. & S.T. 2), in August 1966, scanned a large area of the Moon, the planned orbital height being 46 km (perilune) to 1800 km (apolune). The optical system here is much more sophisticated, and since the requirements are to select a landing site, slope of the surface and presence of protuberances of the order of one metre in height must be known. Stereoscopic viewing of an area of 40,000 km^2 was planned, with a surface resolution of 8 m (Kosofsky 1965).

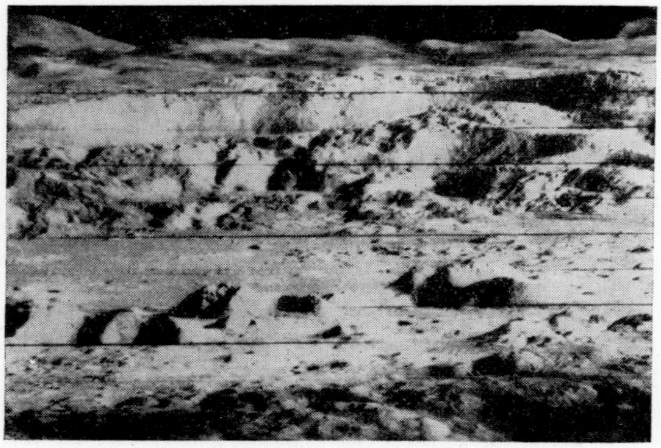

A portion of the first close-up picture of the crater Copernicus, one of the most prominent features on the face of the Moon. Taken at 7·05 p.m. EST, November 23, 1966, by Lunar Orbiter II. Looking due north from the crater's south rim, detail of the central part of Copernicus is shown. Mountains rising from the flat floor of the crater are 1000 ft high with slopes of 30°. The 3000 ft mountain on the horizon is the Gay-Lussac promontory in the Carpathian Mountains. Cliffs on the rim of the crater are 1000 ft high. From the horizon to the base of the Photo is about 150 miles. Lunar Orbiter II was 28·4 miles above the surface of the Moon and about 60 miles due south of Copernicus when this picture was taken. (Reproduced by permission of United Press International.)

Since the velocity of the orbiting capsule is 1·6 km/sec, and an exposure time of 0·04 sec minimum at f/5·6 is dictated by the sensitivity of the high resolution film, surface resolution better than 64 metres requires some form of image movement compensation. The magnitude of movement of the film required for stabilization was determined by an image scanning device which actuated the film movement and controlled the yaw of the spacecraft. Two lens systems, 80 mm f/2·8 and 600 mm f/5·6, recorded on one strip of film allowing a total of 194 exposures. The wide-angle pictures were overlapped so that each part of the surface photographed appeared on at least two frames, allowing stereoscopic viewing of the scene. On-board processing of the film by the Bimat process takes only a few minutes, and the resulting pictures are read out by a flying-spot scan method, giving 17,000 line coverage of each frame.

The first vehicle to make a soft landing on the Moon was the Soviet Luna 9 (February 1966), followed by the American Surveyor I (June 1966). Each sent back details of the terrain with resolution of the order of 1 mm for material near the feet of the craft. Surveyor I sent back 10,000 pictures, apparently by conventional television methods. Luna 9 shifted its position on the lunar surface during the picture taking sequence, enabling the sequence to be viewed stereoscopically (S.F. 1, 1966. S.F. 2, 1966. *Science*, 1966).

Spectroscopy. Several rocket flights have studied solar emission in the I.R., U.V. and soft X-ray regions. The first orbiting astronomical observatory (OAO-1) was a failure; a few rocket-borne spectrograms of stellar u. v. emission from the brightest stars have been published. Plans for a large number of observatories are well advanced.

Stellar spectrograms in the range 2200–4000 Å were obtained by Gemini astronauts, using hand-held spectrographic cameras.

Solar spectroscopy. Several successful rocket flights to study the Sun's corona have been made in British "Skylark" rockets. The nosecone being recoverable, photographic plate recording followed by ground processing has been successful.

The temperature in the immediate neighbourhood of the Sun's surface or photosphere rises from $\sim$5000°K outwards through the chromosphere to $\sim 10^6$°K in the corona (Liller, 1961) in a distance of 40,000 km or an angular distance of one minute of arc. To make useful measurements in this region, the corona is imaged on the slit of a spectrograph with an accuracy of a few seconds of arc. Results (Burton 1965) show that the continuum ends at 2100 Å, below which some 350 emission lines have been detected. Difficulties encountered are in stabilization, and in masking off the photospheric illumination, which is 10^6 times greater than the signal being sought. A minute amount of scattering from the imaging mirror could be very serious. A second experiment in the same payload imaged the Sun's disk (pinhole camera) via a diffraction grating on to the emulsion, with an aluminium foil filter restricting the wave-length coverage between 170 and 400 Å. The pictures were discrete because of the presence of strong emission lines in a weak continuum, and the instrument was capable of producing separated images from emission lines 50 Å apart. The picture obtained from the He 304 Å line was uniformly bright, whereas the $Fe\,IX^+$–$Fe\,XII^+$ picture at 170 Å showed distinct limb brightening in equatorial regions.

Stellar spectroscopy. Published data on stellar U.V. emission refer to a few rocket flights (Stecher and

Milligan 1962), during which some of the brightest early type stars were studied. Since the light levels are very low, long integration times are required, and the major difficulty is in stabilization of the system.

The U.S.A. "O.A.O." series as originally planned was revised in late 1965 (Roman 1966). Descriptions previously published (Rogerson 1963) of the experiments themselves are still correct. A major requirement is for a u. v. Star map, with broad-band data in range 1000–3000 Å of a large number of stars. The "Celescope", in OAO-3, has four telescopes of Schwarzschild type giving $2\frac{1}{2}°$ field, recording on a u.v. sensitive image-tube, the Uvicon. Three filters are used to allow photometry in four bands in this wave-length region, and the map should include stars down to 10th magnitude. The bed-fellow of the Celescope, looking out of the other end of the vehicle, is the University of Wisconsin experiment, which also flew in the unsuccessful OAO-1 (April 1966). Seven instruments are used for stellar and nebular photometry.

OAO-2 and OAO-4 will contain respectively the Goddard Space Flight Centre instrument (1050–4000Å with resolution 2 Å) and the Princeton University Experiment (800–3200 Å, resolution 0·1 Å). Each uses a single 80 cm diameter telescope to image a single star on the slit of a spectrograph.

The European Space Research Organization (ESRO) plans at least two satellites in this field. A stabilized satellite looking radially outwards from the Earth, normal to the Earth-Sun vector, scans the whole sky in six months, and will use four wavebands. The ESRO large astronomical satellite (LAS) payload will be built by the U.K. Atomic Energy Authority. A single 80 cm diameter telescope feeds a concave grating spectrometer, and the spectrum is scanned by a bank of photomultipliers. A resolution of 0·1 Å in the range 900–3300 Å is desired.

Design problems. Telescopes operating down to 900 Å must be reflectors, since no material for use as a refracting element transmits below the cut-off of lithium fluoride at 1050 Å, and even that is not available in large sizes. The coating of evaporated aluminium (Hass and Tousey 1959) has a reflectivity of only 20 per cent at 900 Å, thus dictating a minimum number of reflections. Practical considerations require that for wave-length resolution better than a few angstroms, this minimum number is three—a Cassegrain form of telescope, followed by a concave grating, producing a spectrum along the Rowland circle.

Angular stability of a system analysing one star at a time is determined by the wave-length resolution required, or by the desired efficiency, if a slit is used. For near-normal incidence on the grating, the wave-length change $\delta\lambda$ associated with a maximum change $\delta\theta$ in angle of incidence on the grating (of spacing d and order n) is given by

$$\delta\theta \cong n\delta\lambda/d.$$

If the spectrometer entry slit is of width S, and R is the diameter of the Rowland circle, then $S = R\delta\theta$. For a telescope of diameter D working at focal ratio f, the angular stability required is $\Phi = S/fD$, and finally

$$\Phi fDd = Rn\delta\lambda.$$

Since R and D are usually nearly equal, the result obtained is $\Phi \cong 2n\delta\lambda$, when Φ is in arc sec and $\delta\lambda$ is in angstroms. More precisely, an efficient system with a resolution of 0·1 Å requires stability of 0·1 arc sec. The design of a suitable fine guidance system, and the ability to demonstrate during test that it will function, are both at the very limits of present technology.

Thermal problems. All satellites are subject to temperature variations, since the operating level is determined by heat input from solar radiation, loss to cold space, and internal dissipation, with additional changes due to inclination or orbital night and day. Optical systems are especially vulnerable. As an example, a Cassegrain telescope is particularly sensitive to the positioning of the secondary mirror relative to the primary, and in a large, squat space telescope the tolerance is measured in tens of microns. This is a major problem of thermal design, and requires lagging of the telescope tube. In OAO-4, the mirror separation is determined by rods of fused silica over 2 metres long, which will be placed in position in orbit, after the launch phase.

Precision optical elements should be used at a temperature close to that of manufacture, since surface changes greater than 0·1 micron are important, and even "isotropic" materials show point to point variations in coefficient of linear expansion on this micro-scale. The secondary mirror of a space telescope is particularly subject to variation, being in the most exposed position. The thermal environment of a space telescope is such that the tube and mirrors are colder than expected, since radiation takes place out of the open end. (On ground based instruments this is compensated by radiation in at atmospheric temperature.) Gradients of the order of magnitude 1°C can be expected. Squat telescopes are also exceedingly sensitive to bending, as would be caused by a temperature gradient across the tube.

Finally, photomultiplier detectors should be run cool, to avoid excessive thermal noise.

Launch and weight. Space probes are never as big or as massive as experimenters would like. Launch vibration dictates that edge-held lenses are more likely to fail than uniformly supported mirror systems; this combined with weight restriction sometimes requires that a reflecting telescope (e.g. Mariner IV) be used, although suitable refractors are available. The main mirror of the stellar spectroscopy telescopes, which must be reflectors, are either of fused silica (OAO-4, LAS) or beryllium (OAO-2). Beryllium has a high strength/weight ratio, and a light structure can resist the $1g$ polishing forces and retain its form in zero-g environment. It is not liable to fracture, and its high thermal diffusivity ensures even temperature distribution and rapid recovery from temperature change, compared with silica mirrors; but it does not take a good optical polish, and must be overcoated with amorphous nickel for final figuring before aluminizing. Fine scratches or "sleeks" which cause scattering and consequent inefficiency are particularly difficult to eliminate (Barnes 1966).

Large fused-silica mirrors may be lightened, and their thermal time constants and distortion sensitivity reduced, by a sandwich construction, where two plates are separated by a structure of crossed strips, giving a honeycomb structure with square cells, the whole being subsequently fused. Rigidity is preserved at lower weight.

Space hazards. The remaining hazards are micrometeoroid bombardment, electron and proton bombardment of refracting optics, and reactions caused by long-term outgassing of the spacecraft. Effects of micrometeoroids have been very much less than was originally feared (Wolff 1966). The result of charged particle bombardment of U.V. transmitting optics, especially lithium fluoride, is the formation of centres and loss of transmission at the short wave end.

Testing. Associated optics for pre-launch testing and calibration must be of better quality than the instrument to be tested, and must be immune to distortion. Calibration in the far ultra-violet is difficult because of low light levels and the number of inefficient surface reflections involved. Methods of producing simulated solar flux, in magnitude and spectral density, uniformly over areas of the order of a few square metres have necessitated extensive experiment.

Large high vacuum space environment chambers have used mainly oil diffusion pumps, and the resulting fine film of oil on optical reflecting surfaces is suspect, especially as harmful chemical reactions take place if the oil is cracked by u.v. or other radiation in space.

Bibliography

Aviation Week and Space Technology (1) October 10th, 1966, 66. (2) August 15th, 1966, 28.

Barnes W. P. (1966) *Applied Optics* **5**, 5, May, 701.

Burton W. M. and Wilson R. (1965) *Nature* **207**, July 3rd, 61.

Danielson R. E. *et al.* (1964) *Astron. J.* **69**, 5, June, 344.

Hass G. and Tousey R. *J. Opt. Soc. Amer.*, **49**, 593.

Kosofsky L. J. and Broome G. C. (1965) *Journal SMPTE*, **74**, 9, September, 773.

Leighton R. B. (1966) *Scientific American*, **214**, 4, April, 54.

Liller (Ed.) (1961) *Space Astrophysics*, New York: McGraw Hill.

Lipsky Y. N. (1965) *Sky and Telescope*, **30**, 6, December, 338.

Lowman P. D. and Te Lou Chang (1965) *Scientific Experiments for Manned Orbital Flight, American Astronautical Society*, Science and Technology Series, Vol. 4, 153.

McCall G. J. H. (1966) *Nature* **211**, No. 5056, September 24, 1384.

McClain E. P. (1966) in *Encyclopaedic Dictionary of Physics* (J. Thewlis Ed.), Suppl. Vol. 1, 384, Oxford: Pergamon Press.

McGraw-Hill (1966) *Yearbook of Science and Technology*, "Space", New York.

Martz E. P. (1963) *Applied Optics* **2**, 1, January, 41.

Millburn J. R. and Horton T. (1966) *Spaceflight* **8**, 4, April, 131.

N.A.S.A. (1964) Semi Annual Report to Congress, **12**.

Rogerson J. B. (1963) *Space Sc. Rev.* **2**, 621.

Roman N. (1966) *Spaceflight* **8**, 8, August.

Sagan, C. and Pollack J. B. (1966) *Nature* **212**, No. 5058, October 8th, 117.

Schurmeier H. M. *et al.* (1966) *Scientific American*, January, 52.

Science (1966) **152**, No. 3730, June 24th, 1737.

Spaceflight (1) 8, 9, September (1966) (2) **8**, 4, April (1966), 131.

Space/Aeronautics (1964) **41**, 6, June, 92.

Stecher T. P. and Milligan J. E. (1962) *Astrophys. J.* **136**, 1.

Technology Week, October 17th, 1966.

Wolff C. (1966) *Applied Optics* **5**, May, 701.

R. H. Christie

OPTIMAL CONTROL THEORY

1. Introduction

In every branch of science and engineering there exist systems which are controllable, that is, they can be made to behave in different ways depending on the will of the operator. As an example, the pilot of a jet-propelled aircraft climbing from take-off to cruise can change the flight time by controlling the time history of the elevator position and the thrust setting. As another example, the astronaut of a spaceship being transferred from one orbit to another can change the amount of propellant consumed by simultaneously controlling the time history of the thrust modulus and the thrust direction. As a third example, the ballistician who designs the ogive of an artillery projectile can change the aerodynamic drag by controlling the distribution of longitudinal slopes versus the abscissa.

Every time the operator of a system exerts an option, a choice in the distribution of the quantities controlling the system, he produces a change in the distribution of states occupied by the system and, hence, a change in the final state. Therefore, it is natural to pose the following question: Among all the admissible options, what is the particular one which renders the system optimum? For example, what is the option which minimizes the difference between the final value and the initial value of an arbitrarily specified function of the state of the system? The body of knowledge covering problems of the type indicated above is called the *optimal control theory*. Its main tool is the calculus of variations, in particular, that part of the calculus of variations which deals with the problem of Mayer.

2. *Mayer Problem*

In this section, we consider the set of *state variables* (derivated variables)

$$x_k = x_k(t) \quad k = 1, \ldots, m \tag{1}$$

and *control variables* (non-derivated variables)

$$u_k = u_k(t) \quad k = 1, \ldots, n \tag{2}$$

which satisfy the differential constraints

$$\dot{x}_k - f_k(t, x_1, x_2, \ldots, x_m, u_1, u_2, \ldots, u_n) = 0 \quad k = 1, \ldots, m. \tag{3}$$

We assume that the independent variable and the state variables must satisfy the *separated end conditions*

$$\begin{aligned} \varphi_k(t_i, x_{1_i}, x_{2_i}, \ldots, x_{m_i}) = 0 \quad & k = 1, \ldots, q_i, \\ \psi_k(t_f, x_{1_f}, x_{2_f}, \ldots, x_{m_f}) = 0 \quad & k = 1, \ldots, q_f. \end{aligned} \tag{4}$$

We assume also that the control variables 2 and the derivatives of the state variables 1 are continuous everywhere except at a finite number of *corner points*. With this understanding, we define an *admissible arc* as any set of state variables 1 and control variables 2 which satisfy the differential constraints 3 and the end conditions 4. Then, we formulate the following problem: "In the class of admissible arcs, find that special arc which minimizes the functional

$$I = G_f - G_i, \tag{5}$$

where

$$G = G(t, x_1, x_2, \ldots, x_m) \tag{6}$$

is an arbitrarily specified function of the independent variable and the state variables." In the above relations, the dot sign denotes a derivative with respect to the independent variable t, the subscript i stands for initial point, and the subscript f stands for final point. Furthermore, the numbers m, n, q_i, q_f satisfy the inequalities

$$\begin{aligned} & n \geqq 1, \\ & q_i \leqq m + 1, \\ & q_f \leqq m + 1, \\ & q_i + q_f \leqq 2m + 1. \end{aligned} \tag{7}$$

This problem is a variational problem of the *Mayer type*, which involves one independent variable, $m + n$ dependent variables, m differential constraints, and n *degrees of freedom* (by definition, the number of degrees of freedom of a differential system is the difference of the number of dependent variables and the number of constraining equations). Therefore, the number of degrees of freedom is equal to the number of control variables. The solution of the Mayer problem, the set of functions which renders the functional 5 a minimum, constitutes the *extremal arc*. The latter may or may not include several *subarcs*, depending on whether corner points are or are not present.

A modification of the previous problem arises whenever the *control variables* 2 are required to satisfy a set of *inequality constraints* having the form

$$g_k(t, x_1, x_2, \ldots, x_m, u_1, u_2, \ldots, u_n) \geqq 0 \quad k = 1, \ldots, p. \tag{8}$$

This problem can be reduced to the same mathematical model used in solving the problem where all of the constraints are represented by equalities, if the *real variables*

$$\alpha_k = \alpha_k(t) \quad k = 1, \ldots, p \tag{9}$$

defined by the relationships

$$g_k(t, x_1, x_2, \ldots, x_m, u_1, u_2, \ldots, u_n) - \alpha_k^2 = 0 \quad k = 1, \ldots, p \tag{10}$$

are introduced. Therefore, the new Mayer problem involves one independent variable, $m + n + p$ dependent variables, $m + p$ differential constraints, and n degrees of freedom. Since both the number of equations and the number of variables are increased by p, the number of degrees of freedom is unchanged; hence, any number of inequalities can be imposed on an optimum control problem.

3. *Bolza Problem*

In order to treat the proposed Mayer problem, we introduce the Lagrange multiplier functions

$$\begin{aligned} \lambda_k = \lambda_k(t) \quad & k = 1, \ldots, m, \\ \mu_k = \mu_k(t) \quad & k = 1, \ldots, p, \end{aligned} \tag{11}$$

and rewrite the constraints (3) and (10) in the form

$$\begin{aligned} \lambda_k(\dot{x}_k - f_k) = 0 \quad & k = 1, \ldots, m, \\ \mu_k(g_k - \alpha_k^2) = 0 \quad & k = 1, \ldots, p. \end{aligned} \tag{12}$$

After summing the m equations 12.1 and the p equations 12.2 as follows:

$$\sum_{k=1}^{m} \lambda_k(\dot{x}_k - f_k) + \sum_{k=1}^{p} \mu_k(g_k - \alpha_k^2) = 0 \tag{13}$$

and integrating equation 13 over the interval under consideration, we obtain the relationship

$$\int_{t_i}^{t_f} \left[\sum_{k=1}^{m} \lambda_k(\dot{x}_k - f_k) + \sum_{k=1}^{p} \mu_k(g_k - \alpha_k^2) \right] dt = 0 \tag{14}$$

which is valid regardless of the choice of the Lagrange multiplier functions. Therefore, upon adding equations 5 and 14, we arrive at the functional form

$$I = \int_{t_i}^{t_f} F \, \mathrm{d}t + G_f - G_i, \tag{15}$$

in which the quantity

$$F = \sum_{k=1}^{m} \lambda_k(\dot{x}_k - f_k) + \sum_{k=1}^{p} \mu_k(g_k - \alpha_k^2) \tag{16}$$

is called the *fundamental function*. Clearly, for every admissible arc, the behaviour of the functional 15 is identical with the behaviour of the functional 5 regardless of the choice of the Lagrange multiplier functions. In other words, the problem of minimizing the functional 5 subject to the constraints 3 and 10, is identical with that of minimizing the functional 15 subject to the same constraints. In this way, the treatment of a Mayer problem becomes formally identical with that of a *Bolza problem*.

4. Necessary Conditions

Standard variational methods (see, for instance, Bliss 1946) show that the extremal arc of a Bolza problem must satisfy the following necessary conditions: the Euler equations, the corner condition, the transversality condition, and the Weierstrass condition. Prior to stating these conditions, it is convenient to define the functions

$$A = \sum_{k=1}^{m} \lambda_k \dot{x}_k, \quad B = \sum_{k=1}^{m} \lambda_k f_k$$
$$C = \sum_{k=1}^{p} \mu_k g_k, \quad D = \sum_{k=1}^{p} \mu_k \alpha_k^2 \tag{17}$$

where B denotes the *Hamiltonian* of the problem under consideration. With this understanding, we rewrite the fundamental function (16) in the form

$$F = A - B + C - D \tag{18}$$

and observe that it admits the partial derivatives

$$\begin{aligned} F_{\dot{x}_k} &= A_{\dot{x}_k} = \lambda_k & k &= 1, \ldots, m, \\ F_{x_k} &= C_{x_k} - B_{x_k} & k &= 1, \ldots, m, \\ F_{u_k} &= C_{u_k} - B_{u_k} & k &= 1, \ldots, n, \\ F_{\alpha_k} &= -D_{\alpha_k} = -2\mu_k\alpha_k & k &= 1, \ldots, p. \end{aligned} \tag{19}$$

4.1. Euler equations. From calculus of variations, it is known that the extremal arc of a Bolza problem must satisfy the $m + n + p$ Euler equations

$$\begin{aligned} \mathrm{d}F_{\dot{x}_k}/\mathrm{d}t &= F_{x_k} & k &= 1, \ldots, m, \\ 0 &= F_{u_k} & k &= 1, \ldots, n, \\ 0 &= F_{\alpha_k} & k &= 1, \ldots, p, \end{aligned} \tag{20}$$

which, on account of equations 19, can be rewritten as

$$\begin{aligned} \dot{\lambda}_k &= C_{x_k} - B_{x_k} & k &= 1, \ldots, m, \\ 0 &= C_{u_k} - B_{u_k} & k &= 1, \ldots, n, \\ 0 &= \mu_k \alpha_k & k &= 1, \ldots, p. \end{aligned} \tag{21}$$

The differential system composed of the constraining equations 3 and 10 and the Euler equations 21 includes $2m + n + 2p$ equations and the same number of unknowns; consequently, its solution yields the $m + n + p$ dependent variables and the $m + p$ Lagrange multipliers simultaneously. Generally speaking, approximate methods of integration are needed, since analytical solutions are possible only in special cases. An additional complication arises from the fact that the variational problems of interest in engineering are always of the mixed boundary value type, that is, problems with conditions prescribed in part at the initial point and in part at the final point. Consequently, in the case where closed-form solutions cannot be obtained, trial-and-error techniques must be employed; they consist of guessing the missing initial conditions, integrating numerically the set of Euler equations and constraining equations, and then determining the differences between the resulting final conditions and the specified final conditions. Since these differences are in general not zero, the process must be repeated several times until these differences become sufficiently small; in this connection, it is worth mentioning that analytical schemes have been developed in order to contain the number of iterations within reasonable limits.

Inspection of equations 21.3 shows that the extremal arc includes subarcs internal to the admissible domain defined by

$$\mu_k = 0 \quad k = 1, \ldots, p \tag{22}$$

and subarcs on the boundary of the admissible domain along which

$$\alpha_k = 0 \quad k = 1, \ldots, p, \tag{23}$$

that is,

$$g_k = 0 \quad k = 1, \ldots, p \tag{24}$$

For the subarcs 22, all the multipliers μ_k vanish, and the Euler equations 21 reduce to

$$\begin{aligned} \dot{\lambda}_k + B_{x_k} &= 0 & k &= 1, \ldots, m, \\ B_{u_k} &= 0 & k &= 1, \ldots, n. \end{aligned} \tag{25}$$

For the subarcs 23 and 24, only one of the auxiliary variables α_k (and hence only one of the functions g_k) need vanish.

4.2. First integral. A mathematical consequence of the set of Euler equations is the differential relationship

$$\mathrm{d}(F - \sum_{k=1}^{m} \dot{x}_k F_{\dot{x}_k})/\mathrm{d}t = F_t, \tag{26}$$

whose explicit form is the following:

$$\dot{B} + F_t = 0. \tag{27}$$

Therefore, if the system is *autonomous* (the fundamental function does not contain the independent variable explicitly, $F_t = 0$), the following *first integral* is valid:

$$B = \text{const} \tag{28}$$

and indicates that the Hamiltonian is constant along every subarc composing the extremal arc.

4.3. Corner condition. In the event that the extremal arc is discontinuous (that is, if one or several corner points exist), the corner condition

$$\Delta\left(F - \sum_{k=1}^{m} \dot{x}_k F_{\dot{x}_k}\right)\delta t + \sum_{k=1}^{m} \Delta F_{\dot{x}_k}\delta x_k = 0 \tag{29}$$

must be satisfied at each junction of any two subarcs. In the above relationship, the symbols δt, δx_k denote admissible variations at the corner point and the symbol $\Delta(\ldots)$ denotes the difference of the values taken by the quantity $(\ldots)$ on the two sides of the corner point. On account of equations 19, equation 29 can be rewritten as

$$\Delta B\,\delta t - \sum_{k=1}^{m} \Delta\lambda_k \delta x_k = 0. \tag{30}$$

As a consequence, if the corner point is free, the following relations hold:

$$\Delta B = 0 \tag{31}$$

and

$$\Delta\lambda_k = 0 \quad k = 1, \ldots, m, \tag{32}$$

meaning that the Hamiltonian B and the Lagrange multipliers λ_k are continuous at each corner point. If, in addition, the system is *autonomous*, equations 28 and 31 imply that the integration constant appearing in the first integral 28 has the same value for every subarc composing the extremal arc.

4.4. Transversality condition. The system composed of equations 3, 10, and 21 involves $2m + n + 2p$ relationships, of which $2m$ are first-order differential equations and $n + 2p$ are equations in finite terms. Consequently, $2m + 2$ boundary conditions are needed in order to define one particular solution. Since the problem under consideration is one of the Mayer type, no more than $2m + 1$ boundary conditions may be prescribed *a priori*; otherwise, the value of the functional 5 would be known and the variational problem would cease to exist. The remaining conditions must be determined so that the transversality condition

$$\left[\delta G + \left(F - \sum_{k=1}^{m} \dot{x}_k F_{\dot{x}_k}\right)\delta t + \sum_{k=1}^{m} F_{\dot{x}_k}\,\delta x_k\right]_i^f = 0 \tag{33}$$

is satisfied identically for every set of variations consistent with the separated end conditions 4. On account of equations 19, the transversality condition takes the form

$$\left[\delta G - B\delta t + \sum_{k=1}^{m} \lambda_k \delta x_k\right]_i^f = 0. \tag{34}$$

4.5. Weierstrass condition. After an arc satisfying the Euler equations, the corner condition, and the transversality condition has been determined, the next step is to verify that it yields a *minimum* for the functional 15. In this connection, the necessary condition due to Weierstrass is of considerable assistance. Analytically, this condition states that the functional 15 attains a minimum if the following inequality is satisfied:

$$\Delta F - \sum_{k=1}^{m} F_{\dot{x}_k}\,\Delta\dot{x}_k \geqq 0 \tag{35}$$

for every set of variations of the control variables and the derivatives of the state variables consistent with the constraints (3). Since $\Delta F = 0$, inequality 35 can be rewritten in the form

$$\Delta A \leqq 0 \tag{36}$$

which is equivalent to

$$\Delta B \leqq 0. \tag{37}$$

In the above expression, ΔB denotes the change of the Hamiltonian caused by a change of the control variables for constant values of the state variables and the Lagrange multipliers. Therefore, for the functional 15 to be a minimum, the Hamiltonian B must be a maximum with respect to the control variables at each point of the extremal arc. This statement constitutes *Pontryagin's Maximum Principle* and, as can be seen from the previous considerations, it is a direct consequence of the Weierstrass condition.

Bibliography

BLISS G. A. (1946) *Lectures on the Calculus of Variations*, New York: G. E. Stechert and Company.

PONTRYAGIN L. S., BOLTYANSKII V. G., GAMKRELIDZE R. V., and MISHCHENKO E. F. (1962) *The Mathematical Theory of Optimal Processes*, New York: Wiley.

A. MIELE

OPTIMUM AERODYNAMIC SHAPES

1. Introduction

The determination of optimum aerodynamic shapes has interested the scientific community for centuries. Historically speaking, the first problem of this kind was the study by Sir Isaac Newton of the body of revolution having minimum drag for a given length and diameter. Not only did Newton employ an analytical technique analogous to the modern calculus of variations, but he also postulated a law of resistance which has been recognized to be a good approximation to that of a hypersonic inviscid flow. In the early part of this century, the use of advanced mathematical techniques in the analysis of subsonic and supersonic flows stimulated a renewed interest in optimization problems. In particular, Munk determined the lift distribution which minimizes the induced drag of a subsonic wing having a given span and lift; furthermore, Von Kármán determined the shape of the slender forebody of revolution which minimizes the pressure drag in linearized supersonic flow for a given length and diameter. In more recent times, the advent of jet and rocket engines as aircraft propulsion systems and the parallel increase in flight velocities and altitudes have made it necessary to extend the optimization of aerodynamic shapes to a wider range of Mach and Reynolds numbers, thereby including the hypersonic and free-molecular flow regimes.

Since the distributions of pressure and skin-friction coefficients depend on the flow regime, a single optimum body does not exist; rather, a succession of optimum configurations exists, that is, one for each flow regime and set of free-stream conditions. In addition, the optimum geometry depends on the quantity being extremized (pressure drag, total drag, lift-to-drag ratio, surface-integrated heat transfer rate, sonic boom of an aircraft, thrust of a nozzle) as well as the constraints employed in the optimization process, whether geometric quantities (length, thickness, volume, wetted area, planform area, frontal area) or aerodynamic quantities (lift, pitching moment, position of the centre of pressure).

2. Physical Models

Prior to studying optimum aerodynamic shapes, it is necessary to determine the aerodynamic forces acting over arbitrary shapes. These forces, the drag and the lift, are obtained by integrating the distribution of normal and tangential stresses over the body. In non-dimensional form, normal and tangential stresses are customarily represented in terms of a pressure coefficient C_p and a skin-friction coefficient C_f. The former is the overpressure (local static pressure minus free-stream static pressure) per unit free-stream dynamic pressure; the latter is the tangential stress per unit free-stream dynamic pressure. In turn, the distribution of pressure and skin-friction coefficients is obtained by studying the properties of the flow field over the body.

Ideally, one would like to calculate optimum shapes by employing the Navier–Stokes equations in the entire region surrounding the body. Since this is a formidable task, a more practical approach consists of dividing the flow field into an outer region where the viscosity terms are negligible and an inner region where they are important, that is, a region where the flow is governed by the Euler equations and a region where the flow is governed by the Navier–Stokes equations. Thus, assuming that viscous interaction can be neglected, the pressure distribution is determined by the outer flow while the distribution of skin-friction coefficients is determined by the inner flow.

2.1. Pressure distribution. Depending on the speed regime, several methods can be employed in order to estimate the pressure distribution over a body. Among them, linearized theory, Newtonian theory, and method of characteristics must be mentioned in connection with the continuum flow regime.

For relatively slender shapes in flight at low supersonic Mach numbers, *linearized theory* can be employed. In other words, the non-linear set of equations governing the motion can be replaced by a set which is linear. Because of the linearity, the method of superposition can be employed, and general analytical solutions can be derived for the pressure distribution over either a two-dimensional shape or an axisymmetric shape whose contour is arbitrarily prescribed. For a two-dimensional shape, the local pressure coefficient has the form $C_p \sim \theta$, where θ is the inclination of the tangent to a surface element with respect to the free-stream direction. On the other hand, for an axisymmetric shape, the pressure coefficient no longer depends on the local slope of a surface element, but it is governed by the geometry of the entire body portion preceding that element.

At the other extreme of the velocity spectrum, the hypervelocity regime, *Newtonian theory* can be employed. If the shock wave generated by the body lies so close to the body that it can be regarded as identical with it, the pressure distribution can be determined with the assumption that the particles striking the body conserve the tangential component of their velocity but lose the normal component. For both two-dimensional and axisymmetric shapes, the local pressure coefficient is given by the sine-squared law $C_p = 2 \sin^2 \theta$, which simplifies to $C_p = 2\theta^2$ for slender shapes. This law is also valid for three-dimensional shapes, providing θ is interpreted as the angle which the free-stream velocity forms with the particular tangent which is coplanar with the velocity and the normal to the surface element under consideration.

Finally, if the combination of thickness ratio and Mach number is such that neither linearized theory nor Newtonian theory can be employed, a more precise approach to the determination of the fluid properties is necessary. In this connection, one can employ a pressure coefficient derived from second or higher order approximations to the equations of motion or, where possible, one can use the complete set of equations. If an isoenergetic expansion process is considered (this is the case with a rocket nozzle), the pressure coefficient has the form $C_p = C_p(w)$; that is, it depends only on the local velocity w. On the other hand, if an isoenergetic compression-expansion process is studied (this is the case with the forebody of a fuselage or an artillery projectile), the pressure coefficient has the form $C_p = C_p(w, p_0)$; that is, it depends on the local values of both the velocity w and the stagnation pressure p_0. In turn, the local values of the velocity and the stagnation pressure depend on the geometry of the entire body portion preceding a given surface element and must be determined by solving the partial differential equations governing the flow field within a certain region of interest.

In the previous paragraphs, it was tacitly assumed that the gas is a continuum, that is, the mean free path is small with respect to a characteristic dimension of the body. Whenever the mean free path is large with respect to a characteristic dimension of the body, a *free-molecular flow* takes place. The incident molecules are undisturbed by the presence of the vehicle, that is, the incoming and reflected flows are transparent to each other. For analytical purposes, two idealized models have been employed thus far, and are now illustrated. In the *specular reflection model,* the molecules hitting the surface are reflected optically, which means that the tangential velocity component is unchanged while the normal velocity is reserved. In the *diffuse reflection model,* the molecules hitting the surface are first absorbed and then re-emitted with a Maxwellian velocity distribution corresponding to an equilibrium temperature intermediate between that of the incoming flow and that of the solid surface. Under the hypersonic approximation (that is, if the square of the normal component of the speed ratio is much larger than one), the pressure coefficient is given by $C_p = 4 \sin^2 \theta$ for the specular reflection model and $C_p = 2 \sin^2 \theta + 2k \sin \theta$ for the diffuse reflection model, where k is a constant which depends on the speed ratio and the temperature ratio.

2.2. Skin-friction distribution. For continuum flow, the computation of the distribution of skin-friction coefficients requires the solution of the Navier–Stokes equations in combination with the customary boundary layer simplifications. As far as optimization studies are concerned, the approximation $C_f \sim x^{-\alpha}$ can be employed for slender bodies at low supersonic Mach numbers, where x denotes a coordinate in the undisturbed flow direction and α is an exponent having the typical values $\alpha = 1/2$ for laminar flow and $\alpha = 1/5$ for turbulent flow. At high supersonic Mach numbers, the approximation $C_f \sim p^{1-\alpha} x^{-\alpha}$ can be employed, where p is the static pressure at the outer edge of the boundary layer.

For free-molecular flow, the skin-friction coefficient is given by $C_f = 0$ if the specular reflection model is employed. On the other hand, if the diffuse reflection model is employed in combination with the hypersonic approximation, the skin-friction coefficient is given by $C_f = \sin 2\theta$.

3. Mathematical Models

Regardless of the physical model being considered, the analytical problems occurring in the theory of optimum aerodynamic shapes are problems of the calculus of variations. Generally, they consist of extremizing functionals which involve one or two independent variables and one or several dependent variables. In turn, these variables may be subject to isoperimetric, differential, and inequality constraints.

3.1. One independent variable. The most general problem of the calculus of variations in one independent variable is the problem of Bolza. It consists of extremizing the functional

$$I = \int_{x_i}^{x_f} f(x, y_k, \dot{y}_k)\, dx + g(x_i, y_{ki}, x_f, y_{kf}) \qquad (1)$$

with respect to the set of n functions

$$y_k = y_k(x) \quad k = 1, \ldots, n \qquad (2)$$

which satisfy the p isoperimetric constraints

$$K_j = \int_{x_i}^{x_f} \varphi_j(x, y_k, \dot{y}_k)\, dx + \gamma_j(x_i, y_{ki}, x_f, y_{kf}) \quad j = 1, \ldots, p \qquad (3)$$

the q differential constraints

$$\psi_j(x, y_k, \dot{y}_k) = 0 \quad j = 1, \ldots, q, \qquad (4)$$

the r inequality constraints

$$\omega_j(x, y_k, \dot{y}_k) \geqq 0 \quad j = 1, \ldots, r, \qquad (5)$$

and certain prescribed boundary conditions. In these relations, x denotes the independent variable, y_k the generic dependent variable, and $\dot{y}_k$ the derivative dy_k/dx. The subscripts i, f stand for the initial and final points, respectively; the symbols f, g, φ_j, γ_j, ψ_j, ω_j denote arbitrarily specified functions of the arguments within the parentheses; I is the quantity being extremized, and the symbols K_j denote prescribed constants. The mathematical treatment of this problem can be found, for instance, in Chapters 1 and 2 of Miele (1965). In this reference, it is shown that the extremal arc, the special arc which extremizes the functional 1, must satisfy the following necessary conditions: the Euler equations, the transversality condition, the corner condition, and the Weierstrass condition.

Variational problems with $n = 1$ and $q = 0$ arise, for instance, in the study of the longitudinal contour of two-dimensional wings in linearized supersonic flow, two-dimensional wings in Newtonian hypersonic flow, and axisymmetric bodies in Newtonian hypersonic flow. They also arise in the study of the transversal contour of a conical body in Newtonian hypersonic flow. In a typical case, I is the pressure drag or the total drag; the constants K_j are the enclosed area of a two-dimensional wing, the volume of an axisymmetric body, or the base area of a conical body; and, for hypersonic flow problems, the inequality $\omega \geqq 0$ may represent the requirement that the pressure coefficient be non-negative everywhere.

Variational problems with $n > 1$ and $q \neq 0$ arise, for instance, in the study of the rocket nozzle which produces the maximum thrust for a given length. The thrust, the mass flow, and the length can be expressed as integrals of quantities evaluated along the left-going characteristic line joining the axis of symmetry with the final point. The quantity I is the thrust; the constants K_j are the mass flow and the length; and the constraints $\psi_j = 0$ are the differential equations to be satisfied along the characteristic line, namely, the direction and compatibility conditions.

3.2. Two independent variables. The most general problem of the calculus of variations in two independent variables is the problem of Bolza. It consists of extremizing the functional

$$I = \int\int_S f(x, y, z_k, z_{kx}, z_{ky})\,dx\,dy + \oint_B g(x, y, z_k, \dot{y}, \dot{z}_k)\,dx \tag{6}$$

with respect to the set of n functions

$$z_k = z_k(x,y) \quad k = 1, \ldots, n \tag{7}$$

which satisfy the p isoperimetric constraints

$$K_j = \int\int_S \varphi_j(x, y, z_k, z_{kx}, z_{ky})\,dx\,dy + \oint_B \gamma_j(x, y, z_k, \dot{y}, \dot{z}_k)\,dx \quad j = 1, \ldots, p, \tag{8}$$

the $q = q_1 + q_2$ differential constraints
(within S)

$$\psi_j(x, y, z_k, z_{kx}, z_{ky}) = 0 \quad j = 1, \ldots, q_1,$$

(along B)

$$\tilde{\psi}_j(x, y, z_k, \dot{y}, \dot{z}_k) = 0 \quad j = 1, \ldots, q_2, \tag{9}$$

the $r = r_1 + r_2$ inequality constraints
(within S)

$$\omega_j(x, y, z_k, z_{kx}, z_{ky}) \geqq 0 \quad j = 1, \ldots, r_1,$$

(along B)

$$\tilde{\omega}_j(x, y, z_k, \dot{y}, \dot{z}_k) \geqq 0 \quad j = 1, \ldots, r_2, \tag{10}$$

and certain prescribed boundary conditions. In the surface integrals, x and y are the independent variables; z_k denotes the generic dependent variable, z_{kx} the derivative $\partial z_k/\partial x$ and z_{ky} the derivative $\partial z_k/\partial y$; and the symbol S denotes the domain of integration in the xy-plane. In the line integrals, x is the independent variable; y and z_k denote the dependent variables, $\dot{y}$ and $\dot{z}_k$ denote the derivatives dy/dx and dz_k/dx; and the symbol B denotes the boundary of the domain S. Also, the symbols, f, g, φ_j, γ_j, ψ_j, $\tilde{\psi}_j$, ω_j, $\tilde{\omega}_j$ denote arbitrarily specified functions of the arguments within the parentheses; I is the quantity being extremized, and the symbols K_j denote prescribed constants. The mathematical treatment of this problem can be found, for instance, in Chapters 3 and 4 of Miele (1965). In this reference, it is shown that the extremal surface, the special surface which extremizes the functional 6, must satisfy the following necessary conditions: the Euler equations, the transversality condition, the corner conditions, and the Weierstrass condition.

Variational problems with $n > 1$ and $q \neq 0$ arise, for instance, in the study of the rocket nozzle which produces the maximum thrust for an arbitrary isoperimetric constraint imposed on the contour. Specifically, one deals with the flow properties in a region S limited by a boundary B formed by the nozzle contour, the right-going characteristic line through the initial point, and the left-going characteristic line through the final point. After the thrust and the generalized isoperimetric constraint are expressed as integrals of quantities evaluated along the nozzle contour, the minimal problem can be treated as a Bolza problem, with this understanding: the quantity I is the thrust; the constant K is the value prescribed for the generalized isoperimetric constraint; the constraints $\psi_j = 0$ are the irrotationallity condition and the continuity equation to be satisfied at every point of the region S; and the constraints $\tilde{\psi}_j = 0$ are the tangency condition along the nozzle contour as well as the direction and compatibility conditions along the right-going characteristic line through the initial point and the left-going characteristic line through the final point.

4. *Discussion*

In the previous sections, the physical and mathematical models of interest in optimization studies have been reviewed. Within the frame of these models, considerable research has been performed in recent years on two-dimensional, axisymmetric, and three-dimensional shapes flying at supersonic, hypersonic, and free-molecular flow velocities (for a summary of the state of the art, see Miele 1965). For the hypersonic regime, some of the optimum shapes determined via the calculus of variations have been tested in wind tunnels against a number of comparison shapes and, in every case, they have exhibited superior performance.

While the theory of optimum aerodynamic shapes is only at its beginning, the encouraging results obtained so far make this author feel that, in due time, providing sufficient research effort is expended in this area and providing the present rate of progress is maintained in the design of digital computing machines, the calculus of variations approach may become a fundamental instrument in the design of optimum aerodynamic configurations.

Bibliography

Chernyi G. G. (1961) *Introduction to Hypersonic Flow*, New York: Academic Press.

Ferri A. (1949) *Elements of Aerodynamics of Supersonic Flows*, New York: MacMillan.

Miele A. (Ed.) (1965) *Theory of Optimum Aerodynamic Shapes*, New York: Academic Press.

A. Miele

OPTIMUM FLIGHT TRAJECTORIES

1. Introduction

Flight mechanics is that branch of the aerospace sciences whose purpose is to determine the performance of a vehicle moving under the effect of gravitational forces, aerodynamic forces, and propulsive forces (the term vehicle is employed here in a general sense and, depending on the particular case, may denote aircraft, missile, satellite, or spaceship). A particular, but important, aspect of this study is the analysis of the conditions under which the performance is optimum.

In the years preceding World War II, it was commonly believed among engineers that flight mechanics had reached a conclusive and rather stagnant stage of development. The progress achieved in the last twenty years has disproved this belief. In fact, the advent of jet propulsion systems and the associated increase in flight velocities have generated a wealth of problems in applied mathematics which cannot be handled by conventional methods of performance analysis. Among these problems, those concerned with the optimum flight conditions are discussed here.

Physically speaking, there are two main classes of optimum problems in flight mechanics: problems of quasi-steady flight and problems of non-steady flight, that is, problems where the inertia terms can be neglected in the dynamical equations and problems where they cannot. Regardless of the steadiness or non-steadiness of the motion, the determination of optimum flight programmes requires the study of functional forms which depend on the flight path integrally, rather than locally. Thus, the calculus of variations is of primary importance in flight mechanics, even though there are certain simplified problems of quasi-steady flight which are amenable to treatment by the simpler methods of the theory of maxima and minima. However, since all of the optimum problems of flight mechanics can be handled by means of the calculus of variations, it follows that the most economical and general theory of the optimum flight paths is a variational theory. The results relative to quasi-steady flight can be obtained as a particular case of those pertinent to non-steady flight by letting the acceleration terms appearing in the equations of motion decrease, tending to zero in the limit.

2. Fundamental Equations

Consider a vehicle moving in a great circle plane over a spherical Earth subject to the following forces: the drag D, the lift L, the thrust T, and the weight W which is the product of the instantaneous mass m and the local acceleration of gravity g. The drag and lift have the functional form

$$D = D(h, V, \alpha), \quad L = L(h, V, \alpha), \tag{1}$$

where h is the altitude, V the velocity, and α the angle of attack. The thrust and the mass flow rate of fuel are functions of the following type:

$$T = T(h, V, \pi), \quad \beta = \beta(h, V, \pi), \tag{2}$$

where π denotes the engine control parameter. As an example, π can be identified with the rotor speed of a turbojet engine, the fuel-to-air ratio of a ramjet engine, and the combustion chamber pressure of a rocket engine. The variation of the acceleration of gravity with the altitude h is represented by the inverse square law

$$g = g_0[r_0/(r_0 + h)]^2 \tag{3}$$

in which r_0 denotes the radius of the Earth and g_0 the acceleration of gravity at sea level.

With these considerations in mind, the equations governing the non-steady flight in a great circle plane are given by (see Miele 1962),

$$\begin{aligned} \dot{X} - f_1 &= 0, \\ \dot{h} - f_2 &= 0, \\ \dot{V} - f_3 &= 0, \\ \dot{\gamma} - f_4 &= 0, \\ \dot{m} - f_5 &= 0, \end{aligned} \tag{4}$$

where the functions f_1 through f_5 are defined as

$$\begin{aligned} f_1 &= r_0 V \cos\gamma/(r_0 + h), \\ f_2 &= V \sin\gamma, \\ f_3 &= -g \sin\gamma - D/m + T \cos\varepsilon/m, \\ f_4 &= -g \cos\gamma/V + V \cos\gamma/(r_0 + h) + L/mV \\ &\quad + T \sin\varepsilon/mV \pm 2\omega \cos\varphi, \\ f_5 &= -\beta. \end{aligned} \tag{5}$$

In the above relationships, X denotes a curvilinear coordinate measured on the surface of the Earth, γ the inclination of the trajectory with respect to the local horizon, ε the inclination of the thrust with respect to the velocity, ω the angular velocity of the Earth with respect to the Fixed Stars, φ the smaller of the two angles which the polar axis forms with the perpendicular to the plane of motion, and the dot sign a derivative with respect to time. The sign preceding the Coriolis acceleration term in equation 5·4 is positive for motion in the same sense as that of the Earth's rotation and negative for motion in the opposite sense.

Equations 4.1 and 4.2 are the kinematic relationships in the horizontal and vertical directions. Equations 4.3 and 4.4 are the dynamic relationships on the tangent and the normal to the flight path. And equation 4.5 expresses the principle of conservation of mass. These five equations involve one independent variable, the time t, and the eight dependent variables

$$X, h, V, \gamma, m, \alpha, \varepsilon, \pi. \tag{6}$$

Therefore, the system has $n = 3$ degrees of freedom, as is logical in view of the possibility of controlling the time history of the angle of attack, the thrust direction, and the thrust modulus. Consequently, some optimum requirement can be imposed on the flight trajectory, for instance, that of minimizing the functional

$$I = G_f - G_i, \tag{7}$$

where the subscripts i and f denote the initial and final points and where G is an arbitrarily specified function having the form

$$G = G(t, X, h, V, \gamma, m). \tag{8}$$

This general problem is a variational problem of the Mayer type which includes as a particular case almost every extremal problem of flight mechanics (Miele, in press). Thus, if the quantity to be minimized is the time, we set $G = t$. On the other hand, if the propellant consumption is to be minimized, we set $G = -m$. Finally, if the range is to be maximized, we set $G = -X$.

A modification of the previous problem arises if the engine is in a fixed position with respect to the airframe, that is, if the angle of attack and the thrust inclination must satisfy the relationship

$$\varepsilon - \alpha = \text{const.} \tag{9}$$

The new Mayer problem involves one independent variable, eight dependent variables, six constraining equations, and hence $n = 2$ degrees of freedom.

A further modification arises if one considers the fact that the mass flow rate of fuel is bounded by a lower limit and an upper limit. If it is assumed that the lower limit is ideally zero and the upper limit β_m is a function of the type

$$\beta_m = \beta_m(h, V), \tag{10}$$

equations 4 and 9 must be completed by the two-sided inequality constraint

$$0 \leqq \beta \leqq \beta_m. \tag{11}$$

This problem can be reduced to the same mathematical model used in solving the problem where all of the constraints are represented by equalities, if the *real variable* θ defined by the relationship

$$\beta - \beta_m \sin^2 \theta = 0 \tag{12}$$

is introduced. Since both the number of equations and the number of variables are increased by one, the number of degrees of freedom is unchanged.

3. Necessary Conditions

Standard variational methods (see, for instance, Bliss 1946) show that the extremal arc of a Mayer problem must satisfy the following necessary conditions: the Euler equations, the corner condition, the transversality condition, and the Weierstrass condition. Prior to stating these conditions, it is convenient to define the set of variable Lagrange multipliers

$$\lambda_k = \lambda_k(t) \quad k = 1, \ldots, 5, \qquad \mu = \mu(t), \qquad \nu = \nu(t) \tag{13}$$

and define the fundamental function as the linear combination of the multipliers 13 and the left-hand sides of the constraining equations 4, 9, and 12, as follows:

$$F = \lambda_1(\dot{X} - f_1) + \lambda_2(\dot{h} - f_2) + \lambda_3(\dot{V} - f_3) + \lambda_4(\dot{\gamma} - f_4) + \lambda_5(\dot{m} - f_5) + \mu(\varepsilon - \alpha) + \nu(\beta - \beta_m \sin^2 \theta). \tag{14}$$

Then, we rewrite the fundamental function in the form

$$F = A - B + C \tag{15}$$

where

$$A = \lambda_1\dot{X} + \lambda_2\dot{h} + \lambda_3\dot{V} + \lambda_4\dot{\gamma} + \lambda_5\dot{m}, \quad B = \lambda_1 f_1 + \lambda_2 f_2 + \lambda_3 f_3 + \lambda_4 f_4 + \lambda_5 f_5, \quad C = \mu(\varepsilon - \alpha) + \nu(\beta - \beta_m \sin^2 \theta). \tag{16}$$

Notice that the function A is linear in the derivatives of the unknown functions and the functions B and C do not involve derivatives. The function B is the *Hamiltonian* of the problem under consideration.

3.1. Euler equations. From calculus of variations it is known that the extremal arc of a Mayer problem must satisfy the following set of Euler equations:

$$F_{y_k} - \mathrm{d}F_{\dot{y}_k}/\mathrm{d}t = 0 \quad k = 1, \ldots, 5, \tag{17}$$

$$F_{z_k} = 0 \quad k = 1, \ldots, 4, \tag{18}$$

where the symbols

$$y_1 = X,\ y_2 = h,\ y_3 = V,\ y_4 = \gamma,\ y_5 = m \tag{19}$$

and

$$z_1 = \alpha,\ z_2 = \varepsilon,\ z_3 = \pi,\ z_4 = \theta \tag{20}$$

respectively denote the derivated and non-derivated variables of the problem. In explicit form, the Euler equations are given by

$$\dot{\lambda}_1 = 0, \quad \dot{\lambda}_2 = C_h - B_h, \quad \dot{\lambda}_3 = C_V - B_V, \quad \dot{\lambda}_4 = C_\gamma - B_\gamma, \quad \dot{\lambda}_5 = C_m - B_m \tag{21}$$

and

$$0 = C_\alpha - B_\alpha, \quad 0 = C_\varepsilon - B_\varepsilon, \quad 0 = C_\pi - B_\pi, \quad 0 = C_\theta - B_\theta \tag{22}$$

with the understanding that the subscripts denote partial derivatives.

The differential system composed of the constraining equations 4, 9, and 12 and the Euler equations 21 and 22 includes 16 equations and 16 unknowns; consequently, its solution yields the 9 dependent variables and the 7 Lagrange multipliers simultaneously. Generally speaking, approximate methods of integration are needed, since analytical solutions are possible only in special cases. An additional complication arises from the fact that the variational problems of interest in flight mechanics are always of the mixed boundary value type, that is, problems with conditions prescribed in part at the initial point and in part at the final point. Consequently, in the case where closed-form solutions cannot be obtained, trial-and-error techniques must be employed; they consist of guessing the missing initial conditions, integrating numerically the set of Euler equations and constraining equations, and then determining the differences between the resulting final conditions and the specified final conditions. Since these differences are in general not zero, the process must be repeated several times until they become sufficiently small; in this connection, it is worth mentioning that analytical schemes have been developed in order to contain the number of iterations within reasonable limits.

3.2. Thrust direction programme. If the engine is gimballed ($\mu = 0$), the Euler equations 22.1 and 22.2 are linear and homogeneous in the multipliers λ_3 and λ_4. Consequently, non-trivial solutions exist providing

$$\tan \varepsilon = D_\alpha / L_\alpha. \tag{23}$$

The physical meaning of this result is that the thrust must always be inclined in the direction of the lift. This result is logical since the creation of a normal component of the thrust reduces the amount of aerodynamic lift necessary to perform a certain manoeuvre and, therefore, reduces the induced drag associated with the lift.

3.3. Thrust modulus programme. The explicit form of the Euler equation 22.4

$$\nu \beta_m \sin \theta \cos \theta = 0 \tag{24}$$

shows that the following solutions are possible:

$$\theta = 0, \quad \text{or} \quad \theta = \pi/2, \quad \text{or} \quad \nu = 0, \tag{25}$$

corresponding to

$$\beta = 0, \quad \text{or} \quad \beta = \beta_m, \quad \text{or} \quad 0 < \beta < \beta_m. \tag{26}$$

If the thrust is a monotonically increasing function of the mass flow rate, the optimum thrust programme includes coasting subarcs, maximum thrust subarcs, and variable-thrust subarcs. The total number of these subarcs and their sequence depends on the boundary conditions of the problem.

3.4. First integral. Since the fundamental function F does not contain the independent variable explicitly, the differential system composed of equations 21 and 22 admits the first integral

$$F - \sum_{k=1}^{5} \dot{y}_k F_{\dot{y}_k} = \text{const}, \tag{27}$$

which, in explicit form, is given by

$$B = \text{const} \tag{28}$$

and indicates that the Hamiltonian is constant along every subarc of the extremal trajectory.

3.5. Corner conditions. In the event that the extremal path involves discontinuities in the nonderivated variables 20 and, therefore, discontinuities in the derivatives of the variables 19, the corner conditions

$$\Delta(F - \sum_{k=1}^{5} \dot{y}_k F_{\dot{y}_k}) = 0 \tag{29}$$

and

$$\Delta F_{\dot{y}_k} = 0 \quad k = 1, \ldots, 5 \tag{30}$$

must be satisfied at each point of junction of any two subarcs composing the extremal arc. The explicit form of these conditions is the following:

$$\Delta B = 0 \tag{31}$$

and

$$\Delta \lambda_k = Ok = 1, \ldots, 5 \tag{32}$$

which means that the integration constant of the first integral 28 has the same value for all of the subarcs composing the extremal trajectory and that the transition from one subarc to another occurs without discontinuities in the multipliers λ_k.

3.6. Transversality condition. The system composed of the constraining equations 4, 9, and 12 and the Euler equations 21 and 22 is subject to 12 boundary conditions. Of these, no more than 11 may be assigned *a priori*. The remaining are supplied by the transversality condition

$$\left[\delta G + \left(F - \sum_{k=1}^{5} \dot{y}_k F_{\dot{y}_k}\right) \delta t + \sum_{k=1}^{5} F_{\dot{y}_k} \delta y_k\right]_i^f = 0, \tag{33}$$

which is to be satisfied identically for every admissible system of variations. In explicit form, the transversality condition is given by

$$[\delta G - B\delta t + \lambda_1 \delta X + \lambda_2 \delta h + \lambda_3 \delta V + \lambda_4 \delta \gamma + \lambda_5 \delta m]_i^f = 0. \tag{34}$$

3.7. Weierstrass condition. After an extremal arc has been determined, it is necessary to investigate whether the functional I attains a minimum value. In this connection, the necessary condition due to Weierstrass is of considerable assistance. Analytically, the Weierstrass condition states that the functional I

attains a minimum if the following inequality is satisfied at all points of the extremal arc:

$$\Delta F - \sum_{k=1}^{5} F_{\dot{y}_k} \Delta \dot{y}_k \geqq 0 \tag{35}$$

for all admissible systems of variations. Since $\Delta F = 0$, the Weierstrass condition can be rewritten as

$$\Delta B \leqq 0 \tag{36}$$

where ΔB denotes the change of the Hamiltonian caused by a change of α, ε, π for constant values of the Lagrange multipliers 13.1 and the state variables 19. Thus, for the functional 7 to be a minimum, the Hamiltonian must be a maximum with respect to the angle of attack, the thrust direction, and the thrust control parameter. This statement constitutes the Maximum Principle (Pontryagin *et al.* 1962) for the particular problem under consideration, and, as can be seen from the previous analysis, it is a direct consequence of the Weierstrass condition.

4. Discussion

In the previous sections, the physical and mathematical models of interest in the optimization of trajectories have been reviewed. Within the frame of these models, considerable research has been performed in recent years on turbojet-powered aircraft, ramjet-powered aircraft, and rocket vehicles. A detailed account of this research is beyond the scope of the article; hence, the reader is referred to the specialized literature on the subject (Miele, in press; Lawden 1963).

Bibliography

BLISS G. A. (1946) *Lectures on the Calculus of Variations*, Chicago: The University Press.

LAWDEN D. F. (1963) *Optimal Trajectories for Space Navigation*, London: Butterworths.

MIELE A. (1962) *Flight Mechanics, Vol.* **1**: *Theory of Flight Paths*, Reading, Mass.: Addison-Wesley.

MIELE A. *Flight Mechanics, Vol.* **2**: *Theory of Optimum Flight Paths*, Reading, Mass.: Addison-Wesley (in press).

PONTRYAGIN L. S., BOLTYANSKII V. G., GAMKRELIDZE R. V., and MISHCHENKO E. F. (1962) *The Mathematical Theory of Optimal Processes*, New York: Wiley.

A. MIELE

P

PACKAGE CUSHIONING. The damage that can occur to equipment during transit as a result of shock impact or vibration can introduce serious technical and economic problems. At one time very little attention was paid to providing adequate protection during the packaging stage. In recent years, however, the importance of incorporating an adequate protection system at the outset has been realized and a branch of packaging technology, known as package cushioning, has arisen to deal with the problems in a scientific manner at the design stage.

Before a satisfactory package cushion system can be specified for the protection of any article the expected service life hazards must be known or realistically estimated. From these, the performance requirements for the package cushion may then be defined under two main headings, the so called "static" performance, and dynamic performance.

The static requirements cover the behaviour required of the system under conditions in which the package container is stored for periods of indefinite length in varying climatic conditions. The only load on the package cushion is the applied stress due to the weight of the packaged article. The important criteria are permanent set and creep, both of which are functions of the loading and climatic conditions. This information is important to enable the effects of the static conditions on the stress-strain relationship of the cushion system, and ultimately the dynamic performance, to be assessed.

The dynamic performance requirements cover the behaviour of the package cushion under transit conditions, the important criteria being in this case the effects of bumping, vibrating and dropping of the container, the system being subjected to high rates of strain.

In addition to knowing the conditions to which the container will be subjected, it is important to determine the capability of the article to withstand shock and vibration and this is defined in terms of its fragility. The term "*fragility factor*" is used to define the strength of the equipment to withstand shock and vibration.

Quantitatively the fragility factor is the acceleration required to induce sufficient force to cause failure of the equipment divided by acceleration due to gravity. This results in a "G" value obtained thus:

$$G = a/\boldsymbol{g}$$

where

a = acceleration to induce sufficient force to cause failure of the equipment in feet per second · per second.

$\boldsymbol{g}$ = acceleration due to gravity in feet per second per second.

This factor is very difficult to calculate for any but the simplest pieces of equipment and is generally arrived at by drop test evaluation.

Design. Once the relevant information for both the service hazards and mechanical properties of the packaged article is known, rational design of a suitable protective cushioning system can be initiated.

The methods used for designing a suitable protective system for any given package are based on the application of the laws of classical mechanics to an idealized system representing the package and its contents. A package can be said to consist of four major components illustrated diagramatically in Fig. 1.

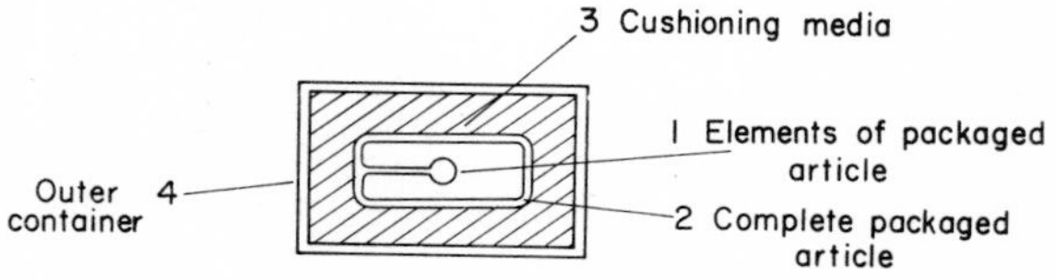

Fig. 1. Representation of a package.

The four major components are (1) the elements of the packaged article which are susceptible to mechanical damage, (2) the complete packaged article, (3) a cushioning medium and (4) the outer container. The complete system is further idealized by considering it to be a "lumped parameter system" where the outer container is considered as a single mass and the cushioning device as a massless spring with friction losses.

The energy characteristics of a cushioning device are dependent upon the load-displacement relationship. It is desirable to have analytical functions to represent the load/displacement characteristics. Ideally, only one family of functions with adjustable parameters to fit all possible shapes of load/displacement curves would be required. However, this is not feasible and the practical shapes are divided into six classes, illustrated in Fig. 2 and defined below:

Class 1—Ideal.

Load/displacement function is

$$P = \text{constant}.$$

Class 2—Linear elasticity.

Load/displacement function is

$$P = Kx$$

where

K = spring rate, pounds per inch.
x = displacement in inches.

Class 3(a)—Cubic elasticity.

Where the slope of the load/displacement function increases with increasing displacement, a suitable load/displacement function is

$$P = K_0x + rx^3$$

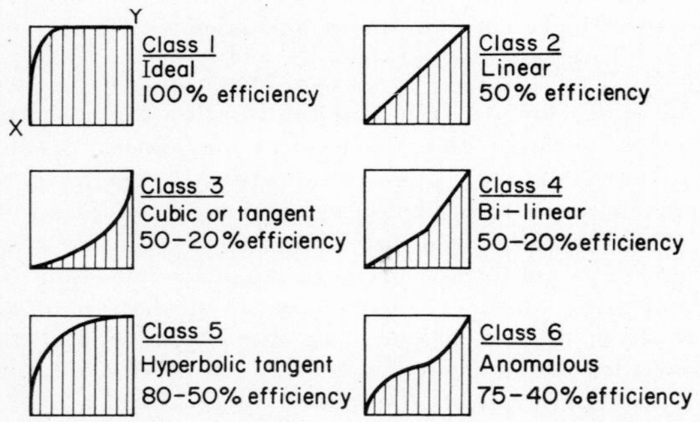

Fig. 2. Compressive stress-strain relationships for package cushioning systems.

K_0 is the initial spring rate of the cushioning and r determines the rate of increase of the spring rate. Cushioning of this type does not "bottom" within the expected range of use.

Class 3(b)—Tangent elasticity.

Cushioning that bottoms, but not abruptly, can be represented by the load/displacement function

$$P = \frac{2K_0d_m}{\pi} \tan \frac{\pi x}{2d_m}$$

K_0 is the initial spring rate and d_m is the maximum available displacement. Figure 2 shows how the cushion stiffness increases as the displacement approaches the maximum and the cushion is effectively solid.

Class 4—Bi-linear elasticity.

This is characterized by a load/displacement curve consisting of two straight line segments, the function being

$$P = K_0x \quad 0 \leqq x \leqq d_s$$

$$P = K_bx - (K_b - K_0)\, d_s x \geqq d_s$$

where K_b is spring rate after bottoming and d_s is displacement at which the spring rate changes from K_0 to K_b.

Class 5—Hyperbolic tangent elasticity.

When the mechanism of the cushioning is such that the maximum force transmitted is limited over a considerable displacement range, i.e. constant force characteristics, the load/displacement function is

$$P = P_0 \tanh (K_0x/P_0)$$

where P_0 is the asymptotic value of the force, and K_0 is the initial spring rate.

Class 6—Anomalous elasticity.

In some instances the load/displacement curve of the cushioning does not match accurately enough any of the above functions, and in these cases a numerical integration can be used.

It should be noted that the function having the closest approximation to the ideal shape where P = constant, resulting in the total area under the load/displacement curve being equal to the maximum energy of the system, is the hyperbolic tangent function.

The finally selected cushioning device for any application and its resulting load/displacement relationship is dependent upon the individual requirements of each package problem. It is, however, rare in practice that a packaging system has linear spring characteristics. The departure from linearity in the majority of cases is possibly due to either non-linear geometry of the system, where, for example, a system of discrete springs is being used, or non-linear characteristics of bulk cushioning materials used in pad form.

The majority of general purpose packaging applications can be solved by designing a cushioning system utilizing flexible materials in pad form.

The flexible materials currently used for packaging include rubberized animal hair; flexible and semi-flexible plastic foams such as polyurethane, polystyrene, polyvinyl chloride and polyethylene; cellular rubbers and other low density, high bulk materials. The compressive stress-strain relationship generally exhibited by materials of this type is that of the tangential form, Fig. 2, Class 3, of relatively low efficiency. The quality, and hence the repeatability of the mechanical properties, is dependent upon the control that can be effected during the manufacturing process, and is more effective for some of these materials than others. This is an extremely important factor in the confidence that can be placed in the design data on these materials.

Recent work in the package cushion materials field indicates that more attention is being given to the manufacture of materials that are more efficient, that is having an anomalous or hyperbolic stress-strain relationship rather than tangential elasticity, and engineered in such a manner to ensure repeatability of the required properties. One example of this is a flexible reinforced plastic honeycomb structure loaded normal to the cell axes as shown in Fig. 3. The structure acts as a buckled strut system under load resulting in a higher efficiency. Control at the manufacturing stage is very effective because the material has been designed as an engineering structure from the outset.

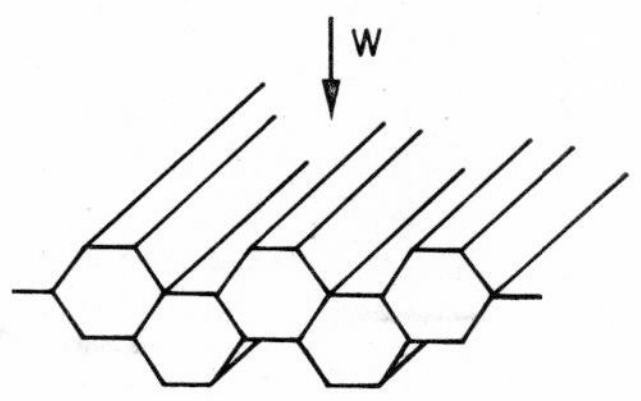

Fig. 3. Hexagonal cell honeycomb loaded normal to cell axes.

Design data for flexible cushioning materials should include both the "static" and dynamic behaviour of the material. The compressive stress-strain curve obtained under static loading conditions, i.e. strain rate $<0{\cdot}5$ inches/minute, provides useful basic information regarding the maximum available deflexion and elasticity, and hence the efficiency, where efficiency is expressed thus:

Efficiency $E =$

$$\frac{\text{Area under curve}}{\text{Maximum Possible Area}} \times 100 \text{ per cent.}$$

Permanent set, creep and the material's sensitivity to change in climatic environment are all relevant data applicable to sound design.

Whilst the static information contributes to better design of a package system it is evident from consideration of the expected service life that the performance under dynamic conditions is more important. The performance characteristics of cushioning materials subjected to rapid, sharp forces of short duration is far different from their performance under static loading.

The dynamic characteristics are obtained from evaluating test specimens in a drop test machine. By varying the dimensions of the specimens, the applied stress and drop height of the impact hammer comprehensive dynamic data can be obtained, which is generally recorded in the form of an acceleration-time pulse by means of a transducer attached to the hammer and connected to an oscilloscope. A typical acceleration-time pulse is shown in Fig. 4 for a material having cubic elasticity. The peak negative acceleration (deceleration) imposed by the cushioning material as the dropped weight slows-up and stops is detailed.

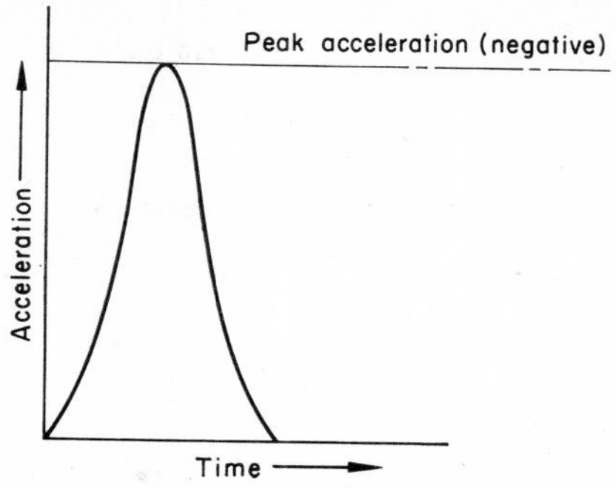

Fig. 4. Typical acceleration-time curve for cushioning with cubic elasticity.

Once the test information is available in the above form the next consideration is the best method of translating and presenting the data in a form best suited to the needs of the packaging designer.

If it is assumed that the complete package is given a constant acceleration by the force of gravity when dropped then at the moment of impact the outer container stops moving; but the packaged article continues to move and compresses the cushion through a distance. During this deflexion the cushioning material absorbs all the kinetic energy (K.E. $= 1/2\, MV^2$) that the article possessed at the moment of impact and at the instant of maximum deflexion, the maximum force and hence maximum deceleration is exerted upon the article.

Based on the above, the following equations can be used to express the relationship of the drop test parameters:

$$f_m = WG_m/A \tag{1}$$

where f_m = maximum dynamic stress in pounds per square inch:

W = weight of article in pounds;
A = bearing area of the article on the cushioning material in square inches;
G_m = maximum number accelerations that the article can stand (fragility factor).

Furthermore:

$$T = f_m h/e_m G_m, \tag{2}$$

where

T = minimum required thickness of cushioning material in inches;
h = drop height, in inches;
e_m = maximum energy absorbed per unit volume of cushioning material, in inch-pounds per cubic inch.

The dimensionless ratio $f_m/e_m = C$ and is known as the cushion factor (3).

By substituting this cushion factor into equation 2,

$$T = \frac{Ch}{G_m} \quad \text{or} \quad C = \frac{G_m T}{h}. \tag{4}$$

Using the above formulae and substituting into equations 1 and 4 the known test data such as W, A, T and h for each specimen of a given density and thickness, then a series of values for f_m and C, i.e. dynamic stress and cushion factor, can be obtained. If these data are then plotted with C as the ordinate and f_m as the abscissa, a cushion factor-dynamic stress curve is obtained. Typical cushion factor-stress curves for two different cushioning materials are shown in Fig. 5 and illustrates the variation between the materials.

This type of presentation was for a long time considered to be the most suitable method of data

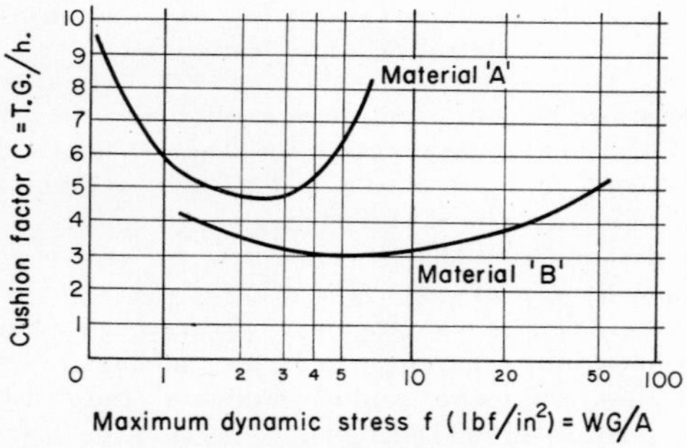

Fig. 5. Typical cushioning factor-dynamic stress curves.

presentation for design use. From a series of such curves for different types of flexible cushioning materials of varying densities the most efficient material could be selected by choosing the one whose cushion factor was at a minimum for a specific dynamic stress. From this figure and knowing the drop height and fragility factor, the minimum thickness of cushioning required can be computed using equation 4. The apparent advantage of this method was that it permitted a single curve for a given density of material to be utilized for several cushion thicknesses or drop heights.

The main disadvantage of this method that has become apparent more recently is that where there is a pneumatic effect contributing to a material's cushioning properties a change in dimensions alters the pneumatic effect and results in a change in the dynamic properties. Because of this it is obvious that data which varies with thickness cannot be used to determine the thickness of cushioning required.

The materials which have this pneumatic component are those whose cushioning performance is contributed to by the effects of air being exhausted from within their cellular structure. The total effect is a function of the type of structure, the size and the straining rate.

A more realistic method of presenting the dynamic data which takes into account the pneumatic effect is that of plotting the variation of the maximum deceleration (G_m) against the static stress (W/A). A curve of this type represents the dynamic compressive response of a material of given density, thickness, type, subjected to a variation in applied stress from a constant drop height. Figure 6 shows typical Peak "G"/Static Stress curves for different thicknesses of two types of flexible cushioning material.

Although this method increases the number of curves to be used, calculations do not have to be made at the test evaluation stage to change the peak "G" value and static load to cushion factor and dynamic stress respectively. The package designer can also read directly the static loading required for any

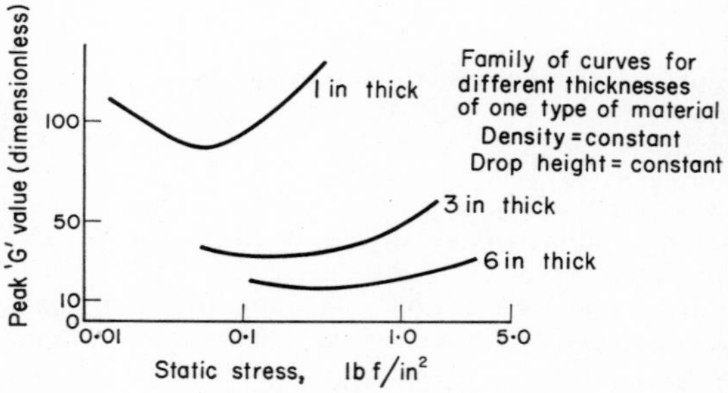

Fig. 6. Typical acceleration-static stress curves for a flexible cushioning material.

value of maximum "G" level and as the basic design information invariably specifies the weight, dimensions and "Fragility Factor" of the article to be packaged, this is obviously an advantage.

When the fragility factor is so low and the drop height is great so that a large deflexion is required to provide the necessary shock attenuation, a discrete spring system such as the tension spring package shown in Fig. 7 may provide the solution. Although each spring under pure tension has a linear characteristic the geometry as shown results in a cubic load/displacement function. Other combinations of

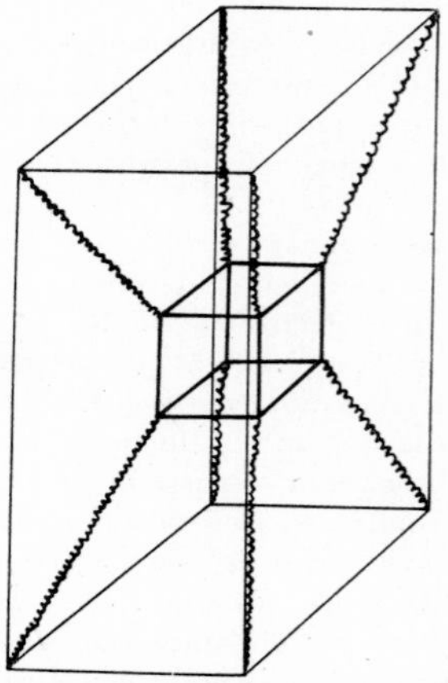

Fig. 7. Diagram of tension spring package.

discrete springs, either metal, rubber or plastic, can be geometrically arranged to provide a solution to many package cushioning problems.

In cases where shock impact forces are high and it is known that only one impact is to be experienced a rigid crushable material in block form can be used very effectively. Materials that have been used successfully for this application include end-grain balsa wood, rigid plastic and metal foams of various types and rigid honeycomb structures manufactured from a variety of materials.

If a piece of rigid honeycomb, as shown in Fig. 8, is crushed parallel to the longitudinal axis of the cells, a compressive stress-strain curve is obtained showing that the material has constant force characteristics approximating to the ideal (Fig. 9). A fairly steep, almost linear, slope is shown until approximately 1·0 per cent strain, when a yield occurs followed by an undulating plateau up to about 75 per cent strain. This indicates very little change in load as deflexion increases. At 75 per cent strain the material becomes solid, owing to excess crushed material packing in the base of the cells, and large load increases are experienced for very little deflexion.

The energy absorption capacity of honeycomb is directly proportional to the volume of honeycomb

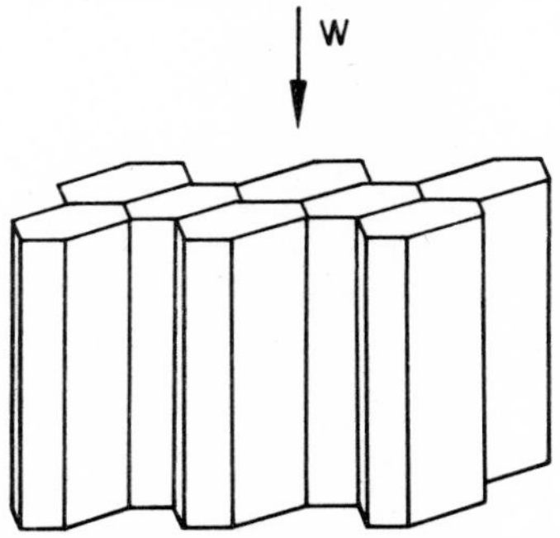

Fig. 8. Rigid honeycomb loaded parallel to cell axes.

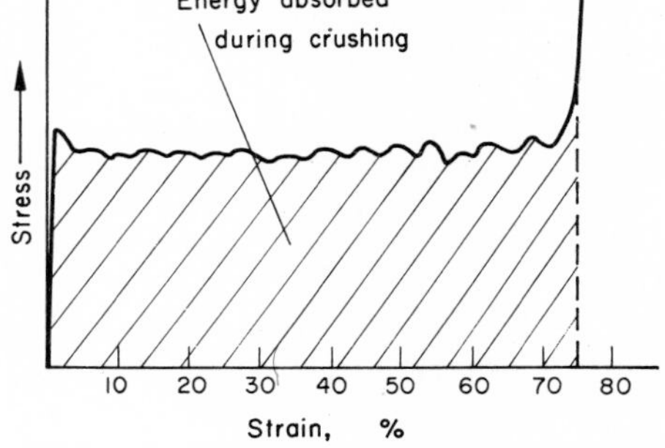

Fig. 9. Compressive stress-strain diagram rigid honeycomb material loaded parallel to cell axes.

under load. It has been shown that about 75 per cent of the thickness is available for the absorption of energy at an almost uniform rate. The energy absorption per square foot of area is therefore

Energy (foot-pounds per square foot) $= 0{\cdot}75\,(144\,f_c)t/12 = 9f_ct$,

where

f_c = crushing strength in pounds per square inch, and
t = thickness in inches.

The crushing strength f_c may be determined at strain rates less than 0·5 inches per minute and utilized for velocities up to 50 feet per second; above 50 feet per second it is considered that the pneumatic effect discussed previously becomes apparent.

When considering the use of rigid honeycomb for a packaging application the stopping distance required can be computed quite simply from the formula:

$$S = V^2/2a.$$

Where

S = stopping distance in feet,
V = impact velocity in feet per second, and,
a = deceleration in feet per second per second.

Knowing the fragility factor of the article $G = a/\boldsymbol{g}$

then $S = V^2/2\boldsymbol{g}G$.

The variables which influence the strength and hence the energy absorption properties are cell size, cell configuration, cell wall thickness and strength of the basic materials.

Suitable applications for these rigid type of energy absorbing materials are for cushioning parachute air drops of supplies, safety helmets and automobile crash protection, amongst many others.

See also: Plastic springs. Rubber springs. Vibration isolation.

Bibliography

Atack E. and Cameron J. B. (1965) The Use of Reinforced Plastics Materials in Energy Absorption Systems, *Trans. Plastics Institute*, **33**, 107, 189, October.

Gigliotti M. E. *Design Criteria for Plastic Package-Cushioning Materials—Plastic Report* **4**, Plastics Technical Evaluation Centre, Picatinny Arsenal, Dover, New Jersey.

Harris C. M. and Crede C. E. *Shock and Vibration Handbook*, Vol. 3, New York: McGraw-Hill.

McCulloch A. H. (1966) Cellular Plastics in Packaging, *Trans. Plastics Institute*. **34**, 111, 147, June.

Mindlin R. D. *Dynamics of Package Cushioning*. Bell Telephone System Technical Publications, Monograph B-1369.

Paine F. A. *Fundamentals of Packaging*, London: Blackie.

E. Atack

PALAEO WIND DIRECTIONS. A quantitative record of wind directions in remote times is available in the geological column. Ash-falls from volcanic eruptions thicken to the lee side and in some cases the bentonite beds so formed from Pleistocene volcanoes have been mapped and the wind direction determined. Prevailing winds inhibit the growth of coral atolls on the windward side through wave action. In certain limestones, even of Cambrian age, the direction of the prevailing wind has been inferred from studies of the colonial corals. The orientation of sand dunes provide the most widespread evidence of wind directions. There are two forms in modern deserts, the longitudinal or seif dune, and the transverse or barchan dune. The former are ridges often tens of miles long in the direction of the wind, the sand being thrown in a forward direction in successive layers on either flank. In the barchan dune which appears to result from steady winds, the sand is blown up the windward slope, which is a few degrees to the horizontal and deposits on the lee side slope where it takes up the angle of repose of about 31°. The lee side slope of the barchan dune forms a crescent in plan. Where there is much sand, the barchan dunes do not form as individuals and the transverse dunes which result have lee side slopes which in plan form wavy lines. In modern deserts, air photographs give evidence of the wind directions which, as the dunes take some tens of years to build up, represent an average over this time, though these may be weighted in favour of the storm winds.

Dunes are preserved in the geological column as aeolian sandstones. These sandstones may be recognized by the extremely well sorted and rounded quartz grains, the frosting of the grains caused by large numbers of impacts and the absence of mica which has been blown away by the wind. Some sandstones which have these characteristics, for example, the St. Peters sandstone of Illinois, U.S.A., are probably desert sands which have been re-deposited after transportation by the water. There are, however, cases for example, the Coconino sandstone of Permian age of the Grand Canyon, in which the dune forms have been preserved. This must have occurred through the lowering of the desert and the planing off by wind or water of the tops of dunes, on which later further aeolian deposition occurs. Perhaps the water table plays a part in determining the level to which the dunes are planed down and the dunes built up on top are sometimes separated from those below by lacustrine deposits. Cementation, which may not be very complete, occurs and the dunes become an aeolian sandstone. A dissection in recent times by river or in quarries reveals the lee side slopes of the barchan dunes as cross strata. These consist of large numbers of laminae making angles up to 32° with the horizontal. These are formed as successive sand deposits of slightly different grain size, colour or packing on the lee side slopes of the barchan dunes. A dissection of a modern barchan dune likewise enables this structure to be seen if the vertical cuts through the dunes are wetted. In an aeolian sandstone the slight differences of resistance to weathering of the different laminae enable them to be clearly visible.

In one set of cross strata the azimuth of lines of greatest slope of the laminae are nearly the same, but differ from those of cross strata belonging to other fossil dunes in the same locality. However, it is found that azimuths form a large number of different dunes in the same aeolian sandstone formation group about a mean. The spread in these azimuths results from the crescent shape of the lee side slopes when seen in plan. The figure shows the mean azimuths of the lee side slopes for a number of localities in the western U.S.A. of Permo-Pennsylvanian sandstones. Each mean is based on observations on about 30 separate dunes; the excellent agreement over these large distances is evidence that this method is capable of determining a significant planetary wind direction.

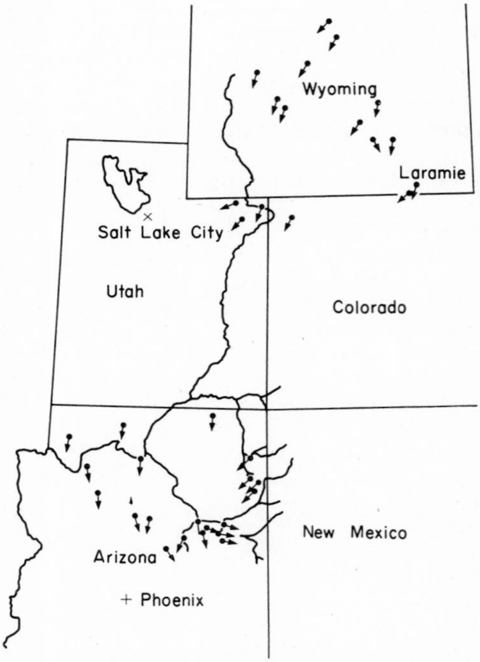

Palaeowind directions for the Western U.S. in later Palaeozoic time (From Opdyke N. D. and Runcorn, S. K. (1960), *Bull. Geol. Soc. Amer.*, **71**, *959*.)

The results of the figure show that the prevailing winds in the western U.S.A. were from the north-east in late Palaeozoic times, considerably different from the westerly prevailing winds of these latitudes today. As the procedures outlined are likely to average the wind directions over a long period of time, perhaps some millions of years, locally produced winds are unlikely to be much in evidence and the consistency of the pattern suggests that this is so.

Aeolian sandstones have been studied in other parts of the world. The New Red sandstones of the English Midlands and Dumfriesshire in Scotland give a consistent easterly wind and the aeolian sandstones of Triassic age in Uruguay and southern Brazil give westerly winds; the former quite different from present prevailing winds, the latter in agreement. The lower latitudes obtained for North America and Europe for the late Palaeozoic from palaeomagnetic observations place these areas in what would now be trade wind belts and the results of the palaeo-wind measurements described above are in reasonable agreement. The latitudes derived for the Triassic for South America from palaeomagnetic observations are not significantly different from the present and again the palaeo-wind directions are as might be expected.

Palaeo-wind directions are an aspect of a new subject, Palaeo-Geophysics, in which measures of planetary physical quantities are recovered from geological deposits for remote times.

S. K. Runcorn

PARTICULATE BEHAVIOUR AND PROPERTIES.

A particle or a *particulate* may be defined as a small entity of matter without regard to its physical state, that is, irrespective of whether it is a solid, a liquid or a gas. Particulates cover a wide range of sizes, the lower limit generally being on the order of only ten Angstroms in diameter and the upper rather loosely defined as that size readily perceptible to the unaided eye or to the touch. This definition results in some overlap, with true solutions and the quantum behaviour of individual atoms and molecules on the small end of the size spectrum and with the continuum mechanics of every-day objects at the large-size limit. Particulate technology then deals with the properties, behaviour, and applications of such small bodies of matter as individual particulates, disperse systems of particulates in various media, and bulk collections in which statistical rather than individual properties are significant. The special characteristic of all particulates and particulate systems is their large ratio of surface to volume with the result that many of their properties are determined by surface effects. This is particularly true for colloidal systems which comprise the extremely small end of the particulate size range, i.e. from about 10 to 10,000 A.

A large number of particulates dispersed in a gaseous medium form an aerosol. Smokes, fumes, fogs, mists, clouds, and hazes are familiar names for various types of aerosols. Condensation of a supersaturated vapour is the most common method for producing aerosols other than carbon smokes, although they may also be formed by mechanical means such as the aerodynamic action of winds on sand or other finely divided matter. The great dust storms of the early 1930's in the American plains states provide an impressive example of mechanically produced aerosol clouds. Dusts consist of solid particles dispersed in a gaseous medium and, for the most part, they are very heterogeneous systems that cover a wide range of size and exhibit poor stability. Smokes and fumes result from incomplete combustion, condensation of supersaturated vapours and, in some cases, from chemical or photochemical reactions. When a gas is produced simultaneously with a particulate cloud, the combination is usually called a fume. Smokes and fumes generally consist of fairly homogeneous systems of particulates ranging in diameter from about 10 microns downward to sub-micron sizes. Mists consist of a dispersion of liquid droplets in a gas. Condensation of supersaturated vapours to relatively large liquid droplets is the usual method for producing mists although they may also be formed by mechanical means.

Dispersions of gases in liquid or solid continuous media are termed foams, and liquid droplets dispersed in other immiscible liquid continuous media are emulsions. Solid particles dispersed in a liquid are called hydrosols when the liquid is water and the solid particles are of colloidal dimensions. In contrast to aerosols which have half-lives measured in hours or days, hydrosols are quite stable for long periods of time. Coarser dispersions of solid particles in liquids wherein the particle dimensions are considerably larger than the usually accepted colloidal range are usually termed slurries. They may be stabilized, however, against coagulation and settling for limited periods of time. Latex and oil-based paints with dispersed fillers or extenders of large particulate size are examples of these latter types of dispersions.

Gels are a rather unusual type of particulate system in that they are formed from two or more continuous phases, one of which is usually a solid. Frequently, only a slight amount of solid phase is required to impart rigidity as, for example, in the case of gelatin desserts where only two to three per cent of gelatin in water imparts the characteristics of a solid. Precipitation- and dispersion-hardened alloys are examples of solid particles dispersed in a continuous solid matrix.

Size determination. If a particulate is spherical or of some other regular geometry there is no ambiguity in specifying a characteristic dimension or "size". However, if a particulate is of irregular shape, as is usually the case, the problem of assigning a characteristic dimension becomes much more arbitrary. Consider, for instance, an acicular particulate such as a fibre which has a very large length-to-diameter ratio. For textile applications, the length, or staple, of the fibre is probably the most meaningful dimension. Consequently, for such applications the fibre length might be reported as its "size" whereas the cross-sectional diameter might be reported as its "size" for some other application. Theoretically, this problem can be alleviated by measuring several dimensions of the particulate and reporting some type of statistical diameter. However, this is a tedious, time-consuming process and is rarely done. Thus one should keep in mind that particulate "size" often refers only to that dimension of interest for a specific application. The most frequently used unit for expressing particulate size is the micron (1×10^{-4} cm).

Direct observation by optical or electron microscope and measurement of individual particulates is the most common technique for size determinations. Graded sieves spanning the range from sub-micron membrane filters to coarse screens are frequently used when separating a solid consisting of a wide range of sizes into several narrower size groups. *Sedimentation methods* of size analysis depend upon the rate of fall of a suspension of particulates through a viscous fluid. Stokes law defines the drag force exerted on a small particulate moving steadily in a viscous fluid and is the basis for most sedimentation techniques. Application of Stokes drag law and the buoyancy-corrected force due to gravity acting on a spherical particulate sedimenting at its terminal velocity v_s gives the following expression for particulate diameter d,

$$d = \sqrt{\left(\frac{18\mu}{(\varrho_p - \varrho_f)\boldsymbol{g}} \frac{x}{t}\right)}$$

where μ and ϱ_f are the viscosity and density, respectively, of fluid; ϱ_p is the particulate density; and x is the vertical distance traversed by the particulate during time t. If the particulate diameter approaches the mean free path of the molecules of the fluid, the particulate velocity increases due to slip between the fluid molecules. When this situation prevails, the terminal settling velocity v_s must be modified by the Cunningham correction

$$v_c = v_s(1 + 2A\lambda/d)$$

where v_c is the corrected terminal velocity, λ is the mean free path of the fluid molecules and A is a constant given by

$$A = 1{\cdot}26 + 0{\cdot}45 \exp(-0{\cdot}54\, d/\lambda).$$

The sedimentation rate of very small particulates is quite low and centrifugal methods are frequently used to shorten the settling time. By determining particulate concentrations at a fixed point at various time intervals, a cumulative size distribution may be obtained. For irregular particulates, the results are usually expressed as equivalent spherical diameters. The initial particulate concentration, the degree of dispersion achieved, and an accurate technique for measuring concentration at some reference point are the three most important factors in sedimentation analysis. Various automatic sedimentation devices are available that shorten considerably the time for analysis and expedite routine analyses.

Sensing-zone methods are now widely used for size distribution analyses. With these methods, a fluid in which particulates are suspended is caused to flow past a sensing device that is designed to monitor some property of the fluid, such as its electrical conductivity. Suspended particulates cause discontinuities in this property, the magnitudes of which are related to some physical parameter of the particulates. Highly refined sensing zone devices have been developed that can monitor electrical conductivity, acoustical properties, or certain optical properties of particulate suspensions. Inherent in most sensing zone methods, however, are certain operational factors that can quite easily lead to erroneous results, and analyses from such sophisticated electronic systems should not be accepted on faith alone

Permeability techniques measure the resistance to flow offered by a packed bed of solid particles. From the measured pressure drop and flow rate encountered under carefully controlled conditions, the surface area of the powder bed can be determined. The average size of the particles that make up the bed can then be deduced from the surface area.

Adsorption methods are also used to calculate the surface area of solid particles. When the surface of a mass of powder particles has been covered by a monolayer of gas molecules, the external surface area of the powder can be calculated from a knowledge of the total amount of gas adsorbed and the effective cross-sectional area of the adsorbed molecules.

A measure of average particle size can be obtained from the degree of line broadening that results when a mass of very fine, crystalline particles are subjected to monochromatic X-radiation.

The latter three methods of size analysis, i.e. *permeametry, adsorption,* and *X-ray line broadening,* yield information on some statistical average diameter but provide no data on the distribution of sizes. Consequently, the derived particle size information from such techniques should be interpreted with care.

Size distribution. Collections of particulates usually consist of a range of sizes, thus such a collection obviously cannot be characterized by any single size. The relative contribution of each size must be specified in the form of a size distribution. This distribution may be defined on a number basis, volume basis or some other basis, and the differences in each are quite important. A size distribution of an atmospheric aerosol, for instance, might indicate on a number basis that most particulates are between 0·01 and 1·0 micron in diameter. A size distribution of the same aerosol on a mass basis would probably indicate that most of the mass was accounted for by particulates from 1 to 5 microns in diameter. The specific application must be the determining factor as to which basis is most significant.

Colloids. Colloidal solutions are stabilized against coagulation and settling by the presence of electrostatic charges on the suspended particulates and/or by adsorption of solvent molecules onto the particulate surfaces. When the stability is largely determined by the adsorption of solvent molecules, the resulting colloid is termed lyophilic. When *colloidal stability* results from electrostatic repulsion, it is called a lyophobic colloid. These two classifications of colloids are, of course, idealizations, and most real colloids exhibit intermediate and overlapping characteristics.

Electrostatic charges on the particulate surfaces of lyophobic colloids may result from adsorption of foreign ions from solution or from ionization of the surface itself. Since the system itself must be electrically neutral, ions of charge opposite to those on the particulate's surface must exist in the solution. These counter-ions, as they are called, actually surround each particulate in a diffuse layer called the *Gouy layer*. Thus, as particulates of lyophobic colloids approach one another they are attracted by the London-Van der Waals forces and repelled by the overlapping clouds of counter-ions and by atomic electrons when they approach to very close ranges. The stability of lyophobic colloids is then a rather complex net result of these competing mechanisms. The *zeta potential* is a measure of the net electrostatic potential and is often used as an index of colloidal stability.

Aerosols, as opposed to colloidal solutions, are inherently unstable and always tend to coagulate and settle from suspension. The rate of coagulation for uncharged, spherical particulates may be predicted from the following relation:

$$\frac{1}{n} - \frac{1}{n_0} = 4\pi DR't$$

where n_0 and n are the number of particles present originally and after time t, respectively; D is the diffusion coefficient; and R' is the effective radius for coagulation which is usually about twice the actual particulate radius. The factor $4\pi DR'$ is frequently replaced by a coagulation constant K.

The presence of electrostatic charges on aerosol particulates does not result in long-term stabilization against coagulation. Although there are conflicting reports in the literature, recent works suggest an increased rate of coagulation in aerosols having particulates carrying opposite charges. Any effect that decreases the probability of collision of particulates will lengthen the life of an aerosol. Consequently, it might be assumed that unipolar charging should stabilize an aerosol in much the same manner as a lyophobic colloid is stabilized. However, it has been found that unipolar charging of aerosols frequently increases the rate of coagulation over that of uncharged aerosols. This effect is apparently a result of the ability of highly charged particulates to induce opposite charges in other particulates which have a lesser charge, or no charge at all. As a result, those particulates with induced charges of opposite sign are attracted more strongly than they are repulsed by the highly charged ones and coagulation results.

The *rheological properties* of particulate systems are of widespread interest in such areas as fluidization, pneumatic transport of powders, pumping and handling of slurries, and the viscoelastic transport properties of certain sols and gels. The flow properties of dispersed particulate systems are influenced by the concentration of particulates, their size and shape, the transport properties of the suspending medium, and interactions among the particulates and between the particulates and the suspending medium. For example, the viscosity of sols that contain fibrous or elongated particulates is usually abnormally high. This is due primarily to the formation of loose structural units of particulates in the sol and the attendant difficulties in disrupting these structures. In contrast, sols that consist principally of globular particulates exhibit much lower viscosities because of the ease with which such particulates can move relative to one another.

Finely divided particulates exhibit marked *diffusional activity* as is evidenced by the well-known Brownian motion. The constant, irregular motion of particulates is a result of collisions with atoms or molecules of the surrounding fluid. This diffusional characteristic causes particulates to strike and adhere to fibres that comprise high-efficiency air filters and is also partially responsible for the spontaneous aggregation and settling of aerosol particulates.

The molecules or atoms on the surface of a particulate experience less total attraction than do those in the interior. Thus the chemical reactivity and other physicochemical properties such as vapour pressure and solubility are more pronounced than for the bulk material. For this sort of reason, it is possible to imagine various systems in which the larger particulates tend to grow at the expense of the smaller ones. Thus differences in vapour pressure cause aerosols composed of liquid droplets or solids of relatively high vapour pressure to become more uniform in size with time.

Many finely divided particulate systems exhibit pronounced *physiological effects*. For example, respiratory disorders due to inhalation and retention of certain materials in the lungs have been recognized for centuries. Pneumoconiosis is the general term that describes a diseased condition resulting from retention of dust particles in the lymph depots of the lungs. Certain types of dusts such as those of the very slightly soluble particles of silica are especially dangerous. Prolonged inhalation of silica particles often results in an uncontrolled proliferation of collagen, a fibrous protein, in the connective tissue of the lymph glands. Many of the dangers of air and water pollution are also a result of the enhanced physiological effects of finely divided particulates.

On the positive side, there are also examples of beneficial effects of certain particulate systems. Many of today's drugs are actually disperse systems wherein the active material exists as the dispersed phase in an emulsion, sol, or coarse slurry. Mass innoculations are sometimes given by exposure to therapeutic aerosols. There is also promise of treatment for certain types of respiratory disorders by administration of aerosols with closely sized particulates. The size range used must contain particulates small enough to circumvent the natural filtering actions of the respiratory process yet not so small as to be removed by diffusional deposition until the desired area of the lungs is reached. Crop spraying and disinfection of large areas may also be effectively accomplished with certain types of aerosols.

Particulates in the atmosphere include clouds and the very fine, suspended particulates that act as condensation nuclei and facilitate cloud formation when the temperature falls below the dew point, i.e at very small supersaturation.

The term "smog" is a combination of the words smoke and fog and originally was used in areas in which large amounts of bituminous coal is burned. When meteorological conditions in such areas are conducive to the formation of fog, they also inhibit the natural removal of suspended coal and soot particles and the result is a dirty, smoky fog. More recently, the term "smog" has also been used to describe a type of urban air pollution that owes many of its properties to photochemical reactions among suspended particulates, hydrocarbon vapours, oxides of nitrogen, ozone and atmospheric oxygen. Fog may also be involved in this "smog" formation but this is not generally the case. Statistical and epidemiological studies have conclusively shown that inhalation of polluted air over extended periods results in severe respiratory disorders and, in some cases, cardio-respiratory insufficiency.

Illumination. Suspended particulates have the capacity to absorb, scatter, and reflect, incident electromagnetic radiation. For example, dust clouds suspended at very high altitudes diffuse sunlight and are visible in good seeing conditions after sunset. Suspended particulates are partly responsible for the colour of the sky and the intensity and duration of twilight. Faraday observed that when a beam of light passes through a colloidal gold sol it appears as a distinct white path in the sol when viewed from the side. This phenomenon is caused by scattering from the dispersed gold particulates. A beam of light passing through a true solution that does not fluoresce cannot be seen when viewed from the side. Colour may also be imparted to a disperse system by selective absorption or by certain types of scattering from suspended particulates.

The *Lambert–Beer law* relates the absorption of light by suspended particulates to the particulate concentration c and the path length x of light through the absorbing media by

$$I = I_0 \exp(-kcx)$$

where I_0 is the incident beam intensity, I is the intensity after path length x, and k is an absorption coefficient characteristic of a given system.

Rayleigh developed a theory for the scattering of light by particulates, the diameters of which are small compared to the wave-length of the incident light. Mie later developed a general theory for the interaction of light and matter which could be applied to particulates comparable in size to, or larger than, the wave-length of the light. When the suspended particulates are comparable in diameter to the wave-length, the intensity of the scattered light is found to depend on the direction from which the observation is made. Mathematical and experimental methods involved in investigations of the optical properties of suspended particulates are quite complex, and further discussion of them is beyond the scope of this article.

Bibliography

CADLE R. D. (1966) *Particle Size*, New York: Reinhold.

DALLA VALLE J. M. (1948) *Micromeritics*, New York: Pitman.

FUCHS K. (1964) *The Mechanics of Aerosols*, New York: Macmillan.

GREEN H. L. and LANE W. R. (1964) *Particulate Clouds: Dusts, Smokes and Mists*, New York: Van Nostrand.

HERDAN G. (1953) *Small Particle Statistics*, New York: Elsevier.

JIRGENSONS B. and STRAUMANIS M. E. (1962) *A Short Textbook of Colloid Chemistry*, New York: Macmillan.

ORR C. JR. (1966) *Particulate Technology*, New York: Macmillan.

ORR C. JR. and DALLA VALLE J. M. (1959) *Fine Particle Measurement*, New York: Macmillan.

J. H. BURSON

PHASE-LOCKED OSCILLATOR. The frequency of a voltage tunable oscillator (v.t.o.) may be stabilized by the use of a phase discriminator to provide a correction voltage dependent on the instantaneous phase difference between the v.t.o. and a reference oscillator. No frequency difference can exist between the reference oscillator and the output frequency of the v.t.o. when the two oscillators are phase locked. The reference oscillator frequency may be selected to be a harmonic of the output of a primary or a secondary frequency standard. Phase locked, laboratory oscillators can assume a short-term frequency stability in excess of $1:10^9$/sec and a long-term stability in excess of $1:10^8$/day.

A tunable phase-locked oscillator can be made by incorporating a number of stable quartz crystals in the reference oscillator. The frequency range over which the frequency of a quartz crystal controlled oscillator can be voltage tuned and the harmonic multiplication factor determines the resonant frequencies of these crystals. The short- and long-term stability is degraded by about a factor of ten using this technique. The output frequency is not strictly continuously variable since the various harmonics of the reference oscillator produce a finite comb of possible output frequencies in the region of interest. Further, any frequency cannot be specified more accurately than its short-term stability.

See also: Microwave frequency stabilizers.

H. A. Buckmaster and J. C. Dering

PHOTOCHROMISM. Photochromism or phototropy can be defined as a reversible colour change induced in a substance by exposure to light radiation, usually in the ultra-violet or the violet end of the visible spectrum. The characteristics of the effect depend on the incident temperature, the spectral distribution and intensity of the exciting light and the presence of any solvent. The phenomenon was first observed by Phipson (1881) and Markwald (1899) who introduced the term "phototropy". When the radiation is in the infra-red region, the term thermochromism is preferred.

Photochromic mechanisms. It is not possible to give a single explanation of photochromism and the following mechanisms are thought to account for some photochromic changes.

(a) Dissociation into ions: The spiro compounds in solution exist in two forms, one of which is coloured. The equilibrium state is shifted by the presence of light.

(b) Dissociation into radicals: The sydnones, a group of stable anhydro compounds some of which are darkened by light, and revert in darkness, on heating, or on the application of an electric field.

(c) Steric isomerism: Anthrones in solution are coloured by ultra-violet radiation.

(d) Oxidation and reduction: Silver halides are reduced to colloidal silver by light. Chlorophyll in an air-free solution of methanol is bleached by light, the effect being reduced by the presence of oxygen.

(e) Excited state population: The aromatic azo compounds are sometimes photochromic with very rapid recovery times.

Applications. The study of photochromism has achieved increasing attention since there are a number of commercial interests. These include the following:

(a) *Sun filters.* Photochromic material on an acetate base or in glass becomes coloured in sunlight and fades in weaker light or in darkness. Some desirable properties of these materials may include (i) rapid rise time in the presence of strong radiation, e.g. for protection from nuclear flash, (ii) a large number of reversals without side effects inhibiting the effect, e.g. for windshields of motor vehicles, and (iii) a neutral tint, e.g. for office windows and sunglasses. Many photochromic substances when excited have a pronounced colour tint.

(*b*) *Data storage.* Suitable materials should darken rapidly when illuminated but bleach slowly or only in the presence of heat. Normal photographic processes are limited by (i) their non-reversible nature, and (ii) projection problems, due to grain size at high magnifications. Photochromic materials offer solutions to these problems.

(*c*) *Q-switches.* A photochromic material which has two stable forms and can be switched from one to the other by radiation of suitable wave-lengths can be used as a bistable logic element. The use of such elements for computing is not at present possible, since a worthwhile number of reversals for this type of service cannot yet be obtained. However, Q switches can be used where a lower number of reversals is acceptable, for example in laser applications. One of their advantages can be that of rapid switching.

Research problems. The regions in which fundamental research is concentrated can be summarized as follows:

(a) The mechanisms which account for photochromism.

(b) The elimination of side effects which produce fatigue and so limit the number of reversals which can be obtained.

(c) Extension of the range of materials which exhibit photochromic behaviour, to increase the spectral range available.

(d) Examination of materials which exhibit rapid rise time of colouration. Sensitivity is usually extremely low compared with photographic materials.

Bibliography

Chalkley L. (1929) *Chem. Rev.* **6**, 217.
Bhatnagar S. S. *et al.* (1938) *J. Indian Chem. Soc.* **15**, 573.
Brown G. H. and Shaw W. G. (1961) *Rev. Pure Appl. Chem.* **11**, 2.
Cohen M. D. *et al.* (1964) *J. Chem. Soc.* **68**, 2041.
Exelby R. and Grinter R. (1965) *Chem. Rev.* 249.

K. J. Dean

PHOTOGRAPHY, STREAK. Streak or smear photography is a technique for continuously recording the position or size of a body or an event with time. In its simplest form time resolution is achieved by moving the image along the film, one dimension of the photographic film or plate accommodates the time axis, the other the movement to be studied.

The important technique of high speed framing photography suffers from the disadvantage of only recording the appearance of an event at discrete intervals of time, between separate frames the object may change and hence no record is made of these changes and also during each frame the record will be of a changing character and again information may be lost in "blurring". An exaggerated example of these disadvantages would be those encountered in photographing a race with an ordinary camera. A record of the part of the race when the film is being wound on will be absent and if the shutter speed is not high enough each individual photograph will be blurred. The framing camera does, however, give a two-dimensional record of the event at some instant in time, providing the exposure time of each frame is short enough. In contrast streak photography records continuously but has the disadvantage of only recording one dimension in space, the other dimension on the film accommodates the time axis. For these reasons it is often desirable to use both framing and streak photography for the study of fast changing events. Frame photographs give a two-dimensional record of the event while streak photographs ensure no loss of information in a particular direction. Combined high speed streak and framing cameras are produced commercially.

Essentially a streak camera consists of a lens which forms a real image of the object on a variable width slit, which is arranged so that the long dimension of the slit selects the direction in the object plane of most interest. This slit image is then swept along the photosensitive material in a direction at right angles to the length of the slit. Many camera arrangements have been used to make streak photographs, the various systems only differing in how the final slit image is swept along the recording surface. They may be conveniently divided into rotating mirror, moving film or drum and electro-optical techniques.

A typical rotating mirror streak camera is shown in the figure. The focal apertures or *f* numbers of objective and relay lenses are usually matched. The long dimension of the slit is perpendicular to the plane of the diagram. The rotating mirror is usually multifaced and made from steel or other metal with a high tensile strength and rigidity to withstand the considerable centrifugal force at high rotation speeds (tensile strength ultimately determines the maximum rotation speed) and to keep distortion of the faces to a minimum. The distortion allowable in streak cameras is less than in framing cameras because the whole surface contributes to the final image whereas in framing cameras an intermediate image is often formed at the mirror surface. Below about 1000 revolutions per second mirrors may be operated at atmospheric pressure using an electric motor drive, at higher speeds air resistance becomes too large and the mirror is run at reduced pressure, the drive mechanism being a compressed air powered turbine. The final moving slit image falls on a photographic film held in an arc with the rotating mirror at its centre. The velocity of the slit image on the photographic film, referred to as the writing speed, is limited by mirror rotation speed, mirror to film distance and the brightness of the object to be photographed, the latter implies that the *f* number of the camera lenses and the photographic film speed must be taken into account.

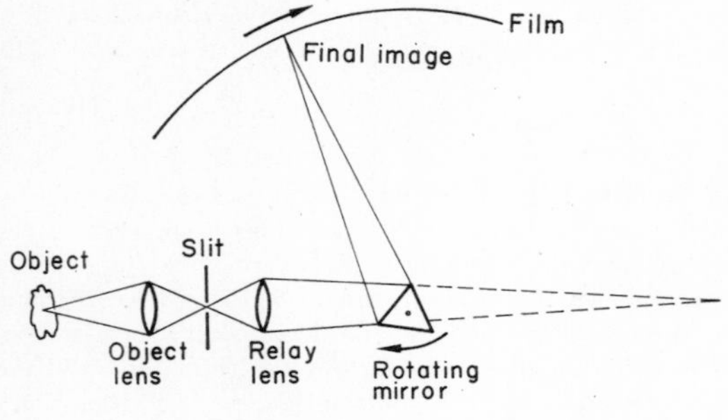

Fig. 1.

Rotating-drum cameras, where the streak function is achieved by moving the film across a stationary image either by mounting the film on the inside of a cylindrical drum or moving the film continuously by some smooth drive mechanism, are of value where synchronization of the event to be photographed and the rotating mirror is difficult. *Drum cameras* are said to have continuous access, meaning that the probability of recording the event is unity. Drum cameras often possess the disadvantage of low writing speeds, but because of their ease of mechanical application and continuous access they form a simple photographic method of *time resolved spectroscopy.* The photographic film, attached to the inside of a drum, is moved in the focal plane of a spectrograph at right angles to the wave-length dispersion. Spatial resolution is completely lost as the entrance slit is limited to a pin hole. A series of point images each corresponding to a spectrum line is streaked along the film. In this case the wave-length and time scales utilize the two dimensions on the film.

The streak photography techniques discussed so far rely on mechanical image displacement. As stated earlier the writing speed is limited by the available light from the object to be photographed or the physical difficulties associated with a long rotating mirror to film distances and high rotation speeds. High streak speeds with objects of low brightness may be achieved with image intensifier tubes. The action of an intensifier tube is to produce an intensified reproduction of the light falling on the photocathode onto a phosphor screen. The photocathode, which is located in the image plane of a camera lens is limited to a narrow strip orientated to lie in the direction of interest in the image plane. A streak effect is obtained by introducing a deflexion plate system into the tube such that the electrons are deflected in a direction at right angles to the strip direction in a similar manner to an oscilloscope time base.

Streak photography finds widest application in the study of explosions and gas discharge physics where the event to be photographed is usually self luminous.

Bibliography

Coleman K. R. (1963) *Rep. Progr. Phys.*, **26**, 269.

Proc. Fifth Congress on High-Speed Photography, 1960, New York: Society of Motion Picture and Television Engineers, 1962.

Proc. Sixth Congress on High-Speed Photography, 1962, Haarlem, Netherlands: Tjeenk Willink, 1964.

D. C. EMMONY

PISTONPHONE. A mechanical device for generating a known sound pressure and used for determining the electroacoustic sensitivity of microphones. A sinusoidally driven piston operates in a cavity of known volume, and the sound pressure thus generated is deduced directly from the dimensions of cavity and piston and the stroke of the piston. At higher frequencies it is valid to assume adiabatic compression within the cavity but at lower frequencies allowance must be made for heat conduction to the walls which remain at constant temperature.

M. E. DELANY

PLASMA DIAGNOSTIC TECHNIQUES. These are the methods by which the behaviour and properties of the plasma are studied experimentally. All information about the plasma at a particular time is contained in the particle distribution function, which is the number and kind of particles in unit volume and unit velocity range at each point. However, this can not be measured directly. In practice certain macroscopic quantities are usually measured, from which the distribution function and its variation with time can be inferred with more or less accuracy, depending on the elegance of the measurement. These are, for example: the density and temperature of various species, including neutrals if present; the macroscopic plasma velocity; plasma pressure; plasma current; magnetic and electric fields; and characteristic dimensions and times such as wave-lengths, density and temperature gradients, relaxation and oscillation periods.

Techniques may be of two kinds. Those in which the signal is produced by the plasma itself may be called passive. The measuring instrument disturbs the plasma very little or not at all. An example is spectroscopy. In the second kind, which may be called active, the signal is the plasma response to an applied perturbation. This may be small enough not to affect the measured parameter, or the effect may be allowed for. An example is microwave refraction. In general active techniques are less desirable, since there is always a danger that the applied perturbation will change the quantity being measured, but they are usually more flexible and give more information than passive techniques. For this reason their use is widespread.

One of the earliest diagnostic tools for laboratory plasmas was a metal probe, used by Langmuir in 1923, and now called the *Langmuir probe*. It is a small sphere, plate or cylinder placed within the plasma at the point of measurement. A voltage (direct, alternating or both) is applied between the probe and some reference electrode, also in contact with the plasma. The current flowing between them is measured. The bias voltage of the probe with respect to the plasma may be positive or negative, when it attracts electrons and positive ions respectively. As the bias voltage is increased the probe current rises at first and then limits at a value called the electron, or ion saturation current. This is a measure of electron density, and the shape of the current-voltage characteristic gives the electron temperature. In practice the interpretation of probe data is not simple. The region between the probe surface and the body of the plasma, called the sheath, has a complicated structure in which the plasma parameters are greatly disturbed. Allowance must be made for this effect to obtain accurate measurements.

Since the early work of Langmuir many refinements to probe diagnostics have been made. For example Sloane and McGregor in 1934 applied an alternating voltage added to the direct bias voltage to measure the shape of the current voltage characteristic more accurately. In 1950 Johnson and Malter used a double probe to reduce the perturbing effect on the plasma. In recent years probe diagnostics have received considerable impetus from their use in rockets and satellites for ionospheric measurements.

Probes are also used to determine the radiofrequency properties of the plasma. The technique consists in measuring the admittance of a probe at a frequency comparable with the plasma frequency. In free space the probe admittance is small and positive. The plasma acts as a dielectric for frequencies near the plasma frequency, and the change in probe admittance reflects the effective dielectric constant of the plasma. The most straightforward example of this type of measurement is that due to Sayers, for measuring the F-region electron density from a satellite, in which the "probe" is simply a parallel plate capacitor with plasma dielectric. A related technique is the resonance rectification effect. A swept radiofrequency voltage covering the frequency range of interest is applied to the probe. The radiofrequency probe current is rectified by the non-linear current-voltage characteristic giving a direct current proportional to the probe admittance. Resonances in the admittance appear as peaks in the rectified probe current. Recently pulsed radiofrequency probes have been used in which the applied voltage excites oscillations in the surrounding plasma which are then detected with a suitable receiver. These depend on the low group velocity of electrostatic plasma waves at certain frequencies: the plasma wave packet remains near to the probe after the exciting voltage is removed.

Probes may also be used passively to detect potential and density fluctuations associated with plasma instabilities and to measure electric fields. An interesting correlation technique makes use of two spaced probes biased to measure ion density. The correlation coefficient between probe signals as a function of their separation is measured electronically, and gives information on the wave number of any waves which may be present. Recently the same technique applied to thermal fluctuations has been used to obtain a value for the diffusion coefficient in a brush discharge plasma.

Small coils of many turns of fine wire are used to measure magnetic field variations. The voltage across the coil, proportional to the rate of change of magnetic field through it, is integrated electronically; usually with a simple RC circuit. The usual method of measuring plasma current is the *Rogowski loop*. A solenoidal coil is bent into a closed loop or torus. It can have a permeable cone, but this is not necessary. The time-integrated voltage across the coil is proportional to the space integral of magnetic induction around the loop, or the current threading it. Often the inductance of the loop itself is used in an LR integrating circuit.

The disadvantage of all kinds of probes placed within the plasma is their disturbing effect on it. Plasma particles reaching the probe surface are absorbed or neutralized and lost. Energetic particles may produce secondary emission giving a spurious probe current, and contaminating the plasma. In the measurement of very tenuous plasmas photoemission currents may be important. Conversely, in hot dense plasmas the flux of energetic particles to its surface may destroy the probe. The main advantage of probes is their ability to sample a small volume, giving good spatial resolution.

Direct mesurement of the velocity distribution in the plasma is obtained by observing particles which arrive at a nearby detector. In the case of a magnetically confined plasma these may be ions and electrons which diffuse from the confinement region or are ejected by an instability. Unfortunately they may not be representative of the whole particle distribution. They are detected by mass spectrometers, electron multipliers, solid state detectors or biased electrodes such as the Faraday cup. Fast electrons from the plasma, which are often produced by instabilities or rapid accelerations, are detected through the X rays which they produce on striking the containing wall. Neutral particles produced by charge-exchange reactions between plasma ions and neutrals are a better diagnostic, since they arise in the relatively undisturbed region in the body of the plasma. If the neutral gas density is not sufficiently high for charge exchange reactions to be frequent cold gas may be purposely injected. This comes into the class of active techniques. Neutral particles are reionized by passing through a gas cell or thin film before detection.

Other active techniques include the use of particle beams. Electron or charged particle beams are deflected by the Lorentz force and electrostatic fields. The deflexion of the beam is a measure of the time average forces acting along its trajectory, giving information on sheath structure, plasma potential and instabilities. Neutral and charged particle beams are used in beam attenuation experiments for measuring scattering and ionization collision cross-sections. Particle beams usually perturb the plasma to a lesser degree than do probes, although they may themselves excite instabilities by resonant interactions with plasma waves. They suffer from the disadvantage that the information gained does not have high spatial resolution.

Diagnostic techniques involving the use of electromagnetic radiation are the most satisfactory where spatial resolution is not required. A plasma interacts strongly with radiation because it contains free electrons. These are accelerated by random electric fields, and by the magnetic field if present, giving bremsstrahlung and cyclotron or synchrotron radiation. They may also recombine with plasma ions to give line spectra. The resulting radiation from the plasma is very complex and the techniques for its observation correspondingly varied. This is almost the only way to study non-terrestrial plasmas. Its power can be judged by the success of the sciences of astronomy and radioastronomy. In particular the techniques of spectroscopy are widely employed for astronomical research and to obtain information on ion species, temperatures and velocities in laboratory plasmas. A detailed discussion of spectroscopic methods will be found elsewhere.

Attention is drawn below to some active techniques using electromagnetic waves. In these cases the plasma electrons are accelerated by the electric vector of the incident wave. The re-radiated wave, to which the heavy plasma ions contribute very little, contains the required information.

Radiation emitted in the direction of the incident wave and at the same frequency interferes with it and gives rise to the phenomenon of refraction. The refractive index of a plasma is in general complex (in the presence of a strong magnetic field plasmas are birefringent). However, if the frequency of the incident wave is much larger than the electron cyclotron frequency, the refraction is simply related to an integral of the electron density along the path of the incident wave. It can be measured by the *Mach-Zehnder refractometer*, or some variant of it. A beam from the source, which may have a frequency from the microwave to the optical band depending on the density of the plasma investigated, is split in two. One beam passes through the plasma, the other through a reference path. They are then made to interfere. The fringe shift in the interference pattern is a measure of the refraction in the plasma. In a more recent method the beam of a laser is passed through the plasma and back again. Interference of the refracted beam in the laser modulates the laser output, which is monitored by a photomultiplier.

Radiation emerging from the plasma in a direction different from the incident wave is said to be scattered. If the plasma were completely uniform there would be no scattered radiation, but any deviation from uniformity such as thermal fluctuations, density variations due to instabilities or mass motion will result in scattering. Suppose the fluctuations in the plasma are fourier analysed into waves with wave number $\mathbf{k}_p$ frequency ω_p. If the incident wave has values $\mathbf{k}_1$, ω_1

the scattered waves will have $\mathbf{k}_2$, ω_2 given by

$$\mathbf{k}_2 = \mathbf{k}_1 \pm \mathbf{k}_p \quad \omega_2 = \omega_1 \pm \omega_p.$$

The dimension $1/k_p$ is called the scattering length in the plasma. When this is much smaller than Debeye length, λ_D, the electrons scatter independently. The frequency spectrum of scattered waves has a width given by the electron velocity distribution. However if $\lambda_D k_p \ll 1$ cooperative effects occur in which the ion motion influences the scattered wave whose bandwidth then gives the ion temperature. The theory of such incoherent scatter was given in connexion with radar backscatter from the ionosphere by Fejer, Dougherty and Farley, and Salpeter in 1961. Incident wave frequencies for this work were about 40 MHz. For laboratory applications at high electron densities optical frequencies are used. Until the advent of the laser no light source was powerful enough to observe this effect. However, the necessary techniques are now developed and laser scattering promises to be a very powerful diagnostic. As well as giving information on ion and electron temperatures it could be used to analyse the structure of short wave-length wave-like instabilities and to observe bulk motions of the plasma which give measurable Doppler shifts. The main difficulty is to eliminate interference from the incident wave.

Not all of the techniques described can be used in a particular case. Plasmas of interest range from the very hot dense plasmas of the laboratory pinch devices, and stellar surfaces to the diffuse weakly ionized plasmas of interstellar space. The technique must very often be adapted to the task in hand. The following bibliography contains more details of individual techniques and includes those relatively new ideas which will be the basis of future developments.

Bibliography

AITKEN K. L. (1967) in *Encyclopaedic Dictionary of Physics* (J. Thewlis Ed.), Suppl. Vol. 2, 47, Oxford: Pergamon Press.

BOURDEAU R. E. (1963) Ionospheric research from space vehicles, *Space Sci. Rev.* **1**, 683, 719.

BROWN S. C. (1962) in *Encyclopaedic Dictionary of Physics* (J. Thewlis Ed.), **5**, 528, Oxford: Pergamon Press.

DOUGHERTY J. P. and FARLEY D. T. (1960) A theory of incoherend scattering of radio waves by a plasma, *Proc. Roy. Soc.* **A 259**, 79.

EMELEUS K. G. (1962, 1967) in *Encyclopaedic Dictionary of Physics* (J. Thewlis Ed.), **5**, 529, Suppl. Vol. 2, 249, Oxford: Pergamon Press.

HEALD M. A. and WHARTON C. B. (1965) *Plasma Diagnostics with Microwaves*, New York: Wiley.

HUDDLESTONE R. H. and LEONARD S. L. (1965) *Plasma Diagnostic Techniques*, New York: Academic Press.

KONSTANTINOV B. R. (1963) *Plasma Diagnostics*, State Publishing House of Literature of Atomic Science and Technology, Moscow.

KRONAST B., RÖHR H., GLOCK E., ZWICKER H. and FÜNFER E. (1966) Measurements of ion and electron temperature in θ-pinch plasma by forward scattering *Phys. Rev. Letts.*, **16**, 1082.

REED T. B. (1966) in *Encyclopaedic Dictionary of Physics* (J. Thewlis Ed.), Suppl. Vol. 1, 242, Oxford: Pergamon Press.

ROBSON A. E. (1962) in *Encyclopaedic Dictionary of Physics* (J. Thewlis Ed.), **7**, 326, Oxford: Pergamon Press.

THOMPSON W. B. (1962, 1967) in *Encyclopaedic Dictionary of Physics* (J. Thewlis Ed.), **5**, 528, Suppl. Vol. 2, 252, Oxford: Pergamon Press.

TSUKISHIMA T. and MCLANE C. K. (1966) Correlation of density fluctuations and diffusion in a plasma, *Phys. Rev. Letts.*, **17**, 900.

J. HUGILL

PLASTIC ANISOTROPY. *1. Introduction.* In a single crystal, variation of the elastic constants with crystallographic direction gives rise to elastic anisotropy (Mason 1961). When deformation extends into the plastic region, the variation of mechanical properties, defined by stress-strain relationships, with direction is known as plastic anisotropy. Plastic anisotropy in a single crystal is due to the limited number of slip and twin systems by which the crystal can deform. The plastic behaviour of a crystal will thus be governed by the disposition of these shear systems relative to the stress axes.

Polycrystalline materials with preferred orientation also exhibit plastic anisotropy arising from the distribution of shear systems in the individual crystals relative to the stress axes. There are other sources of anisotropic behaviour in wrought polycrystalline materials, e.g. a directional microstructure such as the alignment of inclusions and second phase particles with the working direction gives rise to "mechanical fibering"—this is concerned mostly with fracture characteristics. Also reversal of the stress system after moderate plastic deformation often results in a reduction in the flow stress, referred to as the Bauschinger effect. In this article, the main emphasis is on effects arising from the disposition of the shear systems. In favourable cases, this can result in an enhancement of the strength of a material, an effect which has been termed "texture hardening". Besides possible improvements in the mechanical strength of a finished product, advantage can sometimes be taken of plastic anisotropy to increase the limits of formability during processing of metal parts, and this latter aspect has stimulated much interest in the subject in recent years.

2. Assessment of anisotropy. Plastic anisotropy can be revealed by observing the stress-strain characteristics of test pieces cut at various angles to reference directions when the test pieces are subjected to a uniaxial stress. This procedure has some advantages in that the tests are easy to perform under reproducible conditions, and that the relationships between properties and orientation are analysed more readily than when the applied stress is multiaxial. However, the information obtained from uniaxial tests is limited and cannot, in general, be extrapolated to predict the performance of the material under more complex stress conditions.

Much of the interest in plastic anisotropy is concerned with metal in the form of flat sheet or tubes. In the former case variation in the plane of the sheet is readily assessed by tensile tests carried out in various directions relative to the rolling direction—this is referred to as planar anisotropy. It is difficult to make direct strength measurements in the direction normal to the sheet surface and the properties in this direction are often assessed in uniaxial tensile tests by measurements of the ratio of the strain in the width direction to the strain in the thickness direction of a test piece. The strain-ratio, R is given by

$$R = \frac{\ln (W_0/W)}{\ln (T_0/T)} = \frac{\ln (W_0/W)}{\left(\ln \dfrac{WL}{W_0 L_0}\right)} \tag{1}$$

W_0, L_0, T_0 are original dimensions of width, length and thickness, and W, L, T are the corresponding dimensions after an amount of strain which does not exceed the uniform elongation. It is assumed that no change in volume occurs. The average strain-ratio, $\bar{R}$, for tensile tests carried out in several directions in the plane of the sheet is a measure of normal anisotropy, i.e. the difference between the average properties in the plane of the sheet and the property in the direction normal to the sheet surface.

Two types of biaxial test are conventionally carried out on metal sheet. A balanced biaxial tensile stress state occurs at the pole of a disk subjected to bulging by hydraulic pressure. Alternatively, plane-strain tests, in which there is zero strain parallel to one principal direction, can be made in tension or compression.

Uniaxial tests, both in tension and compression, and the above biaxial tests give a fairly complete picture of the yield behaviour and subsequent plastic deformation of metal sheet.

It is easier to obtain a wide range of plane stress states in tubes than in flat sheet. These can be achieved either by axial loading combined with internal or external pressure on the walls of the tube, or by combined axial tension and torsion.

Another method of assessment of plastic anisotropy which can be made on any flat surface involves the use of the Knoop hardness indenter. The indenter is an elongated diamond prism and the length of the long diagonal of the indentation varies with orientation. Most reports of the application of this method refer to single crystals, although polycrystalline specimens have also been examined.

Anisotropy is also manifest in many metal forming operations in which the stress situation is complex. A well-known example is concerned with the production of cups by the deep-drawing of circular blanks. When the metal sheet has planar anisotropy, the rim of the cup is wavy, the undulations being known as ears. Standard earing tests have been proposed to assess directionality in sheet material.

3. Analysis based on crystallographic shear mechanisms. The relationship between crystal orientation and plastic anisotropy is usually approached by consideration of the plastic behaviour of single crystals. When a uniaxial tensile stress, σ, is applied to a crystal, the crystal starts to deform plastically when the applied stress reaches a value X, the yield stress in the x direction (Fig. 1.). At this point, the resolved shear stress on the most favourably oriented slip system is τ_0, where

$$\tau_0 = X \cos \Phi_x \cos \lambda_x \quad \text{(the \textit{Schmid law})} \tag{2}$$

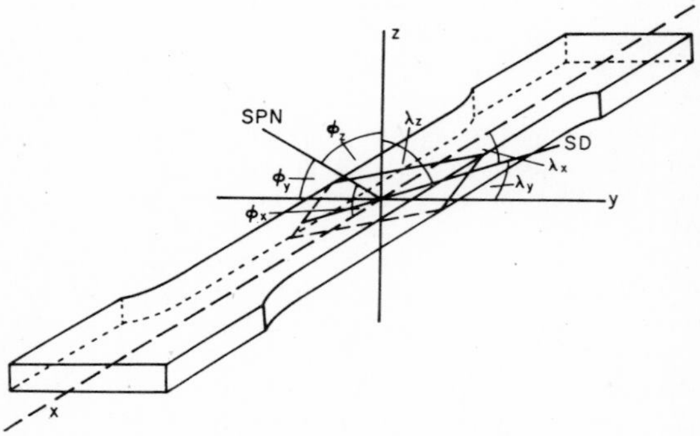

Fig. 1. Flat tensile test piece showing the angles made by the slip plane normal and slip direction with the orthogonal directions corresponding to length, width and thickness.

Φ_x and λ_x being respectively the angles between the slip plane normal and the stress axis, and between the slip direction and the stress axis.

The Schmid orientation factor is often expressed as m, where

$$m = \frac{1}{\cos \Phi \cos \lambda}. \tag{3}$$

From the equivalent work concept

$$dw = \sigma d\varepsilon = \tau\, d\gamma \tag{4}$$

it follows that the tensile strain in the x direction, ε_x, is related to the shear strain, γ, by the same orientation factor m, i.e.

$$\varepsilon_x = \gamma/_m. \tag{5}$$

It also follows that strains can be resolved in any direction, so that the strain ratio, R, mentioned earlier, can be evaluated. Referring to Fig. 1,

$$R = \frac{\text{Width strain}}{\text{Thickness strain}} = \frac{\varepsilon_y}{\varepsilon_z} = \frac{\cos \Phi_y \cos \lambda_y}{\cos \Phi_z \cos \lambda_z} \tag{6}$$

The range of possible values of the orientation factor, m, depends on the number of available slip systems, and high values of m occur only when the slip systems are limited in number as, for example, in hexagonal metals (Fig. 2(a)). For face-centred cubic metals slipping on {111} planes in ⟨110⟩ directions, m varies between a minimum of 2 when the stress axis is close to ⟨123⟩ and a maximum of 3·7 when the stress is applied parallel to ⟨111⟩ (Fig. 2(b)).

(a)

(b)

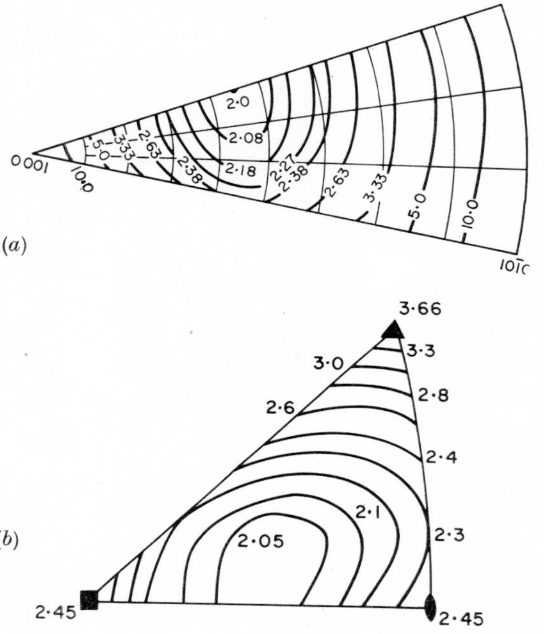

Fig. 2. Values of $m = \dfrac{1}{\cos \Phi \cos \lambda}$ *for slip system on which the shear stress is a maximum (a) Hexagonal crystal (0001)* $\langle 11\bar{2}0 \rangle$ *slip; (b) Cubic crystal {111} ⟨110⟩ or {110} ⟨111⟩* slip.

In polycrystalline aggregates, the situation is complicated by the constraints imposed by neighbouring grains, and it is necessary to make assumptions regarding the uniformity of deformation throughout the aggregate in order to determine an average value of the orientation factor, $\bar{m}$. In a randomly-oriented aggregate of f.c.c. crystals, the value of $\bar{m}$ calculated on the basis that each crystal deforms independently on its system of maximum resolved shear stress is 2·24, the so-called Sachs average. This is the lower limit of $\bar{m}$, but the actual value of $\bar{m}$ must be considerably higher than this value because less favourably oriented systems are forced to operate in order to preserve continuity at the grain boundaries.

In the original theory of G. I. Taylor (1938), it was assumed that each crystal undergoes the same homogeneous strain as the bulk material. This is achieved by the operation of at least five independent slip systems in each crystal. Equations 2, 4 and 5 are thus modified for the polyslip condition to

$$\tau_0 = \frac{1}{M}\,\frac{dw}{d\varepsilon} = \frac{dw}{\Sigma d\gamma} \qquad (7)$$

where M is a generalized Schmid factor relating the applied stress for flow to the basic shear stress for slip.

The choice of the five active systems was based on the "minimum shear" principle, i.e. to produce a given shape change, the shear stress in equation 7 is raised to the critical value for slip by choosing those systems for which $\Sigma d\gamma$ is a minimum. On this basis, the Taylor average, $\bar{M}$, for randomly oriented f.c.c. crystals was calculated to be 3.1.

Bishop and Hill (1951) later postulated a more general theory which did not require the homogeneous strain assumption. The basis of their analysis is the proposal that the actual work done in plastic deformation of a polycrystal is equal to that which would be done if all the crystal underwent the same strain. The operative slip systems were chosen on the "maximum work" principle, i.e. in equation 5, dw is maximized. Both methods finally give equivalent results, but the method of Bishop and Hill is simpler to apply.

From a knowledge of the average orientation factor, $\bar{M}$, it is possible to derive the tensile stress-tensile strain curve for a polycrystalline material from the shear stress-shear strain curve of a single crystal of the same material. The conversion is effected by multiplying the stress axis of the single crystal curve by $\bar{M}$ and dividing the strain axis by the same number.

In order to apply the Taylor and Bishop and Hill theories in this way to derive stress-strain curves, it is implicitly assumed that there is no significant orientation dependence resulting from different rates of work hardening. Experiments with f.c.c. metals subjected to uniaxial stresses, and more recently, to biaxial stresses, have tended to confirm this assumption and the present opinion is that the Bishop and Hill theory is a satisfactory basis for analysis of plastic behaviour of f.c.c. metals.

The above discussion has been concerned with metals deforming by the slip mechanism. In some crystal structures, the alternative shear mechanism of deformation twinning must be taken into account. At normal temperatures and strain rates, twinning is more frequent in hexagonal metals than in cubic metals. An important distinction must be made between slip and twinning, since the latter mode is strongly dependent on whether the applied stress is tensile or compressive in nature. In hexagonal metals with axial rations, c/a, greater than $\sqrt{3}$, twinning on $\{10\bar{1}2\}$ planes is possible when an applied stress nearly parallel to the hexagon axis [0001], is compressive, but not when the stress is tensile. The converse is true for metals with c/a less than $\sqrt{3}$. The mechanical properties of polycrystalline materials are dependent on many other factors apart from orientation, and the combined influence of grain size and orientation on mechanical strength has been assessed. An equation relating the flow strength with these variables has been proposed:

$$\sigma_F = \overline{M}(\tau_0 + \tfrac{1}{2}\overline{M}\tau_c r^{1/2} d^{-1/2}) \quad (8)$$

in which σ_F is the flow strength, d is half the grain diameter, τ_c is a measure of dislocation locking, and r depends on the dislocation arrangement. The large dependence of flow strength on grain size for some hexagonal metals is attributed to large values of $\overline{M}$.

4. *Macroscopic theory of plastic anisotropy.* In this approach, the material undergoing deformation is regarded as an ideal plastic continuum, and no direct account is taken of the crystallographic character of deformation. The mathematical theory was developed independently by Dorn (1949) and Hill (1949), and is based on a criterion of yielding which is a modification of the von Mises law for isotropic materials. The modified yield criterion is

$$F(\sigma_y - \sigma_z)^2 + G(\sigma_z - \sigma_x)^2 + H(\sigma_x - \sigma_y)^2 + 2L\tau_{yz}^2 + 2M\tau_{zx}^2 + 2N\tau_{xy}^2 = 1 \quad (9)$$

F, G, H, L, M, N are parameters characteristic of the state of anisotropy.
$\sigma_x\ \sigma_y\ \sigma_z$ are principal stresses referred to orthogonal axes.
$\tau_{yz}\ \tau_{zx}\ \tau_{xy}$ are shear stresses referred to orthogonal axes.
The stress-strain relationships are

$$\left.\begin{aligned} d\varepsilon_x &= d\lambda[H(\sigma_x - \sigma_y) + G(\sigma_x - \sigma_z)] & d\gamma_{yz} &= d\lambda L\tau_{yz} \\ d\varepsilon_y &= d\lambda[F(\sigma_y - \sigma_z) + H(\sigma_y - \sigma_x)] & d\gamma_{zx} &= d\lambda M\tau_{zx} \\ d\varepsilon_z &= d\lambda[G(\sigma_z - \sigma_x) + F(\sigma_z - \sigma_y)] & d\gamma_{xy} &= d\lambda N\tau_{xy} \end{aligned}\right\} \quad (10)$$

Based on equations 9 and 10, various relationships can be derived which describe the anisotropic characteristics of plastically-deforming materials, e.g. for thin sheet in which the applied stress in the direction perpendicular to the sheet surface is negligible, and with no planar anisotropy, the yield criterion is

$$\sigma_x^2 + \sigma_y^2 - \sigma_x\sigma_y\left(\frac{2R}{R+1}\right) = X^2 = Y^2 \quad (11)$$

The yield locus for different values of R and for different ratios σ_x/σ_y is shown in Fig. 3, and from this figure, it is immediately apparent that texture hardening is effective, as indicated by high values of σ_x/X, for materials having high values of R when the stress system approaches balanced biaxial tension ($\sigma_x = \sigma_y$) and also when the stress state approximates to one of plane strain $\left(\sigma_x/\sigma_y = \dfrac{R}{R+1}\right)$.

The initial assumptions made in the macroscopic theory of plastic anisotropy place severe limits on its general application, e.g. in the present macroscopic theories it is assumed that the stress-strain relationships are similar in tension and compression, although this is certainly not true when mechanical twinning is an important mode of deformation.

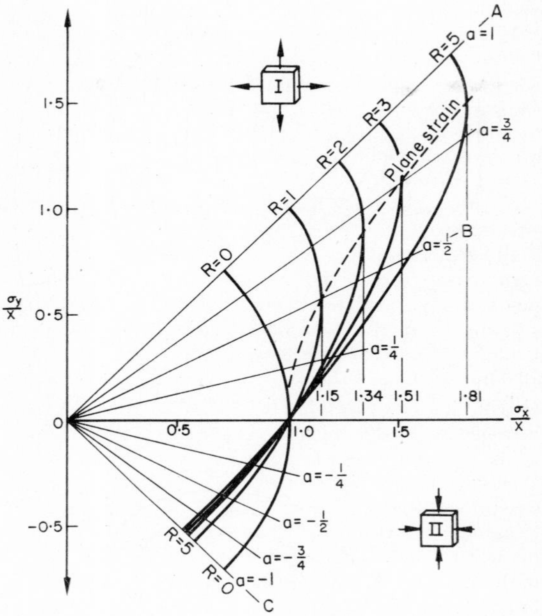

Fig. 3. Yield loci for sheets with textures that are rotationally symmetric about the thickness direction, z. α is the ratio of the applied biaxial stresses, σ_y/σ_x. (After Hosford and Backofen (1962)—Courtesy Syracuse University Press.)

5. *Plastic instability.* Ductile metals tested in uniaxial tension fail when the rate of work hardening is insufficient to compensate for the reduction in area. The instability criterion is expressed mathematically by the relation

$$\frac{d\sigma}{d\varepsilon} = \sigma. \quad (12)$$

It has been assumed that the localized neck formed under these conditions is parallel to a direction of zero strain, but the angle at which the neck forms predicted on this basis is not in good agreement with measured angles. For a sheet subjected to biaxial plane stress, and when both principal strains are tensile everywhere in the plane of the sheet, localized necking cannot occur, and another form of instability called diffuse necking occurs.

Predictions of instability strains from considerations of the disposition of slip systems relative to the stress axis are not possible because present theories of work hardening are not sufficiently quantitative.

Attempts have been made to predict the strain at instability for different plane stress systems and for different types of anisotropy characterized by R values, using the concepts of the macroscopic theory of plastic anisotropy. These analyses are subject to the limitations mentioned at the end of the previous section.

It is in the area concerned with the inter-relation between preferred orientation, grain size, work-hardening behaviour, ductility and fracture characteristics that a great deal more experimental information is required before a satisfactory theoretical interpretation can be contemplated.

6. Magnitude of anisotropic effects. The largest effects of plastic anisotropy occur in metals with limited shear systems. A spectacular example of the high strength attained when normal deformation modes are not available is provided by beryllium single crystals compressed parallel to the hexagonal axis. Under these conditions, the resolved shear stress on all slip systems is zero ($m = \infty$), and twinning on $\{10\bar{1}2\}$ is not possible as this would result in an increase in length in the compression direction. At 77°K, stresses of about 250 tons/in^2, can be attained with negligible plastic deformation, and eventually the crystal shatters explosively.

A high degree of plastic anisotropy is encountered in polycrystalline specimens of hexagonal metals with low axial ratio when they are in the form of sheet or tube with basal planes parallel to the surface of the sheet or tube. In this condition, the material has a high resistance to thinning, resulting in the highest measured values of the strain ratio, $\bar{R}$, in tensile tests. Values of $\bar{R}$ of 4–5 are typical of titanium sheet. Such sheet is strongly "texture hardened" when subjected to biaxial tensile stresses in the plane of the sheet. In contrast with this, another hexagonal metal, zinc, has the lowest measured value of $\bar{R}$ for sheet material —about 0·2.

The range of values of $\bar{R}$ encountered in cubic metals is much less. Face-centred cubic metals have $\bar{R}$ ranging from about 0·3 for cube-texture (100) [001], to a maximum in textures so far attained of just over 1·0. Higher values are sometimes obtained in body-centred cubic metals. Values of $\bar{R}$ in the range 1·4–1·8 obtained in aluminium-killed low carbon steel are associated with significant improvements in deep drawing performance compared with rimming steel, which has $\bar{R}$ values between 1·0 and 1·4. The highest values of $\bar{R}$ in steels are associated with texture components with {111} parallel to the surface, while crystals with {100} parallel to the surface have a strongly depressing effect on $\bar{R}$.

Planar anisotropy is particularly marked in the cube texture, a common recrystallization texture in f.c.c. metals. R varies from 1·0 at 0° and 90° to the rolling direction to zero at 45°, while the ductility in tensile tests varies from about 20 per cent elongation at 0° and 90°, to about 70 per cent at 45°. It is often difficult to produce rolled and annealed metal sheet without planar anisotropy, and some of the consequences of such anisotropy are discussed in the next section.

Texture hardening effects are often more pronounced when the applied stress is multiaxial. In f.c.c. metals, wires with ⟨111⟩ orientation have maximum strength when tested in uniaxial tension. This strength is 50 per cent higher than the weakest direction ⟨100⟩ and is 20 per cent greater than the strength of a randomly oriented f.c.c. wire (Fig. 4.). When tested under a biaxial stress system giving plane strain, the ratio between highest and lowest strengths is approximately two to one.

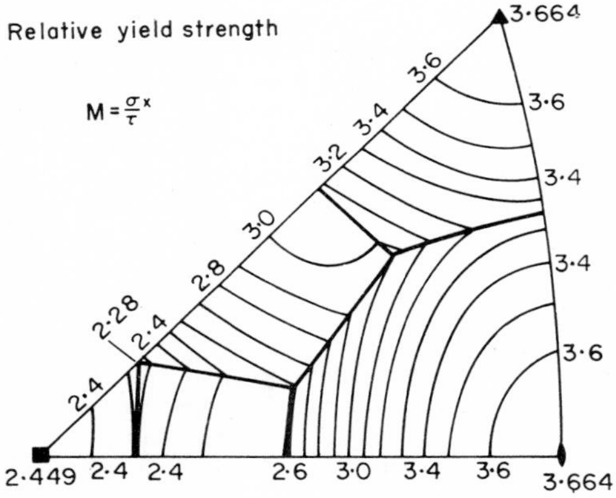

Fig. 4. Values of $\bar{M}$ for axially symmetric flow in cubic crystals slipping on {111} ⟨110⟩ or {110} ⟨111⟩. The figures also represent the relative yield strengths in different directions. (After Hosford and Backofen (1962)—Courtesy Syracuse University Press.)

7. Role of anisotropy in metal forming operations. Metal forming operations are conveniently divided into two groups, drawing and stretching, although in practice, combinations of the two processes are usually encountered. The drawing process is often idealized for research and evaluation purposes by using a circular blank which is pressed through a circular die by a punch to form a cup. When the sheet has planar anisotropy, ears are formed at the top of the cup. The formation of ears in deep-drawn components is generally regarded as a nuisance in production, and considerable efforts have been made to produce sheet which is not prone to this effect.

The most economical manner to achieve this is usually by balancing textures which have different earing tendencies, e.g. in f.c.c. metals, the cube recrystallization textures gives cups with ear peaks at 0 and 90° to the rolling direction, while the "retained rolling" recrystallization texture gives ears at 45° to the rolling direction. When the recrystallized sheet contains about 20 per cent cube texture, the remainder being "retained rolling" texture, approximately flat-topped cups are produced. Analyses of earing behaviour based on crystallographic considerations and also on the macroscopic theory have been made.

The suitability of sheet for deep-drawing can be assessed by the maximum size of blank that can be drawn without failure. This is often expressed as the limiting drawing ratio, i.e. the ratio of maximum blank size to punch diameter. While tool geometry and lubrication conditions are important process factors, it has been demonstrated that so far as the material is concerned, plastic anisotropy is the dominant factor. In order to understand the role of anisotropy, the stress systems operative in different regions of the partly-drawn cup must be examined. In the flange or drawing zone, the stress system is radial tension combined with circumferential compression, and thinning of the sheet is not essential to deformation. At the potential failure site, over the punch profile radius, the stress system is biaxial tension, and the metal cannot deform here without a reduction in thickness. Thus when the metal has a high strength in the through-thickness direction, it will perform well in deep drawing. This condition is associated with a high value of $\bar{R}$ and excellent correlation has been obtained between $\bar{R}$ and the limiting drawing ratio for a wide range of sheet materials.

The influence of plastic anisotropy on formability of metal subjected to biaxial tensile stresses is less well understood at present. Stretching may be effected by a metal tool or by hydraulic bulging and performance is often judged by the depth of bulge attained before fracture develops. It is clear that a high and sustained rate of work hardening is beneficial, but the optimum anisotropic characteristics have not been clearly defined.

Bibliography

BISHOP J. F. W. and HILL R. (1951) *Phil. Mag.* **42**, 414 and 1298.

DORN J. E. (1949) *J. App. Phys.* **20**, 15.

HILL R. (1949) *Proc. Roy. Soc.* (A) **198**, 428.

HOSFORD W. F. and BACKOFEN W. A. (1962) Strength and Plasticity of Textured Metals, *Proc. 9th Sagamore Conference*, 1962, New York: Syracuse University Press.

MASON W. P. (1961) in *Encyclopaedic Dictionary of Physics* (J. Thewlis Ed.) **1**, 193, Oxford: Pergamon Press.

TAYLOR G. I. (1938) *J. Inst. Met.* **62**, 307.

WILSON D. V. (1966) Plastic anisotropy in sheet metals, *J. Inst. Met.* **94**, 84.

W. T. ROBERTS

PLASTIC SPRINGS. Plastic materials are unique in a number of ways and as knowledge of the fundamental behaviour of these materials increases greater use is being made of them for engineering applications.

One of the most important characteristics is their behaviour when strained, the stress-strain relationship being viscoelastic, and not linear elastic, that is, they are load, time and temperature dependent. Where stress is proportional to strain for metallic materials and is independent of rate of strain, the stress-strain relationship of plastic materials is represented by a curve from the origin and is dependent upon the rate of strain, an increase in rate giving an increase in stiffness. The modulus is lower than that for metals and there is a greater deflexion for a given load, indicating high strain energy characteristics shown by a greater area under the stress-strain curve.

This relationship enables plastic materials to be utilized in some ways to a greater advantage than other materials. One particular sphere in which these special properties are utilized very effectively is in energy storage devices such as leaf, helical coil and conical disc springs.

There are further associated advantages that can be obtained from these materials. They are essentially non-corrodible and non-corrosive. The fatigue properties are superior with a considerable improvement in notch sensitivity over a spring steel, and better vibration damping is obtained as a result of the self-damping effects of the viscous property. They are also non-magnetic and they have low electrical and thermal conductivity. The limitations can be classified as lower stress limits, the effects of creep, and temperature sensitivity.

The choice of the most suitable plastic material for any spring design is dependent upon the load system and environmental conditions; therefore each application should be treated individually. Detailed information is always required for spring design whatever the material, but to obtain the best possible use of the material characteristics when considering plastics, more extensive information is required.

The length of time the load will be applied and the number of applications of the load during the spring life should be specified. In addition to the loading conditions, the working temperature range, relative humidity, proximity to chemical substances and any electrical environmental conditions should also be known.

The majority of plastics materials have sufficient resilience to be considered for spring applications in one form or another, but there are a few materials which are outstanding. These are in both of the main plastic materials categories of thermosetting and thermoplastic.

For the more stringent applications it has been found that glass fibre reinforced thermosetting plastics

Fig. 1. Comparison of a reinforced plastic spring material with a spring steel.

	Ultimate flexural strength lbf/in.^{2} × 1000	Flexural modulus lbf/in.^{2} × 10^6	Volume efficiency in lbf/in.^{3}	Flexural fatigue endurance limit lbf/in.^{2} × 1000
Flat spring steel SAE 1060	180	30	60	33
Undirectional reinforced plastic 65% glass 35% epoxide resin	113	47	151	29

can be used very successfully, helical coil and conical disc springs having been produced as well as simple leaf springs.

The most successful composite system of this type is a glass fibre reinforced epoxide resin, where the main advantage over a conventional material such as steel is its high energy absorption capacity per unit volume. This becomes particularly significant where the usable strength of the material is approaching that of steel since the plastic material having a lower modulus of elasticity can deflect further without permanent set.

Ultimate flexural strengths of the order of 110,000 lbf/in^2 and greater have been achieved from glass fibre reinforced epoxide resin laminates. Typical data for leaf springs illustrating this effectively are given in Fig. 1. Since the volume efficiency is a function of the square of the ultimate fibre stress decided by the modulus, i.e. $\propto f^2/E$, the glass fibre reinforced spring is clearly advantageous, being about two and a half times as efficient as the steel spring. Figure 2 shows the variation in f^2/E with temperature for a glass fibre reinforced epoxide resin spring. This indicates that for this particular material system the temperature sensitivity is not a serious problem over the range −40°C–+60°C which covers the operating temperature range of many springs.

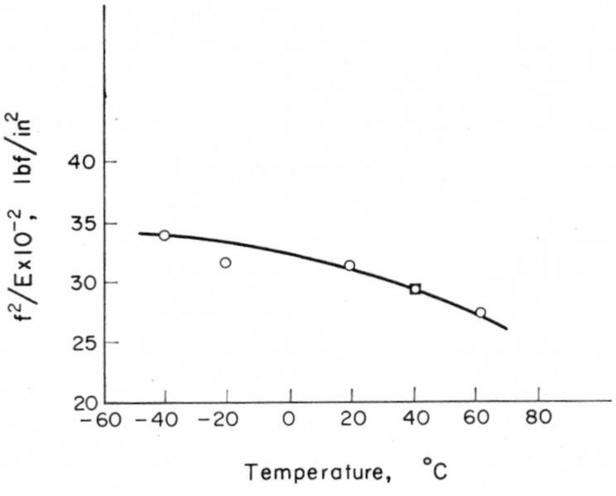

Fig. 2. Variation of f^2/E with temperature glass fabric reinforced epoxide resin laminate. ⊙ *test results;* □ *preconditioned at 40°C, 95 per cent RH for 24 hrs.*

Helical coil springs have been manufactured from a glass fibre reinforced epoxide system having a torsional modulus of rigidity of the order of 1·0 × 10^6 lbf/in^2. Whilst this is poor in comparison with the modulus for spring steel, 11 × 10^6 lbf/in^2, it is adequate to provide many special purpose springs. Some typical results for helical coil springs of approximately 1-in. coil diameter and 3/16 in. wire diameter are shown in Fig. 3.

Glass reinforced epoxide resin laminate has been successfully used for conical disc springs of the Belleville washer type. This laminate system is particularly useful for low load applications where a

Fig. 3. Some test figures on reinforced epoxide helical springs.

Initial		Final*		Energy
Force Lbf.	Energy in. Lbf	Force Lbf	Energy in. Lbf	Available %
28·2	11·0	14·9	3·64	33
36·1	5·6	25·4	2·67	48
43·4	11·72	14·4	1·08	92
40·7	12·3	37·4	5·24	42·6

* Inch points available in static load measurements after being compressed to solid length for a minimum of 13 days at 135°F.

high fatigue resistance is necessary, to a large extent combating the problem of cracking on snap-through. This advantage is very useful where it is necessary to work in a corrosive environment or near electrical, electronic or magnetic components. A high degree of reliability and reproducibility can be achieved. Springs of the Belleville washer type manufactured from glass reinforced epoxide resin have load deflexion curves as shown in Fig. 4 and show that such springs under certain conditions can snap through, i.e. the core will invert, providing a useful source of potential energy.

The superior thermoplastic materials for spring applications are polyacetal, polyamide (nylon), polypropylene, polysulphone and polyphenylene

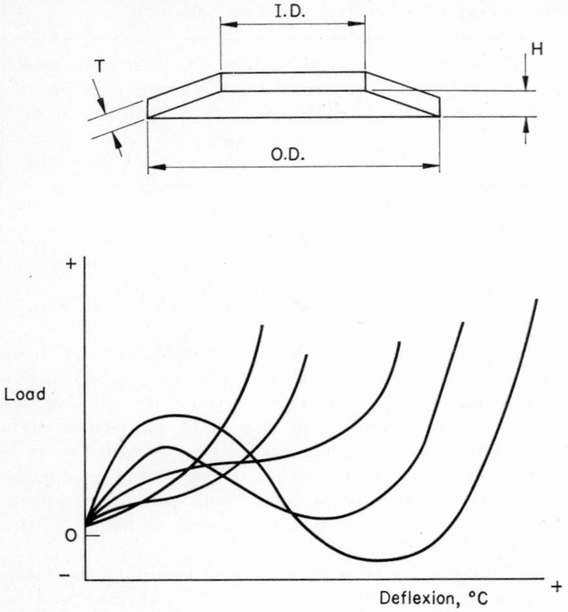

Fig. 4. Load-deflexion curves for conical (Belleville type) springs.

oxide. Polyacetal has the highest permissible fatigue stress, which is a highly desirable characteristic in spring design. These materials are generally better suited for relatively light load spring applications. They have the advantage over the glass reinforced thermosetting materials of being able to be utilized for complex spring forms as well as the more conventional shapes. This is possible because they can be produced by the injection moulding process, which is also a mass production method giving further economic advantages.

The principles used to design plastic springs have been essentially based on the classical linear elastic formulae, sometimes modified empirically, because of the lack of stress-strain-time-temperature data on the basic materials. Although not applicable to viscoelastic materials, linear elastic formulae provide a good approximate solution in certain cases. These are where the stress-strain relationship of the plastics material does not deviate markedly from the linear over the envisaged strain range and the property-temperature spectrum is not critical.

These characteristics are generally exhibited by glass fabric reinforced epoxide resin laminates and confidence in the use of linear elastic, or empirically modified linear elastic formulae, for the design of springs has been proved in practice.

Bending stress. An example of this is the type of flat cantilever spring with clamped ends shown in Fig. 5 where the basic equations are derived thus:

$$M_B = Pl - Pl/2 + W\delta = Pl/2 + W\delta \qquad (1)$$

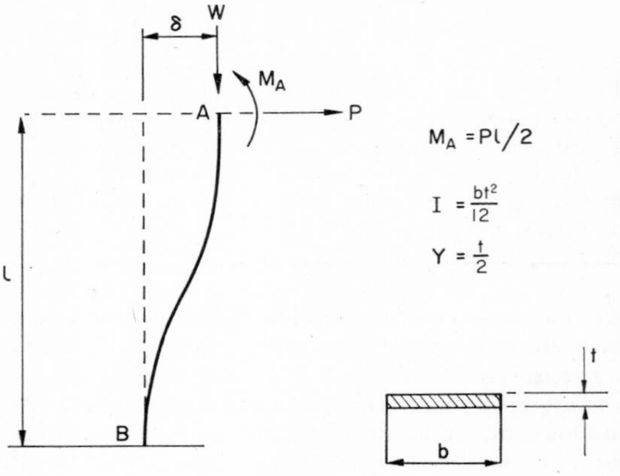

Fig. 5.

Where M_B = Moment at end B.
P = Load in transverse direction.
W = Load in axial direction.
l = Length of Spring.
δ = Deflexion at end A.

$$\delta = Pl^3/3EI - Ml^2/2EI, \qquad (2)$$

where E = flexural modulus of elasticity.
I = moment of inertia of section;

then

$$P = 12EI\delta/l^3 = Ebt^3\delta/l^3. \qquad (3)$$

thus

$$M_B = (6EI/l^2 + W)\delta, \qquad (4)$$

and bending stress $f_B = M_B y/I$.

Where $y = t/2$.

$\therefore\ f_B = (3Et/l^2 + 6W/bt^2)\delta$

for f_B minimum $\mathrm{d}f_B/\mathrm{d}t = 0$.

$\therefore\ \mathrm{d}fB/\mathrm{d}t = 3E\delta/l^2 - 12W6/bt^3 = 0.$

$$\therefore \text{ Optimum thickness } t = \sqrt[3]{\left(\frac{4Wl^2}{Eb}\right)}. \qquad (6)$$

Comparison with a steel spring may be made on the basis of equivalent stiffness where

$$\delta = Pl^3/Ebt^3 \text{ from (2) above.}$$

$\therefore\ \delta$ plastic $= \delta$ steel for equal stiffness.

$\therefore\ Et^3 = nE_s t_s^3,$

where n = number of leaves in the steel spring. Therefore the thickness of plastic laminate for equal stiffness is

$$t = t_s(nE_s/E)^{1/3}.$$

Column buckling. For column buckling thefollowing must hold:

$$W/A = 4\pi^2 E/(l/r)^2 = S_{CR}.$$

Where $r = t/\sqrt{(12)}$ for rectangular beam.

S_{CR} = critical stress.

$$\therefore \; W/A \leq \pi^2 E t^2/3l^2 = S_{CR}.$$

The design of springs utilizing the engineering thermoplastics such as polyacetal, whose properties are much more sensitive to temperature and load duration than the glass reinforced thermosetting plastics, is progressively being made easier by the publication of materials data relating stress-strain-time characteristics at specific temperature levels.

The design data, derived from the basic creep behaviour of the material, can be presented in a number of ways, from the isochronous stress-strain curve as presented diagramatically in Fig. 6, to the three-dimensional envelope diagram shown in Fig. 7. This type of presentation enables the designer

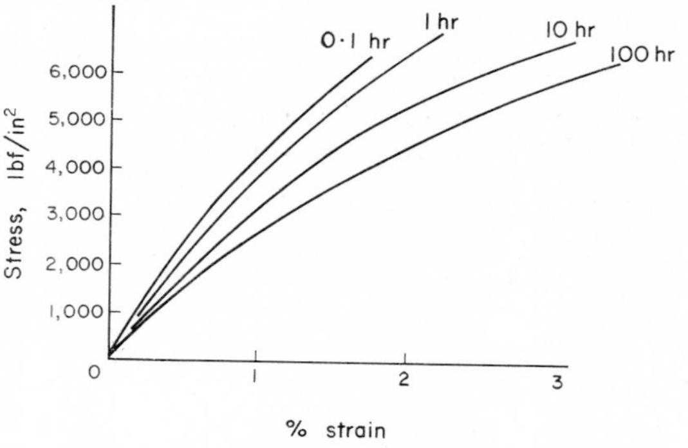

Fig. 6. Isochronous stress-strain curve for acetal resin normal strains, test temperature 23°C.

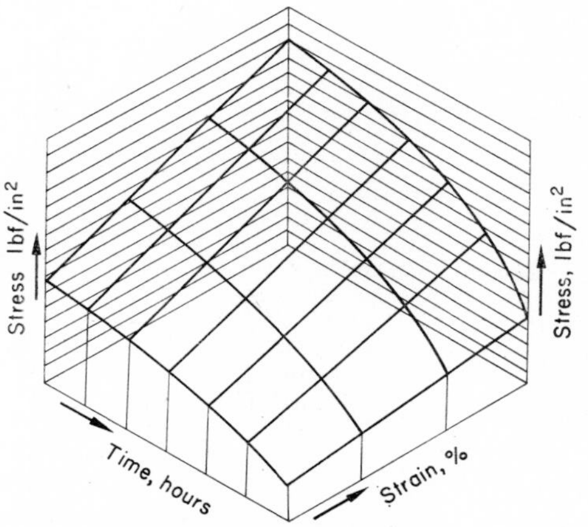

Fig. 7. Three-dimensional stress-strain-time diagram.

to calculate the unknown spring parameters with a greater degree of confidence. This may be exemplified by considering the simple cantilever beam shown in Fig. 8 where all design conditions are known, except the thickness of the beam and the deflexion to give the

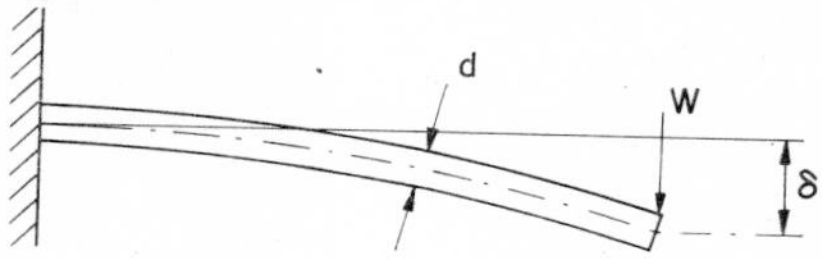

Fig. 8.

required load. Knowing that the selected material is, say, polyacetal, the apparent modulus of elasticity and the maximum permissible stress at the extremes of the working temperature range can be obtained from the material manufacturers' curve shown in Fig. 9. The data plotted relates to a strain level of

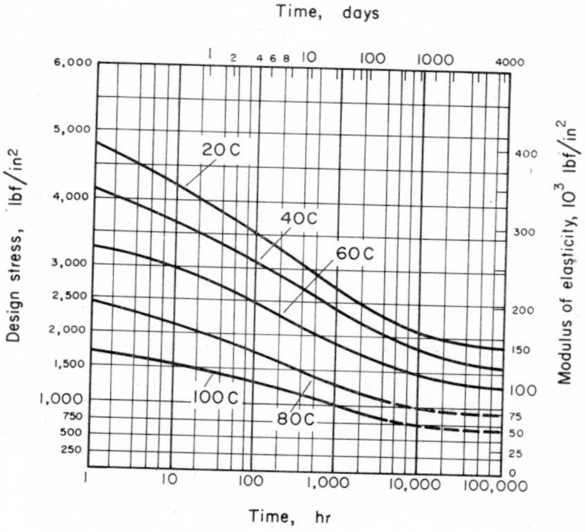

Fig. 9.

1·25 per cent, this being the strain limit above which creep becomes excessive. Once the stress-strain-time relationship is established to meet the design conditions, the unknown parameters, in this case the beam thickness and deflexion for a specified load, can be computed from simple bending theory:

$$\frac{f}{y} = \frac{M}{I} = \frac{E}{R},$$

where f = stress in beam at a distance y from the neutral axis of bending;

M = bending moment;

I = moment of inertia of section;

E = modulus of elasticity;

R = radius of curvature;

and the deflexion formula for a simple cantilever with concentrated end load,

$$\text{deflexion, } \delta = Wl^3/3EI,$$

where W = load;

l = length;

E = modulus of elasticity;

I = moment of inertia of section.

The above calculations are applicable for a beam of uniform cross-section along its length where the stress intensity varies and becomes a maximum at the built-in end. The ideal condition is constancy of stress along the length of the beam which can be achieved by tapering the cross-section. Such a form is easily obtained when using thermoplastic materials, the mould cavity being machined to the required taper. For the ideal condition a new deflexion formula must be derived from first principles to replace the one given above, which becomes invalid.

Where a spring is to operate under dynamic cycling conditions, which can induce complete stress reversal conditions in some cases, it is obviously desirable to cater for the fatigue effects at the design stage. This is, however, very difficult for plastics materials because of the dearth of fatigue endurance figures and the caution with which published information must be regarded due to current test methods and equipment.

The only satisfactory method of evaluating the design of a plastic spring for a given application is to conduct a test programme on a prototype device, simulating as closely as possible the expected service conditions. It is important that the tests should be realistic and great care must be taken in predicting long-term performance from short-term tests, this being particularly important where the spring is to be subjected to a complex stress system and/or dynamic conditions.

Bibliography

Austin C. and Berry J. P. (1966) Design with Plastics as Engineering Materials, *Rubber and Plastics Age*, **47**, (3), March.

E. I. Du Pont de Nemours & Co. (Inc.) *Design and Engineering Data—Plastics Materials.*

Heggernes I. A. (1959) *Product Engineering*, Jan.

Maier K. (1959) *Product Engineering*, August 17th.

Minnesota Mining and Manufacturing Co., Technical Data on "Scotchply".

Morris J. M. (1958) U.S.P. 2,829,881.

Plastics Springs (1956) National Bureau of Standards, *Technical News Bulletin*, Vol. 4, No. 5.

Reinhart F. W. and Newman S. B. (1956) *11th Annual Meeting of Reinforced Plastics Division*, S.P.I.

Reinhart F. W., Llone M. C., Horn L. and George D. A. (1958) U.S.P. 2,852,424.

Sirkin A. L. (1966) *The Design of a Spring Action Polypropylene Hinge*, SPE 22nd Annual Tech. Conf. Vol. 12, March.

Smith T. G. (1965) *The Design and Production of Reinforced Plastic Springs*, Conference on "Engineering Design Problems in Plastics", Jan.

Trainer T. M. (1960) *Product Engineering*, Sept. 12th.

Watson G. (1965) The Design of Beam Springs in Thermoplastic Materials, *Trans. Plast. Inst.*, **33**, No. 106, Aug.

E. Atack

POWDER DIFFRACTOMETRY. Logically the term powder diffractometry would cover all methods of investigating polycrystalline material by diffraction of radiation, and would thus include electron diffraction and neutron diffraction as well as photographic X-ray methods. In practice it is usually restricted to methods involving

(i) partial focusing,
(ii) counter detection of the diffracted radiation, and ordinarily
(iii) X rays.

A Geiger, proportional or scintillation counter may be used, depending on its suitability for the purpose in view. In the future, solid-state detectors might be used. The application of the apparatus is not limited to well-crystallized specimens, and radiation-damaged materials, cold-worked metals, polymers and even liquids may be studied.

The commonest arrangement of a powder diffractometer is shown in Fig. 1. The detector is mounted on a moving arm at a distance from the axis of rotation equal

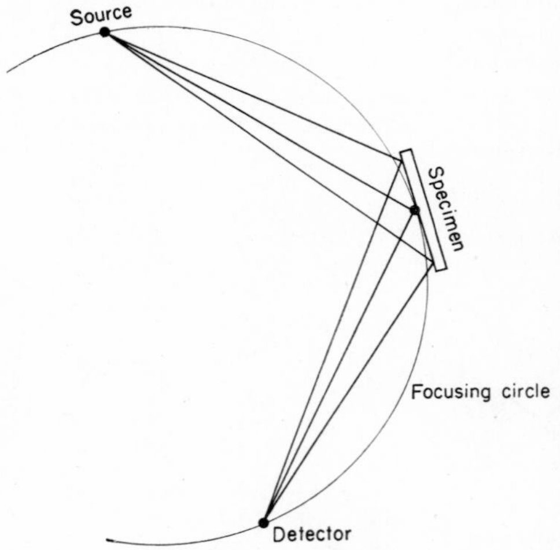

Fig. 1. The principle of a powder diffractometer. The source may be the X-ray tube focus, or the focus of a curved-crystal monochromator. The rays diffracted by the specimen come to an approximate focus (parafocus) at the detector, which is mounted to rotate about the same axis as the specimen.

to the distance from the source of radiation to the axis. The specimen, usually a flat cake of compacted powder, is mounted to rotate about the same axis, but at half the rate. Whatever the position of the detector, arm, therefore, the specimen is in a position to "reflect" radiation from the source to the detector. The 2 : 1 ratio in the angular positions of the specimen and detector is usually maintained by an accurate pair of gears, but a bisecting parallelogram or other device may be used. A considerable measure of focusing is attained in the equatorial plane, the plane of motion of the detector. If the specimen were curved to fit an imaginary circle, called the focusing circle, passing through the source, specimen and detector, all points on its surface would reflect radiation from the source to the detector. This follows from the well-known geometrical theorem that all angles in the same segment of a circle are equal. If Bragg's law is satisfied for one point on the focusing circle it is thus satisfied for all points. In general it is not practicable to curve the specimen appropriately, but a flat cake tangent to the focusing circle is a sufficient approximation for most applications. Methods of preparing flexible specimens and continuously altering their curvature to fit the focusing circle have been devised for special purposes.

Focusing in the axial direction as well as in the equatorial plane would require a specimen curved to fit the surface of a torus obtained by rotating the focusing circle about the chord joining the source to the detector, and the writer is not acquainted with any device capable of producing a continuously variable double curvature in the specimen. In routine applications flat specimens are used, and the divergence of the radiation in the equatorial plane and the axial direction is limited by slits chosen so that the aberrations produced by deviations from the exact focusing conditions are limited to tolerable amounts. The slits, of course, reduce the intensity of the radiation reaching the detector.

Since the ideal of a point source, doubly curved specimen, and point detector cannot be achieved in practice, actual instruments are designed to attain an acceptable compromise between high resolution

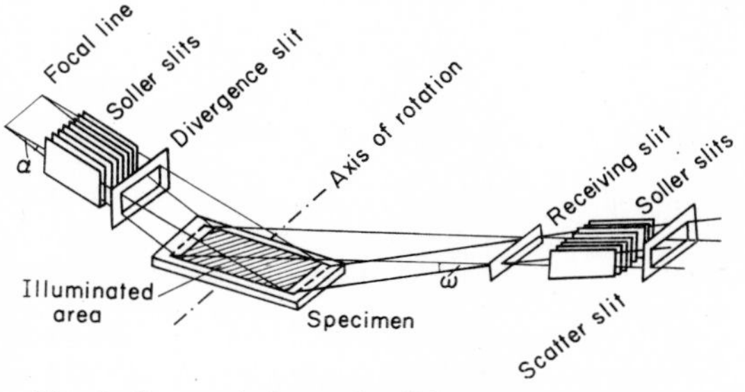

Fig. 2. One typical powder-diffractometer arrangement (diagrammatic, after W. Parrish). The angle of equatorial divergence is α, and of acceptance ω. Axial divergence is limited by the Soller slits.

and high intensity. One common arrangement is shown schematically in Fig. 2. The source is an X-ray tube with a "line" focus of the order of 0·1 cm in width and 1 cm long (different commercial tubes have dimensions varying by a factor of two either way) oriented so that the long dimension is parallel to the diffractometer axis and the short dimension is "seen" at grazing incidence (2–8°) by the specimen. The effective dimensions are thus of the order of 0·01 cm by 1 cm. Axial divergence, that is the angle made by the X rays with the equatorial plane, is limited to a degree or two by Soller slits (Soller 1924). These consist of a set of suitably spaced metal foils set parallel to the equatorial plane, and absorb any ray diverging from the equatorial plane by an angle greater than (spacing of foils)/(length of foils). Soller slits may be placed between the source and the specimen, or between the specimen and the detector, or in both positions. The last arrangement gives the best resolution; if only one set is used it is usually best to place it between the source and the specimen. Divergence in the equatorial plane is limited by a single slit, about 1° being suitable in the lower range of Bragg angles, increasing to 4° in the higher range. The illuminated area of the specimen is about 1 cm by 2 cm, the longer dimension being equatorial, and no ray deviates from the mean ray by more than a degree or two. A receiving slit, with dimensions about the same as the projected size of the focal spot, is placed at the position of approximate focus of the diffracted rays, and the actual detector is placed behind the receiving slit. An acceptance slit, with an angular aperture slightly greater than that of the divergence slit, limits the view of the detector to a region just larger than the illuminated length of the specimen.

As already mentioned, the choice of slit sizes is governed by a compromise between intensity and resolution. Roughly speaking, the total intensity of a line is directly proportional to source width, source length, axial divergence, equatorial divergence, receiving-slit width, and receiving-slit length, whereas the resolution is independent of the lengths of source and receiving-slit, is inversely related to the widths of the source and receiving-slit, and is inversely related to the squares of the equatorial and axial divergences. The effects of the various divergences on the positions and widths of the diffraction maxima, the so-called geometrical aberrations, have been treated in great detail by various authors, and the results have been summarized by Wilson (1963).

X-radiation is not monochromatic, and is analysed into a spectrum by each reflecting plane in the specimen. The overlapping of the spectra from the various reflecting planes forms the recorded diffraction pattern. The crude output of a crystallographic X-ray tube consists of a broad band of white radiation, on which are superposed the emission lines of the characteristic radiation of the target. For the targets in common use (Cu, Ni, Fe, Cr, Mo, etc.) these are the

$K\alpha_1$, $K\alpha_2$ and $K\beta$ lines; occasionally the $K\alpha_3$ group is strong enough to be observed. For special purposes L radiation of, for example, W may be employed. The β radiation can be reduced to negligible intensity by means of a filter or monochromator, but it is not usually practicable to separate the $K\alpha_1$ and $K\alpha_2$ components. If a proportional or scintillation counter is used the counts arising from white radiation can largely be suppressed by incorporating a pulse-height discriminator in the counting circuit, so that the effective spectrum consists of the $K\alpha_1$ and $K\alpha_2$ lines superposed on a small peak of white radiation of wave-length too close to that of $K\alpha$ to be suppressed by the discriminator. Certain physical aberrations arise because of the wave-length spread in the effective spectrum. For example, the long-wave-length components are (in general) more strongly absorbed in the target material, the tube window, the detector window, etc., than are the short-wave-length components, so that the detected radiation has a wave-length distribution different from that of the true emission spectrum. Many factors in the expression for these diffracted intensity are a function of the Bragg angle, and hence differ somewhat for different wave-lengths, leading to a small displacement of the diffraction maximum. Other physical aberrations arise from refraction of the X rays on entering and leaving the powder grains, and from variation with wave-length of the quantum-counting efficiency of the detector. The effect of physical aberrations on the positions of both the peak and the centroid of the recorded diffraction pattern have been discussed by various authors, and the results have been summarized by Wilson (1963, 1965). In general geometrical aberrations affect the peak and the centroid to about the same extent, though the effect on the centroid is easier to calculate. Most physical aberrations affect the peak less than they affect the centroid. Physical aberrations predominate at high angles, geometrical at low and moderate angles, so that it may be found that peaks and centroids as measures of line position have complementary uses.

Each quantum (in neutron or electron apparatus, each particle) absorbed by the detector results in a small burst of current, called a count or a pulse. Suitable electronic circuits, frequently very elaborate, are necessary for the amplification and recording of these pulses. The main components are the following:

(i) For Geiger counters, a quenching circuit to prevent each pulse from triggering off a continuous discharge in the detector.

(ii) A pulse amplifier.

(iii) For proportional and scintillation counters, a pulse-height discriminator. As already mentioned, this circuit rejects pulses whose size differs more than a few percent from that produced by the desired radiation.

(iv) Either an integrating circuit, which averages the rate at which counts are received over a period of the order of seconds, or a scaling circuit, which displays the total number of counts received since the display was last set to zero.

For the less exacting applications of diffractometry an integrating circuit is used. This gives an immediate measure of the intensity diffracted at any angular setting of the diffractometer, and is well adapted for continuous recording of the diffraction pattern. The counter arm is driven at a constant angular velocity (commercial diffractometers frequently provide for speeds of $\frac{1}{4}°$, $\frac{1}{2}°$, 1°, 2° and 4° per minute), and a chart recorder plots counting rate against time. The result is in effect a plot of intensity against Bragg angle, and is immediately useful for such applications as identification and semi-quantitative chemical analysis. By measurement of the line positions such plots may also be used for lattice-parameter measurements, though not of the highest accuracy. To obtain satisfactory plots the scanning speed and the time constant of the integrating circuit must be suitably chosen. Too small a time constant results in a spiky plot, caused by statistical fluctuations in the counting rate. Too large a time constant gives a smooth plot, but with the lines distorted and displaced to lower angles if the diffractometer is scanning from high to low angles, and to higher angles if the scan is in the opposite direction. The slower the scanning speed, the higher the time constant that can be used without undue distortion.

When the highest accuracy is required *step scanning* is used. The diffractometer is set at a particular angle and counts are allowed to accumulate, the counting rate being then given by (counts accumulated)/(time of accumulation). The diffractometer setting is then altered by a small amount (0·01 or 0·02° perhaps) and the process repeated until measurements have been made of the entire pattern, or of that part of it that is of interest for the investigation in hand. There are two main variations of this method, known as fixed-time counting and fixed-count timing. In the first the same counting time is allowed at each step, and the number of counts accumulated is proportional to the intensity of diffraction at that angle. In the second a pre-determined number of counts is accumulated at each step and the intensity of diffraction is inversely proportional to the time required to accumulate them. When it is difficult to stabilize the output of the X-ray tube over long periods a variation of fixed-time counting, called fixed-dose counting, may be used. In this a part of the X-ray output of the tube is fed to a second counter, known as a monitor, and counts are accumulated by the main counter for the time required by the monitor to record a pre-determined number of counts. The monitor beam may be isolated by a dummy diffractometer with a fixed setting operating from one of the other windows of the X-ray tube, or a small part of the actual incident beam may be diverted to the monitor by a slip of mica. By this device the time of counting is automatically adjusted for variations of the tube output.

The actual measurements during step scanning may be displayed in various ways. The simplest is an array of digital counting tubes from which the number of counts accumulated can be read directly and recorded by hand. For extended investigations this is tedious, and it is always liable to error, so that many diffractometers are provided with a device that prints out or punches out the counting rate, in some form or other, on paper tape. The punch-out device is particularly convenient when the measurements are to be subjected to computing procedures, such as routine determinations of the centroid or variance of the line profiles, since the paper tape can be transferred directly to the computer without the risk of human errors in transcription.

The emission of X rays is a random event, with a Poisson distribution in time. The number of counts accumulated in a given time, therefore, is subject to statistical fluctuations, the standard deviation being equal to the square root of the number of counts. The percentage error for fixed-time counting is thus low where the counting rate is high, and high where the counting rate is low. For fixed-count timing, on the other hand, the percentage error is independent of the counting rate. Which procedure is the more advantageous depends on the use to be made of the measurements. For determination of integrated line intensity fixed-count timing wastes a great deal of time, as counting is slow where the counting rate is low, and each step in this region adds but little to the total intensity. For determination of variance, on the other hand, the region of high counting rate near the middle of the line is of less importance than the tails, and time may be saved by going through it quickly. In principle it would be possible to adjust the counting time at each step in order to maximize the accuracy of the particular feature of the line profile under investigation (Wilson, Thomsen and Yap 1965; Wilson 1967).

A powder diffractometer measures a diffraction pattern one line at a time, whereas photographic methods record the entire pattern simultaneously. On the whole, therefore, powder diffractometry is preferred to photography only when accuracy is more important than speed. There is at least one exception to this general statement: in some routine chemical analyses it is necessary only to know the relative intensities of two lines, one from the substance of interest and one from a standard. A single ratio of this kind may be obtainable more quickly as well as more accurately by diffractometry. Some workers, also, prefer an entirely "dry" process and the graphical or numerical presentation of the pattern to the process of developing, fixing and drying and the pictorial presentation of the pattern as provided by photography. However, the real advantages of diffractometry are apparent in the determination of lattice parameters and in the study of line broadening. Since the geometrical and physical aberrations of the diffractometer are well understood, and the line profile is immediately available without the need of microdensitometry, the measured position of each line (centroid) or of the higher-angle lines (peak) can be corrected for aberrations and the corresponding lattice spacing obtained free of systematic error other than that of uncertainty in the wave-length (Delf 1963). Constancy of the lattice parameters derived from different reflections is a test that systematic errors (other than in wave-length) have been eliminated. A more elaborate process of fitting calculated profiles to those observed has been described by Boom (1966). Though developed for microdensitometer traces of photographic powder patterns it seems even better adapted for diffractometer patterns. Perhaps the most characteristic application of powder diffractometry is the study of imperfect crystals. The various types of imperfection—small particle size, internal strain, disorder—all produce broadening of the diffraction lines, and the quantitative measurement of this broadening is more readily undertaken by diffractometry than by photographic methods. The various measures of line breadth (the half width, the integral breadth, the variance, and the Fourier coefficients) are all well adapted to diffractometric determination, and for the last two at any rate punched-tape output and data reduction by computer are highly advantageous. An elementary treatment of line broadening and its interpretation has been given by Wilson (1962).

In this brief article many points of diffractometer design and technique have had to be omitted, and it is not possible to include a comprehensive list of references. Besides the books and papers mentioned in the text, the books by Khejker and Zevin (1963), Klug and Alexander (1954), Parrish (1965), and Peiser, Rooksby and Wilson (1960) may be consulted.

Bibliography

Bacon G. E. (1962) in *Encyclopaedic Dictionary of Physics* (J. Thewlis Ed.), **4**, 818, Oxford: Pergamon Press.

Boom G. (1966) *Accurate Lattice Parameters and the LPC Method*, Groningen: van Denderen.

Delf B. W. (1963) *Brit. J. Appl. Phys.*, **14**, 345.

Khejker D. M. and Zevin L. S. (1963) *Rentgenovskaja Difraktometrija*, Moskva: Fizmatgiz.

Klug H. P. and Alexander L. E. (1954) *X-ray Diffraction Procedures*, New York: Wiley.

Parrish W. (Ed.) (1965) *X-ray Analysis Papers*, Eindhoven: Centrex.

Peiser H. S., Rooksby H. P. and Wilson A. J. C. (Eds.) (1960) *X-ray Diffraction by Polycrystalline Materials*, London: Chapman and Hall.

Soller W. (1924) *Phys. Rev.*, **24**, 158.

Thewlis J. (Ed.) (1962) *Encyclopaedic Dictionary of Physics*, **6**, 545, Oxford: Pergamon Press.

Wilman H. (1961) in *Encyclopaedic Dictionary of Physics* (J. Thewlis Ed.), **2**, 757, Oxford: Pergamon Press.

Wilson A. J. C. (1962) *X-ray Optics*, London: Methuen.
Wilson A. J. C. (1963) *Mathematical Theory of X-ray Powder Diffractometry*, Eindhoven: Philips Technical Library.
Wilson A. J. C. (1965) *Proc. Phys. Soc.*, **85**, 171.
Wilson A. J. C. (1967) *Acta Cryst.* **23**, 888.
Wilson A. J. C., Thomsen J. S. and Yap F. Y. (1965) *Appl. Phys. Letters*, **7**, 163.

A. J. C. Wilson

PROBABILITY, PHILOSOPHICAL ASPECTS OF

1. The Philosophy of Probability

Philosophers of probability agree that their task is to provide an explanation of the meaning, or philosophical analysis, of the term "probability". This task is plainly different from that of the scientist. For although physicists *use* the term "probability" *in*, say, the kinetic theory of gases, and speak perhaps of the probability *of* this theory on the evidence of, for instance, Brownian movement, they do not—*qua* physicists, at any rate—*discuss* the term, or concept, itself.

On the other hand, philosophers of probability disagree about what a philosophical analysis of "probability" ought to be. Some think, like W. Kneale, that it ought to consist in an accurate description of the way in which the concept is actually used, not only by scientists, but also by mathematicians, statisticians, historians, lawyers and plain men. But others think, with R. Carnap, that, since the term as commonly and correctly used is ambiguous, vague and unclear, the philosopher ought to provide an "explication" or "rational reconstruction" of it which, while remaining faithful in essentials to ordinary language, shall nevertheless be unambiguous, clear and precise. The relation between the *explicandum* "probable" and its *explicans* should be like the relation between, e.g., "hot" and "100°C". At the other extreme, yet others hold, with R. von Mises, that the philosopher of probability should follow the lead of the exact scientist by ignoring ordinary language and setting up a definition of "probability" which shall be not only unambiguous and precise but theoretically and practically fruitful. On this view, the relation between "probability" and its philosophical definition should resemble that between, say, "work" in common usage and its definition in mechanics as "product of force and distance".

There is also disagreement about whether "probability" has one meaning or more than one. Most philosophers now believe that it is necessary to distinguish "*statistical* (*or mathematical*) *probability*" from "*inductive probability*" (or "degree of confirmation"). But this has not always been so, notwithstanding that the need for making such a distinction was pointed out over two centuries ago by D. Hume. Plainly, however, philosophical analyses of "probability" will differ according as the author believes that there is only one meaning to be explained, or more than one.

In these circumstances, it is not surprising that no unanimity is to be found in philosophers' opinions about probability. The best course will be, therefore, to sketch the main views which have been taken of the subject.

2. Philosophical Analyses of "Probability"

2.1. Relative frequency theories. According to Aristotle, the probable is that which usually happens. Thus, suppose a bag were to contain 100 balls of which 80 are black; what is the probability that a ball taken at random from the bag is black? The answer is plainly 4/5, so that the probability of the proposition (or event) in question is here identical with the proportion (or relative frequency) of black balls among the balls.

However, probability can be identified with relative frequency only in cases, like the above, in which the population (the balls in the bag) is a finite or "closed" class. When the population is an infinite or "open" class it cannot be so identified, because it is a mathematical convention that the expression "m/n" is meaningless when $n = \infty$. The finite frequency theory has therefore been replaced by the infinite frequency theories of J. Venn, H. Reichenbach and Mises. We will consider the last. What is the probability of throwing heads with this penny? To find out, one must throw it (or make "trials") and record the number of heads thrown ("successes") and the relative frequency of successes among the trials, thus:

Number of trial	1	2	3	4	5	6	7	8	...
Result of trial	T	H	H	T	H	H	H	T	...
Relative frequency of successes	0/1	1/2	2/3	2/4	3/5	4/6	5/7	5/8	...

Then, according to Mises, the probability of a success is the limiting value of the relative frequency of successes, provided that (1) a limit in fact exists (i.e. the sequence in the third row, if continued without end, is convergent); and (2) this infinite sequence is irregular or independent of all possible place-selections. The sequence would be convergent but not irregular if, e.g., every odd-numbered throw was T and every even-numbered throw was H. The point of the irregularity condition is easy to see. For if, in the case just supposed, one is asked the probability of throwing H at the 1 millionth throw, he has grounds for giving two incompatible answers. According to Mises' theory, the answer must be 1/2, since that is the limiting value of relative frequency in the sequence. But the fact that a very large number of observed even-numbered throws have all been H indicates that the answer is 1 or certainty.

Mises' theory has been criticized on various grounds, of which the following is the most significant for present purposes. How, on his theory, is it possible in practice to determine a probability? For one cannot observe an infinite sequence of fractions of relative frequency; empirical statistical sequences are of course finite. One cannot establish *with certainty* that an observed sequence forms part of a convergent infinite sequence. One can say at most that if the observed sequence were continued without end it would *probably* converge upon a certain limit. But the meaning of "probably" in the last sentence is not that defined by Mises' theory, which is "statistical probability", but rather that of "inductive probability". The theory is therefore doubly inadequate. Firstly, it defines only one of the two meanings of "probability". But secondly, even its definition of "statistical probability" cannot be applied in practice without invoking a concept of probability, namely inductive probability, which the theory fails to explain (see Pt. 3).

2.2. Probability and belief. Some think that "degree of probability" is definable in terms of "degree of belief". Hume does so, and interprets "belief" in turn as a sensation, namely a feeling of conviction. In his view, then, when A says "it is very probable that p" (where p is a proposition-variable), this means "I (*sc.* A) feel very sure that p". However, it is a decisive objection against this theory that it involves that probability is subjective, whereas it is generally agreed to be objective. For, on Hume's account, if B replies to A "it is not probable that p", then B is no more *contradicting* A than he is when, if A says "I feel very cold", B replies "I do not feel cold". Yet it is plain that B is in fact contradicting A; so that the proffered analysis must be incorrect.

To escape this objection, other philosophers have defined "probability" in terms, not of *actual* belief, but of *reasonable* belief. B. Russell, for instance, expounds "degree of (inductive) probability" as "degree of credence given by a man who is rational". But the difficulty here is that the notion of "degree of probability" is needed to define the notion of "reasonable belief", so that circularity results if the former is defined in terms of the latter. For the chief criterion of the rationality of a belief is this: A's belief that p is reasonable only if the degree of his belief corresponds to the degree of probability of p.

It is necessary to distinguish the question of the *meaning* of "probability" from the question of the *measure* of probability. The "classical" theorists of probability, J. Bernoulli and P. S. de Laplace, agreed with Hume about the meaning of the term, but propounded a rule for measuring probability which has often been mistaken for a definition of its meaning. According to this rule, one measures a probability by forming a set of equally probable alternatives and taking as the measure the ratio of the "favourable" alternatives to all the alternatives. Thus, if the question is "what is the probability of throwing a number greater than 4 with a true die?", a direct application of the rule gives the answer $2/6 = 1/3$. However, the rule presupposes some method for deciding when alternatives are equiprobable, and the classical theorists proposed for this purpose the Principle of Indifference, according to which alternatives are equiprobable if they are not known not to be. But this leads to paradox. For let the question be "what is the probability of throwing 6 with this die?" The Principle sanctions equally two incompatible answers. One might say that since we have 6 equiprobable alternatives, namely throwing 1 or 2 or ... or 6, of which 1 is favourable, the answer is 1/6. But one might equally well say 1/2, on the ground that we have 2 equiprobable alternatives, namely throwing 6 or not-6.

The distinction between meaning and measure is also important in discussions of the relation between probability and betting-odds. Carnap holds that the meaning of "degree of confirmation (inductive probability)" can be elucidated in terms of the concept of a "fair betting-quotient". The betting-quotient is the ratio $q = u_1/(u_1 + u_2)$, where u_1 and u_2 are the stakes. q is said to be fair if it favours neither party to a betting-contract, i.e. when, given evidence e, betting on a hypothesis h with odds q is as reasonable a choice as betting on not-h with odds $(1 - q)$. On the other hand, F. P. Ramsey and B. de Finetti hold that "degree of probability" means "degree of belief", but that the degree of A's belief that p is measurable by the odds at which A is prepared to bet on p.

Both Carnap's and Ramsey's views involve that degrees of *inductive* probability can be measured by a real number between 0 and 1 inclusive. Carnap and C. G. Hempel hold that a possible form of inductive probability-judgement is "the degree of confirmation of the hypothesis that all the balls in this bag are black on the evidence that all of a sample of balls from this bag are black is 3/5". This is an exemplification of the formula $c(h, e) = r$, where r is any real number between 0 and 1 inclusive. But it seems to others, such as J. M. Keynes, E. Nagel and Kneale, that this is not the case, and that no meaning can be attached to such a statement as the one above. These do not dispute that *statistical* probability can be so measured, but maintain that the fact that inductive probability cannot is precisely one of the main differences between the two sorts of probability.

2.3. Logical theories. Although Keynes sometimes writes as if "degree of probability" is definable in terms of "degree of reasonable belief", his main position is that probability is a logical relation, namely "partial implication", between an hypothesis, h, and the evidence for it, e. Probabilities are accordingly of the form $P(h/e)$, an exemplification of which is "the probability that it will rain on the evidence that the barometer is falling".

However, the only possible interpretation of "e partially implies h" seems to be "when e is true, then h is *usually* true". Partial implication or "probabilification" is thus to be contrasted with plain implication,

$e \supset h$, the meaning of which is "when e is true, then h is *always* true". But in this case Keynes' theory is simply a special form of relative frequency theory, and lies open to the criticism made above (Sec. 2.1). Specifically, it is what may be called, following C. S. Peirce, a relative *truth*-frequency theory. Most frequency theorists equate degree of probability with the *relative frequency* of *events*, e.g. the frequency with which rain has fallen relatively to the frequency with which the barometer has fallen. But it is also possible to equate it with the *relative truth-frequency* of *propositions*, e.g. the frequency with which "it is raining" has been true relatively to the frequency with which "the barometer is falling" has been true.

Carnap holds that inductive probability-judgements are logically true, i.e. that if a statement of the form $c(h, e) = r$ is true at all, then it is an analytic truth. There is, indeed, force in this contention. For consider the judgement "on the evidence that all examined balls taken from this bag are black and very numerous, the probability that all the balls in this bag are black is high". One who thinks, as some do, that the degree of probability of a generalization about a population depends on the size of the examined sample, might well regard this judgement as true by virtue of a connexion of meaning between "inductively very probable" and "confirmed by very numerous instances".

Another logical concept, namely "range", has been invoked to elucidate the meaning of "statistical (or mathematical) probability". It was introduced by J. von Kries and elaborated by F. Waismann, but the latest refinement of it is by Kneale, whose version we shall therefore consider. According to Kneale, the *meaning* of "the probability that a ball taken at random from this bag is black is 4/5" is "the ratio of the range of the attribute being-a-black-ball-in-this-bag to the range of the attribute being-a-ball-in-this-bag is 4/5". Speaking of ratios of ranges presupposes that ranges can be measured, and in this example, where the population is finite, the *measures* of the two ranges are simply the number of black balls in the bag and the number of balls in the bag respectively. By the "range" of an attribute is meant all the "ultimate" alternatives under that attribute. Thus, if there are 100 balls in this bag, then there are under the attribute being-a-ball-in-this-bag 100 ultimate alternatives. Further, these ultimate alternatives are equiprobable. The range theory is in fact an attempt to explain "mathematical probability" in terms of the classical theorists' concept of a set of equiprobable (or equipossible) alternatives, but to provide a more satisfactory test for determining the equiprobability of alternatives than their Principle of Indifference (Sec. 2.2). Similar attempts had been made earlier, e.g. by Keynes, which prescribed that, to be equiprobable, alternatives must be "symmetrical" or "equispecific". Thus, if it is claimed, say, that throwing 6 and throwing not-6 are equiprobable alternatives under the concept of being-a-throw-with-this-die, then the range theorist rejects the claim on the ground that, though throwing 6 is an ultimate alternative, throwing not-6 is not, since there fall under it 5 alternatives which are ultimate.

It is only when the population is finite that ranges can be measured in the simple way just described. When it is infinite, measurement is more difficult. In particular, when the set of equipossible alternatives is infinite because it comprises different possible values of continuous variable magnitudes, the measure of the range is the measure of a region in configuration space. This distinction between the two ways of measuring range corresponds to the distinction which is drawn in the Calculus of Probabilities between "arithmetical" and "geometrical" probability.

But is "ultimacy" in fact a satisfactory test of equiprobability? Consider a loaded die. Throwing 1 or 2 or ... or 6 are all, it would appear, ultimate alternatives under the attribute being-a-throw-with-this-die; yet they are not equiprobable alternatives. There is surely much truth in the widely held opinion that the equiprobability of alternatives must be decided by other means. For instance, one must throw the die several times and note whether each face falls upwards with approximately equal frequency, or perform a simple experiment to determine whether the die's point of balance lies at its geometrical centre.

3. Probability and Induction

"Induction" means "generalization from experience", and we may take for the type of it the inference-pattern

$$m/n \text{ observed } S \text{ are } P$$
$$\therefore\ m/n\ S \text{ are } P;$$

where S and P are term-variables.

Induction is scientifically interesting because, although the logical status of scientific laws is a controversial question, some philosophers believe that these laws are simply comprehensive and important generalizations. By contrast, other philosophers question whether there is any such thing as inductive inference at all. Thus, K. Popper contends that what is mistaken for inductive inference is really hypothetico-deductive inference. This is the process of inventing an hypothesis, deducing from it and additional assumptions some empirical proposition, and testing the hypothesis by observing whether the deduction corresponds with the facts. However, we shall follow majority opinion in accepting that inductive inference does in fact exist.

The central problem of induction is generally considered to be that of justifying it. For, as the inference-pattern above shows, the characteristic feature of inductive inferences, which distinguishes them from deductive inferences, is that the conclusion goes beyond the evidence, or asserts more than the premise. If one has observed only a proper subset of S, what right has one to make an assertion about the whole set S? But students of this problem differ in their interpretations of it. To-day, many think with Reichenbach,

Kneale and R. B. Braithwaite that it is a question of justifying the *process* or "policy" of induction (i.e. inductive inference) by showing, for instance, that it is a "self-correcting" process, or that any alternative policy is self-defeating. Others, however, hold that it is rather a question of justifying the *products* of this process (i.e. inductive conclusions). These maintain further that an inductive conclusion is justified when it is shown to be probable or, in the limiting case, certain. Here, then, is the link between induction and probability.

On this latter view, the problem becomes one of formulating the conditions under which an empirical generalization is probable, i.e. what may be called the determinants of inductive probability. Yet opinions differ about this question too. For instance, some assert that an induction of the form m/n S are P is less probable than one of the form m/n SM are P, where "SM" means "things which are both S and M". The reason they give is that the former asserts more, or is of wider *scope*, than the latter, and is consequently more precarious. The main division of opinion, however, is between those who believe that the determinant of degree of inductive probability is the *number* of the observed confirmatory instances of a generalization (or the size of the sample), and those who believe that the determinant is rather the amount of *variety* among the instances. We shall accordingly examine these positions in turn.

But before doing so, it is desirable to elucidate the relationship between the present inquiry and the task of the philosophy of probability as expounded above (Pt. 1). For it may be objected that exploring the question, *what makes* inductions probable? is distinct from and irrelevant to answering the question, what does "probable", when applied to inductions, *mean*? However, the reply to this objection has been adumbrated already in discussing Carnap's thesis on the analyticity of inductive probability-judgements (Sec. 2.3). It is that the connexion between inductive probability and its determinants is not merely contingent, but rather conceptual. In other words, if the proposition "what is true of any very large sample is very probably true of its population" is true at all, then it is an analytic truth. An investigation into the determinants of inductive probability may therefore legitimately be regarded as forming an essential part of the explanation of the meaning "probability".

Those who consider the number of instances essential base their reasoning on the large number theorems of the Calculus of Probabilities, especially J. Bernoulli's limit theorem. Older writers, notably Laplace, rely also on the inverse probability theorem (T. Bayes' theorem). By "*direct*" *probability* is meant the probability that a random sample of a certain size is approximately statistically similar to its population, and by "*inverse*" *probability* is meant the probability that its population is approximately statistically similar to a random sample of a certain size which has been selected from it. However, Bayes' theorem cannot be used unless the initial probabilities of the alternative possible hypotheses are known. The "initial" probability of an hypothesis is the probability which it possesses independently of any conferred on it by the evidence of observed instances or samples. Laplace met this problem by applying the Principle of Indifference, and so concluding that these initial probabilities are all equal. Thus, suppose the question to be, what proportion in a population of 10 balls in a bag are black? There are 11 possible hypotheses, namely 10 black, 9 black ... 0 black, so that the initial probability of each is 1/11. But we have seen above that the Principle of Indifference is not an acceptable way of determining probabilities (Sec. 2.2). Keynes therefore attempted to surmount the difficulty by his Principle of Limited Independent Variety, which develops an idea of F. Bacon in asserting that any property can enter only into a finite number of combinations. Similarly, Carnap postulates that the number of things in the world is finite and that the number of properties of each thing is also finite. Yet the difficulty remains. Granted that, *if* these principles or postulates are certain or probable, then initial probabilities can be assigned to hypotheses; it is none the less true that there is no reason to think that they *are* certain or probable.

Consequently, modern statisticians have tended to abandon Bayes' theorem. R. A. Fisher, e.g., in his Method of Maximum Likelihood, simply equates the inverse probability (the "likelihood") with the direct probability (the "probability"). Particular interest attaches therefore to an attempt on the problem of induction by a philosopher, D. Williams, who follows contemporary statisticians in dispensing with the inverse probability theorem. Williams claims that, provided only that a sample is large and random, then what is true of it is probably approximately true of its population. An inference from any large random sample to its population goes, he thinks, as follows:

(1) Most large random samples match their populations
(2) This sample is large and random
∴ (3) This sample probably matches its population
∴ (4) Its population probably matches this sample.

Proposition (1) is a truth of applied mathematics, since it contains the empirical terms "sample" and "population"; the corresponding truth of pure mathematics is about "combinations". The verb "match" means "match approximately", not "match exactly". If 1/2 of the balls in the bag (the population) are black, then samples of size 100 which contain, e.g., 51 blacks or 49 blacks count as matching samples. The step from proposition (3) to proposition (4) is justified because "matching", or "being statistically similar to", is a symmetrical relation.

In criticism, we shall concentrate on proposition (2). There is no difficulty about knowing that a sample is large, but how does one know that it is random? For a start, one must know what "*random sample*" means.

The definition is as follows: a sample of given size is random if and only if every possible sample of that size had an equal chance of being selected. That is, "randomness" is defined in terms of "equiprobability". Williams uses the Principle of Indifference, and claims that a sample is random if it is not known not to be random. However, we have seen that this principle does not provide an acceptable test of equiprobability (Sec. 2.2). Statisticians, on the other hand, declare that a sample is random if it has been selected by a "reliable method of random sampling", such as lottery-sampling or sampling by Tippett's Numbers. But their thesis invites two comments. Firstly, this concept of "reliability" is closely related to that of "inductive probability". For "this sample has been selected by a reliable method of random sampling" entails "it is inductively probable that this sample is random". Secondly, how is it established that a particular method of random sampling, say with this lottery, is reliable? It may be said that it is reliable if, when it has been operated a *large number* of times with populations of known compositions, it has selected every possible sample of a given size with approximately equal frequency.

Nevertheless, it may be doubted whether it suffices that the trials should have been numerous. Surely, it is at least equally important that the lottery should have been operated by different persons, on different days, at different temperatures, etc. This is the consideration which has led other students, such as Keynes and Nagel, to conclude that *variety*, not number, of confirming instances is the chief determinant of inductive probability. If this is so, then Williams' theory is exposed to the following objection. It fails to show that an induction is justified provided only that the sample is *numerous* and random, because in practice one can know at best that it is inductively probable that the sample is random from the evidence of *varied* tests on the sampling apparatus (or "method") employed.

It has been said that this methodological dispute between the champions of numerous and of varied samples respectively reflects a corresponding difference in sampling methods. For whereas some pollsters, e.g., rely exclusively on random samples, others use "stratified" samples which are deliberately variegated in respect of sex, age, wealth, etc. Yet the preceding considerations indicate that this account of the situation is oversimplified. For in fact the "random" school rely on variety as well as the "stratified" school, since the concept of an inductive probability which is determined by variety of instances is implicit in their own key concept of a reliable method of random sampling.

Since this article has been largely concerned with the difference between statistical and inductive probability, the reader may naturally wonder whether there are not more than two concepts of probability. In fact, there are good grounds for thinking that there are.

Firstly, some philosophers follow Peirce in distinguishing the inductive probability of *generalizations* from the "abductive probability" of *hypotheses*. This distinction turns on the fact that, whereas the evidence for the former is instantial, the evidence for the latter is not. Compare, for instance, the *law* "all metals are electrical conductors" with the *theoretical principle* "all gases consist of moving molecules". The evidence for the former is a conjunction of observation-reports of the form "this is a metal and an electrical conductor, and that is a metal and an electrical conductor, and ... etc.". But the evidence for the latter cannot consist in observation-reports of the form "this, that and the other samples of gas consist of moving molecules", because gas-molecules cannot be observed directly. It follows that, whatever the determinants of the probability of a theoretical principle may be, they cannot be the variety and number of its confirming instances. Furthermore, since—as we have seen— difference of determinants involves difference of meaning, it also follows that inductive probability and abductive probability are distinct concepts. It would exceed the scope of this article, however, to consider what the determinants of abductive probability actually are.

Secondly, statistical probability and inductive probability are not mutually exclusive, so that there is a mixed concept of probability, the meaning of which is partly statistical and partly inductive. This truth is illustrated by the following inference-pattern:

Most S are very probably P

X is S

$\therefore$ X is probably P;

where it is to be understood that the first premise is a probable generalization-formula. Then, the "probably" in the conclusion is a mixed concept. The statistical constituent of its meaning derives from the quantifier "most" in the first premise, and the inductive constituent of its meaning derives from the modifier "very probably" in the same premise.

Bibliography

Carnap R. (1950) *Logical Foundations of Probability*, Chicago: The University Press.

Day J. P. (1961) *Inductive Probability*, London.

Jeffreys H. (1962) in *Encyclopaedic Dictionary of Physics* (J. Thewlis Ed.) **5**, 658, Oxford: Pergamon Press.

Keynes J. M. (1921) *A Treatise on Probability*, London.

Kneale W. (1949) *Probability and Induction*, Oxford: The University Press.

Mises R. von (1928, English translation 1939), *Probability, Statistics and Truth*, London.

Nagel E. (1939) Principles of the Theory of Probability, *International Encyclopedia of Unified Science*, Vol. 1, No. 6, Chicago: The University Press.

Russell B. (1948) *Human Knowledge: Its Scope and Limits*, Pt. V, London.
Williams D. (1947) *The Ground of Induction*, Cambridge, Mass.: Harvard University Press.

J. P. Day

PROCESS CONTROL, METHODS AND MEASUREMENTS FOR. *Introduction.* Any type of measurement may be utilized to generate a signal to be used for control purposes, but in the case of an actual chemical or physical process the choice of what measurements to make, how to make them, and how to use the values for control purposes will be dictated by various practical considerations. Ideally the process should be well understood so that the variables that need to be controlled can be easily identified and measured; if this is not so it will be wise to make the measuring system flexible so that it can be rearranged if performance is not satisfactory. The nature and finances of the process will also govern the complexity of the control system; whether it is to consist of a few self-contained pressure regulators, or of several hundred control valves with optimum settings calculated by computer. In either case the task of the control system should be closely considered and the economics of more and less complicated schemes carefully evaluated; it is as easy to waste money by over-instrumentation as by under-instrumentation (Eddison considers this point). An equally important consideration is that of maintenance; money spent on instruments will be wasted if they are not properly cared for, and the cost of doing this for a few years may be much larger than that of the instruments (see Carroll 1960).

Control equipment is concerned with obtaining information, transmitting it, and using it to change the operating state of other apparatus. The problem of measuring process variables is a question of mechanical detail which will be mentioned later; the matter of transmitting the measurements and of using them to derive control signals will now be discussed, as this affects the selection of operating medium of the control system and hence the choice of transducer.

Types of process control system. Until ten years ago most process control systems were pneumatically operated. Within the last ten years the use of electronic systems has become increasingly popular, and this has been hastened by the introduction of more reliable solid-state circuits and by the inclusion of data loggers and computers in process control systems. But at present it does not seem likely that pneumatic systems will be completely superseded; most equipment manufacturers still produce both types of apparatus, and each type has particular advantages to recommend it.

In a pneumatic system the measurement and control signals are conveyed as pressures in small tubes. The normal range of signal pressures is 3 to 15 pounds per square inch, and the normal size for tubing is $\frac{1}{4}$ inch o.d. Most commercial process control equipment using pneumatic signals is made to suit these standards, although there is no reason why others should not be used. The tubing is usually of copper or plastic, and large multicore cables containing a dozen or more plastic tubes can be obtained in long lengths; they are no more difficult to install than large electrical cables, though rather bulky. A disadvantage is that there is a lag between the introduction of a pressure change at one end of the tube and its appearance at the other end. With long runs (over 100 metres) the time involved may be several seconds; it depends largely upon the design of transmitters and receivers, and for some purposes is not significant. A major advantage of such a system is safety, as for most processes air is a safer medium than electricity, and even major damage to instruments and signal tubing does not cause a fire risk. Another advantage is that large forces can be obtained by the use of simple diaphragm motors which are very suitable for positioning control valves and other devices. A 10-inch (25 cm) diaphragm supplied with air at 15 pounds per square inch exerts a force of half a ton, needs no maintenance, and lasts for years. A corresponding electrical actuator has many moving parts, needs more attention, and is heavier.

An electrical system transmits information as a current or voltage signal, and unfortunately several standard signals are in use. For Direct Current systems the ranges 0–10 mA, 4–20 mA, and 10–50 mA are all fairly common. There are other ranges, and Alternating Current systems are also used; in some modern process control installations information is transmitted in digital form (certain measuring devices produce a digital output directly and so lend themselves to this type of system—see below). In applications where there is a fire or explosion risk all electrical equipment should be flameproof (enclosed in heavy metal boxes having sealed joints) or intrinsically safe (designed so that the energy level of the circuitry is so low that even the most serious damage or drastic abuse can cause no danger of sparking or overheating). The main advantages of electrical control equipment are accuracy, speed of data transmission, and relative ease of including logging devices or computers in the signal loops. Accuracy and reliability is at present good and is likely to improve considerably in future, with the growing use of integrated circuit components. It is not difficult to convert electric signals to pneumatic form (or vice versa) so that many present-day installations combine the advantages of electrical measurement and control with those of pneumatic operation of the final control element (control valve or motor).

Process controllers. In a conventional process control system the electric or pneumatic signal paths convey information from the point of measurement to a controller. This device compares the measured value with a pre-set desired value, and generates a control signal which is a function of the difference between the two. A simple form of control is achieved by amplifying the error signal and using it as the control output. This is "proportional only" control (so-called because the control signal is always proportional to the error). The degree of amplification used may be expressed as "gain" as in electronic or servo applications, but is normally referred to as "percent proportional band". This quantity is the percentage of scale travel of the measuring instrument that is required to give controller output change of the full range; if a pneumatic temperature controller with a scale range of 0 to 100 degrees gives a change of 3 to 15 pounds per square inch at output when the measured temperature changes by 10 degrees the proportional band is 10 per cent; if this output change occurs for a temperature change of only 2 degrees the proportional band is 2 per cent. Thus "proportional band" gives an inverse representation of amplification, and a small (or narrow) band corresponds to high gain. Most commercial controllers cover a range of about 2 to 300 per cent.

A feature of proportional only control is that such a system cannot maintain zero error for a range of operating conditions. If the liquid level in a cistern is controlled by this means the level must change if the outflow changes; this is because there is a fixed relationship between the level in the tank and the degree of opening of the valve which admits water to it. Thus for a large flow the level must be low, and for a small flow it must be high. For some applications this offset is not a disadvantage, and it can be reduced by increasing the amplification of the control system, but there are usually practical limits to this. Such a system can be simple, cheap, and easy to maintain. More accurate control may be achieved by more complicated controllers. If the "offset" described above cannot be tolerated two-term control may be used; the controller generates a signal proportional to the error and adds to it a term proportional to the time integral of the error. The amplification associated with this integral term is usually represented as the integral action time, and this is the time which must elapse after the application of a step disturbance before the output signal due to integral action exceeds that due to proportional action. The introduction of integral action gives a control system that can always reduce the error to zero after a certain period of time, but it may cause instability and oscillation. This may be remedied by the use of derivative action, in which a third term, proportional to the derivative of error, is added to the controller output. The derivative action time is the time taken for a steadily increasing error to produce equal proportional and derivative components of the control signal; like the integral action time it may range from a few seconds to an hour or more. These quantities are sometimes expressed in other ways (such as reset rate and derivative rate) and the units and definitions used must be carefully checked.

Selection of controllers of this type for a particular duty is often a matter of experience and judgement rather than of calculation. In practice proportional only systems are often used for level control, and three-term controllers are often needed to obtain stable control of temperature regulating systems which include transmission lags; two-term control is used in the majority of cases apart from these. Satisfactory settings of the adjustments of a controller are often selected by experiment, although some theoretical work has provided useful guides for this choice (Caldwell *et al.*). In general the adjustments of process controllers are not accurately calibrated and are intended more for trial and error than for accurate pre-setting; electrical controllers are rather better than pneumatic ones in this respect. There is of course no reason why control actions should not be built up in other ways, but the above methods have become widely accepted in the process control industry, and most of the commercially available controllers are designed along these lines. In spite of this the normal equipment is very flexible, and complicated interconnexions may be made between different control loops. It is also possible to perform quite advanced analog computations. However, large control calculations of this kind may be more easily made if all signals are processed within a digital computer.

Mathematical modelling. Theoretical analysis of process control systems may yield useful results, but accurate treatment is often made difficult or impossible by the large number of non-linearities encountered in problems such as fluid flow and temperature distribution, and by the imposition of non-linear constraints upon some variables or control signals. Calculations on a single problem of this kind are often formidable (even by modern standards) and are frequently handicapped by a shortage of factual information. Because of these factors theoretical results should always be treated with caution and the assumptions upon which they are based should be carefully examined. A great deal of work has been done upon advanced theories of automatic control, but much of it has yet to be proved useful in the field of process control. However, some studies have yielded results of practical interest, and in particular the concept of optimal control seems to hold promise for the future (Noton).

Measurements for control purposes. Because process control instrumentation is used in rapidly expanding industries the techniques and equipment available advance rapidly. The most up to date sources of information are the professional and trade journals which include recent papers and manufacturers' advertisements. There are many publications of this kind, and a study of recent copies of these journals will provide details of solution of practical measurement and control problems too numerous to mention here. The basic quantities that are measured for process control purposes are listed below, and the conventional methods for measurement are mentioned. This is not intended as a list of all possible methods but rather as a review of normal industrial techniques, knowledge of which may facilitate solution of particular problems.

Pressure. Pressure transducers may be of the motion-balance or force-balance type. The former depends upon a curved elastic tube or a spring-opposed diaphragm; the application of internal pressure distorts the tube or deflects the diaphragm and causes movement proportional to the pressure. This motion may be used directly to actuate a valve or to make electrical contacts, or it may be transmitted as an electrical or mechanical signal. The force-balance type of instrument usually incorporates a diphragm whose motion (when pressure is applied) is detected by pneumatic or electrical means and is used to produce an output signal. This signal (air pressure or electric current) is used to generate a force in the instrument which opposes that due to the applied pressure. This opposing force ensures that the deflexion of the diaphragm is always very small, and the use of the feedback principle confers the usual advantages of improved stability and linearity. It is easy to extend this principle to the measurement of differential pressure; in this case the two pressures are applied to opposite sides of the diaphragm and balance is restored by the feedback mechanism. Such a device is known as a d.p. (differential pressure) cell and is a most useful tool for process measurement; its uses include indication of high and low pressures, rates of flow, and liquid levels.

Temperature. Two basic methods are available—mechanical and electrical. The mechanical system comprises a bulb and a length of capilliary tube (usually of steel) connected to a pressure measuring device, and the whole is filled with a liquid, a gas, or an equilibrium mixture of both. The pressure in the system depends on the temperature of the bulb, and the contents of the system can be chosen to give the required range of measurement. Such systems are usually filled, sealed, and calibrated by the makers so that it is difficult to change the range of an existing instrument; however the systems are cheap and reliable (though the capilliary tube may be susceptible to mechanical damage). They have the great advantage that they need no external source of power. The length of the capilliary is usually kept below about 50 feet, but this is only the distance from bulb to pressure transmitter, and the temperature signal may be sent further in pneumatic or electrical form.

Electrical measurements of temperature may be made by thermocouple or resistance thermometers (Golding). The thermocouple method involves accurate measurement of small voltages; the older potentiometric instruments for this purpose are rather cumbersome but modern transistorized d.c. amplifiers can convert a span of a few millivolts into a full-scale output signal, and are very compact. Details of such an installation must be carefully considered; the input impedance of the amplifier must be high, compensating cables of the correct kind must be used in the measuring circuit, and the correct allowance must be made for the temperature at the "cold junction" of the circuit. Many different pairs of metals can be used as thermojunctions, and the choice of appropriate materials is governed by the working temperature range to be covered, the accuracy and repeatability desired, and the money available. Resistance thermometer measurements are usually made in a three-wire circuit so that changes in lead resistance have no effect; as in the case of thermocouples modern transistorized bridge circuits make the instruments simple and compact, and the choice of resistance material is governed by the range and accuracy needed.

Temperature measurements made by radiation instruments can also be used for control, and total radiation pyrometers can be designed to measure the temperature of a small target area some distance away; the newer instruments can be arranged to measure temperatures only a little above ambient. This method and the other two electrical systems mentioned above obviously lend themselves to applications where electrical output signals are needed or can easily be accommodated.

Level. The liquid level in a tank may be measured by a float, whose position is sensed mechanically, electrically, or optically, or by measuring the pressure difference between top and bottom of the tank by means of a differential pressure cell as described above. The force balance principle may be applied to a float mechanism so that the float movement is very small, and the upward force on the float due to its buoyancy in the liquid provides a measure of the level. A modification of this system can be used to measure density of a fluid (the float is then always submerged and the change in upthrust then represents change in density). Another use of such a float is to detect an interface between two liquids of different density.

Flow. The simplest method of flow measurement is to insert a restriction of some kind in the pipe so that a pressure difference is created across it when liquid flows through; this difference may then be measured by a d.p. cell as described above. Several different types of restriction are used for this purpose, including orifice plates, nozzles, Dall tubes, and venturis. Choice must depend upon individual requirements; the two basic differences between the performances of the various devices are their resistance to clogging when solids are present in the fluid, and their degree of pressure recovery. Correct design of a meter of this kind depends upon knowledge of the pressure, temperature, and physical properties of the flowing medium, and flows of unknown or variable composition cannot be measured accurately unless means are available to determine the density of the stream. If pressure and temperature vary at all from the design conditions they must be noted, and a correction factor applied. Spink gives a comprehensive account of all topics relating to design, installation, and use of this type of meter.

Other flowmeters are available in which the fluid is made to drive a light propellor or fan, so that the speed of rotation of this device represents the flow rate. Alternatively the fluid may pass through a positive displacement meter, in which vanes or rotors sweep out a known volume at every turn. It is easy to derive a digital output from both types of meters, by counting turns of the fan or rotor, but any device of this kind must be carefully chosen for its duty, and installed with filters if necessary to guard against the possibility of damage or choking by solids in the fluid. A recent introduction is the electromagnetic flowmeter, which can easily cope with difficult conditions; this measures the voltage induced when a liquid flows through a magnetic field.

Analytical instruments. These are usually tailored to do a particular job, and the service conditions and rough composition of the material to be analysed must be known. Most analysers depend on particular idiosyncrasies of the substance to be measured, although the various types of chromatograph are instruments of more general application. Instruments of this kind often produce output information at intervals rather than continuously, and care must be taken that the frequency of analysis is sufficient for control purposes. Sampled-data control theory can be applied to this problem.

Control valves. The choice of the correct type of valve for a given job can be a difficult problem. The pressure drop across the valve must be sufficient to ensure positive control of flow when it is partly open, but not so large as to cause waste of energy in the system and risk of damage to the valve, and the characteristics (variation of flow with valve opening) must be chosen from several available types to suit individual applications. Some control valve manufacturers provide most useful and comprehensive information on these problems, and will suggest empirical equations that can be used for valve sizing.

Bibliography

CALDWELL, COON, ZOSS (1959) *Frequency Response for Process Control*, New York: McGraw-Hill.

CARROLL (1960) *Industrial Instrument Servicing Handbook*, New York: McGraw-Hill.

CEAGLSKE N. H. (1956) *Automatic Process Control for Chemical Engineers*, New York: Wiley.

EDDISON R. T. (1965) The Meaning and Value of Measurement in Industrial Processes, *Trans. Soc. Instrum. Tech.*, **17**, No. 3, Sept.

GOLDING E. W. (1955) *Electrical Measurements and Measuring Instruments*, London: Pitman.

NOTON A. R. M. (1965) *Introduction to Variational Methods in Control Engineering*, Oxford: Pergamon Press.

POSTLE L. J. (1966) in *Encyclopaedic Dictionary of Physics* (J. Thewlis Ed.), Suppl. Vol. 1, 146, Oxford: Pergamon Press.

SPINK L. K. *Flow Meter Engineering*, U.S.A.: Foxboro company.

BRITISH VALVE MANUFACTURERS' ASSN. (1964) *Valves for the Control of Fluids*, London.

P. H. MYNOTT

PROPAGATION AND RADIATION OF ELECTROMAGNETIC WAVES IN PLASMA. This subject may be studied from the point of view of applications where small amplitude signals propagate through a plasma without producing gross perturbation of the system. With such an emphasis the topics of ionospheric radio wave propagation, plasma diagnostics and microwave electronics become relevant. Situations where large amplitude waves or oscillations arise as a result of the inherently unstable nature of some plasma systems, the waves thus playing an important part in determining the plasma properties, are excluded from the discussion here as more properly coming under the heading of Plasma Physics.

The propagation of small amplitude electromagnetic waves through a plasma is very conveniently represented using an equivalent permittivity describing the medium in a macroscopic sense. In this representation the effects of space charge and current unbalance associated with the passage of an electromagnetic disturbance through the otherwise neutral plasma become included in the permittivity. With a substantial, externally applied magnetic field the plasma exhibits anisotropy and is described as *gyrotropic*. The equivalent permittivity must be written as a *tensor* $\boldsymbol{\varepsilon}$. Problems of wave propagation through a plasma may now be tackled by means of Maxwell's equations as if for a current- and charge-free medium but with an electric constitutive equation

$$\mathbf{D} = \boldsymbol{\varepsilon} \cdot \mathbf{E}. \qquad (1)$$

The magnetic permeability remains a scalar and will generally take the value for free space, $\mu_0 = 4\pi \times 10^{-7}$ henry/metre (we use S.I. units throughout). The actual net charge and current densities may be recovered from

$$\varrho = \varepsilon_0 \nabla \cdot \mathbf{E} \qquad (2)$$

and

$$\mathbf{J} = j\omega(\mathbf{D} - \varepsilon_0 \mathbf{E}) \qquad (3)$$

(the use of phasor quantities, such that a disturbance at frequency ω is represented by the real part of $\mathbf{E} \exp j\omega t$ and so forth, is implicit here). The form taken by $\boldsymbol{\varepsilon}$ depends very much upon the plasma model assumed. It is particularly simple for the linearized theory of a virtually cold, stationary electron plasma neutralized by massive, immobile positive ions and assuming a unique electron collision frequency ν independent of the particle velocity. Fortunately this naive model does represent many practical situations quite well, whilst many features of the simple theory persist for more realistic plasma descriptions.

For the simple model described application of Newton's second law gives an equation in the electron velocity

$$\mathrm{j}\omega m\mathbf{v} + m\nu\mathbf{v} = -e(\mathbf{E} + \mathbf{v} \times \mathbf{B}_0) \qquad (4)$$

where $\mathbf{B}_0$ is the steady magnetic field and the result has been linearized. The electron velocity is related to current by the (linearized) expression

$$\mathbf{J} = -N_0 e\mathbf{v}, \qquad (5)$$

N_0 representing the equilibrium electron density. It is now simple in principle to eliminate $\mathbf{J}$, $\mathbf{v}$ and $\mathbf{D}$ from equations (1), (3), (4) and (5). The result is an identity which, if $\mathbf{E}$ is not to vanish trivially, determines $\boldsymbol{\varepsilon}$. Choosing $\mathbf{B}_0$ to be z-directed the tensor permittivity is found to be

$$\varepsilon = \varepsilon_0 \begin{bmatrix} \varkappa_1 & \varkappa_2 & 0 \\ -\varkappa_2 & \varkappa_1 & 0 \\ 0 & 0 & \varkappa_3 \end{bmatrix}$$

where

$$\varkappa_1 = 1 + \frac{\omega_p^2(\nu + \mathrm{j}\omega)}{\mathrm{j}\omega[(\nu + \mathrm{j}\omega)^2 + \omega_c^2]}$$

$$\varkappa_2 = \frac{\mathrm{j}\omega_p^2\omega_c}{\omega[(\nu + \mathrm{j}\omega)^2 + \omega_c^2]}$$

$$\varkappa_3 = 1 + \frac{\omega_p^2}{\mathrm{j}\omega(\nu + \mathrm{j}\omega)}$$

and the characteristic electron plasma and cyclotron frequencies are defined by $\omega_p^2 = N_0 e^2/m\varepsilon_0$ and $\omega_c = eB_0/m$ respectively.

Forms of $\boldsymbol{\varepsilon}$ for more sophisticated plasma descriptions have been given in the literature. Ultimately the notion of a permittivity associated with the plasma parameters at each point in space becomes insufficient. The properties at all points in space become interrelated because of the relatively unimpeded flow of plasma particles from one point to the next. Thus to describe $\boldsymbol{\varepsilon}$ properly at any given point, the plasma parameters at all surrounding positions and the nature of the plasma boundaries need to be taken into account.

Insofar as the plasma model used to derive $\boldsymbol{\varepsilon}$ is accurate, wave propagation and radiation problems may be dealt with by means of Maxwell's equations in the form

$$\nabla \times \mathbf{E} = -\mathrm{j}\omega\mu_0\mathbf{H} \qquad (6)$$

$$\nabla \times \mathbf{H} = \mathrm{j}\omega\boldsymbol{\varepsilon}\mathbf{E} \qquad (7)$$

subject to whatever boundary conditions the geometry imposes. Evidently with no externally applied magnetic field, when $\boldsymbol{\varepsilon}$ is a scalar, all the results of standard electromagnetic theory for plane waves and for bounded systems such as antennas, waveguides or cavities may be applied directly. Some care needs to be exercised in bounded arrangements with respect to plasma sheath formation, since although such effects arise essentially due to finite plasma temperature, sheaths of appreciable thickness may appear at plasma boundaries for cases where in all other respects the zero temperature model fits very well. In the general case with an external magnetic field the anisotropy introduced requires that fresh thought be brought to bear even in the simplest situations.

The standard analysis for plane waves in ordinary dielectrics shows that with no magnetic field, transverse plane waves propagate in plasma according to the dispersion equation

$$k^2 = \omega^2\mu_0\varepsilon,$$

k corresponding to wave solutions containing the factor $\exp \mathrm{j}(\omega t - \mathbf{k}\cdot\mathbf{r})$. The refractive index n is given by

$$n^2 = k^2/\omega^2\mu_0\varepsilon_0 = \varepsilon/\varepsilon_0.$$

Considering the case with no collisions, when $\varepsilon = \varepsilon_0(1 - \omega_p^2/\omega^2)$, it is evident that there is a cut-off at frequency $\omega = \omega_p$ below which n is purely imaginary and propagation does not occur. Thus free space radiation incident upon a plasma boundary is totally reflected if $\omega < \omega_p$. The effect is exploited in plasma diagnostics for cases in which the plasma frequency, given by $f_p \doteqdot 9N_0$ c/s, is within the accessible microwave spectrum. The cut-off frequency observed when a laboratory plasma is interposed between transmitting and receiving antenna of a microwave arrangement gives f_p directly, whilst phase measurements at higher, non-cut-off frequencies are also easily related to the plasma frequency. There are layers of ionized gas in the upper atmosphere with $f_p \sim 10$ Mc/s and it is easily seen that these can have a marked effect upon short-wave radio propagation.

The more general analysis of waves propagating through a magnetized plasma at some angle θ to the magnetic field, under the idealized plasma assumption discussed earlier, constitutes the basic material of the magneto-ionic theory. This theory is applied successfully to plasma phenomena including radio wave propagation in the ionosphere under the influence of the Earth's magnetic field. Plane wave solutions are assumed and substituted into Maxwell's equations (6) and (7), yielding the Appleton–Hartree equation for the refraction index

$$c^2k^2/\omega^2 = n^2 =$$

$$= 1 - \frac{X}{U - \dfrac{Y^2\sin^2\theta}{2(U - X)} \pm \left[\dfrac{Y^4\sin^4\theta}{4(U-X)^2} + Y^2\cos^2\theta\right]^{\frac{1}{2}}}$$

where (8)

$$X = \omega_p^2/\omega^2, \quad Y = \omega_c/\omega \quad \text{and} \quad U = 1 - \mathrm{j}\nu/\omega.$$

In addition, complicated wave polarizations are required by the theory, these being elliptical in their most general form. The alternative signs in (8) correspond to the elliptically polarized wave vectors rotating in opposite senses. The equation covers a wide variety of situations. The atmospheric radio phenomenon of *whistlers* is explained in the terms of the low frequency limit of (8) with $\theta = 0$. *Faraday rotation* results from natural modes of propagation being elliptically or circularly polarized waves. When a plane polarized wave is launched, its natural mode components propagate at differing phase velocities. This phase difference is equivalent to a rotation of the plane of polarization as the wave progresses away from the launching plane.

The topic of electromagnetic wave propagation in bounded plasma systems finds applications in the studies of waveguides and cavities containing plasma. Again, if there is no magnetic field the permittivity is scalar and we may use the standard results derived for waveguides and cavities containing dielectric material, although in many cases a high degree of sophistication is required for the analysis.

The observation of the resonant frequency shift $\Delta\omega$ and change in Q-factor of a cavity when plasma is introduced offers a useful laboratory method of determining ω_p and ν. The perturbations, if small, may be expressed as

$$\frac{\Delta\omega}{\omega} \doteqdot \frac{1}{2}\,\frac{1}{1+(\nu/\omega)^2}\,\frac{\int\limits_V (\omega_p^2/\omega^2)\,E_0^2\,\mathrm{d}V}{\int\limits_V E_0^2\,\mathrm{d}V}$$

$$\frac{1}{Q_1} - \frac{1}{Q_0} = \Delta\left(\frac{1}{Q}\right)$$

$$\doteqdot \frac{\nu/\omega}{1+(\nu/\omega)^2}\,\frac{\int\limits_V (\omega_p^2/\omega^2)\,E_0^2\,\mathrm{d}V}{\int\limits_V E_0^2\,\mathrm{d}V}$$

where the spatial distribution of electron density and the unperturbed cavity field E_0 are supposed known, whilst the cold electron plasma model without a magnetic field has been assumed. It can be seen that in cases where the plasma density (and hence ω_p^2) is uniform over some part of the cavity, the frequency shift will be directly proportional to this density. The calculation of the perturbations arising when there is a steady, applied magnetic field or for high plasma densities represents a more difficult task which has been carried out in a few cases and is certainly always possible in principle if the simple plasma model may be assumed.

Some attention has been given to the study of wave propagation through plasma-filled waveguide, in attempts to utilize various microwave effects that might be associated with an electronically variable, gyrotropic medium, and the general problems are of some complexity. In many cases the presence of magnetoplasma introduces new modes, not present in ordinary waveguide, with phase velocities small compared with c (i.e. *slow waves*). In such cases the problems may be shown to reduce to an electrostatic form. It is allowable to put $\mathbf{E} = -\nabla\varphi$ and thus to neglect the magnetic field in the first of Maxwell's equations (6). The second of Maxwell's equations (7), taking the divergence of both sides, then yields $\nabla \cdot (\boldsymbol{\varepsilon} \cdot \nabla\varphi) = 0$. For perfectly conducting waveguide walls the appropriate boundary condition is $\varphi = 0$. Often a solution can be obtained in this way whereas a full analysis is intractable or at least exceedingly complicated, and such *quasi-static* solutions have been shown to be in accordance with experiment. It may be noted that the validity of the approximation may be determined from the quasi-static solutions themselves. These will be valid only if they demonstrate slow waves. The quasi-static solutions are in fact related to the evanescent modes in common waveguides, and of course the approximation fails to reproduce the modes corresponding to ordinary waveguide propagation. Application of this method to a circular waveguide filled with plasma, for instance, leads to waves having a dispersion equation,

$$\omega^2/c^2k^2 = \frac{(\omega_c^2 - \omega^2)\,(\omega_p^2 - \omega^2)}{(\omega_p^2 + \omega_c^2 - \omega^2)\,\omega_0^2}$$

where ω_0 is the cut-off frequency for the nth angular and lth radial transverse magnetic (TM_{nl}) mode. This expression evidently yields slow waves if the cut-off frequency ω_0 is sufficiently large.

Bibliography

Allis W. P., Buchsbaum S. J. and Bers A. (1963) *Waves in Anisotropic Plasmas*, Cambridge, Mass.: M.I.T. Press.

Heald M. A. and Wharton C. B. (1965) *Plasma Diagnostics with Microwaves*, New York: Wiley.

Johnson C. C. (1965) *Field and Wave Electrodynamics*, New York: McGraw-Hill.

Ratcliffe J. A. (1959) *The Magneto-ionic Theory and its Applications to the Ionosphere*, Cambridge: The University Press.

Stix T. H. (1962) *The Theory of Plasma Waves*, New York: McGraw-Hill.

Thompson W. B. (1967) in *Encyclopaedic Dictionary of Physics* (J. Thewlis Ed.), Suppl. Vol. 2, 252, Oxford: Pergamon Press.

R. L. Ferrari

PROTECTION OF METALS BY ORGANIC COATINGS. In order for an organic coating to protect a metal, such as iron, from corrosion, it must retard the overall reaction

$$4Fe + 3O_2 + 2H_2O \longrightarrow 2Fe_2O_3H_2O$$

This reaction can be broken down into two reactions, one producing electrons and the other consuming them.

$$4Fe \longrightarrow 4Fe^{++} + 8e \qquad \text{Anodic Reaction}$$

$$2O_2 + 4H_2O + 8e \longrightarrow 8OH^- \text{Cathodic Reaction}$$

or

$$4Fe + 2O_2 + 4H_2O \longrightarrow 4Fe(OH)_2$$

In the presence of oxygen the ferrous hydroxide will be converted into rust, $Fe_2O_3H_2O$.

It follows that when iron rusts, electrons flow in the metal from the anodic to the cathodic regions and this is accompanied by the movement of ions in the solution. This mechanism has been established by Evans and his co-workers, who have shown that, in the case of a number of metals under laboratory conditions, the spatial separation of the anodic and the cathodic zones on the surface was so complete that the current flowing was equivalent to the corrosion.

In order to inhibit corrosion, it is necessary to stop the flow of current. This can be achieved by suppressing the cathodic reaction, or the anodic reaction, or by inserting in the electrolytic path of the corrosion current a high resistance, which reduces the flow of current to a very small value. These three methods have been called cathodic, anodic and resistance inhibition.

If a paint film is to suppress the cathodic reaction, then it must be a bad conductor of electrons, otherwise the cathodic reaction will be transferred from the surface of the metal to the surface of the paint film. Organic polymer films are not in general electronic conductors and conduction in these systems will be taken to be entirely ionic. Under these conditions, the films must have a sufficiently low permeability to either water, or oxygen, so that either are prevented from reaching the metal surface.

The permeability of paint films to water and oxygen has been measured and the quantities which could diffuse through various organic coatings have been calculated, on the assumption that both were consumed as soon as they reached the oxide coated metal. These values were then compared with the quantities consumed by unpainted specimens when exposed to the open air under industrial conditions, and when immersed in sea-water, and it was concluded that organic coatings, even when pigmented, are so permeable to water and oxygen that they cannot inhibit corrosion by preventing water, or oxygen, from reaching the surface of the metal, that is to say, they cannot afford protection by suppressing the cathodic reaction.

The anodic reaction consists of the passage of iron ions from the metallic lattice into solution, with the liberation of electrons. There are two ways in which this reaction can be suppressed:

1. If the electrode potential of iron is made sufficiently negative, positively charged iron ions will not be able to leave the metallic lattice, i.e. cathodic protection.

2. If the surface of the iron becomes covered by a film impervious to ions, then the passage of iron ions into solution will be prevented, i.e. anodic passivation.

In order to make the potential of iron more negative, the iron must receive a continuous supply of electrons. Polymer films do not contain free electrons and there remains the possibility of obtaining them from a pigment. The only pigments which contain free electrons are the metallic ones, and such pigments will protect iron cathodically if the following conditions are fulfilled:

a. The metallic pigment must be of a metal less noble than iron, otherwise the flow of electrons will be in the wrong direction.

b. The pigment particles must be in metallic contact with each other and with the iron, otherwise movement of electrons will not occur.

It has been found that whereas aluminium, magnesium and zinc pigments all fulfil condition "a" only zinc dust fulfils condition "b"; this is probably due to the fact that zinc oxide is an electronic conductor, owing to the presence of excess zinc in the form of interstitial metal. Paints have been made which afford steel cathodic protection, but the dried film may contain as much as 95 per cent by weight of zinc dust.

Anodic passivation may be brought about by certain pigments, which have a limited solubility. Typical examples are zinc chromate and zinc tetrahydroxychromate; these pigments inhibit by virtue of the soluble chromate ion, which repairs and thickens the air-formed oxide film with material containing chromium in the trivalent form.

Certain basic pigments, e.g. red lead, zinc oxide, calcium carbonate, can also modify the anodic reaction. These pigments form soaps when ground in linseed oil and in the presence of water and oxygen the soaps may autoxidize to yield the salts of mono- and dibasic straight chain acids with a chain length of 7–9 carbon atoms. It has been found that the sodium and calcium salts of azelaic and pelargonic acid are inhibitors of corrosion at around pH 4·8, and it has recently been established that inhibition is associated with the formation of complex ferric salts, which reinforce the air-formed oxide film. The lead salts of these acids inhibit at much lower concentrations than the sodium and calcium salts and recent work has indicated that this may be due to the deposition of small quantities of lead on the iron surface; the cathodic reduction of oxygen is easier at these points, consequently the current density is maintained in the region of ferric film formation and the air-formed film becomes reinforced with material of similar composition.

Although certain pigments may inhibit corrosion by reinforcing the air-formed oxide film either with material of similar composition, or a ferric compound, there are many types of protective paint which do not contain inhibitive pigments and clear varnishes have appreciable protective value. It appears that in such cases these products suppress corrosion by virtue of their high electrolytic resistance, which impedes the movement of ions and thereby reduces the corrosion current to a very small value.

An investigation has, therefore, been made of the factors which affect the electrolytic resistance of unpigmented polymer films when immersed in solutions of electrolytes.

In the case of a freshly cast highly cross-linked film it was found that the resistance was higher in concentrated than in dilute solutions of potassium chloride and that both the temperature coefficient of resistance and the diffusion potentials were high. It was concluded that upon immersion water entered the polymer film and clustered round the ionogenic groups attached to the polymer chains; conduction was then by a process of activated diffusion, a high energy of activation being required to move ions from one ionogenic site through the area of low dielectric constant to the next.

When the freshly cast film was less cross-linked, the resistance followed that of the solution in which it was immersed, and the temperature coefficient of resistance and the diffusion potentials were lower. The water uptake was higher, and under these circumstances ions can probably move along paths of relatively high dielectric constant, which may or may not be physical pores in the membrane.

After the film had been in contact with a solution of an electrolyte for some time an ion exchange process was operative, in which potassium ions exchanged with the hydrogen ions, derived from carboxyl groups attached to the polymer network, and since the potassium ions were more easily ionized the resistance fell. The resistance then followed that of the solution in which the film was immersed and the temperature coefficient was similar to that of aqueous conduction.

Although much work has been carried out on pigmented systems, it has usually been of an empirical nature, commercial products, frequently of unknown composition, have been used, consequently very little fundamental information has been obtained. However, recent results obtained by Kumins indicate that, even with synthetic polymers, considerable pigment/polymer interaction may occur and that this may bring about a profound change in the properties of the pigmented film.

Bibliography

EVANS U. R. (1960) *The Corrosion Oxidation of Metals*, London: Arnold.

KUMINS C. A. (1965) *Official Digest*, **37**, 1314.

SCHREIR L. L. (Ed.) (1963) *Corrosion*, Vol. 2, 15.3, London: Newnes.

British Corrosion Journal (1965/66) **1**, 102, 107, 161.

Proc. Third Internat. Congr. on Metallic Corrosion, Moscow, 1966 (to be published).

J. E. O. MAYNE

PROTON RADIOACTIVITY. Proton radioactivity is the emission of protons from nuclei with a measurable delay. In principle at least three varieties are possible, of which only the first has been detected experimentally.

(*1*) *Beta-delayed proton emission* occurs when a nucleus undergoes positive beta decay to a state of the daughter excited by more than its own proton separation energy. Since the actual proton emission is usually relatively prompt, the protons appear to follow the half-life of the parent beta decay. A similar effect occurs for delayed neutron emission, and the proton effect had been long anticipated. Delayed proton events were first detected by Karnaukhov *et al.* in 1962. The first identification of beta-delayed proton emission and of the beta-decaying precursor, ^{25}Si in this case, was made by Barton *et al.* (1963). Figure 1 shows the energy spectrum of protons from ^{25}Si observed with a solid state detector by Hardy and Bell (1965); there is minor contamination from other delayed-proton precursors (^{17}Ne and ^{21}Mg). The peaks are labelled with centre-of-mass energies in MeV. Figure 2 shows the decay scheme corresponding to Fig. 1. (Later work by Reeder *et al.* (1966) has shown considerably more detail in the decay of ^{25}Si.)

In Fig. 2, the precursor ^{25}Si has a T_3 value of $-3/2$ ($Z = 14$, $N = 11$), and hence an isospin T of

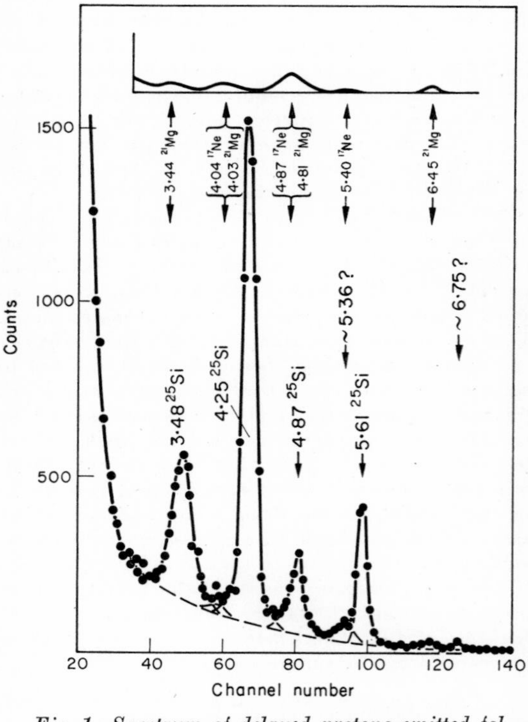

Fig. 1. Spectrum of delayed protons emitted following the decay of ^{25}Si, observed with a silicon detector. The peaks are marked with centre-of-mass energies in MeV. Four peaks are definitely assigned to ^{25}Si and two others are probable; an inset shows possible contamination from protons following ^{17}Ne and ^{21}Mg. (*From Hardy and Bell 1965.*)

3/2. Its strongest decay branch proceeds to a previously-unknown level at 7·91 MeV in ^{25}Al, whose level structure had been relatively well explored. This new level is identified as the lowest $T = 3/2$ level in the $T_3 = -1/2$ nucleus ^{25}Al; it is the analogue state of the ground state of ^{25}Si. On this identification, the beta-decay branch leading to the 7·91 MeV state should be superallowed, and indeed the measured log *ft* value for this transition is 3·0 ± 0·3, in agreement with the theoretical value (Hardy and Margolis 1965). The non-observation of this level in the many previous studies of the levels of ^{25}Al also supports the $T = 3/2$ assignment. If, however, the 7·91 MeV level of ^{25}Al were pure $T = 3/2$, the breakup into the $T = 1/2$ combination ^{24}Mg + p would be forbidden. A measurement of the width of the analogue state would give the degree of $T = 1/2$ admixture in it.

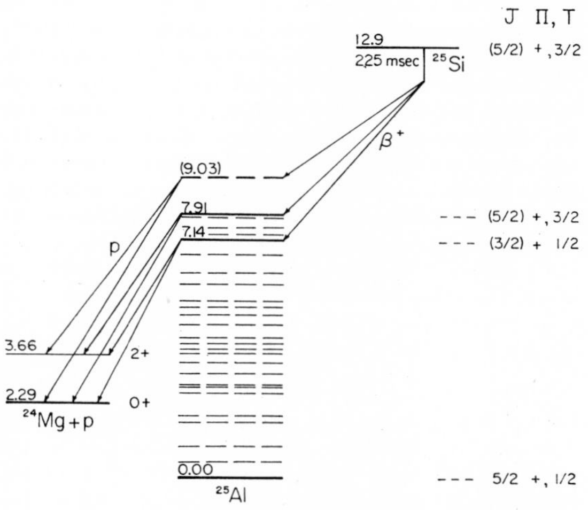

Fig. 2. Partial decay scheme of ^{25}Si to excited states of ^{25}Al, with subsequent proton decay to ^{24}Mg. The excited states of ^{25}Al indicated with broken lines were known from previous work, except that at 9·03 MeV. (From Hardy and Bell 1965.)

A series of similar delayed proton precursors is now known, embracing ^{9}C, ^{13}O, ^{17}Ne, ^{21}Mg, ^{25}Si, ^{29}S, ^{33}A, ^{37}Ca, and ^{41}Ti. All have half-lives of the order of tenths of a second. In the lighter members, the superallowed branch to the analogue state does not appear strongly because the corresponding beta-decay energy is low; starting with ^{21}Mg, however, this branch increasingly dominates the results. Siivola (1965) has also found beta-delayed protons from the decay of ^{108}Te, but no detailed analysis exists. The table gives the main properties of the sequence of delayed-proton cases from ^{9}C to ^{41}Ti; only the more prominent proton groups are included. For the cases ^{21}Mg to ^{37}Ca inclusive, measurements exist (Hardy *et al.* 1966) of the log *ft* values of the superallowed beta decays between the $T = 3/2$ analogue states. The measured values range from 2·9 to 3·6, with errors of about 0·3, so that these beta decays are among the fastest known.

The case of ^{37}Ca has a separate astrophysical interest. The establishment of the ^{37}Ca half-life helps to calibrate the measurement of the flux of solar neutrinos from the yield of the reaction ^{37}Cl(ν, e$^-$)^{37}A, which is the mirror to the beta decay of ^{37}Ca. The delayed-proton results on ^{37}Ca showed that the previous assumptions about the solar neutrino experiment were not seriously in error.

Summary of the known delayed-proton precursors

Nucleus	Q_β (energy of β decay) MeV	$T_{1/2}$ ms	Energies of prominent proton groups, MeV (c.m.) (most intense underlined)
^{9}C	16·5	127	9·25 ($\Gamma = 1·4$) 12·25 ($\Gamma = 0·8$)
^{13}O	18·0	8·7	6·93, 7·55
^{17}Ne	14·7	103	4·04, 4·87, 5·40, 7·39
^{21}Mg	13·1	121	3·44, 4·03, 4·81, 6·45
^{25}Si	12·9	225	1 95, 3·47, 4·25, 4·87 5·61
^{29}S	13·4	195	3·60, 3·86, 5·59
^{33}A	11·9	178	3·26, (3·90), 5·26
^{37}Ca	11·8	173	3·21, 4·12
^{41}Ti	12·2	90	3·13, 3·77, 4·75

(2) *Barrier-delayed proton emission*, similar to ordinary alpha radioactivity, can occur in principle for proton decay energies low enough so that the barrier penetrability is small. The range of appropriate energies is narrow, the absolute energy is very low (a fraction of an MeV), and the process has not been observed.

(3) *Two-proton radioactivity* can occur in principle because owing to the odd-even symmetry effect, a nucleus could be stable against the emission of one proton but unstable against the emission of two protons; cf. double beta decay. Barrier penetration would again provide the mechanism of delay. This process, postulated by Goldanskii (1960), has not been observed.

See also: Delayed neutron emission.

Bibliography

BARTON R. and MCPHERSON R. (1963) *Bull. Am. Phys. Soc.* **8**, 357.
BARTON R. *et al.* (1963) *Can. J. Phys.* **41**, 2007.
GOLDANSKII V. I. (1960) *Nucl. Phys.* **19**, 482.
GOLDANSKII V. I. (1966) *Ann. Rev. Nuc. Sci.* **16**, 1.
HARDY J. C. and BELL R. E. (1965) *Can. J. Phys.* **43**, 1671.
HARDY J. C. and MARGOLIS B. (1965) *Phys. Lett.* **15**, 276.
HARDY J. C., VERRALL R. I. and BELL R. E (1966) *Nucl. Phys.* **81**, 113.
KARNAUKHOV V. A., TER-AKOPIAN G. M. and SUBBOTIN V. G. (1962) Dubna Report P 1072, unpublished.
REEDER P. L. *et al.* (1966) *Phys. Rev.* **147**, 781.
SIIVOLA A. T. (1965) *Phys. Rev. Lett.* **14**, 142.
THEWLIS, J. (Ed.) (1962) *Encyclopaedic Dictionary of Physics*, articles beginning "Proton", **5**, Oxford: Pergamon Press.

R. E. BELL

PULSED NEUTRON RESEARCH. *1. Introduction.* Pulsed neutron research comprises all measurements which entail the generation of a pulse of neutrons of short duration at some stage in the experiment. Three separate, but interrelated, fields of study are pursued by pulsed neutron research, namely, neutron cross-section measurements, measurements of neutron fields and the study of solid and liquid atomic dynamics by the measurement of thermal neutron inelastic scattering. Most of these measurements entail the determination of neutron energy by the time-of-flight technique in which the time taken by neutrons to travel a known distance is measured. In some experiments the decay of the neutron field, generated by the pulses in a given assembly, is observed. In a few experiments, advantage is taken of the short duration of the pulse to distinguish between reactions which occur simultaneously but which are detected at different times after the pulse.

Neutron cross-sections are important in nuclear reactor studies because, when combined with details of reactor geometry and composition, they determine the neutron spectrum (neutron flux per unit energy) in a reactor and, jointly with the spectrum, they determine nuclear reaction rates. They also enable some aspects of nuclear structure to be studied. Thermal neutron inelastic scattering cross-sections are required for the solution of the Boltzmann or neutron balance equation to calculate the neutron spectrum in a given thermal system. Also, because thermal neutron energies (0·025 eV) are about equal to the energy of vibration of atoms in solids and liquids and thermal neutron wave-lengths are comparable with interatomic distances, thermal neutrons can easily exchange energy with atomic lattices and provide, via the inelastic scattering cross-section, a sensitive probe for the study of atomic movement in solids and liquids. Measurements of the neutron field in various media and assemblies enable the methods of solving the neutron diffusion or transport equations as well as the cross-sections used in the calculations to be tested in simple experimental conditions which can be made amenable to mathematical analysis.

2. Principles of time-of-flight technique. The neutron time-of-flight technique enables neutron energies from below thermal neutron energy to several hundred MeV to be determined. The method requires the production of a short burst of neutrons with energies over the entire range to be measured and the subsequent detection of neutrons at a known distance from the source after a measured time lapse. The neutron energy E, usually expressed in eV (1 eV $= 1{\cdot}610^{-12}$ erg), is deduced from the neutron mass m and velocity v from the relation $E = \frac{1}{2}mv^2$. In terms of the flight path length L (metres) and time-of-flight t (μs), the energy is given in eV by

$$E^{1/2} = 72{\cdot}296\ L/t.$$

At energies above about 10 MeV the corresponding relativistic equations must be used. Neutron flight times vary from a few μs to as long as 20 sec depending on the experimental arrangement. For low energy cross-section measurements the neutrons in the burst are produced in some nuclear interaction and typically have energies near the mean fission neutron energy of 2·0 MeV. To provide an adequate number of low energy neutrons, these fast neutrons are slowed down or moderated by collisions in a layer of material containing light nuclei placed near the source. In a single collision with a hydrogen atom the neutron energy E can be reduced to any value from zero to E and the neutrons emerging from a hydrogenous moderator can have any energy below the source energy in a spectrum resembling a Maxwellian distribution at low energy which is joined to the source spectrum by an approximately $1/E$ distribution at intermediate energies. To reduce the loss of neutrons by air scattering, a section of the flight path between the source and the detector is often enclosed in a sealed tube and pumped down to about 1 Torr or filled with helium which has low scattering and absorption cross-sections. The flight path is enclosed in a system of collimators designed to ensure that only neutrons from the neutron source reach the detector. Except in experiments with nuclear explosion neutron sources, data are accumulated from a large number of bursts at rates which vary from one every few seconds to 1 Mc/s.

3. Quality of neutron spectrometers. Two properties which determine the quality of a time-of-flight spectrometer are the neutron intensity, usually expressed as the instantaneous rate of emission of neutrons of all energies from the source during the pulse, and the energy resolution $\Delta E/E$. The energy resolution is given in terms of the error Δt in the time-of-flight and the error ΔL in the flight path by the equation

$$\Delta E/E = 2((\Delta L)^2 + v^2(\Delta t)^2)^{1/2}/L. \quad (1)$$

At low energy where v is small the resolution is governed by ΔL, the error caused by the finite size of the source and detector, while at high energies the resolution is governed by $\Delta t/L$. The timing error Δt depends on the neutron burst width, the timing channel width, the electronic jitter in the detecting and timing equipment and the jitter in the time taken to moderate from the initial energy to energy E. Spectrometers are compared by the ratio $\Delta t/L$ (ns/m) which represents the high energy resolution with the energy dependence removed.

4. Methods of obtaining pulsed neutron sources. Three basically different methods have been used to obtain short bursts of neutrons. These are, by mechanical interruption of a neutron beam by neutron choppers or rotating crystal reflectors, by pulsing particle accelerators and by pulsed reactor techniques and nuclear explosion. The first experiment involving the use of bursts of neutrons was made by Dunning *et al.*, in 1935, who used two disks, carrying radial strips of cadmium, to interrupt the neutrons from a radium-beryllium source encased in a block of paraffin wax.

After the tremendous increase in neutron source strength brought about by the building of nuclear reactors, the combination of nuclear reactor and mechanical chopper developed into a powerful technique for cross-section measurements. The first such experiments were carried out by Fermi using the now famous "Fermi chopper" which consists of alternate laminations of cadmium and aluminium stacked together and enclosed in a steel cylinder. Because of its reliance on cadmium which absorbs neutrons below 0·3 eV this instrument could only be used to interrupt slow neutrons. The most commonly used chopper design nowadays is a variant of the Fermi chopper which is basically a short cylinder or disk, often made of a copper-nickel alloy because of the high neutron scattering cross-section of nickel, with slots cut parallel to one diameter and rotated usually on a vertical axis, in a neutron beam. When the chopper is shut, it scatters neutrons of all energies out of the collimated beam. This design introduced by Seidl *et al.* has given the highest resolution achieved by a chopper of about 5 ns/m obtained using an 18 in. diameter rotor with slits 0·015 in. wide by 1 in. high rotating at 12,000 r.p.m. and a flight path length of 180 m. The energy of neutrons transmitted by a chopper depends on the length and shape of the slots and on the speed of rotation. For straight slots, infinitely fast neutrons are obviously transmitted when the slots are parallel to the neutron beam. Below the "cut-off" energy, however, no neutrons are transmitted; slow neutrons that enter the slots are prevented from emerging because the chopper moves round before these neutrons traverse a diameter. Curved slots can be used to tailor the neutron beam to give a maximum transmission at a chosen energy. When choppers are used to measure neutron spectra, the chopper transmission function must be determined and allowed for.

In measurements of thermal neutron inelastic scattering it is necessary to produce pulses of mono-energetic neutrons. This has been achieved in two ways, by using more than one neutron chopper and by using a rotating crystal monochromator. When a neutron beam passes through two rotors in turn, the first produces a short burst containing neutrons over a wide energy range. Before they reach the second rotor the neutrons spread out in time and only neutrons in a small energy range are transmitted when the second rotor opens. Changing the phasing between the two rotors changes the energy of the neutrons transmitted. In the rotating crystal spectrometer, suggested by Brockhouse, a beam of neutrons is reflected by a rotating crystal on to the scattering sample under investigation. A monochromatic beam of neutrons reaches the sample whenever the crystal is in the correct orientation for Bragg reflection. The energy of the neutrons reflected is changed by moving the sample so as to change the angle between the incident and reflected neutron beams.

Nearly every type of charged particle accelerator has been adapted for use as a pulsed neutron source since the first pulsed accelerator time-of-flight spectrometer, formed by modulating the drive of a cyclotron to produce pulses 4 ms long, was reported by Alvarez in 1938. Both betatrons and synchrocyclotrons, which are intrinsically pulsed accelerators, have been used for pulsed neutron research. The Columbia Nevis synchrocyclotron has been adapted for time-of-flight experiments by using electrostatic deflecting plates to direct the accelerated beam on to a water cooled lead target. The instrument produces 10^{18} neutrons/sec in 20 ns pulses by spallation. The Harwell synchrocyclotron has been similarly adapted to produce 5×10^{18} neutrons/sec in 10 ns pulses or, for high resolution experiments, 3×10^{16} neutrons/sec in 2 ns pulses. In 1949 Cockcroft suggested that a pulsed electron linear accelerator would provide a versatile instrument for producing pulses of neutrons. A pulsed beam of electrons, of about 1 amp peak electron current and pulse width variable from about 10 ns to 2 μs, is stopped in heavy material to produce X rays which then produce neutrons by photodisintegration (γ, n) and, in a fissile target, by photofission. It is possible to direct the electron beam on to a number of targets in turn, arranged so that a number of experiments, using different pulse lengths if necessary, can be carried on simultaneously. At Harwell the primary neutron source is surrounded by a subcritical assembly of ^{235}U, the neutron booster, to increase the neutron output by a factor of 10. Fifteen flight paths radiating from this source can be used at the same time. Van de Graaff machines have been adapted to produce pulses only 1 ns long by top terminal pulsing followed by post acceleration magnetic bunching in which the initial charged particle pulse is compressed in time by causing the early particles to travel further in a magnetic field than the late particles. Neutrons are produced in thin targets by charged particle reactions, such as ^{7}Li(p, n)^{7}Be.

Comparatively inexpensive pulsed neutron sources have been developed for experiments, mainly neutron field measurements, which can be done using less intense sources and long pulse lengths but which require compact flexible sources. Two such sources which produce 14 MeV neutrons from the ^{3}H(d, n)^{4}He reaction have been developed at Karlsruhe. A high voltage pulse is used to extract a deuteron beam, from a plasma formed either in a duoplasmatron or by an intense r.f. field, and to direct it on to a tritium target placed near the exit of the plasma vessel. The neutron flash tube (Neutronenblitzröhre) operated by r.f. gives $2{\cdot}5 \times 10^{14}$ neutrons per second in 2 μs bursts.

At Dubna the pulsed fast neutron reactor IBR has been developed as a pulsed neutron source. The mean reactor power is 6 kW and the peak neutron intensity is $2{\cdot}6 \times 10^{18}$ neutrons/sec for 80 μs pulses at 3·3 pulses per second. The reactor can also be pulsed to a subcritical state and used to multiply the neutrons produced by 30 MeV electrons from a microtron by a factor of 200. Pulses 4 μs wide at 50 pulses per second are produced and a flight path length of 1000 m is used for cross-section measurements.

The most intense terrestrial neutron sources are nuclear explosions. The first experiment to use this source of pulsed neutrons was carried out by Cowan in 1958. A ^{235}U target mounted on a wheel 30 m from a kiloton underground nuclear explosion was rotated at high speed. Sections of the wheel, which has been exposed to neutrons of different energy, were later analysed by radiochemical methods to determine the changes in fission fragment yield with neutron energy. Recently several cross-sections have been measured by detecting neutron reactions in several detectors set up at ground level 187 m above an underground nuclear explosion. High altitude nuclear explosions have also been used for time-of-flight experiments. Albert performed an experiment in 1959 with a path length of 1250 km in which detectors were sent up in a rocket reaching its apogee at the time of the explosion and the data were telemetered back to Earth.

5. *Methods of detecting neutrons.* Neutron detectors are chosen for their efficiency and known spectral response in the case of spectrum, neutron field decay and total cross-section measurements and for their sensitivity to a particular neutron reaction in the case of partial cross-section measurements. Below 100 keV, the most commonly used detectors for spectrum, total cross-section and inelastic scattering cross-section measurements rely on the reactions ^{10}B(n, $\alpha(\gamma)$)^{7}Li and ^{6}Li(n, α)^{3}H. In the reaction between neutrons and ^{10}B, 93 per cent of the interactions give a 0·48 MeV γ ray as well as an α particle. A ^{10}B plug placed in a neutron beam will emit γ rays which can be detected using a NaI(Tl) crystal and photomultiplier. In the proportional counter, neutrons are detected by the intense ionization generated by the α particles from the ^{10}B(n, α) reaction. The ^{6}Li(n, α) reaction is employed in ^{6}Li loaded glass scintillators. At high neutron energy, above 100 keV, detectors sensitive to the proton recoil from hydrogeneous material are used.

Partial cross-section measurements require detectors sensitive to the reaction being measured. In capture cross-section measurements, for instance, the emergent γ ray must be detected. Fission cross-sections can be measured either by detecting the fission fragments using detectors sensitive to the ionization they produce, or by detecting the fission neutrons. Neutron scattering cross-section measurements are made using neutron detectors suitably placed near the scattering sample which is in the neutron beam.

6. *Methods of measuring time of flight.* Pulsed neutron research requires the measurement of neutron flight times ranging from several nanoseconds to a few milliseconds. Two methods of measuring time-of-flight are possible, the analogue method in which something proportional to time-of-flight, such as the length of a trace on an oscilloscope, is recorded and the digital method in which the time of flight is converted to a number which is recorded. Nearly all accelerator and chopper measurements use the digital method which can provide when necessary timing channel widths as short as 1 ns. This method has the advantage of preserving the data in a form readily acceptable for calculation by a digital computer. Experiments using a nuclear explosion source, however, accumulate data at such a high rate all in one burst that only an analogue method can cope with recording the information.

In the digital method two timing pulses are available, one coincident with the neutron burst and one coincident with the time of arrival of neutrons at the detector. The first is used to open a gate which allows timing pulses (typically 1 μs apart) to pass into a binary scaling system until the gate is closed by the detector pulse. The state of the binaries is then either recorded on magnetic tape, or taken as the address of a memory storage location the contents of which are increased by one. In some time-of-flight analysers a condenser is charged linearly, for the time between the two timing pulses, to reach a potential, proportional to time-of-flight, which is then measured by a multichannel pulse height analyser. For fast time-of-flight experiments with 1 ns timing channels, the time is first expanded by a method relying on the rapid linear charging of a condenser during the time-of-flight to be measured, followed by a much slower linear discharging for a time which is measured by time analysers as described above. The output of all time analysers gives the number of events, which is proportional to the neutron interaction rate, in each timing channel.

By contrast, the nuclear explosion time-of-flight technique uses an analogue method in which the reaction rate (e.g. the current from a fission chamber which in these experiments can reach 1 amp) is used to deflect an oscilloscope trace which is swept repetitively every 50 μs. This trace is photographed on 35 mm film travelling at 100 ft/sec. Each detector has its own oscilloscope and camera which record data up to 5 ms after the explosion. Afterwards, so that the data can be processed by a computer, the traces are read and converted to digital form by a projection comparator provided with a card punch.

7. Cross-section measurements. A large fraction of partial and total neutron cross-section measurements of importance in reactor design and of more specific nuclear physics interest are carried out by the time-of-flight technique. Pulses of neutrons are formed as described in section 4, moderated in a layer of hydrogeneous material to produce a wide energy spectrum, and detected at a known distance from the source. If $\Phi(E)\,\mathrm{d}E$ is the flux of neutrons of energy between E and $E + \mathrm{d}E$ at the detector, $\sigma_A(E)$ is the cross-section at energy E for the neutron reaction detected, N_A is the number of atoms of the material in the beam and ε_A is the efficiency for detecting a reaction, then, for a thin sample the observed reaction rate in the energy range E to $E + \mathrm{d}E$ is given by

$$A(E)\,\mathrm{d}E = \varepsilon_A N_A \Phi(E)\,\sigma_A(E)\,\mathrm{d}E. \quad (2)$$

The flux per unit energy, $\Phi(E)$, is normally determined, or eliminated, by a second experiment with a detector employing a known cross-section.

Total neutron cross-sections are determined from the ratio of the count rate from a neutron detector with and without a sample of the material to be measured in the beam. This ratio, known as the "transmission", T, is given by

$$T = \exp(-n\sigma)$$

where n is the number of atoms per unit area in the beam. Accurate values of T can be obtained for values of $n\sigma$ within a certain range and measurements are made for several different sample thicknesses to ensure that $n\sigma$ remains within this range for all values of σ.

Hopkins and Diven have used pulses of monoenergetic neutrons from a Van de Graaff accelerator to measure the ratio, α, of the capture to the fission cross-section for several fissile isotopes. In their experiments the energy of the neutrons was determined by the Van de Graaff energy and the pulsed nature of the source was used to enable γ rays produced in both fission and capture reactions, which were detected almost instantaneously in a large cadmium loaded liquid scintillator, to be distinguished from neutrons, caused by fission reactions only, which were detected after the fission neutrons had slowed down in the scintillator and become captured by the cadmium with the consequent emission of γ rays.

Neutron cross-sections are characterized by Breit–Wigner resonances the distribution of whose parameters and energy spacings throw light on nuclear structure. The determination of these parameters from the measured cross-sections is complicated by the need to remove the effect of Doppler broadening, caused by the thermal motion of the atoms, and the effect of instrumental energy resolution. Once the resonance parameters are known it is possible to calculate the change in the Doppler broadened cross-section with temperature and hence, the Doppler coefficient of nuclear reactor reactivity.

8. Measurements of neutron fields. The pulsed neutron source method has been applied in a variety of ways to the study of the detailed interaction of neutrons with non-multiplying and with multiplying (fissionable) media. The measurements made are either of the decay of the neutron field or of the neutron spectrum in a medium or assembly. The spectrum $\Phi(E)$ is determined from equation 2 by measuring the reaction rate $A(E)$ as a function of time-of-flight with a detector of known spectral response ($\varepsilon_A N_A \sigma_A(E)$). Pulsed source experiments consist of a system, such as a cube of graphite with a source placed inside or on one face, and either a counter in the system which enables the neutron intensity to be measured or a re-entrant hole to allow neutrons from the inner face of the hole to emerge and travel down a flight path for time-of-flight measurement of the spectrum.

8.1. Non-multiplying systems. Under certain conditions, a thermalized neutron field in a moderator will decay exponentially with time and the decay constant α can be measured as a function of the geometric buckling B^2. The α versus B^2 curves are represented by the expansion

$$\alpha = v\overline{\Sigma_a} + D_0 B^2 - CB^4 + FB^6 + \ldots \quad (3)$$

in which v is the neutron velocity, Σ_a the macroscopic absorption cross-section, D_0 the diffusion coefficient and C is known as the diffusion cooling coefficient. In principle an analysis of the α versus B curves yields values for all these parameters and enables the cross-sections to which the parameters are related to be checked.

If the detector is made sensitive to neutrons of a given energy, such as a Sm—Cd shielded Pu fission chamber which detects neutrons only at 0·3 eV, the slowing down time from source energy to this energy can be determined by observing the count rate in such a counter after the neutron burst. The count rate reaches a maximum after a time delay corresponding to the slowing down time.

Bergmann *et al.* have established the relation between slowing down time and neutron energy for a 2 m cube of lead by injecting 1 μs pulses of 14 MeV neutrons into the block and observing the capture γ rays from substances of known cross-section placed in the lead block. In heavy materials the neutron energy spread at any slowing down time is small, being only 11·4 per cent (r.m.s. value) for neutrons of energy below 1 keV in lead. Cross-section measurements with energy resolution of this order are made in the "slowing-down-time spectrometer" by observing reaction rates as a function of slowing down time for materials placed in the lead block.

Beghian has developed a method of studying the diffusion of high energy neutrons by measuring the decay of the high energy neutron field in heavy media. In one experiment a pulsed neutron source from the ^{7}Li(p, n) reaction was used to irradiate one face of an 8-in. cube of lead with 1·24 MeV neutrons. A plastic scintillator on the opposite face was biased so as to detect only neutrons of energy near the source energy. Neutron energy loss by elastic collisions in heavy material is negligible and, in the absence of inelastic scattering, the diffusion of the fast neutrons can be described by a one velocity group theory. After the decay of spatial harmonics, the neutron field decays exponentially with a decay constant given by

$$\alpha = v\Sigma_r + D_0 B^2 + C_T B^4$$

Here Σ_r is a removal cross-section which accounts both for neutron absorption and for inelastic scattering to an energy below the detector bias, D_0 is the diffusion coefficient and C_T represents a transport theory correction to the elementary diffusion theory. Again analysis of the α versus B curve allows these parameters to be found.

8.2. Neutron thermalization and spectra. Pulsed neutron research techniques have been used extensively to probe the methods of calculating neutron spectra in a reactor and to check the cross-sections used in these calculations. Although neutron spectra represent a stage in reactor calculations at which the results can be compared with measurements, few reactors are amenable to direct spectrum measurements. Some measurements of spectra in subcritical assemblies representing specific reactor lattices have been made to obtain spectra which are then used in calculations on reactors with the same lattice structure. The most fruitful research has resulted from measurements of neutron spectra in subcritical or non-multiplying assemblies designed to emphasize some aspect of the neutron balance equation. Three experimental arrangements have been used in making these measurements.

(a) A continuous source of neutrons, normally from a reactor is used to excite the system and the spectrum is measured using a chopper.

(b) In the Poole technique, a pulsed source is used to excite the system and the spectrum of an extracted beam is measured by time-of-flight.

(c) A pulsed source is used to drive the system and a chopper is used to extract a pulse of neutrons for spectrum measurement at a given time after the source pulse. In arrangement (c) the neutron spectrum can be observed as it changes during the neutron slowing down and thermalization time. When a pulse of neutrons is introduced into a system, the time dependent flux $\Phi(E, t)$ is given in the diffusion theory approximation by the neutron balance equation

$$\frac{1}{v}\frac{\partial \Phi(E,t)}{\partial t} = -[\Sigma_a(E) + \Sigma_s(E) + D(E)\,B^2]\,\Phi(E,t) + \int_0^\infty \Sigma(E' \to E)\,\Phi(E',t)\,\mathrm{d}E' + S(E)\,\delta(t) \quad (4)$$

where $\Sigma_a(E)$ and $\Sigma_s(E)$ are the macroscopic absorption and scattering cross-sections and $\Sigma(E' \to E)$ is the energy transfer kernel which represents the probability that a neutron of energy E' will change its energy to E in an inelastic scattering collision. $D(E)$ is the diffusion coefficient, B^2 is the buckling given by

$$\nabla^2\Phi + B^2\Phi = 0$$

and $S(E)\,\delta(t)$ represents the pulsed source. By allowing for the time taken by the neutrons to slow down to a given energy the spectrum measured can be made to yield $\Phi_0(E) = \int_0^\infty \Phi(E, t)\,\mathrm{d}t$ which is the required spectrum for a system with a steady source. In an infinite non-absorbing medium, where $B^2 = \Sigma_a = 0$, $\Phi_0(E)$ is strictly Maxwellian. For finite absorption cross-sections, the spectrum in an infinite medium is sensitive to the energy transfer kernel $\Sigma(E' \to E)$ but not to the angular distribution of the scattered neutrons or the energy dependence of the scattering cross-section Σ_s. Neutron spectra near boundaries are sensitive to the form of the scattering kernel and also to the angular dependence of the scattered neutrons. These facts have been exploited by measuring spectra in uniformly poisoned media and varying the concentration of absorber material (using for instance a solution of cadmium salts in water), by measuring spectra at a boundary formed by a temperature discontinuity and by measuring the spectra near a boundary such as a sheet of cadmium in water. In all experiments care is taken to ensure that the spectrum of the neutrons emerging from the system in the direction of the flight path is representative of, or easily related to, the spectrum in the system.

8.3. Multiplying systems. Pulsed source techniques have been developed for the measurement of the reactivity ϱ of subcritical systems. These techniques were suggested by Von Dardel and first used by Sjöstrand who derived a relation for the reactivity in terms of the prompt and delayed multiplication of a pulsed source. For an experiment in which a subcritical system is pulsed at a rate which is high compared with the delayed neutron decay constants, the delayed neutrons build up to a time independent level which is distinguishable from the prompt neutron distribution. If A_1 is the area under the prompt neutron distribution, then

$$A_1 = S/(1 - k(1 - \beta))$$

where k is the multiplication constant, β the delayed neutron fraction and S is proportional to the source strength. If A_2 is the area under the delayed distribution, then

$$A_1 + A_2 = S/(1 - k)$$

and the reactivity is given by

$$(1 - k)/k = \beta A_1/A_2.$$

This treatment applies only to neutrons in the fundamental spatial mode and methods of circumventing this limitation have been developed by Garelis and Russel with their $k\beta/l$ technique. Here l is the effective prompt neutron lifetime. Again, a time dependent prompt neutron density $N_p(t)$ and a constant delayed neutron density N_d are distinguished and determined experimentally and used to find $k\beta/l$ by an iterative procedure from the equation

$$\int_0^\infty N_p(t) \exp[(k\beta/l)\, t]\, dt - \int_0^\infty N_p(t)\, dt = N_d/R \quad (5)$$

in which R is the pulse repetition frequency.

Simmons and King have obtained the reactivity of a system from the exponential decay constant of prompt neutrons in the fundamental spatial mode.

9. Thermal neutron inelastic scattering. In measurements of thermal neutron scattering, monoenergetic neutrons of energy E and wave number $\mathbf{k}$ are produced and scattered by the sample through an angle θ, to emerge with energy E' and wave number $\mathbf{k}'$. The energy E' is measured by time-of-flight experiment for a large number of values of the angle θ and experiments are often arranged to make measurements at all these angles simultaneously. From these measurements the double differential cross-section $d^2\sigma/d\Omega\, dE$ where Ω is a solid angle, is calculated. Van Hove has shown that this cross-section can be expressed by the equation

$$\frac{d^2\sigma}{d\Omega\, dE} = \frac{\sigma_f}{4\pi}\left(\frac{A+1}{A}\right)^2 \frac{\mathbf{k}}{\mathbf{k}'} S(\mathbf{Q}, \omega) \quad (6)$$

where σ_f is the free atom cross-section, A is the ratio of nuclear to neutron mass, $\mathbf{Q} = \mathbf{k} - \mathbf{k}'$ is proportional to the momentum transfer and $\hbar\omega = (E - E')$ is the energy transfer in a collision. The cross-section thus factorizes into two terms one of which depends on the neutron-nucleus interaction and the other, the so-called scattering law $S(\mathbf{Q}, \omega)$ which is a function of the momentum and energy transfer only and depends solely on the dynamics of the scatterer. Measurements of $d\sigma/d\Omega\, dE$ enable the energy transfer kernel $\Sigma(E' \to E)$ required in the reactor neutron balance equation to be calculated and also enable important information on the atomic dynamics of solids and liquids to be obtained from the scattering law $S(\mathbf{Q}, \omega)$. Van Hove has introduced a space-time correlation function $G(\mathbf{r}, \tau)$ of a system which is defined as

$$G(\mathbf{r}, \tau) = \frac{1}{N} \sum_{m,n} \int d\mathbf{r}' \overline{\delta\{\mathbf{r} + \mathbf{R}_m(0) - \mathbf{r}'\}\, \delta\{\mathbf{r}' - \mathbf{R}_n(\tau)\}} \quad (7)$$

Here N is the number of particles present, $\mathbf{R}_m(0)$ is the position vector of particle m at time $\tau = 0$, $\mathbf{R}_n(\tau)$ is the position vector of particle n at time τ, δ is the Dirac δ-function and the bar denotes a time average. In the classical limit $G(\mathbf{r}, \tau)$ gives the probability that, given a particle at the origin at time zero, there will be a particle (either the same one or a different one) at $\mathbf{r}$ at time τ. The scattering law $S(\mathbf{Q}, \omega)$ is the space and time transform of the correlation function $G(\mathbf{r}, \tau)$, and is given by

$$S(\mathbf{Q}, \omega) = \frac{1}{2\pi\hbar} \int d\mathbf{r}\, d\tau [\exp i(\mathbf{Q} \cdot \mathbf{r} - \omega\tau)]\, G(\mathbf{r}, \tau). \quad (8)$$

Egelstaff has shown that for liquids it is simpler to relate $S(\mathbf{Q}, \omega)$ to a velocity correlation function. The "spectral density" of the motion of a single atom in the system, $z(\omega)$, can be obtained from the expression

$$z(\omega) = \frac{4M\omega}{\hbar} \sinh\left(\frac{\hbar\omega}{2kT}\right) \left[\frac{S(\mathbf{Q}, \omega) \exp[-\hbar\omega/2kT]}{Q^2}\right]_{Q \to 0} \quad (9)$$

For a crystal $z(\omega)$ is equal to the frequency distribution.

Measurements of $z(\omega)$ have been made for a large number of solids and liquids and as an example of the type of conclusions that can be drawn from the experiments we quote a summary, given by Larsson, relating to atomic movements in water:

1. The motion of the scattering centres cannot be described by a gas model or by a simple diffusion model.

2. The diffusion process is probably delayed by a time τ_0 of about 10^{-12} sec.

3. This delay time permits the development of vibratory motions in a relatively fixed or slowly moving position. All vibrations with frequencies $>1/\tau_0$ Are developed and contribute to the frequency spectrum.

4. The structure of the frequency spectrum reveals several details about the molecular and protonic motions.

5. Small energy transfers of the order of 0·6 meV of unexplained origin seem to exist.

Bibliography

BECKURTZ K. H. and WIRTZ W. (1964) *Neutron Physics*, Berlin: Springer-Verlag.

BOBIN K. J. (1961) in *Encyclopaedic Dictionary of Physics* (J. Thewlis Ed.), **4**, 827, Oxford: Pergamon Press.

BROWN B. (1961) in *Encyclopaedic Dictionary of Physics* (J. Thewlis Ed.), **4**, 824, Oxford: Pergamon Press.

EGELSTAFF P. A. (Ed.) (1965) *Thermal Neutron Scattering*, New York: Academic Press.

GREY MORGAN C. (1961) in *Encyclopaedic Dictionary of Physics* (J. Thewlis Ed.), **4**, 826, Oxford: Pergamon Press.

HARVEY J. A. (1961) in *Encyclopaedic Dictionary of Physics* (J. Thewlis Ed.), **4**, 815, 828, Oxford: Pergamon Press.

HAVENS W. W. JR. (1962) in *Encyclopaedic Dictionary of Physics* (J. Thewlis Ed.), **5**, 12, Oxford: Pergamon Press.

HUGHES D. J. (1953) *Pile Neutron Research*, New York Addison-Wesley.

NÈVE DE MÉVERGNIES M., VAN ASSCHE P. and VERVIER J. (Eds.) (1966) *Nuclear Structure Study with Neutrons*, Amsterdam: North-Holland.

PALEVSKY H. (1961) in *Encyclopaedic Dictionary of Physics* (J. Thewlis Ed.) **4**, 817, Oxford: Pergamon Press.

PHILLIPS G. C., MARION J. B. and RISSER J. R. (Eds.) (1963) *Progress in Fast Neutron Physics*, Chicago: The University Press.

Proceedings of AEC-ENEA Seminar on Intense Neutron Sources, Santa Fe (1966). To be published.

Pulsed Neutron Research, Proceedings of a Symposium, Karlsruhe (1965) Vienna, International Atomic Energy Agency.

SAILOR V. L. (1961, 1962) in *Encyclopaedic Dictionary of Physics* (J. Thewlis Ed.), **4**, 827, **5**, 11, Oxford: Pergamon Press.

SPAEPEN J. (Ed.) (1961) *Neutron Time-of-Flight Methods*, Brussels: European Atomic Energy Community (Euratom).

WILLIAMS M. M. R. (1966) *The Slowing Down and Thermalisation of Neutrons*, Amsterdam: North-Holland.

G. D. JAMES

PULSE SHAPE DISCRIMINATION. In many forms of nuclear radiation detector, where the arrival of a charged particle causes an electrical impulse to be generated, it is often possible, by examination of the shape of the pulse, i.e. the variation of the electrical signal with time, to obtain some information on the initiating event, for example on the type of particle entering the sensitive volume of the detector, or on the position at which the event occurs. This technique is generally referred to as Pulse Shape Discrimination.

Scintillation detectors. One of the earliest proposals for pulse shape discrimination was made by Wilkinson (1951), who proposed that by combining two layers of scintillator with different decay times and viewing the combination with a single photomultiplier it would be possible, by separating the two components of the combined light output, to measure the energy deposited in each phosphor. This technique could be used to measure rate of energy loss of the incident particle (dE/dx), and the total energy E. Since the product ($E.dE/dx$) is proportional to the mass of the particle the nature of the incoming particle can be identified. Detectors of this type, sometimes known as "Phoswich" detectors have found a variety of applications. For example, in contamination monitors, by employing a phosphor consisting of silver activated zinc sulphide phosphor powder (decay time about 3 μs) coated on a layer of plastic scintillators (decay time about 3 ns) it is possible simultaneously to survey areas for alpha and beta contamination with a single detector.

Another application (Jones 1960) has employed a plastic phosphor surrounding a caesium iodide scintillator (decay time 0·7 μs for electrons) in order to differentiate between charged particles generated within the caesium iodide by gamma rays, and charged particles entering from outside.

In many phosphors, the mechanism of energy transfer from the primary exciting particle to the luminescent centres is dependent on the excitation density (dE/dx), hence on the mass of the exciting particle. This was described for the thallium activated alkali halide phosphors in 1954 by Eby and Jentscke (1954). In general it is found that for thallium activated NaI, KI, KBr CsI, the light emission from electron excitation takes 10–40 per cent longer than for alpha particle excitation.

Broadly the light output from thallium activated sodium iodide and caesium iodide under alpha particle excitation rises promptly and decays exponentially, but under electron and gamma ray excitation, the light pulse rises more slowly reaching its maximum in about 50 ns in the case of NaI-Tl, 250 ns in CsI-Tl, before decaying exponentially with approximately the same time constant as the alpha particle excitation. These features of the decay have been used to discriminate between electrons or gamma rays and protons, deuterons and alpha particles.

The variation in the shape of light output pulse from organic phosphors irradiated by different particles was first reported by Wright (1956), and the suggestion that this could be applied to the important problem of distinguishing between neutron and gamma ray excitation of hydrogenous organic phosphors was made by Brooks (1956).

The luminescence decay of the organic phosphors can broadly be characterized as having a fast component of a few nanoseconds duration (usually 6 ns in stilbene and 45 ns in anthracene) followed by a slow component persisting up to several microseconds (Owen 1958). Lightly ionizing particles, electrons and gamma rays produce a greater fraction of the faster component, so circuits capable of separating the fast component can identify and reject gamma background in neutron detectors depending on proton recoil scintillation. Of many crystalline organic phosphors, e.g. naphthalene, anthracene, quaterphenyl which show discrimination in this way, stilbene is easily the best. A notable exception is toluene (diphenyl acetylene) which does not show any pulse shape variation with different particles. Some liquid phosphors give good discrimination provided dissolved oxygen is removed either by boiling or by bubbling an inert gas (e.g. argon) through the solution before use. These include p-terphenyl (4 g/l.) or PPO or PBD in xylene or toluene, with or without POPOP as a spectral shifter. Discrimination is also observed in organic mixtures containing methyl borate, e.g. 4 g/l. PBD or PPO + 0·1 g/l. POPOP in a solution of trimethyl borate (47%) xylene (31%) and Naphthalene (22%), although the low intensity of the light flash from the (nα) reaction following neutron capture in the baron, (equivalent in light output to about 40 keV electron energy) makes pulse discrimination difficult owing to the low numbers of photoelectrons carrying the signal information (Jackson and Thomas 1965). Mixtures containing diphenyl hexatriene give no discrimination. Suitable plastic scintillators give pulse shape discrimination between protons and electrons but in general these mixtures have been less attractive than the liquid or solid crystalline phosphors.

Pulse shape discrimination has also been applied to distinguishing between gamma ray and slow neutron induced pulses in lithium-loaded cerium-activated glass phosphors (Coceva 1963). Other systems for neutron/gamma discrimination have employed silver-activated zinc sulphide powder dispersed in a hydrogenous media, for fast neutrons, or in a boron or lithium compound for slow or thermal neutron detection.

A large number of circuits have been developed to examine the pulse shape (Owen 1962); they can be classified into three main groups:

(*a*) *Linear sampling circuits.* In these the initial fast part of the pulse is sampled by passing it through a short time constant filter (lumped CR or distributed line networks), or by means of an active sampler which admits into an integrator the fraction of the pulse arriving during the period for which a gate is opened. The sample is then expanded and shaped and may be compared with the light emitted during the remainder of the pulse, or the total light. A convenient way of comparing the processed sample with the total pulse is simply to subtract one from the other so that the net output is, say, positive or negative according to the type of particle incident. More sophisticated electronic circuits can be employed to develop a pulse whose amplitude is proportional to the ratio between the early and later stages of the pulse.

(*b*) *Non-linear circuits.* Non-linear effects in the photomultiplier or other circuits may be used to separate pulses of different shape. One method is to operate the anode of a photomultiplier at a potential only a few volts above that of the last multiplying surface. During the high current peak of the pulse, space charge between the anode and the last dynode limits the secondary electron current leaving the dynode and the dynode is driven negative by the bombarding primary electrons; during the long tail of the pulse the space charge can be removed by the weak field from the anode and the dynode bchaves normally, becoming positive as more secondary electrons leave than the number of primaries arriving. The voltage between the anode and last dynode is usually chosen so that pulses from the last dynode due to gamma rays striking an organic scintillator are wholly negative while those caused by neutrons are negative initially and then swing positive due to the longer lived component of the light pulse. A simple amplitude discriminator is then all that is required to select the neutrons events.

(*c*) *Zero cross-over discriminators.* The pulse from the photomultiplier is passed through a linear network, e.g. a double differentiating CR network which causes it to swing first positive and then negative. The time before the signal crosses through zero is dependent on the ratio of the light in the early to later stages of the scintillation. By measuring the delay between the initiation of the pulse and its time for zero cross-over discrimination between pulses of different shape is possible.

Pulse shape discrimination can also be used to distinguish between thermally emitted electrons from the photocathode, and pulses originating from light flashes which are of such low intensity that only a few photoelectrons are produced. By accepting only single electron pulses which occur in pairs within the decay time typical of the phosphor, much of the random thermal background can be rejected (Damerell 1962). This technique has been applied to counting low energy beta particles (tritium) incident on silver activated zinc sulphide.

Pulse shape discrimination in proportional counters. The problem of rejecting counts due to a lightly ionizing particle when it is required to detect a heavy ionising is often met in proportional counters. Bennett (1962) describes the use of pulse shape discrimination to reject pulses due to secondary electrons, produced whilst registering pulses produced by proton-recoils from fast neutrons. Proton-recoils from fast neutrons have short ionizing paths, compared with the paths of secondary electrons produced by gamma rays, so that the ionization arrives at the centre wire within a much shorter time-interval and for this reason pulses have a faster rise time. This enables these pulses to be distinguished from gamma ray background.

A similar method can be used to reject carbon recoils in a methane filled proportional counter.

Pulse shape discrimination in semiconductor detectors. The instantaneous value of the current generated in a solid state detector due to equal numbers (N) of positive and negative charges (e) produced within the depletion layer width (W) is given by

$$I(t) = \frac{Ne}{W}[(\mathrm{d}x/\mathrm{d}t)_n + (\mathrm{d}x/\mathrm{d}t)_p]$$

where $\left(\frac{\mathrm{d}x}{\mathrm{d}t}\right)_n$ and $\left(\frac{\mathrm{d}x}{\mathrm{d}t}\right)_p$ are the drift velocities of the negative and positive charge carriers respectively in the direction of the applied electric field. The velocities can be related to the electric field E up to velocities of order 10^6 cm/sec by the relation $\frac{\mathrm{d}x}{\mathrm{d}t} = \mu E$, where μ is the mobility.

As the mobilities of positive and negative carriers are different ($\mu_n = 1300$ and $\mu_p = 500\ \mathrm{cm^2/Vsec}$ in silicon at 300°K), and often, as in p-n junction detectors the electric field is not uniform, the current waveform depends on where the carriers are formed. By examination of the current pulse shape (Legg 1965), or the rise time of the integrated current pulse it is possible to distinguish between, say, alpha particle pulses, which deposit most of their energy within a few microns of the edge of the junction, and say protons or deuterons of equal energy which have a range 50 times longer.

In cases in which the charged particle passes through the depletion layer into the undepleted material, the output pulse consists firstly of a fast component during which electrons and holes are moved out from the depletion layer, followed by a slower pulse which arises from the diffusion of minority carriers produced in the undepleted material which as they reach the depletion layer are swept across it to give an output current. The time t for carriers to diffuse a distance l is given approximately by

$$t = \frac{l^2}{D}$$

where D is the diffusion constant for the minority carrier (D is related to mobility by Einstein's relation $D/\mu = \boldsymbol{k}T/\boldsymbol{e}$ where $\boldsymbol{k}$ is Boltzmann's constant). In silicon, at room temperature, the diffusion time for carriers from 1 mm beyond the depletion layer is of order 10^{-3} sec.

In germanium at liquid nitrogen temperature diffusion across 1 mm takes about 30 μs. The technique has been applied to gamma-ray spectroscopy with germanium. By rejecting those pulses with slow tails, one is rejecting pulses from electrons which have not expended the whole of their energy within the sensitive volume.

Pulse shape discrimination to provide position information. In some forms of detector, the connexion to the distributed detector capacitance can be made by a distributed resistive element. For example, a resistive contact can be made to the depletion layer capacitance of a solid state detector. Signal caused by motion of charges across a section of the capacitor is transferred to the amplifier in effect down a resistor-capacitor transmission line, and will have a fast rise for events occurring at the end close to the amplifier, and a slow rise if they occur at the distant end. Measurement of the rise time will provide information on the position of the initiating event.

The same technique can be applied to proportional counters having a suitable resistive centre wire or cathode. In these applications the rise time of the pulse is converted into a signal whose amplitude depends on position by passing through a differentiating CR circuit. Other forms of position sensitive detectors based on distributed detector capacitance and resistance have been described, but these normally employ two connexions and two amplifiers to develop a total charge or total energy signal.

Conclusion. The addition of positional information or particle identification to rate, time of arrival and energy of the particles incident on a detector provides a valuable facility in many experimental situations. In the past, techniques of amplification, using CR and linear shaping networks, have tended to require that amplifiers be presented with pulses as nearly identical as possible. However, development of faster amplifiers employing active integrators will probably lead to wider use of the information available in pulse shape.

Bibliography

Bennet E. F. (1962) *Rev. Sci. Instrum.* **33**, 1153.
Brooks F. D. (1956) *Progress in Nuclear Phys.* **5**, 251.

Coceva C. (1963) *Nucl. Instrum. and Meth.* **21**, 93.
Damerell C. J. S. (1962) *Nucl. Inst. and Meth.* **15**, 171.
Eby F. G. and Jentscke W. K. (1954) *Phys. Rev.* **96**, 911.
Jackson H. E. and Thomas G. E. (1965) *Rev. Sci. Instrum.* **36**, 419.
Jones F. C. (1960) *IRE Trans. Nucl. Sci.* NS 7, No. 2, 175.
Legg J. C. (1965) *Nuclear Instrum. and Meth.* **36**, 343.
Owen R. B. (1958) *IRE Trans.* NS **3**, 189.
Owen R. B. (1962) *IRE Trans.* NS 9, No. 3, June.
Wilkinson D. H. (1951) *Rev. Sci. Instrum.* **23**, 414.
Wright G. T. (1956) *Proc. Phys. Soc.* (B) **69**, 358.

R. B. Owen

R

RADIATION DAMAGE IN CRYSTALS. *Introduction.* When an untreated crystal, supposedly perfect, is bathed with energetic radiation, defects and imperfections are normally created. The spatial arrangement of defects which remains after the irradiation has ceased is generally described as *radiation* damage, and any physical property change which may result from the introduction of the damage is called a *radiation effect.* Irradiation with energetic particles is a new and very powerful tool. Since a large number of defects can be introduced into a crystal in a reasonably well-controlled manner we consequently have a measure of control over these defect-sensitive properties which may be exploited in various devices.

The nature of the damage which may be produced is directly dependent on the nature of the particle. The distinguishing feature about the neutron, for example, is the fact that it carries no charge so that it suffers no energy loss by Coloumb interaction with the electrons in the solid. In interactions with the nuclei, however, there is an energy transfer and if the total kinetic energy of the participating particles is the same after the event as it was before, such a collision is said to be *elastic.* In an *inelastic* collision some part of the initial kinetic energy is converted into another form, such as electromagnetic radiation, or into energy of excitation of the atom as a whole.

The interactions of charged particles (e.g. electrons, protons, deuterons, fission fragments, etc.) with crystalline matter pose quite a different problem, since both Coulomb interaction and elastic collisions can occur. At high initial particle velocities a large amount of energy is spent in the excitation of electrons bound to lattice atoms or, in a metal, of the more freely moving conduction electrons. As the particle slows down the energy communicated to such electrons assumes less importance and before it comes to rest elastic collisions with lattice atoms are dominant. Gamma irradiation also constitutes a rather special type of bombardment because interaction is mainly with the electrons of the bombarded crystal. The photo- and Compton effects give rise to light and gamma quanta and fast electrons are also produced. When interactions with nuclei do occur they are likely to be followed by gamma-absorption and, subsequently, by the emission of a neutron (the so-called γ-n reaction). If this occurs the atom recoils and for a full explanation of the process of energy dissipation and damage creation the total history of the recoiling atom must also be investigated. In fact, for most types of radiation, in the final analysis we are necessarily intimately concerned with describing what happens when a lattice atom is given a large quantity of energy.

Point defects. The energy which is required permanently to remove a lattice atom from its site is known as the displacement energy or the displacement threshold. The magnitude of this parameter is not only dependent on the direction in the crystal from which it receives a "kick", but also on the material itself. Energies ranging between a few and a hundred electron volts are not uncommon. The displacement threshold, moreover, is not sharp but is "blurred out" by the existence of temperature effects in the crystal and by a complicated dependence on crystallographic direction, though above a certain energy it is always possible for the atom or ion to be ejected by the impinging particle. When this occurs it is described as a *primary* knock-on, and "primaries" can be very energetic indeed. Often they are capable of losing energy themselves by both ionization and further displacement, so that they give rise to secondary, tertiary and higher order displacements. A shower or "cascade" (displacement cascade) of Frenkel defects (vacancy/interstitial pairs) may be deposited in the crystal following the onset of a high number of displacement collisions.

In some circumstances it is possible for a moving atom in a solid to collide with another atom and to replace rather than displace it. Such a replacement collision exhibits many similarities with the phenomenon of diffusion in the solid state and may be regarded as a "dynamic interstitialcy interchange". Furthermore, replacement collisions can have a more profound damaging effect than might be at first imagined, since though the possibility of replacement might not affect the total number of atoms which are finally displaced from lattice sites it must be expected to cause a quite considerable amount of disorder in binary material (an alloy for example).

Ideally what we would really like to be able to predict is the density of Frenkel defects which may be initiated under a given set of irradiation conditions with a known specimen crystal. If the incident flux of bombarding particles is Φ per cm^2 per sec (this is more strictly referred to as the flux density), and τ is the duration of the bombardment, then the number of primary knock-ons produced in a unit volume will be

$$n_p = \Phi\tau n_0 \sigma_d \quad (1)$$

where σ_d is the so-called cross-section for displacement and n_0 is the spatial density of lattice atoms. Now the primary knock-ons are produced with a certain energy spectrum, depending upon the nature of the initial interaction, and so we suppose that any one, in turn, generates on the average $\bar{\nu}$ displacements. The total number of displaced atoms, including the primaries themselves, will therefore be given by

$$n_d = n_p \bar{\nu}. \quad (2)$$

It is a relatively easy matter to obtain values for n_0 and also for σ_d since our theoretical knowledge about the displacement and collision cross-sections is quite profound and based upon firm foundations. There are considerable experimental difficulties in determining the integrated flux $\Phi\tau$, but our knowledge about $\bar{\nu}$ and $\nu(E_p)$, the number of displaced atoms arising on the average from a single knock-on of energy E_p is very poor. The only way to approach the problem is to make drastic assumptions, one of which concerns the nature of the "interaction potential" or the way in which the colliding atoms interact. If they are regarded as behaving like hard spheres then the total number of displacements produced by a primary of energy E_p becomes $E_p/2E_d$, where E_d is the displacement energy, and the analysis of equation (2) is then possible.

Extended defects. The way in which a primary displaced atom interacts with the lattice and loses energy depends on its energy and charge, and on the ratio of its mass to that of the lattice atoms. Since in elemental materials this ratio is one, or nearly so, the energy loss in a collision can be quite considerable. Moreover, if the primary has sufficient energy it will be ionized to some degree and will consequently carry an effective positive charge. It will hence interact in a Coulomb fashion with the electrons of the crystal and will dissipate energy by electron excitation and ionization. As the primary slows down electrons are collected and the effective charge is reduced to zero; the remaining energy is then communicated to the crystal by elastic collisions with lattice atoms. The transition from the regime in which ionization losses are dominant to that in which elastic collisions are of most importance is, of course, a gradual one, though as a useful rule of thumb it may be noted that ionization and electron excitation are only dominant when the energy of the knock-on, in keV, is greater than its atomic weight or mass number A.

The existence of other modes of energy leads to further important concepts of radiation damage physics and to processes by means of which extended defects rather than point defects are created. Consider what happens to the energy which is delivered to a lattice atom in a glancing collision which is *not* followed by displacement. Since the energy Q which the struck atom accepts is less than the displacement energy E_d it clearly cannot be stored as potential energy at some defect, but must be dissipated in another fashion. Evidently we must expect the struck atom to vibrate about its mean position on the lattice and to transfer energy to its neighbours. A very useful way to regard this effect is in terms of the sudden delivery of a pulse of heat to the lattice, which in this example has spherical symmetry. A more energetic *thermal spike* with the same symmetry may be supposed to be initiated by a primary event in which a primary knock-on with an energy of say 300 eV is produced. This spherical thermal spike is then more intense than the former and embraces several dynamic displacement processes, so that its effect may be two-fold. In the first place it may be that some annealing will take place and there may be fewer point defects created than we might have expected. Alternatively it may be that the existence of the temperature pulse is an aid to the clustering of point defects so that extended defects such as small dislocation loops or point defect complexes are created.

A much more vicious thermal spike, this time with cylindrical symmetry, is introduced into the crystal when the bombarding particles are highly energetic and highly charged particles. In the case of fission fragment bombardment several thousand electron volts are delivered to the specimen for every Ångstrom unit of every particle trajectory in the solid. Temperatures as high as several thousand °K endure for times of the order of 10^{-11} sec and certain crystals, particularly the insulators and some of the compound semiconductors, cannot endure this treatment. Along the path of each particle thermally activated Frenkel defects are produced in large numbers, plastic flow and dislocation loop formation occurs, and in some circumstances a transition to an amorphous state ensues.

In the *displacement spike* we have a situation in which it is no longer possible to think of individual unrelated atomic displacements. When the primary knocked-on atom is moving in its parent lattice and its mean free path for displacement collisions λ_d falls below an interatomic distance a large number of neighbouring vacancies are created in a very short time. The atoms which are displaced from nearby sites are forced out and away from the path of the primary so that a multiple displacement of n atoms is produced. This may be thought of as a multiple vacancy of n lattice sites surrounded by n single interstitials. All atoms within one interatomic distance of the trajectory of the primary are displaced as secondaries, and each of these displaces, on the average, one tertiary, so that the result is a shell of interstitials surrounding a core of vacancies. It is not easy to visualize the final effect on the lattice of the initiation of a displacement spike, though there are two distinct dynamic phases. Firstly, the outward motion of the secondary and tertiary displaced atoms must be expected to have some kind of mixing and stirring effect in the lattice. Secondly it is the atoms between the interstitial shell and the void which may be considered to feel the major influence of the temperatures and pressures exerted, so that these are the atoms which are first forced back into the multiple vacancy. Therefore, if some kind of recrystallization does occur it will be preceded by such a turbulence that very few atoms may be considered to remain on, or return to, their original lattice sites. A few Frenkel defects may remain but the great majority of interstitials and vacancies are expected to anneal. Extended defects, such as regions of mismatch, and tangles of dislocation loops, may also be formed.

The introduction of certain refinements into the theory and equations of radiation damage physics leads to the prediction of new phenomena and the explanation of old. Perhaps the most important modern-day development has been the effect of "softening" the interatomic interaction potential and of introducing the lattice structure. If we think of a primary knock-on moving in a crystalline solid in which the size and "sponginess" of the atoms on their sites is a function of the energy of primary it becomes clear that atomic displacement is favoured at the lower energies. Furthermore it will be seen that there are certain crystallographic directions in which the knock-on can move with comparative ease while in other directions it moves towards a seemingly impenetrable wall of atoms. If the energy of the knock-on is not so great then the lattice surrounding the site of a primary collision event is able to impose rigid conditions upon the possible modes of momentum transfer. Thus, when the struck atom cannot penetrate the surrounding lattice it can only transfer energy and momentum to one of its immediate neighbours, and so the possibility of a collision correlation becomes apparent. Sequences of collisions corresponding to an energy pulse can be propagated along close-packed rows of atoms. This is a focusing phenomenon. The bundle of energy which passes down a row of atoms—a *focuson*—is said to have propagated by means of a "*focused collision sequence*". If the close packed row of atoms is bordered by nearby atoms then these also may exert some influence on the behaviour of the focuson, and an assisted focusing process similar to the action of a convergent lens on a beam of light may occur. When the nearby atoms lie close to the "optic" axis then the atomic lens is strong, focusing is enhanced and the effective focal length is small. Moreover, if the focused collision sequence is sufficiently energetic, it can be accompanied by a series of sequential atomic replacements known as a *dynamic crowdion.* A dynamic crowdion loses energy to the surrounding lattice, emitting phonons, and can eventually degrade into a focuson. An interstitial atom is then deposited at a point very distant from the vacancy from which it derived. Yet another possibility is that a displaced atom, instead of propagating as a dynamic crowdion *within* a close-packed row of lattice atoms, can travel *between* two or more such rows so that it is effectively "channelling" in a mass-free zone. Like the focuson the *channelon* is capable of transmitting mass as well as energy away from a primary collision volume.

If we introduce the changes which these new concepts have revealed into the earlier theories then some of the models for the detailed atomic processes have to be modified. The channelling and focusing of mass and energy out of a displacement cascade region makes it certain that a core of vacancies remains behind. The displacement spike therefore leads to the introduction of a *depleted zone* and the interstitials are projected much further from the initial spike than it was hitherto supposed.

Modes of investigation. The influence of point and extended defects is experienced by a crystal lattice at quite some distance from the defects themselves, and because of resulting changes in the lattice potential it proves possible to discover measurable "radiation effects." Investigations of radiation damage in crystals are therefore most often made by "working backwards," as it were, from measurements of the changes induced by bombardment in certain crystal properties (thermal conductivity, specific heat, electrical resistivity, optical absorption, etc.). There are two particular experimental techniques, however, which because of their high magnification and resolution allow an observer to "look inside" a crystal and to examine the arrangement and motion of the constituent atoms. They employ the field-ion and electron microscopes. These instruments are finding increasing use in studies of radiation damage in crystals.

Bibliography

Barnes R. S. (1966) in *Encyclopaedic Dictionary of Physics* (J. Thewlis Ed.) Suppl. Vol. 1, 316, Oxford: Pergamon Press.

Brandon D. G. (1966) in *Encyclopaedic Dictionary of Physics* (J. Thewlis Ed.), Suppl. Vol. 1, 103, Oxford: Pergamon Press.

Chadderton L. T. (1965) *Radiation Damage in Crystals,* London: Methuen.

Clarke F. J. P. (1961) in *Encyclopaedic Dictionary of Physics* (J. Thewlis Ed.), 6, 9, Oxford: Pergamon Press.

Dienes G. J. and and Vineyard G. H. (1957) *Radiation Effects in Solids,* New York: Interscience.

Kiss A. E. (1961) in *Encyclopaedic Dictionary of Physics* (J. Thewlis Ed.), 2, 26, Oxford: Pergamon Press.

Thompson M. W. (1966) in *Encyclopaedic Dictionary of Physics* (J. Thewlis Ed.), Suppl. Vol. 2, 64, 94, 377, Oxford: Pergamon Press.

Thompson W. (1961) in *Encyclopaedic Dictionary of Physics* (J. Thewlis Ed.), 2, 465, Oxford: Pergamon Press.

L. T. Chadderton

RADIATION PROTECTION, PRINCIPLES OF. The fact that ionizing radiations can cause biological damage was realized soon after these radiations were discovered. Unfortunately, the early visible effects of biological damage, e.g. skin reddenning, were often considered as being not serious by many of the early radiation workers; late effects, such as the induction of cancer, brought the realization that biological damage caused by excessive exposure could be very serious indeed.

Although there were early attempts to introduce recommendations concerning the safe use of ionizing radiation rapid progress was not possible until a unit suitable for measuring a quantity of ionizing radiation was devised; this unit, the Roentgen, has been used since 1928. In that year an International Commission on Radiological Protection was formed (although it did not assume this name until 1950), and this body has, from time to time, issued its recommendations concerning maximum radiation doses which it considers to be without significant hazard. These values are now called "maximum permissible doses" and the International Commission on Radiological Protection recommend values not only for radiation workers, but for the general population and a number of so-called "special groups" of persons.

Within recent years the increase in the use of ionizing radiations has meant that the study of methods of achieving protection against such radiations has become very important. Although finally the responsibility for radiation safety is a national one, the International Commission on Radiological Protection have issued recommendations which have found widespread acceptance and form the basis of radiation protection standards in many countries.

The essential problem facing the International Commission on Radiological Protection and national bodies having responsibility concerning radiation protection is to attempt to state "safe" levels of radiation exposure. But this is a complex problem, for it requires an answer to the question: "How safe?". The earliest recommendations adopted the attitude that a "maximum permissible dose" (at that time called a "tolerance dose") was an amount of radiation which would not produce significant damage during the lifetime of the person exposed.

At first sight this seems a satisfactory statement and it is not difficult to specify dose values which conform to this ideal. However, ionizing radiation can cause damage to any cells of the body and therefore the germ cells of individuals may be damaged by these radiations. It is in this regard, the "genetic effect" of radiation, that ionizing radiations present safety problems which are different from those posed by many other hazardous materials which are used in research and in industry. It was necessary for the International Commission on Radiological Protection to take this fact into consideration when formulating its recommendations and for this reason the "permissible dose" is now considered to be:

"... that dose, ... which, in the light of present knowledge, carries a negligible probability of severe somatic or genetic injuries; furthermore it is such a dose that any effects that ensue more frequently are limited to those of a minor nature that would not be considered unacceptable by the exposed individual and by competent medical authorities".

The current recommendations of the International Commission on Radiological Protection list maximum permissible doses; it is of interest to note that, as these values have been revised from time to time, the trend of the recommendations has always been toward lower dose values (Dennis 1966).

Implicit in the definition of "permissible dose" is the fact that all exposure to ionizing radiation causes biological damage.

For this reason the establishment of maximum permissible dose values is said to involve a "philosophy of risk". Simply, this is that although ionizing radiations produce many beneficial effects, e.g. the use of X rays and radioisotopes in medicine, research and industry, their use also causes some damage to persons exposed and so it is necessary to assess the relative benefit and the harm caused by exposure.

The basic problem of radiation protection is to set values of maximum permissible dose so that this damage is acceptable. If by "safe" is meant literally "without any risk of damage" it is clear that such standards are not possible when dealing with exposure to ionizing radiations.

It is the purpose of recommendations concerning radiation protection to ensure that the benefit of using ionizing radiations exceeds the risk to society and to the individual from their use. The risk to society is in the possible genetic damage caused to the population (the "genetic effect" of radiation); the risk to the individual is to his health and well-being, (the "somatic effect" of radiation.)

As a basis of consideration, the International Commission on Radiological Protection and the national protection committees are able to consider the fact that the entire human race is exposed to ionizing radiation from natural background radiation. This exposure is a minimum level below which it is not possible to reduce radiation exposure even if all man-made sources of ionizing radiation were no longer in existence. Typically this represents a dose of about 0·1 rem per year, although the value does vary from place to place geographically. The basic recommendations of radiation protection involve obtaining answers to the following questions:

1. By how much could the radiation exposure be increased without causing unacceptable genetic damage to society?

2. What is a reasonable maximum permissible dose so that significant somatic damage is not caused?

For whole body exposure the limiting factor in determining maximum permissible dose values is the significance of the genetic damage caused by ionizing radiations. In circumstances where this damage is not the limiting factor (e.g. radiation exposure to the extremities) the maximum permissible dose values are based on the somatic effects of ionizing radiation.

Radiation protection authorities assume that the mean age of child-bearing is 30 years. Assuming a value of 0·1 rem per year this represents a genetically significant dose of 3 rem in 30 years on the average. But as it is known that mutations occur spontaneously the question arises: "What fraction of spontaneous mutations is caused by ionizing radiation?".

It is possible, at present, only to estimate this, and the opinion of the International Commission on Radiological Protection is that from 2 to 20 per cent of spontaneous mutations are caused by natural background radiation. To double the number of mutations in a generation is likely to require an average exposure of 50 to 5 times natural background radiation, i.e. between 150 and 15 rem in 30 years. Taking a conservative estimate the International Commission on Radiological Protection have assumed a limit of 5 rem to age 30 years; lower than the more pessimistic of the values estimated above.

The basic principle of radiation protection recommendations is the assumption that if, on the average, the radiation dose to individuals is below this figure, the genetic damage produced will not impose an unacceptable burden on future generations of the society. It is considered that the advantages to be gained from the use of ionizing radiations outweigh the possible damage which would be caused by doubling the mutation rate. At present, average doses to populations in highly developed countries (where ionizing radiations find their most common use) are not near this limit.

Considering an average genetic dose of 5 rem to age 30 years it is apparent that as this is an average figure, a small fraction of the population could be permitted a dose higher than this, provided that the average dose to the remainder of the population is correspondingly reduced. The reason for this is that the total genetic damage can be considered to form a "pool" shared among the total society.

Suppose that the population contains N persons and the average genetic dose to age 30 years has been set, by the reasoning above, at D rem. The total genetic damage will be proportional to DN. Now if a fraction, f, of the population is permitted a higher average dose, say d_1 rem, then the average dose to the remaining fraction $(1 - f)$ must be reduced to say d_2 according to the relation:

$$fNd_1 + (1 - f)\,Nd_2 = DN$$

$$\therefore f\left(\frac{d_1}{D}\right) + (1 - f)\frac{d_2}{D} = 1.$$

The first term on the left-hand side of this equation is that part of the total genetic dose which is apportioned to the fraction, f, of the population. The fraction, f, and the permissible dose, for this group are inversely proportional to one another for a given apportionment of dose to the group. The previous discussion shows that the maximum permissible radiation dose to a group such as (1) could reasonably be greater than that to a group such as (2) providing that the values for (1) are not so great as to produce significant somatic damage.

On the basis of arguments such as these the International Commission on Radiological Protection have set an average weekly whole-body exposure for radiation workers of 100 milli-rem corresponding to 5 rem per year and 60 rem to age 30 years (for it is assumed that radiation workers are not employed in this capacity before age 18). The maximum permissible exposure to members of the general population has been set at one-tenth of the value for radiation workers. In practice the International Commission on Radiological Protection also includes "special groups" of persons whose exposure is likely to exceed that of the general population discussed above.

These ideas explain the apparent paradox contained in radiation protection recommendations that those who are employed as radiation workers (who therefore may receive significant exposure) are permitted to receive a larger exposure than is considered to be acceptable for a member of the general public. The paradox is resolved by considering that, at the low exposure levels now technically achievable in work with ionizing radiation, significant somatic damage should not occur and the genetic damage must be considered to be averaged in the society in future generations.

Bibliography

Recommendations of the International Commission on Radiological Protection (1960) Oxford: Pergamon Press.

Godfrey B. E. (1962) in *Encyclopaedic Dictionary of Physics* (J. Thewlis Ed.), **6**, 11, Oxford: Pergamon Press.

D. J. Rees

RADIOACTIVE AEROSOLS. A radioactive aerosol may be defined as a suspension in air of liquid or solid particles containing radioactive substances. The particle size may vary from 10^{-3} to 50 μm (μm = micrometre or micron = 10^{-4} cm = 10^4 Å). Particles larger than about 50 μm fall out rapidly and are not maintained in the atmosphere by normal turbulent motions. Radioactive vapours are not strictly aerosols, but it is sometimes difficult to be certain whether a particular nuclide is present as vapour or aerosol, and the latter term is therefore sometimes loosely used to include vapours.

There is a distinction between dusts on the one hand and smokes and fumes on the other. Dusts are formed by comminution processes on solids, and usually show a wide range of particle sizes and shapes. When plutonium or uranium is oxidized at moderate temperatures in a stream of air, the oxide is eroded from the surface and a dust is formed (see figure). Smokes or fumes are formed by the condensation from the vapour phase. The basic particles are very small (~0·1 μm), spherical or crystalline in shape, and the aerosol consists of chains or globular aggregates of these particles (Stewart 1963, 1966). A fume is often formed when a metal is machined, since the high but localized excess temperature causes vaporization of metal.

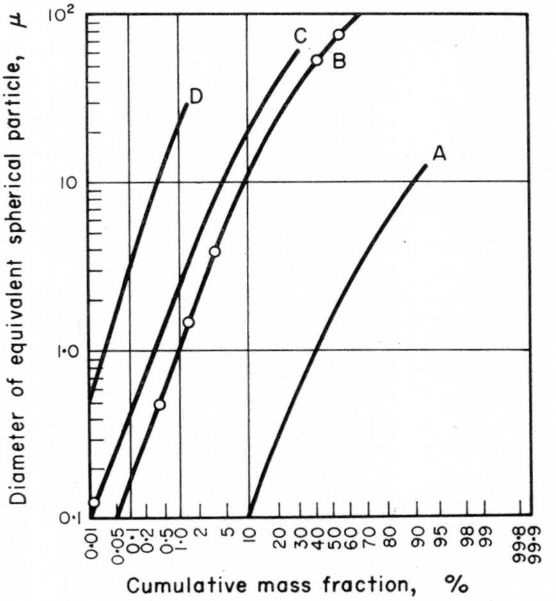

Particle size distributions produced by the oxidation of plutonium under a variety of conditions (Stewart, 1963). Curve A: material airborne under static conditions at all temperatures in air; curve B: oxide formed at 123°C from S a stabilized alloy; curve C: oxide formed at 400°C to 500°C; curve D: oxide formed above ignition point.

The aerosol particle may be wholly formed of radioactive matter, but more commonly a radioactive substance is incorporated in or adsorbed on an aerosol particle of adventitious origin. A monolayer of ^{131}I or other carrier free activity on natural dust will give a highly radioactive aerosol.

Naturally-occurring radioactive aerosols. The noble radioactive gases radon and thoron are exhaled from the Earth's surface to the atmosphere, where they decay and give radioactive daughter products. These products are isotopes of polonium, bismuth and lead, which are solids at ambient temperatures. The decay products are formed as individual ionized atoms, and in dust-free air may remain as a vapour for some time. Normally, however, the decay products become attached to the natural Aitken nuclei, which have diameter 0·01–0·1 μm, and are present in numbers from 10^3 to 10^4 per cm^3 of air in the lower atmosphere. The activity of radon and thoron decay products falls off with height above ground, and is low in the stratosphere and over oceans. It is high in mines, cellars and badly ventilated buildings, especially if the building materials (e.g. granite) contain traces of uranium or thorium.

Radioactive nuclei are also produced by bombardment of atmospheric gases in the stratosphere by cosmic rays. The most notable activity so formed is ^{14}C, which circulates as CO_2 and not as an aerosol, but ^{22}Na, ^{35}S, ^{7}Be, ^{32}P and ^{33}P which are also formed become attached to naturally occurring Aitken nuclei. These activities can be detected in rain, but in amounts very small compared with the decay products of radon and thoron (Junge 1963).

Artificially produced aerosols. The testing of nuclear bombs produces a radioactive aerosol, the physical characteristics of which depend on the power and location of the explosion. If the bomb is detonated near the ground, a large amount of surface material is volatilized into the fire-ball, and the fission products are incorporated into the glass-like particles which are formed when the contents of the fire-ball condense. The larger particles are deposited near the test site, but particles up to about 10 μm diameter can be carried thousands of miles. Stratospheric explosions disseminate the activity into a very large volume of air containing relatively little solid material. The aerosol is very fine (0·01 to 1 μm), and is carried to great heights in the stratosphere, whence it diffuses downwards only slowly. The peak world-wide activity from nuclear explosions was reached some years ago, since when there has been a decline. Even at the peak, the activity of mixed fission products in air at ground level was small compared with the short-lived natural activity.

Measurement of aerosol activity. Airborne activity is usually measured by collecting the aerosol from a known volume of air on a filtering medium or in a bubbler. For health physics measurements of alpha-active dust a high sampling rate (100 l/min) is often required, since the maximum allowable concentration (MAC) is low and the aerosol may consist of only a few relatively large particles per cubic metre of air.

It is usually necessary to allow the radon daughter products to decay before counting the activity. If delay is unacceptable, discrimination between plutonium dust and radon decay products may be made by alpha pulse-height discrimination, analysis of pseudo-coincidences, or mechanical separation of the aerosols (Jehanno *et al.* 1958).

A sampler such as the May Pack (Chamberlain *et al.* 1963) is used for radioiodine, which may be present in air in a variety of gaseous and particulate forms. The pack contains a copper gauze, membrane filter, charcoal loaded filter paper and charcoal pad in series to ensure adequate sampling and enable some discrimination of the physical form to be made.

The distribution of activity by particle size (which is not necessarily the same as the distribution of particle size itself) can be determined approximately by use of the cascade impactor, in which the aerosol is drawn through a series of graded orifices and the particles impacted on slides (Green and Lane 1964). For sub-micron aerosols, some estimate of particle size can be obtained by use of a diffusion battery, in which the aerosol is drawn between closely-set parallel plates. If more detail is required, it is necessary to use the techniques of autoradiography combined with photomicrography (Quon 1959).

Health hazards. The potential danger of breathing radioactive aerosols has been recognized for many years, since the miners in the uranium mines of Joachimstal (Czechoslovakia) suffered from a form of lung cancer, now considered to have been caused by the radon decay products in the lung. Some radioactive substances inhaled as aerosols are rapidly removed from the lung and transferred to other parts of the body. Thus inhalation of radiostrontium results in irradiation of the bone, and inhalation of radioiodine causes irradiation of the thyroid. Insoluble radioactive aerosols are retained in the lung for a period depending on the particle size and many other factors, and cause a localized irradiation of lung tissue. Permissible levels of radioactive substances in air are fixed by the International Commission on Radiological Protection.

Bibliography

CHAMBERLAIN A. C., EGGLETON A. E. J., MEGAW W. J. and MORRIS J. B. (1963) *Reactor Sci. and Tech.* **17**, 519.

GREEN H. L. and LANE W. R. (1964) (2nd Edn.) *Particulate clouds: dusts, smokes and mists*, London: Spon.

International Commission on Radiological Protection Publication No. 2 (1959) *Permissible Dose for Internal Radiation*, Oxford: Pergamon Press.

JEHANNO C., BLANC A., LALLEMANT C. and ROUX G. (1958) *Proc. 2nd United Nations International Conference on Peaceful uses of Atomic Energy*, **23**, 372.

JUNGE C. E. (1963) *Air Chemistry and Radioactivity*, New York: Academic Press.

PEIRSON D. H. (1966) in *Encyclopaedic Dictionary of Physics* (J. Thewlis Ed.), Suppl. Vol. 1, 102, Oxford: Pergamon Press.

QUON J. T. (1959) *J. Am. Ind. Hyg. Assoc.* **20**, 61.

STEWART K. *Progress in Nuclear Energy*, Series IV, Vol. 5 (1963), and Series XII, Vol. 2 (1966).

A. C. CHAMBERLAIN

RADIOACTIVE CHROMIUM IN BIOLOGY AND MEDICINE. Chromium-51 decays primarily by K-capture but approximately 9 per cent of the disintegrations produce X rays of 0·32 MeV; the half life of ^{51}Cr is 27·8 days. This radionuclide has important applications in biology and medicine, for it can be used to measure not only the *blood volume* in a living organism but also the *red blood cell survival time*. The reason for these applications is that ^{51}Cr can conveniently be used to label red blood cells. This is accomplished by the use of a solution of sodium chromate ($Na_2{}^{51}CrO_4$) which is added to a volume of blood taken from the subject. Labelling is accomplished by incubating the red cells with the sodium chromate solution for a few minutes; the cells are washed and centrifuged to remove excess chromium which has not entered the cells. The sodium chromate binds to the haemoglobin molecules in the red cells, is reduced to the trivalent state, and (except for a small elution), remains bound until the red cells die. The ^{51}Cr which is released from a red cell when it dies is then excreted and does not re-label other cells *in vivo*, because in the trivalent state ^{51}Cr does not bind onto red cells; the original tagging was made with ^{51}Cr in the hexavalent state. For blood volume studies in humans the total activity needed is about 30 μCi; for red cell survival studies about 100 μCi is used. Blood volume may be determined by using the dilution principle. First, a known volume (v) of the solution containing the labelled red cells is injected into the subject; some solution is kept to be used as described below. After a few minutes to allow mixing throughout the body a blood sample is withdrawn and the activity of this is measured using a scintillation counter. Let the activity of this sample be S_1 counts per minute. From the solution previously set aside, a volume is taken and used as a standard. The radioactive concentration of this is much greater than in the circulatory blood and so this sample is diluted by a known factor (say f) for convenience. The activity of a sample of this solution is counted as before:
let the activity measured be S_2 counts per minute. The blood volume V will be given by

$$V = \frac{S_2}{S_1} fv.$$

From a measurement of the haematocrit the red cell volume can be determined.

The measurement of blood volume is of considerable value in the management of patients who are in a state of severe shock or dehydration, e.g. after massive haemorrhage, surgical shock or extensive burns. The results of blood volume determination are very useful for determining the amount of intravenous fluids necessary to restore a normal blood volume and so maintain the blood pressure and a normal electrolyte balance.

Red cell survival measurements are carried out by obtaining blood samples from day to day and investigating the reduction in concentration of radioactivity of ^{51}Cr with time. The activity in a given volume of blood decreases with time for three reasons:

(1) the natural decay of the ^{51}Cr
(2) elution of ^{51}Cr for the cells tagged and
(3) the death of the tagged cells.

This activity in a given volume is calculated as a fraction of that of the earliest sample. A graph of this fraction against time results in a curve which when extrapolated to the time axis gives a measure of the red cell survival, typically about 120 days.

The reduction to zero activity corresponds to the death of all the cells which were originally tagged with the ^{51}Cr.

Often the graph is of an exponential form, and when this occurs, a mean cell life (analagous to radioactive mean life) corresponding to the time when 37 per cent of the original activity remains in the blood is used as an index of red cell survival. Corrections for the natural decay of the radioisotope, and for the slow elution of ^{51}Cr from the cells before they die, are necessary so that this graph represents the reduction in activity due to the death of the red cells only. The exact shape of the red cell survival curve has been subject to much investigation (Veall and Vetter 1958; Beirwaltes *et al.* 1957). Red cell survival values are of particular use in the study of certain anaemias, for in these cases a reduced red cell survival time is observed.

Haematocrit. The haematocrit is the fraction of the blood volume which is occupied by blood cells. The term is used synonymously with the phrase "packed cell volume". Experimentally the haematocrit is measured by centrifuging a blood sample (under standard conditions) and directly measuring the fraction of the total volume into which the cells become packed. For highest accuracy a correction may be made to allow for "trapped" blood plasma between the cells. A normal haematocrit is about 45 per cent in humans.

Bibliography

Beirwaltes W., Johnson P. and Solari A. (1957) *Clinical Uses of Radioisotopes,* London: Saunders.

Robertson J. S. (1962) in *Encyclopaedic Dictionary of Physics* (J. Thewlis Ed.), **6**, 50, 51, 56, 57, Oxford: Pergamon Press.

Veall N. and Vetter H. (1958) *Radioisotope Techniques in Clinical Research and Diagnosis,* London: Butterworths.

D. J. Rees

RADIOACTIVITY MEASUREMENTS, LOW LEVEL. Measurements on the radioactivity of materials are of interest in a wide variety of scientific disciplines and over an enormous range of activity level: a nuclear power reactor may contain a total of some 10^8 curie of activity of various kinds and a source used for cancer radiotherapy some 1000 Ci of ^{60}Co, while the occurrence in a particular location of as little as 10^{-15} Ci of ^{232}Th may be of concern to a geochemist (note: 1 curie $= 3{\cdot}7 \times 10^{10}$ disintegrations per second). Not surprisingly, the problems and techniques of measurement differ considerably between the high and low ends of this activity range but, in a great many fields of application, it is only the relatively low activities that are of interest.

Becquerel's observations leading to the discovery of radioactivity in 1896 would probably count today as low level measurements, as, certainly, would J. J. Thomson's observation of the radon content of drinking water in Cambridge in 1902. The real impetus towards the systematic development of low-level techniques and their extension towards an increasingly high degree of sensitivity did not come, however, until the advent, in the 1940's, of nuclear weapons and nuclear power, which brought with them not only a major problem in control of environmental contaminants but also a rapid growth in all fields of applied radioactivity.

One of the major fields of application of low-level radioactivity measurement is in the assessment of human radiation dose. For members of the general population a considerable fraction of such a dose is due to radioactive materials located in the body and it is important to be able to assess this either directly or by measurement of levels of radioactive materials in the diet, in the atmosphere and in the environment generally. This need was highlighted by the problem of weapons fall-out but applies equally to naturally occurring radionuclides. A related problem is the monitoring of individuals who are exposed either occupationally or for medical reasons to radiation from internally deposited radionuclides, and this constitutes a major branch of the subject of Health Physics.

A second important field of application of low-level radioactivity measurement is in the use of radioactive tracers in the study and control of a wide range of physiological, chemical and industrial processes. The need to work with minimal quantities of radioactivity is dictated here primarily by the requirement to minimize the radiation exposure of investigators and, in medical applications, the subjects of investigations.

Radioactive dating constitutes a third field of application: the ability to measure traces of both the natural and artificial radionuclides that are introduced at specific points in naturally occurring environmental processes makes it possible to obtain information on the time scale and dynamics of such processes. Nuclear weapons debris, cosmic-ray induced nuclides and terrestrially originating radon isotopes and their decay products have all been used in studies of bulk movements both of atmospheric air and of ground and ocean water. Relative or absolute concentrations of several of the longer lived natural radionuclides, and particularly ^{14}C, have been shown to provide accurate indications of the age of materials of archaeological and geological interest. As an extension of this field, instruments are now being designed and built with which it is hoped to be able to measure natural radionuclide concentrations in mineral samples retrieved from the surface of the Moon.

A further application of low-level measurement is in the neutron activation analysis technique for estimating concentrations of stable elements in a material by virtue of the radioactivity induced following its exposure to neutrons.

From the foregoing brief review of applications it will be seen that, while the interest in a low-level radioactivity measurement may sometimes be in low absolute activity levels as such, very often it is the mass or volume specific activity that will be the significant parameter. In principle it is usually possible, by chemical or physical means, to concentrate the activity of a low specific activity material. In practice, however, the problems of contamination by chemical reagents and of inadvertent losses, compounded by uncertainty as to the chemical and physical state of the active material of interest, often make it desirable to carry out measurements with the minimum of interference to the sample. Furthermore, in a number of types of application, e.g. human whole body monitoring, it is a basic requirement of the measuring technique that it shall be non-destructive. Techniques for low specific activity measurements are thus of considerable importance.

A fundamental limitation on the sensitivity of a low-level radioactivity counter is imposed by its background counting rate B, which is additive to but statistically independent of the counting rate S due to the source alone. It is important to note, however, that for the low-level situation ($S \ll B$), the figure of merit that is appropriate in any comparison of various possible counting arrangements is the ratio S^2/B. The importance of achieving a high counting efficiency is thus clearly seen.

Background signals may arise from a number of sources, the principle of which are natural external β-γ radiation, cosmic radiation, external man-made radiation, radioactive contamination of the constructional materials of the detector and electronic interference and noise. The relative importance of each of these components, and thus the effectiveness of corresponding measures towards background reduction, will depend on the particular application. In an α-particle detector clear discrimination can usually be provided against less densely ionizing particles and the problem of background reduction reduces essentially to one of eliminating alpha contamination in the counter itself. When detecting β particles and γ rays, externally originating background can be a severe problem and, for the achievement of the lowest possible background, it is frequently necessary to surround detectors with massive shields of very low activity material (e.g. steel, water, chalk, lead) and also to use an anti-coincidence arrangement. In this, the primary detector is partly or entirely surrounded by a second detector whose output is connected in such a way as to block registration of any primary detector pulses that occur in coincidence with a secondary detector pulse; such events being largely (or entirely, in the case of β-counting) due to externally originating background.

The relatively short range of α-particles (typically 10–50 μ in solid materials) is a major factor influencing the design of alpha detectors, particularly for low specific activities where detectors of considerable area may be required in order to observe an adequate mass of source material. In practice, for thick source counting at the lowest levels ($\sim 10^{-13}$ Ci/g), detector areas of 100 cm^2 are necessary while, for energy spectroscopy using sources sufficiently thin to avoid appreciable source self-absorption of particle energy, areas of up to 10,000 cm^2 are used. For thick sources a scintillation counter arrangement based on a thin screen of silver activated ZnS crystals is most commonly employed. To eliminate background due to accumulation of contamination, source materials may be permanently sealed, together with expendable screens, in thin, transparent walled capsules. The radon isotopes constitute an important group of gaseous α emitters, whose estimation is also commonly carried out using ZnS detectors, either directly or following collection of their respective solid, α-emitting decay products. For spectroscopy of thick sources silicon semiconductor detectors provide a resolution of 1 per cent or better but are at present limited in area to a maximum of 10 cm^2. Where larger source areas are to be analysed, gas counters, working in the ionization chamber or proportional region, must be used. Information on spatial distribution of α activity can be obtained by the technique of autoradiography, in which a photographic emulsion is exposed for a suitable period of time in intimate contact with the material or object of interest and is subsequently, following development, scanned microscopically for the characteristic α-particle tracks.

The wide range of maximum energies exhibited by β paricles ($\sim$0·018–3 MeV) entails a corresponding variety in appropriate techniques for their detection. At the higher energies β particles can penetrate a thin window forming one wall of a gas counter (working in either the proportional or Geiger region) and a variety of counters of this type are in common use for work with solid and liquid sources. Counters based on the use of a thin sheet of organic or inorganic (usually CsI) scintillator can also be used for this application and, since they can be windowless, are capable of working to lower beta energies, although they are still subject to the limitation of source self-absorption. Semiconductor devices are coming into widespread use as β detectors and, because of their small physical size, have found application, for example, as implanted detectors for measuring local concentrations of artificially administered β emitters in humans and animals. When suitable shielding is used, background count rates of some 1 count per minute over a 5 cm^2 source (at an efficiency of some 30 per cent for β energies around 1 MeV) can be obtained with scintillation, semiconductor and thin-window gas counters, with further improvement obtainable by the use of anti-coincidence.

An alternative approach to the above types of β-counting arrangement, in all of which there is a spatial separation of counter and source, is to have the source dispersed usually in molecular form, within a detecting medium. This is the basis both of the internal gas counter and of liquid scintillation and Čerenkov counters, and is particularly of importance for work with low energy β emitters such as ^{14}C and ^{3}H. In the internal gas counter, which has been developed very largely for ^{14}C dating work, the specimen is introduced quantitatively in gaseous form (e.g. as $^{14}CO_2$) and forms a constituent of the counting gas. The liquid scintillation counter, in which the specimen is dissolved in a suitable liquid scintillator, is a more recent development but has proved attractive partly due to the readiness with which it can be adapted for routine automatic operation.

Because of the continuous nature of β particle spectra it is not usually possible to obtain a precise indication, in a low activity work, of characteristic β particle energy. A limited amount of such information can, however, be obtained either with scintillation counters or with proportional counters designed to work with internal sources. As in the case of α particles, information on spatial distribution of β emitters can be obtained by autoradiography. A good example of the spatial resolution obtainable with this technique when working with very low energy β emitters is its ability to localize and even to outline the shape of actively metabolizing chromosomes in living cells by virtue of their uptake of substances labelled with ^{3}H.

The achievement of high sensitivity and good resolution in the detection and measurement of γ rays calls for a detector having relatively high photoelectric absorption of the radiation. Scintillation counters using large single crystals of NaI or CsI, activated with thallium, meet this requirement very well and provide overall spectral resolution of some 7–10 per cent. A typical high sensitivity counter consists of a 7·5 cm × 7·5 cm diameter NaI crystal enclosed in a thin metal can and optically coupled to a photomultiplier, the output of which is fed to a single- or multi-channel pulse height analyser. For work with low energy γ rays (below 0·25 MeV) and X rays, thinner crystals may be used, giving lower background without loss of detection efficiency. With adequate shielding such a counter can usefully measure down to levels of 10^{-10} curie of a number of important radionuclides in small sources placed either on the surface of the crystal or in a "well" drilled into it. For very large sources, and particularly for the human body, one or more NaI detectors (often using rather large crystals, e.g. 10 cm × 20 cm diameter) are placed in a standard geometrical relationship to the source or, alternatively, are scanned over its surface in an attempt to determine the internal distribution of activity. With a whole body counter of this type enclosed in a heavily shielded room it is possible to identify and measure total body burdens of such nuclides as ^{40}K, ^{137}Cs and ^{226}Ra at the level of 10^{-8} curie. Whole body counters have also been built employing very large volumes of liquid scintillator, entirely surrounding the subject. The relatively high counting efficiency of this arrangement leads in practice to an order of magnitude improvement in counting time over a NaI counter but has the disadvantage of considerably lower energy resolution.

The recent development of lithium-drifted germanium semiconductors has made available a γ detector of much higher resolution ($\sim$1 per cent) than is provided by the scintillation counter. However, the present upper limit of about 50 cm^3 on the sensitive volume of these detectors constitutes, in practice, a limitation on their usefulness for high sensitivity measurement.

Detection of low levels of γ radioactivity as surface contamination and in mineral deposits is frequently of interest and for this purpose, particularly when a non-analytical but robust and portable instrument is required, a Geiger counter is used. If, however, the measurement is being carried out with the object of assessing radiation exposure resulting from the presence of radioactivity, an ionization chamber is the appropriate instrument and a number of designs, ranging from large, highly sensitive types to small pocket dosimeters, are currently in use. Finally, the blackening of photographic film provides a further valuable means for monitoring γ-ray exposure at moderately low levels.

Bibliography

Abson W. (1966) in *Encyclopaedic Dictionary of Physics* (J. Thewlis Ed.), Suppl. Vol. 1, 139, Oxford: Pergamon Press.

Bobin K. J. (1961) in *Encyclopaedic Dictionary of Physics* (J. Thewlis Ed.), **1**, 353, Oxford: Pergamon Press.

Dennis J. A. (1967) in *Encyclopaedic Dictionary of Physics* (J. Thewlis Ed.), Suppl. Vol. 2, 284, Oxford: Pergamon Press.

I.C.R.U. *Measurement of Low-level Radioactivity* (in prep.), Washington, D.C.: Nat. Bureau of Standards.

I.C.R.U. (1963) *Radioactivity*, Handbook 86, Washington, D.C.: Nat. Bureau of Standards.

Rees D. J. (1967) in *Encyclopaedic Dictionary of Physics* (J. Thewlis Ed.), Suppl. Vol. 2, 289, Oxford: Pergamon Press.

Thewlis J. (Ed.) *Encyclopaedic Dictionary of Physics*, articles beginning "Radioactivity", **6**, Oxford: Pergamon Press.

Watt D. E. and Ramsden D. (1964) *High Sensitivity Counting Techniques*, Oxford: Pergamon Press.

C. R. Hill

RADIOISOTOPE SCANNING. Radioisotope scanning is a technique by which the spatial distribution of radioisotopes can be determined. It is chiefly used in medicine as a diagnostic aid, and its importance is in the fact that the distribution in the body, or within a particular organ of the body, of an administered radioactive material can be visualized, so for example organs concentrating radioisotopes can be delineated or volumes that fail to concentrate radioisotopes be examined. Because there may be a difference in concentration between a tumour and the nearby surrounding normal tissue, radioisotope scanning is often of value in the diagnosis and localization of malignant and benign tumours.

From a purely physical point of view the problem which radioactive isotope scanning attempts to solve is illustrated in a simplified form in Fig. 1. Assume that V is the cross-section of an object, say the human body, in which a smaller volume v is contained. If a suitable radioisotope is given to the patient the concentration in the small volume v may be different from that in the rest of the body, V; the technique of radioisotope scanning attempts to delineate the smaller volume, v, and to determine the distribution of radioactivity within it. It is apparent that this attempt is complicated by the fact that the concentration of radioactivity in the remainder of the body will not, in general, be constant and that the concentrations of the radioisotopes in v and V may both be time dependent; furthermore the concentration in v and V, although different, may be sufficiently close to make certain separation difficult.

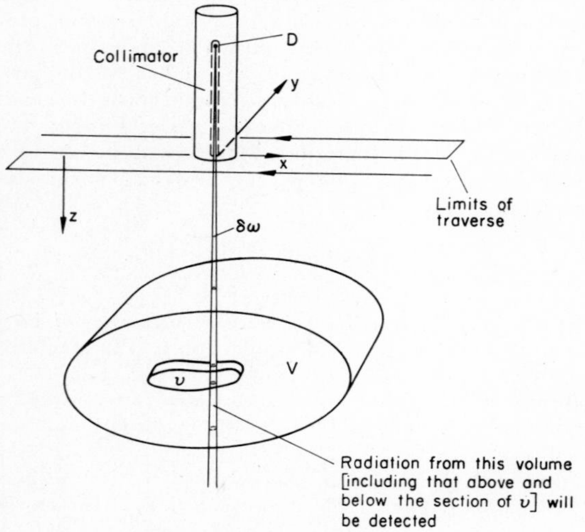

Fig. 1. Principle of radioisotope scanning.

The problems which must be considered in relationship to a scanning system are therefore as follows:

1. What is the smallest volume v that can be detected? This is a problem of resolution of the scanning system.

2. What difference in radioactive content between v and V is significantly detectable? This is a problem of sensitivity of the scanning system.

These problems, of sensitivity and of resolution, are not independent.

In this article interest is in the essential physical problems posed by radioactive scanning and not in the clinical problems involved in interpreting the results obtained. Some of the latter are discussed in International Atomic Energy (1959).

Principles of Radioisotope Scanning

Consider a shielded radiation detector, D with a small collimating aperture as shown in Fig. 1. Let the aperture subtend a small solid angle $\delta\omega$ at the detector. Consider the volume V to be filled with a gamma emitting radioactive material and suppose that the concentration of this material is uniform throughout V, except that V contains a volume v, in which the activity is greater than the remainder. Because the detector is collimated it will detect only the volume of material contained within the solid angle $\delta\omega$; it being assumed here that the shielding material from which the collimator is made is sufficiently thick to cut out all extraneous radiation. If the detector is moved laterally (as shown in Fig. 1) the counting rate will vary as the detector moves across the object scanned. The response of the detector in various positions could be similar to that shown in Fig. 2. The "peak" in this response corresponds to the detection of activity in v as the detector moves across the object. If at the end of traversing the object, the detector is moved a small distance in the y direction and then is returned, parallel to its original direction, a different part of the object will be scanned, and a set of profiles corresponding to each increment of y would be obtained. Such traverses constitute a "scan" of the object. The shape of the object, v, in the x-y plane can be found by this method, but because the counter "sees" a finite solid angle the edge of the object will appear to extend further than it really does. This effect is discussed later in this article.

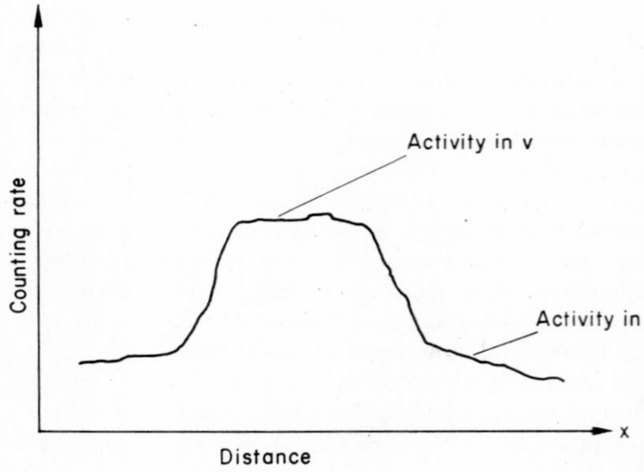

Fig. 2.

The question now arises: In what way does the response of the detector depend on the depth of the object v? A simplified analysis of this problem includes the following assumptions:

1. That the object be sufficiently large that the area seen by the detector is always small compared with the cross-section in the x-y plane.

2. That the radiation be sufficiently penetrating that the absorption of radiation in the object, above v, can be neglected. These assumptions reduce the problem to that illustrated in Fig. 3; the object v can be considered as a thin sheet effectively of infinite extent in the x-y plane. Let the activity per unit volume in v be a μCi/cm³, and let the thickness of the sheet be δz. Suppose that the distance from the detector to sheet is z. The volume δv "seen" by the detector will be given by

$$\delta v = z^2 \, \delta\omega \, \delta z$$

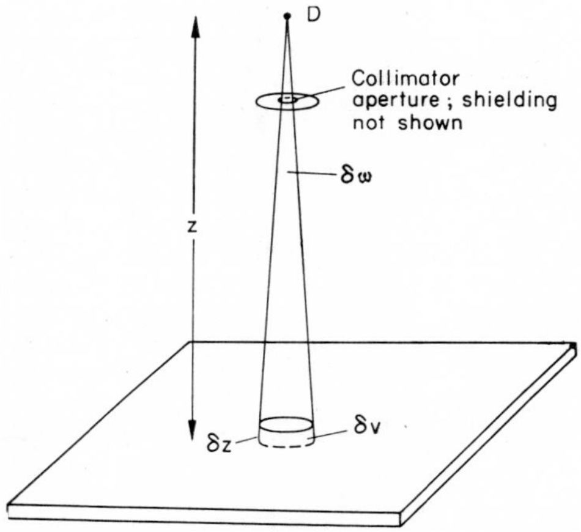

Fig. 3.

and the counting rate δc, observed due to the activity in δv will be

$$\delta c = \frac{K}{z^2} (az^2 \, \delta\omega \, \delta z)$$

K is a proportionality constant; the term z^2 arises in the denominator because the inverse square law is assumed to apply:

$$\therefore \; \delta c = Ka \, \delta\omega \, \delta z.$$

This result shows that for a given detecting system (i.e. K, $\delta\omega$ constant) and if it is assumed that the activity concentration, a, is constant in the sheet then the counting rate due to the thin sheet is independent of the distance of the sheet from the detector.

As any relatively large object can be considered to be composed of a set of thin sheets it can be seen that the response of the scanning system is proportional to the thickness of the object in the z direction but is not dependent on the distance from the detector.

The response of a collimator to a point source of radiation is often considered as being indicative of the performance characteristic of a particular collimator. In practice this response can be measured by moving a point source in a line perpendicular to the axis of the collimator (i.e. in the x direction of Fig. 1) and a "point source response curve" as shown in Fig. 4, can be obtained. The effectiveness of the collimator is often defined in terms of the width of the point source response curve at half maximum; the narrower this width the greater will be the resolution of the collimator. Collimator resolution may be increased by the use of additional shielding and so decreasing the solid angle, $\delta\omega$, for which the detector is sensitive. Thus, the effectiveness of the collimator to distinguish between two sources close together is increased if the collimator solid angle is reduced. On the other hand this has the effect of reducing the counting rate obtained from a

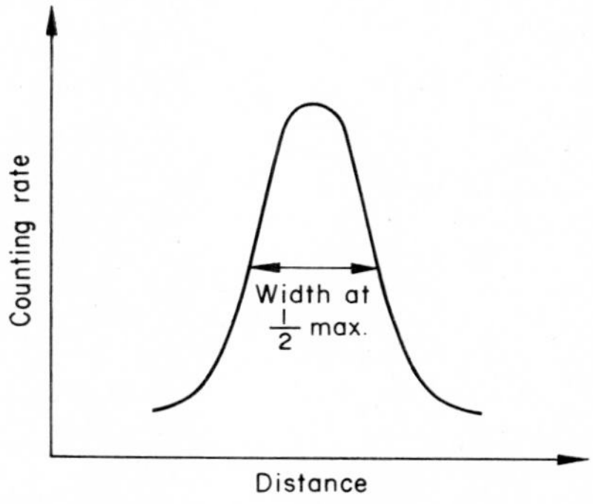

Fig. 4. Point source response curve.

source of given activity, and because of the statistical variation inherent in the radioactive decay process, increasing the resolution of the collimator at the expense of sensitivity of the detector is not necessarily a great advantage. A more comprehensive idea of the performance characteristics of a collimator can be obtained by moving a point source in two dimensions, recording the response of the detector at many positions and joining up points of equal counting rate; the curves obtained are called "iso-response curves", It is unfortunately difficult to quantitate the information from iso-response curves to obtain a figure of merit to compare one collimating and detecting system with another.

Scanning Apparatus

A device which is used to determine variation in concentration of radioisotopes within an object by using the principles described is called a "scanner".

Radioisotope scanners have certain essential features in common; these are illustrated in Fig. 5. The detecting system, consisting of a scintillation counter and collimator, detects radiation emitted from a

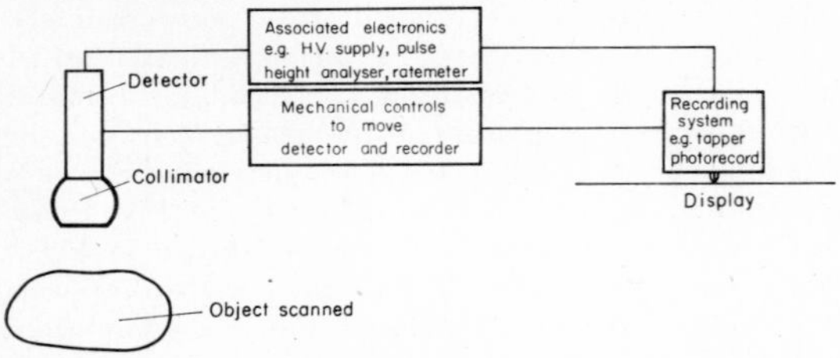

Fig. 5. Essential features of a radioisotope scanner.

limited volume of the object scanned. The counting rate so measured is transferred to a recording device where it is recorded at a position which can be related to the position of the detector. A scanner can be considered to be a device for transferring information from the object scanned to the recording mechanism.

Several different types of scanner have been designed; they may be classified as follows:

1. Profile Scanners
2. Rectilinear Scanners
3. Scanning Cameras

Profile scanners. Profile scanners make use of a collimated detector which can be moved along the body at a set distance from it. The collimator is designed so that the total radioactivity in a narrow section (but the whole width) of the body is measured. The counting rate, measured at various positions along the body, can be plotted against distance along the body, and thus a radioactive "profile" of the body can be constructed (Fig. 6). Peaks in the counting rate are observed where there is selective uptake of the radioisotope. The advantage of this method is that the whole body can be scanned in a short time as a single traverse only is made and although the method measures the total activity in each section and does not locate the activity within the section, any abnormal peak in the profile can be investigated if desired by scanning in two dimensions in this region. The method can also be conveniently applied to make studies of radioisotope turnover, for repeating the profile scan will show the changes in activity which have taken place with time at any site. This method has proved

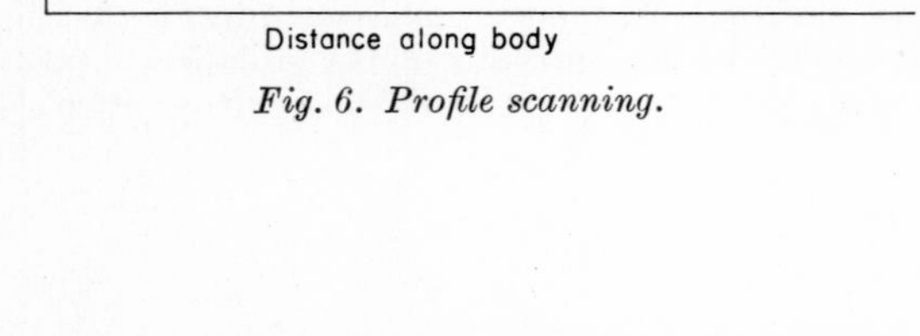

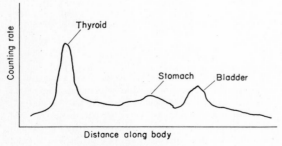

Fig. 6. Profile scanning.

to be of particular value in studying the changes with time, of radioiodine concentration in patients with thyroid cancer.

Rectilinear scanners. A rectilinear scanner makes use of a highly collimated detector so that distribution of the radioactivity within a volume of tissue can be studied in two dimensions. A collimated scintillation counter is moved by a suitable drive mechanism to and fro so that it scans the area of interest. Limit switches terminate the movement at the end of a single line and the counter is moved to an adjacent position and continues to scan along a line parallel to the former line. The output from the scintillation counter is fed to a recorder so that the counting rate at every position occupied by the scintillation counter is recorded. After tracking to and fro over the area desired, a scan will have been recorded which shows the distribution of radioactivity in the object scanned. It is apparent that there are many difficulties in optimizing the many variables. To obtain a clear scan the scintillation counter should be moved as slowly as possible over the object so that the maximum number of counts is recorded at each position of the counter. In practice, for investigations of humans, the rate of travel is determined by the fact that the entire scan must be completed within a reasonable time; furthermore in human applications there are severe restrictions on the activity of a radioactive material which can be administered to a patient and so the counting rate which can be recorded is often quite low. For these reasons a great deal of attention must be paid to increasing the sensitivity of the system so that the maximum number of counts can be recorded at every point. As discussed earlier, sensitivity and resolution are closely related; sensitivity is a property mainly of the scintillation counter, but is also dependent on the design of collimator chosen; resolution is a property mainly of the collimator. The availability of large scintillation crystals has led to the development of the "focused" collimator (Fig. 7). Quite clearly a collimator having many tapered holes focused to a point will result in an increase in sensitivity, for the detector will be more sensitive to radiation emitted from a point source at the focus than will be the case if a single-hole collimator is used. Although collimators of this design are referred to as "focused" collimators, they respond to radiation from points away from the focal point of the collimator. The iso-response curve for a collimator, to the point source, is illustrated in Fig. 7; it will be seen that the response of the collimator is maximum at the "focus" and decreases as the source moves either closer to, or away from the counter. As in the case of a single-hole collimator, it is evident that a focusing collimator will give a response independent of distance if a plane source of uniform activity is used.

Recording systems. The counting rate obtained from the scintillation counter is transferred to the recorder and it is displayed there. In principle any system which

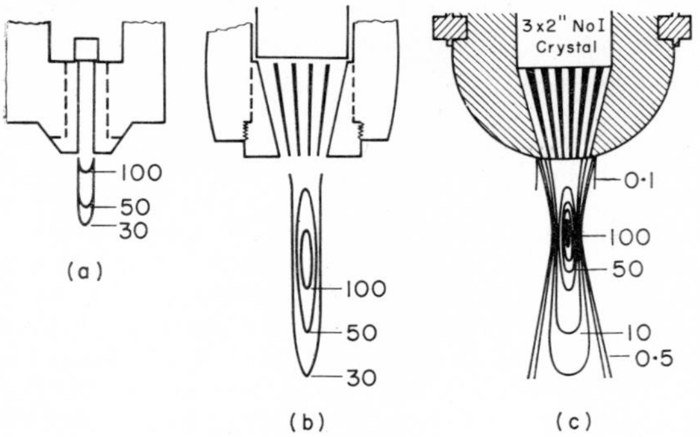

Fig. 7. Isoresponse curves for collimators of various designs. (a) straight ½ in. diameter hole; (b) 19-hole lead collimator; (c) 37-hole gold collimator. These curves illustrate the improved resolution of (c) compared with (b) and (a).

records the counting rate at each point scanned may be used and there are several convenient methods.

For profile scanning, the counting rate may be recorded continually on a chart recorder as the counter traverses the patient.

For rectilinear scanning this system is inconvenient and other methods of recording the response of the detector at each position are commonly used. Most commonly, a tapping device is used, designed so that the rate of tapping corresponds with the counting rate, and each tap can be used to mark a piece of paper. Each mark can represent one pulse detected, or by using scaling circuits several pulses may be needed to activate the tapper once. A set of dots or marks is thus produced, corresponding in position on the paper to the position of the scintillation counter and the separation of these marks is dependent on the counting rate.

Sometimes the record of the scan is obtained by using a tapper which strikes a typewriter ribbon and so makes a permanent record on paper. One elaboration of this method is to mount a set of coloured ribbons on a carriage which moves under the tapping mechanism with the same time-constant as the ratemeter so that as well as the marks on the paper being closer together at higher counting rates the actual colour marked on the paper alters with the counting rate. Such a scan is said to be a "colour scan".

Many commercial scanners make use of light sources, the intensity of which varies with the counting rate, and this is used to expose photographic film. The record obtained in this way is known as a "photoscan".

The "scan" is a visual representation of the isotope distribution in the object studied. Commercial scanners have contrast enhancement devices so that the scan recording mechanism does not record until the counting rate exceeds a certain pre-determined level. The advantage of this is that if there are small differences in activity between two volumes one of them may be suppressed while the other is recorded. Contrast enhancement can give an incorrect impression about the quantitative differences in concentration because it has the effect of recording the scan as an all-or-none situation. Many modern scanners record all the information obtainable from the counting rate on to magnetic tape, and this may be played back with various degrees of background suppression to obtain a scan which gives the desired information.

Scanning cameras. A technical disadvantage of rectilinear scanners is that they collect information on a point-to-point basis over the scanned area; the great majority of the information available from the radioactive decay of the material in the scanned area is lost. Because of the statistical nature of radioactive decay it would be desirable for the collimated scintillation counter to remain at each point for a considerable time to collect enough counts for good statistical accuracy. Unfortunately, this implies a very low scanning speed and often there are practical difficulties which limit the time which can be spent on a particular scan. The scanning camera, or scintillation camera, overcomes this disadvantage to some extent. The principle of a scintillation camera is illustrated in Fig. 8. The detector crystal in this case is a relatively thin, large flat crystal. Above this crystal is placed an array

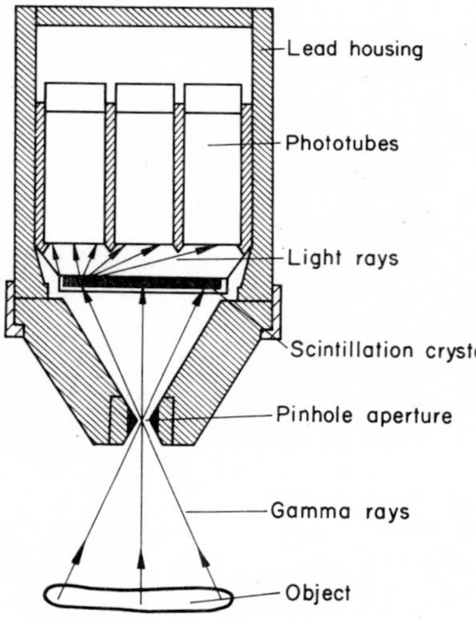

Fig. 8.

of photomultiplier tubes. In its simplest form the collimator is a pinhole aperture. Consider a photon originating from a source which is being scanned which travels through the pinhole and produces a scintillation in the crystal. The position of thi- scintillation will obviously be related to the posis tion at which the photon originated, for the crystal can be considered to be a screen of a pinhole camera. From the photons produced in the crystal, each photomultiplier tube will respond and the size of the pulse which is produced in each of these tubes depends on the location of the interaction in the crystal itself. This can be seen by considering Fig. 9, with an array of seven multiplier tubes. Assume that the pulses produced in the tubes from an interaction occurring at a point (x, y) give pulses of magnitude $p_1, p_2, p_3, p_4, p_5, p_6, p_7$. The relative magnitude of the pulses p_1 to p_7 will depend both on x and on y, and by the use of sum and difference circuits, it is possible to use the magnitude of the pulses to give a value of the co-ordinates, x and y. In addition the total signal $\sum_{n=1}^{n=7} p_n$ depends on the photon energy and this signal may be used to indicate the energy of the radiation which struck the crystal. Consequently the information obtainable from the photomultiplier tubes may be used to produce three separate pieces of information on the display; two give the position and one gives the magnitude of the scintillation pulse. The information concerning the coordinates x and y can be used to control the position of a spot on the screen of a cathode-ray tube, and the total of the pulses can be used to govern the brightness of the spot. The cathode-ray tube therefore will produce flashes of light whose intensity depends on the energy of the radiation emitted, and whose position on the face of the screen depends on the position from which the radiation was emitted

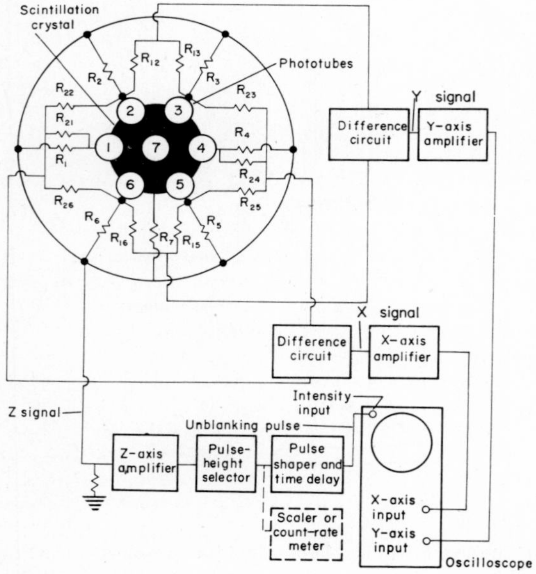

Fig. 9.

in the object scanned. Photographing these flashes results in a scintillation picture being built up within a short time representing the distribution of the radioactive material in the object scanned.

The scanning camera records information from all parts of the area scanned at any one time, and has the advantage that the camera does not move during the time the scan is being taken. Scanning cameras can conveniently be used for dynamic studies of isotope turnover because a complete photograph can be obtained within less than a minute in favourable cases. Like the rectilinear scanner, the camera can make use of multi-aperture collimators or collimators with several pinholes if desired.

Recent developments of scintillation cameras have included the development of a positron camera. The two gamma rays which are produced when a positron is annihilated are emitted in opposite directions, and so, the positron camera uses a scintillation counter positioned opposite to a conventional scintillation camera. A coincidence circuit between these has the effect that only scintillations from the positron emitter are recorded; all other scintillations are not recorded.

Applications

Radioisotope scanning is now one of the most important of the applications of radioisotopes in medicine and is proving a valuable aid in diagnosis of many diseases. Organs which are commonly scanned include the thyroid, liver, brain and kidney. The clinical details and value of this information can be found in the references below. Although radioisotope scanning has been developed chiefly in connexion with medical isotope investigations the principles are applicable to any situation where the distribution of radioactive material in an object is under investigation. Frequently, in purely physical situations, the problems are made much less severe than they are in clinical work because there need not be the limitations either on activity or time which are necessary when human beings are investigated.

Bibliography

Medical Radioisotope Scanning (1959) Proceedings of a Seminar, Vienna: International Atomic Energy Agency.

Medical Radioisotope Scanning (2 Vols.) (1964) Proceedings of a Symposium, Athens 1964. Vienna: International Atomic Energy Agency.

D. J. Rees

RADIOISOTOPE TRACERS IN INDUSTRY. Ionizing radiations can be detected and measured to a useful degree of accuracy at levels corresponding to as little as 10^{-6}–10^{-12} grammes of radioactive emitter; the apparatus and procedures used are often simple by comparison with more "conventional" but much less sensitive techniques not employing radioactivity. Radioisotopes of nearly every element are commercially available, or can be prepared, in a very wide range of chemical and physical forms, and these behave for all practical purposes in precisely the same way as the corresponding inactive materials.

There is usually no need to prepare samples elaborately before measuring their activity and in many cases there is no need to take samples at all; in favourable circumstances measurements can be made through plant or vessel walls without any direct access to the active material. Radioisotopes therefore form almost ideal "labels" for marking many kinds of material and tracing their behaviour and movements in widely varying circumstances. They have proved particularly useful and versatile as tracers for studying industrial processes, installations and products.

General Principles

In any application of radioisotope tracer techniques the tracer must conform in behaviour throughout the investigation as closely as possible with the material it is to follow; ideally it should be identical with it apart from its radioactivity, and it is therefore often best prepared by neutron activation of a sample taken from the bulk. This is not always practicable, in which case a separately prepared tracer must be used consisting of or containing an appropriate radioisotope in a chemical and physical form which will follow the bulk material at least as closely as the investigation requires. The activity of the tracer must be great enough for it to be measured to the required degree of accuracy at the maximum dilution likely to be encountered during the investigation, but not so great as to make it radiologically dangerous or troublesome to handle at any stage, especially after its injection into the system. Satisfactory measurement of the diluted tracer usually requires a count-rate from the tracer of one to ten times the background rate (the latter being reduced by shielding if necessary). At the same time, radiological safety (and the statutory requirements in force to ensure it) require the dose-rate to the operators and the public to be kept within strict limits at all times, and contamination of the product and plant to be reduced to a minimum (see below). The presence of the tracer must have no effect on the materials or the process such as might vitiate the results of the study or harm the product. If the added tracer differs physically or chemically from the bulk, then its specific activity (in curies per gramme) must be as high as is practicable, so that the minimum of foreign material is introduced.

Safety and Legal Requirements

The radioisotope chosen should have the shortest possible half-life consistent with the required accuracy of measurement in the final stages, so that residual activity in plant and product will reach a satisfactorily low level as soon as possible after the investigation is over. A half-life rather longer than the expected duration of the experiment is usually to be preferred for ease of measurement, and a delay-period of 10–20 half-lives afterwards will usually allow activity to fall to a radiologically negligible level. Statutory requirements regarding radiological safety must be complied with before, during and after the investigation, and staff responsible for planning and carrying out the experiment must be aware of the legal as well as the technical aspects of the work. For this reason tracer applications involving radioactivity need to be supervised by trained specialists either from among a firm's own staff or from an outside organization such as a university, a college or a research establishment.

Applications

It is only possible to give here a brief review covering some typical and widely-used applications; these are grouped under the broad headings of plant dynamics, flow-rate measurement, ventilation, filter-testing, process control, investigation of faults, location and measurement of leaks, and research and development.

Plant dynamics. A radioactive tracer of known mass and activity is introduced continuously over a period, or in one or more discrete amounts, into the material entering the plant or process-stage to be studied, and the subsequent distribution of the activity, or its rate of change, is ascertained. In this way it is possible to determine characteristics such as processing or residence times, throughput rate, mixing efficiency, flow patterns and profiles, recycling hehaviour, "dead" volumes and hold-ups, and the diffusion, segregation, carry-over or loss of process material. If the tracer is a γ-emitter it is often feasible to make the measurements through the walls of the containing vessel without sampling or even direct contact with the material; in such a case access to the process stream is only needed at the point of introduction of the tracer. Once the dynamics of a system have been established by a full initial tracer study, and its operation optimized, it is comparatively simple to detect and check deviations from the norm, either by conventional means or by periodic simplified radiotracer studies. Most of the manufacturing and processing industries make use of radiotracers to a greater or less extent, the chemical and petroleum industries being the largest users.

Sodium-24 and bromine-82 are the most widely used isotopes in plant dynamics studies in those cases where a neutron-activated sample of the bulk material or of one of its constituents is not suitable.

Measurement of flow-rate. The velocity or mass-flow rate of fluids of all kinds can be measured by injecting into the system a pulse or stream of radioactive fluid which will mix with the main stream; either the rate of movement of the activity between two points downstream, or its concentration downstream, can be measured and the velocity or the degree of dilution—and hence the volume flow—can be deduced. The velocity technique is suitable for pipes and ducts of known cross-section, and the dilution techniques for channels or pipes of unknown or varying cross-section. Provided that the flow is turbulent and that proper mixing is achieved—an essential condition—the technique, with suitable variations, can be used for measuring flow in liquids, gases or slurries and for checking the efficiency of pumps or turbines. Over the range of flow-rates encountered in most industrial installations measurements can usually be made to within ± 1 per cent. The technique imposes no constraint on the flow being measured, and is particularly useful for calibrating orifice-plates, venturis, and other installed flowmeters. The isotopes most widely used for water flow measurement are sodium-24 and bromine-82 and for air or gases krypton-85 and bromine-82 (as methyl bromide).

Ventilation studies. Similar techniques of injection and subsequent measurement of activity can be applied to the study of air-flow, clearance rates and dead volumes in ventilation systems. Krypton-85 is the tracer most commonly used for this purpose.

Filter testing. The effective pore size of a filter membrane or the integrity or efficiency of a filtration system, can be measured by testing with a suspension of a radioactively labelled particulate solid of appropriate size distribution and other properties. Labelling can be by neutron activation of the particles, or by marking them with a radioactive substance by adsorption or chemical combination. Measurements of activity can be made on the filtrate, the filter cake or the membrane.

Routine control of process or product. Although radioactive tracers are normally used for single investigations, they also find some uses in routine control procedures. The rate at which the inaccessible refractory linings of blast-furnaces are worn away internally is important in their safe and economic operation, but is hard to ascertain by external measurements, and shut-down for internal inspection can be very costly. Radioactive pellets, usually of cobalt-60, are embedded at suitable positions and depths in the lining material, and their presence is periodically checked by external detection of the radiation they emit; absence of a pellet indicates ablation of the lining at that point to at least the depth of the pellet, the latter becoming sufficiently dispersed in the product to present no radiological hazard.

Oil-soluble radioactive tracers have been used to mark the position and spread of the interface between different grades of product in oil pipe-lines, but this is more usually done now by external γ-transmission density-gauging. Pipeline "go-devils" or scrapers can be tagged with pellets of radioactive cobalt or caesium so that if they become jammed they can be located from above ground.

Stock-taking of certain costly materials, such as mercury in electrolytic plants, can be carried out by adding a known mass and activity of a radioisotope (in this case Hg-203) to the bulk; after it has mixed thoroughly the specific activity of a sample gives the dilution ratio and hence the total mass of mercury.

Many manufacturers of branded or high-quality goods have been interested in incorporating radioactive tracers in their products to aid positive identification in case of dispute, or for reasons of security. Although this is possible it is usually more practicable, as well as being radiologically safer, to use an inactive trace material of high neutron capture cross-section and to activate it by neutron bombardment as the occasion arises.

Extremely small air or gas leaks in manufactured articles such as sealed electronic components can be detected by soaking the component in radioactive krypton-85 gas under pressure, then removing and testing for activity; any residual activity will indicate that gas has leaked into the component.

Investigation of Faults

A number of radioisotope tracer techniques have been developed, and are widely used, for locating blockages, for investigating material losses or product contamination, and for locating and measuring leaks. Most of the techniques employed are similar in principle to those already described, but liquid leak location and measurement deserve special mention because of their variety of techniques and their widespread use. In the case of buried pipelines, radioactive liquid can be forced out from the pipe through any leaks into the subsoil, either by filling the pipe with active liquid or by pumping a "plug" of it along the line; the pipe is then flushed clean and any pockets of activity in the surrounding subsoil are located and measured by a detector drawn through on a cable. Alternatively, for long stretches of pipe a detector with a battery-operated recording device can be carried through by a "go-devil" piston; in this case radioactive markers buried at known intervals provide reference points for interpretation of the trace. If leaks are already suspected at known points — e.g. at joints in a newlylaid pipeline — barholes can be prepared and probed for activity.

Where a pipe is buried not more than 3 or 4 ft deep, or embedded in an equivalent amount of concrete or masonry, a sealed tracer can be used (usually a rubber ball fitting the pipe closely and carrying a sodium-24 source); the movement and location of this can be followed easily from the surface; stoppages or significant reductions in speed usually indicate points of leakage. Leaks within a concentric pipe system can be located by injecting a pulse of activity and comparing the times it takes to pass through the system by the correct route and by way of the leak; the size of the leak is found by comparing the amounts of activity travelling by each route. Leaks in engine fuel systems can be detected by adding a soluble β-emitting tracer (e.g. palladium-109) to the fuel, in which it will not be detectable from outside; any leak will carry activity to the surface, where it will remain detectable after evaporation of the fuel.

Research and Development

Many industrial R. and D. problems involving material transfer can be profitably tackled by radiotracer techniques. Lubrication and wear problems are particularly suited to such study because the surfaces involved are often inaccessible and the process of attrition is very slow. By making one of the rubbing surfaces radioactive — e.g. by neutron-activation of a bearing, piston-ring, or cutting-tool — and monitoring the lubricant for radioactive debris, the effect of changing the lubricant or the operating conditions can be determined in hours where it might otherwise take weeks or months.

In metallurgical research radiotracers are used extensively; typical studies include segregation of alloy constituents at grain boundaries, relation between composition and grain-structure, diffusion across grain-boundaries and in bulk metals, action of plating-bath additives, reactions and circulation in the molten pool in arc-welding, and the shape of the solidification profiles in continuously-cast steel and aluminium. In many of these studies autoradiography has proved a valuable ancillary: here the distribution of activity over a surface is revealed by its highly localized effects on a photographic film placed in contact with it.

Radiotracers have proved particularly useful for studying material transfer between surfaces in contact (surface-coatings and their substrates, engine bearings, electrical contacts, tyres, printing type, etc.), carry-over between process stages (dyeing, electroplating, laundering) and the distribution of minor constituents in a product (glue in chip-board, dressings on textile fibres, etc.).

In agriculture and the industries associated with it (e.g. manufacture of fertilizers, feeding stuffs, pesticides, herbicides, veterinary products) radiotracers are widely used in laboratory and field studies. These include investigations into the distribution and behaviour of nutritional and trace elements in the soil and during uptake and metabolism in plants and animals, and, in the case of pesticides etc., into the ultimate distribution of toxic residues. The natural radioactivity of potassium 40 is useful in studies of potassium fertilizers and in the measurement of lean-to-fat ratios in carcase meat or live animals. In the pharmaceutical industry wide use is made of "labelled compounds" containing suitable radioactive atoms in their molecules for studying the biochemistry and metabolism of new products.

See also: Nuclear geophysics.

Bibliography

BAKER P. S. (1961) in *Encyclopaedic Dictionary of Physics* (J. Thewlis Ed.), **4**, 112, Oxford: Pergamon Press.

BRODA E. and SCHOENFELD T. (1966) *Technical Applications of Radioactivity,* Vol. 1, Oxford: Pergamon Press.

ERWALL, L. G., FORSBERG H. G. and LJUNGGREN K. (1964) *Industrial Isotope Techniques,* Copenhagen: Munksgaard.

FAIRES R. A. and PARKS B. H. (1960) (2nd Edn.) *Radioisotope Laboratory Techniques,* London: Newnes.

HELMHOLZ H. R. (1962) in *Encyclopaedic Dictionary of Physics* (J. Thewlis Ed.), **6**, 36, Oxford: Pergamon Press.

INTERNATIONAL ATOMIC ENERGY AGENCY *Symposium on Radioisotope Tracers in Industry and Geophysics,* Prague, November 1966, Proceedings. Vienna, I.A.E.A.

KOHL J., ZENTNER R. D. and LUKENS H. R. (1961) *Radioisotope Applications Engineering,* New York: Van Nostrand.

PUTMAN J. L. (1965) (2nd Edn.) *Isotopes,* London: Penguin Books.

ROGERS G. T. (1966) in *Encyclopaedic Dictionary of Physics* (J. Thewlis Ed.), Suppl. Vol. 1, 15, Oxford: Pergamon Press.

SWAT T. R. (1961) in *Encyclopaedic Dictionary of Physics* (J. Thewlis Ed.), **1**, 633, Oxford: Pergamon Press.

R. M. LONGSTAFF

RADIOMETER. The radiometer is an instrument used to measure radiant power. In general, the radiometer system is composed of three parts: an optical system, a detector (or transducer), and an output system. The optical system may consist of lenses, mirrors, apertures, spectral filters, polarization analyser, etc. Its purpose is to determine the geometry, spectral content, and/or polarization of the radiation beam being measured. The detector (or transducer) element converts the incident electromagnetic power into a more easily measurable physical quantity. Some of the most common types of detectors are: photographic film, photocells, photovoltaic materials, photoconductors, and photoelectric materials. An output system is usually composed of an amplifier section and an indicator system. The amplifier essentially amplifies the signal produced by the detector to values sufficient to operate the output indicator. The indicator simply presents this information in the desired form. Usually, the indicator is a meter, a graph, or oscilloscope.

Bibliography

ATTWOOD W. A. (1962) in *Encyclopaedic Dictionary of Physics* (J. Thewlis Ed.), **6**, 15, Oxford: Pergamon Press.

I. W. GINSBERG and T. LIMPERIS

RADIO NAVIGATION, RECENT DEVELOPMENTS IN. This article is supplementary to that entitled "Navigation-Radio and Radar" in the *Encyclopaedic Dictionary of Physics* (Mills 1962).

The original article is still factually correct but since its publication there has been significant progress towards the achievement of long-range ground-referenced area-coverage navigation systems. In particular, the Decca system (Williams *et al.* 1960) has been developed to provide the Dectra system which is capable of covering an overseas route of many hundreds of miles in length and the present installation does in fact cover the greater part of the North Atlantic Ocean. Moreover, the American v.l.f-based Omega system is now operating to provide a basis for continuing research and development. The experimental transmissions which are provided by these systems offer a valuable research facility and are contributing towards increasing knowledge of propagation characteristics of the Earth-ionosphere medium in the frequency bands 70–120 and 10–14 kHz respectively in which the systems operate.

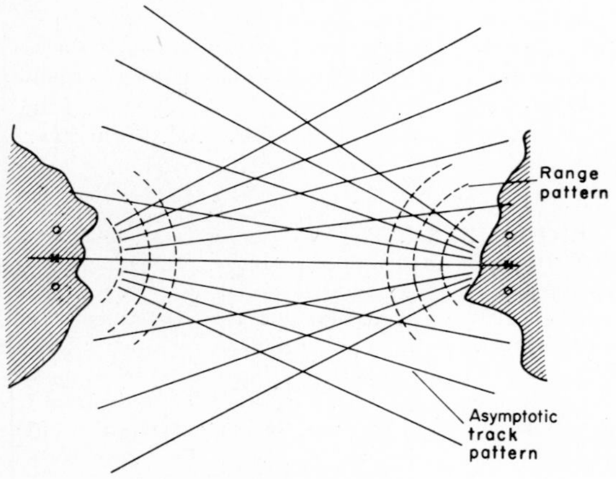

Fig. 1. Principle of Dectra.

Unlike the currently used Loran long-range navigation system, which is a pulsed hyperbolic-lattice system, both Dectra and Omega are C. W. hyperbolic-lattice systems. The trend towards C.W. from pulsed systems is dictated not only by the need to conserve occupancy of the frequency spectrum but also by the fact that at the frequencies at which the systems operate, achievable transmitting aerials are highly reactive and in the case of pulsed systems the mean radiated power is usually limited by voltage breakdown. System performance is determined largely by mean transmitted power and pulsed systems, particularly at v.l.f., would be limited to unacceptably low mean transmitted power.

In the Dectra system, coverage is provided by two pairs of l.f. transmitters, one pair on each side of the Atlantic, the individual stations of the pair being spaced about 100 miles apart along a line approximately normal to the line joining the centres of the two pairs. The operating principle is identical with that of the Decca Navigator System, but the transverse base-length is very short compared with the dimensions of the coverage area, so the position lines which the system defines are, for all practical purposes, over the greater part of the area, defined by the asymptotes of the hyperboli. There are thus two sets of radial position lines which radiate from the mid-points of the transmitter-pairs (Fig. 1).

In principle the pair of hyperbolic-lattices is in itself sufficient to establish a position fix, but in practice the geometry of the intersections is unfavourable to provision of accurate indication of distance along track and, moreover, simultaneous reception from the two transmitter complexes would be essential and this cannot be relied upon in all areas at all times. Consequently, the system is augmented by a "ranging" system which indicates the distance that has been travelled along track.

The ranging system operates by establishing the equivalent of a standing wave pattern in the coverage area so that the navigating aircraft can effectively count the number of cycles traversed in the standing wave. This is achieved by radiating transmissions from the two ends of the system which are phase-locked, one directly and the other through the trans-atlantic propagation path, to a common low frequency which is an integral sub-multiple of the radiated frequencies. The signals, when processed by the receiver, can therefore be treated as common frequency radiations travelling in opposite directions which set up a standing wave.

The ranging system is of course ambiguous but the probability of losing signal for sufficiently long for this ambiguity to be of operational significance is very small on account of the very narrow bandwidth which is used. The hyperbolic-lattice is fundamentally ambiguous but these across-track ambiguities are resolved by a technique similar to that employed in the Decca system (Mills 1962). This ambiguity-resolution technique is very reliable in the area covered by the ground-wave and therefore offers an assurance against setting up an initial error which would be carried throughout the operation. The reliability of this ambiguity-resolution technique is lower in the area which is covered by the sky-wave because the several frequencies upon which the lane-resolution technique depends suffer selective dispersion in the Earth-ionosphere medium. However, there is a further opportunity of establishing that no error has occurred, or of correcting such an error, when the ground-wave coverage area of the distant transmitter complex is approached.

The accuracy of Dectra varies over the area of coverage but even in midatlantic is better than 5 nautical miles except towards the lateral extremities of coverage where it may deteriorate to 10 nautical miles.

The period since the original article was written has been one of intense activity in the investigation of the behaviour of the Earth-ionosphere propagation medium at very low frequencies (v.l.f.) between 10 and 30 kHz, with a view to establishing the feasibility of a v.l.f.-based world-wide area-coverage navigation system.

The fact that v.l.f. waves are propagated over very great distances with low attenuation under all conditions of the ionosphere has been known for many years but experimental work in the late 1950's together with theoretical analysis by Wait (1962) predicted that the phase-length of a long propagation path would be relatively stable over varying ionospheric conditions and, even more important, would be predictable. At the time when these predictions were made relevant research techniques in the form of atomic frequency standards and fast long-range aircraft were becoming available and have subsequently formed the basis of much propagation research work. A summary of those conclusions of the propagation research which are relevant to v.l.f. navigation systems is as follows:

(i) A realizable v.l.f. transmitting aerial would have a very low radiation efficiency (up to 30 per cent) and, being virtually a point source, the radiated energy would excite a variety of modes in the Earth-ionosphere medium.

(ii) The dominant transverse electric mode would be attenuated at a substantially lower rate (of order 2 dB per 1000 kilometres) than the higher modes, especially at the lower frequency (10 kHz) and of the band.

(iii) The effective height of the ionosphere would be about 70 kilometres by day and 90 kilometres by night. Random perturbations are small and a navigation system based on v.l.f. transmissions would not suffer a serious degradation in accuracy in the event of magnetic storms, solar flares, etc.

(iv) Interaction of the Earth's magnetic field with the ionosphere results in a marked non-reciprocity of propagation, especially in equatorial regions, so in some circumstances the signal which has traversed the long path in a west-east direction can arrive at a point some 10,000 kilometres to the west of a transmitter with comparable amplitude to the signal which has traversed the shorter path.

(v) The group velocity of propagation in the dominant mode is most nearly invariant with ionospheric effective height when the frequency is about 11 kHz.

(vi) Attenuation of signals which traverse the polar ice cap greatly exceeds that of signals which are propagated over land or sea.

It follows from the conclusions which have been summarized above that a complex of eight suitably placed transmitters could provide the basis of a world-wide navigation system.

The American "Omega" system is to employ eight stations of which, at the time of writing, four have been installed in Trinidad, Norway, Hawaii and New York State to provide coverage over the greater part of a hemisphere. These stations all radiate signals at frequencies of 10·2 and 13·6 kHz. The stations operate in a time-division multiplex mode and each station in turn radiates, for one second, a signal on each of the two frequencies. The system has substantial development potential and additional frequencies in the 10–14 kHz band may eventually be radiated. The frequencies of all the stations are controlled by atomic oscillators and both frequency- and phase-stability are of a very high order (approximately 1 part in 10^{11} per hour). All the transmitters operate in the same time-reference within a tolerance of about 1 microsecond. The complex of transmitters generates a series of hyperbolic- lattices and the frequencies are such that the inherent ambiguities are spaced at intervals of approximately 8 and 24 nautical miles. Radiation of additional frequencies could of course enable the scale of the ambiguity spacing to be increased. It has been established that when all the transmission-paths are in daylight the accuracy of the system is in the region of 1 nautical mile. The accuracy may be degraded by a factor of 10 or more in the event of some of the transmission paths not being in daylight but, even so, the consistency of the predictable diurnal variations is such that by application of suitable corrections the all-daylight accuracy may still be approached.

(The illustrations, for which Crown Copyright is reserved, are reproduced by permission of Her Majesty's Stationery Office.)

Bibliography

MILLS J. R. (1962) in *Encyclopaedic Dictionary of Physics* (J. Thewlis Ed.), 4, 770.
WAIT J. (1962) *Electromagnetic Waves in Stratified Media*, Oxford: Pergamon Press.
WILLIAMS C. *et al.* (1960) in *Radio Aids to Civil Aviation* (Ed. Hansford), London: Hayward.

S. S. D. JONES

RAMSAUER EFFECT. Gas particles at normal temperature and pressure are in perpetual motion and because of the nearness of adjacent particles, collisions occur. For most of the time the particles move along straight lines but every so often they are deflected by collisions. The distance travelled between two collisions is called the free path and because the free paths vary statistically in length it is useful to consider the mean free path λ_m which can be related to the size or cross-section of the particle in the following way.

Consider only one particle to move whilst the others are at rest. During a mean free path the moving particle will trace out a volume $\lambda_m\theta$ in which θ is the cross-section. For particles of equal size the cross section $\theta = \pi D^2$ in which D is the diameter of the particle as

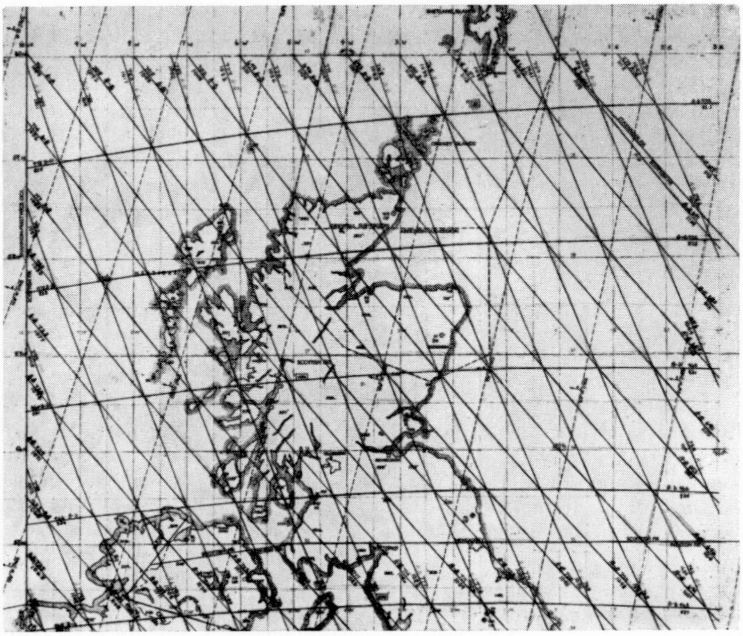

Fig. 2. Typical Omega chart.

illustrated in Fig. 1. Now the average space occupied by each particle is $\frac{1}{N}$ where N is the particle density and is equal to $N_0 p$ where N_0 is the particle density

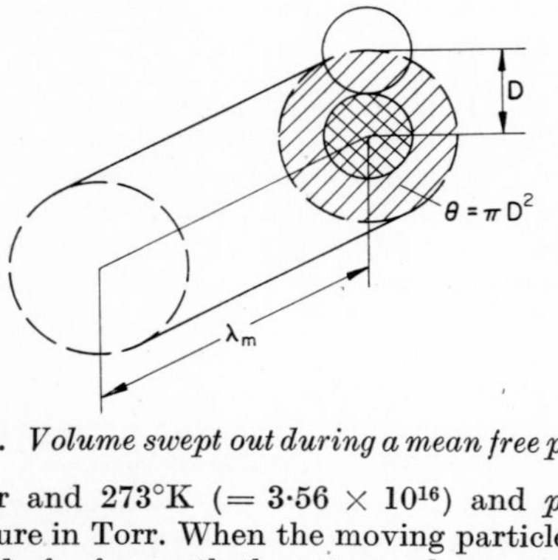

Fig. 1. Volume swept out during a mean free path.

at 1 Torr and 273°K ($= 3{\cdot}56 \times 10^{16}$) and p is the gas pressure in Torr. When the moving particle comes to the end of a free path these two volumes should be equal. The relation is not entirely correct as it was assumed that all the other particles were at rest. Because they actually move as well, the mean free path will be shorter as there is a good chance of some of the particles moving into the swept-out volume. From kinetic theory $\lambda_m = (\sqrt{2}N\theta)^{-1}$.

When free electrons are present in a gas and are in thermal equilibrium, their velocity is much higher for the same energy because of their much smaller mass which is $(1836)^{-1}$ that of a hydrogen atom. Compared with the speed of electrons, the gas particles appear to be at rest and thus the electron mean free path λ_e, will be at least $\sqrt{2}$ greater. In addition, because of their much smaller size, the electrons are able to approach the gas particles to within their periphery and thus the cross section $\theta = \pi D^2/4$. Taking in account the higher speed and smaller size of electrons, the electron mean free path, is given by $\lambda_e = 4\sqrt{2}\lambda_m$. The mean free path or cross-section can be obtained from kinetic theory type measurements and is illustrated in Fig. 2 for He, A and H_2.

In electrical discharges it is unusual for electrons to be in equilibrium with the gas and to remain at room temperature at which the mean energy is only about 0·04 eV. Electric fields are invariably applied between metal electrodes and the electrons are accelerated in the direction of the anode. They gain energy from the field over many free paths and their temperature is soon much higher than that of the surrounding gas. It is not uncommon in Townsend discharges for the electron energy to reach 15 eV or more corresponding to a temperature of some 100,000°K. At these values the interaction of electrons and gas particles is no longer governed by simple kinetic theory principles which assume particles to behave like perfectly elastic billiard balls and thus the cross-section can no longer be expected to remain constant.

It is the great merit of the German Ramsauer that around 1920 he devised an experiment which allowed the total collision cross-section (or mean free path) of

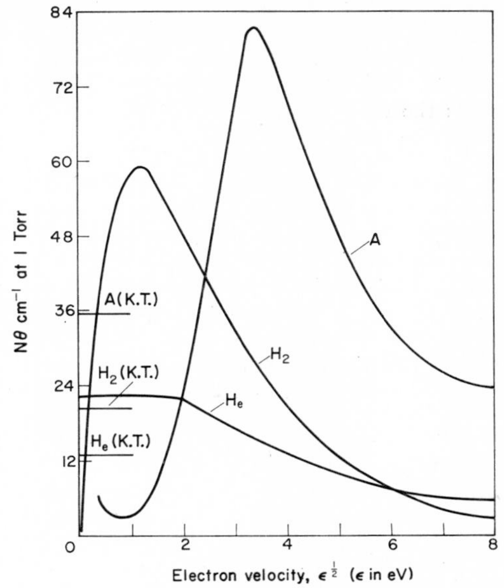

Fig. 2. Electron collision cross-section of helium, argon and hydrogen as a function of electron velocity. Constant values at low velocity are kinetic theory values taken from viscosity measurements.

gas particles towards electrons to be accurately measured as a function of electron velocity. It was found for most gases that as the velocity increased the total collision cross-section gradually decreased as expected from the decrease in time for interaction and that at low velocity the cross-section tended to the kinetic theory value as shown in Fig. 2. An unusual and striking feature is the transparency of argon to low energy electrons. As the energy decreases from 64 eV the cross-section increases in conformity with other gases such as helium and hydrogen but strangely around 12 eV this tendency is reversed and at lower energies the gas becomes more and more transparent to electrons. This transparency of a gas to low energy electrons is called the Ramsauer effect in honour of the man who discovered this phenomenon. The Ramsauer well occurs in all rare gases heavier than argon and in most hydrocarbon gases. It can be satisfactorily be accounted for in terms of wave mechanics.

Due to the Ramsauer effect the drift velocity and the diffusion coefficient tend to remain constant and independent of the electric field or the gas pressure. When impurities are introduced the drift velocity increases as the mean electron energy is reduced towards the Ramsauer well. The Ramsauer effect is responsible for negative resistance characteristics in irradiated thermionic diodes, and inert-gas-filled triodes. Use is made of this in amplifiers and oscillators up to frequencies of some hundreds of Mc/s. Discharges in gases exhibiting the Ramsauer effect are very susceptible to weak magnetic fields whilst at the other end of the scale, very large magnetic fields are required to reach the high magnetic field condition.

An effect analogous to that of Ramsauer in electron collisions has been observed in the total collision cross-section of neutrons in the energy range 0·1–100 MeV.

A. E. D. Heylen

ROAD DESIGN, PHYSICS IN. *Introduction.* At present, roads are designed on the basis of knowledge of the performance of other roads under similar traffic and soil conditions and by testing the soil to determine its ability to support loads (i.e. its bearing capacity). To obtain the information required for future designs, therefore, it is necessary to determine the quality and thickness of each layer in a road, preferably by non-destructive techniques. By making such tests at various ages on the same road, the commencement of changes which might result in failure can be observed. Physical methods are used to obtain this information and to determine the properties of the soil before a road is built upon it.

One of the functions of a road is to spread the dynamic loads imposed by the wheels of moving vehicles so that the stresses communicated to the soil under the road are sufficiently reduced so as not to cause permanent deformation. Theories are being developed for computing the stresses and strains produced in roads by moving vehicles, and are being checked by direct measurement. It is hoped it will eventually be possible to develop a rational system of road design based on the stresses and strains developed in it and the strength of the materials under the appropriate dynamic loading conditions.

Definition of terms. The term "pavement" will be used for the layers of the road above the soil, i.e. the hard crust placed on the soil formation after completion of the earthworks. Its main functions are:

(1) to provide a smooth riding surface,

(2) to distribute the traffic loads over the soil formation sufficiently to prevent the soil being overstressed,

(3) to protect the soil from the adverse effect of weather.

The characteristics of the pavement are dependent not only on the nature of the traffic, but also on the properties of the soil on which the road is built. The main structural element of the pavement is the *base* and its function is to distribute the traffic loads. It is often placed directly on the soil formation (called the "subgrade") but, sometimes, a thin layer of material known as the "*sub-base*" is placed between them. Sub-bases are sometimes used to provide an additional load-spreading layer of somewhat less costly material than the road base or where there is a chance that the soil may work up into the base. The base may be covered with either a concrete or a bituminous surfacing, in which case the pavement is described respectively as "*rigid*" or "*flexible*".

The soil. In preparing the subgrade, the aim is to provide adequate resistance to the action of traffic and weather. The main requirements are:

(1) to obtain adequate compaction of the soil,

(2) to maintain the subgrade in a stable condition at a constant moisture content,

(3) to protect the subgrade from frost.

Before a road can be designed, it is first necessary to identify the type of soil which will lie underneath it. Much can be done by visual inspection, but the engineer needs more information than can be obtained in this way. Borings are made over the site and samples of the soil are taken. The natural moisture content of each sample is measured by weighing it before and after prolonged drying in an oven, and the distribution of particle size in the specimen is also determined. "Gravel", "sand", "silt", and "clay" fractions are recognized as containing particles of decreasing magnitude. The gravel fraction is removed by passing the sample through a No. 7 B.S. sieve. The material passing through the sieve is first treated with hydrogen peroxide to remove organic matter, and with hydrochloric acid to remove carbonates and gypsum. After sieving, the soil is dispersed in a dilute solution of sodium oxalate, which acts as a deflocculating agent, using a high speed stirrer. The material is allowed to settle in a vertical tube, particles of different sizes having different settling velocities, according to Stokes' law. At given time intervals after the initial dispersion samples are extracted by means of a pipette from a given depth below the surface and are dried and weighed. By such sedimentation analysis, it is possible to determine the amount of medium silt (0·02–0·006 mm), fine silt (0·006–0·002 mm) and clay (smaller than 0·002 mm) which the soil contains.

Many of the properties of soils of importance to engineering, such as their strength, compressibility, swelling and shrinkage, are influenced primarily by the clay fraction of the soil. This fraction can have a number of different compositions and these often determine the nature of the soil's behaviour. X-ray diffraction analysis and differential thermal analysis are used to obtain information relating to the composition of the soil. Sealed-off fine focus tubes giving copper and cobalt X radiation are used in conjunction with powder cameras recording low angles of diffraction. The identification of the mineral composition of soil clays is carried out by analysing the diffraction lines obtained for the clay fraction after it has been subjected to various thermal and chemical treatments.

Much information concerning the properties of soil can be obtained from the liquid and plastic limits, which are respectively the moisture contents at which a clay soil passes from the solid to the plastic state and from the plastic to the liquid state. The liquid limit of the soil is defined as the moisture content at which it is sufficiently fluid to flow a specified amount when lightly jarred 25 times in a standard apparatus. The plastic limit is the moisture content at which a thread of soil can be rolled until it is only 3 mm in diameter. The numerical difference between the liquid and plastic limits is termed the "plasticity" index and indicates the range of moisture contents over which the soil is in a plastic condition.

The movement of moisture in soil. In general, the bearing capacity or supporting power of most soils decreases with increase of the moisture content. Adequate compaction of the soil reduces the rate of deformation under load and also the rate of water absorption by the soil. Assuming efficient subgrade drainage, and ignoring the effect of the seasonla wetting and drying of the verges, the subgrade tends to reach an equilibrium moisture content which depends on the level of the water table and the overburden pressure imposed by the road pavement. It is particularly important, therefore, to be able to predict the equilibrium moisture content in any particular set of circumstances because the design thickness of the road must be based on the strength of the soil at this moisture content.

The surface forces by which water is retained in the soil structure are responsible for the pressure reduction (below atmospheric pressure) known as the soil suction or tension. The pressure reduction in a small sample of the soil is measured when the sample is entirely free from externally applied stress. On the pF scale of measuring soil suction, the pF value of the soil moisture is equivalent to the common logarithm of the suction expressed in centimetres of water.

In the ground, each small element of soil is subjected to stress by the surrounding soil and hence indirectly by externally applied loads. This stress may change in the stress-free suction of the moisture in the element, so the pressure of the water in the pores, generally known as the pore water pressure, may be regarded as the algebraic sum of two components: (a) the suction and (b) the effect on the suction of the applied stress. The effect of external stress on pore water pressure is measured directly by measurements on a small unloaded sample. If u is the pressure of the pore water when the sample is loaded, s is the suction in the sample when the soil is free from external loading and P is the applied pressure, then

$$u = s + aP \qquad (1)$$

where a is the fraction of the applied pressure which is effective in changing the pressure of the soil water and ranges between 0 and 1.

Since the value of suction for a soil increases as the soil becomes drier, any local change in moisture content will cause a redistribution of water in the soil. Similarly, the application of a local pressure to a soil mass may produce hydrostatic pressure gradients tending to cause a redistribution of moisture. In either case, equation 1 can be used to calculate the magnitude of the redistribution. In making such cal-

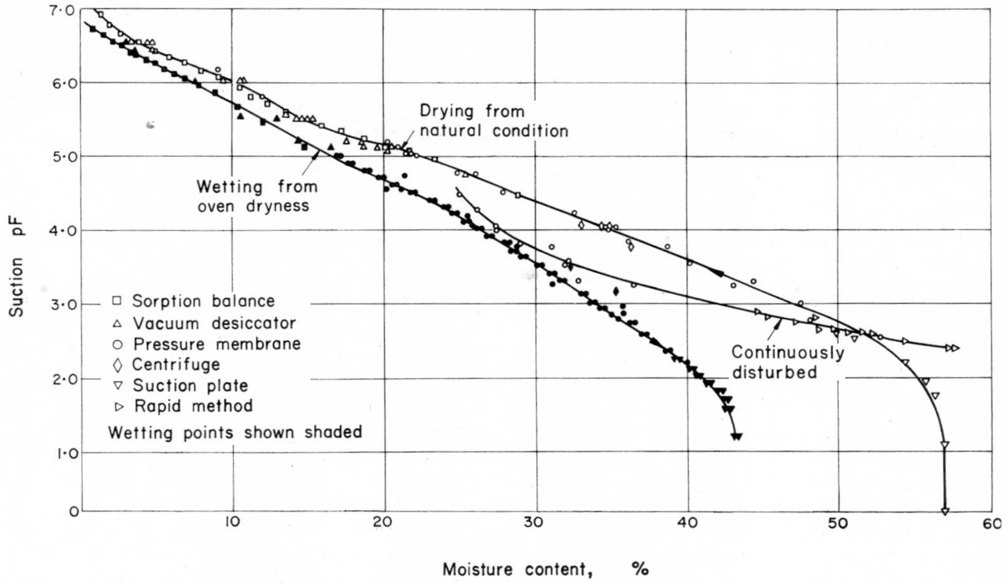

Fig. 1.

culations, the possibility of irreversible changes in the suction of the soil due to changes in the soil structure, i.e. the particle arrangement, must be taken into account.

The migration of water in soil is accompanied by some changes of volume. Volume changes resulting from applied pressure are measured in apparatus in which a sample is subjected to triaxial stresses.

The thermodynamic approach to soil moisture movement can be used to evaluate moisture migration in terms of moisture content on a weight basis if the appropriate relationships between suction and moisture content are known. In one such method, the sample of soil is placed on a thin flat cellulose membrane supported by a porous metal plate filled with water at atmospheric pressure. An air pressure equal to the final suction that is required in the soil sample is applied above the membrane. When moisture exchange through the membrane is complete, the soil is removed for determining its moisture content, strength or other property which is to be related to the moisture suction. Typical results obtained with this apparatus are shown in Fig. 1.

A sorption balance is used for comparatively dry soils. The sample of soil is allowed to reach moisture equilibrium with a known humidity and the equivalent suction is computed from the thermodynamic relation between suction and humidity. A Joly spring balance enables continuous weighing to be made during the progress of equilibration and a series of tests in different humidities can be made without removing the soil from the apparatus.

Because the design thickness of a road depends, among other factors, on the moisture conditions of the soil and the changes likely to occur in the soil after the road has been constructed, considerable effort has been devoted to studying moisture movements in soil. Under impervious pavements, only small seasonal fluctuations of pore pressure and moisture content are observed, except near the edges of the pavement where the soil is influenced by the exposed verges. The ultimate pore water pressure under the impervious surfacing is found to be dependent on the water-table. The method developed for estimating the ultimate moisture-content distribution with depth is found to agree with the measured moisture-content profiles.

Soil compaction. Soil compaction is the process whereby soil is mechanically compressed through a reduction of the air voids. In road construction, good compaction is needed in building embankments and subgrades. In an embankment, subsequent settlement must be minimized to enable a permanent road structure to be placed on it immediately after its completion. Compaction of a subgrade increases its stability and resistance to water absorption. For a constant amount of compaction, all soils have an optimum moisture content at which a maximum dry density is obtainable. At a constant moisture content, increasing amounts of compaction increase the dry density of the soil until the air voids are almost eliminated or the resistance of the soil to further compaction becomes too great. In practice, this means that the first few passes of a compacting machine produce considerable increases of density, but the change becomes less with a greater number of passes until no further change occurs. In controlling embankment construction, therefore, measurements of the density of the soil are made to check that the required degree of compaction has been attained.

Measurements of soil density. A simple, rapid and accurate test is required for measuring the density of soil in embankments. That most commonly employed is the sand replacement method. A hole about 10 cm in diameter is excavated with suitable tools to the depth of the layer being tested, and the weight of soil removed is determined and a moisture content sample taken. Sand graded between No. 25 and No. 52 B.S. sieves is run into the hole from a special bottle and the weight of sand used is determined. The volume of sand required to fill the hole is calculated from the weight of sand used and its bulk density.

Portable radioactive apparatus of the direct transmission type has been developed to measure densities *in situ.* The apparatus has a probe containing a caesium-137 source of gamma radiation situated near to its tip which is inserted into the soil to a depth of 15 cm. The intensity of the radiation emerging from the surface at about 45° to the vertical is measured by a Geiger counter contained within the surface unit. A higher bulk density of the soil results in a greater absorption of gamma radiation and a lower intensity of emergent radiation. It is not always possible to drive in the probe absolutely perpendicular to the surface unit. Errors arising from the variations of path length which thus results can be overcome by having a number of Geiger tubes spaced around the probe. The soil is disturbed by inserting the probe, but this is serious only when the soil contains more than about 35 per cent of material retained on a 1·2 mm sieve. The relation between soil density and intensity of emergent radiation is affected by the type of soil and is not unique. Recent experiments indicate that the effect of soil type can be minimized by better collimation at the detector.

A back-scatter apparatus has also been studied for measuring the density of soil *in situ.* In this, the radioactive source and detector are placed in a surface unit and shielded from each other by lead. A Geiger counter is used to measure the intensity of the emergent gamma radiation. Experiments with beds of soil whose thickness was steadily reduced have indicated that about 80 per cent of the detected radiation is scattered back to the surface by the top 3–5 cm of soil. The amount scattered back by deeper layers diminishes rapidly with increase of depth, but a little comes from depths of the order of 12–15 cm.

Measurement of moisture content in situ. Neutrons are scattered more by hydrogen nuclei than by other elements, and a neutron-scattering device would be expected to be useful to indicate the amount of water contained by a bed of soil. A radium-beryllium neutron source has been tried in a surface back-scatter apparatus but has been found to suffer from all of the difficulties associated with the employment of gamma back-scatter to measure density *in situ.* It gives much more weight to the moisture content of the soil close to the surface and is affected by soil type.

Studies of the variation of moisture content at a depth below the soil surface have been successfully made by employing a deep probe containing a neutron source and a counter.

Development of vibrational methods. The vibrational technique originated in Germany during the years 1938–1939 when the Deutsche Gesellschaft für Bodenmechanik (DEGEBO) developed apparatus for testing soil *in situ.* A mechanical oscillator of the rotary out-of-balance mass type was employed and its vibratory characteristics were studied while operating on many different types of soil. The mechanical properties of the soil were deduced from the results. The resonant frequency of the vibrator varied with the bearing capacity of the soil, but attempts to explain the results in terms of a simple mass and spring system were unsuccessful, largely because the resonant frequency varied with the applied vibratory force.

Since 1948, the development of vibrational testing has been considerably influenced by the work of van der Poel and Nijboer of the Dutch Shell Laboratory at Amsterdam. A machine, known as the "Dutch Shell Vibrator", was devised employing rotating out-of-balance masses to apply vibratory forces of 2 ± 2 tonnes to the road surface at frequencies between 5 and 60 c/s. The peak vibratory force (F_p) developed in the road was calculated from the masses of the machine and the rotary speed, while the peak vibratory displacement (x) was measured under the plate by which the force was applied to the road surface. The stiffness of the construction (S) is given by $S = F_p/x$ which is a function of the mechanical properties and thicknesses of the layers forming the construction, but, unfortunately, it is neither independent of frequency nor force. Measurements on a variety of roads, however, have shown that the value of S will, under certain circumstances, indicate the condition of a flexible road. High values of S indicate a strong construction and low values show the construction to be weak.

Deflexion beam technique. In many countries, a beam apparatus is used to measure the vertical deflexion of the road surface resulting from a loaded lorry wheel moving slowly away from the point at which measurements are made, and the results provide a measure of the stiffness of the road. By correlating values of stiffness with the subsequent performance of different types of road under traffic, it is now possible to assess the condition of many types of road at the time when the measurements are made and to predict their future performance. The beam deflexion technique has the same limitation as the dynamic stiffness obtained by vibration tests in that it cannot provide detailed information about any individual constituent of the road construction. At present, only the wave propagation technique appears capable of indicating which of the layers is changing under traffic or weather.

Measurement and interpretation of surface vibrations on soils and roads. A non-destructive technique which will often provide information about the dynamic modulus of elasticity and the thickness of each layer of the road is the surface wave propagation method. Vibrations are produced at a specific frequency chosen from the range 28–2400 c/s and the wave-length and velocity of the vibrations are measured along the surface of the road. Fractional changes of wave-length are usually measured at successive incremental distances from the vibration generator along a test length of 10–15 cm. If the construction has uniform properties, the average wave-length λ and the average velocity c is derived from the relation

$$c = n\lambda \qquad (2)$$

where c is in m/s, n is the test frequency in c/s and λ is in metres.

To obtain the moduli and thicknesses of all of the layers, it is necessary to measure the average wave-length and velocity of the vibrations at about 30 different frequencies. Analysis of the relation between velocity and wave-length then provides detailed information for the surfacing, base and sub-base. The analysis of the results, for obtaining the elastic properties and thicknesses of the surface and base layers of the motorway type of construction, has now been reduced to a fairly simple curve-fitting procedure which can be carried out in a few minutes with the aid of a slide rule and standard curves or tables.

Variations of the wave-length or velocity at various positions along a long length of road are used to locate areas of lower quality than the remainder. When it is necessary to study only one layer in this way, experiments are first made to determine the frequency providing most information about that particular layer. For cement-bound materials, empirical relations can be determined between the elastic properties and strength so that the velocity measurements can be converted into strengths.

The wave propagation method has been used to observe the compaction of sandy layers under a road by the action of traffic. In one case, the traffic caused a fourfold increase of elastic modulus. It has also been used to detect the onset of cracking in cement-bound layers in bases beneath asphalt surfacings, and to detect stripping of bitumen off bound granular bases. It thus provides much useful information which assists engineers trying to assess the performance of full-scale experimental and other roads under traffic.

Rational design of roads. One of the functions of a road is to spread the dynamic loads imposed by the wheels of moving vehicles so that the stresses communicated to the soil subgrade are sufficiently reduced so as not to cause permanent deformation. Theoretical values of the stresses and deflexions generated in multi-layered elastic systems have shown that the stresses applied to the soil subgrade and the deflexions of the construction under them depend on the relative values of the dynamic elastic moduli of the layers forming the road and their thicknesses. Hitherto the theoretical work has been restricted to two- and three-layer elastic systems, but laboratories in several countries are currently working on a computer programme to deal with an infinite number of layers.

It is hoped it will eventually be possible to develop a rational system of pavement design based on the stresses and strains developed in the road and the strength of the materials under the appropriate dynamic loading conditions. The theoretical computations of stress and strain require knowledge of the elastic moduli of the materials employed. Much information is being obtained from tests made *in situ* by the wave propagation method. The chief difficulty here is that the elastic moduli of bituminous materials increase with increase of the rate of loading and the wave propagation tests which are made at frequencies between 28 and 24,000 c/s are made at a much higher rate of loading than that resulting from the passage of a moving vehicle. Auxiliary experiments are in hand to bridge the gap and consist of repeated flexural tests made on beams of bituminous surfacing materials. Phase sensitive amplifiers are used to measure the components of deflexion in phase and in quadrature with the driving force. It is not yet possible to substitute complex elastic moduli into the equations employed to compute stresses and deflexions, but this will have to be done eventually.

Before elastic theory is accepted for computing the dynamic stresses, strains and deflexions in roads, as many of the results as possible must be checked by direct measurement. Experiments to this end are currently in hand in several European countries. Two types of gauge have been developed, the piezoelectric soil pressure gauge and that having resistance or semiconductor strain gauges attached to a diaphragm. The British piezoelectric soil pressure gauge is shown in Fig. 2 and has the advantage that its very high overall elastic modulus renders its error of registration constant regardless of the soil in which it is employed. The resistive stress gauge works by the deflexion of its diaphragm and is thus less stiff than the piezoelectric gauge with the result that errors of registration often have to be determined for the soil in which it is to be used. Both types of gauge are usually employed with recording oscilloscopes or multi-channel galvanometer recorders, resulting in the subsequent measurement of many oscillograms. Efforts are being made in Britain and Holland to reduce the work involved by recording the signals direct on to magnetic tape for analysis by electronic computer.

The piezoelectric gauges are usually buried at various depths under the road and measurements are made while vehicles of known wheel load are driven along the road at various speeds. The stresses in the soil decrease with increase of vehicle speed as a result of the effect of the changes in the relative elastic moduli of the layers and tend to become approximately constant at speeds above 15 to 20 m.p.h. Attempts are being made to correlate the theoretical values of stress with those measured in this upper range of speeds. Hitherto, comparisons have been made of the load-spreading properties of various types of road and it has been found that good load spreading results from the employment of upper layers on the road having high dynamic elastic moduli. It can be shown theoretically that, when good load spreading is achieved in this way, high tensile strains are developed in the road base in the horizontal direction close to its underside. With some cement-bound bases, there is a risk that such strains will initiate cracks in the material and cause failure.

Because it was found that some road bases failed by cracking under traffic loading conditions, attempts are being made in several countries, notably Germany, Holland and Britain, to measure the dynamic horizontal tensile strains developed in road bases and in bituminous surfacings by moving vehicles. Electrical resistance strain gauges are used, either attached to strips of metallic foil or embedded in prefabricated blocks of the road material. The work is still in progress but comparison between the measured and computed values of strain has been encouraging.

Where concrete roads are concerned, it is necessary to measure the static or slowly varying strains arising from shrinkage and swelling of the material, thermal expansion and contraction and temperature warping. Electrical resistance strain gauges attached to steel strips having enlarged ends have been used for this work but are not very reliable for measurements extending over more than a day at a time because of instability and zero drift. A vibrating wire gauge has been developed which has proved to be more satisfactory. It consists of a length of plated piano wire clamped between enlarged ends and surrounded by a Perspex tube. It is cast into the concrete and, once the material has set, the Perspex tube is of no account. The wire is set into vibration by feeding an electrical impulse to a coil placed opposite its mid-point and the resultant frequency of vibration is measured. When tensile strain is developed in the concrete, the wire is stretched and its natural frequency rises. The gauge is cheap to produce, and this is an important factor when a number have to be cast into a road slab and cannot be recovered.

Temperature measurement in roads. The transient stresses and deflexions produced in flexible roads by moving vehicles vary according to the temperature of the bituminous layers; an increase in temperature lowers the elastic modulus of bituminous material and reduces its load-spreading properties. This weakening of the road may not be important if the high temperature persists for only a short time each year. It is necessary, therefore, to determine the duration of various temperature levels in typical forms of road construction in order to assess the weakening effect of high temperatures and the stiffening effect of low temperatures. Apparatus developed for this purpose records the durations of each of twelve temperature levels in 5°C steps between -10°C and $+50$°C and the durations are read in minutes on electromagnetic counters.

Weighing vehicles in motion. It has been found both experimentally and theoretically that the stresses and strains developed in roads depend on the wheel loads of moving vehicles. It would be expected, therefore, that the performance of a road under traffic would depend not only on the number of vehicles which have travelled along it since it was constructed but also on their wheel loads. Apparatus has been devised to weigh vehicles without affecting the traffic flow. A weighbridge is constructed in the road surface and consists of platforms supported on load cells carrying electrical resistance strain gauges. The electrical signals from these gauges are digitized

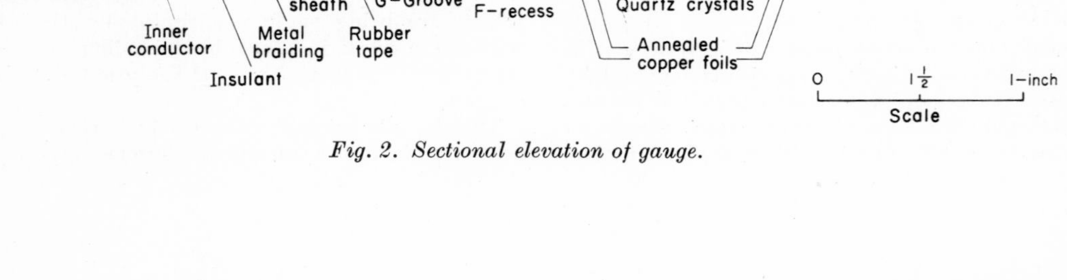

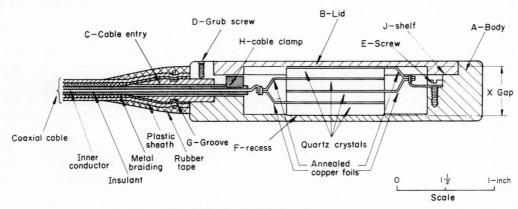

Fig. 2. Sectional elevation of gauge.

electronically and, after classification according to magnitude, are routed to electromagnetic counters which register the number of wheel loads in the groups 300–2000 lb, 2000–4000 lb, 4000–6000 lb, and so on up to 20,000 lb. The counters print out the numbers in each weight group at predetermined intervals, usually hourly. The information being obtained from such counters on trunk roads shows that the number of extremely heavy vehicles is increasing at a greater rate than the traffic and this is very important when designing roads which are expected to have a long life.

Behaviour of road materials under dynamic loading. The development of a rational method of pavement design is complicated by the fact that different types of road fail in different ways. Flexible roads are often considered to have failed when permanent rutting has occurred along one of the wheel tracks. Concrete roads and, occasionally, flexible roads fail when cracking of the surfacing occurs, leading to its break-up and the formation of pot-holes. The cracking of any form of road surface is often accompanied by damage to the foundations resulting from the percolation of water through the surface cracks down to the soil subgrade.

If the stresses applied to the soil subgrade are large, the soil compacts under traffic loading and this eventually leads to damage to the road base and permanent deformation of the road surface. Compaction under traffic leading to rutting may also occur when the sub-base or base of the road consists of unbound aggregate or tar- or bitumen-bound materials which were not adequately rolled when laid. Cement-bound bases and concrete road slabs fail by fatigue when subjected to excessive tensile strains.

Research is taking place in many countries to examine the mode of failure of all types of road, to assess performance under traffic by special machines in which road sections are subjected to between 10^4 and 10^7 passes of a wheel, and by special tests in which lengths of road are subjected to the passage of vehicles of selected wheel load. The condition of the road surface is assessed by measuring its longitudinal profile, the rut depth, and the amount of cracking and patching, and these factors are then combined to provide an index of serviceability. In the U.S.A. some work has been done on the behaviour of soil under repetitive loading, while in the U.S.A., Holland and Britain, some work has been done on the fatigue properties of bituminous materials, largely of the types containing sand or small aggregate used for road surfacings. Work is only just beginning on the materials containing larger stones which are used for road bases. Many of the tests involve either the flexure of beams or plates or rotating cantilever specimens, and the materials are not subjected to the combinations of stress which actually occur in roads. The viscoelastic materials containing tar or bitumen are also sensitive to rate of loading and temperature. These are fields where extensive research is required.

Symposia dealing with the structural design of asphalt pavements and the various aspects of the work described in this note are organized at five-yearly intervals by the University of Michigan and attract workers from all over the world. The first was held at Ann Abor in August 1962, and the second in 1967.

Acknowledgements. The article is contributed by permission of the Director of Road Research. (Crown copyright. Published by permission of the Controller, Her Majesty's Stationery Office.)

Bibliography

International Conference on the Structural Design of Asphalt Pavements, University of Michigan, Ann Arbor, August, 1962, Proceedings.

(1952) *Soil Mechanics for Road Engineers*, London: H.M.S.O.

Croney D., Coleman J. D. and Black W. P. M. (1958) *Studies of the movement and distribution of water in soil in relation to highway design and performance*, American Highway Research Bord, Washington, Special Report No. 40.

Jones R. (1962) *Non-destructive testing of concrete*, Cambridge: The University Press.

Nuclear principles and applications, 1939–1963, American Highway Research Board, Washington, Bibliography No. 41, 1966.

Trott J. J. and Whiffin A. C. (1965) *Measurements of the axle-loads of moving vehicles on Trunk Roads, Roads and Road Construction*, **43**, No. 511, July, 209.

A. C. Whiffin

RUBBER SPRINGS. *1. Introduction.* Rubbers, both natural and synthetic, are polymeric materials which can be greatly elongated, but which rapidly retract to the undeformed state or close to it, upon load removal. This "elastomeric" quality which enables large quantities of strain energy to be stored and regained at will, has distinguished rubber as a spring material for over one hundred years. One of the first vehicle suspensions of natural rubber was a cart spring of 1852. This was made by Stephen Moulton, great grandfather of the designer of the modern Moulton bicycle, which also has rubber springing.

As well as suspensions for road and rail vehicles, present-day rubber spring developments include tyres, anti-vibration mountings, ship fenders, bridge bearings, flexible bearings and couplings, seals and gaskets, snubbers, bump stops and package cushioning. The proper design and use of such springs requires a wide knowledge of the physical and mechanical properties of a number of elastomers, the most important of which is still natural rubber:

2. General physical characteristics of rubber. The following qualities characterize rubber:

1. It has a greater capacity for strain energy storage than steel.
2. It needs no lubrication or maintenance.
3. It can be moulded to shape.
4. It is resilient, but exhibits some hysteresis under both static and dynamic conditions.
5. It impedes the transmission of sound.
6. A range of characteristics can be obtained by suitable compounding, and with the addition of fillers such as carbon black.
7. Because of its flexibility, assembly and service misalignments can be accommodated.
8. It is easily located by means, usually, of chemical bonding to end-plates or directly to the surrounding structure.

   In addition, rubber springs can readily be designed to have:
9. Desired spring rates in more than one direction simultaneously.
10. Inherent stability, which eliminates the need for external supports or guides.

From the above considerations it can be seen that rubber is often preferable to steel as a spring material, particularly in those applications where a measure of energy dissipation, as well as storage, is either necessary or advantageous.

However, rubber may be inferior where high strength and stiffness, or large static deflexions are required, or where extremes of temperature or other unfavourable environmental factors prevail.

Although many synthetics have been produced which have greatly extended the useful range of elastomers in general, few can approach the all-round performance of natural rubber, which possesses the following advantages:

(a) High resilience combined with a small amount of damping.

(b) Low internal heat build-up under dynamic conditions.

(c) High resistance to fatigue, cut growth and tearing.

(d) Low raw material cost allied to ease of processing.

(e) Wide range of operating temperatures.

(f) Less tendency to creep, stress relaxation and permanent deformation under load.

Nevertheless, a number of synthetics have proved to be superior in specialized fields. A short list is given in the table.

3. Tension springs. In engineering applications, the use of rubber under direct tension is avoided where possible because creep and permanent set are usually more pronounced, and there is a tendency to internal cracking in short springs at quite low nominal strains. In addition, there is a greater likelihood of bond failure and it is more difficult to design springs to have fail-safe characteristics. Products like catapults, rubber bands, balloons and diaphragms merely prove the rule that wherever possible, rubber is used in compression, shear, torsional shear, or some combination of these modes of stressing.

4. Static load-deflexion properties of rubber in shear. A typical stress-strain curve for natural rubber in simple shear is given in Fig. 1. The behaviour is linear to well above the normal maximum working deflexion which is equal to about half the rubber thickness. Hence, a straightforward mathematical relationship exists between applied loading and the resultant deflexions, which makes the design of shear springs relatively simple. Another advantage, where soft springs are required, is that like other engineering materials rubber is much less stiff in shear than under direct loading.

Formulae for some springs whose deflexions are due essentially to the rubber being in simple or

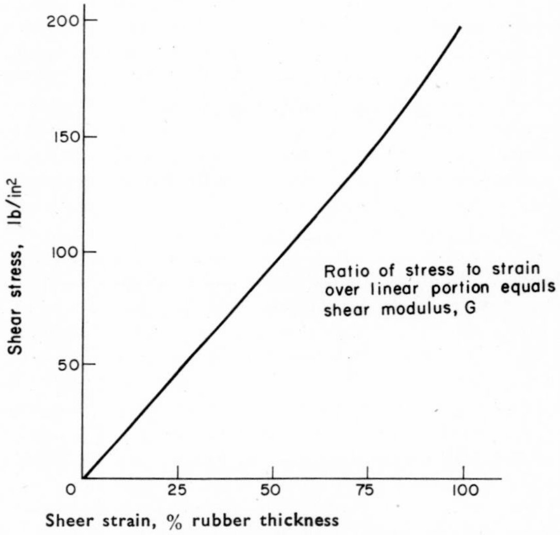

Fig. 1. Stress-Strain curve for ***rubber*** *in simple shear.*

torsional shear, are given in Fig. 2. The rubber is usually chemically bonded to metal end-plates or sleeves, to simplify load application and spring location. In addition, gradual slippage is prevented at the interfaces, and spring performance is thus made more reliable.

5. Static compression-deflexion behaviour. The behaviour of rubber in compression is somewhat difficult to predict because it is essentially non-linear. The principal reason for this is the relative incompressibility of rubber which is of the same order as that of water. If a rubber block is compressed between two parallel plates, as in Fig. 3, its sides expand in order to maintain near constant enclosed volume. As the parallel plates are brought closer together, the lateral deformations of the rubber must accelerate for the volume still to be accommodated, and there is a corresponding increase in the rate of resistance to compression. The stress-strain curve is therefore

Characteristics of some common spring rubbers

Type of rubber	Good properties	Moderate to poor properties
1. Natural	Resilience; strength, creep, fatigue and low temperature resistance	Heat, oil, weather and ozone resistance
2. Butyl	Weather, ozone, chemical and low temperature resistance. Very low gas permeability	Strength and resilience. High inflammability
3. Butadiene-Styrene	Abrasion and ageing resistance better than natural rubber	Oil, tear, weather and ozone resistance
4. Fluorinated	Resistance to heat, fuels, oils, chemicals, ozone and weathering	Poor low temperature resistance. High cost
5. Neoprene	High heat, ageing, weather, fatigue and flame resistance	Moderate oil and chemical resistance
6. Nitrile	Resistance to heat, ageing, oil, petroleum solvents and chemicals	Poor low temperature resistance
7. Polyurethanes	Very high strength, abrasion, tear and ozone resistance. Wide modulus range	Moist heat resistance

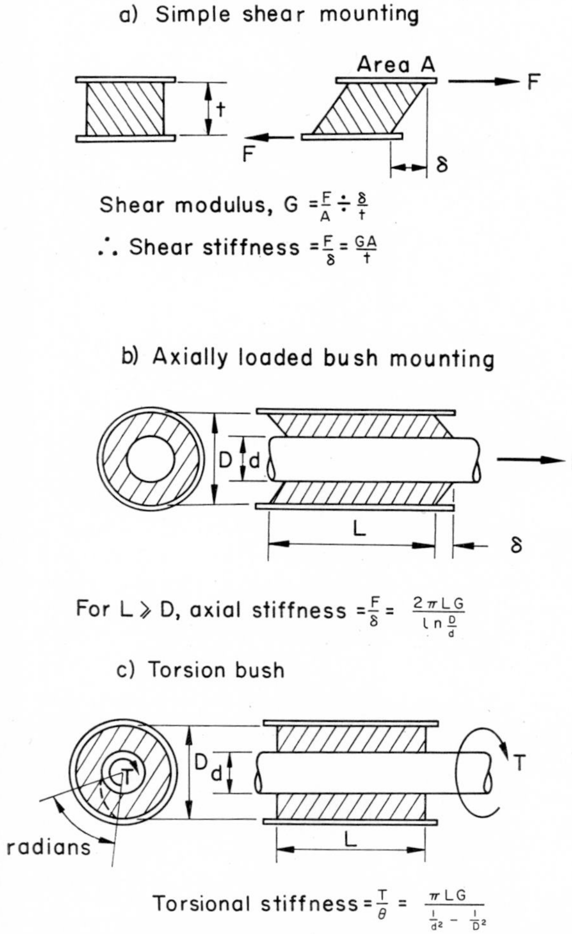

Fig. 2. Springs utilizing rubber in shear. (All the above formulae assume small strains with negligible deflexion due to bending.)

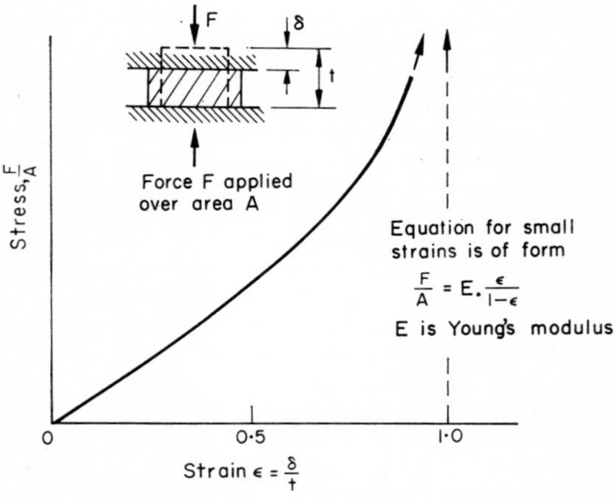

Fig. 3. Constant volume compression of a perfectly lubricated rubber block.

one of increasing slope, and which for perfectly elastic behaviour would be asymptotic to a strain equal to the free height.

The initial stiffness of the rubber is dependent upon the degree to which its sides are free to bulge. If the loaded faces are perfectly lubricated there are no external lateral forces and the load-deflexion equation is of a form such as that in Fig. 3. In real situations, friction or bonding at the loaded faces increase compression stiffness by restricting bulging to an amount dependent upon the relative dimensions of the areas under compression and those where bulging occurs. A convenient indication of this effect is the ratio of the initial area over which compression acts, to that free of external constraints. Sample values of this "shape factor" are given in Fig. 4, and Figs. 5(*a*) and 5(*b*) show how it is included in empirical compression-deflexion formulae.

Sometimes an alternative "shape function" is preferred which takes into account not only di-

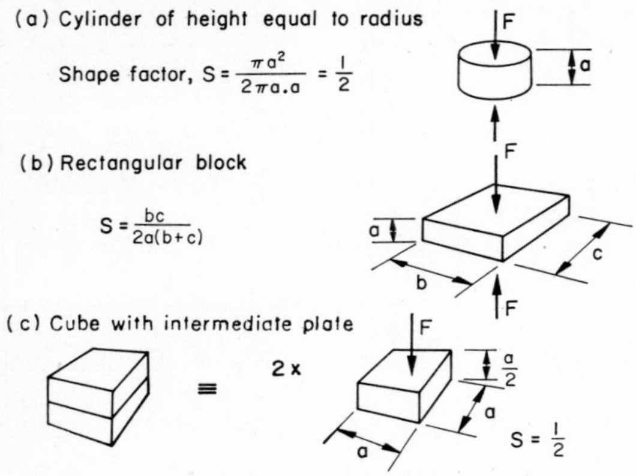

Fig. 4. Shape factors for some simple compression springs.

Limitations to shape factor concept. 1. All loaded faces are perfectly bonded to rigid plates. 2. Unloaded faces are initially normal to loaded faces. 3. Loading is normal to loaded faces. 4. Unloaded faces are free from external constraints. 5. Shape factor must refer to an elementary spring, not one that is really a number of springs in series.

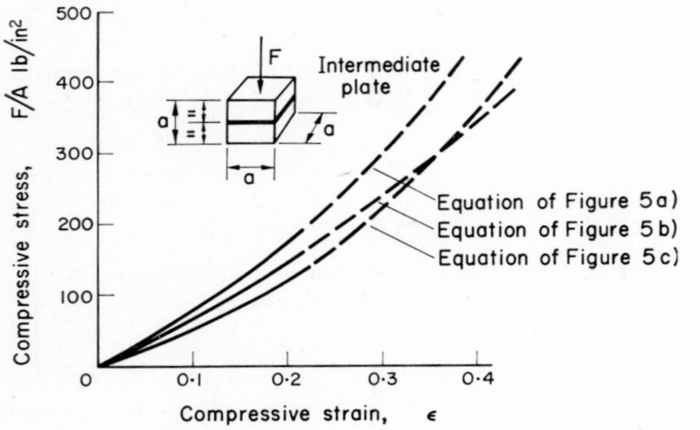

Fig. 6. Prediction of stress-strain behaviour. Consider a cuboid spring with intermediate plate, such as that of Fig. 4c, made of a natural rubber compound of hardness about 55 IRHD. Properties of such a spring are: $S = 0{\cdot}5$, $E = 460$ lb/in², $G = 115$ lb/in², $k = 0{\cdot}64$, $S = 1{\cdot}44$.

mensions, but also less noticeable effects of other parameters on rubber bulging. An example is given in Fig. 5(*c*).

Fig. 5. Empirical formulae for small deflections of rubber compression springs. Unless otherwise stated, symbols are as defined in Figs. 1 to 4.

(*a*) $F/A = E(1 + S)\dfrac{\varepsilon}{1 - \varepsilon}$ (*see ref. 4*);

(*b*) $F/A = E(1 + 2kS^2)\,\varepsilon\,(1 + \varepsilon)$ *where* $0{\cdot}5 < k < 1{\cdot}0$: *k depends on rubber hardness (see ref. 1)*;

(*c*) $F/A = Gs(\lambda^{-2} - \lambda)$ (*see ref. 2*): *s depends upon both spring shape and shear modulus.* $\lambda = 1 - \varepsilon$, *and is the ratio of initial to compressed heights.* (*Note: see also ref. 12.*)

In Fig. 6, the performance of a cylindrical spring is predicted by means of the three equations of Fig. 5. There is some disparity in the curves, but up to the normal maximum working strain of about 0·2, all should predict deflexions which are within ±20 per cent of the observed values. Which formula is most accurate will vary with each spring application according to the experience of the spring designer.

In a number of applications, the rubber is required to deform in both shear and compression, simultaneously. Two such examples, the inclined shear mounting and the radially loaded bush, are shown in Fig. 7.

Fig. 7. Springs under combined loading.

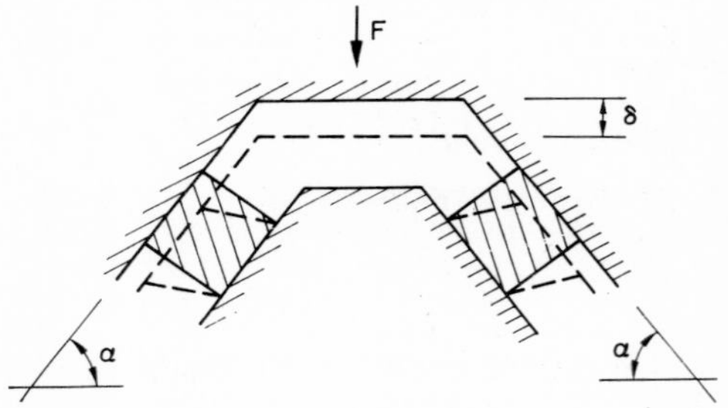

(*a*) *Inclined shear mounting. For small deflexions* $F/\delta = 2(K_c \cos^2\alpha + K_s \sin^2\alpha)$. K_c *and* K_s *are the compression and shear stiffnesses, respectively, for each (identical) rubber unit.*

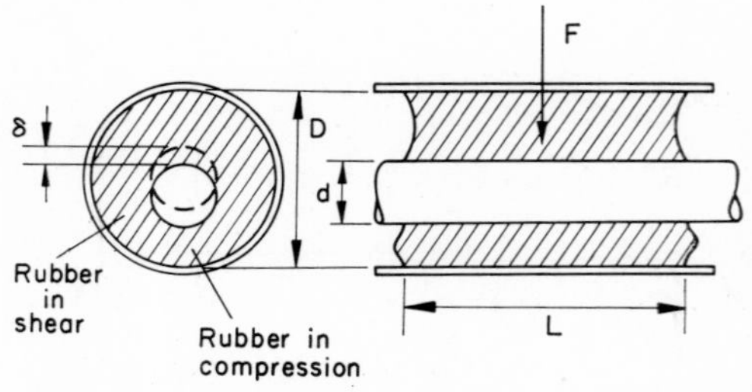

(*b*) *Radially loaded bush. For small deflexions* $F/\delta = \beta LG$. β *is an empirical factor which varies with bush geometry. Values are given in Lindley (1964) and Davey and Payne (1964).*

6. Rubber elastic moduli and their relation to hardness. Since the shear stress-strain curve is essentially linear, the elastic behaviour of rubber in shear over the design range, is defined by the shear or rigidity modulus.

In contrast, the direct stress-strain curve is generally non-linear. However, careful experiment has shown that linearity exists for both tensile and compressive strains of a few per cent, and that the curve is continuous through the point of zero strain. Hence a Young's modulus exists for rubber, and it is usually defined as the slope of the direct stress-strain curve, at the origin (Fig. 8).

In spring design, the Young's and shear moduli are the two most important elastic properties. In the absence of either, the error is not great in assuming that the former is three times the latter, as for small

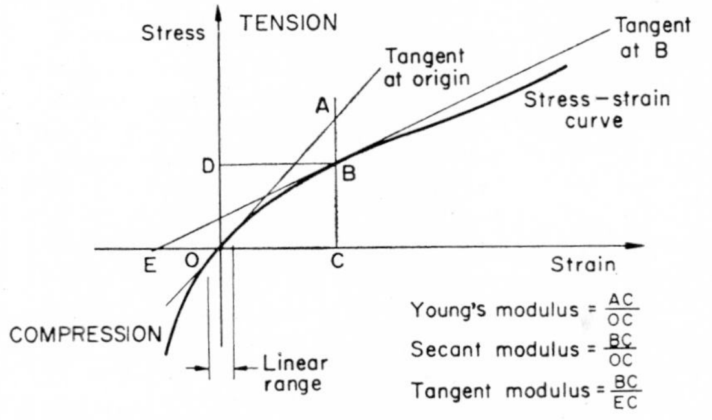

Fig. 8. Rubber elastic moduli. The rubber technologist's modulus at strain C is the value of stress at point D.

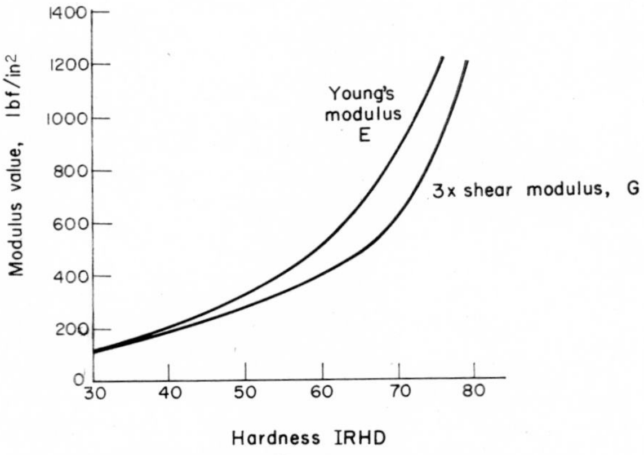

Fig. 9. Relation between elastic moduli and hardness.

strain behaviour of incompressible materials. An indication of the validity of this assumption is given in Fig. 9.

Values of both tangent and secant moduli, which are defined in Fig. 8, give useful information about the shape of the stress-strain curve, but are not essential when using one of the formulae of Fig. 5.

Confusion is sometimes caused by the so-called "rubber technologists modulus" which is simply a reference point on the tensile stress-strain curve. A particular rubber compound may, for example, be said to have a "300 per cent modulus of 1000 lb/in^2". In common with many other oft-quoted physical properties of rubber, it is a value which may be checked quickly by means of a simple standard test, and is thus useful as a measure of quality control. Otherwise, however, it is of little direct significance to the spring designer.

On the other hand, rubber hardness is a useful engineering quantity because it is basically a measure of elastic behaviour, and therefore closely related to Young's modulus. Hardness is also related to many other properties and, since an approximate value is easily obtained, it also is widely used as a quality control.

The hardness reading, obtained from a non-destructive test often with a pocket instrument, is an indication of the surface elastic deformation produced locally by a pre-loaded indentor. The standard unit is the British Standard and International Rubber Hardness Degree (IRHD), although the Shore Durometer "A" scale is often employed. Both methods give similar scale readings for the range of rubbers usually encountered in spring applications, but neither gives any information about the internal condition of the rubber. The relationship between Young's modulus, shear modulus and rubber hardness is given in Fig. 9.

7. *Creep, stress relaxation and set.* The relationship between stress and strain in polymeric materials is time-dependent. The points on a "static" stress-strain curve are obtained under conditions where time effects are restricted to negligible proportions.

If a constant load is applied to a rubber spring, the deflexion will continue to increase. Similarly, the spring force under constant deflexion will decrease with time. These associated phenomena, termed "creep" and "stress relaxation" respectively, have similar characteristics and it is sufficient to discuss only the former.

A typical plot of strain vs time shows a rapid increase in strain for a short while after a constant load is applied, followed by a long period during which strain is approximately proportional to the logarithm of time. Hence, if creep is plotted as a function of logarithm of time as in Fig. 10, the creep rate can be conveniently described by means of the linear part of the curve, and is expressed as percentage creep per decade of time. The reference strain for this purpose is a quasi-instantaneous value usually measured one minute after load application. Typical values for natural spring rubbers are in the range 2–7 per cent per decade, depending on composition.

If at some stage the load is removed, the creep process is reversed, but recovery is incomplete. The residual permanent deformation is known as "set".

When creep is allowed to continue for a very long

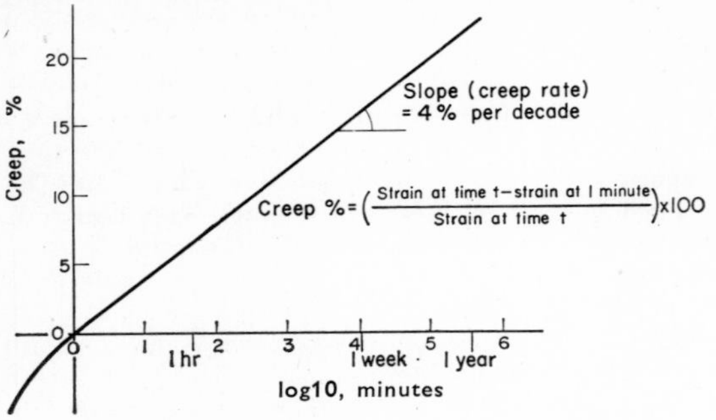

Fig. 10. Deformation vs time relationship (Creep).

period, the rate of strain will eventually increase, and failure will occur. Such "static fatigue" failures are uncommon in practice because the time required is usually far greater than the service life. However, if the rubber is subjected to vibration about a mean applied load, the creep process may be accelerated, particularly if crack growth also occurs because of the continuous flexing process. Damage due to the combined action of creep and cut growth under dynamic conditions, is known as "dynamic fatigue". It is a frequent cause of service failures many of which could be avoided by good design and proper selection of the spring material.

8. Dynamic properties of rubber. Rubbers are poor conductors of heat, and many synthetic elastomers are unsuitable under conditions of continual vibration because they exhibit sufficient damping to cause excessive internal heat generation, which in turn leads to deterioration of spring properties. In large truck tyres, heat build-up must be kept to an absolute minimum. In anti-vibration mountings, a small amount of damping is necessary to reduce vibration amplitudes at system resonance, whilst allowing efficient isolation at higher excitation frequencies. For both these applications, natural rubber is preferred. However, where a synthetic material must be used, the internal temperature rise is limited to an acceptable value by suitable design. Rubber sections are kept thin and in contact with metal plates, so that heat can be more easily dissipated.

The damping, or hysteresis, present in rubber is largely viscous in nature, but care must be used if the theoretical behaviour of spring elements is approximated to mathematically, by means of a linear viscoelastic model such as that in Fig. 11. Values of damping coefficient and spring stiffness are determined by experiment. In Fig. 11 the system response to an harmonically varying force has both in-phase (elastic) and quadrature (viscous) components. Consequently the dynamic modulus is also a complex quantity, and its magnitude is defined as the mean ratio of stress to strain magnitudes during a period of vibration.

Both dynamic modulus and damping vary with excitation frequency, or rate of strain, but for most

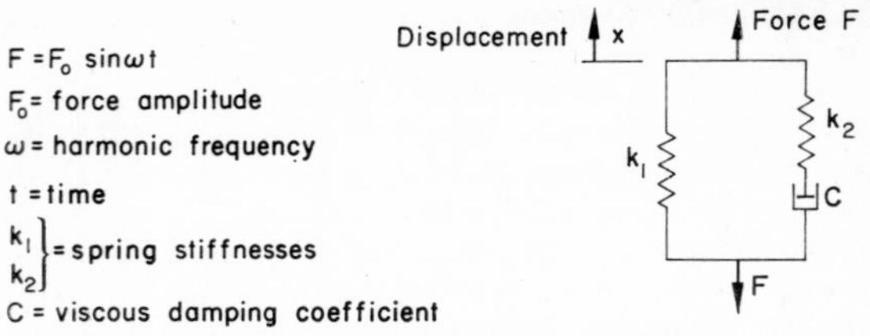

Fig. 11. Typical model for the viscoelastic behaviour of rubbers.

System response,

$$x = \frac{F_0}{k_1 + k_2}\left[\sin \omega t + \frac{k_2 \sin \omega t}{k_1(1 + \omega^2\psi^2)} - \frac{k_2\, \omega \cos \omega t}{k_1\,(1 + \omega^2\psi^2)}\right]$$

$$\textit{where } \psi = c\left(\frac{1}{k_1} + \frac{1}{k_2}\right).$$

rubbers at room temperature, both can be assumed to have reasonably constant values between 10 and 1000 cycles per second. For natural rubber compounds, the dynamic compression modulus varies from about 1·1 to 1·3 times the static value below a hardness of 60 IRHD, to factors of up to 3·0 above 70 IRHD. The viscous damping coefficient is typically 2–20 per cent of critical.

Synthetics may have dynamic moduli of up to an order higher than the static value, with damping two to three times greater than that of natural rubber.

In order to estimate the natural frequency of simple spring mass systems having low or moderate damping, the effects of damping can be ignored. The natural frequency is then expressed as a simple function of the static deflexion, as in Fig. 12.

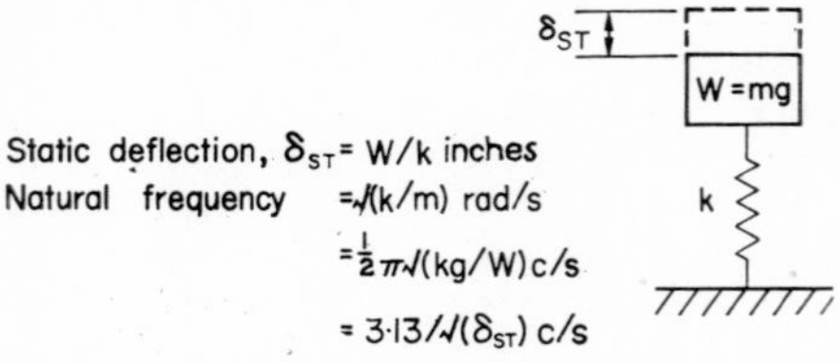

Fig. 12. Natural Frequency—static deflexion relationship for a simple spring-mass system.

9. Environmental effects. The performance of natural rubber springs is dependent on the service environment which is therefore taken into account at the design stage. Natural rubber will perform satisfactorily from −20°C to +60°C, and by careful compounding, this range can extend from −40°C to +100°C. Figure 13 gives an example of the variation of properties. The shear modulus of natural rubber has a minimum value near room temperatures above which the increase in modulus is gradual, and below which the increase is rapid. The working limits are set by the deterioration of other properties such as resilience and recovery at cold temperatures, and strenght and resistance to creep and fatigue under hot conditions.

Sunlight, oxygen and ozone all cause surface deterioration of natural rubber. Minute concentrations of ozone in the atmosphere can cause severe cracking leading to failure in thin unshielded sections. The rubber must be slightly stretched for ozone cracking to occur, but surface tensile strains are often present in shear and compression springs. However, with proper design and compounding, and adequate protective measures, satisfactory service can be obtained.

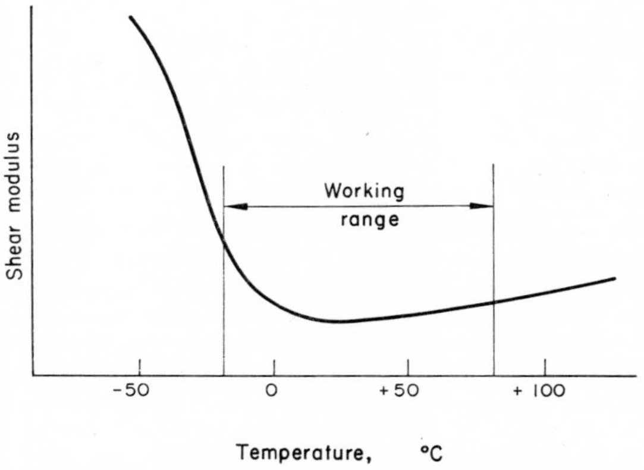

Fig. 13. Characteristic variation of shear modulus with temperature for natural rubber.

Natural rubber has good chemical resistance but should be protected from lubricating oils, degreasing solvents and liquid hydrocarbons generally. These liquids cause rubber to swell by a diffusion process, until it becomes useless. Rubber is attacked chemically by concentrated sulphuric acid, strong oxidizing agents and the halogens.

In environments involving temperature extremes, severe weathering, unfavourable chemical contact or where excessive loadings are encountered, one of the synthetics of the table is usually preferable to natural rubber.

10. Other factors affecting spring design. In order to produce springs requiring a minimum of modification in service, attention to detail at the design stage is all-important. Stress concentrations are minimized both within the material and at bonds. The dimensions of metal, and rubber components ensure stability and, if possible, "fail-safe" functioning. Metals are protected against corrosion, and their stiffness restricts bending deflexions which would otherwise contribute to bond failures. In order to keep production costs to a minimum, consideration is given at an early stage to the method of manufacture. In addition, the flow characteristics associated with each method of moulding produce varyngid egrees of anisotropy in the finished product, which may influence spring performance.

Although when designing a spring, the number of theoretical assumptions appears large, good results are usually obtained in practice from broad simplifying assumptions. Strength, stiffness and simulated service tests are normally performed on a small number of prototype units. Design modifications, which usually entail simple alterations to rubber hardness or compound ingredients, can then be incorporated at a pre-production stage.

11. Advisory organizations. Professional bodies whose advice should be sought on the processing, design, application and testing of rubber springs and spring materials include the following:

(a) The Natural Rubber Producers Research Association (N.R.P.R.A.).

(b) The Rubber and Plastics Research Association (R.A.P.R.A.).

(c) The Spring Manufacturers' Research Association (S.M.R.A.).

(d) The British Standards Institution.

(e) Federation of British Rubber and Allied Manufacturers (F.B.R.A.M.).

(f) Institution of the Rubber Industry (I.R.I.).

Bibliography

AITKEN, D. F. (1960) *Properties of Rubber, Engineering Materials and Design*, November, 676.

DAWSON T. R. and PORRITT B. D. (1935) *Rubber—Physical and Chemical Properties*, Research Association of British Rubber Manufacturers.

DAVEY A. B. and PAYNE A. R. (1964) *Rubber in Engineering Practice*, London: MacLaren.

HARRIS C. M. and CREDE C. E. (1961) *Shock and Vibration Handbook*, New York: McGraw-Hill.

LINDLEY P. B. (1964) *Engineering Design with Natural Rubber*, Natural Rubber Producers' Research Association Technical Bulletin No. 8.

LINDLEY P. B. (1966) Load-Compression Relationships of Rubber Units, *Strain Analysis*, **1**, No. 3.

MCPHERSON A. T. and KLEMIN A. (Eds.) (1956) *Engineering Uses of Rubber*, New York: Rheinhold.

PAYNE A. R. and SCOTT J. R. (1960) *Engineering Design with Rubber*, London: MacLaren.

SNOWDON J. C. (1964) Representation of the Mechanical Damping possessed by Rubberlike Materials and Structures, *Rubber Chem. and Tech.* **37**, No. 2, April–June.

Engineers' Digest (1961) Anti-Vibration Mountings, Survey No. 9, Reprinted from the *Engineers' Digest* of April.

Mechanical Characteristics and Application of Rubber (1964) B. F. Goodrich Industrial Products Co., U.S.A.

Methods of testing Vulcanised Rubber, British Standard Institute Specification B.S. 903.

Rubber in Engineering (1966) N.R.P.R.A. Conference Proceedings, London: McLaren.

Vulkollan—an Engineering Material (1964) Farbenfabriken Bayer AG, July 1964.

N.R.P.R.A. (1965) *List of Publications* 1938–1964.

D. N. COBLEY

S

SCANNING ELECTRON MICROSCOPY. This is a technique of electron microscopy in which a fine electron beam probe of diameter 50–100 A is scanned in a raster across the surface of the specimen. A signal derived from the specimen is used to modulate the brightness of the spot in a cathode-ray display tube, this spot being scanned in a raster in synchronism with the electron probe. The resolution obtained is typically of the order of 100 A, and is therefore inferior to that of a conventional electron microscope. On the other hand, the scanning microscope has four important advantages: (i) the depth of field is *very large*, so that surfaces such as paper or fractured ceramics can be examined directly without the need of making a replica; (ii) it is possible to get a contrast in the image between portions of a specimen surface of different chemical composition; (iii) local magnetic or electric fields in a specimen can be detected. This has led to the use of the instrument for the examination of integrated microelectronic circuits; (iv) access to the specimen is easy because of the relatively large distance (1–2 cm) between the final lens of the electron probe and the specimen surface. The scanning microscope has been developed mainly in the Engineering Department of Cambridge University, and good accounts of it have been given by Smith (1961) and Oatley *et al.* (1965).

In the usual form of the instrument, electrons scattered from the surface of the specimen enter a collector and produce the signal which modulates the brightness of the beam in the display tube. The scattered electrons are of two kinds: secondary electrons, having an energy of a few eV; and reflected electrons, which have energy only a little less than that of the electron probe. By interposing a grid at a suitable potential between the specimen and the collector, it is possible to select one or other of these two kinds of electron. The reflected electrons, because of their high energy, travel in nearly straight lines from specimen to collector; it follows that they cannot reveal any part of the specimen from which there is not a straight line path to the collector. Secondary electrons are not subject to this limitation, since their paths are curved sharply in the electric fields between the specimen and collector. Hence secondary electrons can reveal more detail in a specimen with a rough surface. However, insulating specimens are best examined using the reflected electrons, since they will be less influenced by surface charges produced by the electron probe.

Contrast in the image is due to three factors. The dominant effect is surface topography: a change of a few degrees in the angle between the electron probe and the local surface normal causes a perceptible change in the scattered electron current. Contrast due to the different chemical constitution of different parts of the specimen surface is observed if *reflected* electrons are used, since the number of these electrons scattered increases with the atomic number of the scattering material. Secondary electrons do not show this effect. Images using secondary electrons show contrast due to potential variations across the specimen, and potential variations as low as 0·15 V have been revealed (Thornhill and Mackintosh 1965).

The resolving power of the instrument has been discussed by Smith (1961). The first factor to be considered is the diameter of the electron probe at the specimen. This is determined by spherical aberration in the lens which focuses the beam on to the specimen, and by diffraction at the aperture of this lens. If the semi-angle which this aperture subtends at the specimen has its optimum value, the diameter d of the electron probe is

$$d = AC_s^{1/4}\lambda^{3/4}[(BiT/j) + 1]^{3/8}, \qquad (1)$$

where C_s is the spherical aberration constant of the lens, λ is the electron wave-length, i is the current in the probe, T is the temperature of the electron gun filament, and j is the current density of the thermionic emission from the filament. A and B are constants whose precise value depends on how the contributions of the various aberrations are combined. Approximately, $A = 1$ and $B = 8 \times 10^9$. If only very small probe currents are needed, the first term is negligible, and the expression reduces to that which gives the resolving power of a conventional electron microscope objective. However, because of the particulate nature of an electron beam, the illumination of the specimen is a random process. A mean number n of electrons falling on an element of the specimen surface is associated with a signal-to-noise ratio of $\sqrt{n}$. This must be considerably greater than $1/D$, where D is the threshold brightness contrast, and is defined as $\Delta b/b$ where Δb is the minimum brightness change which it is desired to observe. This consideration leads to equation (1) being modified to

$$d = AC_s^{1/4}\lambda^{3/4}[(CN^2T/D^2tj) + 1]^{3/8}. \qquad (2)$$

In this equation, C is a constant equal to about $1{\cdot}5 \times 10^{-6}$; N is the number of lines in the picture (a square raster is assumed); and t is the recording time in seconds. Unless the recording time exceeds several minutes, the first term is likely to be nearly an order of magnitude greater than the second. This accounts for the relatively poor resolving power of the scanning microscope as compared with a conventional electron microscope.

Other types of scanning microscope. Le Poole (1965) has proposed (but apparently not constructed) an instrument which combines some of the features of a scanning and a conventional microscope. This makes use of the property of a quadrupole lens of forming a *line* image of a point source with extremely small spherical aberration. The specimen is irradiated by a line focus, and the emerging electrons are imaged by a similar quadrupole rotated through 90°. Obviously, good resolution is required only in the direction of the line focus on the specimen, since no electrons hit the specimen outside this focus. A deflecting system scans the line focus across the specimen, and a second deflector in the imaging system produces a corresponding movement of the line image across the fluorescent screen or photographic plate. Theoretically, such an instrument should be capable of a resolution of 1 A.

A. V. Crewe has constructed an instrument for examining transmission specimens of the type used in a conventional electron microscope. The electron gun uses a pointed tungsten wire from which electrons are liberated by field emission. The beam is focused on to the specimen by two quadrupole lenses. Electrons passing through the specimen proceed, through a magnetic spectrometer, to the collector. The use of the spectrometer makes it possible to form an image using only electrons which have lost a precisely defined amount of energy. The instrument produces pictures in which the distribution of any particular material is clearly delineated. The use of an ultra high vacuum system eliminates the usual carbonaceous contamination of the specimen when irradiated by the electron beam.

Bibliography

Le Poole J. B. (1965) *Proceedings of the Second Regional Conference on Electron Microscopy in Far East and Oceania,* Calcutta: Electron Microscopy Society of India, 28

Marton L. (1961) in *Encyclopaedic Dictionary of Physics* (J. Thewlis Ed.), **2**, 795, Oxford: Pergamon Press.

Oatley C. W., Nixon W. C. and Pease R. F. W. (1965) *Adv. Electron. Electron Phys.* **21**, 181.

Smith K. C. A. (1961) *Encyclopedia of Microscopy* (G. L. Clark Ed.), 241, New York: Reinhold.

Thornhill J. W. and Mackintosh I. M. (1965) *Microelectronics and Reliability* 4 (1), 97.

Thornton P. R. (1968) *Scanning Electron Microscopy,* London: Chapman and Hall.

T. B. Rymer

SENSITIVITY IN INDUSTRIAL RADIOGRAPHY. Quantitative information on flaw sensitivity (i.e. the dimensions of a flaw which can be detected when present in a particular thickness of specimen) is not available in any general form. The ability to detect a flaw by radiographic means depends on the nature of the individual flaw, its shape and orientation to the beam of radiation, as well as on the various radiographic parameters. For the calculation of sensitivity some success has been achieved in representing a flaw by some equivalent artificial cavity: thus, a short length of a crack can be represented as a slit-shaped cavity having a known width and a depth. Relationships can then be established between this width and depth, the angle between the slit and the beam of radiation, the various radiographic factors such as unsharpness, film contrast, absorption coefficient, etc., to enable the sizes of slit which can be detected through various thicknesses of material, at various orientations, to be calculated for particular radiographic techniques. These calculated values of artificial-crack sensitivity have been shown to agree reasonably well with experiment. Such values are, however, not used so much as flaw sensitivity data, but as methods of comparing and evaluating different radiographic techniques.

Radiographic sensitivity is more usually measured in terms of some artificial element which does not necessarily bear much resemblance to a flaw in metal. Thus, two of the most widely used methods are:

1. To specify sensitivity in terms of the ability to detect a wire of the same material as the specimen being examined, when the wire is laid on the surface of the specimen remote from the film. The diameter of the thinnest detectable wire is used as a criterion of sensitivity.

2. To use a series of drilled holes in a plate of the same material as the specimen, which is laid on the specimen, and to specify sensitivity in terms of the smallest hole which can be detected on the radiograph.

In one system widely used in Europe, the drilled holes always have depth equal to diameter, so that the plate becomes a step wedge. In several American systems the plate used is a constant thickness and the drill holes are of different diameters.

These devices—sets of wires of different diameters, plates or step-wedges with drilled holes—are called *penetrameters*, or *image-quality indicators* (I.Q.I.). Many patterns other than three described have been proposed, but these three are the best known and are standard patterns in many countries.

The method of specifying sensitivity, as judged with most of these devices, is:

$$\text{Sensitivity} = \frac{\text{thickness of smallest element (wire, drill hole) whose image can be observed on the radiograph}}{\text{thickness of specimen}} \times 100 \text{ per cent}$$

and in quoting values it is necessary to specify the type of I.Q.I. to which the quoted numerical value applies; thus:

percentage wire–I.Q.I. sensitivity
percentage drill-hole — I.Q.I. sensitivity

or to refer to a specific design of I.Q.I., e.g.

(American) A.S.T.M. sensitivity.

Because the sensitivity is expressed in this percentage manner, a smaller numerical value implies that a smaller drill-hole or wire can be detected through a given thickness of specimen. By inference, also, smaller flaws will be detected and the radiograph will therefore have better sensitivity.

Thus 2 per cent sensitivity is not as good as 1 per cent sensitivity if ability to detect small flaws is the criterion of quality. The values of I.Q.I. sensitivity which are obtained in conventional film radiography are usually with the range 0·5–4 per cent and can be found tabulated for a range of techniques in various publications.

The method of specifying sensitivity as a percentage value has led to much confusion and some authorities prefer to quote actual sizes. Thus a radiographic specification might ask for the detection of a 0·2 mm wire through a particular thickness of specimen instead of quoting an acceptable percentage sensitivity.

With the American plaque types of penetrameter it is possible to specify a sensitivity value in terms of either the visibility of the image of the plaque itself or in terms of the drill-holes in the plaque. If judged on the plaque alone it is called *contrast sensitivity* (in Europe this would be called *thickness sensitivity*): if judged on the drill-holes, it is sometimes called "detail sensitivity".

Simple thickness sensitivity can be easily calculated and the formula

$$S = \frac{\Delta x}{x} 100 = \frac{2{\cdot}3\Delta D}{\mu G_D x}\left(1 + \frac{I_s}{I_D}\right) 100$$

is well-established. In this equation Δx is the minimum observable change in thickness (the thinnest detectable plaque thickness laid on the specimen) in a total specimen thickness x; ΔD is the minimum discernible difference in density on the film; μ is the narrow-beam lines absorption coefficient for the radiation used: G_D is the gradient of the film characteristic curve measured at the density of the radiograph, D: I_S/I_D is the ratio of scattered radiation to direct radiation reaching the film at the point where the image is formed, so that the total radiation reaching the film at this point is $(I_S + I_D)$.

In this equation there are parameters depending on the radiation employed, μ, I_S/I_D:

on the radiographic film, G_D;
on the specimen x;
and on the conditions under which the radiograph is examined, ΔD.

There have been several proposals to extend this formula to the calculation of the minimum size of drill-hole or wire which may be detected on a radiograph and to such detail as artificial cracks. The most difficult additional parameters which need to be incorporated are

1. The radiographic unsharpness, which is a complex function of the film characteristics and the radiation source size and
2. The performance of the human eye when viewing very low contrast images of different shapes and sizes.

Methods of measuring and specifying radiographic unsharpness are not yet standardized and its quantitative significance is not fully understood. So far as viewing the radiograph is concerned, film-viewing conditions also are not generally standardized and individuals vary considerably in their ocular performance, which depends both on intrinsic ability and on experience. Experimental values of radiographic sensitivity are therefore liable to exhibit a considerable spread even when the radiographic parameters are fixed.

Although the I.Q.I. sensitivity values are not directly related to flaw sensitivity, there is clearly some relationship, in that an improvement in I.Q.I. sensitivity must also represent an improvement in flaw sensitivity. Drill-hole sensitivity gives a reasonably simple relationship for the ability to detect gas holes and pores in castings or welds, but crack sensitivity depends on the orientation of the crack in addition to the factors affecting I.Q.I. sensitivity and is therefore not related in a simple manner to I.Q.I. sensitivity.

The ability to detect drill-holes or wires clearly cannot be related in a simple manner to the detection of inclusions such as slag in a weld. Thus a 1 per cent I.Q.I. sensitivity does not imply that flaws occupying 1 per cent of the specimen thickness can be detected irrespective of the nature of the flaw.

Bibliography

HALMSHAW R. *et al.* (1966) *Physics of Industrial Radiology*, London: Iliffe.

British Standard BS: 3971 (1966) *Image Quality Indicators.*

R. Halmshaw

SINGLE CRYSTAL DIFFRACTOMETRY. Until recently, nearly all single crystal studies with X rays were carried out by photographic methods. In each of these methods, the X-ray diffraction spectra, or reflexions, are recorded on photographic film, and the intensities of the reflexions are determined by measuring the blackness of the corresponding spots on the film. Provided the phases of the reflexions are known, it is then straightforward to combine the phases and intensities in such a way as to lead to a solution of the crystal structure. Although photographic methods will continue to play an important role in X ray structure determination, their use will decline with the increasing adoption of diffractometer techniques.

A diffractometer (Bacon 1961) is an instrument for recording diffracted X rays (neutrons) by means of a counter detector. Individual X ray photons (neutrons) are detected and counted, so that the diffractometer technique is much more precise and sensitive than one based on the measurement of the blackening of a photographic film. Moreover, the diffractometer can be readily adapted to automatic control, either "off-line" by means of a special-purpose unit, or "on-line" by means of a computer. In the case of a biological compound, there may be many thousands of reflections to be measured, and much of the tedium of carrying out these measurements is avoided by using an automatically controlled diffractometer. The one drawback of the diffractometer is its complexity (and cost). The crystal must be oriented correctly for each reflection in turn, and the detector moved to receive the diffracted radiation: these operations require the independent movement of up to four distinct shafts, whereas in the photographic method a single uniform rotation of one shaft alone may suffice.

The forerunner of the modern single-crystal diffractometer was the ionization spectrometer, developed over fifty years ago by W. H. and W. L. Bragg. The instrument was used by W. H. Bragg as an X ray spectrometer for studying X ray spectra; his son, W. L. Bragg, was more concerned in using it as a diffractometer for the study of crystal structures. A large single crystal was suitably oriented to reflect the X ray beam, and the intensity of the beam was then measured with an ionization chamber. However, because of the difficulty of measuring the very tiny currents from an ionization chamber, the Bragg ionization spectrometer was not really suited to the examination of the structures of small crystals or to the collection of a large amount of intensity data, and during the decade after the 1914–1918 war the ionization spectrometer was gradually replaced in the X ray study of single crystals by various types of X ray photographic camera.

Since the 1939–1945 war, the Geiger counter and, later, the proportional and scintillation counters were developed as reliable quantum detectors. This has been the principal reason for the swing back to the diffractometer method: by replacing the ionization chamber with a quantum detector, the Bragg spectrometer is converted to an extremely sensitive instrument, capable of recording all the quanta diffracted by the sample. Lonsdale and Cochran were amongst the first to exploit the quantum counter technique in the examination of single crystals, and many crystallographers in the decade from 1952 to 1962 began to use single-crystal, hand-operated diffractometers to supplement or to supplant the collection of diffraction data by X ray cameras. Since 1962 there has been further intense activity, especially in Europe and the U.S.A., devoted to the development of automatically-controlled single-crystal diffractometers, and more than ten different kinds of automatic instruments, some manufactured in either an "off-line" or "on-line" version, are commercially available at the time of writing (1967).

It is worth emphasizing that all single-crystal diffractometers are the same, in principle, as the original Bragg ionization spectrometer. The single-crystal sample and the detector must be correctly set for the observation of each individual Bragg reflexion: the intensity of the reflection is then measured by counting the number of diffracted quanta as the crystal rotates uniformly through the reflecting position. The total number of diffracted counts are recorded, the crystal and detector are re-positioned for the next reflection, and the sequence of operations is repeated. The positioning, measuring and recording operations were carried out by hand in the original ionization spectrometer: in the modern instrument they are all performed automatically.

In neutron diffraction there is no practicable photographic alternative to the diffractometer method of data collection, and so the neutron diffractometer has remained as the principal instrument for neutron crystallographic studies. The neutron diffractometer is described in Bacon (1962).

Bibliography

Arndt U. W. and Willis B. T. M. (1966) *Single Crystal Diffractometry*, Cambridge: University Press.

Bacon G. E. (1961, 1962) in *Encyclopaedic Dictionary of Physics* (J. Thewlis Ed.), **2**, 282; **4**, 820, Oxford: Pergamon Press.

B. T. M. Willis

SODIUM COOLING. The core of a fast breeder nuclear reactor, which contains no significant amount of moderator, and in which the fission chain reaction is maintained by fast neutrons with an average energy $\sim$1 MeV, must be small and compact and must have a high power density for the reactor to produce power at an economic price. Typically, the core of a fast reactor producing 1000 MW (Heat) would be a right cylinder 2 m dia. × 1 m high, having an average power density ~500 KW/l, compared with 1 KW/l for the core of a graphite-moderated CO_2-cooled magnox-type reactor, and ~100 KW/l for a pressurized water reactor.

The coolant which takes the heat from this compact, highly rated core to the steam raising plant must have exceptionally good heat transport properties, must be a poor moderator of fast neutrons, and in addition should have a small neutron absorption cross-section, a low melting point, a moderately low density to minimize the pumping power required, and a low vapour pressure at the core outlet temperature, commonly ~650°C, to avoid the inconvenience of a high pressure coolant circuit. Liquid metals, which have high thermal conductivities, are clear candidates for fast reactor coolants, and of those available sodium has the most suitable combination of physical and chemical properties, with the added advantage that it is cheaply and readily available.

Consequently, most fast breeder reactors which have been constructed or which are presently being designed use liquid sodium (melting point 98°C) as the coolant. In one, the Dounreay Fast Reactor, 30 per cent potassium was added to the sodium to produce an alloy (NaK) having a melting point of 42°C. At ambient temperature this alloy forms a sludge, and will not freeze solid if the reactor circuit trace heating fails, but its heat transport and handling properties are inferior to those of sodium, and it is unlikely that it will be used in future power reactors.

At 650°C the thermal conductivity of sodium is 0·145 cal/cm sec °C, and the corresponding Prandtl number $Pr = 5 \times 10^{-3}$, compared with ~1·0 for water and ~0·7 for CO_2. It follows that, in contrast to high Prandtl number fluids which rely primarily on turbulent eddies for their heat transfer, molecular conductivity is of comparable importance to eddy diffusion in the transfer of heat through liquid sodium. Consequently, the heat transfer correlations established for non-metallic fluids do not apply to sodium, and in particular the equivalent diameter concept, which implies that heat transfer is independent of duct geometry is no longer valid and account must be taken of the flow-passage configuration in determining the coolant temperatures.

The maximum allowable power output from a reactor core is commonly limited by the maximum temperature attained by a fuel can anywhere in the core, while the steam conditions at the turbine stop valve are determined by the mixed mean coolant temperature at outlet from the core. For maximum thermal efficiency of the core, the difference between the maximum can temperature and the mixed mean coolant temperature, called the hot spot factor, must be minimized.

A typical fast reactor core consists of fuel pins, ~5 mm diameter, located by grids or fins in a hexagonal or square array having a pitch/diameter ratio ~1·25. The sodium passes axially through the core in the multitude of channels formed by the pins and, because the film temperature drop is small (~10°C), the can temperature is primarily determined by the local coolant temperature, which in turn is mainly dependent on the flow of coolant in each individual channel, and is significantly modified both by the precise shape of the channel and the mixing between adjacent channels. Thus, in a sodium-cooled reactor core, a precise analysis of the flow patterns and coolant mixing in the core is necessary in order to predict the core hot spot factors.

In all sodium-cooled power reactors existing or contemplated, heat is transferred from the core to the steam plant by forced convection of the sodium. Because liquid sodium burns fiercely in air, producing dense caustic fumes, and also reacts violently with water, forming hydrogen and caustic soda, the coolant circuit must be leaktight with an inert cover gas, usually argon above the sodium surface; 18 Cr 8 Ni austenitic stainless steels, which are easily fabricated and which have adequate corrosion resistance in sodium up to 650°C, are the most commonly used circuit materials.

In a typical coolant circuit the sodium enters the core at ~350°C and leaves it at ~650°C, and the circulating pumps are usually placed in the cold leg of the circuit. At 350°C the density of sodium is similar to that of water, and its viscosity is 0·3 cP, which is about one third that of water. It can therefore be circulated round the reactor circuit by mechanical pumps of conventional design, commonly made with a vertical shaft which rises above the sodium surface into the cover gas space and through the top of the reactor vessel via a rubbing face oil-lubricated rotary seal, although pumps in which the shaft is sealed by a plug of frozen sodium have also been built.

Alternatively, by making use of its high electrical conductivity ($5{\cdot}5 \times 10^4$ ohm^{-1} cm^{-1}) sodium can be circulated electromagnetically by pumps which have no mechanical moving parts; they are either of the conduction type, in which an electric current passes through the sodium across the pump duct, interacts with a magnetic field which is normal both to itself and the pump duct, and forces the sodium along the duct, or of the induction type, in which a travelling magnetic field induces eddy currents in the sodium and drags it along the duct. *Electromagnetic pumps* are significantly less efficient (~35%) than conventional mechanical pumps (~80%), but because of their simplicity and reliability they are commonly used in circuits where efficiency is not a primary consideration.

Conventional instruments measuring pressure, temperature, etc. in a sodium coolant circuit are commonly supplemented by additional instruments which make use of the high electrical conductivity of sodium. Flow is determined by measuring the e.m.f. set up across a pipe when the sodium passes through a magnetic field, and the position of the sodium free-surface is measured by detecting the change of inductance of a search coil as it passes through the surface.

When sodium passes through a fast reactor core, the isotope ^{24}Na is produced, which decays with a half-life of 15 h and produces high gamma activity. This necessitates heavy shielding of the sodium coolant circuit, but the short half life permits maintenance of the circuit equipment after a few days have elapsed and the activity has decayed.

Because of the close proximity of the sodium and water circuits in the steam generators there is a significant probability of a leak between the two, which could lead to highly active ^{24}Na entering the steam circuit, and the formation of large quantities of sodium oxide, hydride and hydroxide in the sodium circuit which might lead to circuit blockages. To eliminate this possibility, the steam generators in all existing sodium-cooled reactors are fed by a separate secondary sodium circuit coupled to the primary sodium circuit by sodium/sodium heat exchangers, but the possibility of sodium/water leaks between the secondary sodium circuit and the steam circuit still remains. In some reactors, such as the Dounreay Fast Reactor, another heat transfer substance, in this case copper, is interposed between the sodium and the steam pipes in the steam generator; in other reactors, such as the Enrico Fermi Fast Breeder Reactor, quick-acting valves are incorporated in the secondary sodium circuit so that a leaking steam generator can be quickly isolated to prevent the reaction products reaching the rest of the circuit.

The chemical behaviour of sodium in a coolant circuit is greatly influenced by the presence of impurities, the most important being oxygen, whose solubility in sodium is highly temperature-dependent and varies from $\sim$10 ppm at 150°C to 100 ppm at 300°C and 1000 ppm at $\sim$600°C. The concentration of this dissolved oxygen, which seriously affects the rate of corrosion of 18/8 stainless steels, must be limited to $\sim$15 ppm to obtain acceptable corrosion rates ($\sim 10^{-3}$ cm/year at 650°C), and this is most conveniently achieved by making use of the dependence of its solubility on temperature. A proportion of the sodium coolant flow is directed through a vessel containing a wire mesh filter which is significantly colder than the remainder of the circuit; as the sodium flows slowly through this vessel, which is termed a *Cold trap*, the excess sodium oxide is precipitated, and the sodium leaves the trap with an oxygen content corresponding to the solubility at the trap temperature, typically 150°C to achieve 15 ppm.

The refractory metals (Nb, V, Ti, Ta, Zr) corrode more readily than the austenitic stainless steels in oxygen-contaminated sodium, and when they are present in the sodium circuit, a low oxygen content, typically less than 3 ppm, is necessary; it may be inconvenient to achieve this using a cold trap as the trap temperature required could be so close to the freezing point of sodium that blockages would occur in the cold trap circuit. In these circumstances a *Hot trap* is sometimes used; this comprises a vessel containing a refractory metal foil (e.g. Zr) which, when heated to a temperature greater than the maximum circuit temperature, combines chemically with and acts as a getter for the oxygen.

Hydrogen dissolved in sodium can cause embrittlement of some circuit materials and its concentration is controlled by cold traps in a similar manner to oxygen; calcium, which is a common impurity in sodium as supplied from the manufacturer, combines with dissolved oxygen above about 350°C to form insoluble calcium oxide, which can cause circuit blockages if it is not removed.

Current evidence suggests that carbon is only sparingly soluble (less than 1 ppm) in sodium, but it is certainly present in other forms, either as fine particles in suspension or chemically combined as carbonates, carbides or hydrocarbons, and causes carburization of some circuit materials. It is controlled by the cold trap, although the mechanism of control is not understood.

The impurities present in coolant sodium can be determined by chemical analysis of a sample, but the stringent precautions necessary to prevent contamination of the sample, particularly by oxygen or moisture, are complicated and time consuming, especially if the coolant is active through being irradiated in the reactor core, and consequently simpler methods are used for the routine monitoring of sodium impurities. The instrument most commonly used is the *plugging meter*, in which a proportion of the sodium coolant flow is directed through an orifice or similar constriction whose temperature is varied. When solid impurities start to deposit at the orifice a detectable change occurs in the sodium flow, and the corresponding temperature, termed the plugging temperature, is a measure of the impurity concentration. The instrument responds to all the dissolved impurities, that is, mainly oxygen and hydrogen, but does not discriminate between them; it can either be operated manually or adapted to give a continuous record of the plugging temperature.

Bibliography

GARTH R. C. (1962) in *Encyclopaedic Dictionary of Physics* (J. Thewlis Ed.), **5**, 105, Oxford: Pergamon Press.

WINDSOR H. K. (1962) in *Encyclopaedic Dictionary of Physics* (J. Thewlis Ed.), **5**, 109, Oxford: Pergamon Press.

K. G. EICKHOFF

SOLAR ENERGY, RECENT DEVELOPMENTS IN USE OF. Sun daily showers the Earth with enormous quantities of solar energy, sustains life both directly and indirectly and is the ultimate source of all energy on Earth. Every square metre of the Earth's surface receives energy from the Sun equivalent to 1 kW, power sufficient to run a small fractional horsepower motor, when the conversion efficiency to electrical energy is 10 per cent. The realization that irreplaceable fossil fuels, coal, oil and natural gas, are being rapidly depleted, has accelerated the search for newer and better ways to harness the other sources of energy and among these, solar energy is the most important.

The factors limiting the harnessing of solar energy are its relatively low intensity and intermittent availability. It is not only absent at night, but its intensity also depends on the time of the day, season of the year, weather and location.

Outside the Earth's atmosphere, the intensity of solar radiation received by a surface placed at right angles to the incoming radiation—the solar constant—amounts to 2·0 gcal/cm^2 min. On its passage through the atmosphere, a part of it is reflected back into space, mainly by clouds, another portion is scattered by molecules of dry air, water vapour, carbon dioxide, ozone, etc., while the rest of it reaches the Earth's surface as a beam of direct radiation. A part of the reflected and scattered portions is received back as the incoming diffused radiation, which cannot be concentrated with lenses or mirrors.

Typical energy distribution curves for solar radiation reaching the Earth's surface and that of a perfect radiator or "black body" heated to 6000°C are shown in Fig. 1. Solar radiations of wave-length lower than 3000 Å are completely absorbed by oxygen and ozone

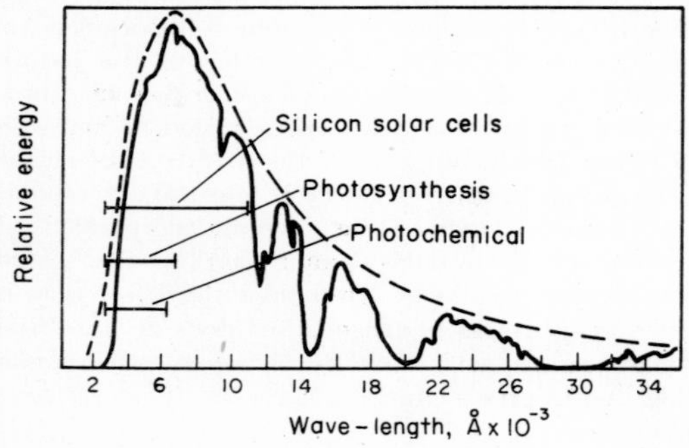

Fig. 1.

in the Earth's atmosphere. Half the solar energy falls in the ultra-violet and visible regions and the other half comprises infra-red radiation. The absorption of energy in the range 3000–6400 Å promotes photochemical reactions, while photosynthesis, the reaction responsible for storage of solar energy in nature operates in the range 3000–7000 Å. Silicon solar cells by absorbing energy in the range 3000–11,000 Å convert it directly into electricity. Appliances, which convert solar radiation into heat, such as flat-plate collectors and focusing reflectors, utilize all the energy in the solar spectrum reaching the Earth's surface.

To assess the performance of a solar equipment, it is necessary to know precisely the intensities of direct and diffuse irradiation of a plane at any orientation at any instant during the year for all places of interest. Several types of instruments are used for measuring and recording these. Several simple and inexpensive intensity measuring instruments have been developed. The *Sol-A-Meter* developed by Yellott utilizes silicon solar cells connected to a galvanometer or milliammeter. Two models, one of recording type and the other of integrating type, are now available in the United States. With suitable calibration, these instruments give results in close agreement with those obtained by the Eppley pyrheliometer.

A set of twelve monthly world maps showing the regional variations in insolation have been constructed independently by the United States Weather Bureau as well as by Black.

The regions with the greatest availability of solar energy lie in two broad bands encircling the Earth between 15° and 35° latitude north and south. Over two-thirds of the area of these countries comprises arid land with nearly 3000 hours of annual bright sunshine and nearly 90 per cent as direct radiation. In these areas direct utilization of solar energy is possible. The equatorial belt between 15°N and 15°S is the next favourable location. Characterized by high humidity, frequent clouds and high proportion of scattered radiation, there are annually about 2300 sunshine hours with very little seasonal variation. Regions between 35° and 45° in the north and in the south are marked by seasonal variations and of low solar intensity during winter months, while regions beyond 45°N and 45°S have low solar energy potential throughout the year. The difficulty of seasonal variation, however, may be overcome largely by constantly orienting the receiving surface to face the Sun.

As a rule, appliances utilizing solar energy make use of a collector or an absorber, the choice being governed by three main factors—conversion efficiency, temperature of operation desired and the capital cost of the equipment. However, photoelectric devices, converting solar energy into electrical energy directly, are an exception. Accordingly, solar operated devices may be classified into (i) low temperature applications below 220°F, where low intensity solar energy as received is used, and (ii) high temperature applications above 220°F employing concentrated energy. Under the first category fall devices for water heating, space heating and cooling, refrigeration, distillation, salt making and liquid concentration, agricultural and industrial drying or dehydration, etc., while the latter comprises solar energy concentrators. A high degree of optical accuracy is required in the construction of solar furnaces used in high temperature material research and chemical and metallurgical processes. Solar cookers and stoves, solar heat engines or motors come next, requiring slightly less degree of optical accuracy.

Solar water heaters. Solar water heating has, by far, been the most successful application of solar energy.

To supply the energy needs for cooking, bathing, washing and other domestic purposes, thousands of solar water heaters are in service in various parts of the world. In Japan alone, over 200,000 units of plastic-pillow type and an equal number of tube-in strip type are in use. Heaters of plastic-pillow type made from polyvinyl chloride plastic sheets are 2 m long, 1 m wide and 10 cm deep and have sufsufficient capacity to dispense with the use of a separate storage tank. Fitted with a transparent canopy and water pipes, it is placed on a level wooden platform in the yard or attached to a roof facing the equator.

Most of the commercial heaters use a series of parallel tubes, whose ends are brazed, soldered or screwed into horizontal header pipes, the lower one for inflow of cold water and the upper one for outflow of heated water. Strips of 24 gauge sheet are soldered in between two successive tubes. An inexpensive construction for parallel channels uses a standard corrugated galvanized iron sheet over a plane galvanized iron sheet. At the points of contact the two sheets are held together with rivets and the edges are lapped and soldered. A typical system as installed for domestic hot water supply in India is shown in Fig. 2.

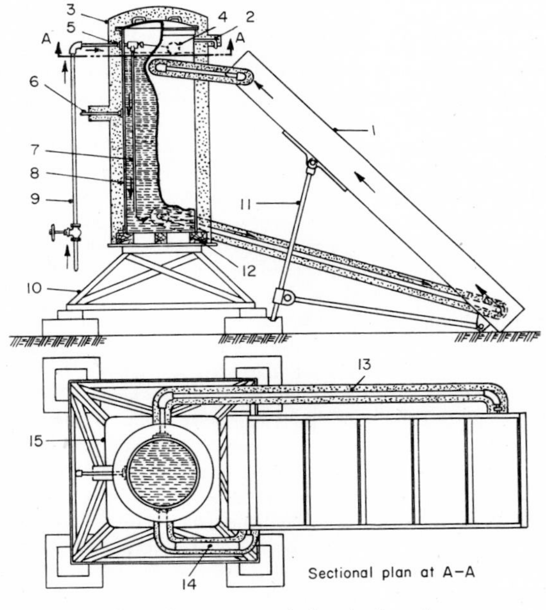

Fig. 2. Complete set-up of the single solar water heater unit. 1, glass cover; 2, overflow; 3, outer cover of water reservoir; 4, float; 5, rockwool; 6, hot water; 7, cold water inlet; 8, wall of water reservoir; 9, connexion to water supply; 10, angle iron stand; 11, support for water heater; 12, wooden base for reservoir; 13, cold water inlet pipe; 14, hot water outlet pipe; and 15, outer frame of the reservoir.

An overall efficiency of about 70 per cent has been realized.

The future of solar water heating is very bright. There is wide scope and need for simplification, standardization and cost reduction.

Storage of solar heat. The diurnal and seasonal variations in solar intensity require that a storage device be used. An effective and efficient storage system can open the way for extensive and more economic utilization of the Sun's heat.

The three methods so far used to store heat are water tanks, pebble beds and chemical systems. Water has a high capacity for heat absorption and is suitable for storing low-temperature heat. Tanks required to store very large quantities of water become expensive and require the use of circulating pumps and electric power. Heat transfer across heat exchangers is efficient and a comparatively small space is required for storage. Pebble beds or rock piles are the cheapest method for storing heat by passing a stream of warm air. Large surface and a zig-zag path through the bed ensure rapid heat exchange and bring about reduction in heat losses from the system. The larger the reservoir the greater is the effective heat storage. They are also suitable for storing heat at high temperatures. Both water tanks and pebble beds are placed underground with proper insulation when large-scale heat storage is contemplated.

A phase change or chemical change involves larger heat effects than those possible with materials retaining sensible heat and, therefore, large heat storage is possible within a given volume. The transition at 90°F of Glauber's salt, $Na_2SO_4 \cdot 10H_2O$, into the dehydrated salt was utilized by Maria Telkes for storing solar heat from hot air or water and recovery of this heat on rehydration of the salt on lowering the temperature. Similarly, disodium phosphate hydrated crystals, $Na_2HPO_4 \cdot 12H_2O$, with a transition temperature of 97°F, have been used for solar heat storage. However, difficulties due to stratification and supercooling of these salts have not yet been fully overcome. Lithium hydride is contemplated for solar energy storage in outer space.

Selective radiation surfaces. Encouraging results have been obtained in the development of black selective coatings for solar absorber surfaces, which possess high absorptivity for sunlight and very low emissivity for long-wave radiation. These selective radiation surfaces, which collect and retain solar radiation, are prepared by depositing very thin films of black oxides or sulphides or other semiconductor compounds that adhere to the bright shining underlying metal surfaces. With a bright metal like aluminium, silver or nickel, covered with a thin coating, the infra-red emissivity is low. The surfaces used by Tabor consisted of cupric oxide on anodized aluminium surface, while Hottel and Unger used cupric oxide deposited on bright shining aluminium. Nickel sulphide and cupric oxide on copper surfaces exhibit lower absorptivity and higher emissivity than cupric oxide deposited on aluminium. On the other hand, surfaces painted flat black or deposited with soot possess high absorptivities in the visible region as well as high emissivity in the infra-red region. The selective radiation properties exhibited depend not only on the special optical properties of thin films, but also on the structure of particles on the surface. These surfaces are not very easy to make in large sizes for use in solar energy collectors.

The use of selective surface as the cover of a flat-plate collector enables it to operate efficiently in the temperature range 250–300°F without optical concentration. Tabor exhibited a flat-plate collector with a selective surface that raised steam in Arizona sunshine. Miromit Sun-Heaters of Israel are now producing selective black surfaces for use in solar water heaters marketed in many countries. Tabor has also developed a solar operated turbogenerator, which uses solar heat collected by large inflated cylindrical solar collectors of plastic film with reflecting surface at the bottom. A similar 600 W turbogenerator developed recently operates with solar energy collected with the help of a number of flat-plate collectors, each one with two transparent covers and one selective black receiver.

Thin electroplated and oxidized coatings possess chemical and mechanical properties that make them suitable for use at high temperatures encountered in solar engines and solar thermoelectric generators. On the other hand, surfaces that absorb little sunlight and emit much infra-red radiation, are needed for preventing overheating of houses and space vehicles. White paints, which reflect sunlight, but exhibit high absorption and emission in the infra-red, are commonly used.

Solar space heating. Space heating with solar energy may potentially be perhaps the most important application of solar energy. Regions, where space heating is needed, have generally low intensity of solar radiation during winter months. Though space heating with solar energy is easily achieved, yet difficulties from the economic and architectural points of view have to be overcome. Solar energy collector is the costliest item of the system and the expenses on heat storage capacity to carry through a week or more of continuous cloudy weather are excessive. Auxiliary heating by fuel is now accepted as a necessity in places with long duration of cloudy days or extremely cold spells. Both solar heating and solar cooling may be accomplished with the same equipment.

In U.S.A. full-scale solar heated houses have been built and operated by Thomason, Löf, Bliss, Telkes, Gardenshire, and the Massachusetts Institute of Technology. The results showed that the Sun can provide two-thirds of the energy needed to supply hot air and hot water for these houses. But at present, the cost of equipment required for the system is higher than can be justified in terms of fuel saving.

The use of a heat pump for solar space heating has not proved to be economically feasible as yet, owing to the relatively high cost of the electricity required for its operation.

In the sub-zero temperature of cold arid regions of Ladakh, there is plenty of sunshine, fuel is scarce and expensive and no electricity is available. Heating of living space was effected by radiative and convective heat losses from the surface of a hot water tank and not by forced circulation of air. An Indian NPL-type solar water heater (Fig. 2) was used to heat water by thermo-siphon action. To increase the overall efficiency of solar energy utilization, the hot water storage reservoir was converted into a heat exchanger-cum-reservoir by providing $\frac{1}{2}$ in. diameter tubes through which air got heated by natural convection; circulation was effected by the "chimney" effect. The latter device of heating air is covered by Indian Pat. No. 99523 by M. L. Khanna and N. M. Singh (assigned to CSIR, New Delhi).

Solar cooling and refrigeration. The use of solar energy for refrigeration and air-conditioning is, next to solar water heating, very promising, because energy is most abundantly available at the time and place when and where the need for cooling is the greatest.

The operation of a vapour compression type of air-conditioner run with mechanical or electrical energy, produced with solar energy, is very expensive. The use of thermoelectric cooling, a hot air engine or small size turbine of high efficiency has been suggested for solar cooling. Triethylene glycol, in an absorption dehumidification system, has been used by Löf to regenerate weak solutions by evaporating off the absorbed moisture with solar energy. For year-round air-conditioning the system was found to be uneconomical as compared to the orthodox air-conditioner. The absorption of moisture in beds of finely divided silica gel was tried to cool a house built in Israel without the use of air blowers. It was found difficult to make the same wall alternatively both an effective collector and rejector of heat during the day.

Absorption refrigeration is the one system which is most suitable for operation with solar energy. Of the various combinations employed, viz. ammonia-water, lithium chloride-water, lithium bromide-water, sodium thiocyanate-ammonia, the ammonia-water system has been found to be the best. The system inherits the property of storing the refrigerant indefinitely until needed. The operation of an intermittent ammonia-water machine with solar heat concentrated with a parabolic reflector has been suggested for low-cost air-conditioning and refrigeration. As yet no house has been built with a closed-cycle absorption-desorption cooling system run on solar energy. A carefully designed solar operated air-conditioning unit using a flat-plate collector and ammonia-water absorption system appears to be economically feasible. Experiments with such a system have been conducted at the Universities of Ceylon and Florida.

Solar drying. Direct exposure drying of materials practised over ages can be accelerated and controlled by using flat-plate collectors. Cheap glass-mirror concentrators have been used for evaporation in Burma and for concentrating palm juice to produce jaggery in India. In the latter case, it has been estimated that the cost of the concentrator is recovered in three tapping seasons of three months each. Prior to retorting, oil shale in Brazil is dried by direct exposure to sunshine, which reduces the cost of production and simultaneously raises product value.

In drying crop products with solar heated air, Buelow reduced drying time and grain spoilage by slightly raising the temperature. He could recover the cost of solar energy collector through power savings in one to five years. Air was heated with solar energy by blowing it along ducts just below the sloping roof of corrugated iron sheet covered with black asphalt paint.

A possible design of a dryer for indoor drying of materials may comprise heating air with solar energy by circulating it with an electrically driven fan through a heat exchanger-cum-reservoir of a solar water heater (Fig. 2). Hot air may be utilized to dry coal fines from coal washers, agricultural products and to cure rubber sheet and timber.

Solar distillation see Khanna (1967).

Salt-making and brine concentration. Salt of very high purity is obtained in large solar evaporating plants and crude salt of less purity on a small scale along the sea coasts throughout the world. However, the efficiency of the solar operated process and the quality of salt produced depend to a large extent on the flatness of the bed, rate of evaporation during crystallization, the average daily precipitation rate, rainfall during summer months, etc.

Tabor used brine solution in a metre deep pond to store large amounts of solar energy. After several days of solar energy absorption by the black bottom of the pond, the temperature of brine in contact with the bed rises to the boiling point of water. Heat losses due to mixing of water in the pond are avoided due to a large difference in density between the lower layer of brine and the upper water layer. Coils of water pipes placed at the bottom of the pond extract heat and low pressure steam is produced for operating a turbine. By storing solar energy in this manner power can be produced day and night. However, there are many difficulties to be overcome before the solar pond can be commercially exploited for power production with solar energy on a large scale.

Solar furnaces. To obtain useful energy outputs at temperatures in excess of 220°F, some form of optical concentration is necessary. The increase in surface area for energy collection has restricted the use of concentrators on account of the high capital cost of equipment concentrating only the direct component of solar radiation. These concentrators are suitable for operation at high-altitude locations and in the tropical arid zones.

Optical concentration may be achieved by the use of lenses or mirrors. The only solar furnace incorporating a number of suitably arranged lenses was built in 1932 at the California Institute of Technology. Three systems of mirrors, namely the paraboloid, the cylindrical parabolic trough and the plane surfaces have been widely used. Their choice is determined on considerations of energy output, operating temperature and capital cost. Concentrator surface may be of glass or metal. In each application, the final choice is a matter of compromise between cost, resistance to corrosion and percentage reflectivity.

With high optical precision, temperatures up to nearly 3000°C can be obtained with solar furnaces, wherein heating is only on the front side and the area of the hot zone is confined to a few cm only. The greatest advantage of solar heating lies in heating being accomplished without contamination from gas, soot, crucible or heater material, without any complication from electrical or magnetic fields, and in oxidizing or reducing atmosphere or vacuum as desired. By changing the position of the material in the focus, temperature can be easily controlled. At the focus, liquid pool or molten cavity is readily formed in the mass. When the cavity is subjected to rotation and centrifugal force, heating occurs throughout the mass.

Large solar furnaces have been built in the United States, Japan, Algiers, France, the Soviet Union and elsewhere. With accurately shaped silvered glass mirrors or highly polished metallic surfaces, energy concentration ratios over 50,000 have been achieved. The 10-foot Convair solar furnace at San Diego is mounted on an equatorial stand and is moved with a synchronous motor to follow the Sun. Alternatively, the parabolic mirror or reflector may be fixed and solar radiation continuously directed upon it by a movable plane mirror or heliostat. The latter system is used in the Quartermaster Solar Furnace of the U.S. Army at Natick, Mass, and the one built by Dr. Felix Trombe at Fort Mont Louis in the French Pyrenees. The guidance system actuates a photocell or bimetallic thermoregulator to operate a motor to turn the large mirror whenever the Sun moves out. Its use has helped W. M. Conn to study new areas of development for the silicate industry, the army to determine the effects of high temperature on building materials and lubricants and Trombe to produce zirconia from zircon and alkali, beryllia from beryl, and tungsten oxide from wolframite respectively.

Relatively cheap reflectors of low optical accuracy have been used in Tashkent, U.S.S.R., to raise steam for central heating of buildings, for operating absorption refrigeration and regenerative distillation equipment. Parabolic aluminium reflectors used as solar cookers are manufactured in India, while cylindrical parabolic reflectors have been used for raising steam in the Soviet Union and for a solar power plant installed by Shuman and Boys in 1912 near Cairo. The use of plane mirrors for concentrating solar energy on a large scale in a steam raising plant was described by V. A. Baum. Nearly 1300 plane mirrors, each of 15 m² area, mounted on carriages running on 23 sets of concentric railway tracks concentrate solar energy around a boiler placed on top of a tower 40 m high. Similarly, the intensity of irradiation of an absorber or a solar stove or oven may be increased by fitting additional plane reflecting surfaces along the sides of the equipment.

Solar cooking. Boiling, baking and frying of food with concentrated sunlight is one of the simplest ways of solar energy utilization. Many designs of solar cookers have been developed. Their economic and social acceptance was studied in low-income rural villages in Mexico and among American Indians in Arizona. The results of field trials reveal that the acceptance of solar cooker depends largely on the cooking and eating habits of the community using it. The saving effected in fuel cost must be balanced against the capital cost and the inconvenience caused by outdoor cooking.

Solar heat engines or motors. Conversion of solar energy into mechanical energy with a heat engine does not present any serious technical problems. The overall efficiency of the system is equal to the product of the efficiencies of the heat collector and the engine. Many large-scale solar power systems have been constructed and operated, but none has proved practical so far.

Parabolic mirrors, mentioned earlier, have been used at the Heliotechnical Laboratory of the U.S.S.R. Academy of Sciences at Tashkent to build two large experimental steam generators. But on account of the high cost of the parabolic reflectors, recent trends are to replace them by flat-plate collectors, each with selective black absorber surface and a transparent cover.

The chances for the development of solar engines utilizing currently well defined cycles appear to be very bright for engines operating on either the vapour-cycle or the hot-air cycle. From the thermodynamic point of view, the working fluid should possess high latent heat, liquid density and vapour pressure at the temperature of engine entry, because it results in low specific consumption and a comparatively cheap and compact engine. Low boiling substances, such as ammonia, ether, sulphur, dioxide, methyl chloride, acetone, methyl and ethyl alcohols, etc. can be used as working fluids for the operation of vapour engines. An Italian firm, SOMOR Ltd. of Milan, manufactures water-pumping machines using sulphur dioxide as the working medium. Sulphur dioxide is vapourized in the long flat-plate collector and the vapours drive the engine, whose power may be used to generate electricity or to pump water. Engines of 0·7, 1·5, 2·5 and 3·5 h.p. are available. A wide range of new fluorinated hydrocarbons developed for use in the refrigeration industry have now become available and remain untapped for vapour engines. They are less toxic, less corrosive and at the same time thermodynamically more suitable than the volatile fluids used so far.

Hot-air engines are quiet in operation, require little maintenance and can be used over a wide range of temperatures. As their performance in actual operations falls short of the theoretical, there is considerable scope for increasing their efficiency by effecting improvements in design, materials, regenerative heat exchangers, etc. High operating efficiencies can be achieved in the Stirling hot-air engine cycles with proper mechanical and thermal development. When a regenerator of 100 per cent efficiency is introduced in the closed-type Stirling cycle, a thermal efficiency equal to the Carnot cycle is made possible. However, in practice engines working on these lines have very low efficiencies. The Phillips Company of Eindhoven, Holland, have developed a closed-cycle hot-air engine employing a thermal regenerator made of a network of very fine wires. In an improved design a transparent quartz window is provided in the cylinder head to focus solar radiation directly inside the hot-air engine. Such an engine operating with focused radiation from an electric lamp was demonstrated at the UN Conference in Rome in 1961.

The use of "solar pond", of large inflated parabolic cylinders with reflecting plastic at the bottom, of flat-plate collectors, each with a selective black absorber anda glass cover, to run a steam turbine have been discussed earlier. "Solar pond" appears to lower the cost of solar power production down to the anticipated values. Whenever it becomes practical, mechanical power production will become more widespread.

High temperature hot-air engines require the use of heat-transfer media to transfer heat from the collector to the engine itself. Liquid metals under development as heat-transfer media for nuclear reactors will probably be able to meet the demands. Work on the development of a 3 kW mercury-vapour turbine system ("Sunflower I") and a 15 kW rubidium-vapour system, both for use in outer space, is under way at the laboratories of the U.S. National Aeronautics and Space Administration. This has simultaneously hastened the development of light-weight large parabolic collectors to produce high temperatures and to open them when in outer space.

The lack of success with solar operated engines is due to such factors as intermittence and low intensity of solar energy itself, inadequate engine development, prohibitive collector costs, competition from I.C. engines, etc. But an urgent need of a small solar-operated engine is felt in the less-developed but sun-rich countries for water pumping, lighting and other purposes not characterized by the exacting demand for continuity and storage.

Direct conversion of solar heat to electricity. Any significant progress in this direction has been hampered by an extremely low conversion efficiency, non-availability of suitable materials, etc. But rapid progress made in solid state physics and the development of semiconductors with special physical properties has considerably changed the situation. Thermoelectronic converters and thermoelectric converters or "thermocouples" used to convert heat directly into electricity require high-temperature heat obtained by concentrating solar radiation with concentrators of high optical precision and with concentration ratios of over 2000. They are also subject to the thermodynamic limitations of the cycle. Both have the advantages of being light in weight, rugged in construction, possibly long life and simple in operation without any moving parts. Their development has been accelerated by space research. Still in the experimental stage, they are suitable for irrigation pumping, communication purposes and various small power uses in remote areas.

Of all the direct converters, thermionic converters are capable of operating at extremely elevated temperatures and rejecting the surplus heat at a lower temperature through radiators of modest dimensions. The introduction of ionized caesium vapour between the anode and cathode to neutralize the space change and to facilitate the flow of electrons and the addition of a third electrode to ionize caesium are two major improvements in the design, which enable operation at about 1100°C with 14 per cent efficiency. A solar thermionic system under development by NASA in the United States would supply 135 watts on Mariner probes and use a solar collector 5 ft in diameter to focus radiation on to a caesium diode converter. However, the defects encountered in operating a large number of these cells in series and parallels make them inferior to the photovoltaic cells, wherein these defects have been completely overcome recently.

With use of highly purified semiconducting materials obtained by zone refining, important advances have been made in obtaining higher efficiencies with thermoelectric devices. U.S. Navy has tested a 250-watt, propane fired thermoelectric generator using lead telluride and germanium bismuth telluride as the two arms. With 450 and 130°C as temperatures of the two junctions, an overall efficiency of 1·5 per cent has been obtained. The higher temperature is limited because of diffusion of the activating elements from the high temperature region to the low temperature region. The life of the unit is limited on account of unequal expansion and contraction of the two arms with alternate day and night cooling. A silicon-germanium system under development at the RCA laboratories appears to promise sufficient lifetime. However, major break-throughs are still awaited in their manufacture, improved physical properties of material and overall reduced costs for economical exploitation.

Solar radiation has a unique quality of having a large part of it coming as light, which can be converted directly into electricity in photoelectric converters (Fig. 1). They are of two types, namely chemical and physical. The chemical photogalvanic cell has not yet become practical. When successful, it would solve problems of storage or intermittence. The physical photovoltaic or photoelectric cell is the Bell Laboratory's solar battery, which represents a major break-through from the feeble currents produced by selenium cells to a miniaturized "power pack" strong enough to operate telephone circuits, short and long-wave transmitters and receivers and even television transmitters far out in space. A solar battery or silicon transistor depends for its operation on either an excess of free electrons or a deficit of electrons in covalent bonds. It is not only more efficient than chemical batteries, but it has also an indefinite life with no parts to wear out, no fluids to replace, no plates to be renewed, no "input" but the Sun itself. With mass production, the initial efficiency of 6 per cent has been raised to about 12 per cent, which is more than half of the theoretical maximum of 22 per cent. Its output may be increased nearly twenty-fold by the use of concentrators. By gridding the cell surface using a precision paraboloidal concentrator and employing a cooling system for the cells, the yields have been raised up to 2 W/cm^2. The production of low cost large area thin films of cadmium sulphide and silicon solar cell is a step further to promote the technology of the direct conversion of solar energy into electrical energy and may bring down the cost within the range of economic feasibility for home use.

Terrestrial applications have so far been confined to some demonstration units, such as unattended relay stations and light houses like the ones in Japan, solar operated telephones, radios, hearing aids, clocks, etc. The untapped applications in space age are communication satellites, manned satellites, manned space platforms, Lunar and Mariner explorations and eventually interplanetary rocket ships.

During the last fifteen years several international symposia have been held in different parts of the world to focus attention on the problems relating to solar energy utilization. The Solar Energy Society encourages the development of solar energy uses by publishing a journal and a news magazine and maintains a complete library and information centre.

Bibliography

Daniels F. (1964) *Direct Use of the Sun's Energy*, New Haven: Yale University Press.

Khanna M. L. (1967) in *Encyclopaedic Dictionary of Physics* (J. Thewlis Ed.), Suppl. Vol. 2, 332, Oxford: Pergamon Press.

Spanides A. G. and Hatzikakidis A. D. (1964) *Solar and Aeolian Energy*, New York: Plenum Press.

U.N. Conference on New Sources of Energy, Rome, 1961, Papers S/1–119 and GR/9–20(S).

Ward G. T. (1961) *Possibilities for the Utilization of Solar Energy in Underdeveloped Rural Areas*, F.A.O., Rome.

Wind Power and Solar Energy (1956) Proceedings of the New Delhi Symposium, Paris, UNESCO.

ZAREM A. M. and ENWAY D. D. (1963) *Introduction to the Utilization of Solar Energy*, New York: McGraw-Hill.

M. L. KHANNA

SOLID FILM LUBRICATION.

Introduction

The ever increasing demands being made on dynamic load-bearing systems in the space, nuclear and aircraft industries are pushing conventional lubricants to their operational limits. Eventually the limitations of these lubricants will present serious obstacles to reliable operation of systems in these fields. For this reason, considerable effort has been made in recent years to develop bearing systems that are not dependent on oils and greases for their lubrication. Instead, these systems utilize solids to furnish the necessary lubrication to load-bearing surfaces having relative motion with respect to one another.

Before focusing attention on the use of solids as lubricants, however, it may be well to examine fluid lubricants in general, briefly discussing their function and the mechanism by which they lubricate.

Conventional Lubricants

An extremely wide variety of materials have been used at one time or another as fluid lubricants, ranging from air, milk and gasoline to molten glass and alkali metals. The most familiar lubricants, however, are the more conventional oils and greases. These lubricants are liquid or semi-liquid materials that are usually obtained from petroleum base stocks. In certain cases, where specific properties are needed such as a low freezing point, the so-called synthetic fluids are used. These include the silicones and polyphenyl ethers among others. In *all* cases, their primary function is to prevent one load-bearing surface from contacting another when relative motion exists between them. The lubricant, of course, also has other functions, such as preventing corrosion, keeping dirt from bearing surfaces, and acting as a coolant. It is emphasized here, however, that their primary purpose is to prevent metal to metal contact.

How is this goal accomplished? Taking the simplest case—that of the journal or sleeve bearing shown in Fig. 1—it is seen that the bearing is made up of a three-component system comprising a shaft, a sleeve, and the lubricant. On shaft rotation, forces are set up in the lubricant film that form a fluid wedge sufficient to literally float the shaft or lift it away from the sleeve. By this mechanism, termed *hydrodynamic lubrication,* metal to metal contact is prevented. Assuming that the bearing has been designed properly, the only resistance to motion in this system is that generated by shearing the oil film between the shaft and sleeve. This resistance to motion is commonly known as *friction.* The magnitude of this frictional force depends on three parameters: namely, lubricant viscosity (Z), shaft speed (N), and load (P). In Fig. 2, a curve plotting the relationship of these three parameters as a function of the *friction coefficient* of a bearing system is shown. Friction coefficient is defined as the ratio of the frictional force in a bearing to the total load carried by the bearing. From the curve, it can be seen that there is an optimum relationship between shaft speed and load and lubricant viscosity that provides a minimum friction coefficient. If shaft speed or lubricant viscosity become too high, more work is required to shear the fluid with a resultant rise in friction coefficient. On the other hand, if these two parameters become too low (or bearing loads excessive) the lubricant film deteriorates until intermittent metal-to-metal contact occurs in the system, again resulting in higher friction coefficients and entry into what is known as the *boundary lubrication* area.

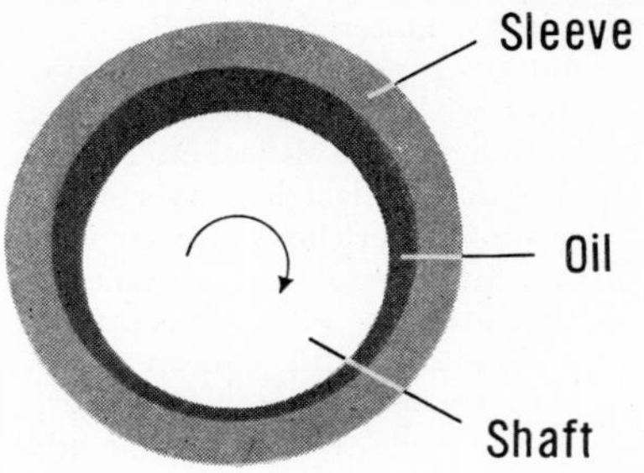

Fig. 1. Journal bearing.

Solid Lubrication

At this point the reader may wonder why an article on solid lubrication is going into this detail regarding the mechanism of fluid lubrication. There are two reasons for the previous discussion. First, many of the terms used above, such as friction force and friction coefficient, are common to both the conventional and the solid lubrication fields. Secondly, the author

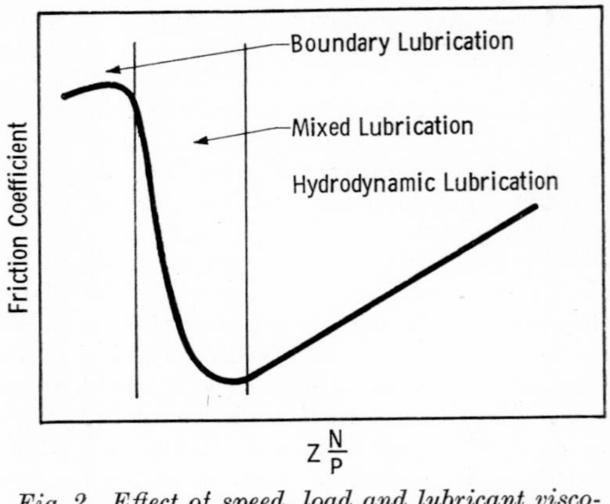

Fig. 2. Effect of speed, load and lubricant viscosity on the frictional behaviour of a journal bearing (Z = viscosity; N = speed; P = load).

believes that the ZN/P curve shown in Fig. 2 offers an excellent example of why an individual should not use a solid as the primary lubricant in a bearing system unless *forced* to do so. Conventional oils and greases are by far the better lubricants and should be used if at all possible. The reason for this is the fact that every solid lubricated bearing operates in the highest portion of the ZN/P curve, or that region which is called the boundary lubrication area. Where friction coefficients of 0·0005–0·001 are common in the hydrodynamic lubrication area, coefficients of 0·1–0·2 are encountered in the boundary lubrication area. In other words, the friction or resistance to motion in a solid lubricated load-bearing system is at least 100 times greater than in an oil lubricated system. From the standpoint of a machine's efficiency alone, it is highly desirable to hold friction coefficients to a minimum.

In view of this disadvantage inherent in all solid lubricated bearing systems, one might ask why they are used at all as lubricants. The answer is simply that quite a large number of applications have come along in the past few years that involve environments that oils and greases cannot tolerate. These are termed hostile environments and include cryogenic and high temperatures, ultra-high vacuums, and high radiation levels. The lowest melting point of any oil now available is about −110°F. If the oil is exposed to lower temperatures it becomes glass-like and immobilizes the bearing. Yet lubrication engineers are currently required to lubricate bearings and seals at liquid hydrogen temperatures, or −400°F. Regarding high temperature applications, the best modern synthetic fluids or greases begin to oxidize severely as temperatures approach 600°F. As illustrated in Fig. 3, however, requirements exist today for lubricants capable of withstanding a 1500°F-oxidizing environment with future requirements approaching 2500°F. Finally, the lubrication engineer is being asked to lubricate bearing systems at ambient pressures less than the vapour pressure of any known oil or grease, particularly at the temperatures where most of these bearing systems function. Prolonged exposure of an oil to such an environment results not only in loss due to evaporation, but also contamination of its environment with oil vapours that proves quite serious in some cases. These, then, are the reasons for the increasing interest in the use of solid lubricants, the physical and chemical characteristics of which are given below.

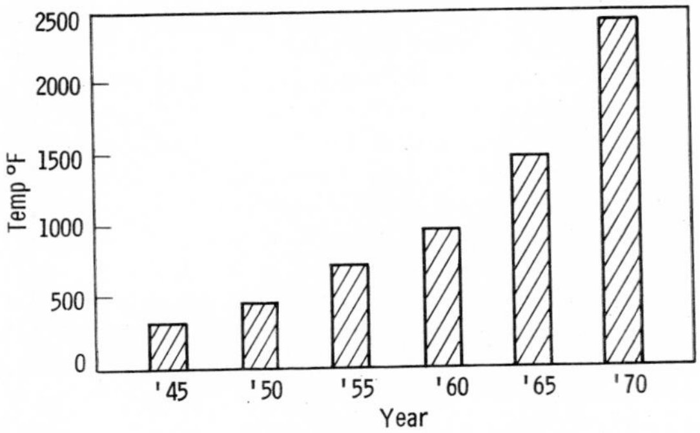

Fig. 3. Past and projected temperature requirements on load bearing systems.

Solid lubrication properties. Probably the best known of all solid lubricants is graphite. A representation of its crystal structure is shown in Fig. 4 and is typical of the great majority of compounds having solid lubricating ability. The material exhibits what is termed a laminar-or layer lattice-structure. This structure is composed of layers of carbon atoms in which the bonds between carbons of a given layer are quite strong, while bonding between layers is of the weak Van der Waals' type. For this reason, layers of graphite can be easily sheared, much like sliding open a deck of cards. It is this ease in shearing that is responsible, at least in part, for the lubricating ability of graphite.

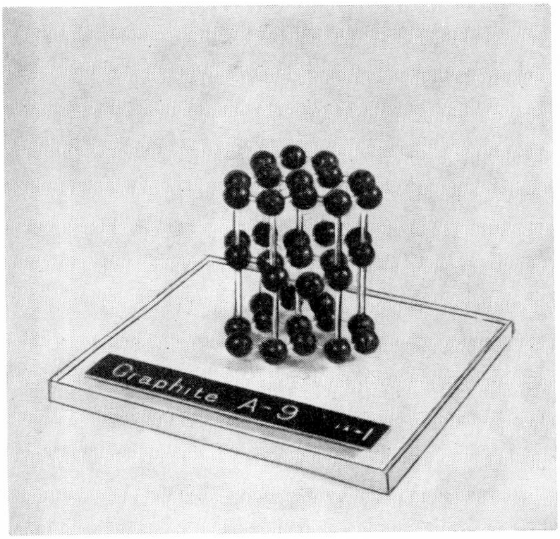

Fig. 4. Model of the crystal structure of graphite.

A relatively inexpensive material, graphite has been used as a lubricant for at least a century. Two outstanding qualities exhibited by graphite are its thermal and oxidative stability. Under inert environments, graphite is capable of withstanding temperatures of 5000°F, while its threshold oxidation temperature is about 750°F (that temperature at which it loses 1 per cent (wt) per 24 hour period). In addition, its excellent chemical inertness enables it to resist attack by such corrosive media as sulphuric acid and highly concentrated caustic solutions. For these reasons its use has met with considerable success and it is widely used in practically every industry. Up until early in the 1940's the explanation for its lubricity was believed to be completely due to the ease in shearing of its layer-lattice structure. During World War II, however, a severe brush-wear problem developed on d.c. machinery in aircraft as their altitude capabilities exceeded about 20,000 ft. By incorporating in the brush small quantities of inorganic materials such as barium fluoride or lead iodide, scientists at the Westinghouse Research Laboratories solved the wear problem, enabling the brush to operate satisfactorily at altitudes substantially higher than 20,000 ft. In later high-vacuum experiments by Savage at the General Electric Laboratories, it became evident that graphite's ability to lubricate depends on the presence of water vapour in its environment. At a level of between 30 and 300 ppm (−50°F dew pt.) it loses all lubricating ability. Since the concentration of water vapour falls below this limit at about 20,000 ft. and is well below the limit in both ultra-high vacuum and cyrogenic environments, it is seen that this deficiency in graphite's lubricating ability is a serious obstacle to its application in these areas.

There is, however, a group of solid lubricants that, while possessing the low shear, layer-lattice structure of graphite, remain independent of water vapour for their lubricating ability. These are the sulphides and selenides of the metals molybdenum, tungsten, tantalum and niobium. Molybdenum disulphide is probably the most familiar of this group. Figure 5 presents a representation of its crystal structure and is typical of all members of this solid lubricant family. It will be noted that the structure of molybdenum disulphide is such that its molecule is again built up of hexagonal layers of atoms. The bonding between the atoms within any given layer as well as between molybdenum and sulphur layers is quite strong. On the other hand, bonding between sulphur layers (called anion pairs) is weak, being of the Van der Waals' type. Through these bounds the compound is cleaved.

Fig. 5. Model of the crystal structure of molybdenum disulphide.

The first patent of the use of MoS_2 as a solid lubricant was awarded to the Westinghouse Electric Corporation in 1942. Strangely enough, this very first application immediately illustrated the fact that the compound's lubricating ability was independent of moisture. At the time it was employed as a lubricant for a rotating anode in a vacuum X-ray tube. In the past fifteen years molybdenum disulphide products have gone through a tremendous growth, until now they are marketed on a world-wide basis. Processed from molybdenite ore, the material is inexpensive and available in large quantities. This fact, coupled with its ability to lubricate in a no-moisture environment, are probably its most significant characteristics. While not as stable chemically as graphite, its physical and chemical characteristics remain quite respectable, with a thermal decomposition temperature above 2000°F and a threshold oxidation temperature of 650°F. It might be well to point out here that improved temperature capabilities have been found with molybdenum disulphide's tungsten counterpart, i.e. tungsten disulphide. This compound has demonstrated resistance to both thermal and oxidative decomposition at temperatures 200 to 300°F higher than molybdenum disulphide.

Rather recent additions to this group of layer-lattice solid lubricants are the diselenides of tungsten and molybdenum, as well as both the disulphides and diselenides of niobium and tantalum. It has been found that not only are these compounds equivalent to molybdenum disulphide in lubricating ability and water vapour independence, but also that the four niobium and tantalum compounds exhibit low electrical resistivities. A comparison of the electrical resistivities (ohm-cm) for molybdenum disulphide, graphite and niobium diselenide, for example, shows them to be 1×10^3, $2{\cdot}5 \times 10^{-3}$ and $0{\cdot}5 \times 10^{-3}$ respectively. One of the immediately obvious applications for these new solid lubricants is in the high vacuum commutation and slip-ring area, for these four compounds retain their lubricating ability despite the absence of water vapor. In addition, a number of programmes are currently underway to evaluate the use of these newer lubricants at temperatures of 1100°F in an air atmosphere.

The introduction of effective solid lubricants with resistance to oxidation at 1100°F would certainly be a major step forward, but it remains far short of the 2500°F future requirement mentioned earlier. Strong candidates for extreme temperature lubrication purposes, however, are the fluorides. In a sense, compounds such as calcium or barium fluoride can be thought of as also having a layer-lattice structure since their molecules are composed of layers of calcium and fluoride atoms. One of the most important points regarding this group of compounds is their excellent oxidation resistance at temperatures up to at least 2000°F, and their ability to retain their lubricating properties at this temperature. A second strong argument on their behalf is their compatibility with alkali metals such as sodium, at temperatures of 1000°F. This, indeed, is an extremely rare and valuable property in the solid lubrication area. A possible limitation to their usefulness at the present time is the tendency for fluoride lubricated load-bearing systems to operate at objectionably high friction coefficients when temperatures drop below 500°F.

Application of solid lubricants. The foregoing discussion has attempted to illustrate that there is available a rather large group of solid lubricants upon which the lubrication engineer can draw to satisfy a wide variety of operating conditions. In addition, concerted efforts continue to develop new materials of improved lubricating ability and chemical stability. Three conditions must be satisfied, however, before these solid lubricants can be successfully applied to load-bearing surfaces. First, a lubricant must be selected that is compatible with the environment in which it operates. Second, a means for introducing the lubricant onto the load-bearing surfaces must be chosen. Finally, a technique must be devised to provide replenishment of the lubricating film as it is lost through wear or ruptured due to shock or an overload condition.

Generally speaking, most of the techniques devised for satisfying these requirements fall into three main categories; namely, solid lubricant dispersions, bonded films, and self-lubricating composites. Illustrations of these three techniques are shown in Fig. 6.

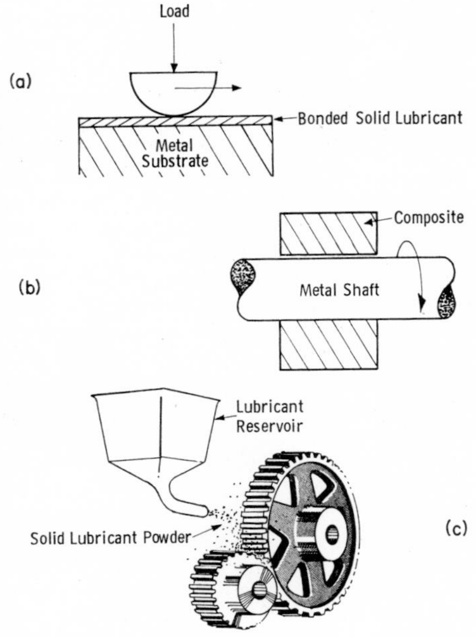

Fig. 6. Techniques for applying solid lubricants (a) bonded film; (b) composites; (c) dispersion.

Dispersion type lubrication. Lubrication by dispersions is usually accomplished by introducing the solid lubricant onto the bearing surfaces by means of a fluid carrier. This carrier can be either a liquid or a gas, with provisions made to meter the lubricant at pre-determined intervals or at continuous, carefully-controlled feeding rates. For example, the powdered solid lubricant may be fed to the bearing system using a carrier gas of air or an inert gas. While this technique requires considerable associated equipment (gas supply, plumbing, and lubricant reservoir) its advantages lie in the fact that the system is of the once-through type. As such, gas not only acts as a coolant for the bearing but also helps clean the bearing or gear of debris, thereby preventing clogging. A second method for introducing a solid lubricant onto a load-bearing surface is to incorporate small percentages of the powdered solid lubricant into an oil or a grease. Recently, the so-called Moly-Greases have received wide-spread attention in the automotive and railroad industry because of the substantial improvement brought about by the solid lubricant in the anti-weld characteristics of conventional lubricants (see Fig. 7).

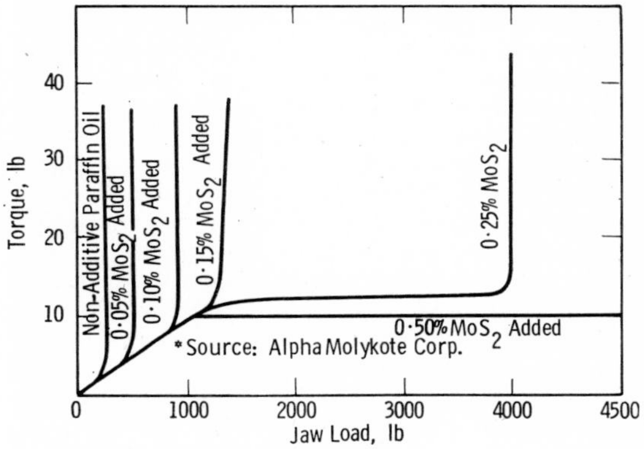

Fig. 7. Effect of MoS_2 additive concentration on anti-weld characteristics of paraffin oil.

Bonded films. Lubrication with bonded films is achieved by coating one of the load-bearing surfaces with a solid lubricant that is held tenaciously to that surface by a bonding agent. While these bonded solid lubricant coatings are hard and wear-resistant, the bearing load must be carried by the surface on which the film is applied. For this reason, the film thickness is quite low, usually not more than 0·0005 in. In most cases, bonded solid film lubricants are used in those applications involving high bearing pressures and low sliding velocities. Under these conditions hydrodynamic lubrication fails and boundary lubrication becomes necessary. Examples of applications where bonded coatings are used successfully are dirty or abrasive environments (mining, road construction), and load-bearing surfaces that are difficult to relubricate, such as hinges and pivots in aerodynamic linkages.

The primary disadvantage of bonded solid lubricant coatings is the fact that lubrication is non-renewable except by complete disassembly of the mechanism. The bonded solid lubricant coating, therefore, is a sacrificial film, so to speak. In other words, the ability of the system to replenish the lubricant lasts only as long as it takes to wear through the film. Once worn through, metal-to-metal contact and bearing failure occurs. The wear life of the film must therefore be carefully correlated to the life desired in the bearing system.

Composite lubrication. In an effort to eliminate the inherent sacrificial nature of the bonded films and the problems associated with it, investigators have been giving increasing attention to solid lubrication by means of composites. The great advantage that self-lubricating composites hold over bonded films is that, rather than relying solely on an initial application of lubricant to a load-bearing system prior to its installation, the bearing system is designed to essentially "lubricate itself" throughout its life. This is accomplished by fabricating one of the bearing components from a mechanically strong material designed to possess self-lubricating and filming characteristics.

Composites are materials composed of one or more lubricants in conjunction with a binder or filler that are mechanically strong enough to act as a load-carrying member of a bearing system. They are usually formed by moulding under pressure and heat, after which they may or may not be sintered, depending on the materials used. Machining and grinding operations are readily performed on them. The two most important characteristics of a good composite are mechanical strength and the ability to supply to a load-bearing surface an adequate amount of lubricant with a minimum amount of wear. The composites are tailor-made, so to speak, to provide these qualities. For every successful composite developed, literally thousands of tests are performed to determine optimum parameters with regard to pressing conditions, particle size, composition and material selection, all of which are critical to the two composite requirements mentioned above.

The three most frequently used types of bearing systems in use today are the ball or roller bearing, the sleeve (journal) bearing, and the gear. The application of the composite lubrication technique to each of the above systems is described below.

Ball bearing lubrication. Ball bearings are composed of an inner and outer race, balls, and a retainer or ball separator. Metal-to-metal contact occurs in this system between the balls and the retainer (sliding) and the balls and the inner and outer races (primarily rolling). It will be noted that the balls are in rolling or sliding contact with every other member of this system. If, therefore, a solid lubricating film is established on these balls and replenished as needed, solid lubrication of the entire system can be achieved through what is called a film transfer mechanism (transference of a solid lubricating film from one metal surface to another as they come into sliding or rolling contact under load). This is exactly what occurs when one replaces the metal retainer of a ball bearing with one fabricated from a composite. As an integral part of the bearing system, the self-lubricating retainer continuously meters minute quantities of lubricant to the balls in sliding contact with it. The balls, in turn, transfer this lubricant to the race grooves in which they are rolling. In this way, all critical load-bearing surfaces are coated with an extremely thin lubricating film that has all the characteristics of the bonded films. A photograph of such a bearing system and the film deposited are shown in Fig. 8. Bearings of this type have operated for thousands of hours under quite difficult conditions with no lubrication other than that furnished by its retainer.

(*a*)

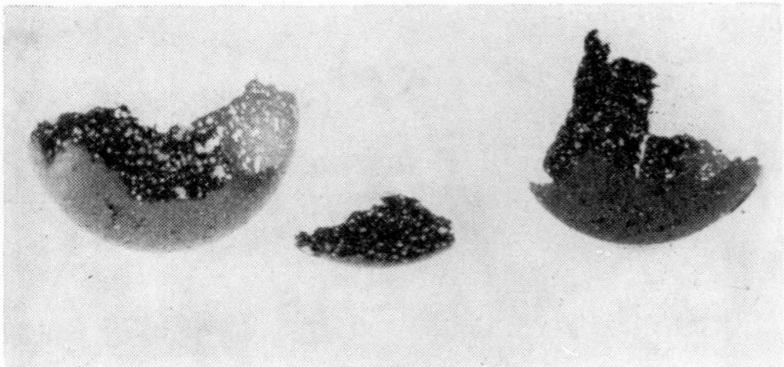

(*b*)

Fig. 8. (*a*) *Glass-MoS_2-filled teflon cage with bearing (800 hrs, 6800 rpm, 10 lb axial load, 40°C).* (*b*) *Dry lubricant film—removed from one ball after 500 hrs operation with teflon cage.*

Journal bearing lubrication. While the application of composite lubrication to journal bearings is somewhat simpler than with ball bearings the results to date have not been as impressive. Composites are usually applied in journal bearings as the sleeve in which the shaft rotates. In most cases, the self-lubricating sleeve is used in conjunction with a metal back-up for additional mechanical strength, and as the shaft slides against the composite, a solid lubricating film is established on its surface. As in conventional lubrication, however, the performance of solid lubricated journal bearings is highly dependent on the clearance between the shaft and the sleeve in which it rotates. As the composite sleeve wears, this critical parameter is changed and the performance of the bearing affected. Another problem common to journal bearing lubrication by composite is the fact that they do not clear themselves of wear debris as readily as ball or roller bearings. This fact, coupled with the close tolerances inherent in the system, results in the bearing literally choking itself with its own wear particles.

Gear lubrication. There are two ways of achieving composite lubrication of gears. The first, and most obvious, is to fabricate one member of a gear pair from the composite. While this approach is by far the simplest, it is limited in the range of operating conditions it can satisfy by the mechanical strength of available composites. Since self-lubricating materials contain lubricants in concentrations as high as 60 per cent in some cases, their mechanical properties cannot be expected to adequately meet some of the extremely high loads imposed upon gears. Load carrying gears fabricated from self-lubricating composites, therefore, are used under relatively mild load and speed conditions. For highly loaded gear applications, a technique utilizing a self-lubricating idler gear is employed. In this case, the load is carried by the metal gears themselves. Meshing with one (or both) of these gears, however, is a second gear fabricated from a self-lubricating composite. Loaded against the metal gear by either springs or a dead weight, the composite "idler" establishes a film of itself on the load-carrying metal gears and in doing so prevents metal-to-metal contact. Furthermore, it offers a readily available supply of solid lubricant for replenishment purposes as the film is worn from the gears.

The success of composite lubricated load bearing systems, particularly in the ball bearing and gear area, has been impressive. Lightly loaded (10 lb) ball bearings have operated for periods of 10,000 hours at room temperature and 7000 rpm with no lubrication other than that received from its retainer. In addition, their load capacity can approach that of conventionally lubricated bearings. For example, solid lubricated ball bearings (35 mm bore) have provided hundreds of hours' operation in ultra-high vacuum environments encompassing liquid nitrogen temperatures to 450°F while submitted to Hertz stresses of 400,000 psi. By utilizing the idler lubrication technique, gear systems have also operated satisfactorily for well over 100 hours at Hertz stresses of 225,000 psi under the conditions mentioned above. Many programmes now under way—sponsored both by private industry and government agencies—are making a concerted effort to not only extend the capabilities of solid lubricated load-bearing systems, but also to better understand the basic mechanisms underlying lubrication with solids.

While it is true that solid lubricants should be used with discretion and in no way will replace conventional lubricants in the foreseeable future, it is also true that the field of solid lubrication is fast developing into a discipline of significant importance that has already made possible a number of major steps forward in the cyrogenic, aeronautical, and space industries.

Bibliography

Bowden F. P. and Tabor D. (1963) *The Friction and Lubrication of Solids,* Part II, Oxford: Clarendon Press.

Fuller D. D. (1956) *Theory and Practice of Lubrication for Engineers,* New York: Wiley.

Gross W. A. (1966) in *Encyclopaedic Dictionary of Physics* (J. Thewlis Ed.), Suppl. Vol. 1, 118, Oxford: Pergamon Press.

Rabinowicz E. (1965) *Friction and Wear of Materials,* New York: Wiley.

Tabor D. (1962) in *Encyclopaedic Dictionary of Physics* (J. Thewlis Ed.), **4**, 360, Oxford: Pergamon Press.

D. J. Boes

SOLID-STATE PLASMA. A solid state plasma is a system of mobile charges which is contained by a fixed matrix of atoms and responds collectively to external stimuli. In contrast to gaseous plasma, the positive and negative mobile charge number density need not be the same since excess charge may be balanced by charges in the matrix; when the number densities are exactly equal, the plasma is called a compensated plasma. Typical solid state plasmas are the one-component plasma of electrons moving in the field of fixed positive ions in a metal, and the two-component plasma of electrons and positive holes in a semiconductor. The high density of mobile charges in a metal ($\sim 10^{22}$ cm^{-3}) results in a Fermi-Dirac distribution of velocities and so the plasma can exhibit quantum effects. In semiconductor plasmas, on the other hand, the density of mobile charges is much less, down to 10^{13} cm^{-3}, and the velocity distribution usually follows a Maxwell–Boltzman law, like that of gaseous plasmas.

Early experiments on solid-state plasmas were made by Ruthemann and Lang in 1948, who measured the transmission of kilovolt range electrons through thin metal films. They found that the electrons lost energy in integral multiples, the value depending on the particular metal; for example, the multiple was 14·7 eV for aluminium and 19·0 eV for beryllium. This was interpreted by Bohm and Pines as showing that the multipoles corresponded to longitudinal resonances at the characteristic frequency of mobile electrons which acted as a plasma in the fixed matrix of atoms of the metal. The frequency, called the *plasma frequency*, is proportional to the square root of the mobile charge density and ranges from the ultra-violet region in metals to microwave frequencies in semiconductors.

Interest in solid-state plasmas has developed with the study of waves and instabilities which can be related to those observed in a gaseous plasma. The solid-state plasma has the great advantage of intrinsic stability because it is contained by a fixed atomic matrix whereas the gaseous plasma is at best only transiently contained by a magnetic field and is prone to spontaneous instabilities arising from, for example, density, temperature and magnetic field gradients, anisotropic velocity distributions, etc. In recent years solid-state plasma research has advanced rapidly with the elimination of scattering due to lattice vibrations by operating at low temperatures, and the application of semiconductor technology to produce very pure solids with a known mobile charge concentration by the addition of controlled impurities.

The study of the dispersion of waves in a solid-state plasma also gives information on the matrix structure because the electric potential due to the fixed charges on the matrix can influence the response of the mobile charges. The response is often expressed by ascribing an effective mass to the mobile charges which in practice range from a few times to one hundredth of the electron mass. The large range of the effective mass enables the dispersion relations to be studied in different regions from experiments using gaseous plasmas; for instance, the mass of the positive holes can be made equal to the mass of the electrons in solid-state plasmas, whereas gaseous plasmas must always have a large ratio of ion to electron mass (at least that of a hydrogen plasma, 1840). When the fixed matrix is anisotropic the dispersion is complicated because it depends on the relative orientation of the matrix, the applied magnetic field and the direction of the propagation of the wave.

Two simple types of transverse wave propagation which have been extensively studied are Alfvén and helicon waves. These are transmitted in the presence of a magnetic field and have frequencies much less than the cyclotron frequency so that the centres of gyration of the mobile charges move with the transverse motion of the magnetic field (that is, the charge motion about the magnetic field lines is adiabatic). The field lines then act as strings with a tension $B^2/4\pi$ dynes/cm^2 where B is the field in gauss, producing a force $B^2/4\pi\ \partial^2 y/\partial x^2$ for displacements of the mobile charges in the y direction transverse to B in the z direction. This is balanced by an inertial force and a Lorentz force on the charge moving with a velocity $\dot{y}$ perpendicular to B giving an equation of motion

$$B^2/4\pi\ \partial^2 y/\partial x^2 = mn\ddot{y} - \sigma B\dot{y}/c$$

where σ is the charge concentration, m the charge mass and n the charge density.

When the inertial term is much greater than the Lorentz term a non-dispersive linearly polarized wave propagates with a velocity $(B^2/4\pi mn)^{1/2}$. Similar waves in the mixture of gaseous plasma and magnetic field in sunspots were originally described by Alfvén. The waves are damped by collisions between the mobile charges and the condition for propagation is that the wave frequency must be greater than the collision frequency. In solid state plasmas the collision frequency is generally greater than 10^{10} sec^{-1} so the wave frequency must be in the microwave region. The further condition that the cyclotron frequency must be greater than the wave frequency means that a magnetic field of kilogauss order must be used. Alfvén waves have been measured to find the effective mass density (mn) and collision times in bismuth which is of particular interest since it has an anisotropic effective mass. They have also been observed in graphite and antimony.

Helicon waves propagate in uncompensated plasmas when the Lorentz term dominates over the inertia term, and the solution of the equation is a circularly polarized dispersive wave $y = y_0 \exp[i(\omega t - kz)]$ with $k^2 = \pm 4\pi\omega\sigma/cB$ which has the distinguishing feature that the velocity (ω/k) is linearity proportional to B for a constant k. This is called a helicon wave because it is circularly polarized and is similar to whistler waves propagated in the gaseous plasma and magnetic field surrounding the Earth. As long as the cyclotron frequency is greater than the collision frequency of the mobile charge in the solid the damping is small irrespective of the wave frequency. These waves can therefore be observed over a wide frequency range and in fact were first studied at 30 cycles per second in a sodium cylinder by Bowers in 1961 yet only a year later were detected at 10^{10} cycles per second in indium antimonide by Libchaber and Veilex. Helicon waves have been used to find the charge density and have accurately shown, for example, that there is one mobile electron per atom in sodium metal. Helicon waves have also been used to measure the Fermi momentum of mobile charge carriers in a metal by adjusting the magnetic field until the Doppler shifted cyclotron resonance is observed (measured be detecting the surface resistance). Since the carriers have a Fermi velocity (v_z) parallel to the magnetic field they experience a Doppler shifted frequency $\omega \pm kv_z$ and the resonance first occurs at a wave frequency much less than the cyclotron frequency, when $kv_z = \omega_c$. From the dispersion relation for helicon waves it follows that the magnetic field at resonance is proportional to the cube root of the frequency and is a simple function of the Fermi momentum of the mobile charges. Taylor has detected the Doppler shifted helicon resonance in sodium at a temperature of 4°K and deduced the radius of the Fermi sphere to be $1{\cdot}09 \times 10^{-8}$ cm, which is within a few per cent of the theoretical value.

In spite of the stabilizing influence of the fixed matrix, instabilities have been generated in solid-state plasmas by causing a current to flow parallel to a magnetic field aligned along the axis of a narrow cylinder of the solid. Larrabee and Steel have observed current oscillations in an indium antimonide crystal with a modulation of up to 70 per cent and a frequency ranging from a few kilocycles to about ten megacycles, depending on the magnetic field and current. Typically fields of a few kilogauss and currents in the milliamp range were used. Devices based on this phenomena are called "*oscillistors*". The effect was explained by Glicksman using a slightly modified form of a theory proposed by Kadomtsev and Nedospasov to account for a similar instability of a positive column of a high current gas discharge plasma in a longitudinal magnetic field. They consider the growth of a screw-shaped perturbation of the current carriers superimposed on the steady-state distribution of density. The applied axial electric field causes charge separation of the positive and negative current carriers which results in radial electric fields and the motion of the carriers in these magnetic and electric fields makes the perturbations grow.

When very high current densities flow, of the order of thousands of amp/cm^2, the relative drift of the mobile charge carriers can cause spontaneous generation of Alfvén or helicon waves. Waves are amplified as they move down the solid when the drift velocity exceeds the phase velocity of the wave. Experiments by Larrabee and Higinbothem using indium antimonide detected 3000 Mc/s waves at a threshold current of around 5000 amp/cm^2 using a magnetic field of 6 kilogauss.

Also using indium antimonide, Glicksman and Steel have found that in the absence of an axial magnetic field the self magnetic field of the current causes the current channel to constrict to a narrow filament in a similar way to the pinch effect in gaseous plasmas. At low currents the plasma is uncompensated and the fixed charges on the matrix prevent pinching, but above a critical current of about 5 amp avalanche breakdown produces a large increase of electrons and holes in equal numbers which rapidly form a pinch. The resistance drops when the avalanche occurs, but does not reach zero because of increased electron/hole scattering in the pinch. If the pinch is inhibited by applying an axial magnetic field the resistance drops much further because the electron/hole scattering is reduced. Since the diffusion time of the axial magnetic field is of the order of 10^{-9} sec or less while the pinching time is 10^{-8} to 10^{-7} sec, the axial magnetic field cannot inhibit the pinch by balancing the self-magnetic field, and it has been suggested that the pinch is destroyed by instabilities.

See also: Helicons.

Bibliography

Chynoweth A. G. and Buchsbaum S. J. (1965) *Physics Today* **18**, 26, November.

Seventh International Conference on the Physics of Semi-conductors (1965) *Plasma Effects in Solids*, Paris: Dunod.

J. A. Reynolds

STORAGE RINGS. *1. Generalities.* An accelerated particle storage ring is a synchrotron-like magnet ring structure of stationary field strength, wherein particles of convenient energy are accumulated. Storage rings are used for experiments with colliding beams. Either one single ring is filled with two counter-rotating intersecting beams of opposite charges or two tangent or interlaced rings are used in the case of identically charged particles.

Storage rings allow for particle collisions at energies in the center of masse that cannot be reached at present by any other means. Generally in the rings, the CM system of colliding particles is the laboratory system. Let $\gamma_{CM} = E/E_0$ be the energy in this system of each particle in each of two colliding beams. A Lorentz transformation shows that one of two colliding particles has then the energy in the system of the other one: $\gamma = 2\gamma_{CM}{}^2 - 1$. This means that the equivalent energy of a particle interacting in the conventional way with an identical one located in a fixed target must be about $2\gamma_{CM}$ times higher than the energy of each particle in a storage device.

The following rings of energy $\geqq 500$ MeV either exist or are in construction or proposed (see Table on p. 390):

The building and operation of storage rings raise problems similar to those for accelerators. However, injection and accumulation techniques are different. The magnet lattice is peculiar. There also exist vacuum problems. mostly in electron rings.

Further, the need to obtain long-lasting beams of very high intensity and density raises new problems, concerning the life-time, stability and convenient size of the beams.

The highest possible interaction rates are wanted. The frequency per interaction region of a given nuclear event is generally expressed as $\dot{n} = L \cdot \sigma$, where σ is the cross-section of the event, and L is called the instrumental "luminosity". An interaction rate of $\dot{n} \approx 1$ per day is considered as a minimum aim for physics experiments.

2. Beam life-time. The beam life-time is limited by several mechanisms and predominantly by interactions of the beam with the residual gas causing, in proton beams, losses mostly by multiple scattering, and in electron beams by single bremsstrahlung energy losses. As a rule, the life-time is in both cases about 10 h at 10^{-9} torr pressure (air equivalent). Pressures smaller than 10^{-10} torr are desirable in the interaction regions in order to obtain a satisfactory ratio of useful events over noise.

Laboratory	Name	Type of particles; Number of rings	One particle energy		State of advancement (1966)
			Lab.	Equiv.	
Orsay LAL	ACO	$\varepsilon^-\varepsilon^+$, 1	500 MeV	980 GeV	exist
Stanford SLAC	PRIST	$\varepsilon^-\varepsilon^-$, 2	500 MeV	980 GeV	exist
Novosibirsk	VEPP-2	$\varepsilon^-\varepsilon^+$, 1	700 MeV	1920 GeV	exist
Frascati, Lab.	ADONE	$\varepsilon^-\varepsilon^+$, 1	1·5 GeV	8800 GeV	under construction
Stanford SLAC		$\varepsilon^-\varepsilon^+$, 1	3 GeV	$3{\cdot}5 \times 10^4$ GeV	project
CERN	CERN-ISR	PP, 2	28 GeV	1690 GeV	project

In electron rings the residual gas pressure depends on gas production by desorption from the chamber walls brought about by the synchrotron radiation of the beam. The gas production rate $\dot{r}$ depends on current, energy and trajectory radius as $\dot{r} \propto IE^{5/2}\varrho^{-1/2}$. In the Stanford ring PRIST, the gas production is, at 300 MeV and after degassing, about 5×10^{-8} torr litre sec^{-1} mA^{-1}, the gas comprising 90% of H_2 and 10% of CO and CO_2.

With electrons the life-time is sometimes further limited by the Touschek effect: energy dispersion in the beam arising from internal Coulomb collisions. In some collisions the energy variation happens to be so large that both the particles involved are thrown out of the stable energy band (R.F. bucket).

The Touschek life time τ is given by

$$\frac{1}{\tau} \approx \frac{\sqrt{\pi} r_e^2 c N}{\delta p_x (\Delta p)^2 k V_p} \left(\ln \frac{1}{\gamma_e (\Delta p / \gamma \delta p_x)^2} - \frac{3}{2} \right)$$

r_e = classical electron radius, N/kV_p = particle density, with $k = RF$ harmonic, $V_p = (4\pi)^{3/2}\,\delta x\,\delta z\,\delta l$, $m_0 c\,\delta p_x$ = radial momentum standard deviation, $m_0 c \Delta_p$ = energy acceptance (central), $\gamma_e = 1{\cdot}78$ (Euler constant).

3. Proton beams. For coasting beams crossing horizontally at angle $2\alpha_h$, the luminosity is obviously

$$L = \frac{N_1}{S} N_2 f,$$

where $S = h(w + \alpha_h 2\pi R)$ is the "apparent beam section" and h, w the height and width of the beams. N_1, N_2 = respective number of particles per beam, f = revolution frequency. The section $h \times w$ depends in proton beams in principle on the original density distribution in phase space of the injected beams. Methods of concentrating ("cooling") of proton beams are under study.

In the planned CERN-ISR proton rings the beam from the existing 28 GeV synchrotron will be injected 400 times in each ring and stacked in the energy phase space, resulting in accumulated beams, characterized each by $N \approx 4 \times 10^{14}$ particles, 20 A equivalent with total momentum dispersion of $\Delta p/p \approx 2\%$ and of dimensions $h \times w = 1 \times 6$ cm^2. Assuming a horizontal crossing angle of $2\alpha_h \approx 15°$, the expected luminosity is $L = 3{\cdot}5 \times 10^{35}$ cm^{-2} day^{-1}.

4. Electron beams. Electrons lose energy by synchrotron radiation. The losses are compensated by RF acceleration. Therefore electron beams are bunched. Natural beam dimensions result as an equilibrium state between a tendency of the beam to shrink by radiation damping and to expand as a consequence of spontaneous photon emissions. The natural repartition function of the beam density is Gaussian with full half-width ($\approx 2{\cdot}36 \times$ standard deviation) typically of order $2{\cdot}36\,\delta x \approx 2$ mm, $2{\cdot}36\,\delta z \approx 1/10$ mm, $2{\cdot}36\,\delta l \approx 30$ cm. The beam size can be enlarged artificially.

The luminosity in an electron beams interaction region is (with k bunches, vertical beam crossing at angle $2\alpha_v$ or head-on collisions):

$$L = \frac{1}{k} \frac{N_1}{S} N_2 f;$$

with $S = 4\pi\,\delta x(\delta z + \alpha_v\,\delta l)$ = "apparent beam section."

The highest possible value of (N/kS) is limited (Amman–Ritson-limit) by beam-beam interaction in the crossing region, resulting in mutual betatron wave number displacements. There is stability only if the small amplitude wave number displacement of the weaker beam (2) by interaction with the stronger one (1) is per period between beam crossings less than

$$\Delta \nu_2 = \frac{2 r_e \beta_v}{\gamma} \frac{N_1}{kS} \leqq 0{\cdot}025.$$

β_v is the vertical Twiss coefficient at the crossing point, r_e = classical electron radius, $\gamma = E/E_0$.

Solving this formula with respect to (N_1/kS) and inserting in the L formula with $N_2 = N_1$ and $N_2 f = I/e$, one gets the maximum luminosity at a given intensity I of each beam:

$$L_M = \frac{1}{e} \frac{\gamma \Delta \nu}{2 r_e \beta_v} I,$$

if the beam section is

$$S = \frac{1}{k} \frac{1}{ef} \frac{2 r_e \beta_v}{\gamma \Delta \nu} I.$$

The formulae show the effect of inserting in the crossing regions special magnet lattices with low β_v. It is possible to make β_v only a few cm large.

There are also mechanisms of unstable collective oscillations, the beam as a whole acting on itself by fields induced by interaction of the beam with the chamber walls, with the cavities, with ions or with the other beam. All these collective instabilities can be controlled by feed-back devices. In some cases they are naturally stable by Landau damping.

In good agreement with these formulae, the Stanford electron–electron rings achieved $L = 1{\cdot}7 \times 10^{33}$ cm^{-2} day^{-1} at 300 MeV with $2{\cdot}36\ \delta x = 3$ mm, $2{\cdot}36\ \delta z = 0{\cdot}8$ mm (artificially enlarged) $N_1 = N_2 = 10^{10}$ (40 mA), and $f = 25 \times 10^6$ s^{-1} (all figures measured in 1966).

The recent Stanford proposal for a 3 GeV electron-positron ring with $I \approx 1$ A, $\beta_v \approx 5$ cm is expected to achieve $L \approx 10^{37}$ cm^{-2} day^{-1}.

Landau-damping. A group of many oscillators of nearly common proper frequency is supposed to be coupled with the environment in such a way that the reaction of the environment in the oscillators induces a growing collective oscillation. However, if the oscillators have sufficiently large frequency spread, Landau has shown that the collective oscillation is damped. (*J. Phys. U.R.S.S.* 11 (1946) page 25).

Twiss coefficient. The Twiss coefficient $\beta(s)$ is a length, a function of azimuthal position s by which the transverse betatron displacement y of the trajectory can be expressed as

$$y(s) = \text{const.}\ \beta^{1/2}(s) \exp\left[\pm i \int (ds/\beta)\right].$$

Bibliography

BLANC-LAPIERRE A. (1966) Les Anneaux de Stockage et de Collision, *Onde Elec.* **46**, 726.

Design Study of Intersecting Storage Rings for the CERN proton synchrotron, CERN report AR/int SG/64.9.

The Frascati Storage Rings, Frascati report LNF 64/65.

International Symposium on Electron and Positron Storage Rings, Orsay-Saclay 1966, Presses Universitaires de France (To be published).

Proceedings Particle Accelerator Conference, Dubna 1963, Frascati 1965 and Washington 1965 (published in *IEEE Trans. Nucl. Sci.* June 1965).

Proposal for a high-energy electron-positron colliding beam storage ring, University of Stanford report, September 1966.

Storage Ring summer study, 1965 on instabilities, Stanford report SLAC 49/1965.

(1946) *J. Phys. U.R.S.S.* **11**, 25.

LAWSON J. D. (1962) in *Encyclopaedic Dictionary of Physics* (J. Thewlis Ed.), **5**, 306, Oxford: Pergamon Press.

H. BRUCK

STRONG INTERACTIONS. The strong interactions are the dominant interactions between elementary particles, and are responsible for nuclear forces, the forces between mesons and nucleons, the production of mesons in particle collisions, and similar phenomena. The range of the forces is short, of the order of 10^{-13} cm, a characteristic length which gives the order of magnitude of nuclear radii. They are one of the three types of interaction among particles, the others being the *weak interactions*, responsible for β-decay, and the well-known electromagnetic interactions. The strengths of these three types may be characterized by the typical time-scales of reactions which they induce. These are of the order of 10^{-10} sec for weak interactions, 10^{-19} sec for electromagnetic interactions, and 10^{-24} sec for strong interactions. Alternatively, one can compare the coupling constants which measure the strengths of the forces. These are of the order of 10^{-7} for weak interactions, 1/137 for electromagnetic interactions, and of the order of 10 for strong interactions.

The elementary particles which are subject to strong interactions are characterized by several quantum numbers, which are conserved in reaction and particle production processes. These quantum numbers are of two kinds, spatial and internal. The spatial quantum numbers are the *spin*, or angular momentum, and the *parity*. The internal ones are baryon number, strangeness, isotopic (or isobaric) spin, and one component of isotopic spin, the electric charge. There are other less well-established internal quantum numbers which will be considered later.

Classification of stable particles. Stable particles are taken to include those which decay by electromagnetic or weak interactions. In fact only one of the strongly interacting elementary particles, the proton, is absolutely stable. The quantum numbers and masses are given in Tables 1 and 2. The baryon number B is a quantum number that is conserved absolutely, and gives a first division of the particles into baryons (with $B = 1$) which all decay eventually to protons, and mesons (with $B = 0$) which all decay to particles (e.g. electrons, photons) which do not interact strongly. All the baryons have spin $\frac{1}{2}$ and the same parity (taken to be positive by convention), and masses of the order of 1 GeV. The mesons all have spin 0 and negative parity.

Table 1. The stable baryons; $B = 1$, parity $+$, spin $\frac{1}{2}$

Particle	Strangeness (S)	Isotopic spin I	I_3	Mass (MeV)
p	0	$\frac{1}{2}$	$+\frac{1}{2}$	938·2
n			$-\frac{1}{2}$	939·1
Λ	-1	0	0	1115·4
Σ^+	-1	1	1	1189·4
Σ^0			0	1192·3
Σ^-			-1	1197·2
Ξ^0	-2	$\frac{1}{2}$	$+\frac{1}{2}$	1314
Ξ^-			$-\frac{1}{2}$	1320·8

Table 2. The stable mesons: $B = 0$, parity —, spin 0

Particle	Strangeness (S)	Isotopic spin I	I_3	Mass (MeV)
π^+			1	139·6
π^0	0	1	0	135·0
π^-			-1	139·6
η	0	0	0	548·8
K^+	$+1$	$\frac{1}{2}$	$+\frac{1}{2}$	493·8
K^0			$-\frac{1}{2}$	497·7
$\overline{K^0}$	-1	$\frac{1}{2}$	$+\frac{1}{2}$	497·7
K			$-\frac{1}{2}$	493·8

Strangeness. S is conserved additively; in other words the total strangeness before and after a reaction must be the same. For example,

$$\pi^- + \mathrm{p} \rightarrow K^0 + \Lambda$$

can occur, but

$$\pi^- + p \rightarrow \overline{K^0} + \Lambda$$

cannot. The concept was introduced to systematize such phenomena. Instead of strangeness, the *hypercharge* Y, defined by $Y = S + B$, is sometimes used. Since B is always conserved, there is no difference in the significance of the rules.

Isotopic spin. This is a more complicated conservation law. With a multiplet of particles of different charges, one can assign a spin, in a purely artificial "isotopic" space, to each particle, so that all members of the multiplet have the same total isotopic spin, but the individual members all have different spin directions. These are characterized by one component of the spin vector, usually taken to be the third, or z-component, just as with ordinary spin. Thus the proton-neutron doublet is taken to have total isotopic spin $I = \frac{1}{2}$; the proton has $I_3 = +\frac{1}{2}$, and the neutron $I_3 = -\frac{1}{2}$. Similarly for the π-meson triplet, $I = 1$, and the three members π^+, π^0, π^-, have $I_3 = +1, 0, -1$, respectively. The analogy with spin in ordinary space is very close. Just as with ordinary spin, it is possible to construct states of more than one particle with definite isotopic spin I and third component I_3.

The conservation law of isotopic spin then states that strong interaction reactions are independent of the direction of the state in isotopic spin space, i.e. that cross-sections will depend only on I and not on I_3. This is also called the *principle of charge independence.* For example, the cross-sections for the reactions

$$p + \eta \rightarrow \Lambda + K^+$$

$$n + \eta \rightarrow \Lambda + K^0$$

will be equal; both η and Λ have $I = 0$, so that the total isotopic spin on each side of the reactions can only be $\frac{1}{2}$. The two reactions represent the states $I_3 = +\frac{1}{2}$, and $I_3 = -\frac{1}{2}$ respectively. When neither particle has $I = 0$, the relations between cross-sections can be more complicated; for example, for π-nucleon scattering

$$\sigma(\pi^+ + p \rightarrow \pi^+ + p) + \sigma(\pi^- + p \rightarrow \pi^- + p)$$
$$= 2\sigma(\pi^0 + p \rightarrow \pi^0 + p) + \sigma(\pi^0 + p \rightarrow \pi^+ + n)$$
$$\sigma(\pi^0 + p \rightarrow \pi^+ + n) = \sigma(\pi^- + p \rightarrow \pi^0 + n).$$

It can be seen from the tables that the relation between I_3 and the charge Q differs for different particles. The rule connecting them is

$$Q = I_3 + \tfrac{1}{2}(B + S) = I_3 + \tfrac{1}{2}Y.$$

Violation of the conservation laws. The conservation laws only hold when the effects of the electromagnetic and weak interactions are neglected. Electromagnetic interactions violate charge independence, but not of course charge conservation. The magnitude of the violation, as can be seen from the tables, is not great; the masses of particles in a multiplet differ by a few per cent at most. Isotopic spin, strangeness, and parity are all violated by the weak interactions, but the dynamical effects are extremely small. Baryon number conservation, like charge conservation, appears to hold absolutely.

Antiparticles. All these particles have their *antiparticles,* which are obtained from them by the operation of charge conjugation. The consequences of this are $B \rightarrow -B$, $Q \rightarrow -Q$, $S \rightarrow -S$, $I_3 \rightarrow -I_3$. For the baryons, this results in a new set of particles with $B = -1$, and negative parity relative to the original particles. For the mesons, which have $B = 0$, it is possible for the antiparticles to be the particles themselves. Thus $\pi^0 \leftrightarrow \pi^0$, $\pi^+ \leftrightarrow \pi^-$, $\eta \leftrightarrow \eta$, $K^+ \leftrightarrow K^-$, $K^0 \leftrightarrow \overline{K^0}$.

Unstable particles. In addition to the stable particles, a considerable number of unstable particles, which might more properly be called resonances, are known, and new ones are continually being discovered. The lifetimes of these particles are so short (of the order of 10^{-23} sec), that there is no possibility of their being seen directly, and they can only be observed by their decay products. The best established mesons are listed in Table 3, and the baryons in Table 4. Many others whose properties are not certainly known have been reported.

The extra quantum number G-parity shown in Table 3 is only of use for mesons (or more complex systems) with $S = 0$. It is defined in terms of isotopic spin and charge conjugation properties, and can be regarded as the equivalent of parity in isotopic spin space. It exists for π-mesons, where it is negative, and η-mesons, where it is positive. For most purposes it can be taken to determine whether a particle can decay, by strong interaction processes, into an even or odd number of π-mesons, according to whether it is positive or negative. Thus the ϱ and f mesons

Table 3. Unstable mesons

Particle	Spin and parity	Strangeness (S)	Isotopic spin (I)	Mass (MeV)	G-parity
$\varrho^{\pm}$, ϱ^0	1−	0	1	765	+
ω	1−	0	0	783	−
K^{*+}, K^{*0}	1−	1	$\frac{1}{2}$	891	
K^{*-}, $\overline{K^{*0}}$	1−	−1	$\frac{1}{2}$		
φ	1−	0	0	1019	−
f	2+	0	0	1253	+
X^0 (or η')	0−	0	0	959	−

decay into two pions and the ω and Φ mesons into three.

In Table 4 only those baryons whose quantum numbers are definitely established are listed. An exception to this is the last entry, Ω^-, which is actually stable under the strong interactions; it is given here for reasons which will appear later. No baryon resonances which have positive strangeness are known.

Table 4. Unstable baryons

Particle	Spin and parity	Strangeness (S)	Isotopic spin (I)	Mass (MeV)
$N^*_{3/2}$	$^3/_2-$	0	$^3/_2$	1236
$N^*_{1/2}$	$^3/_2-$	0	$^1/_2$	1518
$N^*_{1/2}$	$^5/_2+$	0	$^1/_2$	1688
$N^*_{3/2}$	$^7/_2+$	0	$^1/_2$	1924
Y^*_0	$^3/_2-$	−1	0	1405
Y^*_0	$^3/_2-$	−1	0	1520
Y^*_0	$^5/_2+$	−1	0	1815
Y^*_1	$^3/_2+$	−1	1	1385
Y^*_1	$^5/_2-$	−1	1	1765
Y^*_1	$^7/_2+$	−1	1	2065
Ξ^*	$^3/_2+$	−2	$^1/_2$	1530
Ω^-	($^3/_2+$)	−3	(0)	1675

(Data from Rosenfeld *et al.*, *Rev. Mod. Phys.* **37**, 633 (1965).)

Two-particle interactions. The interactions between pairs of particles are often dominated by the resonances described above; in fact, when resonances exist in a particular channel (i.e. with the quantum numbers of a particular net of particles), they can be produced copiously, and often give an almost complete description of the interaction. This is true in many of the meson-meson and meson-baryon channels.

Meson-meson interactions. Since it is not possible to use mesons as targets in experiments, these can only be observed as the interactions between mesons produced in reactions, and only the dominant features are seen. The π-π interactions are dominated by the ϱ and f resonances, the π-K interactions by the K^* and the $K - K$ interactions by the φ. The channels with $S = \pm 2$, involving $K - K$, or $K - K$ interactions, do not appear to have any associated resonances, and there is no significant interaction.

Meson-baryon interactions. Here only the meson-nucleon interactions can be observed directly, since the other baryons cannot be used as targets. The channels with zero or negative strangeness are characterized by series of resonances which again give all the most important features of the interactions. Thus the π-nucleon interaction is dominated at low energies by a large number of N* resonances, four of which are listed in Table 4.

Baryon-baryon interactions. In these channels there is one well-known bound state, the deuteron. No other two-particle bound states or resonances are known. The ordinary nuclei are examples of bound states of many baryons, and there are also known *hypernuclei*, in which a Λ is bound into a nucleus.

The nucleon-nucleon interaction is the best-known of all elementary particle interactions. There are no resonances involved in its description; its main characteristics are a strong attraction which changes to an even stronger repulsion at short distances, and strong spin-dependent forces.

Much less is known of the other baryon-baryon interactions.

Baryon-antibaryon interactions. Again, only the nucleon-antinucleon interactions have been studied. The principal effect is the annihilation of the particle-antiparticle pair, with the production of a number of mesons. The conversion of two baryon masses provides sufficient energy to create at least 13 π-mesons, but frequently only a small number are produced, often in the form of resonant groups.

Further symmetries. The large number of particles which have been found in recent years has stimulated the search for further ways of classifying and systematizing the particles. The most successful, and now generally accepted way of doing this is by the use of the continuous symmetry group $SU(3)$. It is not possible to discuss this fully here, but the train of thought leading to it can be outlined.

The concept of charge independence has been identified with the invariance of the strong interactions under rotations in isotopic spin space. It can also be identified with invariance under unitary transformations in a space of two complex dimensions; these transformations form a continuous group known as $SU(2)$ which is mathematically equivalent to the group of three-dimensional rotations. The extension of this idea to transformations in three dimensions leads to the group $SU(3)$. Using this group, multiplets of particles can be found (corresponding to multiplets of given isotopic spin I in the $SU(2)$ case), which are differentiated by two quantum numbers (instead of the one I_3 in the $SU(2)$ case). These two can be identified with I_3 and the hypercharge Y.

We need only consider three types of multiplet, the singlet, octet, and decuplet, the members of which have the following properties:

Singlet:

$$Y = 0,\ I = 0,\ I_3 = 0$$

Octet:

$$Y = 0,\ I = 1,\ I_3 = \pm 1, 0;\ Y = 0,\ I = 0, I_3 = 0$$

$$Y = \pm 1,\ I = \tfrac{1}{2},\ I_3 = \pm \tfrac{1}{2}.$$

Decuplet:

$$Y = +1, I = {}^3/_2, I_3 = \pm {}^3/_2, \pm {}^1/_2;\ Y = 0, I = 1,$$

$$I_3 = \pm 1;$$

$$Y = -1,\ I = {}^1/_2,\ I_3 = \pm {}^1/_2;\ Y = -2,\ I = 0,$$

$$I_3 = 0.$$

Many of the particles in the tables can easily be seen to fall into octets and decuplets. Thus for the stable mesons, there is an octet:

$$Y = 0,\ I = 1;\ \pi$$

$$Y = 0,\ I = 0;\ \eta$$

$$Y = \pm 1,\ I = \tfrac{1}{2};\ K,\ \overline{K}.$$

The stable baryons belong to another octet:

$$Y = 0,\ I = 1;\ \Sigma$$

$$Y = 0,\ I = 0;\ \Lambda$$

$$Y = 1,\ I = \tfrac{1}{2};\ n, p$$

$$Y = -1,\ I = \tfrac{1}{2};\ \Xi$$

The mass differences in these cases are quite large, especially for the mesons. However, if it is assumed that the symmetry-breaking interaction itself has the octet symmetry, simple formulae relating the masses can be found:

$$m_K^2 = \tfrac{3}{4} m_\eta^2 + \tfrac{1}{4} m_\pi^2$$

$$m_N + m_\Xi = \tfrac{1}{2}(m_\Sigma + 3m_\Lambda).$$

These are satisfied quite well experimentally.

Among the unstable mesons, an octet of 1 — mesons can be found:

$$Y = 0,\ I = 1;\ \varrho$$

$$Y = 0,\ I = 0;\ \omega \text{ or } \varphi$$

$$Y = \pm 1,\ I = \tfrac{1}{2};\ K^*,\ \overline{K^*}.$$

There are two candidates for the $Y = I = 0$ member; the mass formula predicts a mass intermediate between them.

A decuplet of unstable baryons is known:

$$Y = 1,\ I = {}^3/_2 \quad N^*(1236)$$

$$Y = 0,\ I = 1 \quad Y_1^*(1388)$$

$$Y = -1,\ I = {}^1/_2 \quad \Xi^*(1530)$$

$$Y = -2,\ I = 0 \quad \Omega^-(1675).$$

The mass formula here predicts equal spacing for the masses; this is observed, and was in fact used to predict the existence of the Ω particle. A formula for the electromagnetic mass differences of the baryons can also be obtained, and is well confirmed experimentally.

Other multiplets have been suggested, and relations between coupling constants and reaction rates can also be obtained.

See also: Weak interactions.

Bibliography

HAMILTON J. (1959) *Theory of Elementary Particles.* Oxford: The University Press.

KÄLLEN G. (1964) *Elementary Particle Physics*, New York: Addison-Wesley.

SAKURAI J. J. (1964) *Invariance Principles and Elementary Particles*, Princeton: The University Press.

J. K. PERRING

SUPERCONDUCTORS IN POWER ENGINEERING. *The need for superconductors.* The economics of the electrical power industry are dominated by the capital cost, efficiency and reliability of plant; this is true both for the generation and distribution and the utilization of electrical power. Since the second world war there has been a dramatic increase in the rating of single units of plant; turbo-generators, for example have grown from 30 MW to 500 MW and consideration is being given to the construction of 1000 MW units. One of the factors which has allowed such good progress to be made is more efficient cooling; the use of hydrogen gas and water has enabled the ratings to be increased with nothing like the same increase in physical size. The result has been a substantial reduction in capital cost per kilowatt of plant and this has had a marked effect upon the stability of the cost of electricity. How far can the increase in units ratings be pushed, not only for turbo-generators but also for transformers, transmission lines, cables, large motors, etc? The answer to this question is elusive but, relative to the present context, two observations may be made. The weight of equipment is already at the limit for the capacity of the works of some manufacturers and this factor together with the problems of transport, site handling and foundations, must enter the cost equation. The second point is the extent to which conventional cooling techniques may be extended to deal with the losses in larger items of plant; there are a number of factors which must be taken into account in deciding, for example, upon the optimum current density in a machine winding. There may be a need to keep the depth of a slot in an iron core to a minimum, there is a limitation on the water velocity in a hollow conductor because of erosion effects, the power loss in an underground cable is limited by the thermal capacity of the ground and so on.

An alternative approach to the problem of power loss and space limitation is to reduce the losses by reducing the resistance of conductors. Figure 1 illustrates how the resistivity of copper falls when the operating temperature is reduced; the ordinate is shown as the ratio of the resistivity at 20°C (293°K) to

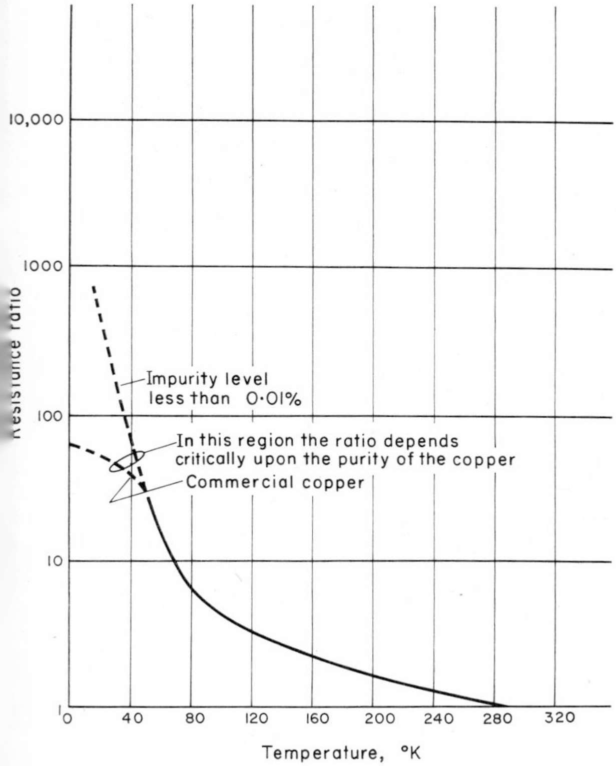

Fig. 1. Resistance ratio $\frac{\varrho_{293}}{\varrho_{T°K}}$ *of copper at low temperatures.*

that at a temperature T°K. The most convenient temperatures to consider for practical applications are 77°K, 22°K and 4·2°K, which correspond to the boiling points of nitrogen, hydrogen and helium. In general it is not economical to use the latter for cooling normal conductors and it is reserved for use with superconductors. A considerable amount of research and development is in progress to employ cryogenic techniques in electrical power equipment, two examples being a transformer with aluminium windings cooled by liquid hydrogen and an underground cable cooled by liquid nitrogen. Let us now consider why the need for superconductors still exists even though practical designs for electrical equipment may be produced using cryogenic techniques.

The power loss in a conductor is ϱJ^2 watts per unit volume where ϱ is the resistivity and J is the current density. Taking a volume of 0·1 m^3 we may draw up a table of power consumption as a function of current density and operating temperature. Further we may take into account the power required by a refrigerator/liquefier to remove the heat dissipated in the conductor at a given operating temperature assuming that only the latent heat of vaporization is added. It will be assumed that cooling water is obtained free of charge and, for the moment, we will neglect the capital cost of the refrigerating system.

The Table (p. 396) gives the results of some calculations of the heating and cooling of copper where P_1 is the power dissipated in the copper and P_2 is the power required by the liquefier.

The values of ϱ are approximate because they depend very much upon the purity of the copper to a large extent.

The Table shows that there is a definite minimum in the total power consumption and that this occurs in the region of the temperature of liquid hydrogen. In the complete analysis the cost of the refrigeration plant must be taken into account together with the cost of the cryostat which is necessary to contain the winding.

The conclusion reached from detailed studies of this nature is that for the special problems which occur for specific applications, such as underground cables in densely populated regions, a good case can be made for a cryogenic installation. However, in general the overall economics are not in favour of cryogenic systems and some other approach is required to solve the problems which arise in the design of large electrical plant. The superconductor provides the means whereby substantial advances may be made in design technique and the present status of the superconductor as an engineering material will now be discussed.

The present status of superconductors. The discovery of superconductivity and the theories which have been formulated to describe the behaviour of superconductors are well covered in the literature. There are some two dozen metallic elements which form a

Coolant	Temperature (°K)	ϱ (ohm-m)	J (amp/m²)	P_1 (kW)	P_2 (kW)	$P_1 + P_2$ (kW)
H_2O	350	$2 \cdot 10^{-8}$	$2 \cdot 10^6$	8·0	0	8·0
			$5 \cdot 10^6$	50·0	0	50·0
			10^7	200·0	0	200·0
N_2	77	$2 \cdot 10^{-9}$	$2 \cdot 10^6$	0·8	10	10·8
			$5 \cdot 10^6$	5·0	60	65·0
			10^7	20·0	240	260·0
H_2	22	$3 \cdot 10^{-11}$	$2 \cdot 10^6$	0·024	5	~5
			$5 \cdot 10^6$	0·150	30	~30
			10^7	0·600	120	~120
H	4.2	10^{-11}	$2 \cdot 10^6$	0·008	10	~10
			$5 \cdot 10^6$	0·050	60	~60
			10^7	0·200	240	~240

class known as Type I (or soft) superconductors; these elements, of which lead and niobium are the most important, are diamagnetic when in the superconducting state. In other words they exclude all flux from the body of the material and the current flows only on the surface. A second class of superconductors (called Type II or hard) are composed of alloys and compounds and are very numerous. The Type II superconductors, of which the alloys niobium-zirconium and niobium-titanium and the intermetallic compound niobium-tin are the most important at present, are not diamagnetic when operating in high magnetic fields. The penetration of flux and current into Type II superconductors has important consequences on their performance; the most important of these being that they are unsuitable for alternating currents but, unlike Type I superconductors, they will operate in very high magnetic fields.

The phenomenon of superconductivity occurs when the temperature is reduced below a critical value, typically about 10°K, and will persist providing that the magnetic field and current density associated with its operation remain within certain limits. The values of critical temperature, magnetic field and current density and their interdependence are different for each superconducting material.

For a Type I superconductor, the relationship between the magnetic field, below which superconductivity can exist, and temperature may be stated approximately:

$$B_c = B_0 \left[1 - \left(\frac{T}{T_c}\right)^2\right]$$

and the values of B_0 and T_c for some superconducting elements are given in the following table.

Element	T_c (°K)	B_0 (W/m²)
Indium	3·4	0·0283
Lead	7·2	0·0803
Niobium	9·5	0·1950
Tantalum	4·5	0·0830

The low maximum magnetic field of Type I superconductors is a severe limitation on their application; in general this class of material is not of great interest to the electrical design engineer.

For d.c. applications, Type II superconductors will operate at very high magnetic field strengths and Fig. 2 shows some characteristic performance curves for Nb-Zr and Nb-Ti. These curves show the rela-

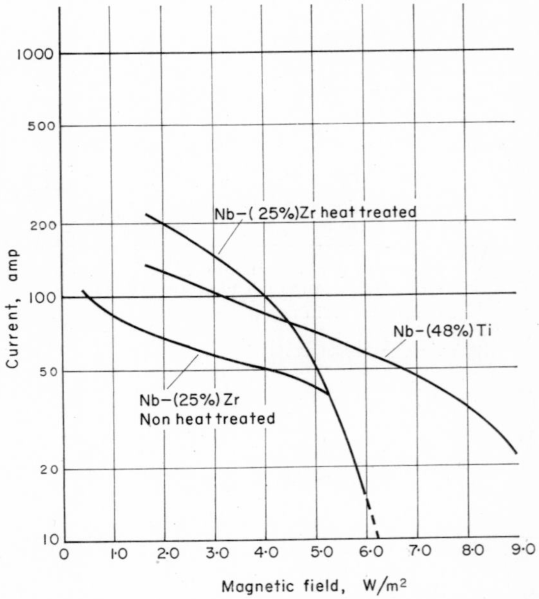

Fig. 2. Typical short sample characteristics for copper plated superconducting wires 0·010 in. diameter.

tionship between current and maximum magnetic fields for short lengths of superconductors at 4·2°K. Points on these curves are obtained experimentally by placing the superconductor in a magnetic field and increasing the current until the material reverts to the normal (non-superconducting) state. Figure 3 shows how the sample length of superconductor is arranged in the magnetic field; the most severe operating condition is at the bottom of the sample where the current and magnetic field are in quadrature. Figure 4 is a photograph showing a typical form of apparatus employed for obtaining the curves of Fig. 2, which are known as the short sample characteristics. There is no precisely defined characteristic

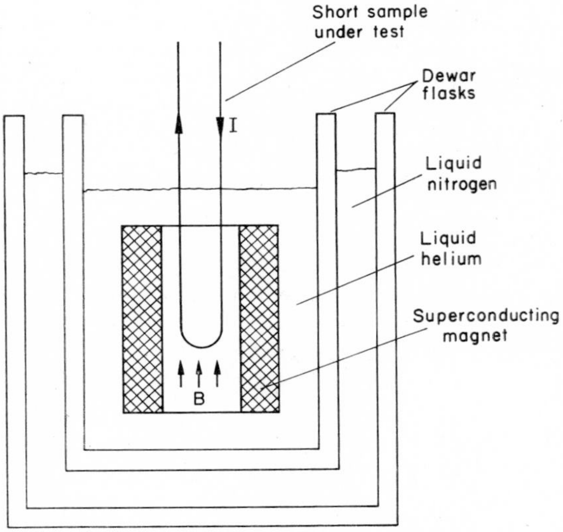

Fig. 3. Diagrammatical representation of apparatus for obtaining short sample characteristics of superconducting wires.

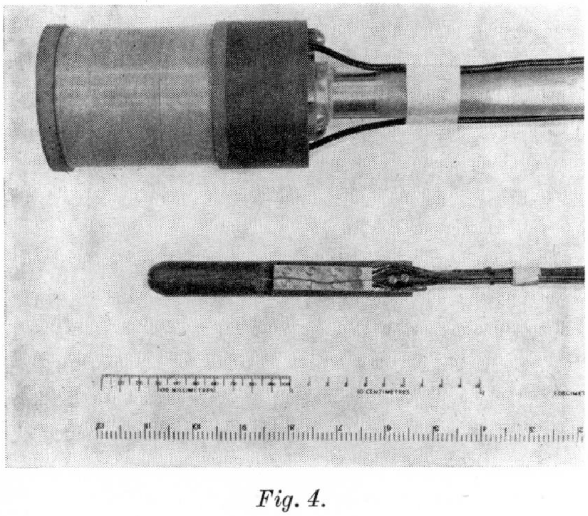

Fig. 4.

for a given type of superconductor; the short sample characteristics may vary considerably depending upon how the raw material was prepared, the level of impurities, the amount of cold working received, what form of heat treatment was applied and so on.

Now, if a magnet coil is wound with a superconductor for which the short sample characteristic is known, a test, when the coil is cooled to the temperature of liquid helium, will reveal a number of alarming facts. It will be discovered that if it is attempted to increase the current too rapidly the coil will quench, that is, it will revert to the normal state. Also it will be found that, however slowly the current is increased, the coil will not reach the design performance if this was based upon the short sample characteristic. The cause of these effects is associated with the manner in which the magnetic flux penetrates into the superconductor. Consider the situation where some current is flowing in the body of the superconductor and where, therefore, some flux has already penetrated; if the current is now increased more flux must pass into the superconductor and it must cross the current path. To achieve the latter some work must be performed and heat will be dissipated, if a total flux of Φ webers crosses a current i amp, the total work is Φi joules. The rate of working depends, of course, upon the velocity with which the flux moves.

Now if the flux moved gradually into the superconductor it is possible that the heat could be transferred into the surrounding helium bath without raising the temperature of the superconductor above its critical value. However, this is not always the case and the flux moves erratically through the superconductor; it tends to become attached or "pinned" at metallurgical dislocations in the material (called pinning centres) until, with an increase in flux level, it suddenly moves to another pinning centre. The sudden dissipation of energy caused by these flux jumps is often sufficient to raise the temperature above the critical value and, in a small region, superconductivity vanishes. The small region now has a high electrical resistance and the current quickly raises the temperature and the normal region propagates until the winding is quenched; the reason for the degraded performance of the coil is now apparent. Since the superconductor cannot be made perfectly homogeneous it is better to provide as many metallurgical dislocations as possible so that the distance between pinning centres is reduced and the magnitude of the energy dissipated by the flux jump is reduced.

The magnitude of a critical avalanche of flux depends upon a large number of factors including the size of the region in which it occurs relative to the diameter of the wire, the specific heat, the critical temperature, the current flowing in the wire and the rate at which heat is dispersed in the wire and its surroundings. The phenomenon is a transient and must be accommodated by restricting the temperature rise and decreasing the proportionate area of the normal region generated. Figure 5 shows the effect on the short sample characteristic of copper plating the superconductor and the explanation of the increased performance is as follows. Liquid helium has a low latent heat of vaporization and if the heat flux through the surface of the superconductor is greater than

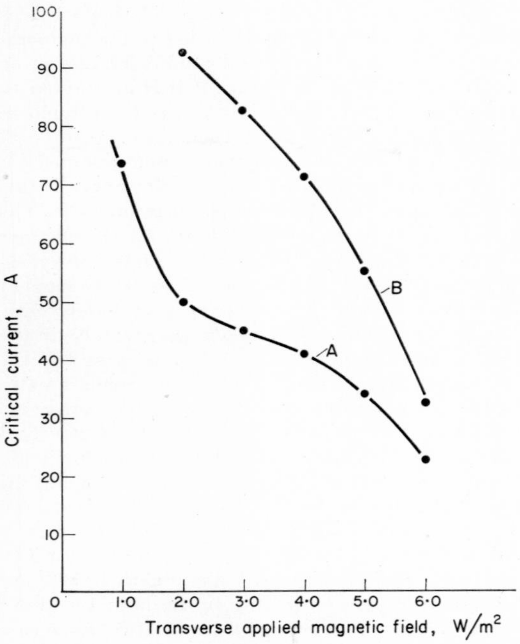

Fig. 5. Short sample characteristics of Nb-(25%) Zr wires 0·010 in. diameter (a) bare wire; (b) with 0·001 in. electrodeposited copper.

about 0·5 W/cm², film boiling occurs and a high temperature difference results. The effect of the copper is to disperse the heat over a large area to reduce the heat flux due to the affected region and the heat transfer stays in the nucleate boiling regime where the wire is at almost the same temperature as the liquid helium. The proportionate area of the normal region may be reduced by placing a number of wires in parallel. The two steps described, i.e. copper plating and using a number of superconductors in parallel leads to the next advance which has been made, the fully stabilized superconductor.

Consider a superconducting wire in good thermal and electrical contact with a substantial quantity of normal material. The current is flowing in the superconductor until some perturbation in field and current causes a region of the superconductor to become normal. The current is able to move out of the superconductor into the copper and back into the superconductor and thus by-pass the normal region of the superconductor.

The detailed calculations for producing a fully stabilized superconductor are well covered in the literature (Stekly and Zar 1965) but a brief outline of the procedure is given below.

The limiting stable current for a copper clad superconductor, for which any small normal region in the winding will collapse spontaneously when the perturbation is removed is determined by equating the heat generated per unit length of stabilized superconductor to the heat transferred to the liquid helium:

$$J^2 \varrho A = qP\alpha.$$

In this expression J is the current density in the normal material, e.g. copper, ϱ is the resistivity and A is the cross-sectional area of the normal material. The parameter P is the perimeter of the heat transfer surface and α is the fraction of the surface which is cooled. The heat flux, q, is a function of the temperature difference between the conductor and the helium bath and also of the mode of heat transfer. The value of q depends, for example, upon whether there is nucleate or film boiling, whether there is forced cooling and upon the overall geometry. Normalizing the relationship between A and P by putting

$$k = \frac{P\alpha}{\sqrt{A}}$$

and noting that the current in the conductor, I, is equal to JA we obtain

$$J = \left(\frac{qk}{\varrho}\right)^{2/3} I^{-1/3}$$

and therefore the maximum current density, Jm, for stability is given by

$$Jm < \left(\frac{qk}{p}\right)^{2/3} I^{-1/3}.$$

It must be noted that is a simplified solution since in addition to an analysis of the value of q in relation to the critical temperature of the superconductor and mode of cooling, the resistivity of the normal material will change according to the magnetic field strength (magnetoresistance effect). It should also be noted that the critical temperature of the superconductor is a function of the magnetic field strength. However, these effects do not affect the nature of the solution and the engineer is well trained to the iterative type of solution which is the most expedient for taking everything into account.

The value of Jm obtained from the above calculation provides the critical current, I_c, for the superconducting wire (or wires, since it is advantageous to put a number of thin wires, typically 0·010″ diameter, into a composite conductor). If the critical current is exceeded the normal region in the superconductor will persist and if it is increased to a value I_P, the normal region will propagate along the superconductor. It may be shown (Williams 1965) that I_P, the minimum propagating current is related to the critical current I_c by the following expression:

$$\frac{I_P}{I_c} = \frac{A}{2\alpha}\left\{-1 + \sqrt{\left(1 + \frac{8a}{A}\right)}\right\}$$

where a is the cross-sectional area of the superconductor.

For a current between the values I_P and I_c a voltage will appear across the ends of the winding because of the energy dissipation which is occuring. A classical experiment (Stekly and Zar 1965) was performed, to illustrate the stability of a superconductor, in which the voltage across a superconducting winding was measured at different current values. When I_c was exceeded a voltage appeared and remained under steady state conditions; the voltage disappeared when the current was reduced.

Another important aspect of the design of superconducting magnets is protection against the sudden release of the stored inductive energy. As an introduction to this topic, it is useful to discuss the various methods of operating a superconducting magnet. When the current in a magnet coil has reached its operating value no further supply of energy is required and it is quite feasible to short-circuit the winding with a superconducting link and the current will persist in complete isolation. To achieve this result the shorting link, known as the persistent current switch, is made non-superconducting during charging by energizing a small heater which maintains the temperature above the critical value. At the required coil current the heater is de-energized and the, then superconducting, link takes the current and the coil current supply may be disconnected. An alternative method is to provide a short circuit to the winding using normal material, e.g. copper; the resistance of the switch may be made very small but, of course, there will be energy dissipation and the current will decay albeit with a very long time constant. A third method is not to disconnect the power supply at all but allow it to feed a steady current into the winding.

The stability for a superconductor will be destroyed if the supply of liquid helium is lost; for example a cooling channel may become blocked and cause a region of the winding to become warm. The resistivity of a superconductor in the normal state (i.e. non-superconducting) is very high depending upon the temperatures but typically about 20×10^{-8} ohm-metres. The resistivity of copper at low temperatures is very dependent upon the level of impurities which it contains but may be of the order of 2×10^{-6} ohm metres and is certainly two orders of magnitude lower than the "normal" superconductor. If a warm spot develops in a non-stabilized superconductor the heating is intense and the normal region quickly propagates throughout the winding; depending upon the size of the magnet and its operating conditions the superconducting winding could be severely damaged. A warm spot in a stabilized superconductor, however, results in the current being carried in the stabilizing material which we have assumed to be copper. If cooling is lost the copper will warm up and for unit mass the energy balance is given by

$$J^2 \varrho \, dt = s\delta \, d\theta .$$

Clearly the current must be made to decay as rapidly as possible, firstly, by removing the power supply, if this is still connected and secondly by causing the current to flow in a resistor external to the winding. Let us assume, for the moment, that we have achieved these two objectives; the energy source is that stored inductively ($\frac{1}{2}Li^2$) and the dissipation is in the affected region of the winding and the external resistance, R. The value of R must not be so large that the voltage across the winding exceeds the electrical insulation level, but it must be sufficient to prevent the temperature of the winding rising above a specified or optimized level. The analysis to determine optimum current density in the copper, maximum voltage level, winding temperature rise, etc. is complex but reasonably straightforward for a given shape and size of magnet.

A disadvantage of the persistent current switch method of operating is now apparent; to put an external resistor in series with the winding, the switch heater must first be energized and the switch link must then develop sufficient voltage to drive the current into the external resistor. The external switch method may be employed in a reasonable manner by placing the resistor across the switch contacts; however, the use of a dissipative switch element under healthy operating conditions is not desirable for many applications. A suitable method of operating the magnet is shown in Fig. 6 and consists of a power supply which is permanently connected to the winding

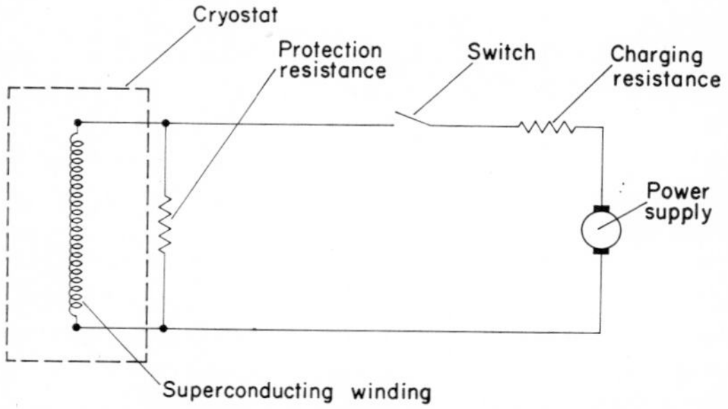

Fig. 6. Method of protecting a superconducting winding.

through a switch and an external resistor permanently connected to the winding terminals. With the advent of a fault condition the switch is opened and the current is forced into the resistor. Figure 7 shows the voltage obtained from the terminals of a superconducting winding with a 1 ohm resistor across the terminal when the switch was opened. The coil inductance was two henries and the current was 200 As;

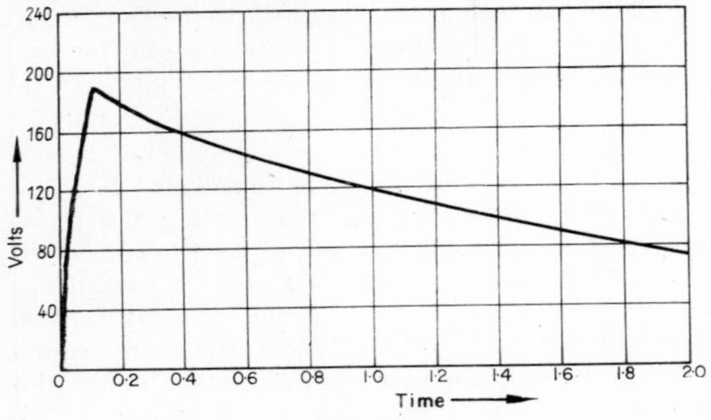

Fig. 7.

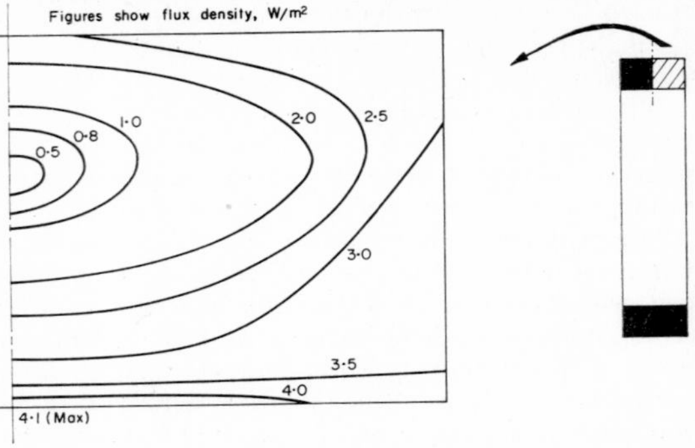

Fig. 8. Magnetic flux distribution inside a superconducting winding.

it is seen that the maximum voltage is nearly 200 V. Very little liquid helium boil off resulted to show that the protection system worked efficiently.

Applications. The development of the stabilized d.c superconductor enables a magnet to be designed to give a guaranteed performance. Consider the design of a winding of 2 metres mean diameter with a maximum magnetic field strength of 4·0 W/m^2; suitable parameters are $1{\cdot}9 \times 10^6$ amp turns and a coil cross-section of 0·27 m long by 0·135 m radial width. These figures give an overall current density of 5200 amp/cm^2. The magnetic field distribution through the winding, Fig. 8, shows the position of near zero magnetic field and the equipotentials. Of particular interest is the contour for 3·5 W/m^2; since this is reasonably flat it is possible to consider a change in the current density of the superconductor. All the turns of the winding are in series and consist of a number of strands of superconductor in a matrix of copper. Assuming that the current per turn is 500 A and that the superconductor is niobium-zirconium, the short sample characteristic for wires of 0·010 in. diameter shows that at 4·1 W/m^2 seven wires are required but at 3·5 W/m^2 only five wires are necessary. A saving in the quantity of superconductor may

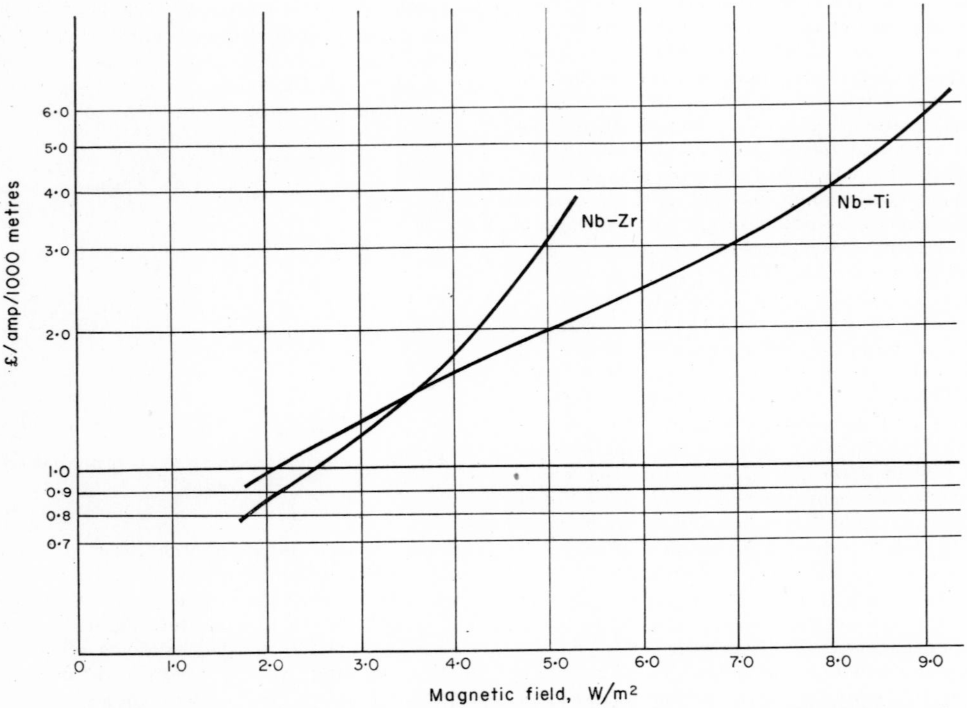

Fig. 9. Typical curves of the cost of superconductors as a function of magnetic field strength.

thus be achieved by reducing the number of superconducting wires in the composite conductor after part of the coil has been wound.

The cost of a superconductor is most conveniently expressed in £ per amp per 1000 m (or some other length) and Fig. 9 shows some typical curves of this parameter as a function of magnetic field strength. The penalty which is incurred in designing for higher magnetic fields is clearly shown up in Fig. 9. Consider, however, the enormous advantages to be obtained by using superconductors; a large winding with a magnetic field of 6·0 W/m^2 would require of the order of 30 MW of power if wound with normal materials but with a superconducting winding the power requirement is only about 15 kW for the helium refrigerator.

At the present time large magnets are being designed for MHD (magnetohydrodynamic) methods of direct energy conversion; in this system of power generation the electrical output is proportional to the square of the magnetic field strength (for closed cycle MHD systems the output may depend upon a higher power of the field strength) and it will be necessary to employ superconductors to produce an economically viable design. In the field of high energy nuclear particle research, superconducting magnets are being designed for bubble chambers and beam bending. A major advance in the application of superconductors was announced in February, 1968, by International Research & Development Co. Ltd., Newcastle upon Tyne; this company has tested a small superconducting homopolar motor and is proceeding to construct a motor of 3250 H.P. at 200 r. p. m. This machine will employ $5\frac{1}{4}$ tons of niobium-titanium copper composite superconductor and the economics of this new class of machine are very attractive. It is estimated that large slow speed D. C. motors as required, for example, in a steel rolling mill, will be substantially cheaper than conventional motors even when the cost of the refrigerator is taken into account. The power to weight ratio is also greatly improved and this development is the first major industrial application for superconductors. In the U.S.A. a small superconducting alternator has been constructed by Dynatech Corporation of Boston which employs niobium-tin tape but full details of the A. C. losses have not been revealed. The development of an A. C. superconducting power cable was announced in 1967, by B. I. C. C. Ltd.; this employs niobium on a substrate of aluminium on nickel-iron; the current carrying capacity of the superconductor in the form of a thin walled tube is about 375 A/cm and the A. C. loss is about $5 \mu W/cm^2$.

It is clear that the application of superconductors to electrical power equipment is receiving serious attention by industry and the state of the art is developing rapidly. The situation in respect of alternating current applications, particularly where a high working level of magnetic field is required, is not as advanced as for direct current equipment but there is no doubt that substantial progress will be achieved over the next few years.

In giving consideration to the use of superconductors it is vitally important to adopt the systems approach, i.e. a superconducting winding may not be considered in isolation from the design of the associated cryogenics. Furthermore, an individual piece of superconducting machinery should not be considered out of the context of the system in which it is going to operate; it may, for example, be impossible to justify the cost of a single superconducting motor or generator but taken in the context of several machines located together perhaps with a superconducting cable or transformer, there may well be a substantial cost incentive.

Bibliography

Thewlis J. (Ed.) (1962, 1966) *Encyclopaedic Dictionary of Physics*, articles beginning "Superconducting", "Superconductors", 7, Suppl. Vol. 1, Oxford: Pergamon Press.

A. D. Appleton

T

THERMAL CONDUCTION IN SEMICONDUCTORS. Studies of the conduction of heat in semiconductors have proved to be exceptionally interesting for a number of reasons. In the first place, some semiconductors have been prepared in the form of large single crystals having particularly small impurity concentrations; the processes of heat conduction in such materials are much less complex than in ordinary substances. Secondly, there are mechanisms of heat transfer in semiconductors, associated with the presence of an energy gap, that are not found at all in metals and that are negligibly small in electrical insulators. Finally, certain semiconductors have potential application in thermoelectric energy convertors, the efficiency of which depends on the materials having high thermal resistivities. There has, thus, been an intensive search for semiconductors of low thermal conductivity, as well as attempts to lower the thermal conductivity of known materials.

Lattice component of the thermal conductivity. Heat can be conducted through a solid by any system of mobile particles or waves. In semiconductors, one is principally concerned with heat transfer through the lattice vibrations (as in electrical insulators) and through the quasi-free electrons (as in a metal). Thus, the total thermal conductivity $\varkappa$ is the sum of a lattice component $\varkappa_L$ and an electronic component $\varkappa_e$:

$$\varkappa = \varkappa_L + \varkappa_e. \quad (1)$$

It is the lattice component that is discussed first.

Most semiconductors display an inverse variation of lattice thermal conductivity with absolute temperature T (Eucken's law), for T larger than the Debye temperature, this behaviour being typical of dielectric crystals. Eucken's law holds when scattering of phonons by other phonons in 3-phonon Umklapp processes is predominant; a stronger temperature variation associated with 4-phonon scattering processes has sometimes been invoked to explain experimental results at elevated temperatures.

At low temperatures, in pure semiconducting crystals, the mean free path of the phonons becomes limited by the specimen boundaries, and the thermal conductivity and specific heat show the same dependence on temperature (a T^3 law, or something close to it, is followed). It is interesting to note that boundary scattering effects on both the thermal conductivity and phonon-drag Seebeck coefficient have been observed on the same samples of germanium. Because of their longer wave-length, the phonons responsible for the drag effects have longer free path lengths in large crystals than have the heat-conduction phonons; thus boundary scattering produces an effect on the Seebeck coefficient at temperatures which are too high for the thermal conductivity to be affected.

A common feature seen in even the purest of semiconductor crystals, is the persistence of Eucken's law to well below the Debye temperature, whereas Peierls' theory suggests that an exponential temperature dependence of the lattice conductivity should be found in this range. The reason for this behaviour lies in the fact that even the purest elements are usually made up of a number of different isotopes each of different atomic weight; the phonon free path length becomes limited by mass-defect scattering as Umklapp scattering ceases to be effective. The validity of this explanation was demonstrated by Geballe and Hull in 1958 in an experiment in which a crystal consisting almost entirely of the isotope ^{74}Ge was compared with a chemically pure sample of germanium containing the usual mixture of isotopes. The isotopically enriched sample had more than three times the thermal conductivity of the other at temperatures below about 50°K.

Much work has been devoted to the study of the lattice thermal conductivity in solid solutions between elemental or compound semiconductors. This follows the suggestion by A. F. Ioffe that disturbances in the short-range order should lead to strong scattering of the short-wave-length phonons that are primarily responsible for the conduction of heat, whereas there should be much weaker scattering of the charge carriers in view of their much longer wave-length. The importance of this idea lies in the fact that the ratio of carrier mobility to lattice thermal conductivity is one of the overriding factors in the selection of thermoelectric materials. In certain solid solutions, e.g. Bi_2Te_3—Sb_2Te_3 and Ge—Si, the reduction of the thermal conductivity is accounted for reasonably well by the fluctuations in density arising from the different masses of the unit cells. On the other hand, in other systems, such as PbTe—PbSe and InAs—InP, the scattering of phonons is much stronger and it must be supposed that there are substantial fluctuations in the elastic constants too.

Electronic thermal conductivity in extrinsic conductors. The theory of metals shows that the electronic thermal conductivity of a degenerate conductor should be equal to $(\pi^2/3)\,(\boldsymbol{K}/\boldsymbol{e})^2\,\sigma T$, where σ is the electrical conductivity, irrespective of the nature of the scattering of the charge carriers (provided that each scattering event is effective in transferring momentum as well as energy). On the other hand, the quantity $\varkappa_e/\sigma T$, known as the Lorenz number L, becomes dependent on the scattering law for a non-degenerate system of carriers. In particular, for a quadratic density-of-states function and a scattering law of the form $\tau \propto E^\lambda$, where τ is the relaxation time, E is energy and λ is a constant, the Lorenz number of a non-degenerate semiconductor is $(5/2+\lambda)\,(\boldsymbol{K}/\boldsymbol{e})^2$. Thus, for acoustic–mode lattice scattering L is $2(\boldsymbol{k}/\boldsymbol{e})^2$, and for the Conwell–Weisskopf form of ionized impurity scattering L is $4(\boldsymbol{k}/\boldsymbol{e})^2$. The Lorenz number becomes modified in the region of partial degeneracy, depending then on both the scattering parameter and the Fermi energy.

A simple calculation shows that the electronic thermal conductivity is negligible compared with the lattice component in those semiconductors which have electrical resistivities of the order of 1 ohm-cm or more. However, when the electrical resistivity is of the order of 10^{-3} ohm-cm the two components become of comparable magnitude. It is then often important to be able to separate the two contributions to the thermal conductivity. The most generally applicable method involves the measurement of the electrical conductivity and the calculation of the Lorenz number from the scattering parameter λ (obtained from the variation of carrier mobility with temperature, or from the longitudinal or transverse Nernst coefficients) and the Fermi energy (obtained from the Seebeck coefficient). The lattice thermal conductivity for zero doping level can be estimated from measurements of the total thermal conductivity on a range of samples having different electrical conductivities; the plot of $\varkappa$ against σ is merely extrapolated back to zero electrical conductivity. Yet another method is available for semiconductors in which the mobility μ of the charge carriers is large; then, on applying a magnetic field H that is so large that $(\mu H)^2 \gg 1$, the electronic thermal conductivity can be made to vanish.

Mixed and intrinsic semiconductors. The thermal conductivity is always defined for the condition that the flow of electric current should be zero. Thus, in an extrinsic conductor the transfer of heat is accomplished by the flow of more energetic carriers in one direction and the flow of less energetic carriers in the other direction, the overall charge flow being zero. However, if both electrons and holes are present it is possible for there to be substantial flows of each type of carrier in the same direction; because electrons and holes have opposite charges these flows need not lead to any electric current. In effect, the electrons and holes transport their potential energy, whereas if one type of carrier is present only the kinetic energy can be carried.

The thermal conductivity of a mixed semiconductor is easily found by solving equations for the electric and thermal currents in terms of the gradient of the Fermi energy and the temperature, setting the flows of electrons and holes equal to one another. It is found that

$$\varkappa_e = \varkappa_n + \varkappa_p + \frac{\sigma_n \sigma_p}{(\sigma_n + \sigma_p)}\, T(\alpha_p - \alpha_n)^2 \qquad (2)$$

where α represents Seebeck coefficient and the subscripts n and p refer to partial coefficients for electrons and holes respectively. Some idea of the magnitude of the third term on the right-hand side of equation 2 can be obtained by assuming non-degenerate statistics to be applicable and by supposing that the scattering parameter λ is equal for both electrons and holes. Then

$$\varkappa_e = \left\{\left(\lambda + \frac{5}{2}\right) + \frac{\sigma_n \sigma_p}{(\sigma_n + \sigma_p)^2}\left(\frac{E_g}{\boldsymbol{k}T} + 5 + 2\lambda\right)^2\right\}\left(\frac{\boldsymbol{k}}{\boldsymbol{e}}\right)^2 \sigma T \qquad (3)$$

where E_g is the energy gap. For the particular case of an intrinsic semiconductor with comparable electron and hole mobilities and with a scattering parameter λ equal to $-\frac{1}{2}$, the Lorenz number becomes $\{2 + (E_g/\boldsymbol{k}T + 4)^2/4\}\,(\boldsymbol{K}/\boldsymbol{e})^2$ which can, of course, be very much larger than the value given by the Wiedemann–Franz law. For example, bismuth telluride has an energy gap equal to about 6 $\boldsymbol{k}T$ at room temperature leading to a Lorenz number, according to equation 3, of about $27(\boldsymbol{K}/\boldsymbol{e})^2$; the high electronic thermal conductivity corresponding to this value of L has indeed been observed for intrinsic samples of the compound.

The figure shows a schematic plot of thermal conductivity against electrical conductivity for a semiconductor. The large Lorenz number in the intrinsic and mixed samples is shown in region (1). The constant Lorenz number in the extrinsic non-degenerate range (2) depends on the value of the scattering parameter, whereas in the fully degenerate region (4) L is constant but independent of the scattering law. Region (3) represents the transition from non-degenerate statistics. It is supposed that the lattice thermal conductivity shows some dependence on the doping level.

Other mechanisms of heat transfer. The mechanisms of heat transfer, that have been discussed so far, by no means exhaust the possibilities that exist. For example, excitons (electron-hole pairs that can move through the lattice although remaining bound to one another) can transport their energy of formation in much the same way as quasi-free electron-hole pairs.

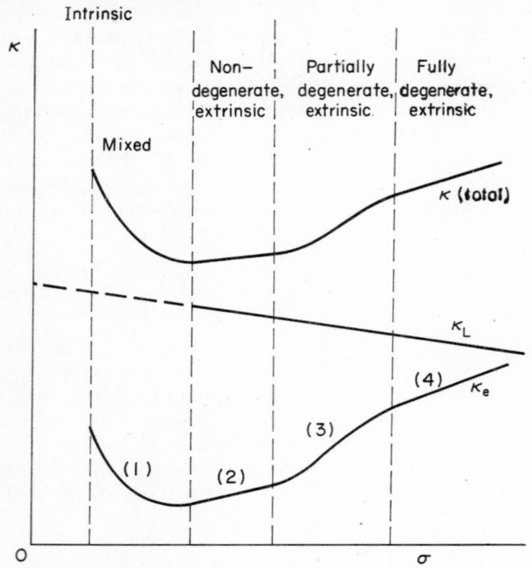

Schematic plot of thermal conductivity against electrical conductivity at a constant temperature for variously doped samples of a given semi-conductor.

However, heat conduction by excitons remains to be demonstrated conclusively in any particular case. There is rather more evidence for heat transfer by radiation in pure semiconductors; it must be remembered that absorption of photons of energy less than that of the band gap is small, provided that there are not too many free carriers. Genzel has shown that the effective increase in thermal conductivity by radiative transfer is

$$\varkappa_R = \frac{16}{3}\,\frac{n^2 s T^3}{a}, \tag{4}$$

where s is Stefan's constant, n is the refractive index and a is the absorption coefficient, assuming that a single value for this quantity can be defined. The application of this equation is complicated by the fact that the absorption coefficient is quite different for photons having energies greater than or less than the gap width. Nevertheless, Devyatkova has demonstrated the validity of Genzel's formula for pure tellurium.

Techniques of measurement. Most of the methods of measurement of thermal conductivity that are used for metals or insulators are applicable for semiconductors. Here, mention will be made only of one or two techniques that have special relevance to semiconducting materials.

At elevated temperatures the thermal conductivity of a semiconductor is often determined from the thermal diffusivity which is itself measured by Angström's method. It will be recalled that, in this method, temperature waves are passed along the sample, and the relative amplitude and phase of the wave are found as a function of distance. The method is important since it allows the surface losses (which become greater at high temperatures) to be eliminated. One difficulty lies in the attachment of thermometers to the sample but, if the sample is a semiconductor, it can be used as its own thermometer. The semiconductor itself forms one branch of a thermocouple, the other branch being a thin metallic wire joined to the surface.

If a semiconductor has a reasonably large thermoelectric figure of merit it is possible to utilize *Harman's* elegant *method* of measuring its thermal conductivity. In this method an electric current I is passed through a sample which is suspended in an adiabatic enclosure. A temperature difference ΔT is established by the Peltier effect; at equilibrium

$$\pi I = K \Delta T \tag{5}$$

where π is the Peltier coefficient and K is the thermal conductance. The Peltier coefficient (which, according to Kelvin's laws is equal to αT) is found either in a separate measurement or from the difference between the adiabatic and isothermal electric conductivities. Because no heat source is needed in Harman's method, the losses are smaller than in a conventional method.

Bibliography

APPEL J. (1960) Thermal Conductivity of Semiconductors, *Progress in Semiconductors*, **5**, 141.

COEKIN J. A. (1966) in *Encyclopaedic Dictionary of Physics* (J. Thewlis Ed.), Suppl. Vol. 1, 286, Oxford: Pergamon Press.

DRABBLE J. R. and GOLDSMID H. J. (1961) *Thermal Conduction in Semiconductors*, Oxford: Pergamon Press.

HARRIS D. J. (1962) in *Encyclopaedic Dictionary of Physics* (J. Thewlis Ed.), **6**, 447, Oxford: Pergamon Press.

IOFFE A. F. (1956) Heat Transfer in Semiconductors, *Canad. J. Phys.*, **34**, 1342.

H. J. GOLDSMID

THERMAL IMAGING. High-temperature research invariably requires the creation of an environment which is capable of reaching the desired high temperature, is free of contamination, is closely controlled, and can permit measurements and observation of the test object. Various devices have been available to create such a high-temperature environment, e.g. resistance heating, induction heating, electron bombardment, plasma heating, lasers. An alternative approach, thermal imaging—the concentration of radiant energy by means of an optical system from a source of a desired spectral distribution— has found unique applications in high-temperature research and testing, ranging from materials research to the simulation of the thermal effect of nuclear fireballs. The laboratory techniques are also useful for testing instrumentation and developing procedures applicable to large systems, such as solar furnaces and solar collectors for producing power in space.

Techniques. The usefulness of thermal imagers was already recognized at the beginning of the seventeenth century when a metal paraboloidal mirror was cast to concentrate the rays of the Sun to heat ceramic materials. Concentrators consisting of an assembly of small flat mirrors and large glass lenses were also employed in the construction of solar furnaces. Lavoisier devised a solar furnace using a lens to carry out experiments in a controlled high-temperature environment. A number of solar-powered devices requiring paraboloidal concentrators was built in the late nineteenth century.

Interest in solar furnaces was revived by the emerging interest in high-temperature phenomena. Large solar furnace installations are now in operation in France, Japan, the U.S.S.R. and the United States.

Recent technological advances have required the attainment of high temperatures to develop instruments capable of measuring such temperatures, withstanding their effects, and obtaining data on physical and chemical properties of materials at these temperatures. Although other heating techniques have been developed which can provide higher temperatures, the applications of thermal imagers became more widespread as the following unique characteristics of this technique became better known:

1. An object can be heated in a controlled atmosphere ranging from vacuum to high pressure, or the type of gas can be quickly changed while a test is in progress.
2. Only a small portion of a sample material is heated; therefore, the sample can be arranged to form its own crucible, thus eliminating undesirable reactions at high temperature between sample and container.
3. The electrical conductivity, magnetic properties, or composition of the sample do not interfere with the heating of the material.
4. Heating is carried out only by radiation; therefore, the possibility of reactions between the sample and the furnace atmosphere or the heating medium is eliminated.
5. The sample can be heated and cooled in a programmed sequence by controlling the radiation flux impinging on the sample.

In a laboratory thermal-imaging apparatus, only a small area of the sample is exposed to high temperatures; therefore, the sample is limited in size since large-temperature gradients will be set up unless an optical system is used to provide a more uniform distribution of the radiant flux. Thermal imaging techniques are particularly applicable to materials which must be heated on one surface only, and where an incremental heat input is desired. Although considerable experience has been accumulated on instrumentation and techniques, the experimenter must use ingenuity when calibrations, measurements, and other than routine tests have to be performed, and where special sample mounts, instrumentation and procedures are required.

Optical systems. The selection of an optical system for a thermal imager is governed by the characteristics of the radiation source, the required level of radiation flux, and the flux distribution required at the sample. Of course, no optical system can be devised to concentrate energy from any source so effectively that the temperature of the sample will exceed the temperature of the source. Generally, paraboloidal mirrors consisting either of a single mirror or multiple flat mirror elements have been used to concentrate solar radiation. The mirror may either track the Sun directly, or, particularly in large installations, it may be stationary and receive the Sun's radiation via one or more heliostats which track the Sun. If the mirror tracks the Sun, the sample placed at the focal point of the paraboloid is subjected to continual motion; therefore, a stationary mirror and a fixed position for the sample at the paraboloid's focal plane are preferred. With large solar concentrators, wind loads on the structure can cause vibrations and distortions, and reduce the level of the radiation flux reaching the sample. With electrically powered radiation sources, thermal imagers can be used in a great variety of optical systems. The most common optical systems employed in laboratory thermal imagers are shown in Fig. 1. The single ellipsoidal mirror causes the smallest reflection loss, with the sample receiving radiation over a small solid angle only. However, the irradiated area is magnified, thus causing a lower flux level. The image area of the two-mirror ellipsoidal system is equal to that of the source, with the sample receiving radiation over a large solid angle. The flux levels near the crossover point in this optical system are low; therefore, devices can be installed at the crossover point to modulate, chop, or filter the radiation. A watercooled flat mirror can be inserted next to the crossover point so that the optical axis of the second ellipsoidal mirror is at right angles to the optical axis of the source mirror. This arrangement is particularly useful when molten samples must be observed. However, in a two-mirror paraboloidal system, the flux cannot be easily modulated, because of the large area of the radiation beam.

When the radiation source and its image are located within the optical system, the size of the source and its support must be small enough to prevent shadowing of the radiation. The shadowing can be eliminated by a refractive optical system; however, in such a system energy is lost by absorption as it travels through the lenses. Furthermore, the angle of pickup between the source and the lens is small.

In a compound reflective optical system, two paraboloidal mirrors are mounted face to face so that the

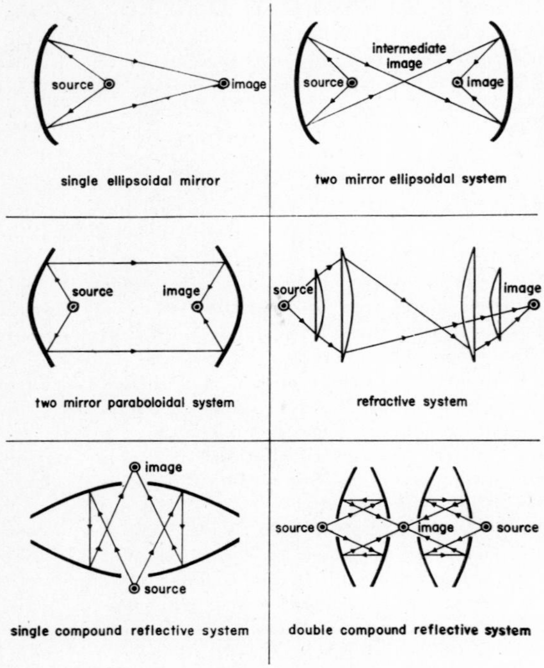

1

Fig. 1. Reflecting and refracting optical systems for thermal imagers with electrical radiation sources.

focal point of each falls at the vertex of the other. Holes at the vertices permit the radiation to enter and leave the system; the dimensions of the holes depend on the type of radiation source and on the rim diameter of the mirror. In this optical system, neither the source mechanism nor the sample casts a shadow.

A further extension of this optical system is a double compound reflective system, where the sample is placed between two sets of paraboloidal mirrors, so that it is heated at both sides. This method of heating reduces cooling by radiation because most of the sample's volume is immersed in the radiant beam.

An optical element which can be combined with any optical system to produce spatial uniformity of the radiation flux at the sample is the *flux redistributor*. Figure 2 shows a flux redistributor which consists of a square light pipe, with highly reflective inside surfaces, placed in the focal plane of the optical system. The spatially non-uniform energy which enters the flux redistributor is reflected by the inside surfaces and emerges uniformly distributed over an area equal to the cross-section of the redistributor. The overall efficiency of this optical element depends on the reflectivity of the materials used for the inside surfaces; a redistributor with high-quality specular reflecting surfaces transmits about 50 per cent of the incoming energy.

Concentrators for thermal imagers have generally been made of glass. Metal mirrors produced by stamping, spinning, or molding do not have the optical quality required. Electrolytically deposited metal mirrors are too costly for most applications. Glass mirrors are usually produced in a mold by slumping, and are subsequently ground and polished. When the back of a mirror consists of aluminized or silvered reflecting surfaces, a loss in energy is introduced. When such coatings are applied to the front surface, the energy loss is reduced; however, a coating of silicon monoxide is usually applied to protect the reflecting surface from oxidation and erosion.

Sources of radiant energy. The characteristics of the radiation source are of crucial importance in achieving the desired high temperatures; they determine both the maximum flux obtainable and the spatial distribution of this flux. The Sun provides continuous radiation flux with excellent time-independent uniformity. However, such energy is available only during

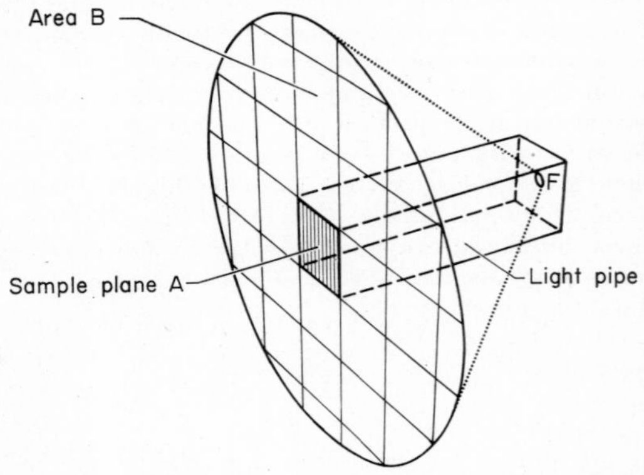

Fig. 2. Principle of Operation of flux redistributor—with redistributor radiation diverges from focal place F to illuminate area B. The redistributor superimposes the subdivided area B over its cross-sectional area A. Non-uniformities as a result of shadowing are reduced by fraction A/B.

a portion of the daylight hours, and only when the atmosphere is clear enough to reduce radiation dispersion. High-intensity carbon arcs are among the most widely used radiation sources for thermal imagers. The conventional carbon arcs have the characteristic tail flame caused by natural convection; high-intensity Gretener arcs can be operated at about 300 amperes. In the latter arc, the anode is surrounded by air jets, which concentrate the gaseous material in the form of a cylinder of brilliantly glowing gases, while also cooling the anode surfaces. Variations in electrode composition, anode creater and cathode tip configuration, and the quantity of vaporized material and solid particles in the arc gap cause high-frequency instabilities in the arc. Inaccuracies in the electrode deef mechanism and compositional changes along the electrode can cause longer-term variations. An automatic electrode feed mechanism can be used to overcome the limitations in operating time dictated by electrode length.

Plasma devices capable of generating plasma temperatures of up to 7000°K have been developed. While the radiation from a carbon arc closely resembles the radiation from a black body, infra-red radiation is greatly reduced in a plasma device. Two types of plasma devices have been developed. In one, two electrodes are aligned on the axis of the unit, and a cool gas continually flows around the electrode to create a vortex-stabilized radiation source. In the other, a ring-shaped anode surrounds the cathode to form a volume of plasma which can radiate over most of its volume.

The power inputs of plasma radiation sources, ranging from 5 to 150 kW, are useful where a large area has to be irradiated, as, for example, in a solar simulator. Where continuous operation is required (e.g. in crystal growth or zone refining), high-pressure compact arc lamps with non-consumable electrodes of tungsten operating in a xenon or mercury pressurized atmosphere are used. For example, a 10-kW d.c. xenon lamp with an arc length of 8 mm is rated for several hundreds hours of operation. In an arc lamp, the radiation has to pass through an enclosing fused-silica envelope. To increase the useful operating life of high-power input lamps, this envelope must be cooled by forced convection. Because the arc lamp radiates over its entire volume, to utilize most of the radiation an auxiliary reflector is placed behind the lamp.

Where heating of the sample is desired only for very short periods (e.g. in ignition studies), very high heat fluxes can be obtained from a high-intensity spark. The spark is powered by a capacitance-inductance circuit, and discharged in from 5 to 1000 microseconds. The radiation flux corresponding to these times varies from 250,000 to 10,000 cal/cm^2 sec.

Electrical-resistance-heated radiation sources, e.g. tungsten filaments within a lamp, or graphite resistors, have found application where the flux stability has to be very high, continuous operation is desired, and close control over the radiation flux has to be achieved. A graphite resistor is capable of providing uniformly high temperatures, up to about 2500°C, over a large area. A conical graphite resistor can be heated in air without an enclosure within a double protective cone of argon and helium; the argon forms the outer sheath, and helium forms the inner sheath to protect the carbon from oxidation. Alternatively, a pyrex hemisphere can be placed over a graphite resistor and pressurized with dry argon.

Instrumentation. In most experiments it is of utmost importance to "*characterize*" the flux reaching the sample and to measure the temperature of the sample surface. A wide variety of instruments has been developed to perform these measurements.

For steady-state measurements a commonly used instrument is the *Gardon foil radiometer* (see Fig. 3). In this radiometer, a thin circular foil of constantan (about 1 mm in diameter) is attached at its outer edge to a water-cooled block. A fine copper thermocouple wire is attached to the centre of the foil disk. The radiation falling on the disk is absorbed in a blackened layer and flows radially to the outer edge of the disk. The flux intercepted by the disk is a function of the radial temperature gradient measured by the thermocouple, which is calibrated with an absolute calorimeter.

The absolute black-body cavity calorimeter is a steady-state device which can measure high flux levels. Figure 4 shows a typical calorimeter. This calorimeter consists of a blackened water-cooled cavity which collects the radiation passing through the image area. The flux is determined by measuring the flow rate and the temperature rise of the cooling water. A calorimeter for transient flux measurements consists of a blackened copper disk with a thermocouple attached

Optical system	Radiation source	Peak flux density cal/cm^2-sec	Equivalent black body temperature °K	Duty cycle
Compound reflective	Resistance heated carbon temperature—2350°K	20	1950	$>$ 100 hr
Compound reflective	Blown arc—300 amp	150	3400	5 min
Double paraboloid	Blown arc—500 amp	300	3900	5 min
Double ellipsoid	Blown arc—153 amp	350	4000	20 min
Single paraboloid	Sun	350	4000	6 hr
Double paraboloid	10 kW Xenon compact arc lamp	420	4200	$<$ 100 hr
Double ellipsoid	Blown arc—300 amp	456	4380	5 min
Double paraboloid	Pressurized blown arc 30 psi—1300 amp	650	4700	5 sec

to measure the temperature rise as the radiation is absorbed by the disk. From the temperature rise, the mass of the disk, and its surface absorptivity, the flux can be calculated after corrections for re-radiation by areas of the disks not exposed to radiation and after correction for heat loss from the disk by convection.

A radiation probe consisting of a quartz *light-pipe* within a small stainless steel tube can be used to scan the flux distribution over the image area. The sensing element is a photomultiplier tube placed at the end of the light pipe and calibrated at known flux levels. Filters can be interposed to provide information on the spectral distribution of the radiation. A qualitative indication of the flux distribution over the image area can be obtained with a chemically treated paper which discolours on heating.

The temperature of the sample is usually measured by optical means because it exceeds the limitations of thermocouple alloys. To measure sample surface temperatures, the reflected radiation must be separated from the radiation emitted by the surface. To measure the emitted radiation only, it is necessary either to interrupt the incident radiation with a chopper, or place in its path a series of filters to exclude a portion of the radiation spectrum.

Figure 5 shows a pyrometer which combined with a chopper alternately measures the radiation emitted from a point on the sample, the reflected radiation from the same point, and which can be used to measure the radiation illuminating the sample at that point. To measure emission and reflectance of the sample, the instrument first measures emitted plus reflected light. Then the radiation illuminating the sample is briefly interrupted so that only the radiation emitted by the sample is measured. The difference between the emitted plus reflected radiation and the emitted radiation is used to calculate the reflected radiation. The measurement of the incident radiation is based on the reflectance of a "white sample" placed at the sample position. A stable reference lamp checks drifts in the instrument.

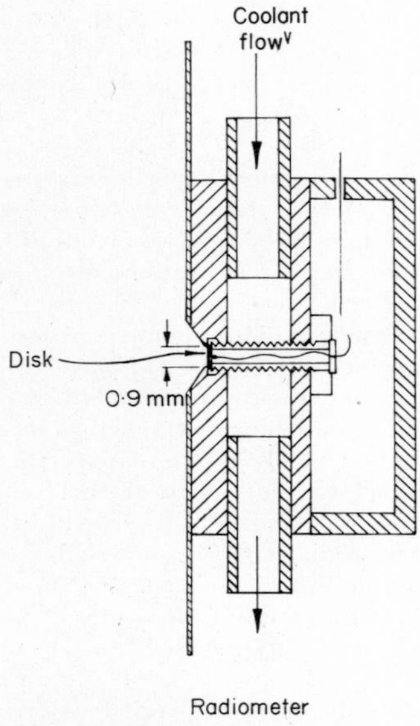

Fig. 3. Gardon foil radiometer.

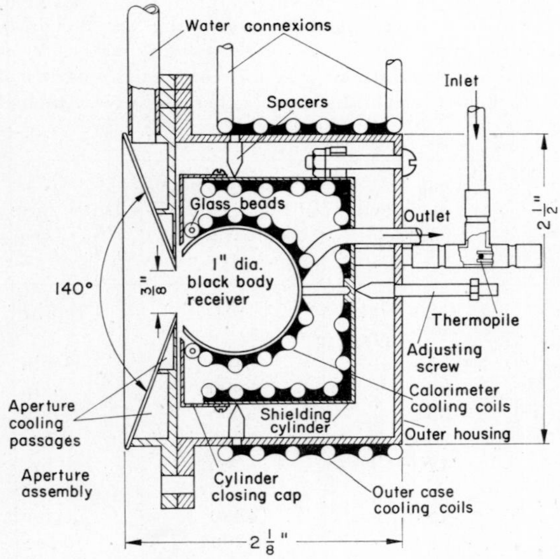

Fig. 4. Absolute water cooled cavity calorimeter.

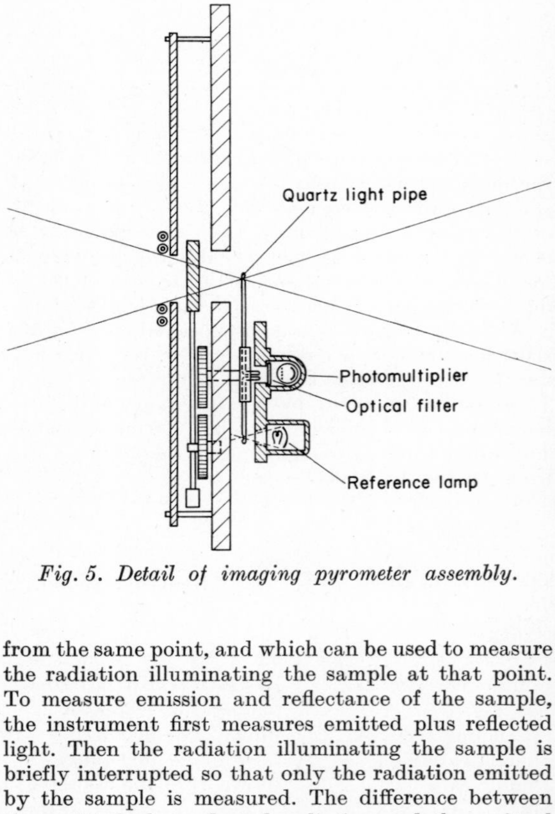

Fig. 5. Detail of imaging pyrometer assembly.

Another instrument to measure the temperature of a sample employs an adjustable reflecting system to collect the radiation from the sample. Part of the time this radiation is directed to a detector which has been previously calibrated against a black body. The remainder of the time the radiation is collected by a monochromator. The dispersed radiation can be received by one or more detectors for the ultra-violet, visible, and infra-red regions of the spectrum. By calibrating these detectors with standard sources, the absolute spectral radiance of the sample can be determined. Uncertainty in the emittance is the greatest cause of error in this spectroradiometric method. To overcome this limitation, an ultra-violet radiometer has been developed which measures the radiation from the sample at sufficiently short wave-length that uncertainties in sample emittance have relatively little effect on the calculated temperatures.

Applications. The applications of thermal-imaging techniques range over the major scientific disciplines. In most cases such techniques have been developed to meet specific experimental requirements which could not be met by other approaches. These techniques have been oriented towards producing the high temperatures required for the following purposes:

1. Preparation of materials, e.g. crystal growth, fusion of materials, and sintering.
2. Measurement of material properties, e.g. thermal conductivities, thermal diffusivity, electrical resistance, thermal expansion, thermal shock, heat capacity, melting point, vaporization and phase changes, and resistance to ablation.
3. Studies of the effects of thermal radiation, e.g. on various materials, on biological tissues, and on retinas.
4. Analysis of chemical reactions, e.g. between various solids, gases and liquids, during ignition of solid propellants.
5. Test of the performance of radiant energy conversion devices, e.g. solar cells; thermionic diodes.

As evidenced by the variety of these applications, thermal-imaging techniques have a wide range of usefulness. Most important, these techniques can be used to obtain rapidly qualitative information at high temperatures prior to embarking on an experimental programme requiring elaborate apparatus.

Bibliography

Burda E. J. (Ed.) (1955) *Bibliography of Applied Solar Energy Research,* Tempe, Arizona: Solar Energy Society.

Glaser P. E. and Walker R. F. (Eds.) (1964) *Thermal Imaging Techniques,* New York: Plenum Press.

Laszlo T. S. (1965) Image Furnace Techniques, *Techniques of Inorganic Chemistry,* Volume V, New York: Interscience.

P. E. Glaser

THERMODYNAMICS OF SURFACES. *Introduction.* When an interface between a liquid and a gas phase changes its extent and its shape but the 3-phase line (if any), in which the gas, the liquid and the solid meet, does not shift, then the variation of the free (Helmholtz) energy of the system ($F = U - TS$, see Klemperer 1961) must be expressed by two terms, namely

$$dF = \gamma\, dA + C\, d\varkappa. \qquad (1)$$

In this equation, γ is the interfacial (or surface) tension, A is the area of the interface, $\varkappa$ is an abbreviation for $(1/R_1) + (1/R_2)$; R_1 and R_2 are the two principal radii of curvature of the interface, and C is a proportionality factor which has no simple physical meaning (such as γ or pressure p have) but depends on the system.

The term $C\, d\varkappa$ exists because the pressure at the convex side of the curved surface (between two fluids) is by $\gamma[(1/R_1) + (1/R_2)]$ smaller than at the point across the interface, i.e. at the concave side (Laplace 1805). Thus, every change of $\varkappa$ changes the pressure difference between the two phases.

Consider the simple system illustrated in Fig. 1. Liquid (L) occupies the central part of a vessel, the rest of which is filled with the saturated vapour (V) of the liquid; P is a piston. If the piston is pushed in and the liquid pellet is displaced to the left without changing its shape, the work performed by the piston is zero; friction is, of course, neglected. However, if the meniscus is deformed from a plane (thin solid line in Fig. 2) to a spherical segment (gravitation being disregarded) indicated by the interrupted curve, the work of the piston is equal to the right-hand terms of equation (1). If two systems of the type of Fig. 1 differ from each other only by the inner radius of the tube in which the meniscus is deformed, and if the piston is pushed in so as to make the increase dA in the meniscus area equal in both systems, then the term $C\, d\varkappa$ of (1) will be greater for the narrower tube (because

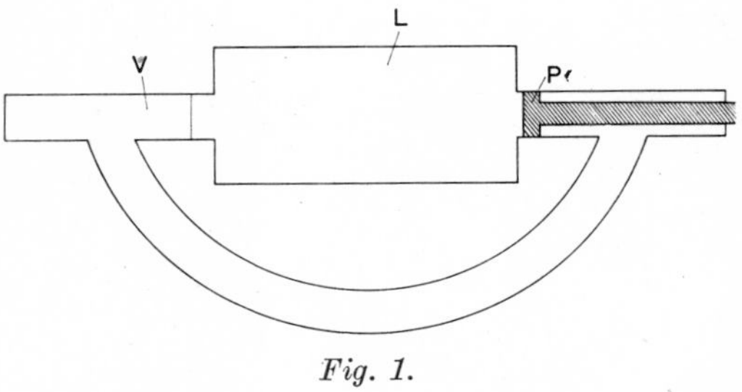

Fig. 1.

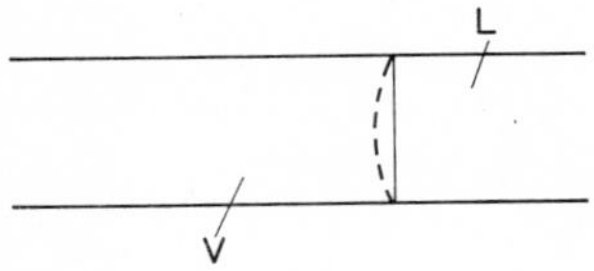

Fig. 2.

the radii of curvature will be shorter in it); hence, dF also will be greater.

If a drop of volume v_2 is in equilibrium with volume v_1 of its saturated vapour whose pressure is p and if the two possible events are a change in v_1 and a change in the curvature $\varkappa$ and, consequently, in the internal pressure $\gamma\varkappa$ of the drop, then $dF = -p\, dv_1 + v_2\gamma\, d\varkappa$. As dF is a complete differential, $\gamma(\partial v_2/\partial v_1)_K = -(\partial p/\partial \varkappa)v_1$. The ratio $(\partial v_2/\partial v_1)_\varkappa$ is equal to $-\varrho_1/\varrho_2$, ϱ_1 and ϱ_2 being, respectively, density of the vapour and of the liquid. Integration affords

$$p - p_0 = (\varrho_1/\varrho_2)\,\gamma\varkappa = (\varrho_1/\varrho_2)\,\gamma[(1/R_1) + (1/R_2)], \quad (2)$$

where p_0 is the vapour pressure above a plane liquid surface (for which $\varkappa$ is zero). This is the equation derived by William Thomson (Lord Kelvin) for the dependence of vapour pressure above a meniscus on the curvature of the latter.

Surface properties. In many instances, the properties of the surface layer are of greater interest than the properties of a system. Hence, curvature $\varkappa$ is supposed to remain constant during all the transformations considered.

(1) Let entropy S and surface area A be the two independent variables. An alteration in the internal energy is then

$$dU = T\, dS + \gamma\, dA; \quad (3)$$

T is absolute temperature. Thus,

$$(\partial T/\partial A)_S = (\partial \gamma/\partial S)_A. \quad (4)$$

If entropy is varied by adding the amount of heat $c_A\, dT$ to the system, c_A being the heat capacity at a constant area (and also constant volume, constant composition, etc.), equation (4) is transformed into

$$(\partial T/\partial A)_S = (T/c_A)\,(\partial \gamma/\partial T)_A. \quad (5)$$

As surface tension decreases with increasing temperature for all one-component and the majority of many component systems, formula (5) shows that an extension of the surface without any heat supply results in a cooling of the system.

If equation 3 is combined with the general equation $dF \equiv d(U - TS) = dU - T\, dS - S\, dT$, it is found that

$$dF = -S\, dT + \gamma\, dA. \quad (6)$$

Thus,

$$(\partial \gamma/\partial T)_A = -(\partial S/\partial A)_T. \quad (7)$$

As $(\partial \gamma/\partial T)_A$ usually is negative, the entropy of the system, as a rule, increases with the interfacial area at a constant temperature. Sometimes, the ratio $(\partial S/\partial A)_T$ is designated as the specific surface entropy; its dimension is gsec^{-2} degree.

When $dF = \gamma\, dA$ and as, from equation 7, $dS = -(\partial \gamma/\partial T)_A\, dA$, it follows that, at a constant temperature,

$$dU = \left(\gamma - T\frac{\partial \gamma}{\partial T}\right) dA. \quad (8)$$

The expression $\gamma - T(\partial \gamma/\partial T)_A$ often is referred to as the specific total surface energy; its dimension is gsec^{-2}. Again, as $(\partial \gamma/\partial T)_A$ is negative for the majority of liquids, the total energy is greater than the free energy per unit area (i.e. γ); it is also, far from the critical temperature, generally less temperature-dependent than γ.

Expansion of the interface by dA raises the internal energy of the system by dU of equation 8. If this energy is increased also by adding so much heat to the system as to raise its temperature by dT, then

$$dU = c_A dT + \left(\gamma - T\frac{\partial \gamma}{\partial T}\right)_A. \quad (9)$$

Hence,

$$(\partial c_A/\partial A)_T = -T(\partial^2 \gamma/\partial T^2)_A; \quad (10)$$

the heat capacity of the system (at constant area, volume, etc.) increases with the area if $(\partial^2\gamma/\partial T^2)_A$ is negative and is smaller the greater the area if $(\partial^2\gamma/\partial T^2)_A$ is positive. Usually, $\partial^2\gamma/\partial T^2$ is very small; thus, the heat capacity is almost independent of the interface area or, in other words, practically no heat is needed to raise the temperature of the surface as such.

(2) If area and volume are the two independent variables and the process occurs isothermally and at constant curvature, then

$$dF = -p\, dv + \gamma\, dA. \quad (11)$$

Thus,

$$(\partial \gamma/\partial v)_A = -(\partial p/\partial A)_V. \quad (12)$$

In a one-component system, i.e. one consisting of a pure liquid and its saturated vapour, both terms of equation 12 are equal to zero. However, if two substances are present and p represents the partial vapour pressure of one of them, equation 12 is a form of the Gibbs adsorption equation. If v is the volume of the saturated vapour and this volume can be varied by adding more vapour from the outside, then

$$(\partial \gamma/\partial v)_A = (\partial p/\partial A)_V. \quad (12a)$$

For instance, let mercury and ethyl ether be the two ingredients; the liquid phase is mercury (saturated with ether but still practically pure) and the gas phase is ether (saturated with mercury vapour but still very little different from ether alone). In this case p may be taken as the total vapour pressure. If, at a constant A, a volume dv of ether is added to the gas phase, the surface tension of the liquid decreases, and the derivative $\partial\gamma/\partial v$ is equal to the variation of the gas pressure p occurring when the area increases by dA at a constant volume. If p increases when A contracts, it is said that ether, which previously was adsorbed on the area dA, became free and raised the amount of ether in the vapour phase.

Another expression which perhaps is even easier to interpret is obtained when a differential of the function $F - pv$ (which, of course, depends only on the state of the system) is formed. With reference to (12a),

$$d(F - pv) = -v\,dp + \gamma\,dA. \quad (13)$$

Thus,

$$(\partial\gamma/\partial p)_A = -(\partial v/\partial A)_p. \quad (14)$$

As the tension of the mercury-ether interface decreases when the pressure of ether vapour increases, i.e. the left-hand term is negative, an additional volume of ether must be introduced to keep the pressure constant when A increases. This additional amount is said to be adsorbed on the new area dA.

Adsorption. As would be expected, equations 12a and 14 give no information on the mechanism of adsorption or on the whereabouts of the ether molecules which were present in volume dv and, according to (14), did not contribute to the gas pressure p. Attempts have been made to determine the location of the adsorbed material in a more definite manner but they are not convincing and certainly do not belong to the science of thermodynamics (Bikerman, 1961).

The quantity p in equations 12a and 14 may mean also osmotic pressure. The most general form of the Gibbs adsorption equation is $\Gamma_1 = -d\gamma/d\mu$; Γ_1 is the amount of substance No. 1 adsorbed per unit surface, expressed in g-molecules per cm^2 if the chemical potential μ_1 of this substance is given in ergs per g-molecule.

Mixtures. Equations 12 and $(\partial\gamma/\partial p)_A = (\partial v/\partial A)_p$ must be valid also for systems of two mutually saturated liquids, A and B; in this instance p means the external pressure exerted on the system, v the total volume and γ the interfacial tension acting in the boundary of A and B. If, for instance, this tension is lowered by enhanced pressure, then the combined volume of the two phases decreases when the interfacial area increases at a constant pressure. Apparently, the predicted effect has not yet been confirmed by experiment. Presumably, it can be accounted for by the effect of pressure on the mutual miscibility of A and B and by the effect of this miscibility on the interfacial tension.

Electrostatics. The fundamental relation between surface and electrostatic energies was derived by Gabriel Lippmann in 1875. If the interface between two fluids can not only be extended or contracted but also electrically charged by transferring charge de from a source whose potential is ψ volts greater than that of the interface, then $dF = \gamma\,dA + \psi\,de$. As also $F - \psi e$ is a function of the state only,

$$d(F - \psi e) = \gamma\,dA - e\,d\psi \quad (15)$$

is a complete differential. Hence,

$$(\partial\gamma/\partial\psi)_A = -(\partial e/\partial A)_\psi. \quad (16)$$

A charge de must be imparted to the interface when the area of the latter increases to keep the potential difference constant. Thus it may be said that $(\partial e/\partial A)_\psi$ is the surface density σ of the charge and the derivative of γ by the potential is equal to it. If (16) is differentiated with respect to ψ at a constant A, relation

$$(\partial^2\gamma/\partial\psi^2)_A = -(\partial\sigma/\partial\psi)_A \quad (17)$$

is obtained. The ratio $(\partial\sigma/\partial\psi)_A$ is the differential capacity of the interface (per unit area); the ratio σ/ψ would be the integral capacity. Equation 16 and 17 are essential in the study of electrocapillarity.

Surface reactions. Let $A + B - C + D$ represent a homogeneous chemical reaction. In the absence of surface effects the equilibrium constant $[C]\,[D]/[A]\,[B]$ may be designated by the letter K; $[A]$ etc are the activities of the four components. If a shift of the equilibrium causes a change in the surface tension γ of the liquid mixture, then in the presence of an interface A_0 (per g-molecule), the constant becomes $K + \Delta K$. The variation in the chemical free energy associated with this change is $\boldsymbol{R}T \ln (K + \Delta K/K)$ per g-molecule, i.e. approximately $\boldsymbol{R}T\,dK/K$; $\boldsymbol{R}$ is the gas constant. This variation must be equal to the (negative) change in the free surface energy, i.e. $-A_0(\partial\gamma/\partial K)\,dK$. Hence,

$$-(\partial K/\partial\gamma) = A_0 K/\mathbf{R}T. \quad (18)$$

If γ increases with K, i.e. with the relative concentration of C and D, then K is smaller the greater A_0, i.e. an increase of the surface area lowers the mole fractions of C and D and raises those of A and B. In other words, the equilibrium is shifted toward formation of more surface-active substances when surface area is more extensive. The integrated form of (18), that is

$$\ln (K/K_0) = (A_0/\boldsymbol{R}T)\,(\gamma_0 - \gamma) \quad (19)$$

also is used; K_0 is the equilibrium constant in the absence of an interface, and γ_0 is the surface tension of this equilibrium mixture; K and γ are the analogous quantities when the mixture has a surface A_0 against the vapour phase.

Surface phase theories. The above 19 relations were derived by the simplest and most reliable methods known in thermodynamics. Many papers and books have been published in which surface layers were treated as surface phases. For instance, the accumulation of surface-active ingredients near the surface, which is the physical phenomenon underlying equation 14, is accounted for by the greater solubility of these ingredients in the surface as compared with the bulk phase. This treatment is not recommended, as real fluid phases in Gibbs' sense are homogeneous in both physical and chemical aspects, while inhomogeneity is a most conspicuous property of surface layers.

If surface phases existed, a system consisting of a liquid and its saturated vapour would have contained three phases, in contradiction to the phase rule. In reality, the phase rule is correct as long as plane interfaces between pure liquids and their vapours are the only interfaces present. If the boundaries between two phases are or may be curved, an additional degree of freedom is available. Thus, a pure liquid and its vapour, as long as $\varkappa = 0$, can exist, at a given temperature, only under one definite pressure. A look at Fig. 1 shows that equilibrium is possible when the pressure on the piston P varies within a wide range.

When pressure increases, $\varkappa$ increases also in conformity with Laplace's formula, and the increase in $\varkappa$ gives rise to a higher vapour pressure as indicated in equation 2. If the system consists of n independent components, φ phases and s interfaces and if the curvature of each interface can be varied in an independent manner, then the number of the degrees of freedom is $n - \varphi + s + 2$.

Solid surfaces. The formulae of this article are not applicable to solid surfaces because neither A nor $\varkappa$ in equation 1 are independent variables in a solid. It is impossible to stretch or shrink the surface area of a solid body without affecting its bulk energy (usually, the strain energy); the latter is altered also when the curvature of the body varies. This was clear already to Gibbs. However, Gibbs believed that surface effects could be experimentally separated from bulk effects also in solids provided that no visible deformation of the solid was involved, when, for instance, a solid bar was gradually immersed in, or withdrawn from, a liquid; in this phenomenon, a liquid-solid interface is gradually supplanted by a solid-vapour interface, or vice versa. It appears, however, that the situation is more complex. The first difficulty has to do with the hysteresis of wetting. In the mind experiment to which Fig. 1 and Fig. 2 refer, the angle which the liquid-vapour interface forms with the wall varies when the piston is pushed in or out; theoretically, this angle should be constant.

Several other difficulties which must be overcome before a thermodynamic treatment of solid surfaces is possible have been pointed out recently (Bikerman 1965).

Bibliography

Bikerman J. J. (1961) *Contributions to the Thermodynamics of Surfaces,* Cambridge, Mass.

Bikerman J. J. (1965) *Physica Status Solidi* **10**, 3.

Gibbs J. W. (1961) *Thermodynamics,* New York: Dover.

Klemperer W. (1961) in *Encyclopaedic Dictionary of Physics* (J. Thewlis Ed.), **3**, 293, Oxford: Pergamon Press.

J. J. Bikerman

THERMOGRAPH, INFRA-RED. The thermograph is a thermal imaging device, which converts an infra-red image into a visible image. The visible image is produced by the selective evaporation of an oil layer from a film or membrane upon which is present an infra-red image. Basically, the thermograph is a chamber containing oil vapour and source of that vapour. One wall of the chamber has a window through which may be viewed the above mentioned membrane which is part of another wall. The infra-red image of the object under consideration (which is outside the chamber) is placed upon the membrane either through contact with the object or by means of an optical system. The image selectively heats the membrane in a pattern similar to that image. The layer of oil coating the membrane has a thickness dependent on the temperature at that point, warmer areas having less oil. When illuminated with white light this oil layer produces a coloured interference pattern which is viewed through the window. The thickness, and therefore the colour, of the layer is related to the distribution of intensity in the infra-red image.

I. W. Ginsberg and T. Limperis

THIN FILMS, OPTICAL PROPERTIES OF. *Single films.* The optical behaviour of a thin film may be readily calculated, either by summing multiple-reflected waves or as a boundary-value problem, and is conveniently described in terms of Fresnel coefficients. These are defined as the amplitude reflection and transmission coefficients at the interfaces and, for the situation of Fig. 1 and for normal incidence, are given by

$$r_1 = \frac{n_0 - n_1}{n_0 + n_1}, \quad t_1 = \frac{2n_1}{n_1 + n_0}$$

$$r_2 = \frac{n_1 - n_2}{n_1 + n_2}, \quad t_2 = \frac{2n_2}{n_2 + n_1}$$

(Note that the direction of incidence is important. For reflection in medium n_1 at interface n_0/n_1, the Fresnel coefficient is *minus* r_1 as given above).

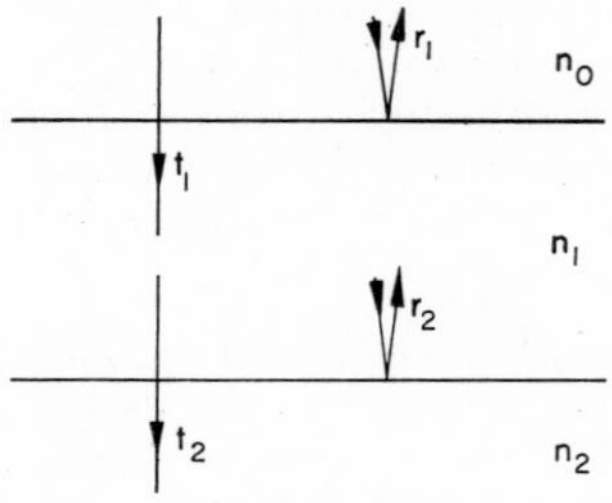

Fig. 1.

The case of non-normal incidence may be dealt with by replacing the refractive indices in the above expressions by effective values in the following way:

(a) Electric vector of lightwave parallel to plane of incidence:

Replace n_r by $n_r \cos \varphi_r$, where φ_r is the angle between the light ray and normal in medium r ($r = 0, 1, 2$),

(b) Electric vector perpendicular to plane of incidence: Replace n_r by $n_r/\cos \varphi_r$.

For a film of index n_1 and thickness d and for light of wavenumber $\nu = 1/\lambda$ the reflected amplitude R and transmitted amplitude T are given by

$$R = \frac{r_1 + r_2 \exp(-2i\delta_1)}{1 + r_1 r_2 \exp(-2i\delta_1)},$$

$$T = \frac{t_1 t_2 \exp(-i\delta_1)}{1 + r_1 r_2 \exp(-2i\delta_1)}$$

where $\delta_1 = 2\pi\nu n_1 d_1$. For non-normal incidence, the Fresnel coefficients are modified as above and the phase thickness term is written $\delta_1 = 2\pi\nu n_1 d_1 \cos\varphi_1$. The energy reflectance and transmittance (defined as the ratio of transmitted to incident flux per unit area normal to the beam) are given by

$$\mathbf{R} = \frac{r_1^2 + 2r_1r_2\cos 2\delta_1 + r_2^2}{1 + 2r_1r_2\cos 2\delta_1 + r_1^2r_2^2}$$

$$= \frac{(n_0^2 + n_1^2)(n_1^2 + n_2^2) - 4n_0n_1^2n_2 + (n_0^2 - n_1^2)(n_1^2 - n_2^2)\cos 2\delta_1}{(n_0^2 + n_1^2)(n_1^2 + n_2^2) + 4n_0n_1^2n_2 + (n_0^2 - n_1^2)(n_1^2 - n_2^2)\cos 2\delta_1}$$

$$\mathbf{T} = \frac{n_2}{n_0}\frac{t_1^2t_2^2}{1 + 2r_1r_2\cos 2\delta_1 + r_1^2r_2^2}$$

$$= \frac{8n_0n_1^2n_2}{(n_0^2 + n_1^2)(n_1^2 + n_2^2) + 4n_0n_1^2n_2 + (n_0^2 - n_1^2)(n_1^2 - n_2^2)\cos 2\delta_1}.$$

The variation of R and T with film thickness is shown in Fig. 2 for various refractive indices and a substrate index of 1·52. The maxima and minima in the reflectances (measurement of which enables the index of the film to be found) are

$n_0 \lesseqgtr n_1 \lesseqgtr n_2$

$$\mathbf{R}_{\min} = \left(\frac{n_1^2 - n_0n_2}{n_1^2 + n_0n_2}\right)^2, \quad \mathbf{R}_{\max} = \left(\frac{n_2 - n_0}{n_2 + n_0}\right)^2$$

$n_0 \lesseqgtr n_1 \gtreqless n_2$

$$\mathbf{R}_{\min} = \left(\frac{n_2 - n_0}{n_2 + n_0}\right)^2, \quad \mathbf{R}_{\max} = \left(\frac{n_1^2 - n_0n_2}{n_1^2 + n_0n_2}\right)^2.$$

If the film index is intermediate between those of the adjoining media, minimum reflectance occurs for optical thicknesses which are odd multiples of a quarter-wave-length and maxima for multiples of a half-wave-length. The reverse applies when $n_0 \lesseqgtr n_1 \gtreqless n_2$.

Figure 2 may be regarded as a spectrophotometric curve of a film if variation of the indices with wavelength is ignored. It shows how, when films are used for antireflecting or high-reflecting purposes, the reflectance changes away from the central wavelength. Thus an antireflecting film designed for zero reflectance at the centre of the visible spectrum with a glass of index 1·52 reflects approximately 0·1 per cent

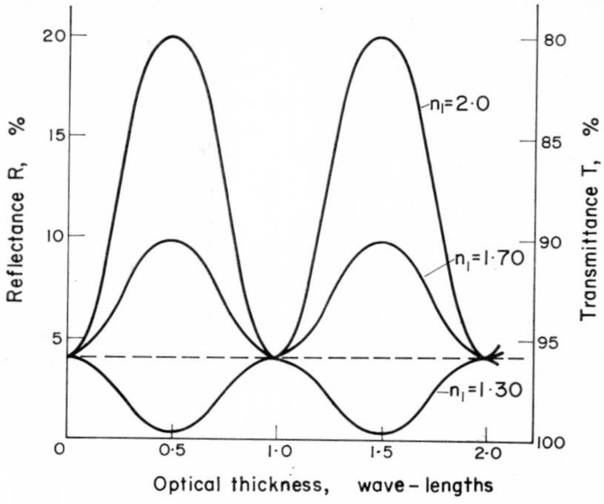

Fig. 2.

at the two ends of the visible region. Better performance may be obtained using multiple layers, as indicated below.

The phase change suffered by a beam on reflection from a film may be calculated from the amplitude reflectance, expressed in the form $\varrho \exp i\Delta$. Zero phase change occurs for optical thicknesses equal to a quarter-wave-length. The form of the phase-change vs. thickness curve for a MgF_2 film ($n_1 = 1·38$) on glass ($n_2 = 1·52$) in air is shown in Fig. 3.

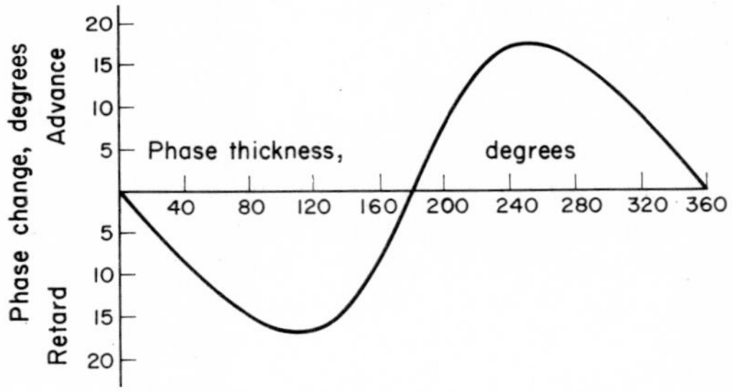

Fig. 3.

The above remarks apply to non-absorbing films. The effect of absorption may be taken into account by substituting a complex index ($n_1 - ik_1$) for n_1 in the expressions for Fresnel coefficients and for R, T. For small values of k_1 (say less than $\sim$0·1, maxima and minima are seen in the curves of reflection vs. thickness, but with damping. For large values of k_1, typical of metals in the visible region, the amplitudes of multiple-reflected beams are so small that no interference effects are observed and a monotonic reflectance/transmittance vs. thickness curve is observed.

With absorbing media, a highly complex phase change behaviour is observed. The variation of Δ with film thickness depends on the value of k_1 and may exhibit a retardation (small k_1) or an advance (large k_1). The behaviour is illustrated in Fig. 4. Since the optical constants of many metals are highly dispersive, the phase change behaviour will likewise be highly wave-length-dependent, a factor to be noted in cases where metal films are used for interferometry.

Although many metal films, when suitably prepared, show a general behaviour in agreement with that expected from calculation using the equations

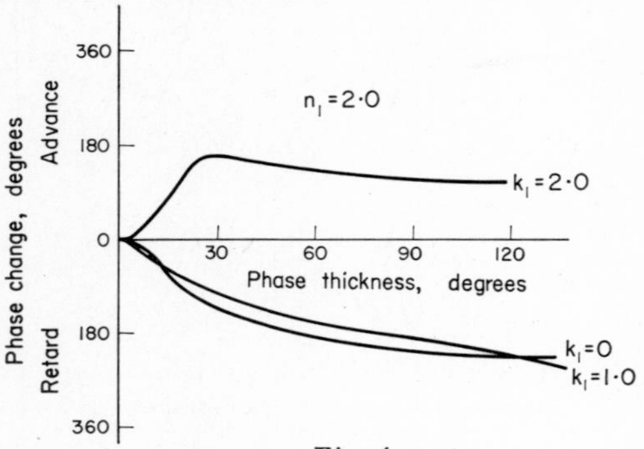

Fig. 4.

above, agreement in detail is rare. Such films often exhibit an apparent dependence of optical "constants" on thickness, an effect which may be accounted for in terms of the discontinuous structure of the films. (The expressions given above assume a homogeneous, isotropic medium with parallel, mathematically plane boundaries.) Closer agreement with experiment is obtained if the "film" is treated as a distribution of conducting spheres in a dielectric medium, using the theory of Maxwell-Garnett. Further generalizations to distributions of ellipsoidal particles have been made. On the experimental side, effort has been directed, using processes such as epitaxy, towards preparing films which are homogeneous.

Multilayer systems. The properties of multiple systems of films, especially of dielectric materials, are of considerable interest and of some importance in scientific and technological fields. Some of the more useful characteristics are given below.

The reflecting and transmitting properties of a stack of parallel-sided slabs may be readily calculated as a boundary-value problem. The results are conveniently expressed in terms of 2×2 matrices characterizing each film of the stack. The total electric and magnetic field strengths in the incident medium (n_0) are given in terms of those in the mth medium by the relation

$$\begin{pmatrix} E_0 \\ H_0 \end{pmatrix} = M_1 M_2 \dots M_m \begin{pmatrix} E_m \\ H_m \end{pmatrix}$$

where

$$M_r = \begin{pmatrix} \cos \delta_r & \dfrac{i \sin \delta_r}{n_r} \\ i n_r \sin \delta_r & \cos \delta_r \end{pmatrix}$$

and $\delta_r = 2\pi \nu n_r d_r$ for normal incidence. (The case of non-normal incidence is dealt with as discussed above). Since the total fields E, H are sums of positive- and negative-going waves, the reflectance and transmittance of a stack is readily found.

One of the most useful multilayer systems is the multilayer stack, consisting of a series of layers of equal optical thickness and with alternating high and low refractive indices. Such a system has a broad reflectance band centred on the wave-length λ_0 for which the optical thicknesses of the layers are $\lambda_0/4$. The reflectance band has the form in Fig. 5. The bandwidth of such a stack depends on the refractive indices, being given by

$$\Delta\sigma = \sigma_0 \text{ arc sin} \left(\frac{n_L - n_H}{n_L + n_H} \right).$$

The reflectance region becomes flatter and higher as the number of layers is increased. For a stack consisting of an even number $2s$ of layers of indices n_L and n_H on a substrate of index n_s, (with a layer n_L

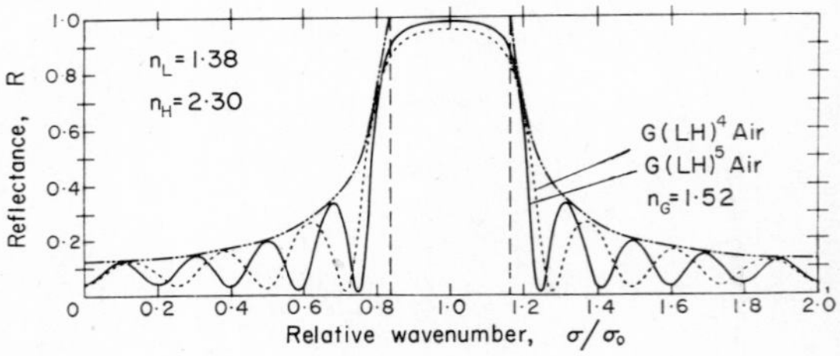

Fig. 5.

next to the substrate) the reflectance $\mathbf{R}_{2s}$ is given by

$$\mathbf{R}_{2s} = \left(\frac{n_s f - n_0}{n_s f + n_0} \right)^2$$

where n_0 is the medium of incidence and $f = (n_H/n_L)^{2s}$. The transmittance is simply $1 - \mathbf{R}_{2s}$. For an odd number $(2s + 1)$ of layers, starting and finishing with a high index layer, $\mathbf{R}_{2s+1}$ is given by

$$\mathbf{R}_{2s+1} = \left(\frac{n_H^2 f - n_0 n_s}{n_H^2 f + n_0 n_s} \right)^2.$$

Since present techniques enable systems of fifty or more layers to be formed, very high reflectances are obtainable by this means. Since all the layers used are dielectrics, absorption is negligible and the value of $\mathbf{R} + \mathbf{T}$ for the stack approaches closely to unity. Such stacks are invaluable in high-Q optical resonators. Note must be taken of the variation with wave-number of the phase change on reflection at such a stack (Fig. 6) which shows a considerably higher dispersion than is found with the metal layers hitherto used in interferometry. The limitation in bandwidth, imposed by available refractive indices for such a stack may be overcome by the use of other combina-

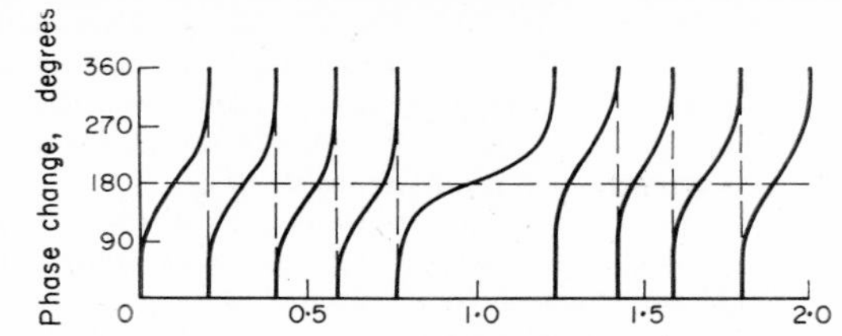

Fig. 6.

tions of thicknesses. Figure 7 shows the reflectance band calculated for a staggered multilayer stack, in which the layer thicknesses form a geometrical progression.

The use of two high-reflecting stacks, with a half-wave layer sandwiched between them yields a Fabry–Perot type interference filter, displaying a narrow transmission band at λ_0 and a rejection region on either

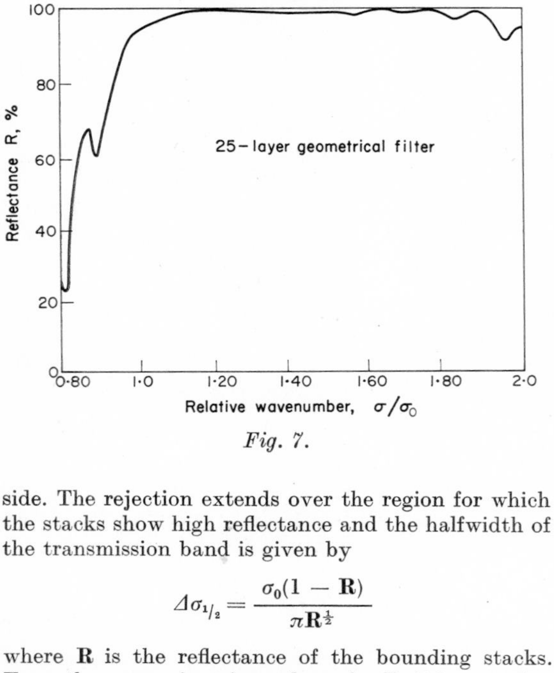

Fig. 7.

side. The rejection extends over the region for which the stacks show high reflectance and the halfwidth of the transmission band is given by

$$\Delta\sigma_{1/2} = \frac{\sigma_0(1 - \mathbf{R})}{\pi \mathbf{R}^{\frac{1}{2}}}$$

where **R** is the reflectance of the bounding stacks. From the expression given above for **R**, it is seen that bandwidths of the order of a few Angström units are quite feasible with this type of filter. Since there are no absorbing elements, the theoretical peak transmittance is 100 per cent (for a filter with identical media on either side). Filters may be made with properties approaching the theoretical performance.

An alternative potentially powerful system displaying a very narrow passband is the frustrated total reflection filter, employing a pair of prisms (Fig. 8) on the hypotenuse of each of which a layer is deposited such that, when the prisms are united about a spacer

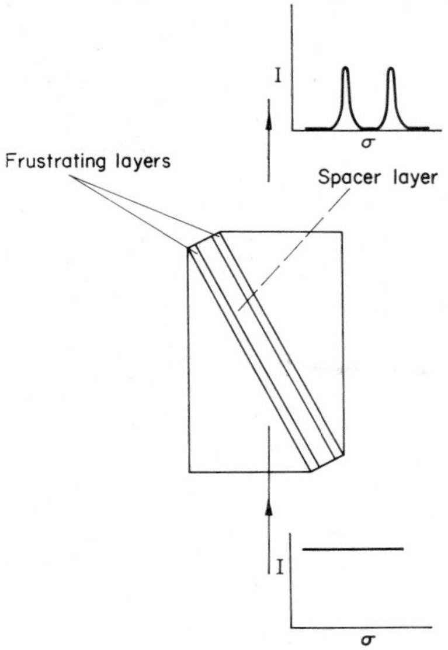

Fig. 8.

layer, tunnelling through the frustrating layer yields a high-reflecting, theoretically lossless, system. The reflectance is governed simply by the thickness of the frustrating layer. Since non-normal incidence is essential and since the effective optical thickness of the spacer layer (including the path equivalents of phase changes on reflection) depends on the state of polarization of the incident light, then this filter possesses two passbands (but see below).

In addition to their use in high-reflecting systems and filters, multilayers may be used to give anti-reflecting coatings over a broader band than is obtainable with a single layer. A typical three-layer system is shown in Fig. 9.

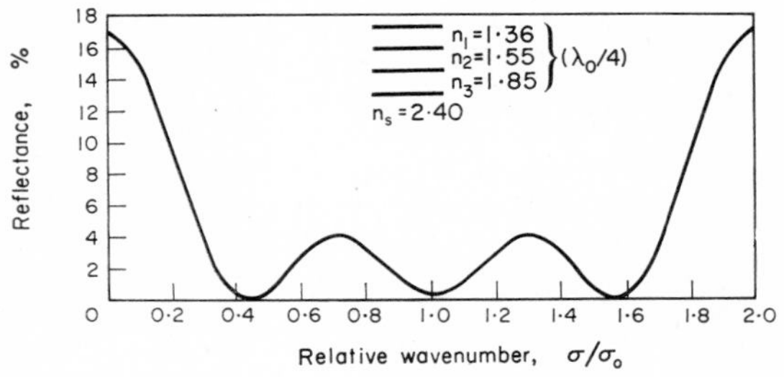

Fig. 9.

Anisotropic films. The extension of electromagnetic theory to the case of birefringent films is straightforward, if somewhat laborious. Such films have not been dealt with experimentally to any great extent, partly from difficulties in forming such films. An ingenious application of birefringent films is that in which the frustrated total reflection filter is made with a birefringent spacer layer, so chosen that the differential phase change on reflection at the frustrating layers is just compensated by the optical path difference for the two components due to the birefringence. The two passbands of the isotropic spacer case are thus brought together.

The optical behaviour of films of ferromagnetic materials cannot be described by the expressions given above. This is because such films transmit and reflect incident plane-polarized light as elliptically polarized, with the major axis of the ellipse inclined to the original direction of polarization. Such media are described as *gyrotropic* and may be characterized phenomenologically by a skew-symmetric tensor permeability and/or permittivity, of the form

$$[\mu] \equiv \begin{bmatrix} \mu & -ip\mu & 0 \\ ip\mu & \mu & 0 \\ 0 & 0 & \mu' \end{bmatrix} \quad [\varepsilon] \equiv \begin{bmatrix} \varepsilon & -iq\varepsilon & 0 \\ iq\varepsilon & \varepsilon & 0 \\ 0 & 0 & \varepsilon' \end{bmatrix}$$

where p and q are the gyromagnetic and gyroelectric constants.

The effects on the optical behaviour of films of skew symmetry in the μ and ε tensors are best discussed by considering separately the three following cases:

(1) Magnetization (or electrification) axis normal to the plane of the film (polar case);
(2) Axis parallel to film surface and parallel to the plane of incidence (longitudinal case);
(3) Axis parallel to film surface and perpendicular to the plane of incidence (transverse case).

The transverse case displays the simplest behaviour: an incident ray produces a single refracted ray which behaves normally on reflection within the film. In the longitudinal case, double refraction occurs and at non-normal incidence, two refracted rays appear, again showing normal behaviour. In the polar case, double refraction again occurs but in this case the behaviour of rays within the film is abnormal inasmuch as the angle of incidence is *not* equal to the angle of reflection. Pairs of rays within the medium couple in the manner indicated in Fig. 10.

The reflectance and transmittance of films of gyrotropic media may be calculated in a similar fashion to that for normal films, with the substitution of the appropriate tensor form in Maxwell's equations.

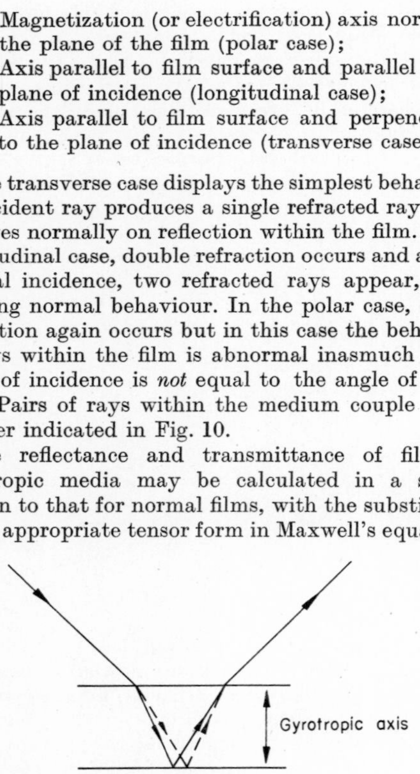

Fig. 10.

The general case is rather cumbersome but simplification is possible for $p, q \ll 1$, a condition found to be amply fulfilled by known materials. The propagation matrix for the film is now 4×4, owing to the fact that the s- and p-components of the magnetic (electric) vectors are no longer independent—the value of each component at any point depends on *both* components at neighbouring points. Thus the components of E, H in medium 2 are related to those in medium 0 by a relation of the type

$$\begin{pmatrix} E_{0p} \\ H_{0p} \\ E_{0s} \\ H_{0s} \end{pmatrix} = C \begin{pmatrix} E_{2p} \\ H_{2p} \\ E_{2s} \\ H_{2s} \end{pmatrix}$$

where C is a 4×4 matrix whose off-diagonal elements depend on the gyrotropic constants.

The forms of C required for computation of the optical behaviour of such films are given below (a, b, c). These expressions are given for the gyromagnetic case and are valid subject to the condition $p \ll 1$. Corresponding expressions may readily be written for the gyroelectric case, with $q \ll 1$.

(*a*) *Polar case.*

$$C_p \equiv \begin{pmatrix} \cos\delta & (\mathrm{i}\sin\delta)/m & \mathrm{i}p\delta^2/2 & 0 \\ \mathrm{i}m\sin\delta & \cos\delta & mp\delta & -\mathrm{i}p\delta^2/2 \\ -\mathrm{i}p\delta^2/2 & 0 & \cos\delta & -(\mathrm{i}\sin\delta)/m \\ mp\delta & \mathrm{i}p\delta^2/2 & -\mathrm{i}m\sin\delta & \cos\delta \end{pmatrix}$$

where δ is the phase thickness of the film and m the wave impedance, given by $\mu/n\sqrt{(\mu_0\varepsilon_0)}$.

(*b*) *Longitudinal case.*

$$C_l = \begin{pmatrix} \cos\delta & -(\mathrm{i}\sin\delta)/m\gamma & 0 & \mathrm{i}p\gamma^2\delta^2/m \\ -\mathrm{i}m\gamma\sin\delta & \cos\delta & -\mathrm{i}mp\delta^2\gamma^2 & -p\gamma\delta \\ -p\gamma\delta & \mathrm{i}p\delta^2 & \cos\delta & (\mathrm{i}\gamma\sin\delta)/m \\ -\mathrm{i}mp\delta^2 & 0 & (\mathrm{i}m\sin\delta)/\gamma & \cos\delta \end{pmatrix}$$

where γ is the cosine of the angle between the direction of the ray and the gyro axis.

(*c*) *Transverse case.* This is in effect an isotropic case, with no coupling between perpendicular components of the propagating modes, so that 2×2 matrices, as for the isotropic case, suffice for description.

Bibliography

ANDERS H. (1965) *Dünne Schichten für die Optik*, Stuttgart: Wissenschaftliche Verlagsgesellschaft.

HASS G. and THUN R. (Eds.) (1963, 1964) *Physics of Thin Films*, Vols. 1 and 2, New York: Academic Press.

HEAVENS O. S. (1966) *Optical Properties of Thin Solid Films*, New York: Dover.
SOKOLOV A. V. (1961) *Opticheski Svoistva Metallov*, Moscow; *Scripta Technica* (1966).
THEWLIS J. (Ed.) (1962) *Encyclopaedic Dictionary of Physics*, articles beginning "Thin films", Suppl. Vol. 2, Oxford: Pergamon Press.
VASICEK A. (1960) *Optics of Thin Films*, Amsterdam: North Holland.
WOLF E. (Ed.) (1963–5) *Progress in Optics*, Vols. 2, 4 and 5. Amsterdam: North Holland.

O. S. HEAVENS

THIN FILMS; SUPERCONDUCTING. Superconductors have two characteristic lengths associated with them, a penetration depth λ (Mackinnon 1962), and a coherence length ξ (Enderby 1962). For a Type I superconductor, $\lambda \sim 500$Å, and $\xi \sim 5000$Å. When the thickness of a superconducting film is of the same order, or less than one of these lengths, deviations from bulk behaviour occur, and this is what we take as a definition of thin film behaviour.

Enhanced critical field. When a superconducting film is of the same order of thickness as the penetration depth, it is no longer correct to assume $\mathbf{B} = 0$ within the film, i.e. the film will no longer show a complete Meissner effect (Enderby 1962). By allowing magnetic flux into the specimen, the magnetization becomes less than that for a perfect diamagnetic:

$$(m_v)_{\mathrm{T.F}} < (m_v)_{\mathrm{BULK}} = -\frac{H_0}{4\pi},$$

where H_0 is the applied field.

The free energy difference between the normal and superconducting phases is given by

$$G_N(0) - G_S(0) = V \int_0^{H_c} m_v \mathrm{d}H \tag{1}$$

where $G_N(0)$ and $G_S(0)$ are the free energies in the normal and superconducting states in zero applied field. If we assume that this free energy difference between the two phases is the same for a thin film as for bulk, then if $(m_v)_{\mathrm{T.F.}} < (m_v)_{\mathrm{BULK}}$, the integral has to be taken to higher fields, H_s, for the equation to be satisfied. Thus the critical field for a thin specimen is raised above that for bulk material.

The simplest description can be given in terms of London's theory, which assumes that the penetration depth is independent of magnetic field. The solution of London's equations for a film of thickness $2a$ in a uniform field, H_0, applied parallel to the surface, shows that at a distance z from the median plane the field is given by

$$H = H_0 \cosh\left(\frac{z}{\lambda_L}\right) \Big/ \cosh\left(\frac{a}{\lambda_L}\right) \tag{2}$$

and

$$m_v = -\frac{H_0}{4\pi}\left(1 - \frac{\lambda_L}{a}\tanh\frac{a}{\lambda_L}\right). \tag{3}$$

Where λ_L, the London penetration depth is given by

$$\lambda_L = (mc^2/4\pi n_s e^2)^{1/2} \tag{4}$$

and n_s = number of "superconducting electrons". Substituting for m_v in equation 1, from equation 3,

$$G_N(0) - G_S(0) = -V \int_0^{H_c} \left[\frac{H_0}{4\pi}\right] \mathrm{d}H$$

$$= -V\left(1 - \frac{\lambda_L}{a}\tanh\frac{a}{\lambda_L}\right)\int_0^{H_s}\left[\frac{H_0}{4\pi}\right]\mathrm{d}H. \tag{5}$$

Where H_s is the enhanced critical field of the thin film.

Thus

$$H_s^2 = H_c^2 \Big/ \left(1 - \frac{\lambda_L}{a}\tanh\frac{a}{\lambda_L}\right). \tag{6}$$

For

$$a \gg \lambda_L, \quad H_s = H_c\left(1 + \frac{\lambda_L}{2a}\right). \tag{7}$$

For

$$a \ll \lambda_L, \quad H_s = \sqrt{3}\,\lambda_L H_c/a. \tag{8}$$

In the defining relation for the London penetration depth, equation 4, the only temperature dependent term is n_s. By fitting their two fluid theory to a T^3 dependence of specific heat, Gorter and Casimir showed

$$n_s = 1 - \left(\frac{T}{T_c}\right)^4$$

and hence

$$\left(\frac{\lambda_L(T)}{\lambda_L(0)}\right)^2 = \frac{1}{1 - \left(\frac{T}{T_c}\right)^4},$$

where $\lambda_L(0) = (mc^2/4\pi n_0 e^2)^{\frac{1}{2}}$ is the value of λ_L at 0°K (n_0 here is the effective free electron density).

Thus as the temperature is raised towards the transition temperature the penetration depth increases rapidly and consequently from equation 8, so does H_s.

The coherence length. The London treatment of enhanced critical fields for thin films has to be modified when account is taken of the coherence length. Pippard (1950) introduced the concept of a coherence length and subsequently modified London's treatment with the *Pippard non-local theory* (1953).

In this theory the number of superelectrons is allowed to vary spatially, and the penetration depth becomes a function of the coherence length. Pippard assumed that in a pure bulk superconductor the wave function describing the superconducting ground state extends over a distance $\xi(0)$. (Thus rapid variations in the number of superelectrons cannot occur in distances smaller than this.) In specimens in which the electronic mean free path is l, the coherence length is reduced to $\xi(l)$, and $\xi(l) \to \xi(0)$ as $l \to \infty$ and $\xi(l) \to l$ as $l \to 0$. In a bulk metal, the electronic mean free path is limited by a combination of phonon and impurity scattering. In a thin film, scattering of electrons at the surfaces of the films also becomes important, and will constitute the main limitation to the mean free path in very thin films.

Experimentally measured values of the penetration depth did not agree very well with theoretical values calculated from equation 4. This follows from the assumption in the London model that the number of "superconducting electrons" remains constant when the magnetic field is applied. If we allow n_s to vary over the coherence length, the penetration depth in a bulk sample will clearly be higher than the London value. Thus if we measure the penetration depth in a pure bulk specimen we obtain a value λ_0. Pippard showed that for a thin film of this same material in which the mean free path is l, the effective penetration depth is given by

$$\lambda = \lambda_0 \sqrt{\left(\frac{\xi(0)}{\xi(l)}\right)}.$$

When this value of λ is used in equations **7** and **8**, good agreement is obtained with experiment.

Ginzburg–Landau theory. The Pippard non-local theory is not altogether satisfactory theoretically as it modifies London's equations at a late stage and the concept of a coherence length is not built in to the theory at its outset. A much more satisfactory way of treating a superconductor in a magnetic field is given by the Ginzburg–Landau theory (1950).

In this theory a parameter ψ is introduced and ψ is a kind of wave function normalized so that $|\psi|^2 = n_s$ represents the density of superconducting electrons. They allow ψ to vary spatially and include a contribution to the free energy associated with the gradient in ψ. Minimization of this more complicated expression again leads to the idea that n_s can only vary fairly slowly with position, in a similar way to Pippard's theory. The solutions of the Ginzburg–Landau equations are normally expressed in terms of a parameter $\varkappa$ and Gorkov has been able to show from first principles

$$\varkappa \sim 0{\cdot}96 \frac{\lambda_0}{\xi_0}$$

where λ_0 is the empirical penetration depth in a bulk superconductor.

The value of the enhanced critical field of thin films is higher than that deduced from London's theory because the penetration depth becomes a function of magnetic field, i.e. as H increases, λ increases and hence H_s increases. Lynton provides a good treatment of the G. L. Theory and some of its important results. For a film of thickness $2a \leqq \lambda_0$ of a pure, elemental, type I superconductor, i.e. $\varkappa$ small,

$$H_s/H_c = \sqrt{6}\lambda_0/a$$

Magnetic field dependence of energy gap. Using the G.L. equations it is possible to calculate how the order parameter ψ, varies inside a thin film subjected to a magnetic field. From the definition of coherence length or in G.L. terms,

$$\frac{\lambda_0}{\varkappa},$$

it follows that if the film thickness $2a \ll \xi$, ψ must be uniform through the thickness of the film, and for very thin films, with $\varkappa = 0$, Ginzburg (1958) finds

$$\frac{\psi^2(H_e)}{\psi^2(0)} = \left[1 - \left(\frac{H_e}{H_s}\right)^2\right] \bigg/ \left[1 - \frac{4}{5}\left(\frac{2a}{\lambda_0}\right)^2\right].$$

Thus as $H_e \to H_s$, the ratio $\psi^2(H_e)/\psi^2(0) \to 0$, i.e. the value of $\psi^2(H_e)$ just before the transition to normality (critical value) is zero. There is a thickness $2a = \sqrt{5}\lambda_0$, below which there is a second order transition and above which there is a first order transition, characteristic of bulk material.

Gorkov (1960) has shown that it is possible to identify the order parameter ψ of the G.L. Theory, with the energy gap, ε, of the Bardeen, Cooper and Schrieffer (1957) (B.C.S.) microscopic theory. The energy gaps in superconductors have been measured by a variety of techniques, far infra-red absorption, ultrasonic attenuation and more recently by electron tunnelling. Merservey and Douglass (1964) have used electron tunnelling to measure the magnetic field dependence of the energy gaps of aluminium films from $T = T_c$ to $T = 0{\cdot}7\,T_c$ on films ranging in thickness from 420 to 9850Å. For $\frac{2a}{\lambda} > 1$ they find agreement with the theory but for $\frac{2a}{\lambda} < 1$ they observe deviations from the theory which they suggest may result from the assumption that ψ is spatially invariant. They suggest that periodic "vortex" solutions as obtained by Abrikosov for bulk superconductors where $\varkappa > 1/\sqrt{2}$ (Type II superconductors) may be the stable configuration.

Bardeen (1962) has calculated the critical currents and critical fields for thin films on the basis of the microscopic theory. He shows that for any thickness of film below a reduced temperature of 0·325 there should be a first order transition from the superconducting to the normal phase. In this paper Bardeen also gives expressions for the critical current densities in thin films under a variety of conditions.

Proximity effect. The proximity effect is closely related to the idea of coherence length. A thin film of superconductor is evaporated onto some substrate, and subsequently a very thin layer of normal metal is deposited on top of, and in direct contact with, the first film. The wave function describing the Cooper pairs extends as we have seen over a coherence length, a distance of $\sim 10^{-4}$ cm, and so it seems reasonable that the Cooper pairs (the superelectrons of the two fluid theory), could extend into the normal region, in which the interaction between electrons is not attractive. (Further evidence for the plausibility of this argument comes from the *Josephson effect*, where Cooper pairs tunnel through an oxide separating two superconductors.) As a result of this "diffusion" the ground state of the bimetallic layer can be characterized by some average value of the energy gap, and transition temperature. Thus the normal metal may, under suitable conditions, be made a superconductor, while in the superconductor the energy gap is reduced. Theoretical treatments of this effect have been given by Cooper (1961), Parmenter (1960) and de Gennes (1964) and a phenomenological theory by Douglass (1962).

Extremely elaborate precautions have to be taken during the preparation of suitable samples, to prevent diffusion or alloying of the two metals. Experimental results have been reported by Adkins (1966), Fischer (1966), Hauser (1966) and others. Adkins, using an electron tunnelling technique, found that a well defined gap-like structure was found to be induced in the normal film by its proximity to the superconductor. By varying the normal film thickness he was able to show that initially the gap in the normal metal decreased exponentially with distance from the interface. At larger distances from the interface the energy gap decreased more slowly. From the rate at which the gap falls off with increasing normal metal film thickness it is possible to obtain an idea of how nearly the normal metal is to being a superconductor. If the gap falls off rapidly one can say that the interaction between electrons via a virtual phonon (the B.C.S. mechanism) is repulsive and one can attribute a negative transition temperature to such a material. Adkins was able to show that if copper is a superconductor, then its transition temperature is less than 20μ°K.

Ginzburg superconductivity. Ginzburg (1964) has proposed that a new type of surface superconductivity may be observable in sufficiently thin films. The magnitude of the B.C.S. energy gap is a function of the phonon spectrum and density of states at the Fermi surface. Ginzburg argues that the surface of a crystal may differ appreciably from the bulk in two important respects. Firstly the density of states may vary, for example by suitably doping a semiconductor a sufficiently large density of surface states may be produced to form a two-dimensional metal. Secondly, the phonon spectrum will be different at the surface, and so it is conceivable that within the surface layer there may be an attractive interaction and hence a two-dimensional superconducting surface may result. In order to study this effect it is necessary to use very thin films so that the energy gap in the surface layer is not suppressed by the normal metal bulk, i.e. by the proximity effect. Possible experimental detection of this effect has been reported by Strongin *et al.* (1965).

Energy gap anisotropy. The B.C.S. theory is based on an idealized metal with a spherical Fermi surface. In any real superconductor such a model is at best a good approximation. Thus in a real metal one can visualize that different sheets of the Fermi surface have different energy gaps, and that the energy gap on a particular sheet may vary with direction. Anderson (1959) suggested that in order to observe this "anisotropy" of the energy gap the mean free path for electrons must be greater than the coherence length. Thus if the mean free path is limited by reducing the film thickness then there will be a film thickness below which energy gap anisotropy is smeared out and we measure an energy gap which is direction independent.

Energy gap anisotropy in lead was reported by Townsend and Sutton (1962) and Rochlin and Douglass (1965). A theoretical treatment of anisotropy in lead has been given by Bennett (1965).

We can take advantage of this sharpening up of energy gaps in films $2a < \xi$ in tunnelling experiments, i.e. when trying to measure fine structure in a given superconductor, we can remove all structure in the reference superconductor simply by making it thin.

Real films and thin films. The ideal thin film would consist of a very thin section of a perfect single crystal. Experimentally the films can only be produced by evaporation or vapour deposition and the situation immediately becomes non-ideal. The films have to be deposited on substrates and consequently differential thermal contraction between the film and the substrate will, on cooling, cause distortion of the lattice and a change in transition temperature T_c. Some improvement can be obtained by attempting to match the expansion coefficients of the film to the substrate.

To avoid contamination with gases during evaporation, films should be prepared in ultra-high vacuum conditions, and if possible measurements made on the films without exposing them to the atmosphere.

For very thin films it cannot be assumed that the film is parallel sided, and for the thinnest films (<100Å thick) the film is not even continuous but consists of isolated "islands" of metal.

It is therefore vitally important to be absolutely sure of the structure, purity and nature of films used in superconducting experiments before concluding too much from experimental results.

Bibliography

Adkins C. J. (1966) LT 10 (Moscow), To be published.
Anderson P. W. (1959) *J. Phys. Chem. Solids* **11**, 26.

BARDEEN J. (1962) *Rev. Mod. Phys.* **34**, 667.
BARDEEN J., COOPER L. N. and SCHRIEFFER J. R. (1957) *Phys. Rev.* **108**, 1175.
BENNETT A. J. (1965) *Phys. Rev.* **140**, A 1902.
COOPER L. N. (1961) *Phys. Rev. Letters* **6**, 698.
DE GENNES P. G. (1964) *Rev. Mod. Phys.* **36**, 225.
DOUGLASS D. H. (1962) *Phys. Rev. Letters* **9**, 155.
ENDERBY J. (1962) in *Encyclopaedic Dictionary of Physics* (J. Thewlis Ed.), **7**, 105, 108, Oxford: Pergamon Press.
FISCHER G. and KLEIN. LT 10 (Moscow), To be published.
GINZBURG V. L. (1958) *Soviet Physics, J.E.T.P.* **7**, 78.
GINZBURG V. L. (1964) *Phys. Letters* **13**, 101.
GINZBURG V. L. and LANDAU L. D. (1950) *J.E.T.P., U.S.S.R.* **20**, 1064.
GORKOV L. P. (1960) *Soviet Physics, J.E.T.P.* **9**, 1364; **10**, 593, 998.
HAUSER J. J. LT 10 (Moscow), To be published.
LONDON F. (1961) *Superfluids*, Vol. 1, New York: Dover.
LYNTON E. A. (1962) *Superconductivity*, London: Methuen.
MACKINNON L. (1962) in *Encyclopaedic Dictionary of Physics* (J. Thewlis Ed.), **7**, 109, Oxford: Pergamon Press.
MERSERVEY R. and DOUGLASS D. H. (1964) *Phys. Rev.* **135**, A 24.
PARMENTER R. H. (1960) *Phys. Rev.*, **118**, 1173.
PIPPARD A. B. (1950) *Proc. Roy. Soc.* A **203**, 210.
PIPPARD A. B. (1953) *Proc. Roy. Soc.* A **216**, 547.
ROCHLIN G. I. and DOUGLASS D. H. (1965) *Phys. Rev. Letters* **16**, 359.
SHOENBERG D. (1952) *Superconductivity*, Cambridge: The University Press.
STRONGIN M., KAMMERER O. F. and PASKIN A. (1965) *Phys. Rev. Letters* **14**, 949.
THEWLIS J. (Ed.) (1967) *Encyclopaedic Dictionary of Physics*, Suppl. Vol. 2, articles beginning "Thin film(s)", Oxford: Pergamon Press.
TOWNSEND P. and SUTTON J. (1962) *Phys. Rev.* **128**, 591.

P. TOWNSEND

TIME-OF-FLIGHT DRIFT VELOCITY MEASUREMENTS. The most accurate method of measuring the drift velocity of ions or electrons is to determine the time taken for an electron or ion pulse to travel a known distance in a uniform electric field. The pulse of charged particles is usually produced by photoemission or by an electrical "shutter" which is opened for a short time interval. In order to obtain accurate measurements it is necessary to ensure that the effect of any distortion of the uniform electric field caused by the operation of electrical shutters is negligible. Attention should also be given to the effects of diffusion. Under certain circumstances errors greater than 15 per cent can be incurred in electron drift velocity measurements if the effects of diffusion are ignored.

Of the many methods devised, those most commonly used are discussed under the following headings:

(*a*) *Methods which use electrical shutters.* In Fig. 1, S is an ion source and S_1 and S_2 are electrical shutters, a distance d apart. The shutters are operated in phase and are opened at regular time intervals. The current transmitted to the collector C when plotted as a function of the frequency at which the shutters are

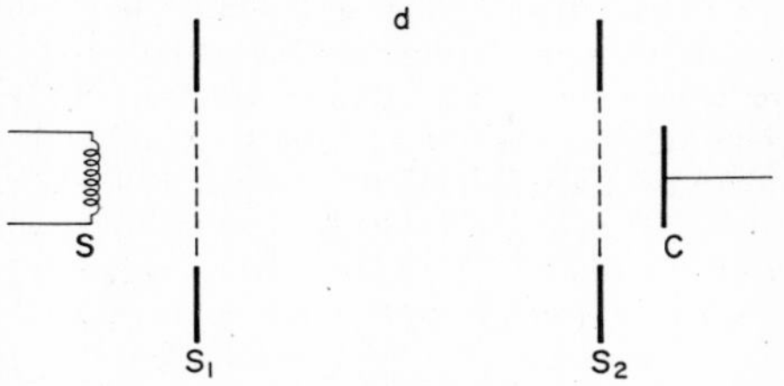

Fig. 1.

opened, exhibits a series of maxima and minima, the maxima occurring at integral multiples of the frequency f_0 at which the first maximum occurs. The drift velocity is given by $W = f_0\, d$ where d is the distance between the shutters.

The shutters may be of two forms. In the "four-gauze method" (Tyndall and Powell), each shutter consists of two closely spaced grids or gauzes. A d.c. retarding field is placed between the grids of the shutter, the ions being transmitted when the field produced by an alternating potential applied to the grids exceeds the retarding field. This method has been used extensively for measurements of the drift velocity of both positive and negative ions. An alternative form of shutter consists of a series of parallel, coplanar wires, alternate wires being connected together (Bradbury and Nielsen). When an alternating potential is applied between the two sets of wires electrons are transmitted every half cycle, the drift velocity being given by $W = 2f_0d$. This method can be used for both ions and electrons and is capable of high accuracy (error $< 1\%$). A typical current frequency curve is shown for potassium ions in hydrogen (Fig. 2). Both types of shutter can be operated by voltage pulses instead of a voltage which varies sinusoidally.

In an alternative method (Pack and Phelps) of measuring electron drift velocities electron pulses are produced not by the operation of a shutter but by photoelectric emission from a cathode illuminated by a pulsed source of ultra-violet radiation. The electron pulses travel to a shutter which is maintained opaque to electrons by a d.c. biasing potential. The potential is removed for a short time at varying time intervals after the flash of ultra-violet radiation and from the time at which the maximum transmitted current occurs the transit time of an electron pulse from the cathode to the shutter is determined. The same procedure is adopted using a second shutter, a further distance from the cathode than the first. The difference between the transit times to the two shutters enables the drift velocity to be determined independent of any end effects at the cathode. The same apparatus can be used in a way which eliminates distortion of the uniform electric field caused by the d.c. bias on the shutters. In the zero-bias method the shutter is maintained open and closed only for a short time interval. The electric field is therefore maintained uniform throughout the transit of the pulse. The transit time is determined from the time delay at which a minimum occurs in the measured current waveform.

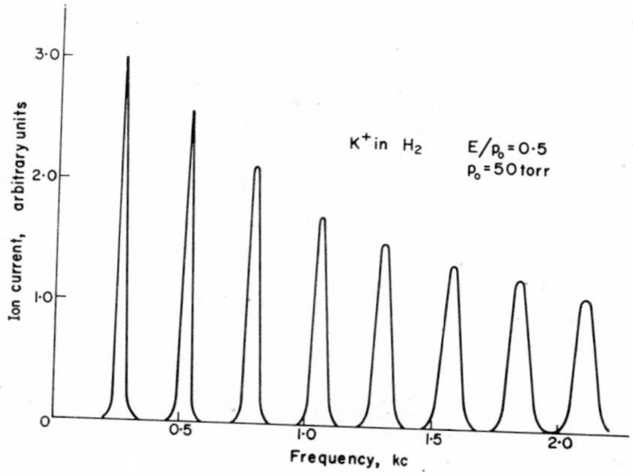

Fig. 2.

A method of determining both the drift velocity of negative ions formed by attachment and the attachment coefficient has been devised by Doehring. Shutters S_1 and S_2 are of the double gauze type, the shutters being operated by square wave pulses of variable duration, amplitude and repetition. When shutter S_1 is opened for a short time a group of electrons is transmitted and, as they traverse the region between S_1 and S_2, negative ions are formed by attachment along their path. Due to the drift velocity of the electrons being approximately 1000 times that of the negative ions, the ions may be considered to have remained stationary until the electrons have been collected. The ion distribution left in the space between S_1 and S_2 then drifts with a velocity W_i towards the collector. Shutter 2 is opened for a short time interval at a time t with respect to the opening of shutter S_1 and the current received by the collector measured as a function of the delay time t. A typical curve is shown by the full line in Fig. 3. The dashed curve is the theoretical curve, assuming no diffusion and an infinitely narrow ion pulse width. It can be shown that the current received by the collector is given by $I = I_0 e^{\alpha W_i t}$ where α is the number of attachments formed per cm of path in the direction of the electric field, and hence on a log-linear plot the linear portion of the ion current versus delay time curve has a slope given by αW_i. The drift velocity W_i is found from the transit time t_i.

Instead of the double-grid type of shutter, those of the form used by Bradbury and Nielsen have also been employed in this type of measurement. Chanin, Phelps and Biondi have used a modified form of Doehring's method in which the electron pulses were produced not by a shutter S_1 but by a pulsed source of ultra-violet radiation. Shutter S_2 was of the Bradbury–Nielsen type.

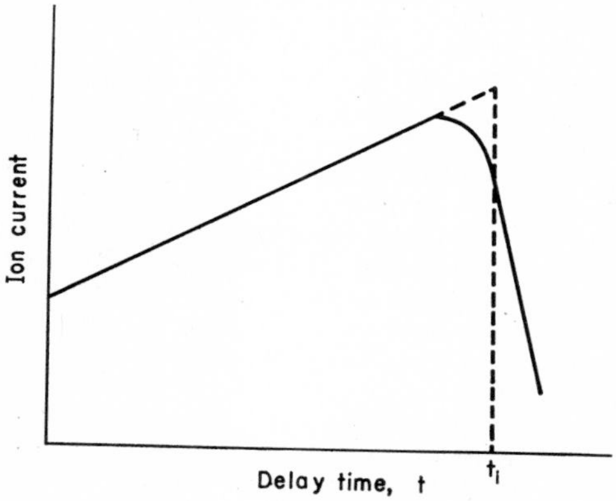

Fig. 3.

(*b*) *Observations of current transients.* In the method devised by Hornbeck a uniform electric field is established between an anode and cathode, and a flash of ultra-violet radiation used to produce a pulse of electrons at the cathode. As the electron pulse moves to the anode a Townsend avalanche is produced, the current transient being observed oscillographically. The transient which contains of a positive ion current component and a component due to secondary electron emission is analysed and the transit time of the ions determined. The method has also been used to measure electron drift velocities at low values of E/p where no ionization occurs (E is the electric field strength and p the gas pressure). Hornbeck's method is limited to measurements of ion drift velocity at high values of E/p. A modified form (Chanin and Biondi) can be used to measure ion drift velocities at low values of E/p. The cathode is replaced by a grid, the ions being formed in a discharge produced by a high voltage pulse of short duration between an auxiliary electrode and the cathode. The ion current transient is observed for two different anodes to cathode spacings in order to avoid end effects, and the drift velocity determined from the difference in transit times.

(*c*) *The use of ionization chambers.* The electron drift velocity can be measured by observing the electron current transient after electrons have been produced throughout an ionization chamber by a flash of

X rays (Hudson). In the method of Bortner, Hurst and Stone a pulse of electrons is produced by a stream of α particles, the time of formation of the pulse being recorded by the α particles activating a Geiger-counter. The time of arrival of the pulse after it has travelled a known distance is recorded by a second counter.

(*d*) *Observation of transit times of single particles.* The transit times for single electrons to traverse a known distance are measured and the drift velocity derived from the distribution of transit times. In the method of Hurst, O'Kelly, Wagner and Stockdale a Geiger–Muller counter is used to observe the arrival of the electrons. In a later modification of this method the detector used was an electron multiplier, differential pumping being employed. A similar method (Martin, Barnes, Keller and McDaniel) has been used to measure the drift velocity of positive ions. The ions are released in short pulses from a moveable ion source. The transit time of the pulse is found from the distribution of ion transit times for two positions of the ion source and hence the drift velocity determined.

(*e*) *Observations on electron avalanches.* Electron drift velocities at high values E/p can be measured by the observation of an electron avalanche initiated by photoelectrons emitted from the cathode. Observations can be made by means of a cloud chamber (Raether) which enables the spatial distribution of the avalanche to be determined, or by observing the current transient oscillographically and analysing its form.

Bibliography

DAVIES D. E. (1961) *in Encyclopaedic Dictionary of Physics* (J. Thewlis Ed.), **2**, 523, Oxford: Pergamon Press.

LOEB L. B. (1955) *Basic Processes of Gaseous Electronics,* Berkeley and Los Angeles: University of California Press.

McDANIEL E. W. (1964) *Collision Phenomena in Ionized Gases,* New York: Wiley.

M. T. ELFORD

TOWNSEND ENERGY FACTOR: TOWNSEND ENERGY RATIO. When an assembly of ions drifts and diffuses through a gas in the presence of an electric field the mean energy of the ions and of the gas molecules is nearly equal provided the ratio E/N of electric field strength to gas number density is sufficiently small. In these circumstances Townsend showed that the following equation relates the drift velocity W and diffusion coefficient D of the ions to the electric field strength, the gas temperature T and Boltzmann's constant $\boldsymbol{k}$:

$$W/D = E\boldsymbol{e}/\boldsymbol{k}T. \tag{1}$$

In general the situation is rarely as straightforward as this for electrons. Like the ions, the electrons acquire energy in addition to their thermal energy as they drift in the electric field but, because of the large discrepancy between the mass of the electron and the mass of even the lightest molecule, the transfer of excess energy from electrons to gas molecules is usually a very inefficient process. As a result, although it is possible, when E/N is very small, for the energy loss to balance the energy gain without a significant rise occurring in the mean electron energy, more usually a steady state is attained only when the mean electron energy is considerably in excess of the mean molecular energy. If the electron mass and speed are denoted by $\boldsymbol{m}$ and c respectively then $^1/_2\,\boldsymbol{m}\overline{c^2} > {}^3/_2\,\boldsymbol{k}T$ and the Townsend energy factor k_T is defined as the ratio $^1/_2\boldsymbol{m}c^2/(^3/_2\,\boldsymbol{k}T)$, that is

$$\frac{1}{2}\,\boldsymbol{m}\overline{c^2} = k_T\,\frac{3}{2}\,\boldsymbol{k}T.$$

In typical laboratory experiments the magnitude of k_T can range from one to several hundred.

When $k_T > 1$, equation 1 no longer applies but in special cases the equation can be simply modified. For example, when the electron energy distribution function is Maxwellian it can be shown that equation 1 can be replaced by

$$W/D = E\boldsymbol{e}/k_T\boldsymbol{k}T \tag{2}$$

where k_T is the ratio of electron to gas temperature. In these circumstances the use of equation 2 forms the basis of a convenient and accurate method of determining the electron temperature since the ratio W/D can be measured directly in a comparatively straightforward experiment (see Dutton 1961).

Such circumstances are, however, the exception. For instance, when elastic collisions alone determine the energy distribution, the distribution function is Maxwellian only when the collision frequency for momentum transfer ν_m is independent of electron energy. In the majority of cases the distribution is non-Maxwellian and it can be shown that equation 2 is no longer valid. Nevertheless the measurement of the ratio W/D can still be used to estimate the mean electron energy. (The use of the concept of electron temperature has now, strictly speaking, to be discarded.)

When formulae for the transport coefficients W and D are derived in terms of the electron speed c and the momentum transfer cross-section $q_m(c)$ the following formula for W/D is obtained:

$$W/D = E(\boldsymbol{e}/\boldsymbol{m})\left[\overline{c^{-2}\frac{d}{dc}(c^2/q_m)}\right] \Big/ [\overline{c/q_m}] \tag{3}$$

where the barred terms represent averages over all the speeds in the distribution. From equation 3 it follows that

$$\left.\begin{aligned} W/D &= (Ee/m\overline{c^2})\,[\overline{c^2}]\,\overline{\left[c^{-2}\frac{d}{dc}(c^2/q_m)\right]}\Big/\overline{[c/q_m]} \\ &= (Ee/k_T 3kT)\,[\overline{c^2}]\,\overline{\left[c^{-2}\frac{d}{dc}(c^2/q_m)\right]}\Big/\overline{[c/q_m]} \\ &= Ee/(kT(k_T/F)) \end{aligned}\right\} \quad (4)$$

where

$$F = [\overline{c^2}]\,\overline{\left[c^{-2}\frac{d}{dc}(c^2/q_m)\right]}/3[\overline{(c/q_m)}].$$

Since the factor F is a dimensionless quantity whose value is often close to unity, equation 4 shows that an approximate value of k_T can always be calculated from measured values of W/D but that equation 2 is valid only in special situations. (Using the formula for F, it can be shown that $F = 1$ for *any* form of the energy distribution function provided that ν_m is constant. Furthermore, the formula shows that, if the energy distribution function is Maxwellian, $F = 1$ for any energy dependence of q_m. These conditions rarely apply.)

With the realization of this fact, a further coefficient, the *Townsend energy ratio* k_1, was defined such that $k_1 = k_T/F$. Thus analogous to equation 2 we have

$$W/D = Ee/k_1 kT \quad (5)$$

and the energy ratio k_1 is therefore directly related to the ratio W/D. Unlike W/D, k_1 is a function of E/N only and not of E/N and N. This parameter is thus frequently used to summarize the results of experimental determinations of W/D since the measurements are usually made at a number of gas pressures. Its use has the additional advantage that the mean electron energy can be quickly estimated.

It is to be noted that the relation of k_1 to W/D carries no assumption as to the magnitude of the factor F; k_1 is, therefore, like W/D, an experimentally determined macroscopic property of an electron swarm. By contrast k_T, and hence the mean electron energy, can only be calculated from k_1 (or W/D) provided the energy dependence of q_m is known together with the form of the electron energy distribution function.

An alternative parameter, *the characteristic energy* ε_K is frequently used in place of k_1. If $\mu = W/E$, equation 5 may be written as

$$D/\mu = k_1(kT/e).$$

The parameter $\varepsilon_K = De/\mu$ thus estimates the mean electron energy directly rather than the ratio of this energy to the mean molecular energy. It is important to note that $(3/2)\,\varepsilon_K$ rather than ε_K is approximately equal to the mean energy since $(3/2)\,\varepsilon_K = (3/2) \times (De/\mu) = \frac{1}{2}mc^2$ when $F = 1$.

Bibliography

Dutton J. (1961) in *Encyclopaedic Dictionary of Physics* (J. Thewlis Ed.), 2, 403, Oxford: Pergamon Press.

Harrison J. A. (1967) in *Encyclopaedic Dictionary of Physics* (J. Thewlis Ed.), Suppl. Vol. 2, 407, Oxford: Pergamon Press.

Huxley L. G. H. and Crompton R. W. (1962) in *Atomic and Molecular Processes* (Ed. D. R. Bates), New York: Academic Press.

Loeb L. B. (1955) *Basic Processes of Gaseous Electronics*, California: The University Press.

McDaniel E. W. (1964) *Collision Phenomena in Ionized Gases*, New York: Wiley.

R. W. Crompton

TRANSFORMATIONS IN METALS AND ALLOYS, THEORY OF. A phase transformation may be produced by altering the external constraints on an assembly of atoms or molecules in such a way that the initial configuration becomes less stable (i.e. has a higher free energy) than some other configuration. The difference in chemical free energies is loosely described as the "driving force" for the transformation. This article outlines the general theory of the mechanism of phase transformations in metallic systems; more detailed descriptions of some particular transformations have been given in the separate articles listed at the end. The atomic mechanisms clearly also apply to phase changes in many nonmetallic materials, and to some extensive re-arrangements in the solid state which do not involve a change in the phase constitution of the assembly. These changes, which may be included in a wider definition of transformations, are characterized by driving forces of different origin; for example, the strain energy of a deformed structure may lead to recrystallization, or the internal surface energy of a fine-grained structure may lead to grain growth.

The net driving force is determined by macroscopic parameters appropriate to relatively large volumes of the phases involved, but the initiation of transformation is dependent on the effects of small fluctuations from the original condition, and is also usually very sensitive to the types and densities of defects present in the structure. Most phase transformations begin from discrete centres, a process known as nucleation, and the new regions then grow outwards into the surrounding material. Regions of discontinuity (macroscopic surfaces) are thus necessarily introduced into the assembly during transformation, even though the initial and final states may be single phase.

Growth of new phase(s). Transformations are usually classified by reference to the growth process, which may generally be more readily related than the nucleation stage to observable features of the reaction. Two main categories have been distinguished, namely nucleation and growth transformations and martensitic transformations. This terminology is not very satisfactory, since nucleation is involved in all cases, and alternative names such as diffusional, diffusionless, shear-type, etc. have been applied to the various transformations. These are also partially misleading, but a recent descriptive classification that has found some acceptance is to divide transformations into "*civilian*" and "*military*" types. In a nucleation and growth, or civilian, reaction, the atoms move independently of each other during the growth process, and the motion of an individual atom is nearly random. The atomic re-arrangement is much more orderly in a martensitic, or military, transformation, and in general each atom retains the same neighbours throughout the change. The division into two groups is not rigid, and information of intermediate types are also known.

In principle, a civilian transformation proceeds to completion under fixed external constraints, and the transforming regions do not normally undergo substantial changes of shape. Since the movements of individual atoms are dependent on local fluctuations in thermal energy, however, the rate of reaction varies markedly with temperature. Factors which affect the velocity of an interphase boundary include the atomic process by which the atoms are transferred across the boundary from one phase to the other, the supply or removal of the latent heat of the transformation, and the diffusion of matter if composition changes are required. The energy transfer conditions are often the primary factor governing growth rate during solidification or melting, but are seldom important in the normal experimental conditions used to investigate solid state transformations. When there is no composition change in the reaction, the velocity is then determined entirely by the interface processes, and this is sometimes referred to as interface-controlled growth. The other extreme is diffusion-controlled growth, in which an individual region grows or shrinks at a rate which is limited only by the time taken for atoms of a particular component to arrive at, or be removed from, the vicinity of the interface. In intermediate cases, the driving force of the reaction must be divided between the interface and diffusion processes, but one or other of these is likely to be predominant in determining the overall rate.

The kinetic features of a martensitic reaction are normally quite different from those of a nucleation and growth transformation, except for circumstances in which the overall martensitic reaction rate is governed by thermal nucleation. A martensitic transformation is characterized by the existence of a mode of growth which is not dependent on thermal activation, so that the interface velocity is very rapid and does not vary with temperature, even at temperatures near 0°K. On the other hand, the transforming regions normally undergo a large change of shape, the main component of which is a shear on the habit plane of the plate-shaped or lenticular crystals, and this change of shape has to be accommodated in the surrounding matrix. The associated strain energy must be supplied by the energy of the transformation, with the result that at a fixed temperature (driving force) growth may cease when only a fraction of the assembly has been transformed. Thus in a martensitic change, the amount of transformation may be a function of temperature, although the rate of transformation is not; such a reaction is said to be athermal. If the growth of the plates is halted before the yield stress of parent or product phases is reached, there is a state of thermoelastic equilibrium between the chemical driving force of the transformation and the opposing strain energy of the shape change. This equilibrium can be displaced by altering the temperature or external mechanical stresses, causing the plates to grow or shrink. More usually, growth of a plate ceases when plastic deformation destroys its coherency with the matrix, and further transformation under increased driving force then requires the formation of new crystals of the product phase.

Lattice correspondence. The theoretical concept which distinguishes military from civilian transformations is the postulate of a lattice correspondence. Let the initial and final structures each have primitive unit cells containing only one atom and suppose the atoms are labelled in some way. Then a lattice correspondence implies that a direction defined in one structure by a particular set of labelled atoms becomes a *corresponding* vector of the other structure, defined by the same set of atoms, though with a different spacing. Similarly, a labelled plane of one structure becomes a *corresponding* plane of the other structure. The relationship between the structures is described mathematically as an affine transformation, and the change of phase may be considered as an effective deformation of one structure into the other.

The existence of a lattice correspondence cannot be proved directly, but may be inferred from observations of the change of shape of a transformed region. As shown in Fig. 1, a volume of crystal transforming from one structure to the other under-

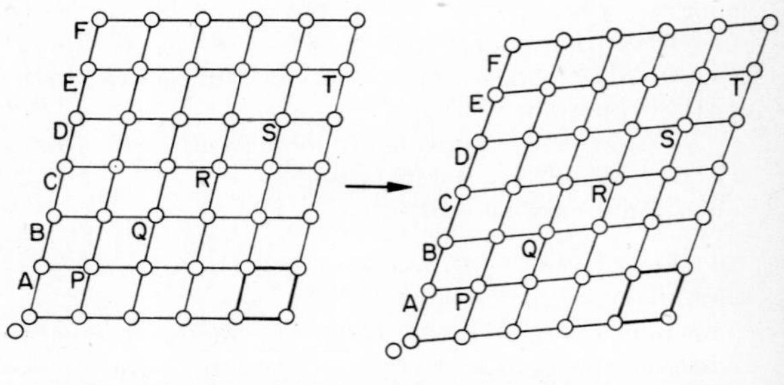

Fig. 1. Simple illustration of lattice correspondence. Labelled atoms define corresponding lattice vectors; corresponding unit cells are indicated by heavy lines.

goes a change of shape which is the same as the change of shape of corresponding unit cells. In a civilian transformation, there is no correspondence and the labelled atoms of one structure occupy a random set of sites in the other structure. There need thus be no net change of shape, other than that required by the change in volume per atom.

The concept of a lattice correspondence may also be applied to transformations in which one or both structures have primitive unit cells containing more than one atom. It is then necessary to select sets of equivalent lattice points in the two structures in such a way that these define unit cells containing the same number of atoms, and the correspondence then relates directions, planes and unit cells specified by an imaginary labelling of these equivalent lattice points. The complete structural change requires additional movements of atoms within the unit cell; these displacements are called shuffles, and do not contribute to the overall shape change.

The structure of the interface is clearly relevant to the existence of a correspondence. Boundaries in the solid state may be classified as coherent, semi-coherent and incoherent. A fully coherent interface must be a plane of matching which is common to the lattice of the two structures. Rows and planes of lattice points are continuous across this boundary, but have different orientations in the two crystals. Coherent twin structures are of this type, but two crystals of different structure do not generally have a plane (rational or irrational) of exact matching. Fully coherent interphase boundaries are thus rare when the crystals are reasonably large. If one crystal is very small, two structures which nearly fit together may be elastically strained into coherence.

If the transition region between the structures is narrow and very disordered, so that there is no continuity of lattice rows and planes across the interface, the boundary is said to be incoherent. The structure is not known in detail, but presumably resembles that of a high angle grain boundary. The boundaries in most civilian transformations are incoherent when the growing crystals are large enough to be visible under an optical microscope.

The semi-coherent boundary may similarly be compared to a low-angle grain boundary. The lattices are elastically strained into coherence over local regions of the boundary, but periodic discontinuities are necessary to correct accumulating misfit. These discontinuities have many of the formal properties of dislocations, except that they are constrained to remain in the boundary, and the net Burgers vector of the dislocations crossing any unit vector in the boundary can be defined in either of the lattices.

When an incoherent boundary is displaced, the atoms crossing the boundary from one plane to the other must move independently, and there can be no lattice correspondence. The atoms could also move independently with a coherent boundary, but it is also possible for such a boundary to be displaced in such a way that a lattice correspondence is maintained. The change of shape in the volume swept by the boundary is then equal to the change of shape of a unit cell, as illustrated in Fig. 1.

For a semi-coherent boundary, the change in shape of the macroscopic volume swept has to be distinguished from the change of shape of a unit cell. The correspondence may be maintained in the regions between discontinuities in the boundary, but the overall shape change is the net effect of the lattice deformation inherent in the correspondence and the "lattice invariant deformation" produced by the migration of the discontinuities in the boundary. This is the basis of the formal theory of martensite crystallography, discussed below.

Thermally-activated, interface-controlled growth. In civilian growth processes, atoms cross the interface independently of each other, and in so doing must move relative to their neighbours through distances at least of the order of an interatomic distance. A distinction is made in the theory between interfaces which are so disordered on an atomic scale that simultaneous growth is possible at all points on the surface, and interfaces which are stepped or facetted on an atomic scale. In the latter case, atoms are added to the growing phase only along the step edges, which thus move laterally across the interface as the crystal grows.

The surfaces of a crystal growing from the vapour phase always consist of stepped-sections of close-packed planes. It follows that high-index surfaces, which contain high densities of steps will grow most rapidly, and in so doing will remove themselves from the crystal habit. Thus the crystal should soon be bounded only by close-packed planes, and if there are no internal defects these planes should be atomically flat even at quite high temperatures. Further growth of such a perfect crystal takes place by the successive nucleation of each atomic layer, and this is a very slow process at normal supersaturations. The difficulty is avoided in real crystals by Frank's dislocation theory of crystal growth, according to which the emergent dislocation lines provide non-destructible growth steps on the close-packed surfaces.

The situation for liquid-solid interfaces and for some solid-solid interfaces is less certain. The calculation of the disorder, or surface roughness, is a complex problem, and has been only partially solved. It is usually considered that the disorder is related to the thickness of the transition region between the structures, continuous growth being possible for a diffuse interface, but not for a sharp interface. A thermodynamic theory due to R. W. Cahn, however, introduces the idea of steps even in a diffuse interface. According to this theory, continuous growth is always possible at sufficiently high values of the driving force, but stepped growth is required when the driving force is less than some critical value. For diffuse interfaces, this value is so low that continuous growth is possible in all circumstances, whereas for sharp interfaces the driving force may never become large enough for continuous growth.

A plausible model for continuous growth in condensed phases is to suppose that each atom has to surmount a free energy barrier of height Δg^* in crossing the boundary. If the difference in free energy per atom of the two phases (the driving force) is $\Delta g^{\alpha\beta}$, a simple application of rate theory gives the velocity, v, of the interface as

$$v = \delta\nu \exp(-\Delta g^*/\boldsymbol{k}T)\,[1 - \exp(-\Delta g^{\alpha\beta}/\boldsymbol{k}T)] \quad (1)$$

$$\simeq (\delta\nu/\boldsymbol{k})\,(\Delta g^{\alpha\beta}/T)\exp(-\Delta g^*/\boldsymbol{k}T) \quad (1\,\text{a})$$

where δ is the atomic jump distance across the interface, and ν is the atomic vibration frequency, assumed equal in the two phases. Equation 1a is valid if $\Delta g^{\alpha\beta} \ll \boldsymbol{k}T$, and shows that the growth velocity is directly proportional to the difference of free energy of the phases in the immediate vicinity of the boundary. The derivation is strictly applicable only to the motion of a planar interface, and the change in surface free energy has to be included to give the net driving force for the migration of a curved interface. This is usually unimportant, except for very small particles, but it becomes important in the theory of grain growth or particle coarsening, when the surface free energy provides the only driving force.

The kinetics of interface-controlled growth by a lateral step mechanism obviously differ from those of continuous growth, but it is difficult to formulate a general theory since there are so many possibilities. A simple assumption is that the growth rate is given by the product of the right-hand side of equation 1 and the fraction of atomic sites which are situated at steps. For spiral steps centred on dislocations, the minimum radius of the step depends on the driving force, so that the step length per unit area of interface is proportional to $\Delta g^{\alpha\beta}$. This gives a growth velocity v which varies as $(\Delta g^{\alpha\beta})^2$, in contrast to the linear variation for continuous growth. It is frequently convenient to regard the two growth velocities as proportional to ΔT and $(\Delta T)^2$ respectively, where ΔT is the difference in the growth temperature and the temperature at which the phases have equal free energies.

The activation energy for growth, Δg^*, may be approximately equal to the activation energy for atomic migration within the parent phase, or may be less than this if the interface structure is disordered. For an incoherent interface, Δg^* may reasonably be expected to approximate to the activation energy for grain boundary diffusion, rather than that for lattice diffusion.

Diffusion-controlled growth. In any growth problem in which long-range transport of matter is involved, the diffusion equation

$$\partial c/\partial t = D\nabla^2 c \quad (2)$$

where the concentration, c, of a given component is a function of time, t, and position vector, $\mathbf{x}$, must be solved for appropriate initial and boundary conditions. The diffusion coefficient D is usually taken to be independent of c in order to make the problems tractable, but this assumption is far from valid.

Consider the growth of an isolated β particle of solute concentration c^β from an α matrix of initial solute concentration c^m (continuous precipitation). If the growth rate is limited by diffusion, the concentration in the matrix immediately adjacent to the growing particle will differ infinitesimally from the equilibrium α concentration, c^α and this provides one boundary condition. The other boundary condition is simply $c = c^m$ as $\mathbf{x} \to \infty$ in an infinite matrix, but is very complex when an array of β particles grows from one α region and begins to compete for the available solute.

Equating the diffusion flux across the boundary to the excess solute in the volume swept by the boundary gives the boundary velocity as

$$v = (\mathrm{d}r/\mathrm{d}t)_{r_0} = \{D/(c^\beta - c^\alpha)\}\,(\partial c/\partial r)_{r_0} \quad (3)$$

where r is a co-ordinate normal to an element of the boundary and $r = r_0$ is the position of the boundary at any time. The concentration gradient may be written rather loosely as $\Delta c/y^D$ where Δc is the concentration difference between regions of the α phase near to and far from the boundary, and y^D is the effective diffusion distance. Thus if y^D is constant, the growth is linear with $r_0 \propto t$, but if $y^D \propto r$, the growth is parabolic and $r_0 \propto t^{1/2}$. A parabolic law is predicted for a spherical particle growing under diffusion control, for which an exact solution of the diffusion equation 2 can readily be obtained. Exact solutions for ellipsoidal particles of general shape were first given by Ham, who found that shape preserving solutions exist for all ellipsoids, so that the volume is proportional to $t^{3/2}$ and the eccentricity of a plate or needle-shaped particle does not change with time. This result depends on assumptions that the particle grows from a very small size, that the concentration is uniform over the surface, and that atoms remain in the sites at which they reach the β crystal. The result was contrary to earlier assumptions that plate and needle particles would have constant y^D (and thus grow linearly) in one or two dimensions, giving volumes proportional to $t^{5/2}$ and t^2 respectively. An unsolved problem, however, is the stability of the interface to perturbations. Recent work has shown that a fluctuation in interface shape should lead to a breakdown of smooth surface into a dendritic form, and although the curvature of the surface imposes some stability because of the effect of a curved surface on the composition, this is restricted to particles smaller than about seven times the critical nucleus radius. Experimentally, particles of very much larger size are observed to maintain smooth shapes, and it is possible that the growth in such cases is partially controlled by interface processes.

The other main growth problem for an isolated particle under diffusion control is that of a needle or plate growing inwards from a grain boundary. The simplest treatment of the edgewise growth of a plate without thickening, is known as the *Zener-Hillert model*. The radius of curvature, r, at the tip of the plate must exceed a critical value r_c at which the net driving force becomes zero because of the surface free energy created. It is assumed that Δc defined above is reduced by $(1 - r_c/r)$, and that the actual radius of the tip is that which maximizes the growth rate, and putting the diffusion distance $y^D = Cr$ it is readily seen that this maximum radius is $r = 2r_c$. The growth rate is then

$$v = \frac{D}{2Cr_c} \frac{c^m - c^\alpha}{c^\beta - c^\alpha}.$$

Since r_c is readily shown to be inversely proportional to the chemical free energy released per unit volume, this model gives a growth velocity proportional to the driving force. A more elaborate calculation made by Hillert gives $C \approx \frac{1}{2}$ for a plate, and the maximum growth rate for a needle is $\sim 1{\cdot}5$ times larger than that of a plate.

The difficulty with the Zener–Hillert model is that the boundary conditions of the diffusion problem are not in fact consistent with a steady state solution to the diffusion equation. An exact solution for the problem of a particle growing linearly under diffusion control in one direction and having the general shape of an elliptical paraboloid may be obtained, provided that the effect of a curved surface on the composition in equilibrium is neglected. This effect is emphasized in the Zener–Hillert treatment, and is the basis of the classical theory of nucleation, but its significance is less certain in growth problems. The exact treatment of the diffusion problem neglecting curvature originates from a theory of dendritic growth by Papapetrou and has been given in its most advanced form by Horvay and Cahn. The theory results in a value for the product of tip radius and velocity, but some other condition is needed to specify both quantities. The difficulty of using assumptions such as the Gibbs–Thomson condition is that they modify the boundary conditions of constant composition utilized in the solution of the diffusion equation. Horvay and Cahn believe that this condition is a good approximation and will give the correct velocity if the tip geometry is determined experimentally. A notable result of the theory is then that the velocity is given by an expression of form:

$$v = (\Delta T)^z / r \qquad (4)$$

where the supercooling is ΔT and z is a slowly varying function of ΔT. For a needle of circular cross-section, z is about 1·2, and for a flat platelet, z is about 2. This result is contrary to the linear dependence of the Zener–Hillert model.

The growth problem just mentioned is formally identical with that of the growth of a solid dendrite from the liquid phase, when growth is controlled by heat transport. Indeed all transformations controlled by diffusion of matter have formal analogues in which diffusion of energy is rate controlling.

There is one other important form of diffusion controlled growth, namely the formation of a lamellar aggregate of α and β phases from a γ phase (eutectoidal decomposition) or from a supersaturated α phase (discontinuous or cellular precipitation). In this case, the mean composition of the duplex product is equal to that of the parent phase, so that there is a steady state solution to the growth problem and a linear growth law. The diffusion transport is essentially parallel to the interface, rather than normal to it, and the extent of the diffusion field is independent of the position of the interface.

In most theories, it is assumed that the rate-limiting process is the segregation of solute by volume diffusion in the matrix. The situation is shown schematically in Fig. 2, which also serves to define some solute concentrations used in the following discussion. With an origin in the moving boundary and co-ordinates x_1 and x_2 normal to the boundary and in the boundary normal to the lamellae, the growth equation is

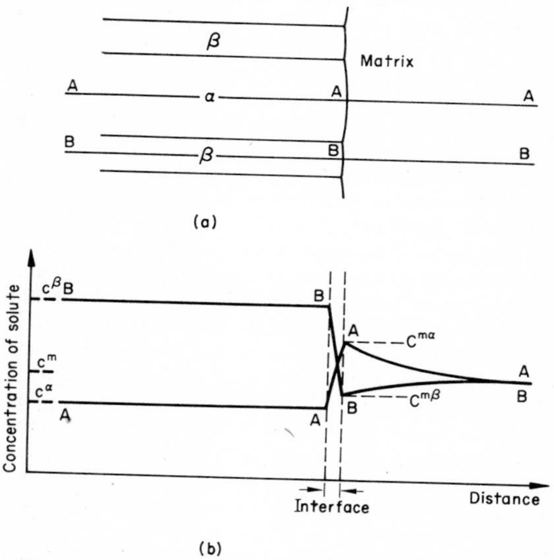

Fig. 2. To illustrate the growth of a laminar two-phase aggregate. In (b) the solute concentration along the lines A–A–A and B–B–B of (a) is plotted.

$$D\{(\partial^2 c/\partial x_1^2) + (\partial^2 c/\partial x_2^2)\} + v(\partial c/\partial x_1) = 0. \qquad (5)$$

An exact solution of (5) may be written in the form of an infinite series, but the boundary conditions are not sufficient to enable the constants to be evaluated. Approximate solutions are written in the form

$$v = \alpha\, D/y \qquad (6)$$

where α is a rather complex homogeneous function of the various compositions in Fig. 2.

The interlamellar spacing y is a parameter in the diffusion problem, and the question of what determines its value has next to be answered. In the Zener–Hillert theories, equation 6 is multiplied by $(1 - y_c/y)$ where y_c is the critical value of y at which growth ceases because all the chemical free energy is transformed into interfacial free energy. If the growth rate is then maximized, this gives $y = 2y_c$, and implies that $y \propto \Delta T$. Experimental results are often consistent with this prediction, but the slope of the line gives unreasonably high values for the interfacial free energy of the $\alpha - \beta$ interfaces.

In general, it is clear that the observed spacing y is usually considerably larger than the Zener prediction for both eutectoidal decomposition and discontinuous precipitation. Various possible reasons for this include growth controlled by processes other than diffusion, selection of the spacing by some principle other than that of maximum growth rate, and formation of a product with non-equilibrium segregation of solute. There is little doubt, however, that the most important effect is frequently that diffusion takes place predominantly along the incoherent boundary between parent phase and duplex product. This boundary is particularly effective since it sweeps the matrix and is favourably orientated to produce the required segregation parallel to itself.

A simple modification of the theory to allow for boundary segregation is to replace D/y by $D^B\delta/y^2$ where D^B is the boundary diffusion coefficient and the factor δ/y allows approximately for the reduced cross-section of the diffusion path. A complete theory, however, requires more radical modification, since there must be a composition gradient along the boundary, and it is therefore impossible to achieve equilibrium segregation at any finite growth rate. The diffusion equation in fact provides a relation between three quantities, the growth rate v, the spacing y and the degree of segregation achieved. Within the thermodynamic limitation that the free energy must decrease, the boundary could move slowly for a fixed spacing, producing near-equilibrium segregation, or more rapidly, producing less separation. This applies equally to segregation by diffusion in the volume of the decomposing phase, and it may be incorrect to assume that such diffusion produces equilibrium segregation.

Clearly further assumptions are necessary to determine the unknown variables and provide a complete theory of growth. Cahn assumed that the boundary mobility is also important, i.e. that the growth rate is both diffusion- and interface-controlled. This fixes both the degree of segregation and the growth rate for a fixed spacing. The spacing itself in Cahn's theory is obtained by maximizing the net free energy decrease, which also maximizes the growth rate. There is reason to believe that a correct growth condition should be obtained by considering the stability of the lamellar array against fluctuations which tend to either increase or decrease the spacing.

The permissible segregation achieved in a eutectoidal reaction where neither phase can approximate to the matrix composition is never far removed from the equilibrium segregation, but wide variations are possible in discontinuous precipitation. According to Cahn's theory, a small ratio of boundary mobility to diffusion coefficient results in a large spacing, and almost all the driving force is used to move the rather sluggish boundary. The rapid diffusion gives near equilibrium segregation, even with large y. When the mobility is relatively much larger, the spacing is smaller and most of the chemical energy released is converted into $\alpha - \beta$ interface energy. In precipitation reactions, the change in spacing is lessened because of smaller segregation, but in eutectoidal reactions a very small spacing is possible, approaching a limit of $1{\cdot}25\, y_c$ in contrast to the Zener spacing of $2y_c$.

Crystallography of martensite. Martensite was originally the name given to the product of the rapid phase transformation in quenched steels, but is now used for all transformations of this type. The interface velocity can be of the order of one-third the velocity of sound, even at temperatures approaching 0°K, so that the growth is not thermally activated. A martensite transformation is thus characterized by an interface structure which permits this rapid growth, and the interface is sometimes described as glissile.

The theory of martensite crystallography depends on the postulate of a local lattice correspondence and the experimental observation that a transforming region changes shape. The change of shape (Fig. 3) is

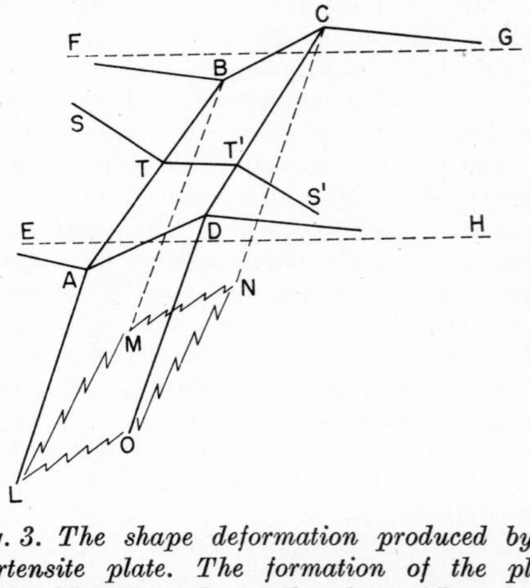

Fig. 3. The shape deformation produced by a martensite plate. The formation of the plate ABCDOLMN produces tilts along AB and CD where the habit plane intersects the free surface, and deviates a straight scratch into STT'S'.

an invariant plane strain, and may be regarded as the combination of a simple shear on the invariant plane plus an expansion or contraction normal to this plane. The martensite crystals usually have the form of thin plates (lenticular shaped when the crystal is constrained by the surrounding matrix) and the habit plane of the plates is the invariant plane of the shape change.

If the shape change is applied to a unit cell of the parent structure, it will not usually produce a unit cell of the product. This is a necessary result, since, as already pointed out, two structures cannot in general have a completely coherent or matching interface. The martensite interface is thus semi-coherent, the relations between lattices being specified by the lattice correspondence, and the overall shape change being the net result of the lattice and lattice invariant deformations. This is illustrated schematically in Fig. 4.

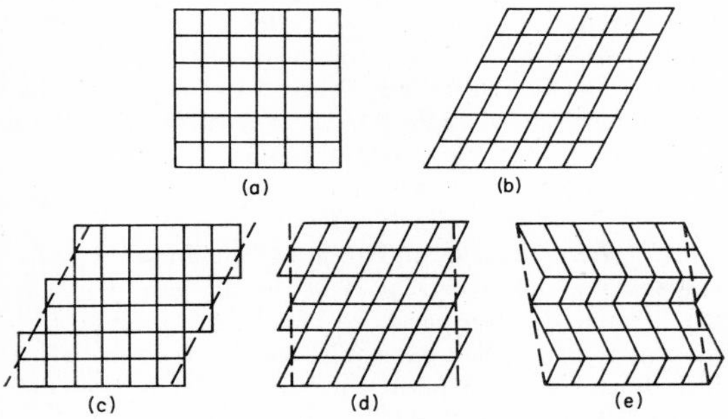

Fig. 4. Types of deformation in martensitic transformations. (a) Original undeformed crystal; (b) Lattice deformation; (c) Lattice invariant (or dislocation) deformation by slip; (d) Lattice deformation and lattice invariant deformation combining to give zero total shape deformation; (e) shape deformation caused by combining two different lattice deformations.

A lattice invariant deformation produced by slip involves the motion of interface dislocations in slip planes which meet edge to edge in the interface (Fig. 5(a)). The lattice invariant deformation is thus a simple shear, and since it produces no change in vectors in the direction of the dislocation lines, this direction is an invariant line in the lattice deformation.

Figure 4 also shows that the total shape deformation may be varied by formation of a duplex product in which different regions have different lattice correspondences with the parent phase. The interfaces introduced into the product in this way clearly must have low surface free energies, which means in practice that they must be coherent twin boundaries. So far as the macroscopic effects are concerned, the twinned product has a shape change which is obtained by adding a simple shear to the lattice deformation in region 1 of the product; this effective lattice invariant deformation is given by the twinning shear multiplied by the volume fraction of region 2. The invariant line of the lattice deformation is now the intersection of the twin boundaries and the habit plane (Fig. 5(b)).

The complete theory may be summarized as follows. Given the two structures, the simplest lattice correspondence is selected, usually by inspection. This determines the pure deformation inherent in the lattice relations, but the whole lattice deformation also

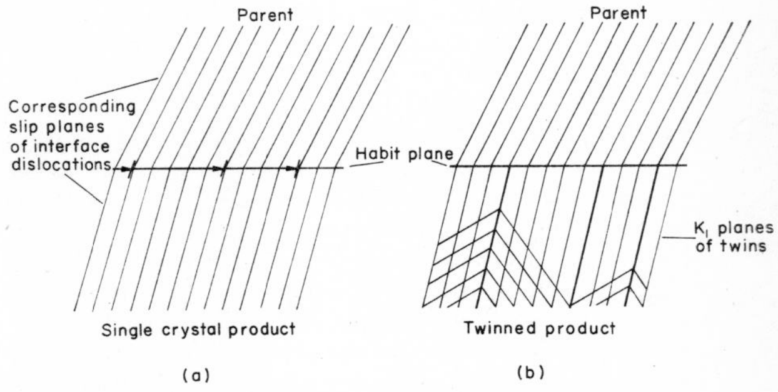

Fig. 5. Section through martensite interface normal to the invariant line of the lattice deformation. (a) Dislocation interface; (b) Twin interface.

includes a rotation, unknown at this stage. The elements of the lattice invariant deformation (or equivalently the second lattice correspondence in the twinning case) are not specified, but the choice is usually restricted to slip or twinning elements of either parent or product structures. The amount of shear (i.e. the dislocation spacing, or the volume ratio of the twins) is then determined by the condition that the shape deformation shall have two undistorted planes. Finally the rotation is determined by the condition that one of these planes shall be returned to its original orientation, giving an invariant plane shape deformation. The theory predicts the habit plane, the magnitude of the shape change and the orientation relations between the lattices. In general one set of input data and assumptions about the elements of the lattice invariant deformation lead to four non-equivalent solutions of the crystallographic problem, but in some cases these reduce to two or to one solution.

This theory was developed independently by Wechsler, Lieberman and Read and by Bowles and Mackenzie, following earlier work by Greninger and Troiano and others. It has been very successful in predicting, for example, the irrational habit planes of type {3, 10, 15} often found in steels. Moreover, thin foil transmission electron microscopy has shown that many martensite plates do consist of stacks of fine twins, exactly as assumed in the theory. Some transformations have crystallographic features which are not predicted by the simple theory, but which can be explained if a small relaxation of the invariant plane strain condition is permitted, as in the Bowles–Mackenzie theory. Even with this degree of freedom, however, some crystallographic relations are not fully explained at present; it is likely that these depend on the matrix constraints on the shape deformation which are not included in the formal geometrical theory of matching. It should perhaps be emphasized that the various stages of the calculation above represent convenient mathematical factorizations and not physically separable atomic movements. At the interface, each atom of the parent effectively passes into the nearest available site of the product, and moves relative to its neighbours only by a small fraction of an interatomic distance.

Nucleation. Gibbs distinguished two kinds of fluctuation within a metastable phase, namely an appreciable atomic re-arrangement confined to a very small volume, and a very small re-arrangement spread over a large volume. The first type of fluctuation is considered in the classical theory of nucleation; the second type is only possible in certain restricted circumstances and leads to quasi-homogeneous transformation.

The classical theory of nucleation was initially developed by Volmer and by Becker and Döring for the formation of liquid droplets from a supersaturated vapour, but with appropriate modifications may be applied to all transformations. The change in free energy on forming a nucleus of given size first increases and then decreases as the size increases; this is conventionally expressed by dividing the free energy change into a negative term proportional to the volume and a positive term proportional to the surface area. The division is really arbitrary, although convenient, and the surface free energy, for example, is not necessarily to be identified with the macroscopic parameter appropriate to larger surfaces.

The nucleation rate is dependent on the maximum increase in free energy at some critical nucleus size, and the variation in free energy with nucleus shape has also to be considered. In the solid state, the optimum shape depends not only on the surface energy but on the elastic strain energy produced by the constraints of the surrounding matrix, and further variables are the degrees of coherency of the nucleus and matrix. A coherent particle in general will have a low surface energy but a higher strain energy than an incoherent particle, and since the strain energy for fixed shape varies as the volume, there will be a tendency for very small particles to be coherent, but to become incoherent at some critical size. Finally the variation of free energy with the composition of the nucleus has to be considered in many transformations, and it cannot necessarily be assumed that the nucleus has the equilibrium composition of the new phase.

The theory of nucleation shows that the nucleation rate along any path depends on $\exp(-\Delta G_c/\boldsymbol{k}T)$ where ΔG_c is the maximum increase in free energy along that path. Because of the form of this expression, the minimum value of ΔG_c when alternative paths are considered will give a nucleation rate far exceeding the others. Thus the nucleation rate depends on the value ΔG_c which is a saddle point in the energy field. This is the minimum energy barrier which permits continuous increase in the size of the nucleus.

The rate of nucleation is never constant, but in the early stages of the reaction a quasi-steady state distribution of embryos of various sizes may be established, and this leads to a nearly constant nucleation rate per unit volume of these assembly. This steady state rate of homogeneous nucleation depends on the encounter rate of atoms and nuclei as well as on the exponential term. In a gaseous phase, it is thus proportional to the surface area of a critical-sized nucleus and to the collision rate of vapour atoms, whereas in condensed states of matter the nucleation rate may be written

$$I = C \exp(-\Delta g^*/\boldsymbol{k}T) \exp(-\Delta G_c/\boldsymbol{k}T) \qquad (7)$$

where Δg^* is the energy for the movement of atoms across the interface and C contains the total number of atoms in the assembly and the atomic vibration frequency as well as other dimensionless combinations of the quantities involved. When long-range diffusion is involved in the nucleation, Δg^* is replaced by the activation energy for such diffusion.

For a transformation which takes place on cooling, equation 7 predicts the familiar C-shaped T-T-T diagram (Christian 1962). The first exponential term, representing the growth part of the process, decreases with decreasing temperature, since Δg^* is nearly independent of temperature. The second term increases with decreasing temperature, since ΔG_c is inversely proportional to $(\Delta T)^2$ where the supercooling, ΔT, is assumed proportional to the driving force. The result of the two opposing factors is a maximum nucleation rate and hence transformation rate at some finite supercooling. If the high temperature phase can be cooled sufficiently rapidly through this "nose" of the C-curve, transformation may be prevented almost indefinitely at lower temperatures.

In practice, nucleation does not usually occur homogeneously but is catalysed by impurity particles or by grain boundaries, dislocations, and other imperfections in the solid state. The theory of heterogeneous nucleation depends on a reduction of ΔG_c by utilizing some of the energy already frozen into the structure, for example the surface energy of an impurity or a grain boundary, or the strain energy of a dislocation. A reduction in ΔG_c usually means a faster overall nucleation rate, but not invariably so, since the pre-exponential term is now proportional to the number of atoms in contact with the catalyst, rather than to the number of atoms in the whole assembly.

Gibbs's second type of fluctuation is only possible if a continuous range of structures can form between two phases, with effectively a diffuse interface. An example in the solid state is the establishment of periodic composition fluctuations of relatively long wave-length, which increase gradually in amplitude until ultimately they lead to the formation of two solid solutions of the same crystal structure but differing composition. This process is called *spinodal decomposition,* and is only possible when the second derivative of the free energy with respect to composition is negative, i.e. effectively when the diffusion coefficient is negative. This type of transformation does not represent nucleation in the normal sense, and the original phase is unstable rather than metastable to the fluctuations. The decomposition, however, takes place at a finite rate because of the energy barrier to atomic migration. There are relatively few solid state transformations which can begin in this homogeneous fashion; apart from spinodal decomposition, the most likely examples are some forms of order-disorder transformation.

The above nucleation processes are all thermally activated, and in some martensitic transformations thermal nucleation is clearly involved since isothermal transformation characteristics of the C-type are observed. The condition leading to the transition from thermally activated growth in the nucleation stage to athermal martensitic growth is not clear, and the classical nucleation theory is also difficult to apply to martensite since the large strain energy implies an improbably large free energy barrier, ΔG_c.

Two kinds of athermal nucleation process have been discussed. In the first of these, a distribution of embryos which is subcritical at one temperature becomes supercritical by rapid quenching to another temperature. This can occur in all types of transformation. A related way of avoiding nucleation at a particular transformation temperature is to give a short inoculating treatment at some temperature where nucleation is rapid, and then to up-quench to the required transformation temperature.

The second kind of athermal nucleation is the possibility that in some martensitic transformations, existing lattice defects can lead to the spontaneous formation of regions of product phase. This is likely to happen for example in martensitic transformations between f.c.c. and h.c.p. phases. In this special case, a stacking fault in one phase, formed by dissociation of a lattice dislocation, is already a two-dimensional nucleus of the other phase.

Kinetics of nucleation and growth reactions. In a homogeneous reaction, the volume transforming in any small time interval may be taken to be proportional to the volume remaining untransformed at the beginning of this interval, and this leads to a first order rate process. In a nucleation and growth reaction, the situation is more complex. At time t, the volume of an individual region of a new phase originating at time τ may be written $\eta v_1 v_2 v_3 \ (t - \tau)^3$ where η is a shape factor and v_1, v_2, v_3 are growth velocities (assumed constant) for three mutually perpendicular directions. If the nucleation rate per unit volume is I', the volume fraction of the assembly transformed at time t becomes

$$f(t) = v_1 v_2 v_3 \int_{\tau=0}^{\tau=t} I'(t - \tau)^3 \, d\tau \qquad (8)$$

provided that the interference which transformed regions cause to one another by impingement is neglected. This is only possible at the beginning of the transformation, but Johnson and Mehl, and Avrami showed how impingement can be treated accurately, and the volume fraction transformed then becomes

$$f(t) = 1 - \exp\left[\eta v_1 v_2 v_3 \int_0^t I'(t - \tau)^3 \, d\tau\right]. \qquad (9)$$

Equation 9 can be integrated only by making assumptions about the variation of I' with time. If I' is constant

$$f(t) = 1 - \exp(-\eta v_1 v_2 v_3 t^4/4). \qquad (10)$$

In general I' may either decrease or increase with time, and this leads to changes in the time exponent of the growth law. A general formulation of equation 10, suggested by Avrami, is

$$f(t) = 1 - \exp(-kt^n) \qquad (11)$$

where $3 \leqq n \leqq 4$ for a three-dimensional growth process in which I' is either constant or some decreasing function of time. Various other possibilities exist, for example if growth is constrained by a free surface in one dimension (a thin sheet) or two dimensions (a wire) n is decreased by 1 and 2 respectively.

Isothermal transformation curves typically have sigmoidal shapes as predicted by equation 11 and the value of n may be determined experimentally. It was once believed that this gave information about the growth process and the shape of the particles, but this is not generally true unless there is independent evidence of the time dependence of the nucleation rate.

The situation is more complex when heterogeneous nucleation, e.g. on grain boundaries, is considered. The first treatment of this was given by Johnson and Mehl, but a much more complete theory has been developed by Cahn. The outstanding feature of any form of heterogeneous nucleation is that the transformed product is concentrated near the nucleating regions, and ultimately covers them completely, a situation referred to by Cahn as "nucleation site saturation". Before saturation occurs, the fact that nuclei are confined to particular sites scarcely affects the overall transformation rate, and equation 9 applies if I' is the nucleation rate per unit volume of the assembly due to the heterogeneous sites. After site saturation occurs, the nucleation rate is effectively zero, and the transformation law is effectively just a growth law. For randomly distributed sites, this gives $n = 3$ as already noted; for grain boundary sites, growth is one-dimensional and $n = 1$.

The above development considered linear growth, and modifications are necessary for diffusion-controlled parabolic growth. The volume of a transformed region is now proportional to $(t - \tau)^{3/2}$ for three-dimensional growth, but in a process such as continuous precipitation the fraction transformed has to be treated more carefully, since complete transformation does not correspond to the whole assembly having changed phase. If $f(t)$ is now defined as the ratio of the volume of β at time t to the equilibrium volume of β at time τ, an uncritical application of the previous theory would give an equation of the form of (9) with a homogeneous function of the concentrations c^{α}, c^{β}, c^{m} included in the exponential term, and $(t - \tau)^3$ replaced by $(t - \tau)^{5/2}$. Thus it is often stated as an approximation that Avrami's equation is valid also for this case, with $n = 5/2$ for constant nucleation rate and $n = 3/2$ for early site saturation of randomly distributed heterogeneous nuclei. However, this is strictly incorrect. The form of equation 11 depends on the geometrical treatment of impingement. In a diffusion controlled reaction, two neighbouring β particles do not cease to grow because they impinge upon one another, but because all the excess solute has been drained from the region between them. The stage of competition for solute is sometimes called soft impingement, and the rather complex theory has been given by Zener and Wert and by Ham.

Some particular transformations. Crystal growth from the vapour and solidification are dealt with in extensive separate articles, as also are many solid state transformations. It may be convenient to assemble here, however, a few remarks about particular transformations to which the above theory applies.

Melting. The absence of superheating when a solid melts is believed to be due to the free surface of the solid which acts as a perfect nucleation catalyst ($\Delta G_c = 0$) for the formation of liquid.

Polymorphic transitions. In elements, polymorphic transformations occur by nucleation and interface-controlled growth at temperatures where atomic mobility is adequate. If the equilibrium transformation temperature is sufficiently low, only martensitic type changes are possible (e.g. cobalt, lithium, sodium) and complete transformation is then sometimes not achieved. Elements such as iron or titanium transform in a civilian manner if cooled moderately slowly, but the transformation becomes martensitic at higher cooling rates.

Massive transformations. This term originated from the "block" microstructure in some transformed alloys, and has inadvertently been applied to two quite different types of transformation. The first of these is a civilian transformation in alloys, essentially equivalent to a polymorphic transition in an element. The compositions of the product and parent are identical, since the rapid growth mechanism leaves no time for diffusion, but the boundaries are incoherent, and new regions grow across the prior grain boundaries of the parent phase. The second use of the term is for a type of martensitic transformation in which individual plates grow in parallel, self-accommodating groups. Careful microscopy is necessary to reveal the individual transformed regions, rather than the groups, and a distinction has sometimes been made between this so-called massive martensite and normal or acicular martensite. The two types of massive transformation appear to have nothing in common other than the production of a new phase of identical composition, and a superficially similar microstructure.

Precipitation. Separate consideration is usually given to continuous, discontinuous and low-temperature precipitation. Continuous precipitation means the formation of new β particles in a supersaturated α' matrix, and their subsequent growth until an equilibrium $\alpha + \beta$ array is obtained. The distribution of the precipitates may be random or may be localized, at the grain boundaries for example, but the final α grains have the same orientations and essentially the same sizes and shapes as the original α' grains. Discontinuous precipitation in contrast implies the formation of a duplex $\alpha + \beta$ array, which then grows and consumes the α' matrix.

It follows from the above discussion that the growth rate is likely to be larger for discontinuous precipitation, but nucleation of the duplex product is more difficult. This is supported by the experimental observation that continuous precipitation occurs mainly under conditions where nucleation is difficult (low driving force, or large misfit energies between phases) whilst the reverse is true for discontinuous precipitation.

Low temperature precipitation processes usually begin with the formation of solute atom clusters, which are known as Guinier–Preston zones when they have reached a size sufficient to give X-ray diffraction effects. The sequence of zones, intermediate coherent precipitates and final equilibrium precipitates involved in low temperature ageing reactions is frequently very complex. The transformations are greatly influenced by excess concentrations of vacancies retained in quenching from the solution temperature.

Eutectoidal transformations. The duplex laminar product in a eutectoidal decomposition or discontinuous precipitation does not consist of separate parallel plates, but of two continuously connected phases. At high temperatures, where nucleation is difficult, pearlite grows from a few centres as approximately spherical group nodules, each containing many "colonies" or cells of parallel plates. At lower temperatures, site saturation at grain boundaries occurs early in the reaction, and the pearlite then grows inwards from these boundaries. In the upper temperature range, the nucleation rate can be independently measured, and it is found to increase with time. Whilst this may be partially a transient in establishing a quasi-steady state nucleation rate, the main effect is probably due to a two-stage nucleation process. The rate of nucleation of the second phase of the duplex product may be dependent on the surface area of the previously formed nucleus of the first phase.

Recrystallization. Formally this is an interface-controlled process of growth from small nuclei. The available driving force, however, is so small that the classical theory of nucleation is probably inapplicable. Recrystallization in moderately deformed materials has been shown to begin by the bulging of an existing grain boundary separating local regions of different strain energy. In other cases, recrystallization probably originates from the growth of favourable regions which have polygonized. Thus, although an operational nucleation rate may be defined, the process is mainly one of growth, with an initially increasing growth rate.

Military transformations. Various transformations with characteristics differing from those of martensite nevertheless appear to include a lattice correspondence, since a shape change is observed. The formation of bainite in steels involves long-range diffusion to produce ferrite and cementite, and is plausibly explained as a martensitic transformation for which there is insufficient driving force unless diffusion also occurs. The carbon atoms behave as civilians and diffuse rapidly, but the iron atoms are relatively immobile so that a military type transformation is required to produce ferrite from austenite. In lower bainite, a supersaturated ferrite is produced first, and carbide is precipitated subsequently within this phase. In upper bainite, ferrite of new equilibrium composition forms, enriching the austenite until carbide is precipitated. This theory of bainite implies that the growth rate is diffusion controlled, but there is also some evidence for a rival hypothesis that it is interface controlled.

Other examples of military transformations are certain order-disorder changes, to which the theory of martensite crystallography applies, even though atomic interchanges must be involved in ordering. A generalization of the idea of a military transformation requires that some of the atoms move only a few interatomic distances relative to their neighbours during the period of growth. In a martensitic transformation, a correspondence is maintained, and relative movements of neighbours are less than an interatomic distance. In a military order-disorder transformation, the restriction is relaxed, but the movements of an individual atom must still be much smaller than the dimensions of a product crystal, otherwise the correspondence would be destroyed because of the favourable reduction in strain energy. Finally in a bainite transformation, only some of the atoms maintain a correspondence. Clearly there may also be intermediate type transformations in which a correspondence is partially destroyed during growth by civilian-type atom migrations.

Mechanical twinning. This may formally be regarded as a type of martensitic transformation in which the only driving force comes from external stress. The reverse of this statement is also true, i.e. martensite formation is often a mode of plastic deformation. The crystallography and physics of twin formation have many features in common with martensite, including the existence of irrational habit planes in type II twins. However, twins can always form fully coherent interfaces, so that some of the complexities of the theory of the crystallography of martensite are avoided.

Bibliography

Burke J. (1965) *The Kinetics of Phase Transformation in Metals*, Oxford: Pergamon Press.

Christian J. W. (1961) in: *Encyclopaedic Dictionary of Physics* (J. Thewlis Ed.), **3**, 655, **7**, 405, Oxford: Pergamon Press.

Christian J. W. (1965) *The Theory of Transformations in Metals and Alloys*, Oxford: Pergamon Press.

Christian J. W. (1965) in: *Physical Metallurgy* (R. W. Cahn Ed.), 443.

Clifton D. F. (1962) in: *Encyclopaedic Dictionary of Physics* (J. Thewlis Ed.), **4**, 497, Oxford: Pergamon Press.

Fine M. E. (1964) *Introduction to Phase Transformations in Condensed Systems*, New York: MacMillan.

Fletcher N. H. (1961) in: *Encyclopaedic Dictionary of Physics* (J. Thewlis Ed.), **3**, 300, Oxford: Pergamon Press.

Frank F. C. (1961) in: *Encyclopaedic Dictionary of Physics* (J. Thewlis Ed.), **2**, 208, Oxford: Pergamon Press.

Smallman R. E. (1961) in: *Encyclopaedic Dictionary of Physics* (J. Thewlis Ed.), **1**, 98, Oxford: Pergamon Press.

Smoluchowski R. (1961) in: *Encyclopaedic Dictionary of Physics* (J. Thewlis Ed.), **2**, 384, Oxford: Pergamon Press.

Turnbull D. (1962) in: *Encyclopaedic Dictionary of Physics* (J. Thewlis Ed.), **5**, 163, 627, Oxford: Pergamon Press.

Wayman C. M. (1964) *Introduction to the Crystallography of Martensitic Transformations*, New York: MacMillan.

J. W. Christian

U

ULTRA HIGH-SPEED RADIOGRAPHY. In conventional radiography exposure-times are usually measured in seconds or minutes, with exposure-times down to 2 milliseconds being used in diagnostic medical radiography to stop motion. To obtain sharp radiographs of a high-speed projectile or an explosive event, however, exposure-times of 1 microsecond or less are necessary. This is the technique of *ultra-high-speed radiography*, or *flash radiography*. Between these methods and conventional radiography, systems of cine-fluoroscopy usually using some form of X-ray image intensifier system, are available; these are at present limited to frame rates of about 100 frames/second.

The use of X-rays is often the only method suitable for observing explosive phenomena, as the details are generally obscured by smoke or flame, which limits conventional high-speed photography, or the phenomena take place in an opaque container. Flash radiography methods are therefore widely used in armaments research and some recent developments of equipment have not been published because of the classified nature of the applications.

The problem of generating a pulse of X rays containing sufficient energy to produce a record, within an exposure-time of 10^{-7}–10^{-6} sec, is considerable. The X-ray pulse must be produced consistently, with predictability of time of production so as to enable synchronization with the event under examination to be possible. Desirable but less essential features of the equipment are that the X ray tube has a small focal spot and also has a reasonable life in terms of the number of flashes possible before tube failure; the latter is a serious practical problem as the peak anode dissipation during pulses is typically a few hundred megawatts.

The first really practical flash X-ray tube was described by Slack and Ehrke in 1941 and consisted of a plate cathode surrounded by a trigger electrode and a plate anode, in a vacuum envelope. It was intended to be used with a Marx impulse generator at 180–300 kV. Concurrently, Schaaffs and others developed a flash X-ray tube operating on the same principle, but using a conical anode so as to obtain an effective small focus. With both these types of tube, there was no control of exposure time, which was around 1 to 2 microseconds. These tubes use a field-emission-initiated vacuum arc and operation tends to be inconsistent. Later versions of this type of tube reduced exposure times to the order of 10^{-7} sec and focal spot sizes to about 1 mm.

In 1957 Tsukerman and Manakora described a simple two-electrode flash X-ray tube which could be operated at voltages up to 2 MV, and similar tubes have been built and used in the U.K. In these tubes the voltage impulse from a Marx generator is switched across the gap between a conical anode and a ring cathode. Later versions used ultra-short square-wave pulses.

More recently Dyke and associates have developed new types of flash X-ray tube, making use of field emission effects from a needle-point or a wire. These are available for operation at voltages up to 2,000,000 volts, the exposure times may be controlled between 10^{-6}–10^{-8} sec and so far as commercial equipment is concerned, have superseded all other types. In research establishments, however, "home-made" equipment is still used and superior performances to those of commercial apparatus have been claimed.

Electron linear accelerators used to generate X-rays have also been used to a limited extent for flash radiography. By electronic switching of the microwave energy carrying the electrons, a single pulse or short train of pulses of X rays can be generated. Usually high energy accelerators, e.g. 30 MeV, are used, so as to obtain more X-ray energy in each pulse, and pulse lengths as short as 10^{-7} sec have been reported with an interval between pulses, in a train of nine pulses, of 3×10^{-6} sec.

Typical performances for flash X-ray equipment are given in the table.

A typical *field-emission X-ray tube* has several comb-shaped arrays of needle points as electron sources, arranged around a conical anode. This shape of anode permits the dissipation of large amounts of heat while acting as a small X-ray source. Field emission from a needle point takes advantage of the intense field which surrounds a sharply pointed conductor and so enables the voltage required to be kept to a reasonable level; a typical cathode has a hemispherical tip of radius from 10^{-4} to 10^{-6} cm, and such a single point can emit a current pulse of 140 amperes when pulsed for 1×10^{-8} sec at 300 kV, so that currents of 2000–5000 amperes pass through flash X-ray tubes. Modern field-emission X-ray tubes are sealed off and are remarkably compact, e.g. 6″ long overall for use

Peak voltage (kV)	Pulse length (nano sec)	Output of X rays (mr/pulse at 1 m)	Focus size (mm)
400	45*	60	4 × 0·5
2000	30**	700	5 × 5
3000	10	2000	—

* Pulse rise time 4×10^{-9} sec.

** Capable of producing radiographs through 10 cm steel at 1 m source-to-film distance.

at 100 kV. The generators usually use a Marx surge circuit in which electrical energy is stored in capacitors and then discharged through the X-ray tube after conversion into a square-wave pulse; delay lines may be used instead of conventional condensers. To eliminate atmospheric effects such as dust and humidity, the generator is usually sealed and pressurized.

Recording in ultra high-speed radiography is almost always on X-ray film used in conjunction with high-speed fluorescent (salt) intensifying screens. Such screens limit the resolution of the image, but are essential because of the small amount of X-ray energy available.

Most applications employ a single pulse of X-rays to produce a single radiograph, with the generation of the pulse timed to coincide with some critical aspect of the event under examination, but a limited amount of sequential radiography is also possible. This can be achieved in a number of ways:

(1) By operating several X-ray tubes from a single generator. One X-ray tube is required for each radiograph in the sequence, but the interval between pulses can be as short as 10^{-6}–10^{-7} sec.

(2) Repeated pulses through a single X-ray tube. Rates up to 12,000 frames/sec have been achieved, but this technique can only be used when the image of different frames can be projected on to different parts of the recording film (or array of films), i.e. for objects such as a travelling projectile.

(3) Repeated pulses through a single X-ray tube, when the interval between pulses is long enough to move the recording film between pulses. The film can be on a rotating drum, revolving at a suitable speed to match the pulse interval to be used, but again the total number of frames possible is limited by the length of film round one circumference of the drum.

As already stated, most of the applications to date have been to explosive and ballistic phenomena in armaments research, but applications to shock waves by detecting the large density change in a shock front, the study of transient effects in radiation damage studies, the study of purely mechanical processes, and some biodynamic applications, have also been reported.

Bibliography

Dyke W. P. and Dolan W. W. (1956) *Adv. Electron. and Elec. Phys.* **8**, 90.

Schaaffs W. (1954) *Vakuum Techn.* **3**, 31.

Slack C. M. and Ehrke L. F. (1941) *J. Appl Phys.* **12**, 165.

Stenzel A. and Thomer G. (1961) *J.S.M.P.T.E.* **70**, 18.

Tsukermann V. A. and Manakora M. A. (1957) *Soviet Phys.* (*Techn. Phys.*) **2**, 2.

R. Halmshaw

UNDERWATER ACOUSTICS, NON-LINEARITY IN.

1. Introduction

The set of differential equations describing the propagation of acoustic waves in a compressible fluid can be written in the following form neglecting viscous effects:

Force equation $$\varrho \frac{du}{dt} + \nabla p = 0$$

Continuity equation $$\frac{\partial \varrho}{\partial t} + \nabla \cdot (\varrho u) = 0$$

Equation of state $$p = p(\varrho).$$

Here, p is the differential pressure, ϱ is the density of the medium and u is the particle velocity; Eulerian co-ordinate system is used.

The non-linear nature of the resulting wave-equation has been known to the investigators for a long time. Rayleigh (1877) and Lamb (1931) describe the earlier efforts to find the complete solution to these non-linear equations. More recently Beyer (1963) and Blackstock (1962) have reviewed some of the earlier attempts which have contributed significantly to the present-day thinking on the subject.

The main object of most of the investigations appears to have been a study of the distortion of a plane wave as it progresses through the medium. A general solution of the non-linear differential equations was found to be difficult and some limited solutions have been proposed for specific cases. The simplest case studied was that of an acoustic plane wave which is sinusoidal at the origin. As such a wave progresses, the phases associated with higher particle velocities travel faster than those associated with lower particle velocities, and thus a progressive bunching of the phases occur, resulting in the triangulation of the waveform. This effect, which was predicted as early as the middle of the 19th century, was demonstrated experimentally by Burov and Krasil'nikov (1958) and by Ryan *et al.* (1962). For further information on the subject the reader is referred to Beyer (1963) and the references given therein.

The present-day acoustic technology is based on the linear-approximation in the solution of the acoustic wave equations. For example, if the variation in the density of the medium due to the propagating acoustic wave is negligible, then the wave equation can be expressed in the approximate form

$$\nabla^2 \varrho + \frac{\partial \varrho}{\partial t} = 0$$

This is now a linear differential equation and if a sinusoidal plane wave of frequency ω (propagating in the x direction) is assumed to exist at the boundary $x = 0$, then the solution of the wave equation yields (making allowance now for absorption of the waves)

$$u(x, t) = U_0 \exp(-\alpha_0 x) \cos(\omega t - kx),$$

which indicates a single-frequency plane wave being attenuated at a rate determined by the absorption coefficient α_0 where k is the wave-number and U_0 the amplitude of u at $x = 0$.

In the case of an acoustic wave of finite-amplitude, the attenuation mechanism is further complicated because of the progressive distortion of the wave. A wave which is initially sinusoidal becomes distorted and, consequently, the fundamental frequency component of the intensity is reduced with distance, not only because it is absorbed, but also because it is (in part) converted into waves at harmonic frequencies. Further, as the absorption coefficient α_0 has a higher value at the harmonic frequencies, the rate of attenuation (with distance) of the total acoustic intensity in the wave is also increased.

The investigation of the finite amplitude effects in underwater propagation has been accelerated in recent years mainly as a consequence of increased use of high intensity acoustic waves in underwater technology.

If more than one acoustic wave is present in a given part of the medium, non-linear interaction between these waves produces intermodulation products. A quasi linear analysis of this effect was provided by Westervelt (1957), who applied Lighthill's (1952) results to the analysis of first-order interaction between acoustic waves. The basic assumption in this analysis is that the intermodulation products propagate in a linear manner. Hence, if a source function can be defined mathematically for the interaction products at all points of the region of interaction of the primary sound waves, then the velocity potential at any point can be calculated in a closed form by evaluating Kirchoff's integral over the region in which the source-function exists. Westervelt's expression for the source-density, q, at the first-order interaction frequencies, produced by the non-linear interaction of two plane waves intersecting one another at an angle β, was given in the form

$$q = (1/\varrho_0^2 c_0^4) \frac{\partial}{\partial t} (p_i^2) (\cos \beta + B/2A) \qquad (1)$$

where

p_i is the total pressure due to the primary waves (as the interaction components are assumed to propagate linearly, only the terms appropriate to the particular interaction component within the expression for p_i^2 need be considered)

B/A, the so-called parameter of non-linearity of the medium, has a value between 5 and 7 for water (see Beyer (1960), for example), and ϱ_0 and c_0 are, respectively, the density of and the velocity of sound propagation in an undisturbed medium. The parameters B and A are coefficients in the first two terms of a series expansion of the equation of state in ascending powers of $(\varrho - \varrho_0)/\varrho_0$ in the form

$$p = A(\varrho - \varrho_0)/\varrho_0 + (B/2)(\varrho - \varrho_0)^2/\varrho_0^2 + \cdots$$

In the particular case of coincident plane waves (i.e. $\beta = 0$) equation (1) reduces to

$$q = (1/\varrho_0^2 c_0^4) \frac{\partial}{\partial t} (p_i^2) (1 + B/2A). \qquad (2)$$

If the medium is non-dispersive, the spatial rate of change of the phase of the source-function is such that the interaction-frequency components add co-phasally only along the direction of propagation of the primary waves. This effect is similar to radiation from an end-fire array formed by virtual sources at the interaction frequencies (Westervelt 1960). This leads to the possibility of obtaining highly directional beams even at low difference frequencies, using relatively small transducers. The properties of these virtual arrays are discussed in more detail in section 2.

Some practical examples of possible applications of the end-fire array principle in the field of underwater acoustics are evaluated (and some considerations pertinent to such applications are discussed) in section 3.

Theoretical work points to the possible use of non-linear interaction between acoustic waves in reception and processing of acoustic signals. A short discussion of these is included in section 4.

2. *End-fire array of Virtual Sources*

Interaction between coincident plane waves. If the two primary waves are launched from the same transducer and the waves are both planar and well collimated, the total differential pressure at a point

(x, y, z) due to the primary waves can be expressed as

$$p_i = P_1 \exp(-\alpha_1 x) \cos(\omega_1 t - k_1 x) + P_2 \exp(-\alpha_2 x) \cos(\omega_2 t - k_2 x) \quad (3)$$

when

$$|y| \leqq b \quad \text{and} \quad |z| \leqq d,$$

and $p_i = 0$ when the point (x, y, z) is outside the interaction volume. In general, as waves of finite-amplitude are being considered, α_1 and α_2 are functions of x and the initial pressure amplitudes. But, for the moment, α_1 and α_2 will be considered to have the values appropriate to waves of infinitesimal amplitude; the finite-amplitude effects in this connexion will be considered at a later section.

The source-density function depends upon p_i^2; therefore it will have components at the frequencies $2\omega_1$, $2\omega_2$, $\omega_1 \pm \omega_2$. Only the effects due to the difference-frequency component in the source function will be considered in the remainder of this section.

The difference-frequency component in the source-function will arise from the term

$$P_1 P_2 \exp(-\overline{\alpha_1 + \alpha_2} x) \cos(\Omega t - Kx) \quad (4)$$

where $\Omega = \omega_1 - \omega_2$ and $K = k_1 - k_2$.

Substituting this for p_i^2 in the expression for the source-function, the difference-frequency component of the latter is obtained:

$$\text{(i.e.)} \quad q = -(P_1 P_2 \Omega / \varrho_0^2 c_0^4)(1 + B/2A) \times \exp(-\overline{\alpha_1 + \alpha_2} \cdot x) \sin(\Omega t - Kx) \quad (5)$$

Thus, a cross-sectional wafer at x and of thickness δx would include a uniform distribution of co-phasal virtual sources. Hence, radiation from such a wafer would be identical to that from a uniformly insonified aperture of the same shape and size. As the cross-section consists of a rectangle $2b \times 2d$, the radiated field at a point in the far-field of the aperture is proportional to (in spherical co-ordinates)

$$(2d)(2b) \frac{\sin(dK \cos\gamma)}{(dk \cos\gamma)} \frac{\sin(bK \sin\gamma \sin\theta)}{(bK \sin\theta \sin\gamma)}.$$

Having thus isolated the aperture effect due to the finite cross-section of the array of virtual sources, the integration with respect to x to obtain the overall scattered pressure reduces to that obtaining in an end-fire array in which the amplitude of the source-function is decaying exponentially with increasing x. This last is the case studied by Westervelt and by Bellin and Beyer (1962).

If, for simplicity, the observer is constrained to move on the plane $z = 0$ in the far-field [(i.e.) $r \gg x$, thus giving $R \doteqdot x \cos\theta + r$ when $\gamma = \pi/2$], then $\gamma = \pi/2$. The amplitude of the scattered pressure field can then be found by integration with respect to x to obtain the total effect of all the wafers of thickness δx and is given by

$$P(R, \theta) = \frac{P_1 P_2 \Omega^2 (2b) \cdot (2d)}{4\pi \varrho_0 c_0^4 R} \exp(-\alpha \cdot R) \times (1 + B/2A) \frac{\sin(bK \sin\theta)}{(bK \sin\theta)} \times [N^2 + 4K^2 \sin^4(\theta/2)]^{-1/2} \quad (6)$$

where $N = \alpha_1 + \alpha_2 - \alpha$.

If the observer is on the axis of symmetry of the transducer (i.e. $\theta = 0$) the amplitude of the differential pressure at the difference frequency will be

$$P(R, O) = \frac{P_1 P_2 \Omega^2 \cdot S}{4\pi \varrho_0 \mathrm{Co}^4 RN} \exp(-\alpha R)(1 + B/2A), \quad (7)$$

where $S = 2b \times 2d$ and is the cross-sectional area of the column.

As the primary waves are assumed to be planar and collimated, the acoustic powers in the primary waves at $x = 0$ will be, respectively,

$$W_1 = SP_1^2/(2\varrho_0 c_0) \quad \text{and} \quad W_2 = SP_2^2/(2\varrho_0 c_0).$$

Substituting for P_1 and P_2 in equation 7, one obtains for the axial pressure amplitude

$$P(R, O) = \frac{\sqrt{(W_1 W_2)} \cdot \Omega^2}{2\pi \mathrm{Co}^3 RN} \exp(-\alpha R)(1 + B/2A). \quad (8)$$

From equations 6 and 7 a directivity-function $D(\theta)$ can be defined as

$$D(\theta) = P(R, \theta)/P(R, O) = \frac{\sin(bK \sin\theta)}{bK \sin\theta} \times \{1 + [(2K/N) \sin^2(\theta)/2)]^2\}^{-1/2}. \quad (9)$$

$D(\theta)$ gives the two-dimensional pattern of the array of the virtual sources, and consists of two factors; one being that due to the finite aperture formed by the cross-section of the column and the other due to the end-fire array with exponential taper.

If $bK \ll 1$, the aperture factor is unity. Then, the $3dB$ points of the directivity pattern will occur when

$$(2K/N) \sin^2(\theta/2) = 1. \quad (10)$$

The value of θ satisfying this equation will be denoted by θ_d.

For propagation in water at the frequencies of general interest, $(N/2K) \ll 1$. Then, from equation 10, the $3dB$ beam-width of the end-fire array with small cross-sectional dimensions would be given by

$$2\theta_d = 4\sqrt{(N/2K)}. \quad (11)$$

It is worth noting that the factor in the expression for $D(\theta)$ due to the end-fire effects does not have any maxima other than at $\theta = 0$ for $0 \leqq \theta \leqq 2\pi$, and is a minimum when $\sin(\theta/2) = 1$, (i.e.) when $\theta = \pi$.

In the general case, the directivity pattern for the end-fire effects is modified by an aperture factor which, in the case of a circular cross-section, would have been of the form $2J_1(M)/M$, where J_1 is the Bessel function of order 1. The effect of the aperture factor will be to increase the directionality of the difference-frequency waves.

An effective source-strength can be calculated for the end-fire array at the difference frequency, by putting $R = 1$ in equation 8. Assuming, further, that $W_1 = W_2 = \frac{1}{2}W_0$ where W_0 is the total acoustic power transmitted, the r.m.s. pressure at the difference-frequency referred to 1 m distance can be calculated for water. This calculation gives

$$\overline{P}(1, 0) = 4{\cdot}1 \times 10^{-2} W_0(f_1 - f_2)/(2\theta_d)_0^2 \ \mu \text{ bars},$$

where $(2\theta_d)_0$ is the value of $2\theta_d$ in degrees.

The predicted values given above are in agreement with the results of laboratory-scale experiments (Berktay and Smith 1965).

Cylindrical waves. Non-linear interaction between spherical waves and that between cylindrical waves has been considered by many authors. However, a complete study of the problem is complicated. A simplified version of the problem is studied in Berktay (1965), using a procedure similar to that used in the previous subsection; an outline of the results obtained there is given below.

The primary waves are now assumed to consist of two cylindrically spreading waves (at frequencies of ω_1 and ω_2) confined between the planes $z = \pm 1$ and within the half-planes including the z-axis and making angles $\pm\psi_1$ with the x-axis. It is further assumed that the radial power flow is uniformly distributed over the spreading angle of $2\psi_1$. (In a practical case, this could be approximated to by assuming $2\psi_1$ to be the 3 dB beamwidth of the fan-beam.) The 3 dB beamwidth at the difference frequency, $2\Phi_d$, was calculated in a normalized form (as Φ_d/θ_d) as a function of the normalized variable $\psi_d = \psi_1/\theta_d$. Similarly, the variation of the axial pressure at the difference-frequency (normalized with respect to the value obtaining for $\psi_d = 0$, which corresponds to the plane wave case) was calculated as a function of ψ_d. The results were given in graphical form.

A study of these results shows that, provided the sector angle $2\psi_1$ is not greater than $2\theta_d$, the difference frequency effects produced by the interaction of cylindrical primary waves will be the same as if the primary waves were planar. If, however, the value of ψ_d is made greater than unity (either by increasing the sector-angle $2\psi_1$ *or* by reducing $2\theta_d$, say by increasing the difference frequency), then a loss of directional gain can be observed in the form of a reduced axial pressure and an increased beamwidth.

Experimental results have shown very good agreement with theory (Berktay and Smith 1965). One feature of these results is the constancy of the beamwidth over a very wide range of frequencies. The reason for this can be seen from the theoretical curves given in Berktay (1965a). For values of ψ_d between 1·25 and 5 the graph of Φ_d/θ_d vs. ψ_d can be approximated to by the straight-line

$$\Phi_d/\theta_d = 0{\cdot}85\ \psi_d \quad (\text{i.e.} = 0{\cdot}85\ \psi_1/\theta_d).$$

$$\therefore\ 2\Phi_d = 0{\cdot}85\ (2\psi_1), \text{ provided } 1{\cdot}25 \leqq \psi_d \leqq 5.$$

One of the applications being considered for non-linear acoustic interactions is the production of wide-band signals with constant beamwidths, based on the ideas outlined here.

Spherical waves. The primary waves are now assumed to be confined and uniform within a conical pencil beam of width $2\psi_1$. Because of the complexity of the computations involved, results were computed only for the axial pressure (again normalized with respect to that for plane waves, i.e. for $\psi_d = 0$) as a function of $\psi_d = 2\psi_1/2\theta_d$. These results indicate that the end-fire array gain associated with the interaction of plane waves can be realized provided the spreading angle of the spherical waves $(2\psi_1)$ is not greater than $2\theta_d$.

Deflexion of the difference-frequency beam. If the two primary beams are deflected simultaneously and in such a way as to remain coincident, the virtual end-fire array will have been rotated in space, thus producing the deflexion of the difference-frequency beam. The primary beams can be deflected electronically by using phased arrays, thus producing an electronically-deflected difference-frequency beam.

Some preliminary experimental results supporting this view have been obtained using two transducer elements each having the dimensions of 25 mm × 5 mm and resonant at about 2·5 MHz.

3. *Considerations Regarding Practical Underwater Applications of the Virtual End-fire Array*

Some fundamental aspects of the virtual end-fire array principle are discussed in the previous sections. The main features of this phenomenon which may be of interest in practical applications in underwater acoustics can be summarized as follows:

(i) Highly directional low-frequency acoustic waves can be obtained with relatively small transducers.

(ii) For a given transducer Q, the difference-frequency waves will have wider bandwidths than would be available if the same frequency were being transmitted directly from a resonant transducer, as the transducers are resonant at the higher primary frequencies.

(iii) Difference-frequency waves can be produced which have a constant beamwidth over very wide frequency-bands (e.g. greater than 2 octaves).

(iv) The wide bandwidth capability of the difference-frequency beam which can be deflected mechanically or electronically makes it possible to think in terms of practicable transmit-beam scanning sonar systems.

As opposed to these favourable features, the main disadvantage of the non-linear conversion process is its low efficiency.

Some possible applications of the virtual end-fire arrays and some factors affecting the use of such devices in practice are discussed and evaluated in the literature (Becktay 1965d, 1967a) where it is shown that in spite of the disadvantage of the low conversion efficiency, in some applications the non-linear process can provide distinct technical and/or economical advantages.

The effects of increased absorption at finite amplitudes. Equations 5 and 8 show that the source-strength and the conversion efficiency, respectively, of a non-linear transmitting device increases rapidly with the acoustic power in the primary waves. Therefore, for a given transducer area, the usefulness of such a device is increased by raising the transmitted acoustic intensity. As the primary frequencies can be much higher than the difference frequency, the cavitation-threshold is higher and the intensity of the acoustic waves can be raised accordingly as compared with the intensity which would be permissible in a system transmitting the low-frequency waves directly. But, the rate of attenuation of acoustic waves of high intensities is not constant and is a rather complicated function of distance, intensity and frequency (Blackstock 1964). Therefore, the difference-frequency effects would be expected to be dependent upon the intensity of the primary waves.

Some experiments have been made to assess the effects of the increased acoustic intensity on the pressure and the beamwidth at the difference frequency. Preliminary results obtained with cylindrically spreading primary waves have shown no marked variation in pressure magnitudes from those predicted by using the equation 8.

4. Parametric Amplification due to Acoustic Non-linearities

One form of the effects arising from energy transfer to new frequencies through non-linear interactions of acoustic waves was discussed in the preceding sections. The possibility of obtaining parametric amplification through such energy transfer has been investigated in two forms: up-converter and travelling-wave type parametric amplification (Berktay 1965b 1967b). In each case a low-energy signal wave of frequency ω_2 is assumed to interact with a locally generated collimated plane wave ("pump") of high intensity and at a higher frequency ω_1. The various frequency components are assumed to be picked up at a distant point along the "pump" column. In the travelling-wave parametric amplifier (T.W.P.A.) an enhancement of the power at the signal frequency is looked for. In the up-converter type of parametric amplifier, on the other hand, the available power at an intermodulation frequency (Ω) is compared with that at the signal frequency.

A quasi-linear analysis shows that travelling-wave parametric amplification at the signal frequency cannot be obtained unless the component wave at the sum-frequency $\omega_1 + \omega_2$ is suppressed. (This corresponds to the suppression of the upper-idler in an electromagnetic travelling-wave parametric amplifier).

Up-converter type of parametric amplification of about 18 dB has recently been obtained in laboratory experiments in the Electronic and Electrical Engineering Department of the University of Birmingham, using a pump wave of moderate intensity. The results are in good agreement with the theory. Possible applications of this phenomenon to the reception of acoustic waves in underwater applications is proceeding.

5. Conclusions

Finite-amplitude effects in underwater acoustics have been known over a long period of time. However, the realization that noticeable non-linear effects can be obtained at moderate acoustic intensities is more recent. Theoretical work and laboratory-scale experiments suggest that some aspects of these non-linear effects can be put to good use in underwater technology in the form of more economical transducer units for highly directional sonar or communication systems.

Bibliography

Bellin J. L. S. and R. T. Beyer (1962) Experimental investigation of an end-fire array, *J. Acoust. Soc. Amer.* **34**, 1051.

Berktay H. O. (1965a) Possible exploitation of non-linear acoustics in underwater transmitting applications, *J. of Sound and Vibration*, **2**, 435.

Berktay H. O. (1965b) Parametric amplification by the use of acoustic non-linearities and some possible applications, *J. of Sound and Vibration*, **2**, 462.

Berktay H. O. (1967a) Some proposals for underwater transmitting applications of non-linear acoustics *J. of Sound and Vibration*, **6**, 244.

Berktay H. O. (1967b) A study of the travelling-wave parametric amplification mechanism in non-linear acoustics, *J. of Sound and Vibration*, **5**, 155.

Berktay H. O. and Smith B. V. (1965) *Electronic Letters*, **1**, 6; (1965) *Electronic Letters*, **1**, 202.

Beyer R. T. (1960) Parameters of non-linearity in fluids, *J. Acoust. Soc. Amer.*

Beyer R. T. (1965) Non-linear acoustics, *Physical Acoustics* (Ed. W. P. Mason) Volume II, Part B, New York: Academic Press.

Blackstock D. T. (1962) Propagation of plane sound waves of finite-amplitude in nondissipative fluids, *J. Acoust. Soc. Amer.* **34**, 9; (1964) Thermoviscous attenuation of plane, periodic, finite-amplitude sound waves, *J. Acoust. Soc. Amer.* **36**, 534.

Burov V. A. and Krasil'nikov V. A. (1958) Direct observation of the deformation of intense ultrasonic waves in liquids, *Doklady Akad. Nauk SSR*, **118**, 920 (translation in *Sov. Phys. Doklady*, **3**, 173).

LAMB H. (1931) *The Dynamical Theory of Sound*, New York: Dover (2nd edition), 1960 re-issue.

LIGHTHILL M. J. (1952) On sound generated aerodynamically, *Proc. Roy. Soc.* A**211**, 564.

RAYLEIGH J. W. S. (1877) *The Theory of Sound* (two volumes), New York: Dover (2nd Edn.), 1945 re-issue.

RYAN R. P., LUTSCH A. G. and BEYER R. T. (1962) Measurement of the distortion of finite ultrasonic waves in liquids by a pulse method, *J. Acoust. Soc. Amer.* **34**, 31.

WESTERVELT P. J. (1957) Scattering of sound by sound, *J. Acoust. Soc. Amer.* **29**, 199.

WESTERVELT P. J. (1960) Parametric end-fire array, *J. Acoust. Soc. Amer.* **32**, 934A.

H. O. BERKTAY

V

VAN DE GRAAFF SUPERVOLTAGE X-RAY GENERATOR. *Introduction.* The Van de Graaff supervoltage X-ray generator is a special adaptation of the Van de Graaff particle accelerator (Buechner 1962) and is routinely used in hospitals and industries as a flexible source of penetrating X rays for clinical cancer therapy and for non-destructive testing by radiographic techniques. Its usage in these organizations is similar to that of the betatron, the electron linear accelerator, and the resonant transformer. In recent years, radioactive sources of gamma rays, such as cobalt-60, have similarly been applied.

Principles of operation. The X rays from this type of machine are produced by the impingement of 1- or 2-MeV electron beams on gold or tungsten targets. The 1- or 2-MV constant potential of a Van de Graaff electrostatic generator provides a uniform electric field along a sealed-off, multiple-electrode acceleration tube, to impel the electrons to their final energy. Figure 1 is a schematic diagram of a typical Van de Graaff supervoltage X ray generator.

The d.c. electron current in the acceleration tube is of the order of 0·25 mA, and X ray outputs up to 85 R/min at a metre's distance are routine with the 2-MeV model. Of particular value is the small electron focal spot at the X-ray-producing target, from which the penetrating X rays are propagated preferentially in the forward direction. The "point source" of X rays makes it possible to collimate the X-ray beam with very little penumbra and to make radiographs of sharp definition, even under adverse geometrical conditions. The electron beam is focused electrostatically at the cathode end of the acceleration tube, for focal spots of 2–3 mm in diameter. Additional electron focusing is achieved by a solenoidal magnet, after the acceleration process, to achieve focal spots of 1 mm or less.

Applications. Over forty 2-MeV Van de Graaff X-ray generators are in clinical use today, typified by the sketch of Fig. 2. In some hospitals, up to fifty patients can be treated per day with these machines.

About 35 industrial Van de Graaffs are used in foundries or welding shops, for the non-destructive inspection of heavy metal casting and fabrications. Steel sections up to 8 inches can be examined on a production basis, to determine their soundness and

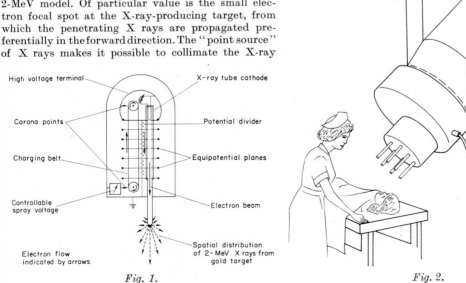

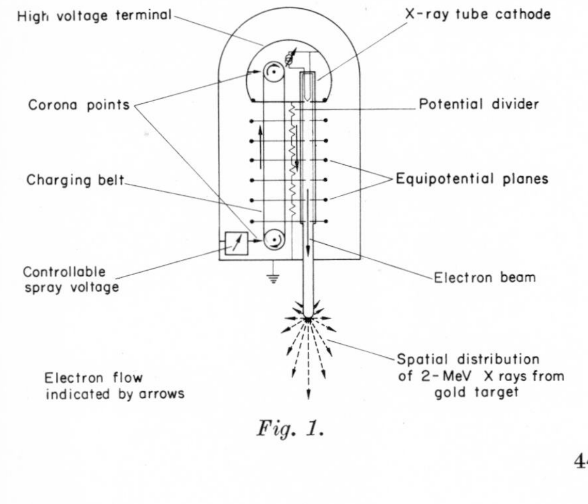

Fig. 1.

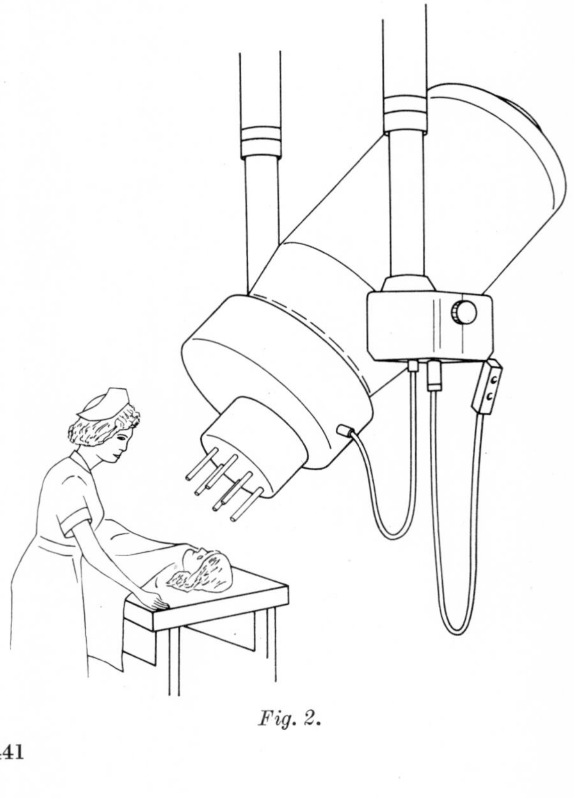

Fig. 2.

safety. Solid plastic propellant motors for missiles up to 30 inches in cross-section can also be inspected for their internal integrity. Figure 3 is a sketch of a 1-MeV Van de Graaff in an unusual mounting, for the simultaneous radiography of the seams of complex weldments.

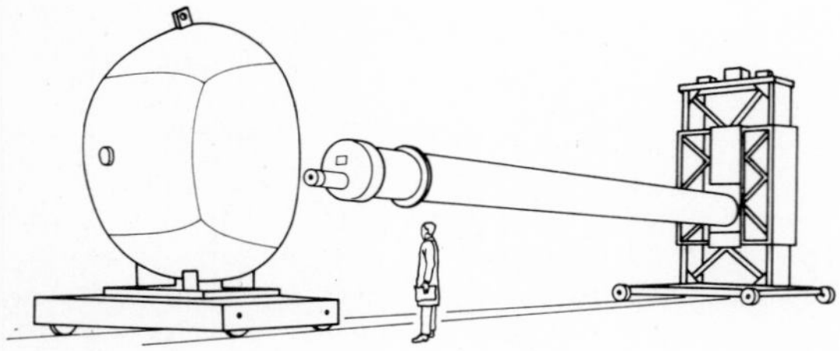

Fig. 3.

Historical background. The great expense and relative unavailability of radium gamma-ray sources in the early 1930's led many radiologists, physicists, and engineers to devise electrical systems for producing X rays with ionizing and penetrating characteristics similar to radium emanations. The first 1-MeV supervoltage X-ray generator for cancer therapy was constructed in 1937 by John G. Trump and Robert J. Van de Graaff, both professors at Massachusetts Institute of Technology, for the Huntington Memorial Hospital in Boston. This generator operated in open air and required a considerable space to prevent terminal spark-over. Figure 4 is a sketch of this pioneering X-ray generator.

Subsequent Van de Graaffs were insulated by compressed gases to reduce the overall dimensions, and two such units were built at M.I.T. for hospital use. During World War II, five 2-MeV Van de Graaff X-ray generators were designed and developed at M.I.T. for the nondestructive examination of ordnance materiel. Since the war, almost all Van de Graaffs for hospital and industry have been commercially manufactured.

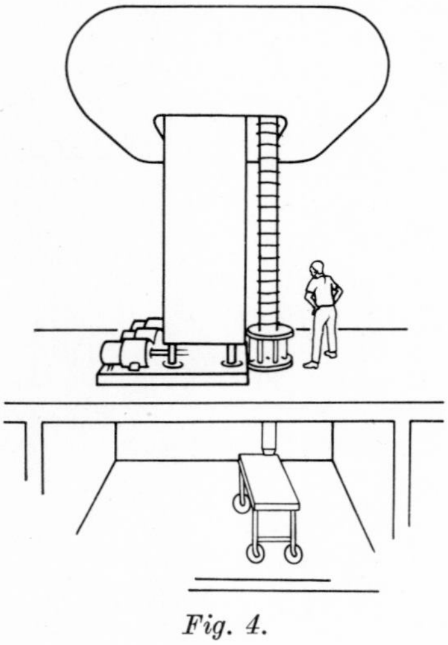

Fig. 4.

Bibliography

BUECHNER W. W. (1962) in *Encyclopaedic Dictionary of Physics* (J. Thewlis Ed.), **7**, 580, Oxford: Pergamon Press.

BURILL E. A. (1967) in *Encyclopaedic Dictionary of Physics* (J. Thewlis Ed.), Suppl. Vol. **2**, 374, Oxford: Pergamon Press.

E. A. BURRILL

VIBRATION ISOLATION. *1. Introduction.* The term "vibration isolation" means the reduction to an acceptable level, of the transmission of unwanted vibratory effects from one structure to another. Isolation is effected by means of one or more resilient elements inserted between the source of vibration, and the structure to which the source is necessarily attached. The resilient elements, called "vibration isolators" are fundamentally springs made of materials such as steel, felt, cork, rubbers and plastics. They can be produced as pads, mats and carpets covering a relatively large contact area, or as suspensions or mountings positioned at discrete points of the structure.

Two basic situations can be distinguished. The reduction of vibration passing from machinery to support structure is termed "active" isolation. An example is the lessening of out-of-balance forces transmitted from factory machines to the foundations. Passive isolation is the protection of equipment from vibrations occurring in the surroundings. A good illustration is the cushioning of automobile passengers from road surface irregularities.

The principles of isolation are most conveniently shown by analysis of a simple one degree of freedom translational system. The results obtained are readily adaptable to most real situations and, in particular, may be applied to torsional systems by direct analogy.

2. Active isolation. Although damping is not necessary for isolation to be achieved, a certain amount is often desirable. A simple idealized isolator whose behaviour is in many ways characteristic of real isolators, is the linear viscous damper system of Fig. 1. The rigid mass, *m*, represents a source of vibration

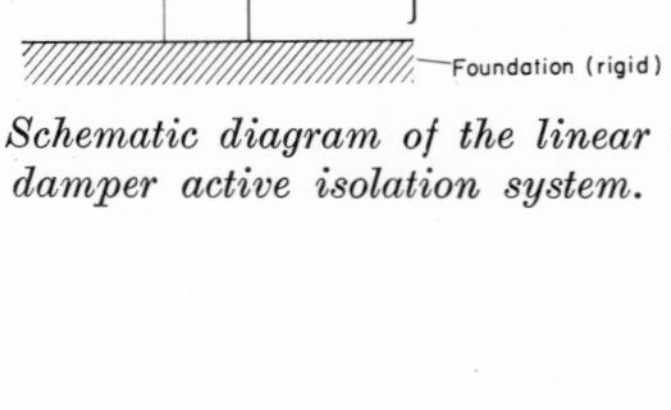

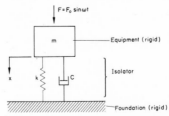

Fig. 1. Schematic diagram of the linear viscous damper active isolation system.

which is isolated from a rigid foundation by means of a massless linear spring of stiffness k. The dashpot of damping coefficient c, is placed in parallel with the spring in order to give reduced response amplitudes near resonance.

The vibration source is assumed to excite the system with an harmonic force of amplitude, F_0, and constant circular frequency, w. If at time t, the mass is displaced a distance x, from its static equilibrium position, the equation of motion is

$$m\ddot{x} + c\dot{x} + kx = F_0 \sin wt.$$

The system response is given by the general solution of this equation, which is

$$x = Ae^{-\frac{ct}{2m}} \sin(\omega_d t + B) + [c^2\omega^2 + (k - m\omega^2)^2]^{-1/2} F_0 \sin(\omega t - \Phi).$$

The first term represents the response of the unforced system, i.e. when F_0 equals zero. It has been assumed that damping is light enough for this free response to be oscillatory. A and B are constants determined by values of x and $\dot{x}$ at time t equal to zero, and the damped natural frequency is given by

$$\omega_d^2 = \frac{k}{m} - \left(\frac{c}{2m}\right)^2.$$

Two important properties are obtained from this last expression. By putting the damping, c, equal to zero, the undamped natural frequency, ω_n, is found to be equal to $\sqrt{k/m}$. On the other hand, if the damping is increased to that ω_d is equal to zero, the critical damping value, c_c, is obtained at which the free response just becomes non-oscillatory,

i. e. $$c_c = 2\sqrt{(km)}.$$

It can be seen that the contribution of the free vibration to the total system response is of transient importance. It decays exponentially with time leaving the steady state response

$$x = \frac{F_0}{k}\left[\left\{1 - \left(\frac{\omega}{\omega_n}\right)^2\right\}^2 + \left(2\frac{c}{c_c}\frac{\omega}{\omega_n}\right)^2\right]^{-\frac{1}{2}} \sin(\omega t - \Phi)$$

where the phase angle Φ is given by

$$\tan\Phi = \frac{c\omega}{k - m\omega^2} = \frac{2c_c/c\,\omega/\omega_n}{1 - (\omega/\omega_n)^2}.$$

The displacement amplitude, x_0, is therefore given by

$$\frac{x_0}{F_0/k} = [\{1 - (\omega/\omega_n)^2\}^2 + (2c/c_c\omega/\omega_n)^2]^{-1/2}.$$

This expression, called the "motion response", gives the dimensionless ratio of forced displacement amplitude to the spring deflexion that would occur if F_0 were applied statically. It is useful in determining clearance and space requirements of an isolator. A graphical representation is given in Fig. 2.

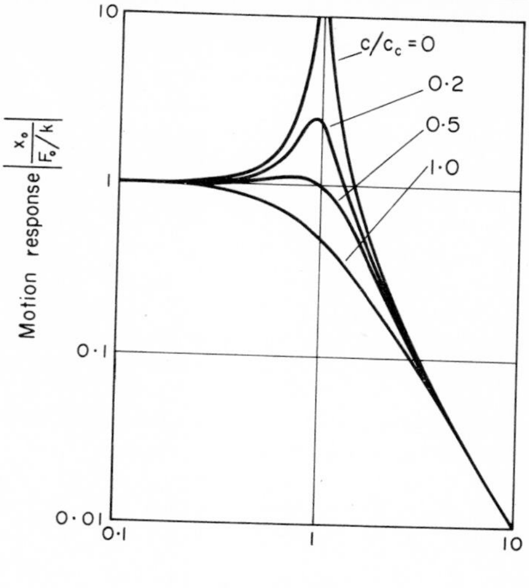

Fig. 2. Motion response of the isolation system of Fig. 1.

The instantaneous force, P, transmitted to the foundation is

$$P = kx + c\dot{x} = F_0\left[\left\{1 - \left(\frac{\omega}{\omega_n}\right)^2\right\}^2 + (2c/c_c\,\omega/\omega_n)^2\right]^{-\frac{1}{2}} \left[\sin(\omega t - \Phi) + \frac{c\omega}{k}\cos(\omega t - \Phi)\right].$$

The force amplitude, P_0, which is easily obtained from this equation, can now be written as follows:

$$\frac{P_0}{F_0} = \left[\frac{1 + (2c/c_c\,\omega/\omega_n)^2}{\{1 - (\omega/\omega_n)^2\}^2 + (2c/c_c\,\omega/\omega_n)^2}\right].$$

This ratio of amplitudes of force transmitted to force applied, is called the "force" or "absolute transmissibility", T_A. It indicates the effectiveness of the system in reducing the magnitude of transmitted vibratory forces, and is therefore the most important single property of the isolator.

For the particular case of zero damping, force transmissibility and motion response reduce to

$$\left(\frac{x_0}{F_0/k}\right)_{c=0} = (T_A)_{c=0} = \frac{1}{1 - (\omega/\omega_n)^2}.$$

Figure 3 shows the variations of transmissibility with frequency, for values of damping from zero up to the critical. The behaviour is summarized as follows:

(a) Statically, the force transmissibility equals unity, which is obvious from considerations of equilibrium.

(b) For frequencies between zero and $\sqrt{2}$ times the undamped natural frequency, the isolator actually magnifies the applied forces, particularly near resonance.

(c) For frequency ratios above $\sqrt{2}$, the vibratory forces are reduced on transmission, and transmissibility decreases with increase in excitation frequency. This is the fundamental principle of vibration isolation.

(d) For frequency ratios greater than unity, the transmissibility is negative, because the transmitted force becomes 180 degrees out of phase with the applied force as the excitation frequency passes through resonance. It is therefore usual to plot transmissibility as an absolute value, as in Fig. 3. This remark is equally true of Fig. 2.

(e) Damping reduces transmissibility near system resonance, but increases transmissibility, i.e. decreases isolation, at frequency ratios greater than $\sqrt{2}$.

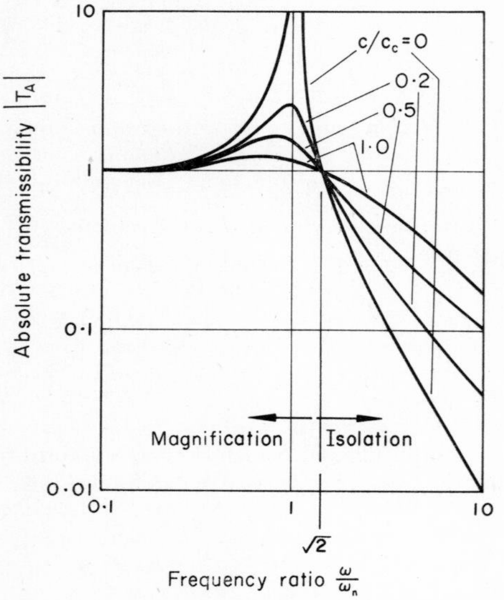

Fig. 3. Absolute transmissibility of the isolator of Fig. 1.

(f) As in Fig. 2, there is a slight shift towards zero of the peak frequency as damping is increased. However, for small damping, the damped natural frequency, and the frequencies of peak transmissibility and peak motion response, can be assumed equal for practical purposes, to the undamped natural frequency. Actually, for damping of about 10 per cent of critical, each of these three frequencies is about one per cent less than ω_n.

Although it is never intended for isolators to function continually at resonance, they may sometimes be required to operate for short periods under resonant conditions. For example, when starting up a piece of machinery the excitation frequency will pass through the resonant range before steady running speed is reached. One of the main uses of damping is to control vibration amplitudes under such conditions. A rough idea of the relationship between peak vibration amplitudes and damping ratio can easily be obtained. By assuming that damping is small, and hence that peak frequencies are approximately equal to the undamped natural frequency, previous formulae reduce to the following:

$$\left(\frac{x_0}{F_0/k}\right)_{\max} = (T_A)_{\max} = \frac{1}{2c/c_c}.$$

It should be noted that these peak values apply only to the fully developed resonant state. Actual amplitudes can be kept to a lower level by quickly traversing the resonant range. The selected value of damping should be compatible with the desired level of isolation. In practice, steady state transmissibilities of from 0·5 to 0·1 are usually achieved.

3. Passive isolation. The "passive" isolation situation is shown diagrammatically in Fig. 4 where the previous spring/mass/damper system is excited by a sinusoidal motion of amplitude, μ_0, applied to the rigid foundation. If at time, t, the mass is displaced a distance, x, from its static equilibrium position, the equation of motion is

$$\text{i.e.}\quad m\ddot{x} + c(\dot{x} - \dot{u}) + k(x - u) = 0,$$

$$m\ddot{\delta} + c\dot{\delta} + k\delta = -m\ddot{u} = m\omega^2 u_0 \sin \omega t,$$

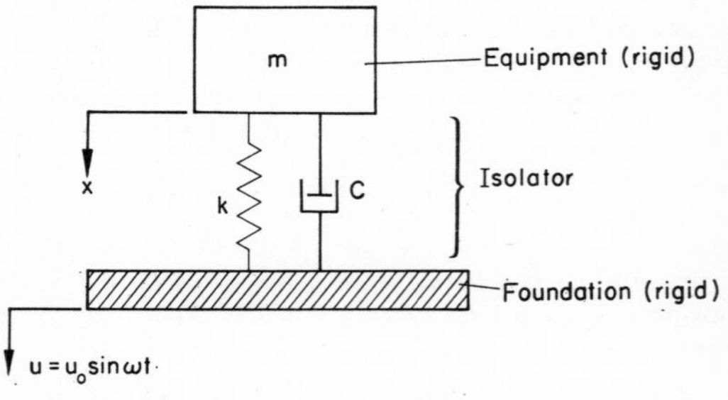

Fig. 4. Schematic diagram of the linear viscous damper passive isolation system.

where δ is the relative displacement $(x - u)$, between the mass and the foundation. This equation is now similar to that in the active isolation case. The steady state relative displacement amplitude is therefore given by

$$\frac{\delta_0}{u_0} = \left(\frac{\omega}{\omega_n}\right)^2 \left[\left\{1 - \left(\frac{\omega}{\omega_n}\right)^2\right\}^2 + (2c/c_c\omega/\omega_n)^2\right]^{-\frac{1}{2}}.$$

This ratio relates the relative displacement amplitude to that of the applied motion, and is called the "relative transmissibility", T_R, of the isolator. Like the motion response of the active isolation case, it is useful in determining clearance and space requirements. Curves of relative transmissibility are given in Fig. 5, where the following points are noted:

(a) At low frequencies the mass tends to follow the motion of the base, so that low relative transmissibilities result.

(b) At high frequencies, the mass tends to remain stationary, and values therefore tend towards unity.

(c) In the region of resonance the peak transmissibility is again given by the approximate relationship

$$\left(\frac{\delta_0}{u_0}\right)_{\max} = \frac{1}{2c/c_c}.$$

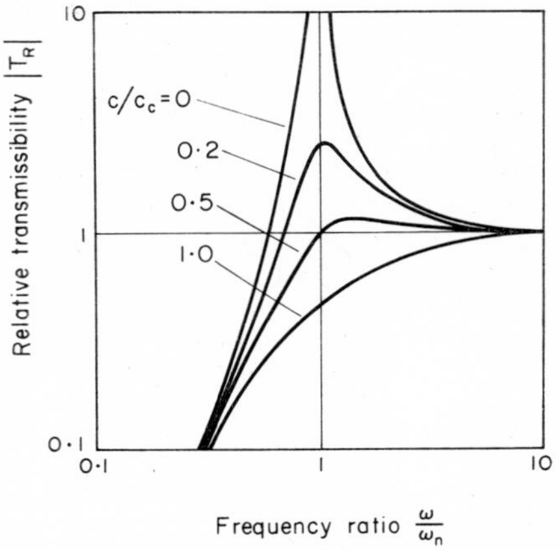

Fig. 5. Relative transmissibility of the isolator of Fig. 4.

As for the active isolation case, this equation only applies where damping is light, and where the resonant condition is allowed to build up. It should be noted that, in contrast to the active isolation case, there is a slight shift of peak frequency away from zero as damping is increased.

The equation of motion can also be written

$$m\ddot{x} + c\dot{x} + kx = c\dot{u} + ku$$
$$= u_0(k^2 + c^2\omega^2)^{1/2} \sin(\omega t + \psi).$$

where $\tan \psi = c\omega/k$.

The steady state solution is

$$x = \left[\frac{k^2 + c^2\omega^2}{(k - m\omega^2)^2 + c^2\omega^2}\right]^{\frac{1}{2}} u_0 \sin(\omega t + \psi - \Phi)$$

and the response amplitude, x_0, is given by

$$\frac{x_0}{u_0} = \left[\frac{1 + (2c/c_c\,\omega/\omega_n)^2}{\{1 - (\omega/\omega_n)^2\}^2 + (2c/c_c\,\omega/\omega_n)^2}\right]^{\frac{1}{2}}.$$

This ratio represents the extent to which the foundation vibration amplitude is modified due to transmission through the isolator and is called the "displacement transmissibility". It is remarkable that displacement and force transmissibilities are identical for a linear system, and in this case each is also known as the absolute transmissibility of the isolator. Figure 3 and all previous discussion of force transmissibility also, therefore, apply to displacement transmissibility.

4. *General remarks concerning real situations.* The isolation principles characteristic of the one degree of freedom case are also generally valid for multi-degree of freedom systems. However, in the latter case several modes of vibration can be forced by the excitation which it is desired to isolate. Each mode will, in general, have a different natural frequency whose value will depend not only upon mass, but also upon the way in which mass is distributed, and the number, arrangement and stiffnesses of the spring mountings.

The transmissibility curve will contain a number of peaks corresponding to the number of modes excited. In order to achieve isolation, the highest frequency at which a peak occurs must be arranged to fall well below the applied steady state forcing frequency. It is sometimes possible to make this task easier by modifying the mounting arrangement so as to reduce the bandwidth containing the resonance peaks. Methods of achieving this reduction include limiting the number of modes which it is possible for the applied vibration to excite, by employing guides to prevent motion in unwanted directions. Alternatively, is should be ensured that as many modes as possible can be excited quite independently. These "uncoupled" modes can then be arranged to have natural frequencies which are equal or close to each other. The method of uncoupling modes consists of suitable positioning and orientation of isolators, with respect to planes of symmetry containing the direction of loading and the centre of mass of the equipment.

In practice, the assumptions of rigid equipment and foundation structures, and massless springs, generally give good results for fairly low forcing frequencies. However, they can lead to large errors for sufficiently high frequency excitation.

The non-rigid structure will effectively increase the number of degrees of freedom of the combined equipment/isolator/foundation system and one or more secondary peaks will occur in the transmissibility curve, as in Fig. 6.

Again, while vibration would be transmitted instantaneously through massless springs, the velocity of propagation in real isolators is finite. Hence standing waves will tend to occur at high forcing frequencies when the length of the spring in the direction of propagation just contains one or more half-wave-lengths of the excitation frequency. A secondary peak in the transmissibility curve will be associated with each order resonance of the standing wave system, as in Fig. 7.

Since transmissibility may be significantly increased in the region of secondary peaks, whatever their origin, it is necessary to ensure that they do

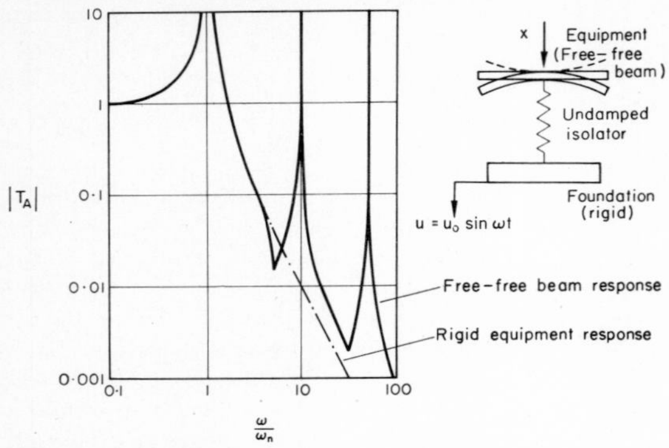

Fig. 6. Effect on transmissibility of non-rigid structure.

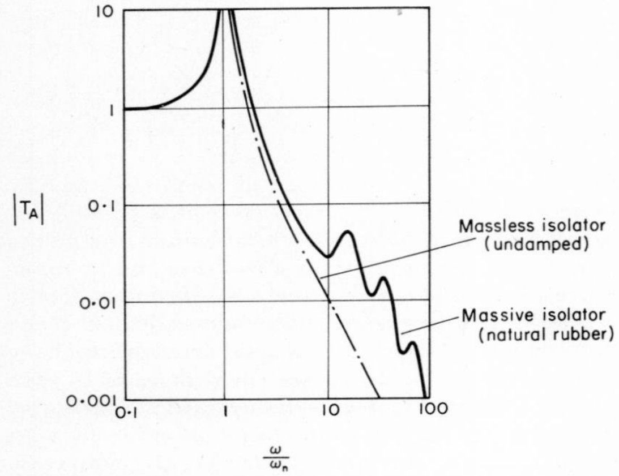

Fig. 7. Increase in transmissibility due to isolator wave effects.

not coincide with the design steady forcing frequency. Changes of mounting geometry, stiffnesses and damping, and stiffening of connected structures may be required together with experimental verification, before satisfactory isolation is achieved, especially in difficult applications. Choice of spring materials will depend upon a large number of economic, processing, performance, and environmental considerations. The advantages and limitations of the available materials is well documented in the literature, to which reference should be made before a design is finalized.

See also: Plastic springs. Rubber springs.

Bibliography

BERANEK L. L. (Ed.) (1960) *Noise Reduction*, New York: McGraw-Hill. (1966) *How to Analyse Vibratory Systems*, Design Monograph on Vibration Control, Lord Manufacturing Co. U.S.A.

DAVEY A. B. and PAYNE A. R. (1964) *Rubber in Engineering Practice*, London: Maclaren.

DEN HARTOG J. P. (1956) *Mechanical Vibrations*, New York: McGraw-Hill. (1961) Anti-Vibration Mountings, Engineers' Digest Survey No. 9, reprinted from the *Engineers' Digest*, April 1961.

GROOTENHUIS P. (1963) Low Frequency Vibration Isolation whith Air Bellows Mountings, *Society of Environmental Engineers, Symposium Proceedings*, London: Kenneth Mason Publications.

HARINGX J. A. (1949) *On Highly Compressible Helical Springs and Rubber Rods, and their Application for Vibration-free Mountings*, Phillips Research Reports, Eindhoven.

HARRIS C. M. (Ed.) (1957) *Handbook of Noise Control*, New York: McGraw-Hill.

HARRIS C. M. and CREDE C. E. (1961) *Shock and Vibration Handbook*, New York: McGraw-Hill.

PAYNE A. R. (1963) Properties and Uses of Viscoelastic Materials, *Society of Environmental Engineers Symposium Proceedings*, London: Kenneth. Mason Publications.

SNOWDON J. C. (1958) The Choice of Resilient Material for Anti-Vibration Mountings, *Brit. J. Appl. Phys.* **9**, Dec.

SNOWDON J. C. (1964) Representation of the Mechanical Damping Possessed by Rubberlike Materials and Structures, *Rubber Chemistry and Technology*, April–June.

D. N. COBLEY

W

WEAK INTERACTIONS. It has long been known that, as well as the very fast interactions between nucleons, there exists a much slower type of interaction, manifested in beta decay, which determines the rate of many natural processes, for example stellar evolution. When muons, mesons, and the whole family of elementary particles known as baryons were discovered, it became evident that slow decay processes occurred with all these particles, and the notion emerged of weak interaction forces, distinct from the strong interaction forces to which most of these particles are also subject.

Thus four basic types of interaction are now recognized in nature, as follows. Although their relative strengths cannot readily be compared in terms of dimensionless quantities, an approximate idea of their magnitudes is indicated:

Interaction	*Relative strength*
strong	~ 1
electromagnetic	1/137
weak	$\sim 10^{-14}$
gravitational	$\sim 10^{-39}$

Whereas the electromagnetic and gravitational are long-range, and the strong (nuclear) forces have a range of the order of nuclear dimensions ($\sim 10^{-13}$ cm), the weak forces appear to be extremely short-range.

Present understanding of weak interaction processes is still far from perfect although, following the discovery of parity violation, there has been intensive interest in the subject over the past decade. New discoveries and theories continue to be announced, and a review is likely to be to some extent out of date by the time it is printed. In this article the main emphasis is placed on well-established facts and concepts, with briefer reference to some of the recent developments, which may be subject to revision as further experimental and theoretical discoveries emerge. Credits are in general not given here but may be found in the detailed reviews listed in the bibliography.

Classification. In addition to nuclear beta decay, close on forty different weak interaction processes amongst the elementary particles are at present known. For the purpose of categorizing these different interactions, elementary particles may be classified as follows:

leptons:	ν_e ν_μ	
	e^- μ^-	
mesons:	π^- π^0 π^+	$(S = 0)$
	K^0 K^+	$(S = 1)$
baryons:	n p	$(S = 0)$
	Λ^0	$(S = -1)$
	Σ^- Σ^0 Σ^+	$(S = -1)$
	Ξ^- Ξ^0	$(S = -2)$
	Ω^-	$(S = -3)$

The antiparticles are similarly classified, viz: leptons, $\bar{\nu}_e$, $\bar{\nu}_\mu$, e^+, μ^+; mesons $\overline{K}^\circ$, K^-, etc., with the same pions. The distinction between ν_e and ν_μ arises from the experimental discovery by Danby *et al.* that the neutrinos associated with muons are not the same as those associated with electrons, although the nature of the difference is not yet understood. The symbol S given in parentheses denotes the quantity strangeness, attributable to mesons and baryons according to the definition $S = Y - B$, where Y, the hypercharge, is twice the average charge of a multiplet (e.g. 0 for pions, $+1$ for nucleons, etc.) and B is the baryon number ($+1$ for baryons, -1 for antibaryons, 0 for mesons).

All the particles listed above (except Σ^0) are known to participate in weak interactions. The mesons and baryons (known collectively as hadrons) are subject also to strong interactions, whereas the leptons are not. Hadrons, muons, and electrons undergo electromagnetic interactions as well, but neutrinos experience only the weak interaction. The neutrinos are generally assumed to have zero mass.

Presently known weak interaction processes can be broadly classified as follows:

(1) Purely leptonic processes,

e.g. $\mu^+ \rightarrow e^+ + \nu_e + \bar{\nu}_\mu$

(2) Semileptonic processes (involving hadrons and leptons),

(a) strangeness conserving ($\Delta S = 0$),

e.g. $n \rightarrow p + e^- + \bar{\nu}_e$

$\pi^+ \rightarrow \mu^+ + \nu_\mu$

(b) strangeness non-conserving ($\Delta S \neq 0$)

e.g. $\Lambda^0 \rightarrow p + e^- + \bar{\nu}_e$

$K^+ \rightarrow \mu^+ + \nu_\mu$

(3) Non-leptonic processes (all $\Delta S \neq 0$),

e.g. $K^0 \rightarrow \pi^+ + \pi^-$

$\Lambda^0 \rightarrow p + \pi^-$.

In the weak interactions of class (1) no strong forces are involved, but for the classes (2) and (3) effects of strong forces have also to be considered. In strong interactions strangeness is always conserved, but in weak interactions this is not necessarily so. The distinction in the class (2) above between decays with $\Delta S = 0$ and those with $\Delta S \neq 0$ is made because it is known, from observations of relative decay probabilities (e.g. for π^+ or $K^+ \rightarrow \mu^+ + \nu_\mu$), that the coupling strengths for strangeness non-conserving decays are an order of magnitude less than for those in which $\Delta S = 0$. Processes in class (3), which do not include leptons, always involve a change of strangeness. In all established cases $|\Delta S| = 1$.

The effects of strong forces in the cases (2) and (3) above make difficult the extraction of data on the weak interaction contributions to the processes. The overall strength of a particular interaction is modified by the strong forces, the coupling constant being said to be *renormalized*. There are also electromagnetic effects. Consequently with the present imperfect state of knowledge of nuclear forces it has not yet been possible to establish an exact or universal weak interaction theory. This is particularly so as regards non-leptonic decays, where experimental data are far from complete and theories are correspondingly speculative.

Basic concepts. Most of the established ideas about weak interactions stem from data on the purely leptonic decays and on the extensively studied beta decay processes. The concepts are as follows:

A weak interaction involves two pairs of fermions (particles of spin 1/2). usually represented as a contact interaction. Thus in muon decay we have two lepton pairs:

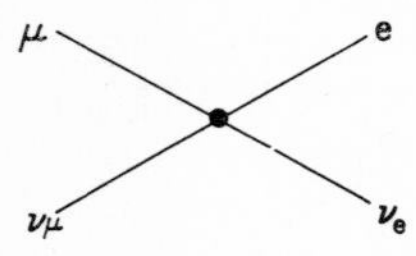

Fig. 1.

the designation particle or antiparticle for each participant being determined by conservation laws discussed below.

Similarly beta-decay processes are represented by the interaction of nucleon and lepton pairs:

Fig. 2.

These include decay of the neutron (free or in a complex nucleus): $n \rightarrow p + e^- + \bar{\nu}_e$; proton decay in a nucleus: $p \rightarrow n + e^+ + \nu_e$; and orbital electron capture: $e^- + p \rightarrow n + \nu_e$. The same diagram represents neutrino capture processes, such as $\bar{\nu}_e + p \rightarrow n + e^+$ observed with the aid of the antineutrino flux from a nuclear reactor.

A third type of process involving leptons and nucleons is muon capture ($\mu^- + p \rightarrow n + \nu_\mu$):

Fig. 3.

and this diagram also represents the experimentally observed capture processes $\nu_\mu + n \rightarrow p + \mu^-$ and $\bar{\nu}_\mu + p \rightarrow n + \mu^+$. These latter experiments, using high energy machines to produce neutrinos and antineutrinos from π^+ and π^- decays, were in fact those which established the distinction between electron-associated and muon-associated neutrinos, since it was shown that e^- and e^+ production did not occur from these neutrinos.

For cases like the pion decay $\pi^+ \rightarrow \mu^+ + \nu_\mu$, the four-fermion weak interaction is visualized as arising when the pion is dissociated into a nucleon plus antinucleon pair, thus:

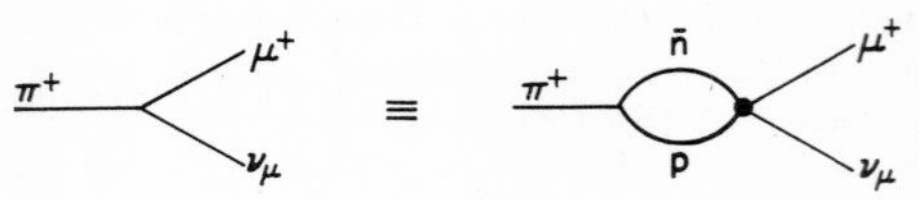

Fig. 4.

In non-leptonic decays the equivalent diagrams are even more complicated.

From a series of key experiments on beta decay the following results have been obtained:

(i) parity is not conserved in the weak interaction;
(ii) neutrinos are fully polarized, the ν_e spinning left-handedly about its direction of motion and the $\bar{\nu}_e$ right-handedly;
(iii) electrons have the maximum polarization ($+v/\mathbf{c}$);

(iv) lepton number (l_e) is conserved, as well as charge and baryon number [$l_e = +1\ (-1)$ for $e^-(e^+)$ and for $\nu_e(\bar{\nu}_e)$; $l_e = 0$ for baryons].

In muon decay, and other leptonic interactions, similar principles are found to apply. For muon decay the lepton conservation rule must be more fully expressed because of the distinction between the two kinds of neutrinos. For muons and their associated neutrinos we define the lepton number $l_\mu = +1(-1)$ for $\mu^-(\mu^+)$ and for $\nu_\mu(\bar{\nu}_\mu)$, and $l_\mu = 0$ for electrons and their neutrinos. Likewise $l_e = 0$ for the muonic particles. Then both l_e and l_μ are separately conserved. We thus find

$$\mu^- \to e^- + \bar{\nu}_e + \nu_\mu$$

and

$$\mu^+ \to e^+ + \nu_e + \bar{\nu}_\mu .$$

Accordingly the decay $\mu \to e + \gamma$, which is not observed, is not expected to occur.

Strange particle decays. The decays of strange particles are not so well understood, but a number of selection rules have been discovered, to which there are at present no established exceptions. These are as follows:

(i) the total strangeness S of hadrons in any weak interaction can only change by an amount $\Delta S = 0$ or ± 1;

(ii) in strangeness-changing leptonic reactions the change ΔQ in the total charge of the hadrons has the same sign as ΔS, i.e. $\Delta Q = +\Delta S$;

(iii) in any strangeness-changing reaction the total isotopic spin of the hadrons changes by one half, i.e. $|\Delta I| = \frac{1}{2}$, subject to the effect of electromagnetic interactions;

(iv) in strangeness-conserving reactions $|\Delta I| = 1$.

Basic theories. The theory of beta decay developed by Fermi in 1934 was based on the idea of interacting currents, by analogy with the treatment of the electromagnetic interaction, and this same idea is still used today. The theory has been modified to incorporate the deductions from the parity-violation and associated beta-decay experiments, including the inference that there are only two basic types of interaction, the vector (**V**) and axial vector (**A**), in opposite phase.

For the simplest case, that of muon decay, the Hamiltonian is represented as the interaction of two leptonic currents:

$$H_\mu = G/\sqrt{2}\, (j_\lambda^\mu)^\dagger\, (j_\lambda^e) + \text{h.c.}$$

the coupling constant G determining the strength of the interaction. The lepton currents are expressed in terms of Dirac wave functions and the operators γ_λ, $\lambda = 1, 2, 3, 4$, and $\gamma_5 = \gamma_1\gamma_2\gamma_3\gamma_4$:

$$j_\lambda^e = i\bar{\psi}_e \gamma_\lambda (1 + \gamma_5)\, \psi_{\nu_e}$$

and similarly for j_λ^μ.

For beta decay the Hamiltonian represents the interaction of baryon and lepton currents:

$$H_\beta = G/\sqrt{2}\, (J_\lambda)^\dagger\, (j_\lambda^e) + \text{h.c.}$$

where

$$J_\lambda = i\bar{\psi}_p \gamma_\lambda (g_V - g_A\gamma_5)\, \psi_n .$$

The two terms here, representing the vector and axial vector components, have additional constants g_V and g_A to allow for renormalization of the coupling constant G due to the effects of strong forces, so that

$$G_\beta^V = g_V G \quad \text{and} \quad G_\beta^A = g_A G .$$

The coupling constant $G_\mu(=G)$ for muon decay is regarded as basic since no strong forces are involved in this case.

The magnitude of G_μ is determined from accurate measurements of the muon mass and lifetime and is found, after a small radiative correction, to be

$$G_\mu = (1{\cdot}4350 \pm 0{\cdot}0011) \times 10^{-49} \text{ erg cm}^3 .$$

Values of the vector and axial vector coupling constants in beta decay (G_β^V and G_β^A) are determined from measurements of the comparative half-lives (ft values) of pure vector (Fermi) decays between $J = 0^+$, $T = 1$ isobaric analogue states, and of neutron decay which involves both vector and axial-vector (Gamow–Teller) transitions with known matrix elements. From these measurements it has been found that

$$G_\beta^V \simeq G_\mu ; \text{ in fact } (G_\mu - G_\beta^V)/G_\mu = (2{\cdot}02 \pm 0{\cdot}25) \text{ per cent,}$$

and

$$G_\beta^A = -(1{\cdot}18 \pm 0{\cdot}02)\, G_\beta^V ; \text{ phase difference } 180° \pm 8° .$$

Thus the axial vector interaction apparently involves some renormalization due to strong forces, but the vector interaction does not, to within 2 per cent.

Conserved currents. The similarity, both in structure and coupling strengths, of the muon and beta-decay interactions, and likewise muon capture, has led to the idea of a universal Fermi interaction, for strangeness-conserving processes at least. However, a question arises as to why the *vector* coupling constant for beta decay, where virtual pion emission would be expected to produce some renormalization effects, is so nearly equal to that for muon decay, which is without pionic effects. An explanation is provided by the hypothesis of a *conserved vector current* (CVC), which effectively says that virtual pions contribute an additional current component which maintains the beta interaction strength, so that the total vector current is conserved, and $G_\beta^V = G_\mu$ exactly. Thus, by analogy with the conservation of electric current,

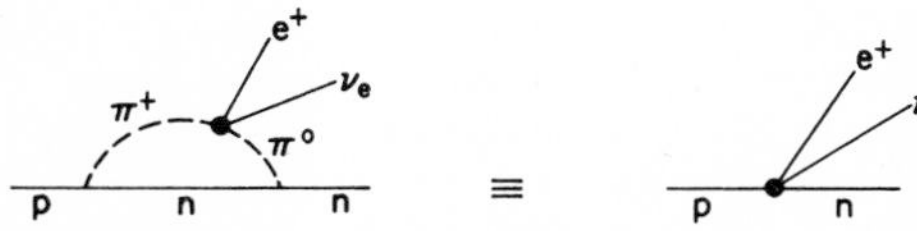

Fig. 5.

The observed rate for $\pi^+ \rightarrow \pi^0 + e^+ + \nu_e$, and a number of other test experiments, support (though they do not prove) the conserved vector current hypothesis. The observed residual 2 per cent discrepancy between G_β^V and G_μ has caused some concern, but a possible explanation is provided by the theory discussed in the next section.

The case of the axial vector current is more complicated. A renormalization factor of 1·18 for the axial vector coupling constant has already been mentioned, and theoretical arguments have shown that absolute conservation of the axial vector current would be contrary to experimental observation, particularly of the leptonic decay of pions. However, some successes have been achieved with the theory of a so-called partially conserved axial vector current (PCAC) in which the divergence of the current is expressed in terms of a pion field operator. Calculations by Adler and by Weisberger of the expected renormalization factor for the axial vector coupling constant in beta decay (which they are able, with some assumptions, to express in terms of pion-nucleon scattering cross sections) give the magnitudes 1·24 and 1·16 respectively, in good agreement with the experimental value.

Further developments and conjectures. (*a*) *Theoretical extensions.* In order to embrace in the theory the strange particle decays, leptonic and non-leptonic, a universal weak interaction Hamiltonian has been envisaged, of the form:

$$H = G/\sqrt{2}\,(j_\lambda^e + j_\lambda^\mu + J_\lambda + S_\lambda)^\dagger \times (j_\lambda^e + j_\lambda^\mu + J_\lambda + S_\lambda) + \text{h.c.}$$

where, in addition to the lepton currents and the strangeness-conserving baryon current J_λ, a strangeness non-conserving component S_λ has also been introduced. As well as providing for all known interactions, this predicts, from the square terms, processes as yet unknown, except possibly the effect of a $J_\lambda^\dagger J_\lambda$ term in producing a small parity violating component in nuclear reactions. (This latter effect, evidenced for example by a mixed $E_1 + M_1$ gamma-ray transition, is not yet confirmed.) The main objection to the above current-current representation is that it does not account for the very much weaker coupling strengths of $|\Delta S| = 1$ decays, nor for all the observed selection rules.

A more promising approach for the *leptonic* decays is that due to Cabibbo. Following up the observation of SU_3 symmetry properties for the strongly interacting particles (see *Strong interactions*), he has postulated that the total weak interaction current $\mathscr{J}_\lambda$, including both J_λ and S_λ components, transforms according to the eightfold representation of SU_3, the vector and axial vector parts of $\mathscr{J}_\lambda$ each having an octet of currents.

From this assumption follow the selection rules $\Delta S = +\Delta Q$ and $|\Delta I| = \frac{1}{2}$ for $|\Delta S| = 1$, as well as $|\Delta I| = 1$ for $\Delta S = 0$. Furthermore, the total baryon current interacting with the lepton current then becomes

$$\mathscr{J}_\lambda = aJ_\lambda + bS_\lambda$$

where a and b are constants. A relatively small value of b could account for the comparative weakness of strangeness changing decays.

The further postulate has been made that a, b satisfy $a^2 + b^2 = 1$, so that they can be represented as $\cos\theta$ and $\sin\theta$ respectively, with θ an adjustable parameter. Thus, separating out the vector and axial vector components, one has

$$\mathscr{J}_\lambda^V + \mathscr{J}_\lambda^A = \cos\theta\,(J_\lambda^V + J_\lambda^A) + \sin\theta\,(S_\lambda^V + S_\lambda^A),$$

(although the value of θ may not necessarily be the same for the vector and axial vector parts).

In the spirit of the conserved vector current hypothesis, applied to $\mathscr{J}_\lambda^V$, one then excepts the following vector coupling strengths:

with G for muon decay,
$G\cos\theta_V$ for $\Delta S = 0$ decays,
$G\sin\theta_V$ for $|\Delta S| = 1$ decays.

A value of θ_V can be determined from a comparison of the decay rates of $|\Delta S| = 1$ and $\Delta S = 0$ decays, for example

$$\frac{R(K^+ \rightarrow \pi^0 + e^+ + \nu_e)}{R(\pi^+ \rightarrow \pi^0 + e^+ + \nu_e)} = \frac{1}{2}\tan^2\theta_V$$

whence $\theta_V = 0{\cdot}22$. This result can be compared with the value inferred from the experimental ratio of G_β^V (for beta decay) to G_μ, discussed earlier. Putting $G_\beta^V = G_\mu \cos\theta_V$ gives $\theta_V = 0{\cdot}21$. The agreement appears to give satisfactory support to the basic hypothesis, although there are some possible corrections to be considered. The value obtained for θ_A appears to be somewhat greater than for θ_V.

(*b*) *An intermediate boson.* It has been conjectured that the weak interaction field is mediated by a quantum, termed the intermediate vector boson (W, a particle of spin 1), in the same way that the electromagnetic field is mediated by the photon. If such were the case the diagram for beta decay, for example, would be represented thus:

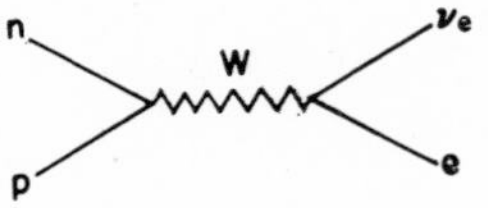

Fig. 6.

Attempts are being made to obtain experimental evidence for the existence of such a boson, produced by high energy neutrino interactions, but no definite success has yet been achieved. The lower limit to the mass (m_W) of such a particle stands at present at about

1800 MeV. This means that the range of weak interactions ($\sim h/m_W\mathbf{c}$) is certainly extremely short, and theoretical formulae based on the assumption of a point interaction are a good approximation, particularly in low momentum transfer reactions. However, the discovery of the W boson would be a great stimulus to unifying the theory of weak interactions.

(c) *CP invariance.* Until recently it was supposed that all weak interactions were invariant under the combined operation of charge conjugation C (interchanging particles and antiparticles) and the parity operator P, i.e. e^- (L.H.) $\xrightarrow{CP}$ e^+ (R.H.) etc. This is theoretically desirable in order to maintain invariance both under time reversal T and the combination CPT. However, some CP violation has now been observed in the case of K° decays. There are two modes of K° decay: one, known as K_1^0, into two pions, with lifetime $\sim 10^{-10}$ sec; the other K_2^0, into three pions, or leptonic, with lifetime $\sim 10^{-8}$ sec. CP invariance implies $K_2^0 \nrightarrow 2\pi$. However, Christenson and others have found that the long-lived component, K_2^0 although decaying mainly to $\pi^+ + \pi^- + \pi^0$ or by leptonic modes, has a 0·2 per cent probability of the two pion process $K_2^0 \rightarrow \pi^+ + \pi^-$. This could imply a small CP violation in the weak interaction. On the other hand, since both strong and electromagnetic interactions are also involved in the K_2^0 decay, it is possible that the violation applies to one of these, or even to an unknown type of force. This question remains to be resolved.

See also: Strong interactions.

Bibliography

GOEBEL C. J. (1962) in *Encyclopaedic Dictionary of Physics* (J. Thewlis Ed.), **4**, 578, Oxford: Pergamon Press.

LEE T. D. and WU C. S. (1965) *Ann. Rev. Nucl. Sci.* **15**, 381.

ROMAN P. (1961) in *Encyclopaedic Dictionary of Physics* (J. Thewlis Ed.), **3**, 85, Oxford: Pergamon Press.

SCHOPPER H. F. (1966) *Weak Interactions and Nuclear Beta Decay*, Amsterdam: North-Holland.

SOPER J. M. (1961) in *Encyclopaedic Dictionary of Physics* (J. Thewlis Ed.), **1**, 394, Oxford: Pergamon Press.

TAYLOR J. C. (1964) *Rep. Progr. Phys.* **27**, 407.

J. M. FREEMAN

X

X-RAY ASTRONOMY. *Introduction.* The possibility that the Sun might be an emitter of radiations in the far ultra-violet and soft X-ray region with an intensity far greater than could be expected from a black-body with a temperature equal to that of the solar photosphere (~6000°K) was first suggested independently by Hulburt and Vegard in 1938. They were led to this conclusion by consideration of possible means of production of the observed ionization in the lower regions of the ionosphere, and this seemed to suggest a radiation capable of penetrating to a height of 70–80 km and yet not penetrating appreciably to lower levels. Radiation with $\lambda \sim 10$ Å and greater seemed to answer this need. The first experimental demonstration that such radiation was emitted by the Sun was probably obtained by Burnight who in 1948 sent up a piece of photographic film suitably protected from visible and u.v. light in a V-2 rocket and found blackening of the film indicative of soft X-ray irradiation. This result was confirmed and placed beyond doubt in 1949 by Friedman, Lichtman and Byram of the U.S. Naval Research Laboratory (N.R.L.) who used Geiger counters as X-ray detectors.

All the earlier work was concerned with solar X-ray physics, but in 1962 a non-solar X-ray source was discovered by the American Science and Engineering Inc. group (A.S. & E.). This was followed shortly after by the discovery of further sources of a cosmic nature thus leading to further developments in this field. It is interesting to note that so far all observations in the field of *cosmic* X rays have been made from rockets. Bearing in mind that the useful observational life of a rocket is of the order of three or four minutes this means that the large amount of information now available has been obtained through a total observing time of the order of two hours!

During the past decade an increasing number of groups has become interested in the field in America, in Europe and in Japan. In the U.K. a group at University College London and one at Leicester University have been engaged in the field both separately and in collaboration.

Techniques of X-ray astronomy. Although X rays of energy greater than 10 keV (~1 Å) penetrate sufficiently far into the Earth's atmosphere for it to be possible to observe them from high altitude balloon flights, the lower energy radiation (4–10 Å) is so strongly absorbed that observations must be carried out at altitudes where the air density is negligible (~200 km) and the bulk of the work has been carried out with rocket or satellite borne equipment. The means of detection of the X rays so far employed are:

(*1*) *Photographic.* This is of value when the equipment is recoverable and has been used considerably in rocket experiments. The photographic film must be protected from visible and u.v. radiation by suitable filters, e.g. Al, Be foils or plastic film coated with Al. By using filters of different materials and different thicknesses a rough indication of the spectral energy distributions may be arrived at, and by using calibrated photographic film absolute intensities may be measured. Photographic methods are not suitable for the very low intensity radiations from cosmic sources.

(*2*) *Gas counters.* These may be of the Geiger type or the proportional type. The latter if used in conjunction with provision for pulse height analysis into several channels gives better information about spectral distribution than can be obtained with filters. The energy resolution falls off, however, with decreasing photon energy and is very poor above 40 Å. Gas counters with very large windows have been used for cosmic X-ray work.

(*3*) *Ionization chambers.* These are of use when the radiation intensity is sufficient to give measurable steady ionization currents and have been employed on many occasions. Energy resolution is absent and recourse must be had to filters for this purpose.

(*4*) *Photomultipliers.* Recently attention has been paid to photomultipliers of the channel type. Being windowless they are particularly useful for longer wave-lengths (>40 Å) where window absorptions become particularly troublesome. They yield no new information about spectral distribution.

(*5*) *Scintillation counters.* These are useful only for high photon energies and have been employed only to a limited extent.

In the field of solar X rays, photographs giving the distribution of X-ray sources over the Sun's disk have been obtained by using pin-hole cameras, and cameras with imaging devices based on grazing incidence reflection optics.

Recently greatly improved knowledge of the Sun's X-ray spectrum has been obtained from rockets carrying Bragg crystal spectrometers. These employ suitable crystals in conjunction with gas counters.

Solar X rays. Most of the work carried out since 1949 has been concerned with the Sun's X-ray emission. Measurements of the intensity of this emission are now sufficiently extensive to show that although extremely variable, on the average the X-ray intensity follows the 11-year solar cycle, being a maximum at sunspot maximum. The short period variations appear to be connected closely with various types of solar activity particularly with plages and flares (*vide infra*).

Earlier work succeeded only in measuring the radiation integrated over the whole solar disk with a limited energy resolution into rather broad energy bands. These earlier results showed that the X-ray intensity at the top of the Earth's atmosphere varied over the range 0·1 to 1·0 erg/cm²/sec in the wave-length band 10–100 Å where the bulk of the energy seems to lie. The high absorbability of radiation in this wave-length band indicates that the source must be in a medium of very low density—most likely in the corona or at most the upper chromosphere. A coronal origin was confirmed by estimates of the temperature of the source based on observations of the spectral distribution. Treating this as a continuous spectrum emitted by a "grey" body this gives temperatures in the range $0{\cdot}5\text{–}2 \times 10^6$°K, consistent with the temperatures needed to account for the coronal spectrum lines identified by Edlen as being emitted by highly ionized atoms such as Fe X and Fe XIV. This is confirmed also by photographs of the Sun taken in X rays (see accompanying photograph) which show brightening of emissions at the Sun's limb in contradistinction to the limb darkening observed in the visible light coming from optically denser regions.

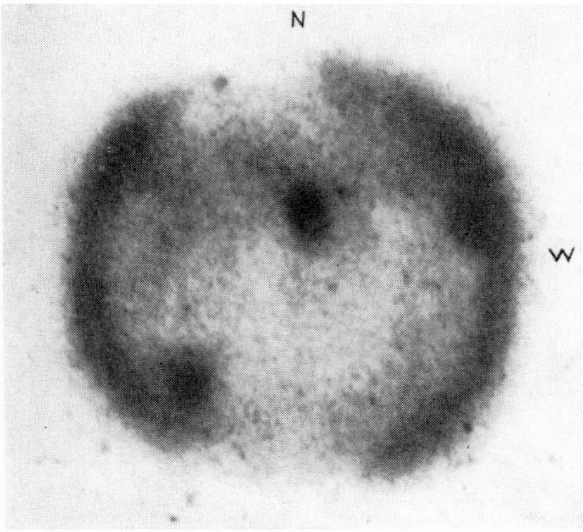

Solar X-ray image from Skylark 306, flown October 20th 1965. Camera resolution 2·5 arc min and wave-length band 44–50 Å. (Leicester University Space Research Group.)

Spatial distribution of sources. Information about this is afforded by X-ray photographs such as the pin-hole photograph reproduced here. The first of such photographs was taken in 1963 by the N.R.L. group and has been followed more recently by pin-hole photographs taken by the Leicester group and focused image photographs by the A.S. & E. and NASA groups. The bulk of the emission appears to come from a few regions of high intensity. These are associated in all cases with optically observed plages and show close correlation with the spatial distribution of radio emissions. Close study of such photographs suggests that the "hot spots" from which X rays arise are probably very small in extent (<1 arc min.). This is confirmed by observations made during the eclipse of the Sun in May 1966 when a very sharp cut-off in X-ray intensity was observed as the Moon occulted an active "hot-spot". That the "quiet" corona is still an emitter is indicated, however, by the presence of the limb brightening. The lowest intensity so far measured is one of the order of 0·01–0·1 ergs/cm²/sec measured by the N.R.L. group for the wave-length range 10–60 Å. The Sun was very "quiet" at the time. Pin-hole photographs taken through different filters indicate that the limb-brightening depends on the wave-length, being about 90 per cent higher over the range 40–50 Å than for wave-lengths $\sim$20 Å. It is supposed that a strong O VII emission at 21·5 Å may be optically thick on the limb, since O is a fairly abundant element and at a temperature of $1\text{–}2 \times 10^6$°K it would be largely in an O VII state.

Spectral distribution. Earlier work making use of filters or pulse height analysis resulted in an analysis into broad bands. These results gave a spectrum having in general a negligible intensity at c. 4 Å and rising steadily with the wave-length. No maximum has so far been observed. During a strong solar activity (e.g. a flare) this spectrum may be profoundly changed. Under these circumstances there is an enormous increase in intensity at the short wave-lengths amounting to several orders of magnitude. At longer wave-lengths the increase is much less, however, and it seems likely that for $\lambda > 40$ Å the increase may be very small.

Lately it has become possible to get more detailed spectral information by flying crystal spectrometers. This has been done by the N.R.L. group and also by a group at Leicester University. These experiments have revealed a large number of lines in the range 10–25 Å. Many of these can be identified with lines from known atoms in high ionization states, such as O VIII, Ne IX, Ne X and Fe XVII. Comparison of the Ne IX and Ne X lines suggests that they originate in active regions with an effective temperature $\sim$3–4 $\times$ 10^6 °K. It is not impossible that lines of shorter wave-length arising from even more highly ionized atoms may exist in which case the active region temperatures may be even higher.

Temporal variation. In addition to the solar-cycle variation, "spot" measurements of intensity from rockets indicate considerable variations in the integrated intensity. These variations are largely determined by variations in plage regions. These variations appear to be relatively slow ($\sim$ days or longer). In addition very rapid variations ($\sim$ minutes) occur. These are associated with solar flares and in general require observations from satellites. Such variations were observed for example from the Ariel I satellite in 1963. Here large increases and changes in the spectral distribution were observed during periods when visual observation showed the occurrence of flares. In some cases after an initial increase of X-ray intensity of moderate amount there followed a rapid and intense burst of X rays. A close correlation between such bursts and D-region disturbances was noticed. Much more monitoring of the Sun from this point of view is needed, particularly in the short wave-length region.

It will be seen that further observations in this field and in particular of the spatial and temporal distribution of solar X rays is likely to make a vital contribution to solar physics. Of particular importance will be the high resolution spectroscopy of individual coronal regions, yielding information on local temperatures and particle densities.

Cosmic X rays. A fairly active Sun gives a radiation intensity above the Earth's atmosphere of the order of 1 erg/cm^2/sec in the wave-length range 1–100 Å, or a photon count rate of the order of 10^8–10^9/cm^2/sec. If Sirius emitted X rays with a similar intensity the count rate in the vicinity of the Earth would be of the order of 10^{-3}/cm^2/sec, or less than 100 per day, or taking account of absorption by interstellar matter probably considerably less even than this. Such a count rate would be very difficult to detect against a background of other types of radiation, and even in 1962 seemed to suggest that galactic X-ray astronomy would not be possible until there were available detecting systems of large collection area and stabilized satellites capable of high pointing accuracy so that they could be oriented for long periods towards the possible galactic sources. It must be counted a major discovery therefore that in 1962 Giacconi, Gursky, Paolini and Rossi of A.S. & E., using a rocket only partially stabilized, while attempting unsuccessfully to observe X rays from the Moon, observed X rays of measurable intensity in the general direction of the galactic centre, the angle of survey in this experiment being very broad. This surprising result was verified by Bowyer, Byram, Chubb and Friedman in 1963 using detectors with a field of view restricted to 10° carried in an Aerobee rocket. They placed the source found by Giacconi *et al.* some little way from the galactic centre in the direction of the constellation of Scorpius and in the approximate position R.A. $16^h\,15^m$, Decl. $-15°$. At the same time they found a weaker source in Taurus in the direction of the Crab nebula. The X-ray intensity from these sources is remarkably high. That from the Scorpius source for $\lambda < 8$ Å amounts to 5×10^{-7} erg/cm^2/sec, equivalent in optical wave-lengths to the light flux from a third magnitude star. Since then, a number of surveys have been carried out and most of the sky covered and at present some 25 cosmic X-ray sources of intensity measurable by existing techniques have been detected. It is possible that at least one of these is extragalactic (*vide infra*). Assuming most of the sources so far detected are within 3000 pc, Friedman estimates the galaxy to contain some 600 such sources. These surveys have seen increasing precision in determining the position of the sources and very recently Giacconi *et al.* have succeeded in placing the position of the Scorpius source to within 20″. These results have been achieved by using large aperture ($>$500 cm^2) gas counters combined with "honey-comb" collimators, and more recently (Giacconi *et al.*) with "modulation collimators" consisting of a series of wire grids. Some information about spectral distributions of the X rays from the stronger sources has been gained by using gas counters having different gas fillings and different window materials. Proportional counters and (for X rays of higher energy) scintillation counters have also been used for this purpose.

The sources are now designated by the letters X or XR prefixed by the abbreviation for the constellation and followed by a number indicating its relative strength in that constellation. Thus the Scorpius source is Sco XR-1, the Taurus source Tau XR-1 etc. At present only one X-ray source has been indubitably identified with an optically known object (namely the Crab nebula) although recently it seems likely that Sco XR-1 has been identified optically with fair certainty. A few other sources can be identified with likely objects, but not with great certainty owing to the poor angular resolution ($\sim$1–2°) of the X-ray detector systems. We give now particulars of a few of the more important and more interesting sources known at present.

1. Scorpius XR-1. This is far and away the most intense source known at the time of writing and was the first to be discovered. Its position has been determined with increasing accuracy through successive observations by various groups of workers. The latest and most precise is that of Gursky, Giacconi, Gorenstein and Waters of A.S. & E. in association with Oda, Bradt, Carmire and Sreekantan of M.I.T.

(1966) who consider the uncertainties in position to be $\pm 4^s$ in R.A. and $\pm 30''$ in decl. Very recently these groups in association with Sandage and Osmer of Mt. Wilson and Palomar Observatories, and Oda, Osawa and Jugaku of the Institute of Space Science and the Tokyo Astronomical Observatory consider that they have succeeded in identifying the source optically with a star of Mag. 13, possessing an ultra violet excess ($B - V = +0{\cdot}23$, $U - B = -0{\cdot}88$). The object varies rapidly and irregularly in brightness and is considered to have some of the properties of an old nova. Assuming it to be such they estimate 250 pc as a likely value for its distance. Previously Manley (1966) assuming the X rays to arise from a synchrotron type of emission and the source to have an angular size of $1''$ put an upper limit to the distance of 9 kpc. Various other estimates of its distance have been made depending on various theories of the nature of the source but all agree that it is a galactic object.

Studies of the X-ray spectrum of Sco XR-1 have been made in the range 1–40 keV (12–0·3 Å) by Graber, Hill, Seward and Toor (1966) and to 50 keV by Peterson and Jacobson. The former found a spectral distribution which among other possibilities can be regarded as arising from a thermal bremsstrahling emission from a hot optically thin gas at a temperature $\sim 3 \times 10^7$ °K. The energy distribution curve reaches a maximum at approximately 1·5 keV ($\sim$9·0 Å). It is uncertain at present whether the drop in intensity at wave-lengths longer than this is instrumental or not, but assuming it to be real Felton and Gould (*Phys. Rev.* 1966) show that interstellar absorption could account for it. The extension to 50 keV by Peterson and Jacobson is in fair agreement with the results of Graber *et al.*, where the observations overlap, but the former find above 35 keV a "non-thermal" tail, which could (although considered unlikely) be due to a line emission at 42 keV superimposed on an exponential continuum.

2. Taurus XR-1 (Crab nebula). This, the second strongest X-ray source known has been identified beyond doubt with the Crab Nebula by Chubb, Byram, Bowyer and Friedman (1964) who made measurements of the X-ray intensity in the direction of the Crab Nebula while the latter was being occulted by the Moon. The results not merely showed that the Crab was the source of the X-ray emission but showed also that the X rays come not from a single point source in the nebula but from an extended region on the visible disk of the object. The spectral distribution of energy has been measured by Graber *et al.* (loc. cit.) who found it to be of a different character from that of Sco XR-1 and consistent with the idea that it is a synchrotron radiation. This is in harmony with the known synchrotron nature of the continuum in the visible spectrum of the Crab Nebula, and the idea is generally accepted, although it encounters some difficulties in view of the high electron energies needed ($\sim 10^{14}$ eV) and the relatively short life-time of such electrons emitting X rays at the observed rates.

3. The Cygnus sources. Cyg XR-1 was discovered by Bowyer *et al.* in their survey of 1964. Since then Byram, Chubb and Friedman have discovered some 8 or 9 sources in Cygnus Cyg XR-1 is of great interest as Friedman considers that there is positive evidence that over a period of approximately 8 months its X-ray intensity has dropped by nearly 75 per cent, the first example of cosmic X-ray variability so far discovered. Measurements of the spectral distribution have been made by Graber *et al.* The results are rather indeterminate but this seems to be due to their being derived from the combined effects of Cyg XR-1 and other nearby sources.

4. Cassiopea A. An X-ray source in the direction of Cas A was found by Byram, Chubb and Friedman in their survey of April 1965 and they have tentatively identified the source with this body. Cas A is the brightest radio source in the Galaxy and is usually regarded as a supernova remnant. The emission of X rays from this body, if confirmed, is obviously of great interest.

5. Centaurus XR-2. By a fortunate seres of circumstances six surveys which included the Centaurus region were carried out between October 1965 and September 1967. The first of these detected no source stronger than background, while three in April at nearly weekly intervals found a strong source—on April 4th nearly as strong as Sco XR-1 (Harries, McCracken, Francey and Fenton) on 10th April stronger than Sco XR-1 (Cooke, Pounds, Stewardson and Adams, A.P.J. Letters, January 1968), and on 20th April about half the strength of Sco XR-1 (Harries *et al*). On May 18th 1967 Chodil, Mark, Rodrigues and Swift found the intensity to be $\sim 10\%$ of that on May 10th while by September 28th the same observers found it indistinguishable from backgroud. This is a second example of a variable X-ray source and Chodil *et al.* have sought to interpret it in terms of a nova model.

6. Extragalactic Sources. X-rays from discrete extragalactic sources have been occasionally reported, but at the time of writing only one of these appears to be fairly certain, namely M 87. This was reported by Friedman *et al.* (1967) and apparently confirmed by Bradt *et al.* (1968).

M 87 is an elliptical nebula in the Virgo cluster, characterized by a jet apparently emitted from its centre region. It is also a radio emitter albeit some 1000 times weaker than Cyg A. Again the identification if confirmed, is of great interest.

Reports of X-rays from Cyg A and 3C273 are still subject to some doubt, while an X-ray emission in the energy range 30–100 kev from the direction of the Coma cluster of galaxies also awaits confirmation.

No X-ray sources other than those given above have so far been identified even tentatively with known optical objects, so that theories as to the nature of the sources and the mechanism of X-ray production are necessarily still largely speculative. Of theories at present current the following are worth noting: (1) The galactic sources are supernova remnants. This is supported by the fact that the Crab Nebula is known to be such, and that the Cas A source is likely to be. Braes and Hovenier have sought support for the association of X-ray sources with old supernovae by pointing out that a number of the observed X-ray sources are located close to known OB associations, three of these being between such associations and the "runaway" stars which originated from them. Confirmation of this, however, would depend much on greater precision in the determination of the positions of the sources.

(2) The sources are old novae. This has been suggested by many writers. Present evidence based on the optical properties of Sco XR-1 and on recent work of McCracken (*vide infra*) suggest that this particular source may be a nova, and not a supernova and although it is almost certain that some sources at least do not fall into this category, a number of sources may be of this type. (3) The sources are Neutron stars. That there might exist highly condensed stars constituted mainly of neutron gas was first suggested by Zwicky in 1934. The theory of such stars has recently been taken up and extended. The extremely small size of such a star would make it virtually invisible optically, while its very high temperature would result in its maximum emission intensity lying in the soft X-ray region. There is, however, still some uncertainty as to how long a neutron star would remain in an X-ray emittive condition and if this should turn out to be too small $\sim$ decades) it would make it unlikely that any appreciable number of the known X-ray sources are of this type. The matter is still *sub judice*. (4) The sources may be protostars. This possibility has also been considered, particularly by Manley (but mainly in connexion with Sco XR-1, which is most likely invalidated by the recent optical identification of this source). (5) Flare stars. The possibility that flare stars could be sources of X rays has been suggested, and worked out, particularly by Gurzadyan. Such sources would necessarily be variable.

Since two of the three suspected extragalactic sources are known radio sources any theory of their X-ray emission is probably linked to theories of the radio emissions. At present "synchrotron" theories are favoured. More recently K. G. McCracken has published the results of some observations made in the hard X-ray region (20–60 keV) by balloon-borne equipment. He finds the Cygnus complex of sources to form the brightest object in the sky in this X-ray region, and combining these results with those already known for the region 2–10 Å shows that the photon energy distribution is of the form $(h\nu)^{-1\cdot7}$. Upper limits to hard X-ray fluxes from Sco XR-1, Cyg XR-2, Oph XR-1 and Ser XR-1 were also established. The Sco XR-1 source was found to be very soft and differences between this body and bodies such as Tau XR-1 or Cyg (XR-1 + A) seems also to exclude the possibility of Sco XR-1 being a supernova remnant.

The interstellar medium. Owing to the pronounced dependence of absorption on wave-length in the soft X-ray region, absorption by interstellar matter plays an even more important role in X-ray astronomy than it does in Optical Astronomy. It seems likely from current estimates of the density of the interstellar medium that radiations in the range 100–1000 Å from all cosmic sources will suffer complete extinction through absorption by the medium and that only near and intense sources would be able to give measurable intensity for $\lambda > 50$ A. Calculations of the possible effects of the interstellar medium based on estimates of the mean density of interstellar matter and on the probable relative abundance of the elements have been made by Strom and Strom and further extended by Fenton and Gould. The latter apply their results to the observed radiation from Sco XR-1 and show that the spectral peak at 1·5 keV can be interpreted as being due to absorption (*vide supra*). They pay some attention also to intergalactic absorption but find that in view of the uncertainty in the X-ray data no firm conclusion can be drawn.

An interesting possible application of X-ray methods is to the determination of the absolute amount of a given atomic species between a source and the Earth by the measurement of the magnitude of the absorption discontinuities in continuous X-ray spectra. The K edges of O and Ne suggest themselves in this respect.

The X-ray background. In addition to the discrete sources described above most observers agree in finding a general background of soft X-rays. For $\lambda \sim 1 - 10$ Å the background radiation has an intensity $\sim 10^{-8}$ erg/cm^2/sec/ster, and appears to be isotropic. Recently Bowyer, Field and Mark (*Nature*, **217**, p. 32, 1968) report measurements of a background X-ray flux in the range 44–70 Å. This component of the background appears to have a maximum in the direction of the galactic pole. Current opinion at present is that the X-ray background generally is of extragalactic origin. How it is produced is still an open question. The possible cosmological significance of this has been discussed by Rees, Sciama and Setti (*Nature*, **217**, p. 326, 1968).

Bibliography

The subject of X-ray astronomy is so new that no books devoted to it have yet appeared. Short review articles have been published by K. A. Pounds, *X-ray Focus*, Vol. 6, No. 3, 1965, by T. A. Chubb, *Endeavour*, Vol. 25, No. 96 and by R. J. Gould, *American Journal of Physics*, Vol. 35, p. 376, 1967 while an account of the subject as known in 1964 will be found in Chapter 7 of *Space Physics* by Sir Harrie Massey, 1964. For more detailed accounts the original papers should be consulted. A number of these are quoted in the above mentioned articles. Reference should be made also to the *Proceedings of the Annual COSPAR Symposia* while the following papers contain references to recent and earlier work.

BYRAM E. T. *et al.* (1966) *Science*, **152**, 66.
GRADER R. J. *et al.* (1966) *Science*, **152**, 1499.
GURSKY H. *et al.* (1966) *Astrophys. J.*, **144**, No. 3, 1249.
MANDEL'STAM S. L. (1965) *Space Sci. Rev.* **4**, 587.
RUSSELL P. C. and POUNDS K. A. (1966) *Nature*, **209**, No. 5022, 490.

E. A. STEWARDSON

INDEX

QUEEN MARY
COLLEGE
LIBRARY

ENCYCLOPAEDIC DICTIONARY OF PHYSICS

SCOPE OF THE DICTIONARY

For convenience in planning, and to provide a framework on which the Dictionary could be erected, physics and its related subjects have been divided into upwards of sixty sections. The sections are listed below, but, as the Dictionary is arranged alphabetically, they do not appear as sections in the completed work.

Acoustics
Astronomy
Astrophysics
Atomic and molecular beams
Atomic and nuclear structure
Biophysics
Cathode rays
Chemical analysis
Chemical reactions, phenomena and processes
Chemical substances
Colloids
Cosmic rays
Counters and discharge tubes
Crystallography
Dielectrics
Elasticity and strength of materials
Electrical conduction and currents
Electrical discharges
Electrical measurements
Electrochemistry
Electromagnetism and electrodynamics
Electrostatics
Engineering metrology
General mechanics
Geodesy
Geomagnetism
Geophysics
Heat
Hospital and medical physics
Industrial processes
Ionization
Isotopes
Laboratory apparatus
Low-temperature physics
Magnetic effects
Magnetism
Mathematics
Mechanics of fluids
Mechanics of gases
Mechanics of solids
Mesons
Meteorology
Molecular structure
Molecular theory of gases
Molecular theory of liquids
Neutron physics
Nuclear reactions
Optics
Particle accelerators
Phase equilibria
Photochemistry and radiation chemistry
Photography
Physical metallurgy
Physical metrology
Positive rays
Radar
Radiation
Radioactivity
Reactor physics
Rheology
Solid-state theory
Spectra
Structure of solids
Thermionics
Thermodynamics
Vacuum Physics
X rays